STUDENT'S SOLUTIONS MANUAL

MATH MADE VISIBLE, LLC

INTERMEDIATE ALGEBRA: GRAPHS AND MODELS

FOURTH EDITION

Marvin L. Bittinger

Indiana University Purdue University Indianapolis

David J. Ellenbogen

Community College of Vermont

Barbara L. Johnson

Indiana University Purdue University Indianapolis

016

PEARSON

Boston Columbus Indianapolis New York San Francisco Upper Saddle River
Amsterdam Cape Town Dubai London Madrid Milan Munich Paris Montreal Toronto
Delhi Mexico City Sao Paulo Sydney Hong Kong Seoul Singapore Taipei Tokyo

Copyright © 2012, 2008, 2004 Pearson Education, Inc.
Publishing as Pearson, 75 Arlington Street, Boston, MA 02116.

ISBN-13: 978-0-321-72577-6
ISBN-10: 0-321-72577-8

5

www.pearsonhighered.com

CONTENTS

Chapter 1
Basics of Algebra and Graphing

1. constant

3. value

5. division

7. rational

9. terminating

11. We substitute 5 for b and 7 for h and multiply:

$$a = \frac{1}{2} \cdot b \cdot h = \frac{1}{2} \cdot 5 \cdot 7 = 17.5 \ \text{ft}^2$$

13. We substitute 7 for b and 3.2 for h and multiply.

$$A = \frac{1}{2} \cdot b \cdot h = \frac{1}{2}(7)(3.2) = 11.2 \ \text{m}^2$$

15. $3(x-7)+2 = 3(10-7)+2$
$= 3(3)+2$
$= 9+2$
$= 11$

17. $12+3(n+2)^2 = 12+3(1+2)^2$
$= 12+3(3)^2$
$= 12+3(9)$
$= 12+27$
$= 39$

19. $7x+y = 7 \cdot 3 + 4$
$= 21+4$
$= 25$

21. $2c \div 3b = 2 \cdot 6 \div 3 \cdot 2$
$= 12 \div 3 \cdot 2$
$= 4 \cdot 2$
$= 8$

23. $25+r^2-s = 25+3^2-7$
$= 25+9-7$
$= 34-7$
$= 27$

25. $3n^2p - 3pn^2 = 3 \cdot 5^2 \cdot 9 - 3 \cdot 9 \cdot 5^2$

Observe that $3 \cdot 5^2 \cdot 9$ and $3 \cdot 9 \cdot 5^2$ represent the same number, so their difference is 0.

27. $5x \div (2+x-y) = 5 \cdot 6 \div (2+6-2)$
$= 5 \cdot 6 \div (8-2)$
$= 5 \cdot 6 \div 6$
$= 30 \div 6$
$= 5$

29. $\left[10-(a-b)\right]^2 = \left[10-(7-2)\right]^2$
$= \left[10-5\right]^2$
$= 5^2$
$= 25$

31. $\left[5(r+s)\right]^2 = \left[5(1+2)\right]^2$
$= \left[5 \cdot 3\right]^2$
$= 15^2$
$= 225$

33. $m^2 - \left[2(m-n)\right]^2 = 7^2 - \left[2(7-5)\right]^2$
$= 49 - \left[2 \cdot 2\right]^2$
$= 49 - 4^2$
$= 49 - 16$
$= 33$

35. $(r-s)^2 - 3(2r-s) = (11-3)^2 - 3(2 \cdot 11 - 3)$
$= 8^2 - 3(22-3)$
$= 8^2 - 3(22-3)$
$= 64 - 3(19)$
$= 64 - 57$
$= 7$

37.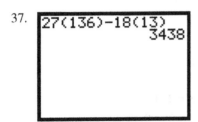

39.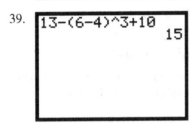

41.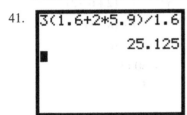

43.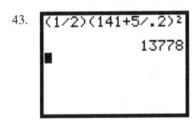

45. a) We replace y with 7.

$$\frac{12-y \;=\; 5}{12-7 \;\; ? \;\; 5}$$

$$\quad\;\; 5 \;=\; 5 \;\; \text{TRUE}$$

Since $5 = 5$ is true, 7 is a solution of the equation.

 b) We replace y with 5.

$$\frac{12-y \;=\; 5}{12-5 \;\; ? \;\; 5}$$

$$\quad\;\; 7 \;=\; 5 \;\; \text{FALSE}$$

Since $7 = 5$ is false, 5 is not a solution of the equation.

 c) We replace y with 12.

$$\frac{12-y \;=\; 5}{12-12 \;\; ? \;\; 5}$$

$$\quad\;\; 0 \;=\; 5 \;\; \text{FALSE}$$

Since $0 = 5$ is false, 12 is not a solution of the equation.

47. a) We replace x with 0.

$$\frac{5-x \;\le\; 2}{5-0 \;\; ? \;\; 2}$$

$$\quad\;\; 5 \;\le\; 2 \;\; \text{FALSE}$$

Since $5 \le 2$ is false, 0 is not a solution of the inequality.

 b) We replace x with 4.

$$\frac{5-x \;\le\; 2}{5-4 \;\; ? \;\; 2}$$

$$\quad\;\; 1 \;\le\; 2 \;\; \text{TRUE}$$

Since $1 \le 2$ is true, 4 is a solution of the inequality.

 c) We replace x with 3.

$$\frac{5-x \;\le\; 2}{5-3 \;\; ? \;\; 2}$$

$$\quad\;\; 2 \;\le\; 2 \;\; \text{TRUE}$$

Since $2 \le 2$ is true, 3 is a solution of the inequality.

49. a) We replace m with 6.

$$\frac{3m-8 \;<\; 13}{3\cdot 6-8 \;\; ? \;\; 13}$$

$$18-8 \;\; ? \;\; 13$$

$$\quad\;\; 10 \;<\; 13 \;\; \text{TRUE}$$

Since $10 < 13$ is true, 6 is a solution of the inequality.

 b) We replace m with 7.

$$\frac{3m-8 \;<\; 13}{3\cdot 7-8 \;\; ? \;\; 13}$$

$$21-8 \;\; ? \;\; 13$$

$$\quad\;\; 13 \;<\; 13 \;\; \text{FALSE}$$

Since $13 < 13$ is false, 7 is not a solution of the inequality.

 c) We replace m with 9.

$$\frac{3m-8 \;<\; 13}{3\cdot 9-8 \;\; ? \;\; 13}$$

$$27-8 \;\; ? \;\; 13$$

$$\quad\;\; 19 \;<\; 13 \;\; \text{FALSE}$$

Since $19 < 13$ is false, 9 is not a solution of the inequality.

51. $\{a, l, g, e, b, r\}$ or $\{a, b, e, g, l, r\}$

53. $\{1, 3, 5, 7, ...\}$

55. $\{5, 10, 15, 20, ...\}$

57. $\{x \mid x \text{ is an odd number between 10 and 20}\}$

59. $\{x \mid x \text{ is a whole number less than 5}\}$

61. $\{x \mid x \text{ is an odd number between 10 and 20}\}$

63.
$$\overset{\sqrt{5}}{\underset{-4 \; -2 \;\; 0 \;\; 2 \;\; 4}{\longleftrightarrow}}$$

65.
$$\overset{-\sqrt{22}}{\underset{-4 \; -2 \;\; 0 \;\; 2 \;\; 4}{\longleftrightarrow}}$$

67. a) $0, 6$
    b) $-3, 0, 6$
    c) $-8.7, -3, 0, \frac{2}{3}, 6$
    d) $\sqrt{7}$
    e) $-8.7, -3, 0, \frac{2}{3}, \sqrt{7}, 6$

69. a) $0, \frac{30}{5}$
    b) $-17, 0, \frac{30}{5}$
    c) $-17, -4.\overline{13}, 0, \frac{5}{4}, \frac{30}{5}$
    d) $\sqrt{77}$
    e) $-17, -4.\overline{13}, 0, \frac{5}{4}, \frac{30}{5}, \sqrt{77}$

71. Since 5.1 is not a natural number, the statement is false.

73. Since every member of the set of whole numbers is also a member of the set of integers, the statement is true.

75. Since $\frac{2}{3}$ is not an irrational number, the statement is false.

77. Since $\sqrt{10}$ is a real number, the statement is true.

79. Since some members of the set of integers are not natural numbers, the statement is true.

81. Since some members of the set of rational numbers are not integers, the statement is false.

83. *Thinking and Writing Exercise.*

85. *Thinking and Writing Exercise.*

87. The only whole number that is not also a natural number is 0. Using roster notation to name the set, we have $\{0\}$.

89. $\{5, 10, 15, 20, \ldots\}$

91. $\{1, 3, 5, 7, \ldots\}$

93. Recall from geometry that when a right triangle has legs of length 2 and 3, the length of the hypotenuse is $\sqrt{2^2 + 3^2} = \sqrt{4 + 9} = \sqrt{13}$. We draw such a triangle:

Exercise Set 1.2

1. True

3. True

5. False

7. False

9. True

11. $|-9| = 9$ -9 is 9 units from 0.

13. $|6| = 6$ 6 is 6 units from 0.

15. $|-6.2| = 6.2$ -6.2 is 6.2 units from 0.

17. $|0| = 0$ 0 is 0 units from 0.

19. $\left|1\frac{7}{8}\right| = 1\frac{7}{8}$ $1\frac{7}{8}$ is $1\frac{7}{8}$ units from 0

21. $|-4.21| = 4.21$ -4.21 is 4.21 units from 0.

23. $-6 \le -2$ is true because -6 is left of -2.

25. $-9 > 1$ is false since -9 is left of 1.

27. $3 \geq -5$ is true since -5 is left of 3.

29. $-8 < -3$ is true since -8 is left of -3.

31. $-4 \geq -4$ is true since $-4 = -4$.

33. $-5 < -5$ is false since -5 is not left of -5.

35. $3 + 9 = 12$

    Two positive numbers: Add the numbers, getting 12. The answer is positive.

37. $(-3) + (-9) = -12$

    Two negative numbers: Add the absolute values, 3 and 9, getting 12. The answer is negative, -12.

39. $-3.9 + 2.7 = -1.2$

    A negative and a positive number: The absolute values are 3.9 and 2.7. Subtract 2.7 from 3.9 getting 1.2. Since the negative number is farther from 0, the result is negative, -1.2.

41. $\frac{2}{7} + \left(-\frac{3}{5}\right) = -\frac{11}{35}$

    A positive and a negative number: The absolute values are $\frac{10}{35}$ and $\frac{21}{35}$. Subtract $\frac{10}{35}$ from $\frac{21}{35}$ getting $\frac{11}{35}$. The negative number is farther from 0, so the answer is negative, $-\frac{11}{35}$.

43. $-3.26 + (-5.8) = -9.06$

    Both numbers are negative. Add the absolute values getting 9.06. The answer is negative, -9.06.

45. $-\frac{1}{9} + \frac{2}{3} = -\frac{1}{9} + \frac{6}{9} = \frac{5}{9}$

    A negative and a positive number: The absolute values are $\frac{1}{9}$ and $\frac{6}{9}$. Subtract $\frac{1}{9}$ from $\frac{6}{9}$ getting $\frac{5}{9}$. The positive number is farther from 0, so the answer is positive, $\frac{5}{9}$.

47. $0 + (-4.5) = -4.5$

    Any number added to zero is itself, so the answer is -4.5.

49. $-7.24 + 7.24 = 0$

    A negative and a positive number: The numbers have the same absolute value, 7.24, so the answer is 0.

51. $15.9 + (-22.3) = -6.4$

    A positive and a negative number: The absolute values are 15.9 and 22.3. Subtract 15.9 from 22.3 getting 6.4. Since the negative number is farther from 0, the answer is negative, -6.4.

53. The opposite of 3.14 is -3.14 because $3.14 + (-3.14) = 0$.

55. The opposite of -138 is 138 because $-138 + 138 = 0$.

57. The opposite of 0 is 0 because $0 + 0 = 0$.

59. If $x = 9$, then $-x = -9$.

61. If $x = -\frac{1}{10}$, then $-x = -\left(-\frac{1}{10}\right) = \frac{1}{10}$.

63. If $x = -4.67$, then $-x = -(-4.67) = 4.67$.

65. If $x = 0$, then $-x = -0 = 0$.

67. $8 - 5 = 8 + (-5) = 3$

69. $5 - 8 = 5 + (-8) = -3$

71. $-5 - (-12) = -5 + 12 = 7$

73. $-5 - 14 = -5 + (-14) = -19$

75. $2.7 - 5.8 = 2.7 + (-5.8) = -3.1$

77. $-\frac{3}{5} - \frac{1}{2} = -\frac{6}{10} + \left(-\frac{5}{10}\right) = -\frac{11}{10}$

79. $-3.9 - (-3.9) = -3.9 + (-3.9) = 0$

81. $0 - (-5.37) = 0 + 5.37 = 5.37$

83. $(-5)6 = -30$

Two numbers with unlike signs: Multiply their absolute values, getting 30. The answer is -30.

85. $(-4)(-9) = 36$

Two number with the same signs: Multiply their absolute values, getting 36. The answer is positive, 36.

87. $4.2(-5) = -21$

Two numbers with unlike signs: Multiply their absolute values, getting 21. The answer is negative, -21.

89. $\frac{3}{7}(-1) = -\frac{3}{7}$

Two numbers with unlike signs: Multiply their absolute values, getting $\frac{3}{7}$. The answer is negative, $-\frac{3}{7}$.

91. $(-17.45) \cdot 0 = 0$

The product of any number and 0 is 0.

93. $-\frac{2}{3}\left(\frac{3}{4}\right) = -\frac{1}{2}$

Two numbers with unlike signs: Multiply their absolute values, getting $\frac{1}{2}$. The answer is negative, $-\frac{1}{2}$.

95. $\frac{-10}{-2} = 5$

Two numbers with the same signs: Divide their absolute values. The answer is positive, 5.

97. $\frac{-100}{20} = -5$

Two numbers with unlike signs: Divide their absolute values. The answer is negative, -5.

99. $\frac{73}{-1} = -73$

Two numbers with unlike signs: Divide their absolute values. The answer is negative, -73.

101. $\frac{0}{-7} = 0$

Two numbers with unlike signs: Divide their absolute values. The answer is -0, or 0. Zero divided by any quantity equal 0.

103. The reciprocal of 4 is $\frac{1}{4}$, because $4 \cdot \frac{1}{4} = 1$.

105. The reciprocal of $-\frac{5}{7}$ is $-\frac{7}{5}$, because $-\frac{5}{7} \cdot \left(-\frac{7}{5}\right) = 1$.

107. The reciprocal of $\frac{1}{8}$ is 8, because $\frac{1}{8} \cdot 8 = 1$.

109. $\frac{2}{3} \div \frac{4}{5} = \frac{2}{3} \cdot \frac{5}{4} = \frac{10}{12}$, or $\frac{5}{6}$.

111. $-\frac{3}{5} \div \frac{1}{2} = -\frac{3}{5} \cdot \frac{2}{1} = -\frac{6}{5}$

113. $\left(-\frac{2}{9}\right) \div (-8) = \left(-\frac{2}{9}\right)\left(-\frac{1}{8}\right) = \frac{2}{72}$, or $\frac{1}{36}$

115. $-\frac{12}{7} \div \left(-\frac{12}{7}\right) = -\frac{12}{7} \cdot \left(-\frac{7}{12}\right) = \frac{84}{84}$, or 1

117. $-10^2 = -10 \cdot 10 = -100$

119. $-(-3)^2 = -(-3)(-3) = -9$

121. $(2-5)^2 = (-3)^2 = (-3)(-3) = 9$

123. $9 - \left(8 - 3 \cdot 2^3\right) = 9 - (8 - 3 \cdot 8)$
$$= 9 - (8 - 24)$$
$$= 9 - (-16)$$
$$= 9 + 16$$
$$= 25$$

125. $\dfrac{5 \cdot 2 - 4^2}{27 - 2^4} = \dfrac{5 \cdot 2 - 16}{27 - 16}$
$$= \dfrac{10 - 16}{27 - 16}$$
$$= -\dfrac{6}{11}$$

127. $\dfrac{3^4 - (5 - 3)^4}{8 - 2^3} = \dfrac{3^4 - (2)^4}{8 - 8}$
$$= \dfrac{81 - 16}{0}$$
$$= \dfrac{65}{0}$$

Since division by 0 is undefined, the expression is undefined.

129. $\dfrac{(2-3)^3 - 5|2-4|}{7-2\cdot 5^2} = \dfrac{(-1)^3 - 5|-2|}{7-2\cdot 25}$

$= \dfrac{-1-5(2)}{7-50}$

$= \dfrac{-1-10}{-43}$

$= \dfrac{-11}{-43}$

$= \dfrac{11}{43}$

131. $\left|2^2-7\right|^3 + 4 = \left|4-7\right|^3 + 4$

$= \left|-3\right|^3 + 4$

$= (3)^3 + 4$

$= 27 + 4$

$= 31$

133. $32 - (-5)^2 + 15 \div (-3)\cdot 2$

$= 32 - 25 + 15 \div (-3)\cdot 2$

$= 32 - 25 - 5\cdot 2$

$= 32 - 25 - 10$

$= 7 - 10$

$= -3$

135. (a)

137. (d)

139.
```
((6-11)²-abs(7-4
²))/(4-(-2)^3)
          1.333333333
Ans▶Frac
               4/3
```

141.
```
(2(3-4²)²-2²)/((
-1)^4/abs(3*5-5²
))
               3340
■
```

143.
```
17-√(11-(3+4))/(
-5-(-6))²
               15
```

145. *Thinking and Writing Exercise.*

147. Substitute and carry out the indicated operations.

$2(x+5) = 2(3+5) = 2\cdot 8 = 16$

$2x+10 = 2\cdot 3 + 10 = 6+10 = 16$

149. *Thinking and Writing Exercise.*

151. $(8-5)^3 + 9 = 36$

153. $5\cdot 2^3 \div (3-4)^4 = 40$

155. Any value of a such that $a \le -6.2$ satisfies the given condition. The largest of these values is –6.2.

Exercise Set 1.3

---

1. Commutative law

3. Commutative law

5. Distributive law

7. Using the commutative law of addition:
    $3 + 4a = 4a + 3$
    Using the commutative law of multiplication:
    $3 + 4a = 3 + a\cdot 4$
    Other answers are possible.

9. Using the commutative law of multiplication:
    $(7x)y = y(7x)$, or $(7x)y = (x\cdot 7)y$
    Other answers are possible.

11. Using the associative law of multiplication:
    $(3x)y = 3(xy)$

13. Using the associative law of addition:
    $x + (2y+5) = (x+2y) + 5$

15. Using the commutative law of addition:
$2+(t+6)=(t+6)+2$
Using the associative law of addition:
$2+(t+6)=(2+t)+6$
Other answers are possible.

17. Using the commutative law of multiplication:
$(3a)\cdot7=7\cdot(3a)$
Using the associative law of multiplication:
$(3a)\cdot7=3(a\cdot7)$
Other answers are possible.

19. $(5+x)+2=(x+5)+2$ commutative law
of addition
$\quad\quad\quad\quad=x+(5+2)$ associative law
of adddition
$\quad\quad\quad\quad=x+7$ simplifying

21. $(m\cdot3)7=m(3\cdot7)$ associative law
of multiplication
$\quad\quad\quad=m(21)$ simplifying
$\quad\quad\quad=21m$ commutative law
of multiplication

23. Using the distributive law, we have:
$7(t+2)=7t+14$

25. Using the distributive law, we have:
$4(x-y)=4x-4y$

27. Using the distributive law, we have:
$-5(2a+3b)=-10a-15b$

29. Using the distributive law, we have:
$9a(b-c+d)=9ab-9ac+9ad$

31. $5x+50=5(x+10)$

33. $9p-3=3\cdot3p-3\cdot1=3(3p-1)$

35. $7x-21y+14z=7(x-3y+2z)$

37. $255-34b=17(15-2b)$

39. $xy+x=x(y+1)$

41. $4x-5y+3=4x+(-5y)+3$
The terms are: $4x$, $-5y$, and 3.

43. $x^2-6x-7=x^2+(-6x)+(-7)$
The terms are: x^2, $-6x$, and -7.

45. $3x+7x=(3+7)x=10x$

47. $9t^2+t^2=(9+1)t^2=10t^2$

49. $12a-a=(12-1)a=11a$

51. $n-8n=(1-8)n=-7n$

53. $5x-3x+8x=(5-3+8)x=10x$

55. $4x-2x^2+3x$
$=4x+3x-2x^2$
$=(4+3)x-2x^2$
$=7x-2x^2$

57. $6a+7a^2-a+4a^2$
$=6a-a+7a^2+4a^2$
$=(6-1)a+(7+4)a^2$
$=5a+11a^2$

59. $4x-7+18x+25$
$=4x+18x-7+25$
$=(4+18)x+(-7+25)$
$=22x+18$

61. $-7t^2+3t+5t^3-t^3+2t^2-t$
$=-7t^2+2t^2+3t-t+5t^3-t^3$
$=(-7+2)t^2+(3-1)t+(5-1)t^3$
$=-5t^2+2t+4t^3$

63. $2x+3(5x-7)=2x+15x-21$
$\quad\quad\quad\quad\quad\quad=(2+15)x-21$
$\quad\quad\quad\quad\quad\quad=17x-21$

65. $7a-(2a+5y)=7a-2a-5y$
$\quad\quad\quad\quad\quad\quad=(7-2)a-5y$
$\quad\quad\quad\quad\quad\quad=5a-5y$

67. $m - (m - 1) = m - m + 1$
$\qquad = (1 - 1)m + 1$
$\qquad = 0 \cdot m + 1$
$\qquad = 1$

69. $3d - 7c - (5c - 2d) = 3d - 7c - 5c + 2d$
$\qquad = (-7 - 5)c + (3 + 2)d$
$\qquad = (-12)c + 5d$
$\qquad = -12c + 5d$

71. $2(x - 3) + 4(7 - x) = 2x - 6 + 28 - 4x$
$\qquad = (2 - 4)x + (-6 + 28)$
$\qquad = -2x + 22$

73. $3p - 4 - 2(p + 6) = 3p - 4 - 2p - 12$
$\qquad = (3 - 2)p + (-4 - 12)$
$\qquad = 1p + (-16)$
$\qquad = p - 16$

75. $x + 3[2x + 4(1 - x)] = x + 3[2x + 4 - 4x]$
$\qquad = x + 3[(2 - 4)x + 4]$
$\qquad = x + 3[-2x + 4]$
$\qquad = x - 6x + 12$
$\qquad = (1 - 6)x + 12$
$\qquad = -5x + 12$

77. $-2(a - 5) - [7 - 3(2a - 5)]$
$= -2a + 10 - [7 - 6a + 15]$
$= -2a + 10 - [-6a + 22]$
$= -2a + 10 + 6a - 22$
$= (-2 + 6)a + (10 - 22)$
$= 4a - 12$

79. $5\{-2a + 3[4 - 2(3a + 5)]\}$
$= 5\{-2a + 3[4 - 6a - 10]\}$
$= 5\{-2a + 3[-6a - 6]\}$
$= 5\{-2a - 18a - 18\}$
$= 5\{(-2 - 18)a - 18\}$
$= 5\{-20a - 18\}$
$= -100a - 90$

81. $2y + \{7[3(2y - 5) - (8y + 7)] + 9\}$
$= 2y + \{7[6y - 15 - 8y - 7] + 9\}$
$= 2y + \{7[6y - 8y - 15 - 7] + 9\}$
$= 2y + \{7[-2y - 22] + 9\}$
$= 2y + \{-14y - 154 + 9\}$
$= 2y + \{-14y - 145\}$
$= 2y - 14y - 145$
$= -12y - 145$

83.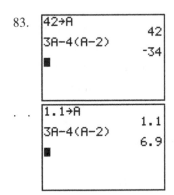

85.

87. Substitute each value in the first expression.
$$2x-3(x+5)=2(1.3)-3(1.3+5)$$
$$=2(1.3)-3(6.3)$$
$$=2.6-18.9$$
$$=-16.3$$
$$2x-3(x+5)=2(-3)-3(-3+5)$$
$$=2(-3)-3(2)$$
$$=-6-6$$
$$=-12$$
$$2x-3(x+5)=2\cdot0-3(0+5)$$
$$=2\cdot0-3(5)$$
$$=0-15$$
$$=-15$$
Now substitute each value in the second expression.
$$-x+15=-1.3+15=13.7$$
$$-x+15=-(-3)+15=3+15=18$$
$$-x+15=-0+15=15$$
Since the values of the expressions differ when one or more of the replacements are the same, the expressions are not equivalent.

89. Substitute each value in the first expression.
$$4(x+3)=4(1.3+3)=4(4.3)=17.2$$
$$4(x+3)=4(-3+3)=4\cdot0=0$$
$$4(x+3)=4(0+3)=4\cdot3=12$$
Now substitute each value in the second expression.
$$4x+12=4(1.3)+12=5.2+12=17.2$$
$$4x+12=4(-3)+12=-12+12=0$$
$$4x+12=4\cdot0+12=0+12=12$$
Since the values of the expressions are the same for three replacements, we can be reasonably certain that the expressions are equivalent.

91. *Thinking and Writing Exercise.*

93. $8-(-3)=8+3=11$

95. The opposite of -35 is 35, because $-35+35=0$.

97. *Thinking and Writing Exercise.*

99. $11(a-3)+12a-\{6[4(3b-7)-(9b+10)]+11\}$
$$=11(a-3)+12a-\{6[12b-28-9b-10]+11\}$$
$$=11(a-3)+12a-\{6[3b-38]+11\}$$
$$=11(a-3)+12a-\{18b-228+11\}$$
$$=11(a-3)+12a-18b+228-11$$
$$=11a-33+12a-18b+228-11$$
$$=23a-18b+184$$

101. $z-\{2z+[3z-(4z+5z)-6z]+7z\}-8z$
$$=z-\{2z+[3z-9z-6z]+7z\}-8z$$
$$=z-\{2z-12z+7z\}-8z$$
$$=z-\{-3z\}-8z$$
$$=z+3z-8z$$
$$=-4z$$

103. $x-\{x+1-[x+2-(x-3-\{x+4-$
$$[x-5+(x-6)]\})]\}$$
$$=x-\{x+1-[x+2-9x-3-\{x+4-$$
$$[x-5+x-6]\})]\}$$
$$=x-\{x+1-\{x+2-9x-3-\{x+4-$$
$$[2x-11]\})]\}$$
$$=x-\{x+1-[x+2-9x-3-\{x+4-$$
$$2x+11\})]\}$$
$$=x-\{x+1-[x+2-(x-3-\{-x+15\})]\}$$
$$=x-\{x+1-[x+2-(x-3+x-15)]\}$$
$$=x-\{x+1-[x+2-(2x-18)]\}$$
$$=x-\{x+1-[x+2-2x+18]\}$$
$$=x-\{x+1-[-x+20]\}$$
$$=x-\{x+1+x-20\}$$
$$=x-\{2x-19\}$$
$$=x-2x+19\}$$
$$=-x+19$$

105. *Thinking and Writing Exercise.*

Exercise Set 1.4

1. The Power Rule

3. Raising a Product to a Power

5. The Product Rule

7. Raising a Quotient to a Power

9. The Quotient Rule

11. Positive power of 10

13. Negative power of 10

15. Positive power of 10

17. $5^6 \cdot 5^4 = 5^{6+4} = 5^{10}$

19. $m^9 \cdot m^0 = m^{9+0} = m^9$

21. $6x^5 \cdot 3x^2 = 6 \cdot 3 \cdot x^5 \cdot x^2 = 18x^{5+2} = 18x^7$

23. $\left(-2m^4\right)\left(-8m^9\right) =$
     $(-2)(-8)m^4 \cdot m^9 = 16m^{4+9} = 16m^{13}$

25. $\left(x^3 y^4\right)\left(x^7 y^6 z^0\right) =$
     $\left(x^3 x^7\right)\left(y^4 y^6\right)\left(z^0\right) = x^{3+7} y^{4+6} \cdot 1 = x^{10} y^{10}$

27. $\dfrac{a^9}{a^3} = a^{9-3} = a^6$

29. $\dfrac{12t^7}{4t^2} = \dfrac{12}{4} \cdot t^{7-2} = 3t^5$

31. $\dfrac{m^7 n^9}{m^2 n^5} = m^{7-2} n^{9-5} = m^5 n^4$

33. $\dfrac{32x^8 y^5}{8x^2 y} = \dfrac{32}{8} \cdot x^{8-2} y^{5-1} = 4x^6 y^4$

35. $\dfrac{28x^{10} y^9 z^8}{-7x^2 y^3 z^2} = \dfrac{28}{-7} \cdot x^{10-2} y^{9-3} z^{8-2} = -4x^8 y^6 z^6$

37. $-x^0 = -(-2)^0 = -1(1) = -1$

39. $(4x)^0 = \left(4(-2)\right)^0 = (-8)^0 = 1$

41. $(-2)^4 = (-2)(-2)(-2)(-2) = 16$

43. $-2^4 = -2 \cdot 2 \cdot 2 \cdot 2 = -16$

45. $(-4)^{-2} = \dfrac{1}{(-4)^2} = \dfrac{1}{(-4)(-4)} = \dfrac{1}{16}$

47. $-4^{-2} = -\dfrac{1}{4^2} = -\dfrac{1}{16}$

49. $a^{-3} = \dfrac{1}{a^3}$

51. $\dfrac{1}{5^{-3}} = 5^3 = 125$

53. $-6x^{-1} = -6 \cdot \dfrac{1}{x} = -\dfrac{6}{x}$

55. $3a^8 b^{-6} = 3a^8 \cdot \dfrac{1}{b^6} = \dfrac{3a^8}{b^6}$

57. $\dfrac{z^{-4}}{3x^5} = \dfrac{1}{3x^5} \cdot \dfrac{1}{z^4} = \dfrac{1}{3x^5 z^4}$

59. $\dfrac{ab^{-1}}{c^{-1}} = \dfrac{ac}{b}$

61. $\dfrac{4a^{-3} bc^{-1}}{d^{-6} f^2} = \dfrac{4bd^6}{a^3 cf^2}$

63. $\dfrac{1}{x^3} = x^{-3}$

65. $6^8 = \dfrac{1}{6^{-8}}$

67. $4x^2 = 4 \cdot \dfrac{1}{x^{-2}} = \dfrac{4}{x^{-2}}$

69. $\dfrac{1}{(5y)^3} = (5y)^{-3}$

71. $8^{-2} \cdot 8^{-4} = 8^{-2+(-4)} = 8^{-6}$, or $\dfrac{1}{8^6}$

73. $b \cdot b^{-5} = b^{1+(-5)} = b^{-4}$, or $\dfrac{1}{b^4}$

75. $a^{-5} \cdot a^4 \cdot a^2 = a^{-5+4+2} = a^1 = a$

77. $\left(-7x^4 y^{-5}\right)\left(-5x^{-6} y^8\right) = (-7)(-5)x^{4+(-6)} y^{-5+8}$
$$= 35x^{-2} y^3$$
or, $\dfrac{35y^3}{x^2}$

79. $\left(5a^{-2} b^{-3}\right)\left(2a^{-4} b\right) = 5 \cdot 2 a^{-2+(-4)} b^{-3+1}$
$$= 10a^{-6} b^{-2}$$
or, $\dfrac{10}{a^6 b^2}$

81. $\dfrac{10^{-3}}{10^6} = 10^{-3-6} = 10^{-9}$, or $\dfrac{1}{10^9}$

83. $\dfrac{2^{-7}}{2^{-5}} = 2^{-7-(-5)} = 2^{-7+5} = 2^{-2}$, or $\dfrac{1}{2^2}$, or $\dfrac{1}{4}$

85. $\dfrac{y^4}{y^{-5}} = y^{4-(-5)} = y^{4+5} = y^9$

87. $\dfrac{24a^5 b^3}{-8a^4 b} = \dfrac{24}{-8} a^{5-4} b^{3-1} = -3a^1 b^2 = -3ab^2$

89. $\dfrac{15m^5 n^3}{10m^{10} n^{-4}} = \dfrac{15}{10} m^{5-10} n^{3-(-4)} = \dfrac{3}{2} m^{-5} n^7$,
or $\dfrac{3n^7}{2m^5}$

91. $\dfrac{-6x^{-2} y^4 z^8}{-24x^{-5} y^6 z^{-3}} = \dfrac{-6}{-24} x^{-2-(-5)} y^{4-6} z^{8-(-3)}$
$$= \dfrac{1}{4} x^3 y^{-2} z^{11}, \text{ or } \dfrac{x^3 z^{11}}{4y^2}$$

93. $\left(x^4\right)^3 = x^{4 \cdot 3} = x^{12}$

95. $\left(9^3\right)^{-4} = 9^{3 \cdot (-4)} = 9^{-12}$, or $\dfrac{1}{9^{12}}$

97. $\left(t^{-8}\right)^{-5} = t^{-8 \cdot (-5)} = t^{40}$

99. $(-5xy)^2 = (-5)^2 x^2 y^2 = 25x^2 y^2$

101. $\left(-2a^{-2} b\right)^{-3} = (-2)^{-3} (a^{-2})^{-3} b^{-3}$
$$= \dfrac{1}{(-2)^3} a^{(-2)(-3)} b^{-3}$$
$$= \dfrac{1}{-8} a^6 b^{-3} = -\dfrac{1}{8} a^6 b^{-3}, \text{ or } -\dfrac{a^6}{8b^3}$$

103. $\left(\dfrac{m^2 n^{-1}}{4}\right)^3 = \dfrac{m^{2 \cdot 3} n^{-1 \cdot 3}}{4^3} = \dfrac{m^6 n^{-3}}{64}, \text{ or } \dfrac{m^6}{64n^3}$

105. $\dfrac{\left(2a^3\right)^3 4a^{-3}}{\left(a^2\right)^5} = \dfrac{2^3 \cdot a^9 \cdot 4 \cdot a^{-3}}{a^{10}} = \dfrac{2^3 \cdot 4 \cdot a^9 a^{-3}}{a^{10}}$
$$= \dfrac{8 \cdot 4 \cdot a^{9+(-3)}}{a^{10}} = \dfrac{32a^6}{a^{10}}$$
$$= 32a^{6-10} = 32a^{-4}, \text{ or } \dfrac{32}{a^4}$$

107. $\left(8x^{-3} y^2\right)^{-4} \left(8x^{-3} y^2\right)^4 = \left(8x^{-3} y^2\right)^{-4+4}$
$$= \left(8x^{-3} y^2\right)^0 = 1$$

109. $\dfrac{\left(3x^3 y^4\right)^3}{6xy^3} = \dfrac{3^3 x^9 y^{12}}{6xy^3} = \dfrac{27x^9 y^{12}}{6xy^3}$
$$= \dfrac{9}{2} x^{9-1} y^{12-3} = \dfrac{9x^8 y^9}{2}$$

111. $\left(\dfrac{-4x^4 y^{-2}}{5x^{-1} y^4}\right)^{-4} = \left(\dfrac{-4x^{4-(-1)} y^{-2-4}}{5}\right)^{-4}$
$$= \left(\dfrac{-4x^5 y^{-6}}{5}\right)^{-4} = \left(\dfrac{5}{-4x^5 y^{-6}}\right)^4$$
$$= \left(\dfrac{5y^6}{-4x^5}\right)^4 = \dfrac{5^4 y^{24}}{(-4)^4 x^{20}}$$
$$= \dfrac{625y^{24}}{256x^{20}}$$

113. $\left(\dfrac{4a^3 b^{-9}}{6a^{-2} b^5}\right)^0 = 1$

115. $\left(\dfrac{21x^5 y^{-7}}{14x^{-2}y^{-6}}\right)^0 = 1$ because any nonzero

number raised to the power 0 is 1.

117. $-8^4 = -4096$

119. $(-2)^{-4} = 0.0625$

121. $3^4 5^{-3} = 0.648$

123. $4 \times 10^{-4} = \dfrac{4}{10^4} = \dfrac{4}{10,000} = 0.0004$

125. $6.73 \times 10^8 = 673,000,000$

Moving the decimal point 8 places to the right.

127. $8.923 \times 10^{-10} = 0.0000000008923$

Moving the decimal point 10 places to the left.

129. $9.03 \times 10^{10} = 90,300,000,000$

Moving the decimal point 10 places to the right.

131. $47,000,000,000$

$= \dfrac{47,000,000,000}{10^{10}} \times 10^{10} = 4.7 \times 10^{10}$

133. 0.000000016

$= \dfrac{0.000000016}{10^8} \times 10^8 = \dfrac{1.6}{10^8} = 1.6 \times 10^{-8}$

135. $407,000,000,000$

$= \dfrac{407,000,000,000}{10^{11}} \times 10^{11} = 4.07 \times 10^{11}$

137. 0.000000603

$= \dfrac{0.000000603}{10^7} \times 10^7 = 6.03 \times 10^{-7}$

139. 5.02×10^{18}

141. -3.05×10^{-10}

143. $(2.3 \times 10^6)(4.2 \times 10^{-11})$

$= (2.3 \times 4.2)(10^6 \times 10^{-11})$

$= 9.7 \times 10^{-5}$

145. $(2.34 \times 10^{-8})(5.7 \times 10^{-4})$

$= 13.34 \times 10^{-12}$

$= 1.3 \times 10^{-11}$

147. $(2.0 \times 10^6)(3.02 \times 10^{-6})$

$= 6.04 \times 10^0$

$= 6.0$

149. $\dfrac{5.1 \times 10^6}{3.4 \times 10^3} = \dfrac{5.1}{3.4} \times \dfrac{10^6}{10^3} = 1.5 \times 10^3$

151. $\dfrac{7.5 \times 10^{-9}}{2.5 \times 10^{-4}} \times \dfrac{7.5}{2.5} \times \dfrac{10^{-9}}{10^{-4}} = 3 \times 10^{-5}$

153. $\dfrac{1.23 \times 10^8}{6.87 \times 10^{-13}} = \dfrac{1.23}{6.87} \times \dfrac{10^8}{10^{-13}}$

$= 0.1790393 \times 10^{21}$

$= 1.79 \times 10^{20}$

155. $5.9 \times 10^{23} + 6.3 \times 10^{23}$

$= (5.9 + 6.3) \times 10^{23}$

$= 12.2 \times 10^{23} = 1.22 \times 10^{24}$

157. *Thinking and Writing Exercise.*

159. $\dfrac{1}{2}x - 7 = \dfrac{1}{2}(10) - 7 = 5 - 7 = -2$

161. $x^2 - 1 = (-1)^2 - 1 = 1 - 1 = 0$

163. *Thinking and Writing Exercise.*

165. $\dfrac{12a^{x-2}}{3a^{2x+2}} = \dfrac{12}{3}a^{x-2-(2x+2)} = 4a^{x-2-2x-2}$

$= 4a^{-x-4} = 4a^{-(x+4)} = \dfrac{4}{a^{x+4}}$

167. $\left(3^{a+2}\right)^a = 3^{(a+2)a} = 3^{a^2+2a}$

169. $\dfrac{4x^{2a+3}y^{2b-1}}{2x^{a+1}y^{b+1}} = \dfrac{4}{2} \cdot x^{(2a+3)-(a+1)}y^{(2b-1)-(b+1)}$

$= 2x^{2a+3-a-1}y^{2b-1-b-1} = 2x^{a+2}y^{b-2}$

171. The larger number is the one in which the power of ten has the larger exponent. Since -90 is large than -91, $8 \cdot 10^{-90}$ is larger than $9 \cdot 10^{-91}$. We find the difference:

$8 \cdot 10^{-90} - 9 \cdot 10^{-91} = 10^{-90}\left(8 - 9 \cdot 10^{-1}\right)$

$= 10^{-90}(8 - 0.9)$

$= 7.1 \times 10^{-90}$

173. $(4096)^{0.05}(4096)^{0.2} = 4096^{0.25}$

$= \left(2^{12}\right)^{0.25}$

$= 2^3$

$= 8$

175. 0.125, $\frac{1}{8}$, 1.25×10^{-1}

Mid-Chapter Review

Guided Solutions:

1. $-3 - (2 \cdot 3^2 - 100 \div 4) = -3 - (2 \cdot 9 - 100 \div 4)$

$= -3 - (18 - 25)$

$= -3 - (-7)$

$= 4$

2. $\dfrac{-8xy^4z^{-3}}{2x^2yz^{-10}} = -4x^{-1}y^3z^7 = -\dfrac{4y^3z^7}{x}$

Mixed Review:

1. $10x \div 2y = 10(4) \div 2(5)$

$= 40 \div 2(5)$

$= 20(5) = 100$

2. Replace x with 4.

$\begin{array}{ccc} 12 - x & \le & 8 \\ \hline 12 - 4 & \le & 8 \quad ? \\ 8 & \le & 8 \quad \text{TRUE} \end{array}$

Since $8 \le 8$ is true, 4 is a solution to the inequality.

3. $|123| = 123$

4. $-x = -(-30) = 30$

5. The reciprocal of $-\frac{1}{3}$ is $-\frac{3}{1} = -3$ since $-\frac{3}{1} \cdot (-3) = 1$.

6. Using the associative law of addition:

$x + (2x + 4) = (x + 2x) + 4$

$= (1 + 2)x + 4$

$= 3x + 4$

7. $6x - 9y + 15 = 3 \cdot 2x - 3 \cdot 3y + 3 \cdot 5$

$= 3(2x - 3y + 5)$

8. $-15 + .01 = -(15 - 0.1) = -14.9$

9. $-14 - 17 = (-14) + (-17) = -31$

10. $(-3)(-13) = 39$

11. $\frac{2}{3} \div \left(-\frac{5}{6}\right) = \frac{2}{3} \cdot \left(-\frac{6}{5}\right) = -\frac{2 \cdot 6}{3 \cdot 5} = -\frac{4}{5}$

12. $\dfrac{3 - 2(1 - 5)}{2^3 - 3^2} = \dfrac{3 - 2(-4)}{8 - 9} = \dfrac{3 - (-8)}{-1} = \dfrac{11}{-1} = -11$

13. $\left(-3x^4\right)\left(5x^{-3}\right) = (-3) \cdot 5x^{4+(-3)} = -15x$

14. $-10^4 = -10 \cdot 10 \cdot 10 \cdot 10 = -10{,}000$

15. $\dfrac{12x^8}{6x^3} = \dfrac{2x^{8-3}}{1} = 2x^5$

16. $\left(-3a^5b^{-3}\right)^{-2} = (-3)^{-2}a^{5(-2)}b^{(-3)(-2)}$

$= \dfrac{1}{(-3)^2}a^{-10}b^6$

$= \dfrac{1}{9} \cdot \dfrac{1}{a^{10}} \cdot b^6$

$= \dfrac{b^6}{9a^{10}}$

17. $\left(\dfrac{7x^0 y^{-1}}{2y^{-12}}\right)^{-1} = \left(\dfrac{7\cdot 1\cdot y^{-1-(-12)}}{2}\right)^{-1} = \left(\dfrac{7y^{11}}{2}\right)^{-1}$

$\quad\quad = \left(\dfrac{2}{7y^{11}}\right)^1 = \dfrac{2}{7y^{11}}$

18. $5x - x^2 - 6x - 4x^2 = (5-6)x + (-1-4)x^2$
$\quad\quad\quad = (-1)x + (-5)x^2$
$\quad\quad\quad = -5x^2 - x$

19. $3m - (2m - 16) = 3m - 2m + 16$
$\quad\quad\quad = (3-2)m + 16$
$\quad\quad\quad = m + 16$

20. $-(a-3) - 2[3 - (5a + 7)]$
$\quad = -a + 3 - 2[3 - 5a - 7]$
$\quad = -a + 3 - 2[-5a - 4]$
$\quad = -a + 3 + 10a + 8$
$\quad = 9a + 11$

Exercise Set 1.5

1. axes

3. third

5. solutions

7. A is 5 units right of the origin and 3 units up, so its coordinates are $(5,3)$.

B is 4 units left of the origin and 3 units up, so its coordinates are $(-4,3)$.

C is 0 units right or left of the origin and 2 units up, so its coordinates are $(0,2)$

D is 2 units left of the origin and 3 units down, so its coordinates are $(-2,-3)$.

E is 4 units right of the origin and 2 units down, so its coordinates are $(4,-2)$.

F is 5 units left of the origin and 0 units up or down, so its coordinates are $(-5,0)$.

9.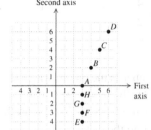

$A(3,0)$ is 3 units right and 0 units up or down.

$B(4,2)$ is 4 units right and 2 units up.

$C(5,4)$ is 5 units right and 4 units up.

$D(6,6)$ is 6 units right and 6 units up.

$E(3,-4)$ is 3 units right and 4 units down.

$F(3,-3)$ is 3 units right and 3 units down.

$G(3,-2)$ is 3 units right and 2 units down.

$H(3,-1)$ is 3 units right and 1 unit down.

11.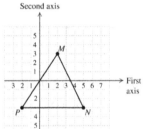

A triangle is formed. The area of a triangle is found by using the formula $A = \dfrac{1}{2}bh$. In this triangle, the base and height are 7 units and 6 units, respectively.

$A = \dfrac{1}{2}bh = \dfrac{1}{2}\cdot 7\cdot 6 = \dfrac{42}{2} = 21$ square units.

13.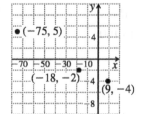

15.

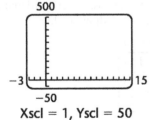

17. The *x*-coordinate is positive, and the
    y-coordinate is negative, so the point $(7, -2)$
    is in quadrant IV.

19. Both coordinates are negative, so the point
    $(-4, -3)$ is in quadrant III.

21. The *x*-coordinate is 0, so the point
    $(0, -3)$ is on the *y*-axis.

23. The *x*-coordinate is negative, and the
    y-coordinate is positive, so the point
    $(-4.9, 8.3)$ is in quadrant II.

25. The *y*-coordinate is 0, so the point
    $\left(-\frac{5}{2}, 0\right)$ is on the *x*-axis.

27. Both coordinates are positive, so the point
    $(160, 2)$ is in quadrant I.

29.

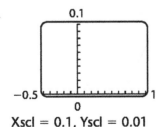

Xscl = 1, Yscl = 50

31.

0.1

−0.5 ⌊⌊⌊⌊⌊⌊⌊⌊⌊⌊⌊⌋ 1

0

Xscl = 0.1, Yscl = 0.01

33. Substituting 1 for *x* and −1 for *y* (alphabetical
    order of variables).

$$\begin{array}{rcl} y & = & 3x - 4 \\ \hline -1 & ? & 3 \cdot 1 - 4 \\ -1 & ? & 3 - 4 \\ -1 & ? & -1 \end{array}$$

    Since $-1 = -1$ is true, $(1, -1)$ is a solution of
    $y = 3x - 4$.

35. Substituting 2 for *s* and 4 for *t* (alphabetical
    order of variables).

$$\begin{array}{rcl} 5s - t & = & 8 \\ \hline 5 \cdot 2 - 4 & ? & 8 \\ 10 - 4 & ? & 8 \\ 6 & ? & 8 \end{array}$$

    Since $6 = 8$, is false $(2, 4)$ is not a solution
    of $5s - t = 8$.

37. Substituting 0 for *a* and $\frac{3}{5}$ for *b* (alphabetical
    order of variables).

$$\begin{array}{rcl} 6a + 5b & = & 3 \\ \hline 6 \cdot 0 + 5 \cdot \frac{3}{5} & ? & 3 \\ 0 + 3 & ? & 3 \\ 3 & ? & 3 \end{array}$$

    Since $3 = 3$ is true, $\left(0, \frac{3}{5}\right)$ is a solution of
    $6a + 5b = 3$.

39. Substituting 2 for *r* and −1 for *s* (alphabetical
    order of variables).

$$\begin{array}{rcl} 4r - 2s & = & 10 \\ \hline 4 \cdot 2 - 2(-1) & ? & 10 \\ 8 + 2 & ? & 10 \\ 10 & ? & 10 \end{array}$$

    Since $10 = 10$ is true, $(2, -1)$ is a solution of
    $4r - 2s = 10$.

41. Substituting 6 for *r* and −2 for *s* (alphabetical
    order of variables).

$$\begin{array}{rcl} r - s & = & 4 \\ \hline 6 - (-2) & ? & 4 \\ 6 + 2 & ? & 4 \\ 8 & ? & 4 \end{array}$$

    Since $8 = 4$ is false, $(6, -2)$ is not a
    solution to $r - s = 4$.

43. Substituting 3 for x and -1 for y (alphabetical order of variables).

$$\begin{array}{ccc} y & = & 3x^2 \\ \hline -1 & ? & 3(3)^2 \\ -1 & ? & 3\cdot 9 \\ -1 & ? & 27 \end{array}$$

Since $-1 = 27$ is false, $(3,-1)$ is not a solution of $y = 3x^2$.

45. Substituting -2 for x and 9 for y (alphabetical order of variables).

$$\begin{array}{ccc} x^3 + y & = & 1 \\ \hline (-2)^3 + 9 & ? & 1 \\ -8 + 9 & ? & 1 \\ 1 & ? & 1 \end{array}$$

Since $1 = 1$ is true, $(-2, 9)$ is a solution to $x^3 + y = 1$.

47. $y = -x$

To find an ordered pair, we choose any number for x and then determine y by substitution. We choose several ordered pairs, plot them, and draw the line.

x	y	(x, y)
1	-1	$(1,-1)$
2	-2	$(2,-2)$
-1	1	$(-1,1)$

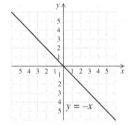

49. $y = x + 4$

To find an ordered pair, we choose any number for x and then determine y by substitution. We choose several ordered pairs, plot them, and draw the line.

x	y	(x, y)
0	4	$(0,4)$
-2	2	$(-2,2)$
-4	0	$(-4,0)$

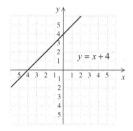

51. $y = 3x - 1$

To find an ordered pair, we choose any number for x and then determine y by substitution. We choose several ordered pairs, plot them, and draw the line.

x	y	(x, y)
0	-1	$(0,-1)$
2	5	$(2,5)$
-1	-4	$(-1,-4)$

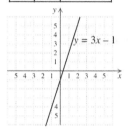

53. $y = -2x + 3$

To find an ordered pair, we choose any number for x and then determine y by substitution. We choose several ordered pairs, plot them, and draw the line.

x	y	(x, y)
-1	5	$(-1,5)$
1	1	$(1,1)$
3	-3	$(3,-3)$

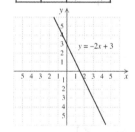

55. $y + 2x = 3$

To find an ordered pair, we choose any number for x and then determine y by substitution. We choose several ordered pairs, plot them, and draw the line.

x	y	(x, y)
0	3	$(0, 3)$
-1	5	$(-1, 5)$
4	-5	$(4, -5)$

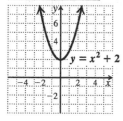

57. $y = x^2 + 2$ is a nonlinear equation, so make a table of several values, some positive and some negative.

x	y	(x, y)
-2	6	$(-2, 6)$
-1	3	$(-1, 3)$
0	2	$(0, 2)$
1	3	$(1, 3)$
2	6	$(2, 6)$

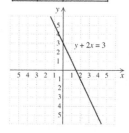

59. $y = -\dfrac{3}{2}x + 1$

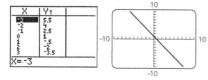

61. $y = \dfrac{3}{4}x + 1$

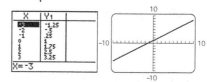

63. $y = -x^2$

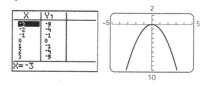

65. $y = |x| + 2$

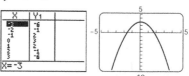

67. $y = 3 - x^2$

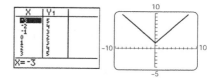

69. $y = x^3$

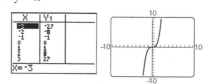

71. Only window (b) shows where the graph crosses the x-axis and y-axis.

73. Only window (a) shows where the graph crosses the x-axis and y-axis and the vertex.

75. Window (b) shows where the graph crosses the x-axis and y-axis best.

77. The equations in the odd-numbered exercises 47–55 , and 59, 61, and 71 have graphs that are straight lines, so they are linear equations.

79. *Thinking and Writing exercise.*

81. We replace x with 5.

$$3x - 5 = 10$$

$$\begin{array}{ccc} \hline 3 \cdot 5 - 5 & ? & 10 \\ 15 - 5 & ? & 10 \\ 10 & ? & 10 \end{array}$$

Since $10 = 10$, is true, 5 is a solution of the equation.

83. We replace x with 10.

$$6 - \tfrac{1}{2}x = 11$$

$$\begin{array}{ccc} \hline 6 - \tfrac{1}{2}(10) & ? & 11 \\ 6 - 5 & ? & 11 \\ 1 & ? & 11 \end{array}$$

Since $1 = 11$ is false, 10 is not a solution of the equation.

85.
$$\begin{aligned}
3x - 2(x - 7) + 5 &= 3(-1) - 2(-1 - 7) + 5 \\
&= -3 - 2(-8) + 5 \\
&= -3 + 16 + 5 \\
&= 18
\end{aligned}$$

87. *Thinking and Writing exercise.*

89. *Thinking and Writing exercise.*

91. Substitute $-\dfrac{1}{3}$ for x and $\dfrac{1}{4}$ for y in each equation.

a)
$$-\frac{3}{2}x - 3y = -\frac{1}{4}$$

$$\begin{array}{ccc} \hline -\dfrac{3}{2}\left(-\dfrac{1}{3}\right) - 3\left(\dfrac{1}{4}\right) & ? & -\dfrac{1}{4} \\[2mm] \dfrac{1}{2} - \dfrac{3}{4} & ? & -\dfrac{1}{4} \\[2mm] -\dfrac{1}{4} & ? & -\dfrac{1}{4} \end{array}$$

Since $-\dfrac{1}{4} = -\dfrac{1}{4}$ is true, $\left(-\dfrac{1}{3}, \dfrac{1}{4}\right)$ is a solution of the equation.

b)
$$8y - 15x = \frac{7}{2}$$

$$\begin{array}{ccc} \hline 8\left(\dfrac{1}{4}\right) - 15\left(-\dfrac{1}{3}\right) & ? & \dfrac{7}{2} \\[2mm] 2 + 5 & ? & \dfrac{7}{2} \\[2mm] 7 & ? & \dfrac{7}{2} \end{array}$$

Since $7 = \dfrac{7}{2}$ is false, $\left(-\dfrac{1}{3}, \dfrac{1}{4}\right)$ is not a solution of the equation.

c)
$$0.16y = -0.09x + 0.1$$

$$\begin{array}{ccc} \hline 0.16\left(\dfrac{1}{4}\right) & ? & -0.09\left(-\dfrac{1}{3}\right) + 0.1 \\[2mm] 0.04 & ? & 0.03 + 0.1 \\[2mm] 0.04 & ? & 0.13 \end{array}$$

Since $0.04 = 0.13$ is false, $\left(-\dfrac{1}{3}, \dfrac{1}{4}\right)$ is not a solution of the equation.

d)
$$2(-y + 2) - \frac{1}{4}(3x - 1) = 4$$

$$\begin{array}{ccc} \hline 2\left(-\dfrac{1}{4} + 2\right) - \dfrac{1}{4}\left[3\left(-\dfrac{1}{3}\right) - 1\right] & ? & 4 \\[2mm] 2\left(\dfrac{7}{4}\right) - \dfrac{1}{4}(-2) & ? & 4 \\[2mm] \dfrac{7}{2} + \dfrac{1}{2} & ? & 4 \\[2mm] \dfrac{8}{2} & ? & 4 \\[2mm] 4 & ? & 4 \end{array}$$

Since $4 = 4$ is true, $\left(-\dfrac{1}{3}, \dfrac{1}{4}\right)$ is a solution of the equation.

93. Plot $(-10, -2)$, $(-3, 4)$, and $(6, 4)$, and sketch a parallelogram.

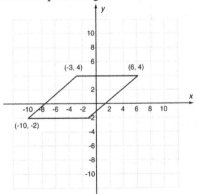

Since $(6, 4)$ is 9 units directly to the right of $(-3, 4)$, a fourth vertex could lie 9 units directly to the right of $(-10, -2)$. Then its coordinates are $(-10 + 9, -2)$, or $(-1, -2)$.

If we connect the points in a different order, we get a second parallelogram.

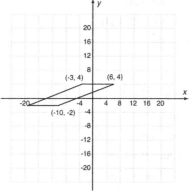

Since $(-3, 4)$ is 9 units directly to the left of $(6, 4)$, a fourth vertex could lie 9 units directly to the left of $(-10, -2)$. Then its coordinates are $(-10-9, -2)$, or $(-19, -2)$. If we connect the points in yet a different order, we get a third parallelogram.

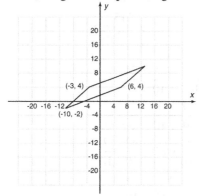

Since $(6, 4)$ lies 16 units directly to the right of and 6 units above $(-10, -2)$, a fourth vertex could lie 16 units to the right of and 6 units above $(-3, 4)$. Its coordinates are $(-3+16, 4+6)$, or $(13, 10)$.

95. Equations a), c), and d) appear to be linear.

Exercise Set 1.6

---

1. linear

3. identity

5. circumference

7. $A = bh$

9. subscripts

11. The equation $t + 5 = 11$ is true only when $t = 6$. Similarly, $3t = 18$ is true only when $t = 6$. Since both equations have the same solution, they are equivalent.

13. The equation $12 - x = 3$ is true only when $x = 9$. Similarly, $2x = 10$ is true only when $x = 5$. Since both equations do not have the same solution, they are not equivalent.

15. Zero is a solution of $5x = 2x$. However, since $\dfrac{4}{x} = 3$ is not defined for $x = 0$, the equations $5x = 2x$ and $\dfrac{4}{x} = 3$ are not equivalent.

17.
$$x - 2.9 = 13.4$$
$$x - 2.9 + 2.9 = 13.4 + 2.9$$
$$x - 0 = 16.3$$
$$x = 16.3$$

19.
$$8t = 72$$
$$\frac{1}{8} \cdot 8t = \frac{1}{8} \cdot 72$$
$$t = 9$$

21.
$$\frac{2}{3}x = 30$$
$$\frac{3}{2} \cdot \frac{2}{3}x = \frac{3}{2} \cdot 30$$
$$x = 45$$

23.
$$4x - 12 = 60$$
$$4x - 12 + 12 = 60 + 12$$
$$4x + 0 = 72$$
$$4x = 72$$
$$\frac{1}{4} \cdot 4x = \frac{1}{4} \cdot 72$$
$$x = 18$$

25. $\dfrac{3}{5}n + 2 = 17$

$\dfrac{3}{5}n + 2 - 2 = 17 - 2$

$\dfrac{3}{5}n + 0 = 15$

$\dfrac{3}{5}n = 15$

$\dfrac{5}{3} \cdot \dfrac{3}{5}n = \dfrac{5}{3} \cdot 15$

$n = 25$

27. $6x + 3x = 54$

$(6 + 3)x = 54$

$9x = 54$

$\dfrac{1}{9} \cdot 9x = \dfrac{1}{9} \cdot 54$

$x = 6$

29. $x + 0.06x = 57.24$

$(1 + 0.06)x = 57.24$

$1.06x = 57.24$

$\dfrac{1}{1.06} \cdot 1.06x = \dfrac{1}{1.06} \cdot 57.24$

$x = 54$

31. $4(t - 3) - t = 6$

$4t - 12 - t = 6$

$3t - 12 = 6$

$3t - 12 + 12 = 6 + 12$

$3t = 18$

$\dfrac{1}{3} \cdot 3t = \dfrac{1}{3} \cdot 18$

$t = 6$

33. $3(x + 4) = 7x$

$3x + 12 = 7x$

$3x - 7x + 12 - 12 = 7x - 7x - 12$

$-4x = -12$

$\left(\dfrac{1}{4}\right)(-4x) = \left(\dfrac{1}{4}\right)(-12)$

$-x = -3$

$x = 3$

35. $70 = 10(3t - 2)$

$70 = 30t - 20$

$70 + 20 = 30t - 20 + 20$

$90 = 30t$

$3 = t$

37. $1.8(n - 2) = 9$

$1.8n - 3.6 = 9$

$1.8n - 3.6 + 3.6 = 9 + 3.6$

$1.8n + 0 = 12.6$

$1.8n = 12.6$

$n = \dfrac{12.6}{1.8} = 7$

39. $5y - (2y - 10) = 25$

$5y - 2y + 10 = 25$

$5y - 2y + 10 - 10 = 25 - 10$

$3y = 15$

$y = \dfrac{15}{3} = 5$

41. $\dfrac{9}{10}y - \dfrac{7}{10} = \dfrac{21}{5}$

$\dfrac{9}{10}y - \dfrac{7}{10} + \dfrac{7}{10} = \dfrac{21}{5} + \dfrac{7}{10}$

$\dfrac{9}{10}y + 0 = \dfrac{42}{10} + \dfrac{7}{10}$

$\dfrac{9}{10}y = \dfrac{49}{10}$

$\dfrac{10}{9} \cdot \left(\dfrac{9}{10}y\right) = \dfrac{10}{9} \cdot \dfrac{49}{10}$

$y = \dfrac{49}{9}$

43. $7r - 2 + 5r = 6r + 6 - 4r$

$12r - 2 = 2r + 6$

$12r - 2r - 2 + 2 = 2r - 2r + 6 + 2$

$10r = 8$

$r = \dfrac{8}{10} = \dfrac{4}{5}$

45. $\frac{2}{3}(x-2)-1=\frac{1}{4}(x-3)$

$\frac{2}{3}x-\frac{4}{3}-1=\frac{1}{4}x-\frac{3}{4}$

$\frac{2}{3}x-\frac{1}{4}x=-\frac{3}{4}+\frac{4}{3}+1$

$\left(\frac{2}{3}-\frac{1}{4}\right)x=-\frac{3}{4}+\frac{4}{3}+1$

$\left(\frac{8}{12}-\frac{3}{12}\right)x=-\frac{9}{12}+\frac{16}{12}+\frac{12}{12}$

$\frac{5}{12}x=\frac{19}{12}$

$\frac{12}{5}\cdot\left(\frac{5}{12}x\right)=\frac{12}{5}\cdot\frac{19}{12}$

$x=\frac{19}{5}$

47. $2(t-5)-3(2t-7)=12-5(3t+1)$

$2t-10-6t+21=12-15t-5$

$-4t+11=7-15t$

$-4t+15t=7-11$

$11t=-4$

$t=-\frac{4}{11}$

49. $3x-2(1-x)=4(3x-5)+6(3-2x)$

$3x-2+2x=12x-20+18-12x$

$5x-2=-2$

$5x=-2+2$

$5x=0$

$x=\frac{0}{5}=0$

51. $5+2(x-3)=2\left[5-4(x+2)\right]$

$5+2x-6=2\left[5-4x-8\right]$

$5+2x-6=10-8x-16$

$2x+8x=10-16-5+6$

$10x=-5$

$x=\frac{-5}{10}=-\frac{1}{2}$

53. $7x-2-3x=4x$

$7x-4x-2+2-3x=4x-4x+2$

$0x+0=2$

$0=2$

Since the original equation is equivalent to $0=2$, which is false for any choice of x, the equation has no solution. There is no choice of x that will solve $7x-2-3x=4x$. The equation is a contradiction and the solution set is $\varnothing$.

55. $2+9x=3(4x+1)-1$

$2+9x=12x+3-1$

$2+9x=12x+2$

$0=3x$

$0=x$

There is one solution, 0. For other choices of x, the equation is false. This is a conditional equation since it can be true or false depending on the replacement for x.

57. $-9t+2=-9t-7\left(6\div2(49)+8\right)$

$-9t+2=-9t-7\left(3(49)+8\right)$

$-9t+2=-9t-7\left(147+8\right)$

$-9t+2=-9t-7\left(155\right)$

$-9t+2=-9t-1085$

Since the original equation is equivalent to $-9t+2=-9t+1085$, which is false for any choice of t, the equation has no solution. There is no choice of t that will solve the original equation. The equation is a contradiction, and the solution set is $\varnothing$.

59. $2\{9-3[-2x-4]\}=12x+42$

$2\{9+6x+12\}=12x+42$

$2\{6x+21\}=12x+42$

$12x+42=12x+42$

This is true regardless of what x is replaced with, so all real numbers are solutions. This equation is an identity since it is true for all replacements. The solution set is $\mathbb{R}$.

61. $E = wA$

We want A alone so we divide both sides of the equation by w.

$$\frac{E}{w} = \frac{wA}{w}$$

$$\frac{E}{w} = A$$

63. $F = ma$

We want a alone so we divide both sides of the equation by m.

$$\frac{F}{m} = \frac{ma}{m}$$

$$\frac{F}{m} = a$$

65. $V = lwh$

We want h alone so we divide both sides of the equation by lw.

$$\frac{V}{lw} = \frac{lwh}{lw}$$

$$\frac{V}{lw} = h$$

67. $L = \dfrac{k}{d^2}$

We want k alone so we multiply both sides of the equation by d^2.

$$L \cdot d^2 = \frac{k}{d^2} \cdot d^2$$

$$Ld^2 = k$$

69. $G = w + 150n$

We want n alone so we add $-w$ to both sides of the equation and then divide both sides of the equation by 150.

$$G = w + 150n$$

$$G - w = w - w + 150n$$

$$G - w = 150n$$

$$\frac{G - w}{150} = \frac{150n}{150}$$

$$\frac{G - w}{150} = n$$

71. $2w + 2h + l = p$

We want l alone so we add $-2w$ and $-2h$ to both sides of the equation.

$$2w + 2h + l = p$$

$$2w + 2h - 2w - 2h + l = p - 2w - 2h$$

$$l = p - 2w - 2h$$

73. $Ax + By = C$

We want y alone, so we first add $-Ax$ to both sides of the equation and then divide both sides of the equation by B.

$$Ax + By = C$$

$$Ax - Ax + By = C - Ax$$

$$By = C - Ax$$

$$\frac{By}{B} = \frac{C - Ax}{B}$$

$$y = \frac{C - Ax}{B}$$

75. $C = \dfrac{5}{9}(F - 32)$

We want F alone, so we first multiply both sides of the equation by $\dfrac{9}{5}$ and then add 32 to both sides of the equation.

$$C = \frac{5}{9}(F - 32)$$

$$\frac{9}{5} \cdot C = \frac{9}{5} \cdot \frac{5}{9}(F - 32)$$

$$\frac{9}{5}C = 1 \cdot (F - 32)$$

$$\frac{9}{5}C = F - 32$$

$$\frac{9}{5}C + 32 = F - 32 + 32$$

$$\frac{9}{5}C + 32 = F$$

77. $V = \dfrac{4}{3}\pi r^3$

We want r^3 alone, so first multiply both sides of the equation by $\frac{3}{4}$ and then divide both sides of the equation by π.

$$\frac{3}{4}V = \frac{3}{4}\cdot\frac{4}{3}\pi r^3$$

$$\frac{3}{4}V = \pi r^3$$

$$\frac{1}{\pi}\cdot\frac{3}{4}V = \frac{1}{\pi}\cdot\pi r^3$$

$$\frac{1}{\pi}\cdot\frac{3}{4}V = r^3, \text{ or } r^3 = \frac{3V}{4\pi}$$

79. $np + nm = t$

We want n alone, so first factor n out of the two terms on the left side of the equation. Then divide both sides of the equation by the second factor.

$$np + nm = t$$

$$n(p+m) = t$$

$$\frac{n(p+m)}{(p+m)} = \frac{t}{(p+m)}$$

$$n = \frac{t}{p+m}$$

81. $uv + wv = x$

We want v alone, so first factor v out of the two terms on the left side of the equation. Then divide both sides of the equation by the second factor.

$$uv + wv = x$$

$$v(u+w) = x$$

$$\frac{v(u+w)}{(u+w)} = \frac{x}{(u+w)}$$

$$v = \frac{x}{u+w}$$

83. $A = \dfrac{q_1 + q_2 + q_3}{n}$

We want n alone, which appears in the denominator, so first multiply both sides of the equation by n to clear fractions. Then divide both sides of the equation by A.

$$A\cdot n = \frac{q_1 + q_2 + q_3}{n}\cdot n$$

$$An = q_1 + q_2 + q_3$$

$$\frac{An}{A} = \frac{q_1 + q_2 + q_3}{A}$$

$$n = \frac{q_1 + q_2 + q_3}{A}$$

85. $v = \dfrac{d_2 - d_1}{t}$

We want t alone so we multiply both sides of the equation by t and then divide both sides of the equation by v.

$$v = \frac{d_2 - d_1}{t}$$

$$t\cdot v = t\cdot\left(\frac{d_2 - d_1}{t}\right)$$

$$tv = d_2 - d_1$$

$$\frac{tv}{v} = \frac{d_2 - d_1}{v}$$

$$t = \frac{d_2 - d_1}{v}$$

87. $v = \dfrac{d_2 - d_1}{t}$

We want d_1 alone so we first multiply both sides of the equation by t and then add $-tv$ and d_1 to both sides of the equation.

$$v = \frac{d_2 - d_1}{t}$$

$$t\cdot v = t\left(\frac{d_2 - d_1}{t}\right)$$

$$tv = d_2 - d_1$$

$$tv - tv + d_1 = d_2 - tv - d_1 + d_1$$

$$d_1 = d_2 - tv$$

89. $r - m = mnp$

We want m alone so first add m to both sides of the equation to get all the m-terms on the same side. Then use the distributive law to factor m out of the two terms on the right side of the equation and divide both sides of the equation by the second factor.

$$m + r - m = m + mnp$$
$$r = m + mnp$$
$$r = m(1 + np)$$
$$\frac{r}{(1 + np)} = \frac{m(1 + np)}{(1 + np)}$$
$$\frac{r}{1 + np} = m$$

91. $y + ac^2 = ab$

We want a alone so first subtract ac^2 from both sides of the equation to get all the a-terms on the same side. Then use the distributive law to factor a out of the two terms on the right side of the equation and divide both sides of the equation by the second factor.

$$y + ac^2 - ac^2 = ab - ac^2$$
$$y = ab - ac^2$$
$$y = a(b - c^2)$$
$$\frac{y}{(b - c^2)} = \frac{a(b - c^2)}{(b - c^2)}$$
$$\frac{y}{b - c^2} = a$$

93. $2x - y = 1$
$$2x - y + y = 1 + y$$
$$2x = 1 + y$$
$$2x - 1 = 1 + y - 1$$
$$2x - 1 = y$$
$$y = 2x - 1$$

95. $2x + 5y = 10$
$$2x - 2x + 5y = 10 - 2x$$
$$5y = 10 - 2x$$
$$\frac{5y}{5} = \frac{10 - 2x}{5}$$
$$y = \frac{10 - 2x}{5}$$

Since $\dfrac{10 - 2x}{5} = \dfrac{10}{5} - \dfrac{2x}{5} = 2 - \dfrac{2}{5}x$

$y = 2 - \dfrac{2}{5}x$, or $y = -\dfrac{2}{5}x + 2$ are also correct.

97. $4x - 3y = 6$
$$4x - 3y + 3y = 6 + 3y$$
$$4x = 6 + 3y$$
$$4x - 6 = 6 + 3y - 6$$
$$4x - 6 = 3y$$
$$y = \frac{4x - 6}{3}$$

Since $y = \dfrac{4x - 6}{3} = \dfrac{4x}{3} - \dfrac{6}{3} = \dfrac{4}{3}x - 2$,

$y = \dfrac{4}{3}x - 2$ is also correct.

99. $x = y - 7$
$$x + 7 = y - 7 + 7$$
$$x + 7 = y$$

101. $y - 3(x + 2) = 4 + 2y$
$$y - 3x - 6 = 4 + 2y$$
$$y - y - 3x - 6 - 4 = 4 - 4 + 2y - y$$
$$-3x - 10 = y$$

103. $4y + x^2 = x + 1$
$$4y + x^2 - x^2 = -x^2 + x + 1$$
$$4y = -x^2 + x + 1$$
$$\frac{4y}{4} = \frac{-x^2 + x + 1}{4}$$
$$y = \frac{-x^2 + x + 1}{4}$$

105. The formula for simple interest is
$A = P + Prt$. Solving this equation for r, we have:

$$A = P + Prt$$
$$A - P = P - P + Prt$$
$$A - P = Prt$$
$$\frac{A - P}{Pt} = \frac{Prt}{Pt}$$
$$\frac{A - P}{Pt} = r$$

Substituting 2756 (or 2600 + 156) for A, 2600 for P and 0.5 or (6/12) for t, we have:

$$r = \frac{A - P}{Pt} = \frac{2756 - 2600}{2600 \cdot (0.5)} = \frac{156}{1300} = 0.12$$

So, the interest rate must be 0.12 or 12%.

107. The formula for the area of a parallelogram is
$A = b \cdot h$. Solving this formula for h, we have:

$$A = b \cdot h$$
$$\frac{A}{b} = \frac{bh}{b}$$
$$\frac{A}{b} = h$$

Substituting 78 for A and 13 for b, we have:

$$h = \frac{A}{b} = \frac{78}{13} = 6$$

So, the height of the parallelogram is 6 cm.

109. The Thurnau model is given by
$P = 9.337da - 299$. Solving this equation for d, we have:

$$P = 9.337da - 299$$
$$P + 299 = 9.337da - 299 + 299$$
$$P + 299 = 9.337da$$
$$\frac{P + 299}{9.337a} = d$$

Substituting $P = 1614$ and $a = 24.1$, we have

$$d = \frac{P + 299}{9.337a} = \frac{1614 + 299}{9.337 \cdot 24.1}$$
$$= \frac{1913}{225.0217} = 8.5014$$

So, the diameter of the fetal head is about 8.5 cm.

111. The first step in solving the equation
$2x + 5 = -3$ would be to isolate the term $2x$ on the left side of the equation. Pat should begin by adding –5 to both sides of the equation which would give:
$2x + 5 - 5 = -3 - 5$ or $2x = -8$.
The final step would be to multiply both sides of the equation by $\frac{1}{2}$ resulting in $x = -4$.

113. To put 0.00031 into scientific notation, the decimal must be moved 4 places to the right. Therefore, $0.00031 = 3.1 \cdot 10^{-4}$.

115. The exponent of positive 8 tells us to move the decimal point in 4.6 eight places to the right. Therefore, $4.6 \cdot 10^8 = 460,000,000$.

117. $(8.3 \cdot 10^{12})(4.7 \cdot 10^{18}) = (8.3 \cdot 4.7) \cdot 10^{12+18}$
$$= (39.01) \cdot 10^{30}$$
$$= (3.901 \cdot 10^{-1}) \cdot 10^{30}$$
$$= 3.901 \cdot 10^{29}$$

119. When $c = 0$, $ac = bc$ is equivalent to $a \cdot 0 = b \cdot 0$. No matter what values are substituted for a and b, this is a true statement. For example, $7 \cdot 0 = 5 \cdot 0$. However, the fact that $7 \cdot 0 = 5 \cdot 0$ does not imply that $7 = 5$.

121. $4.23x - 17.898 = -1.65x - 42.454$
$$4.23x + 1.65x - 17.898 + 17.898 =$$
$$-1.65x + 1.65x - 42.454 + 17.898$$
$$5.88x = -24.556$$
$$\frac{5.88x}{5.88} = \frac{-24.556}{5.88}$$
$$x = -4.176190476$$

123. $8x - \{3x - [2x - (5x - (7x - 1))]\} = 8x + 7$
$$8x - \{3x - [2x - (5x - 7x + 1)]\} = 8x + 7$$
$$8x - \{3x - [2x - 5x + 7x - 1]\} = 8x + 7$$
$$8x - \{3x - 2x + 5x - 7x + 1\} = 8x + 7$$
$$8x - 3x + 2x - 5x + 7x - 1 = 8x + 7$$
$$9x - 1 = 8x + 7$$
$$x = 8$$

125.
$$17 - 3\{5 + 2[x - 2]\} + 4\{x - 3(x + 7)\}$$
$$= 9\{x + 3[2 + 3(4 - x)]\}$$
$$17 - 3\{5 + 2x - 4\} + 4\{x - 3x - 21\}$$
$$= 9\{x + 3[2 + 12 - 3x]\}$$
$$17 - 3\{2x + 1\} + 4\{-2x - 21\}$$
$$= 9\{x + 3[-3x + 14]\}$$
$$17 - 6x - 3 - 8x - 84 = 9\{x - 9x + 42\}$$
$$17 - 6x - 3 - 8x - 84 = 9x - 81x + 378$$
$$-14x - 70 = -72x + 378$$
$$58x = 448$$
$$x = \frac{448}{58} = \frac{224}{29}$$

127. Answers vary.

$\dfrac{x+3}{2} = y - 6$ It would be preferable to clear fractions in this case by multiplying both equations by 2 resulting in

$$\frac{x+3}{2} = y - 6$$
$$2\left(\frac{x+3}{2}\right) = 2(y - 6)$$
$$x + 3 = 2y - 12$$

Now, the addition principle could be used to continue to solve for either variable.

129.
$$A = 4lw + w^2$$
$$A - w^2 = 4lw$$
$$\frac{A - w^2}{4w} = l$$

131.
$$\frac{b}{a - b} = c$$
$$b = c(a - b)$$
$$b = ac - bc$$
$$b + bc = ac$$
$$b(1 + c) = ac$$
$$b = \frac{ac}{(1 + c)}$$

133.
$$s + \frac{s + t}{s - t} = \frac{1}{t} + \frac{s + t}{s - t}$$
$$s = \frac{1}{t}$$
$$st = 1$$
$$t = \frac{1}{s}$$

135.
$$y - 2(x^2 + y) = 0$$
$$y - 2x^2 - 2y = 0$$
$$-2x^2 = y$$

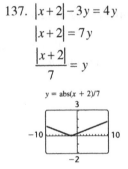

137.
$$|x + 2| - 3y = 4y$$
$$|x + 2| = 7y$$
$$\frac{|x + 2|}{7} = y$$

Exercise Set 1.7

1. Expression

3. Equation

5. Equation

7. Expression

9. Equation

11. Expression

13. Six less than some number

    Let n represent the number. Then we have $n - 6$.

15. Twelve times a number

    Let t represent the number. Then we have $12t$.

17. Sixty-five percent of some number.

Let x represent the number. Then we have

$0.65x$, or $\dfrac{65}{100}x$.

19. Nine more than twice a number

Let y represent the number. Then we have
$2y+9$.

21. Eight more than ten percent of some number

Let s represent the number. Then we have
$0.1s+8$, or $\frac{10}{100}s+8$.

23. One less than the difference of two numbers

Let m and n represent the numbers. Then we
have $m-n-1$.

25. Ninety miles per every four gallons of gas

We have $90 \div 4$, or $\dfrac{90}{4}$.

27. *Familiarize*. The problem asks for two
numbers that sum to 91. It states that one of
the numbers is 9 more than the other.
Translate. Let $x =$ the smaller number. Then
$x + 9 =$ the larger number. The equation for the
sum is:

First number	plus	second number	is	91.
↓	↓	↓	↓	↓
x	$+$	$(x+9)$	$=$	91

29. *Familiarize*. The problem states that Stella can
kayak at a speed of 3.5 mph in calm water. If
Stella kayaks with the current, the current's
speed must be added to her speed in calm
water to determine her speed in the current. If
Stella kayaks against the current, the current's
speed must be subtracted from her speed in
calm water to determine her speed in the
current. Use the formula
$d = rt$ (distance = speed × time)
for an object moving at constant speed.
Translate. Let $t =$ the time, in hours, it takes
Stella to make the 6 mile trip. Since Stella is
kayaking against the current, her speed in the
current equals her speed in calm water minus
the speed of the current
$= 3.5$ mph $- 1.9$ mph $= 1.6$ mph. Therefore,

distance	=	speed	×	time
↓	↓	↓	↓	↓
6 mi	=	1.6 mi/h	×	t

31. *Familiarize*. The problem asks for the measures
of three angles in a triangle. It states that the
measures are consecutive integers. Consecutive
integers are one unit apart. Use the fact that the
three angle measures of a triangle sum to 180°.
Translate. Let $x =$ the degree measure of the
smallest angle. Then $x + 1° =$ the degree
measure of the middle angle, and $(x + 1°) + 1°$
$= x + 2° =$ the degree measure of the largest
angle. The equation for the sum of the three
measures is:

First	plus	second	plus	third	is	180°.
↓	↓	↓	↓	↓	↓	↓
x	$+$	$x+1°$	$+$	$x+2°$	$=$	$180°$

33. *Familiarize*. The problem gives the length and
speed of an escalator, and Dominik's speed on
the escalator. To determine Dominik's speed
toward the top of the escalator the speed of the
escalator must be added to Dominik's speed
on the escalator. Use the formula
$d = rt$ (distance = speed × time)
for an object moving at constant speed.
Translate. Let $t =$ the time, in minutes, it takes
Dominik to get to the top of the 230 ft
escalator. Dominik's speed equals his speed on
the escalator plus the speed of the escalator
$= 100$ ft/min $+ 90$ ft/min $= 190$ ft/min.
Therefore,

distance	=	speed	×	time
↓	↓	↓	↓	↓
230 ft	=	190 ft/min	×	t

35. *Familiarize*. The problem states that the cost of flash drives to customers is the wholesale cost to the seller, plus a given percent of the wholesale cost, plus a fixed cost. The problem then gives the selling price of the flash drives and asks for the wholesale price. Use the equations: mark-up = percent mark-up × wholesale cost + fixed cost, and selling cost = wholesale cost + mark-up

Translate. Let w = the wholesale cost of a flash drive. The associated mark-up is then:

50% of w + $1.50

$= 50\% \cdot w + \$1.50 = 0.50w + \$1.50$

The equation used to determine the wholesale cost to the seller is:

		wholesale		
selling price	is	cost	plus	mark-up
↓	↓	↓	↓	↓
$22.50	=	w	+	$0.50w + \$1.50$

37. *Familiarize*. The problem gives the starting and final altitudes for a plane that is instructed to increase its altitude. The problem also gives the rate at which the plane ascends, and the time it takes to make the ascent is asked for. Use the starting and final altitudes of the plane to determine how far it climbed, and the equation, distance = rate × time, to determine how long it takes to reach its final altitude.

Translate. Let t = the time, in minutes, required for the plane to reach 29,000 ft. The distance it must climb is

$29,000 \text{ ft} - 8000 \text{ ft} = 21,000 \text{ ft}$

The equation needed to determine the time to complete its ascent is:

distance				
climbed	is	rate	times	time
↓	↓	↓	↓	↓
21,000 ft	=	3500 ft/min	×	t

39. *Familiarize*. The problem asks for the measures of three angles in a triangle. It gives the measures of the first and third angles in terms of the measure of the second. The three angle measures of a triangle sum to 180°.

Translate. Since the measures of the first and third angles are defined in terms of the second, let x = the degree measure of the second angle. Then $4x$ = the degree measure of the first angle, and $2x + 5°$ = the degree measure of the third angle. The equation for the sum of the three measures is:

First plus second plus third is 180°.

↓	↓	↓	↓	↓	↓	↓
$4x$	+	x	+	$2x+5°$	=	$180°$

41. *Familiarize*. The problem asks for three consecutive odd integers, and the sum of the first, twice the second, and three times the third is given. Consecutive odd integers are two units apart.

Translate. Let x = the smallest number. Then $x + 2$ = the middle number, and $(x + 2) + 2 = x + 4$ = the largest number. The equation for the sum of the first number, twice the second, and three times the third is:

first		twice the		three times		
number	plus	second	plus	the third	is	70
↓	↓	↓	↓	↓	↓	↓
x	+	$2(x+2)$	+	$3(x+4)$	=	70

43. *Familiarize*. The problem gives the length of a steel rod and asks how it may be cut so that when the two resulting pieces are bent into equilateral triangles, the side lengths of the larger triangle are twice the side lengths of the smaller one. Note that if the side lengths of the larger triangle are twice the side lengths of the smaller one, then the rod used to make the larger triangle must be twice as long as the one used to make the smaller triangle. Also the lengths of the two resulting rods must sum to the length of the original single rod.

Translate. Let x = the length of the rod, in cm, used to make the smaller triangle. Then $2x$ = the length of the rod used to make the larger triangle. The equation for determining the length of each rod is:

length of		length of		
smaller piece	plus	larger piece	is	90 cm.
↓	↓	↓	↓	↓
x	+	$2x$	=	90 cm

45. *Familiarize*. The problem states the average number of calls per shift for Stockton rescue crews is 3. The number of calls Cody received on each of four shifts is given, and the question asks for the number of calls he must receive on his fifth shift for his shift average to equal the crew average. Cody's average number of calls per shift equals the total number of calls he receives divided by the number of shifts he works.

***Translate*.** Let x = the number of calls Cody receives on his fifth shift. The equation needed to determine x is:

$$\underbrace{\text{Cody's average}} \quad \underset{\downarrow}{\text{equals}} \quad \underbrace{\text{crew average.}}$$

$$\underset{\dfrac{5+2+1+3+x}{5}}{\downarrow} \quad \underset{=}{\downarrow} \quad \underset{3}{\downarrow}$$

47. *Familiarize*. The price Tess paid for her calculator is given. The problem states that this is \$13 less than the price Tony paid for his calculator. The question asks for the amount Tony paid. To find the price Tony paid, use the fact that the difference between the larger price (Tony's) and the smaller price (Tess's) is \$13.

***Translate*.** Let p = the price Tony paid for his calculator. Then:

$$\underbrace{\substack{\text{price Tony} \\ \text{paid}}} \quad \text{minus} \quad \underbrace{\substack{\text{price Tess} \\ \text{paid}}} \quad \text{is} \quad \$13$$

$$\underset{p}{\downarrow} \quad \underset{-}{\downarrow} \quad \underset{84}{\downarrow} \quad \underset{=}{\downarrow} \ \underset{13}{\downarrow}$$

***Carry out*.** Solve the equation.

$$p - 84 = 13$$
$$p - 84 + 84 = 13 + 84$$
$$p = 97$$

***Check*.** $97 - 84 = 13$
$$13 = 13$$

The answer checks.

***State*.** Tony paid \$97 for his calculator.

49. *Familiarize*. The monthly cost of day care in Boston is given. The problem then states that the cost of day care in Boston is eleven-fourths the cost in Billings, and asks for the monthly cost of day care in Billings. Eleven-fourths "of" a number is eleven-fourths "times" that number.

***Translate*.** Let c = the monthly cost of day care

in Billings. Then:

$$\underbrace{\text{cost in Boston}} \quad \text{is} \quad \dfrac{11}{4} \quad \text{of} \quad \underbrace{\text{cost in Billings}}$$

$$\underset{1089}{\downarrow} \quad \underset{=}{\downarrow} \ \underset{\dfrac{11}{4}}{\downarrow} \ \underset{\cdot}{\downarrow} \quad \underset{c}{\downarrow}$$

***Carry out*.** Solve the equation.

$$1089 = \dfrac{11}{4}c$$
$$\dfrac{4}{11} \cdot 1089 = \dfrac{4}{11} \cdot \dfrac{11}{4}c$$
$$396 = c$$

***Check*.** $\dfrac{11}{4} \cdot 396 = 1089$
$$1089 = 1089$$

The answer checks.

***State*.** The monthly cost of day care in Billings is \$396.

51. *Familiarize*. The problem states that 40 seniors were photographed and that Robbin took 8 fewer pictures than Michelle, or, equivalently, that Michelle took 8 more pictures than Robbin. The problem then asks for the number of seniors Robbin photographed.

Use the fact that the number of seniors Michelle photographed is the number Robbin photographed plus 8, and that the total number of photographs taken is the sum of how many each girl took.

***Translate*.** Let n = the number of seniors Robbin photographed. Then $n + 8$ = the number of seniors Michelle photographed. Then:

$$\underbrace{\substack{\text{number Robbin} \\ \text{photographed}}} \quad \text{plus} \quad \underbrace{\substack{\text{number Michelle} \\ \text{photographed}}} \quad \text{is} \quad 40$$

$$\underset{n}{\downarrow} \quad \underset{+}{\downarrow} \quad \underset{n+8}{\downarrow} \quad \underset{=}{\downarrow} \ \underset{40}{\downarrow}$$

***Carry out*.** Solve the equation.

$$n + n + 8 = 40$$
$$2n + 8 = 40$$
$$2n + 8 - 8 = 40 - 8$$
$$2n = 32$$
$$\dfrac{1}{2} \cdot 2n = \dfrac{1}{2} \cdot 32$$
$$n = 16$$

Check. $16 + 16 + 8 = 40$

$$32 + 8 = 40$$
$$40 = 40$$

The answer checks.

State. Robbin photographed 16 seniors.

53. **Familiarize.** The problem gives the perimeter of a rectangular mirror and states that its length is 3 times its width.

Use the fact that the perimeter P of a rectangle, of length l and width w, is given by the formula: $P = 2l + 2w$

Translate. Let $w =$ the width, in cm, of the mirror. Then $3w =$ its length. Therefore

$$P = 2l + 2w$$
$$120 = 2(3w) + 2w$$

Carry out. Solve the equation.

$$120 = 2(3w) + 2w$$
$$120 = 6w + 2w$$
$$120 = 8w$$
$$w = \frac{120}{8} = 15$$
$$l = 3w = 3 \cdot 15 = 45$$

Check. $120 = 2 \cdot 45 + 2 \cdot 15$

$$120 = 90 + 30$$
$$120 = 120$$

The values for l and w check.

State. The mirror is 45 cm long and 15 cm wide.

55. **Familiarize.** The problem gives the perimeter of a rectangular greenhouse and states that its width is one-fourth its length.

Use the fact that the perimeter P of a rectangle, of length l and width w, is given by the formula: $P = 2l + 2w$

Translate. Let $l =$ the length, in m, of the greenhouse. Then $\frac{1}{4}l =$ its width. Therefore

$$P = 2l + 2w$$
$$130 = 2l + 2\left(\tfrac{1}{4}l\right)$$

Carry out. Solve the equation.

$$130 = 2l + 2\left(\tfrac{1}{4}l\right)$$
$$130 = 2l + \tfrac{1}{2}l$$
$$2 \cdot 130 = 2 \cdot 2l + 2 \cdot \tfrac{1}{2}l$$
$$260 = 4l + l$$
$$260 = 5l$$

$$l = \frac{260}{5} = 52$$
$$w = \tfrac{1}{4}l = \tfrac{1}{4}(52) = 13$$

Check. $130 = 2 \cdot 52 + 2 \cdot 13$

$$130 = 104 + 26$$
$$130 = 130$$

The values for l and w check.

State. The greenhouse is 52 m long and 13 m wide.

57. The **Familiarize** and **Translate** steps were done in Exercise 29.

distance	=	speed	×	time
↓	↓	↓	↓	↓
6 mi	=	1.6 mi/h	×	t

Carry out. Solve the equation.

$$6 = 1.6t$$
$$t = \frac{6}{1.6} = 3.75$$

Check. $(1.6 \text{ mi/h})(3.75 \text{ h}) = 6 \text{ mi}$

The answer checks.

State. Stella takes 3.75 hr to paddle 6 mi.

59. The **Familiarize** and **Translate** steps were done in Exercise 39.

First	plus	second	plus	third	is	180°.
↓	↓	↓	↓	↓	↓	↓
$4x$	+	x	+	$2x + 5°$	=	180°

Carry out. Solve the equation.

$$4x + x + (2x + 5) = 180$$
$$7x + 5 = 180$$
$$7x = 180 - 5$$
$$7x = 175$$
$$\frac{1}{7} \cdot 7x = \frac{1}{7} \cdot 175$$
$$x = 25$$
$$4x = 4 \cdot 25 = 100$$
$$2x + 5 = 2 \cdot 25 + 5 = 55$$

Check. $100° + 25° + 55° = 180°$.

The answers check.

State. The angle measures are 100°, 25°, and 55°.

61. The *Familiarize* and *Translate* steps were done in Exercise 41.

$$\underbrace{\text{first number}}_{} \quad \text{plus} \quad \underbrace{\text{twice the second}}_{} \quad \text{plus} \quad \underbrace{\text{three times the third}}_{} \quad \text{is} \quad 70$$

$$\downarrow \qquad \downarrow \qquad \downarrow \qquad \downarrow \qquad \downarrow \qquad \downarrow \ \downarrow$$

$$x \quad + \quad 2(x+2) \quad + \quad 3(x+4) \quad = 70$$

Carry out. Solve the equation.

$$x + 2(x+2) + 3(x+4) = 70$$
$$x + 2x + 4 + 3x + 12 = 70$$
$$6x + 16 = 70$$
$$6x = 54$$
$$x = \frac{54}{6} = 9$$
$$x + 2 = 11$$
$$x + 4 = 13$$

Check. $70 = 9 + 2 \cdot 11 + 3 \cdot 13$
$$70 = 9 + 22 + 39$$
$$70 = 31 + 39$$
$$70 = 70$$

The answers check.

State. The three numbers are 9, 11, and 13.

63. The *Familiarize* and *Translate* steps were done in Exercise 35.

$$\underbrace{\text{selling price}}_{} \quad \text{is} \quad \underbrace{\text{wholesale cost}}_{} \quad \text{plus} \quad \underbrace{\text{mark-up}}_{}$$

$$\downarrow \qquad \downarrow \qquad \downarrow \qquad \downarrow \qquad \downarrow$$

$$\$22.50 \quad = \quad w \quad + \quad 0.50w + \$1.50$$

Carry out. Solve the equation.

$$22.50 = w + 0.50w + 1.50$$
$$22.5 = 1.5w + 1.5$$
$$22.5 - 1.5 = 1.5w$$
$$21 = 1.5w$$
$$w = \frac{21}{1.5} = 14$$

Check. $22.50 = 14 + 0.50 \cdot 14 + 1.50$
$$22.50 = 14 + 7 + 1.50$$
$$22.50 = 21 + 1.50$$
$$22.50 = 22.50$$

The answer checks.

State. The wholesale price was \$14.

65. *Familiarize.* We need to divide to determine the average number of megabytes of information generated per person.

Translate. We divide 5 exabytes, or 5×10^{12} megabytes by the worldwide population of 6.3 billion.

$$\frac{5 \times 10^{12}}{6.3 \times 10^9}$$

Carry out. Calculate and write scientific notation for the answer.

$$\frac{5 \times 10^{12}}{6.3 \times 10^9} = \frac{50 \times 10^{11}}{6.3 \times 10^9}$$
$$= \frac{50}{6.3} \times 10^{11-9}$$
$$= 7.936507937 \times 10^2$$
$$\approx 8 \times 10^2$$

Check. We reverse the process (multiply) to check our answer.

$$(8 \times 10^2)(6.3 \times 10^9) = (8 \times 6.3)[10^{2+9}]$$
$$= 50.4 \times 10^{11}$$
$$\approx 5 \times 10^{12}$$

Reminder: We used an approximate value to write the answer with the correct number of significant digits. Our answer checks.

State. The average number of megabytes of information per person is 8×10^2 megabytes.

67. 121 million $\cdot$ 5.8 hours

$$= 121 \cdot 10^6 \cdot 5.8 \ \cancel{\text{hr}} \cdot \left(\frac{60 \text{ min}}{1 \ \cancel{\text{hr}}} \right)$$
$$= 121 \cdot 10^6 \cdot 5.8 \ \cdot 60 \text{ min} = 42,108 \cdot 10^6 \text{ min}$$
$$= 4.2108 \cdot 10^4 \cdot 10^6 \text{ min} = 4.2108 \cdot 10^{10} \text{ min}$$

69. *Familiarize.* We will divide the diameter by the light year to determine n, the number of light years.

Translate. $n = \dfrac{5.88 \times 10^{17}}{5.88 \times 10^{12}}$

Carry out. We solve the equation.

$$n = \frac{5.88 \times 10^{17}}{5.88 \times 10^{12}}$$
$$= \frac{5.88}{5.88} \times 10^{17-12}$$
$$= 1 \times 10^5$$

Check: $(1 \times 10^5)(5.88 \times 10^{12}) = 5.88 \times 10^{17}$
Our answer checks.

State. It is 1×10^5 light years from one end of the Milky Way to the other end.

71. The cable is like a long cylinder. Let $l =$ the length of a cylinder and $d =$ its diameter. Then the volume of the cylinder is given by the formula: $\pi r^2 l = \pi \left(\dfrac{d}{2} \right)^2 l = \dfrac{\pi d^2 l}{4}$.

In order to calculate a volume, make sure that the units for d and l match. The problem gives l in km and d in cm. Converting both to meters will give an answer in units of m^3.

$$125 \text{ km} = 125 \cdot 10^3 \text{ m} = 1.25 \cdot 10^2 \cdot 10^3 \text{ m}$$
$$= 1.25 \cdot 10^5 \text{ m}$$

and $0.6 \text{ cm} = 0.6 \cdot 10^{-2} \text{ m} = 6 \cdot 10^{-1} \cdot 10^{-2} \text{ m}$
$$= 6 \cdot 10^{-3} \text{ m}$$

Therefore the volume of cable is:

$$\frac{\pi d^2 l}{4} = \frac{\pi \left(6 \cdot 10^{-3} \text{ m} \right)^2 \left(1.25 \cdot 10^5 \text{ m} \right)}{4}$$
$$= \frac{\pi \left(36 \cdot 10^{-6} \text{ m}^2 \right) \left(1.25 \cdot 10^5 \text{ m} \right)}{4}$$
$$= \frac{\pi \left(36 \cdot 1.25 \cdot 10^{-6} \cdot 10^5 \right)}{4} \text{ m}^3$$
$$= \frac{\pi \left(45 \cdot 10^{-1} \right)}{4} \text{ m}^3 = \frac{4.5 \pi}{4} \text{ m}^3$$
$$= 1.125 \pi \text{ m}^3 \approx 3.5 \text{ m}^3$$

73. *Familiarize.* There are 10 million bacteria per square centimeter of coral in a coral reef. The reefs near the Hawaiian Islands cover 14,000 square kilometers.
Translate. We will multiply after converting the given information to the same units of area. Since

$1 \text{ m}^2 = 100 \text{ cm} \times 100 \text{ cm} = 10^4 \text{ cm}^2$ and

$1 \text{ km}^2 = 1000 \text{ m} \times 1000 \text{ m} = 10^6 \text{ m}^2$, we

simply multiply $\left(10 \times 10^6 \right) \cdot 10^4 \cdot 10^6$ to

convert cm^2 to km^2, obtaining 10×10^{16}

bacteria per km^2.
Carry out. $10 \times 10^{16} \cdot 14,000$
$$= 1 \times 10^{17} \cdot 1.4 \times 10^4$$
$$= 1.4 \times 10^{17+4}$$
$$= 1.4 \times 10^{21}$$
Check. We reverse the process to check our answer.

$1.4 \times 10^{21} \div 1.4 \times 10^4 = 1 \times 10^{21-4} = 1 \times 10^{17}$

which is 10×10^{16} bacteria per km^2.
Our answer checks.
State. 1.4×10^{21} bacteria are in Hawaii's coral reef.

75. *Familiarize.* Each side of the house is covered with a. 4-ft high-sheet of 8-mil plastic. These dimensions remain constant for all of the sheets/sides which are rectangular prisms. The volume of a rectangular prism is the area of the base times the height of the prism, or $v = l \cdot w \cdot h$. Two sides have length of 24-ft, and the other two sides have length of 32-ft. $1 \text{ ft} = 12$ inches
Translate. We will convert the measurement in feet to inches.

$24 \text{ ft.} = 24 \cdot 12 = 288 \text{ in.}$

$32 \text{ ft.} = 32 \cdot 12 = 384 \text{ in.}$

$4 \text{ ft.} = 4 \cdot 12 \ = 48 \text{ in.}$

Substitute these values into the volume formula.
Note: 8-mil = .008
$V = 2(288 \times 48 \times .008) + 2(384 \times 48 \times .008)$
Carry out. We solve the equation
$V = 2(288 \times 48 \times .008) + 2(384 \times 48 \times .008)$
$$= 2(110.592) + 2(147.456)$$
$$= 221.184 + 294.912$$
$$= 516.096$$
$$= 5.16096 \times 10^2 \text{ in}^3.$$
$$\approx 5 \times 10^2 \text{ in.}^3, \quad \text{or} \quad 3 \times 10^{-1} \text{ ft}^3$$
Note: $1 \text{ in.}^3 = \left(\dfrac{1}{12} \right)^3 \text{ ft}^3, \quad \text{or} \quad \dfrac{1}{1728} \text{ ft}^3$

Check. We recalculate to check our solution. The answer checks.
State. The volume of plastic used is $5 \times 10^2 \text{ in}^3$, or $3 \times 10^{-1} \text{ ft}^3$.

77. Locate the point directly above 225. Then estimate its second coordinate by moving horizontally from the point to the vertical axis. The rate is about 75 heart attacks per 10,000 men.

79. Locate the section of the curve above May 2009. Note that the year labels shown are under September of the corresponding year and, therefore, the months before September appear to the left of the year label. Then use the vertical axis of the graph to estimate the price of gas at that time. The price was about $2.40 per gallon.

81. The highest point on the graph occurs halfway between May 2008 and September 2008. Since the months are May, June, July, August, September, halfway between would be in July. Prices were highest in July 2008.

83.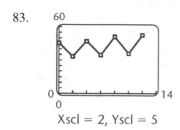
Xscl = 2, Yscl = 5

85. Replace f with y and t with x and graph
$y = -\dfrac{1}{100}x + 2.3$ in an appropriate window.
One good choice is $[0,50,0,3]$ with Xscl = 5. Note that $2005 - 1990 = 15$, so 2005 is 15 years after 1990. Using the Value feature from the CALC menu, we see that $y = 2.15$ when $x = 15$.

$y = -1/100x + 2.3$

X = 15   Y = 2.15
Xscl = 5

Thus, there were 2.15 million farms in the United States in 2005.

87. Replace H with y and graph
$y = -0.7x^3 + 12.1x^2 - 60.5x + 870$ in an
appropriate window. One good choice is
$[0,8,0,1000]$ with Yscl = 100. Using the
TRACE feature, we find that $x \approx 1.6$ and
$x \approx 7.1$ years after 2000 when $y \approx 800$. This
corresponds to the years 2001 and 2007,
respectively.

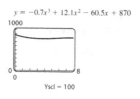
$y = -0.7x^3 + 12.1x^2 - 60.5x + 870$

Yscl = 100

89. *Thinking and Writing Exercise.*

91. $5t - 7 = 5(10) - 7 = 50 - 7 = 43$

93. $(3-x)^2(1-2x)^3 = \left(3-\tfrac{1}{2}\right)^2\left(1-2\left(\tfrac{1}{2}\right)\right)^3$
$= \left(3-\tfrac{1}{2}\right)^2(1-1)^3$
$= \left(3-\tfrac{1}{2}\right)^2 \cdot 0^3$
$= \left(3-\tfrac{1}{2}\right)^2 \cdot 0 = 0$

95. $\dfrac{2x+3}{x-4} = \dfrac{2\cdot 0+3}{0-4} = \dfrac{3}{-4} = -\dfrac{3}{4}$

97. $x + 4 = 0$
$x + 4 - 4 = 0 - 4$
$x = -4$

99. $1 - 2x = 0$
$1 - 2x + 2x = 0 + 2x$
$1 = 2x$
$\dfrac{1}{2}\cdot 1 = \dfrac{1}{2}\cdot 2x$
$\dfrac{1}{2} = x$

101. Although both graphs show some peaks and valleys, the graph in Exercise 84 for Pacific typhoons appears more constant in that its fluctuations are centered around a value of 13 or 14. The graph corresponding to Atlantic hurricanes shows an overall decline over these years.

103. Let a and b represent the numbers. The quotient of the sum of the two numbers and their difference would then be expressed as
$\dfrac{a+b}{a-b}$.

105. Let r and s represent the numbers. Half of the difference of the squares of the two numbers would then be expressed as $\frac{1}{2}\left(r^2 - s^2\right)$ or

$$\frac{r^2 - s^2}{2}.$$

107. The current average is
$$\frac{83 + 91 + 78 + 81}{4} = \frac{333}{4} = 83.25.$$ Let x
represent the score on Tico's next text. After taking the next test, the new average must be 85.25. So. Solving for x, we have:

$$\frac{333 + x}{5} = 85.25$$
$$333 + x = 5\left(85.25\right)$$
$$333 + x = 426.25$$
$$x = 426.25 - 333$$
$$x = 93.25$$

So, Tico must score 10 points, or 93.25 – 83.25 to raise his average 2 points.

109. Let $x =$ the worth of the house in 2007.
The house increased in value by 6% from 2007 to 2008, so its value, in terms of x, at the end of 2008 would have been
$x + 0.06x = 1.06x$.
The house increased in value by 2% from 2008 to 2009, so its value, in terms of x, at the end of 2009 would have been

$$1.06x + 0.02(1.06x)$$
$$= 1.06x + 0.0212x$$
$$= 1.0812x.$$

The house decreased in value by 1% from 2009 to 2010, so its value, in terms of x, at the end of 2010 would have been
$$1.0812x - 0.01(1.0812x)$$
$$= 1.0812x - 0.010812x$$
$$= 1.070388x.$$
Since the house sold for \$117,743 in 2010, $1.070388x = 117743$.
Solving for x:
$1.070388x = 117,743$

$$x = \frac{117,743}{1.070388} \approx \$110,000$$

The house had a value of approximately \$110,000 in 2007.

111. a) Graph IV seems most appropriate for this situation. It reflects 10 minutes on the local streets at 10 to 20 mph. This is followed by 20 minutes from time 10 to 30 minutes of driving at approximately 65 mph, followed by 5 minutes from time 30 to 35 minutes of driving at speeds of approximately 15 mph.

b) Graph III seems most appropriate for this situation. It reflects driving 10 minutes to the train at speeds of 30 to 20 mph, followed by a 45 mph train ride of 20 minutes from time 10 to 30 minutes, followed by a 5 minute walk from time 30 to 35 minutes of approximately 5 mph.

c) Graph I seems most appropriate for this situation. It reflects a 10 minute walk of approximately 5 mph, followed by a bus ride of 20 minutes from time 10 to 30 minutes at a speed of about 45 mph, followed by a 5 minute walk from time 30 to 35 minutes at a speed of about 5 mph.

d) Graph II seems most appropriate for this situation. It reflects a 10 minute wait for the school bus, followed by a 20 minute ride from time 10 to 30 minutes with intermittent starts and stops to pick up students with speeds ranging from 15 to 35 mph, followed by a 5 minute walk from time 30 to 35 at a speed of about 5 mph.

Chapter 1 Study Summary

1. $3 + 5a - b = 3 + 5 \cdot 6 - 10$
    $= 3 + 30 - 10$
    $= 23$

2. $|167| = 167$

3. $-15 + (-10) + 20 = -25 + 20 = -5$

4. $7 - (-7) = 7 + 7 = 14$

5. $-2(-15) = 2 \cdot 15 = 30$

6. $10 \div (-2.5) = -\dfrac{10}{2.5} = -4$

7. $6 + 10n = 10n + 6$

8. $3(ab) = (3a)b$

9. $10(5m + 9n + 1) = 10 \cdot 5m + 10 \cdot 9n + 10 \cdot 1$
$$= 50m + 90n + 10$$

10. $26x + 13 = 13 \cdot 2x + 13 \cdot 1 = 13(2x + 1)$

11. $3c + d - 10c - 2 + 8d = (3 - 10)c + (1 + 8)d - 2$
$$= -7c + 9d - 2$$

12. $6^1 = 6$

13. $(-5)^0 = 1$

14. $x^5 x^{11} = x^{5+11} = x^{16}$

15. $\dfrac{8^9}{8^2} = 8^{9-2} = 8^7$

16. $\left(y^5\right)^3 = y^{5 \cdot 3} = y^{15}$

17. $\left(x^3 y\right)^{10} = \left(x^3\right)^{10} \cdot y^{10} = x^{3 \cdot 10} y^{10} = x^{30} y^{10}$

18. $\left(\dfrac{x^2}{7}\right)^5 = \dfrac{\left(x^2\right)^5}{7^5} = \dfrac{x^{2 \cdot 5}}{7^5} = \dfrac{x^{10}}{7^5}$

19. $10^{-1} = \dfrac{1}{10^1} = \dfrac{1}{10}$

20. $\dfrac{x^{-1}}{y^{-3}} = \dfrac{y^3}{x^1} = \dfrac{y^3}{x}$

21. We need to move the decimal point 4 places to the right. The exponent needed for the power of 10 is therefore -4.
$$0.000904 = 9.04 \cdot 10^{-4}$$

22. The exponent of positive 5 tells of we need to move the decimal point 5 places to the right.
$$6.9 \cdot 10^5 = 690,000$$

23. The graph of $y = 2x + 1$ is a line. Make a table of values, plot the points, and draw a line through them.

x	$y = 2x + 1$	(x, y)
-2	$2(-2) + 1 = -3$	$(-2, -3)$
0	$2 \cdot 0 + 1 = 1$	$(0, 1)$
2	$2 \cdot 2 + 1 = 5$	$(2, 5)$

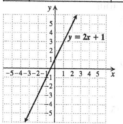

24. $4(x - 3) - (x + 1) = 5$
$$4x - 12 - x - 1 = 5$$
$$3x - 13 = 5$$
$$3x - 13 + 13 = 5 + 13$$
$$3x = 18$$
$$\frac{1}{3} \cdot 3x = \frac{1}{3} \cdot 18$$
$$x = 6$$

25. Use the distributive law to factor the y out of both terms on the left side of the equation. Then divide both sides of the equation by the second factor to get y by itself.
$$xy - 3y = w$$
$$y(x - 3) = w$$
$$\frac{y(x - 3)}{(x - 3)} = \frac{w}{(x - 3)}$$
$$y = \frac{w}{x - 3}$$

26. **Familiarize.** Deborah took two bicycle tours. One tour was 25 miles longer than the other, and she rode a total of 120 miles.
Translate. Let x = the distance of the shorter tour. Then $x + 25$ = the distance of the longer tour, and $x + (x + 25) = 120$.

Carry out. Solve the equation in the previous step for x and give the values of x and $x + 25$, the lengths of each tour.

$$x + (x + 25) = 120$$
$$2x + 25 = 120$$
$$2x + 25 - 25 = 120 - 25$$
$$2x = 95$$
$$\frac{1}{2} \cdot 2x = \frac{1}{2} \cdot 95$$
$$x = 47.5, \text{ or } 47\frac{1}{2}$$
$$x + 25 = 47.5 + 25 = 72.5, \text{ or } 72\frac{1}{2}$$

Check. We check that the two values obtained have a sum of 120 and a difference of 25.

$47.5 + 72.5 = 120 \quad \sqrt{}$
$72.5 - 47.5 = 25 \quad \sqrt{}$

The two values check.
State. The lengths of the two tours were 47.5 mi and 72.5 mi.

Chapter 1 Review Exercises

1. e

2. g

3. j

4. a

5. i

6. b

7. f

8. c

9. d

10. h

11. Substituting $x = -2$, $y = 3$, and $z = -5$ into the expression:

$$
\begin{aligned}
7x^2 - 5y \div zx &= 7(-2)^2 - 5(3) \div (-5)(-2) \\
&= 7(4) - 5(3) \div (-5)(-2) \\
&= 28 - 15 \div (-5)(-2) \\
&= 28 - (-3)(-2) \\
&= 28 - 6 \\
&= 22
\end{aligned}
$$

12. The set containing the first five odd natural numbers can be expressed using roster notation as $\{1, 3, 5, 7, 9\}$ and using set-builder notation as $\{x \mid x$ is an odd natural number between 0 and 10$\}$.

13. The area of a triangle is given by the formula $A = \dfrac{1}{2}bh$, where b is the base and h is the height. So, the area of a triangle with base of 50 cm and height of 70 cm would be

$$A = \frac{1}{2}bh = \frac{1}{2}(50)(70) = \frac{1}{2}(3500) = 1750 \text{ cm}^2.$$

14. a) We replace x with 7.

$$
\begin{array}{rcl}
10 - 3x &=& 1 \\
\hline
10 - 3(7) &?& 1 \\
10 - 21 &?& 1 \\
-11 &=& 1 \text{ False}
\end{array}
$$

Since $-11 = 1$ is false, 7 is not a solution of the equation.

 b) We replace x with 3.

$$
\begin{array}{rcl}
10 - 3x &=& 1 \\
\hline
10 - 3(3) &?& 1 \\
10 - 9 &?& 1 \\
1 &=& 1 \text{ True}
\end{array}
$$

Since $1 = 1$ is true, 3 is a solution of the equation.

15. a) We replace a with 0.

$$
\begin{array}{rcl}
5a + 2 &\leq& 7 \\
\hline
5(0) + 2 &?& 7 \\
0 + 2 &?& 7 \\
2 &\leq& 7 \text{ True}
\end{array}
$$

Since $2 \leq 7$ is true, 0 is a solution of the inequality.

b)   We replace a with 1.

$$\frac{5a+2 \quad \le \quad 7}{5(1)+2 \quad ? \quad 7}$$

$$5+2 \quad ? \quad 7$$

$$7 \quad \le \quad 7 \quad \text{True}$$

Since $7 \le 7$ is true, 1 is a solution of the inequality.

16. $|-19| = 19$

17. $|4.09| = 4.09$

18. $|0| = 0$

19. $-6.5 + (-3.7) = -6.5 - 3.7 = -10.2$

20. $\left(-\frac{2}{5}\right) + \frac{1}{3} = \left(-\frac{2}{5} \cdot \frac{3}{3}\right) + \frac{1}{3} \cdot \frac{5}{5}$

$\qquad = \left(-\frac{6}{15}\right) + \frac{5}{15} = \frac{(-6)+5}{15}$

$\qquad = \frac{-1}{15} = -\frac{1}{15}$

21. $10 + (-5.6) = 4.4$

22. $-13 - 12 = -25$

23. $-\frac{2}{3} - \left(-\frac{1}{2}\right) = -\frac{4}{6} - \left(-\frac{3}{6}\right) = -\frac{4}{6} + \frac{3}{6} = -\frac{1}{6}$

24. $12.5 - 17.9 = -5.4$

25. $(-4.2)(-3) = 12.6$

26. $\left(-\frac{2}{3}\right)\left(\frac{5}{8}\right) = -\frac{10}{24} = -\frac{5}{12}$

27. $\dfrac{72.8}{-8} = -9.1$

28. $-7 \div \dfrac{4}{3} = -7 \cdot \dfrac{3}{4} = -\dfrac{7}{1} \cdot \dfrac{3}{4} = -\dfrac{21}{4}$

29. If $a = -4.01$, then $-a = -(-4.01) = 4.01$.

30. $9 + a = a + 9$

31. $7y = 7 \cdot y = y \cdot 7$

32. $5x + y = x \cdot 5 + y$

$\qquad = y + 5x$

33. $(4 + a) + b = 4 + (a + b)$

34. $(xy)3 = x(y \cdot 3)$

35. $28 - 14mn + 7m = 7 \cdot 4 - 7 \cdot 2mn + 7 \cdot m$

$\qquad = 7(4 - 2mn + m)$

36. $3x^3 - 6x^2 + x^3 + 5 = 3x^3 + x^3 - 6x^2 + 5$

$\qquad = 4x^3 - 6x^2 + 5$

37. $7x - 4\left[2x + 3(5 - 4x)\right]$

$\qquad = 7x - 4\left[2x + 15 - 12x\right]$

$\qquad = 7x - 4\left[-10x + 15\right]$

$\qquad = 7x + 40x - 60$

$\qquad = 47x - 60$

38. $\left(5a^2b^7\right)\left(-2a^3b\right) = (5)(-2)a^{2+3}b^{7+1}$

$\qquad = -10a^5b^8$

39. $\dfrac{12x^3y^8}{3x^2y^2} = \dfrac{12}{3} \cdot x^{3-2}y^{8-2}$

$\qquad = 4xy^6$

40. For $a = -8$.

a) $a^0 = (-8)^0 = 1$

b) $a^2 = (-8)^2 = (-8)(-8) = 64$

c) $-a^2 = -64$ (Use the result from part b.)

41. $3^{-4} \cdot 3^7 = 3^{-4+7} = 3^3$, or 27

42. $\left(5a^2\right)^3 = 5^3\left(a^2\right)^3 = 125a^6$

43. $\left(-2a^{-3}b^2\right)^{-3} = (-2)^{-3}\left(a^{-3}\right)^{-3}\left(b^2\right)^{-3}$

$\qquad = \left(\dfrac{1}{-2}\right)^3 a^9 b^{-6} = -\dfrac{a^9}{8b^6}$

44. $\left(\dfrac{x^2y^3}{z^4}\right)^{-2} = \left(\dfrac{z^4}{x^2y^3}\right)^2 = \dfrac{\left(z^4\right)^2}{\left(x^2y^3\right)^2} = \dfrac{z^8}{x^4y^6}$

45. $\left(\dfrac{2a^{-2}b}{4a^3b^{-3}}\right)^4 = \dfrac{\left(2a^{-2}b\right)^4}{\left(4a^3b^{-3}\right)^4} = \dfrac{2^4a^{-8}b^4}{4^4a^{12}b^{-12}}$

$= \dfrac{16a^{-8}b^4}{256a^{12}b^{-12}} = \dfrac{b^4b^{12}}{16a^{12}a^8} = \dfrac{b^{16}}{16a^{20}}$

46. $\dfrac{7(5-2\cdot3)-3^2}{4^2-3^2} = \dfrac{7(5-6)-9}{16-9}$

$= \dfrac{7(-1)-9}{7} = \dfrac{-7-9}{7} = -\dfrac{16}{7}$

47. $1-(2-5)^2+5\div10\cdot4^2 = 1-(-3)^2+5\div10\cdot4^2$

$= 1-9+5\div10\cdot4^2$

$= 1-9+5\div10\cdot16$

$= 1-9+0.5\cdot16$

$= 1-9+8 = -8+8$

$= 0$

48. $0.000000103 = \dfrac{0.000000103\times10^7}{10^7}$

$= \dfrac{1.03}{10^7} = 1.03\times10^{-7}$

49. $30,860,000,000,000$

$= \dfrac{30,860,000,000,000\times10^{-13}}{10^{-13}}$

$= \dfrac{3.086}{10^{-13}} = 3.086\times10^{13}$

50. The correct answer should have 2 significant digits.

$\left(8.7\times10^{-9}\right)\times\left(4.3\times10^{15}\right)$

$= \left(8.7\times4.3\right)\times\left(10^{-9}\times10^{15}\right)$

$= 37.41\times10^{-9+15} = 37.41\times10^6$

$= \dfrac{37.41\times10^{-1}}{10^{-1}}\times10^6$

$= \dfrac{3.741}{10^{-1}}\times10^6 = 3.741\times10^{1+6}$

$= 3.7\times10^7$ to 2 significant digits.

51. The correct answer should have 2 significant digits.

$\dfrac{1.2\times10^{-12}}{1.5\times10^{-7}} = \dfrac{1.2}{1.5}\times\dfrac{10^{-12}}{10^{-7}} = 0.80\times10^{-12+7}$

$= 0.80\times10^{-5} = \dfrac{0.80\times10}{10}\times10^{-5}$

$= \dfrac{8.0}{10}\times10^{-5} = 8.0\times10^{-1-5}$

$= 8.0\times10^{-6}$

52. Substituting 3 for p and 7 for q (alphabetical order of variables).

$\begin{array}{rcl}
4p-q & = & 5 \\
\hline
4(3)-7 & ? & 5 \\
12-7 & ? & 5 \\
5 & = & 5 \quad \text{True}
\end{array}$

Since $5=5$, is true, $(3,7)$ is a solution of $4p-q=5$.

53. Substituting –2 for x and 4 for y (alphabetical order of variables).

$\begin{array}{rcl}
x-2y & = & 12 \\
\hline
-2-2(4) & ? & 12 \\
-2-8 & ? & 12 \\
-10 & = & 12 \quad \text{False}
\end{array}$

Since $-10=12$, is false, $(-2,4)$ is not a solution of $x-2y=12$.

54.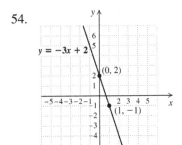

55.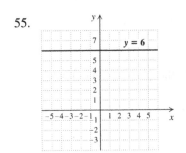

56.

x	y_1
-3	3
-2	1
-1	-1
0	-3
1	-1
2	1
3	3

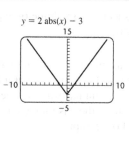

57. $3(t+1)-t=4$

$3t+3-t=4$

$2t+3=4$

$2t+3-3=4-3$

$2t=1$

$\frac{1}{2}\cdot 2t=\frac{1}{2}\cdot 1$

$t=\frac{1}{2}$

58. $\frac{2}{3}n-\frac{5}{6}=\frac{8}{3}$

$6\cdot\frac{2}{3}n-6\cdot\frac{5}{6}=6\cdot\frac{8}{3}$

$2\cdot 2n-1\cdot 5=2\cdot 8$

$4n-5=16$

$4n-5+5=16+5$

$4n=21$

$\frac{1}{4}\cdot 4n=\frac{1}{4}\cdot 21$

$n=\frac{21}{4}$

59. $-9x+4(2x-3)=5(2x-3)+7$

$-9x+8x-12=10x-15+7$

$-9x+8x-12-10x+12=$
$\qquad\qquad 10x-15+7-10x+12$

$-11x=4$

$x=-\frac{4}{11}$

60. $3(x-4)+2=x+2(x-5)$

$3x-12+2=x+2x-10$

$3x-10=3x-10$

$0=0$

Since $0=0$ is true for all x, this is an identity and the solution set is $\mathbb{R}$, the set of all real numbers.

61. $5t-(7-t)=4t+2(9+t)$

$5t-7+t=4t+18+2t$

$6t-7=6t+18$

$6t-6t=18+7$

$0=25$

Since $0=25$ is a contradiction, the solution set is the empty set, $\varnothing$.

62. $P=\dfrac{m}{S}$

$P\cdot S=\dfrac{m}{S}\cdot S$

$PS=m$

63. $c=mx-rx$

$c=(m-r)x$

$\dfrac{c}{m-r}=\dfrac{(m-r)}{(m-r)}x$

$\dfrac{c}{m-r}=x$

64. Let x represent the number.

$\underbrace{\text{Fifteen}}\ \underbrace{\text{more than}}\ \underbrace{\text{twice a number}}\ \underbrace{\text{is}}\ \underbrace{21}$
$\quad\downarrow\qquad\quad\downarrow\qquad\qquad\downarrow\qquad\quad\downarrow\quad\ \downarrow$
$\quad 15\qquad\ +\qquad\qquad 2x\qquad\ \ =\quad 21$

or $2x+15=21$

65. Let x represent the other number. Then the smaller number is $x-19$. Since the sum of the number is 115, we have: $x+(x-19)=115$. Solving for x:

$x+(x-19)=115$

$x+x-19=115$

$2x=115+19$

$2x=134$

$x=67$

Since the smaller number is $x-19$, it is $67-19=48$.

66. Let x represent the second angle of the triangle. Then the first angle measures $3x$ and the third angle measures $2x$. The sum of the measures of the three angles must equal $180°$, so we have:
$$3x + x + 2x = 180.$$
Solving for x, we have
$$3x + x + 2x = 180$$
$$6x = 180$$
$$x = 30.$$
So, the first angle measures $3x$, or $90°$, the second angle measures $30°$, and the third angle measures $2x$, or $60°$.

67. Let $h =$ the height of the candle in cm. The candle is in the shape of a cylinder with height h and radius 3.5 cm. The volume of the candle is 538.51 cm^3. The formula for the volume V of a cylinder is $V = \pi r^2 h$.

Substituting 538.51 cm^3 for V, 3.14 for π, and 3.5 cm for r, we have
$$538.51 \text{ cm}^3 = \pi (3.5 \text{ cm})^2 h.$$
Solve for h,
$$538.51 \text{ cm}^3 = (3.14)(3.5 \text{ cm})^2 h$$
$$538.51 \text{ cm}^3 = (3.14)(12.25 \text{ cm}^2)h$$
$$\frac{538.51 \text{ cm}^3}{12.25 \text{ cm}^2 (3.14)} = h$$
$$14 \text{ cm} = h$$
The height of the candle is 14 cm.

68. Let V represent the volume of the plastic shrink wrap in mm^3. The sheet is in the shape of a rectangle which has volume $V = lwh$, where l represents the length, w represents the width, and h represent the height.
If the length of the sheet is 79 m, then the length is 79,000 mm since one meter is 1000 mm. If the width of the sheet is 1.2 m, then the width is 1200 mm. Assuming the height of the sheet is represented by its thickness, the height is 0.00015 mm.
So, the volume of the sheet is given by:
$$V = lwh$$
$$= (79000)(1200)(0.00015)$$
$$= 14220$$
$$= 1.4220 \times 10^4$$

The sheet has a volume of approximately 1.4×10^4 mm^3.

69. The gas mileage is given by the vertical height of the graph. The graph takes on its highest point halfway between 40 and 50. Therefore, the gas mileage is highest at a speed of 45 mph.

70. The graph appears to be at a height of 7mpg above 30.

71.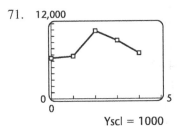

72. To write an equation that has no solution, begin with a simple equation that is false for any value of x, such as $x = x + 1$. Then add or multiply by the same quantities on both sides of the equation to construct a more complicated equation with no solution.

73. Use the distributive law to rewrite a sum of like terms such as a single term by first writing the sum as a product. For example, $2a + 5a = (2 + 5)a = 7a$.

74. If the odor of gasoline is detected at 3 parts per billion, then there must be 3 parts gasoline per 1,000,000,000 parts air. We have $\frac{3}{1,000,000,000} = \frac{3}{10^9} = 3 \times 10^{-9}$. To convert this number to percent, we must multiply by 100, so
$$3 \times 10^{-9} \times 100 = 3 \times 10^{-9} \times 10^2$$
$$= 3 \times 10^{-9+2} = 3 \times 10^{-7}$$
$$= 0.0000003$$
So, the percent of air occupied by the gasoline is 0.0000003%.

75. Substituting $a = 3$, $b = -2$, and $c = -4$, we have:

$$a + b\left(c - a^2\right)^0 + \left(abc\right)^{-1}$$
$$= 3 + (-2)\left(-4 - 3^2\right)^0 + \left[(3)(-2)(-4)\right]^{-1}$$
$$= 3 + (-2)(-13)^0 + [24]^{-1}$$
$$= 3 + (-2)(1) + \frac{1}{24}$$
$$= 3 - 2 + \frac{1}{24}$$
$$= \frac{72 - 48 + 1}{24}$$
$$= \frac{25}{24}$$

76. The area of the 17-in. diameter pizza is:

$A = \pi r^2 \approx 3.14(8.5)^2 \approx 226.865$ in.^{2}.

To determine the cost per square inch, we divide the cost by the area,

$$\frac{1100}{226.865} = 4.8487.$$

The area of the 13-in. diameter pizza is:

$A = \pi r^2 \approx (3.14)(6.5)^2 \approx 132.665$ in.^{2}

To determine the cost per square inch, we divide the cost by the area,

$$\frac{800}{132.665} = 6.0302$$

The 17-in. pizza is a better deal. It costs about 5¢ per square inch; the 13-in. pizza costs about 6¢ per square inch.

77. Let x represent the length of one side of the cube. The surface area of the cube is given by the formula $S = 6x^2$. The volume of the cube is given by the formula $V = x^3$.

Since the surface area is 486 cm^2, we have $S = 6x^2 = 486$. Solving for x:

$$6x^2 = 486$$
$$x^2 = 81$$
$$x = 9$$

So, the length of one side of the cube is 9 cm. Substituting $x = 9$ in the formula for volume, we have $V = x^3 = 9^3 = 729$.

So, the volume of the cube is 729 cm^3.

78.
$$m = \frac{x}{y - z}$$
$$m(y - z) = \frac{x}{y - z}(y - z)$$
$$my - mz = x$$
$$my - x = mz$$
$$\frac{my - x}{m} = z$$

So, $z = \dfrac{my - x}{m}$, or $z = y - \dfrac{x}{m}$.

79.
$$\frac{\left(3^{-2}\right)^a \cdot \left(3^b\right)^{-2a}}{\left(3^{-2}\right)^b \cdot \left(9^{-b}\right)^{-3a}}$$
$$= \left(3^{-2}\right)^{a-b}\left(3^b\right)^{-2a}\left(9^{-b}\right)^{3a}$$
$$= \left(3^{-2}\right)^{a-b}\left(3^b\right)^{-2a}\left[\left(3^2\right)^{-b}\right]^{3a}$$
$$= \left(3^{-2}\right)^{a-b}\left(3^b\right)^{-2a}\left[3^{-2b}\right]^{3a}$$
$$= 3^{-2a+2b}\left(3^{-2ab}\right)\left(3^{-6ab}\right)$$
$$= 3^{-2a+2b+(-2ab)+(-6ab)}$$
$$= 3^{-2a+2b-8ab}$$

80. Let x represent the average on the first text. His overall average after taking the first text will be $\dfrac{4(82.5) + 3x}{7} = 85$. Since test scores count 3 times as much as a single quiz score, we divide by 7 because there have been 4 quizzes already, and the single test counts as 3 quiz scores, making a total of 7 scores. Solving for x, we have:

$$\frac{4(82.5) + 3x}{7} = 85$$
$$4(82.5) + 3x = 7(85)$$
$$330 + 3x = 595$$
$$3x = 595 - 330$$
$$3x = 265$$
$$x = \frac{265}{3} = 88.\overline{3}$$

Ray needs to get a score of $88.\overline{3}$ on his first test to raise his average to 85.

81. In order for the equation to be an identity, it must simplify to a statement that is true for all values of x. Let k represent the quantity in the blank.

$$5x - 7(x+3) - 4 = 2(7-x) + k$$
$$5x - 7x - 21 - 4 = 14 - 2x + k$$
$$-2x - 25 = 14 - 2x + k$$
$$-25 = 14 + k$$
$$-25 - 14 = k$$
$$-39 = k$$

So, the quantity in the blank must be -39 in order for the equation to become an identity.

82. In order to produce a contradiction, we must find the value k such that the equation simplifies to a false statement for all values of x.

$$20 - 7\left[3(2x+4) - 10\right] = 9 - 2(x-5) + k$$
$$20 - 7\left[6x + 12 - 10\right] = 9 - 2x + 10 + k$$
$$20 - 7\left[6x + 2\right] = 19 - 2x + k$$
$$20 - 42x - 14 = 19 - 2x + k$$
$$6 - 42x = 19 - 2x + k$$
$$6 - 40x = 19 + k$$

To produce a contradiction, k must equal $-40x$.

83. $a \cdot 2 + cb + cd + ad = ad + a \cdot 2 + cb + cd$
$$= a(d+2) + c(b+d)$$

84. Answers may vary. We need to find x such that x is between $\frac{1}{2}$ and $\frac{3}{4}$, where x is irrational. Note that the irrational number $\sqrt{5} \approx 2.236068$.

Therefore, $2 < \sqrt{5} < 3$

$$\frac{2}{4} < \frac{\sqrt{5}}{4} < \frac{3}{4}$$
$$\frac{1}{2} < \frac{\sqrt{5}}{4} < \frac{3}{4}$$

So, $\dfrac{\sqrt{5}}{4}$ satisfies the required condition.

Chapter 1 Test

1. Substitute $a = -2$, $b = 6$, and $c = 3$.
$$a^3 - 5b + b \div ac = (-2)^3 - 5(6) + 6 \div (-2)(3)$$
$$= -8 - 5(6) + 6 \div (-2)(3)$$
$$= -8 - 30 + 6 \div (-2)(3)$$
$$= -8 - 30 - 3(3)$$
$$= -8 - 30 - 9$$
$$-47$$

2. The area of a triangle is given by $A = \frac{1}{2}bh$, where b is the base and h is the height. Substituting $b = 7.8$ m and $h = 46.5$ m :
$$A = \tfrac{1}{2}bh = \tfrac{1}{2}(7.8 \text{ m})(46.5 \text{ m}) = 181.35 \text{ m}^2$$

3. a) Let $x = 1$.

$14 - 5x$		4
$14 - 5(1)$	?	4
$14 - 5$	?	4
9	$=$	4 False

Since $9 = 4$ is false, $x = 1$ is not a solution of the equation.

   b) Let $x = 2$.

$14 - 5x$		4
$14 - 5(2)$	?	4
$14 - 10$	?	4
4	$=$	4 True

Since $4 = 4$ is true, $x = 2$ is a solution of the equation.

   c) Let $x = 0$.

$14 - 5x$		4
$14 - 5(0)$	?	4
$14 - 0$	?	4
14	$=$	4 False

Since $14 = 4$ is false, $x = 0$ is not a solution of the equation.

4. $-25 + (-16) = -41$

5. $-10.5 + 6.8 = -3.7$

6. $\frac{1}{3} + \left(-\frac{1}{2}\right) = \frac{1}{3} - \frac{1}{2} = \frac{1}{3} \cdot \frac{2}{2} - \frac{1}{2} \cdot \frac{3}{3}$
$$= \frac{2}{6} - \frac{3}{6} = \frac{2-3}{6}$$
$$= -\frac{1}{6}$$

7. $29.5 - 43.7 = -14.2$

8. $-17.8 - 25.4 = -43.2$

9. $-6.4(5.3) = -33.92$

10. $-\frac{7}{3} - \left(-\frac{3}{4}\right) = -\frac{28}{12} - \left(-\frac{9}{12}\right) = -\frac{28}{12} + \frac{9}{12} = -\frac{19}{12}$

11. $-\frac{2}{7}\left(-\frac{5}{14}\right) = \frac{10}{98} = \frac{5}{49}$

12. $\frac{-42.6}{-7.1} = 6$

13. $\frac{2}{5} \div \left(-\frac{3}{10}\right) = \frac{2}{5}\left(-\frac{10}{3}\right) = -\frac{20}{15} = -\frac{4}{3}$

14. $5 + (1-3)^2 - 7 \div 2^2 \cdot 6$

    $= 5 + (-2)^2 - 7 \div 4 \cdot 6$

    $= 5 + 4 - 7 \div 4 \cdot 6$

    $= 5 + 4 - \frac{7}{4} \cdot 6$

    $= 5 + 4 - \frac{42}{4}$

    $= 5 + 4 - \frac{21}{2}$

    $= 9 - \frac{21}{2}$

    $= \frac{18}{2} - \frac{21}{2}$

    $= -\frac{3}{2}$

15. $3 + x = x + 3$

16. $4y - 10 - 7y - 19 = -3y - 29$

17. $9x - 3(2x-5) - 7 = 9x - 6x + 15 - 7$

    $= 3x + 8$

18. $(12x^{-4}y^{-7})(-6x^{-6}y) = (12)(-6)x^{-4}x^{-6}y^{-7}y$

    $= -72x^{-4+(-6)}y^{-7+(1)}$

    $= -72x^{-10}y^{-6}$

    $= -\dfrac{72}{x^{10}y^6}$

19. $-3^{-2} = -(3^{-2}) = -\left(\frac{1}{3^2}\right) = -\frac{1}{9}$

20. $\left(-5x^{-1}y^3\right)^3 = (-5)^3 x^{3(-1)} \cdot y^{3(3)}$

    $= -125x^{-3}y^9$

    $= -\dfrac{125y^9}{x^3}$

21. $\left(\dfrac{2x^3y^{-6}}{-4y^{-2}}\right)^{-2} = \left(-\dfrac{x^3}{2y^{-2-(-6)}}\right)^{-2}$

    $= \left(-\dfrac{x^3}{2y^4}\right)^{-2}$

    $= \left(-\dfrac{2y^4}{x^3}\right)^{2}$

    $= (-1)^2 \dfrac{2^2 y^{2 \cdot 4}}{x^{2 \cdot 3}}$

    $= \dfrac{4y^8}{x^6}$

22. $\left(5x^3 y\right)^0 = 1$

23. Each factor has 3 significant digits so the answer should also have 3.

    $\left(9.05 \times 10^{-3}\right)\left(2.22 \times 10^{-5}\right)$

    $= (9.05)(2.22) \times 10^{-3+(-5)}$

    $= 20.091 \times 10^{-8}$

    $= 2.01 \times 10^{-7}$ to 3 significant digits

24. Each number in the quotient has 2 significant digits, so the answer should also have 2.

    $\dfrac{5.6 \times 10^7}{2.8 \times 10^{-3}} = \dfrac{5.6}{2.8} \times 10^{7+3} = 2.0 \times 10^{10}$

25. Let $p = 1$ and $q = -4$.

$-2p + 5q$		$=$	18
$-2(1) + 5(-4)$		?	18
$-2 - 20$		?	18
-22		$=$	18 False

    Since $-22 = 18$ is false, $(1, -4)$ is not a solution of the equation.

26.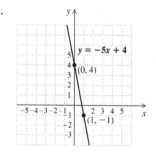

27.

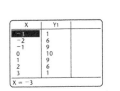   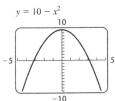

28. $10x - 7 = 38x + 49$

$10x - 10x - 7 - 49 = 38x - 10x + 49 - 49$

$-56 = 28x$

$-2 = x$

29. $13t - (5 - 2t) = 5(3t - 1)$

$13t - 5 + 2t = 15t - 5$

$15t - 5 = 15t - 5$

$0 = 0$

Since $0 = 0$ is true for all values of t, this equation is an identity and the solution set is the set of all real numbers, $\mathbb{R}$.

30. $2p = sp + t$

$2p - sp = sp + t - sp$

$2p - sp = t$

$p(2 - s) = t$

$\dfrac{p(2 - s)}{2 - s} = \dfrac{t}{2 - s}$

$p = \dfrac{t}{2 - s}$

31. If we let m and n represent the two numbers, then the product of the two numbers is mn and four less than the product is: $mn - 4$.

32. Let x represent Linda's score on her sixth test. Her average is computed by adding the six scores and dividing the sum by 6. Since this average must equal 85, we have the equation:

$$\frac{84 + 80 + 76 + 96 + 80 + x}{6} = 85$$

Solving for x:

$$\frac{84 + 80 + 76 + 96 + 80 + x}{6} = 85$$

$$84 + 80 + 76 + 96 + 80 + x = 85(6)$$

$$416 + x = 510$$

$$x = 510 - 416$$

$$x = 94$$

Linda must score 94 on her sixth test to have an overall average of 85.

33. Let x, $x + 2$, and $x + 4$ equal the three consecutive odd integers. The sum of four times the first, three times the second, and two times the third must equal 167 can be expressed as

$$4x + 3(x + 2) + 2(x + 4) = 167.$$

Solving for x, we have:

$$4x + 3(x + 2) + 2(x + 4) = 167$$

$$4x + 3x + 6 + 2x + 8 = 167$$

$$9x + 14 = 167$$

$$9x = 153$$

$$x = 17$$

So, the three consecutive odd integers are 17, 19, and 21.

34. Divide the mas of the alpha particle by the mass of the neutrino to determine how many neutrinos it takes to get the same mass as the alpha particle.

$$\frac{3.62 \cdot 10^{-27} \text{ kg}}{1.8 \cdot 10^{-36} \text{ kg}} = \frac{3.62}{1.8} \cdot 10^{-27 - (-36)} = 2.0 \times 10^{9}$$

Note that the answer was given to two significant digits, because the mass of the neutrino was only given to two significant digits. So, 2.0×10^{9} neutrinos are equal to the mass of one alpha particle.

35. $\left(4x^{3a} y^{b+1}\right)^{2c} = (4)^{2c} \cdot \left(x^{3a}\right)^{2c} \cdot \left(y^{b+1}\right)^{2c}$

$= \left(4^{2}\right)^{c} \cdot x^{(3a)(2c)} \cdot y^{(b+1)(2c)}$

$= 16^{c} x^{6ac} y^{2bc + 2c}$

36. $\dfrac{-27a^{x+1}}{3a^{x-2}} = \left(\dfrac{-27}{3}\right) \cdot a^{(x+1)-(x-2)}$

$= -9a^{x+1-x+2} = -9a^{3}$

37.
$$-\frac{5x+2}{x+10} = 1$$

$$-\frac{5x+2}{x+10} \cdot (x+10) = 1 \cdot (x+10)$$

$$-(5x+2) = x+10$$

$$-5x-2 = x+10$$

$$-5x-2+5x = x+10+5x$$

$$-2 = 6x+10$$

$$-2-10 = 6x+10-10$$

$$-12 = 6x$$

$$-12 \cdot \frac{1}{6} = \frac{1}{6} \cdot 6x$$

$$x = -2$$

Chapter 2

Functions, Linear Equations, and Models

Exercise Set 2.1

1. correspondence

3. domain

5. horizontal

7. "f of 3," "f at 3," or "the value of f at 3"

9. The correspondence is a function because each member of the domain corresponds to just one member of the range.

11. The correspondence is a function because each girl's age corresponds to just one weight.

13. The correspondence is not a function because one member of the domain, 2008, corresponds to three musicians and another, 2009, corresponds to two musicians.

15. The correspondence is a function because each predator corresponds to just one prey.

17. The correspondence is a function because each USB flash drive would have only one storage capacity.

19. The correspondence is a function because each team member would have only one number on his or her uniform.

21. a) The domain is the set of all x-values. It is $\{-3, -2, 0, 4\}$.
    b) The range is the set of all y-values. It is $\{-10, 3, 5, 9\}$.
    c) The correspondence is a function.

23. a) The domain is the set of all x-values. It is $\{1, 2, 3, 4, 5\}$.
    b) The range is the set of all y-values. It is $\{1\}$.
    c) The correspondence is a function.

25. a) The domain is the set of all x-values. It is $\{-2, 3, 4\}$.
    b) The range is the set of all y-values. It is $\{-8, -2, 4, 5\}$.
    c) The correspondence is not a function.

27. a) Locate 1 on the horizontal axis, and then find the point on the graph for which 1 is the first coordinate. From that point, look to the vertical axis to find the corresponding y-coordinate, –2. Thus, $f(1) = -2$.
    b) The domain is the set of all x-values in the graph. It is $\{x \mid -2 \le x \le 5\}$.
    c) To determine which member(s) of the domain are paired with 2, locate 2 on the vertical axis. From there look left and right on the graph to find any points for which 2 is the second coordinate. One such point exists. Its first coordinate is 4. Thus, the x-value for which $f(x) = 2$ is 4.
    d) The range is the set of all y-values in the graph. It is $\{y \mid -3 \le y \le 4\}$.

29. a) Locate 1 on the horizontal axis, and then find the point on the graph for which 1 is the first coordinate. From that point, look to the vertical axis to find the corresponding y-coordinate, –2. Thus, $f(1) = -2$.
    b) The domain is the set of all x-values in the graph. It is $\{x \mid -4 \le x \le 2\}$.
    c) To determine which member(s) of the domain are paired with 2, locate 2 on the vertical axis. From there look left and right on the graph to find any points for which 2 is the second coordinate. One such point exists. Its first coordinate is –2. Thus, the x-value for which $f(x) = 2$ is –2.
    d) The range is the set of all y-values in the graph. It is $\{y \mid -3 \le y \le 3\}$.

31. a) Locate 1 on the horizontal axis, and then find the point on the graph for which 1 is the first coordinate. From that point, look to the vertical axis to find the corresponding y-coordinate, 3. Thus, $f(1) = 3$.

   b) The domain is the set of all x-values in the graph. It is $\{x \mid -4 \le x \le 3\}$.

   c) To determine which member(s) of the domain are paired with 2, locate 2 on the vertical axis. From there look left and right on the graph to find any points for which 2 is the second coordinate. One such point exists. Its first coordinate is -3. Thus, the x-value for which $f(x) = 2$ is -3.

   d) The range is the set of all y-values in the graph. It is $\{y \mid -2 \le y \le 5\}$.

33. a) Locate 1 on the horizontal axis, and then find the point on the graph for which 1 is the first coordinate. From that point, look to the vertical axis to find the corresponding y-coordinate, 1. Thus, $f(1) = 1$.

   b) The domain is the set of all x-values in the graph. It is $\{x \mid -3, -1, 1, 3, 5\}$.

   c) To determine which member(s) of the domain are paired with 2, locate 2 on the vertical axis. From there look left and right on the graph to find any points for which 2 is the second coordinate. One such point exists. Its first coordinate is 3. Thus, the x-value for which $f(x) = 2$ is 3.

   d) The range is the set of all y-values in the graph. It is $\{y \mid -1, 0, 1, 2, 3\}$.

35. a) Locate 1 on the horizontal axis, and then find the point on the graph for which 1 is the first coordinate. From that point, look to the vertical axis to find the corresponding y-coordinate, 4. Thus, $f(1) = 4$.

   b) The domain is the set of all x-values in the graph. It is $\{x \mid -3 \le x \le 4\}$.

   c) To determine which member(s) of the domain are paired with 2, locate 2 on the vertical axis. From there look left and right on the graph to find any points for which 2 is the second coordinate. There are two such points. They are $(-1, 2)$ and $(3, 2)$. Thus, the x-values for which $f(x) = 2$ are -1 and 3.

   d) The range is the set of all y-values in the graph. It is $\{y \mid -4 \le y \le 5\}$.

37. a) Locate 1 on the horizontal axis, and then find the point on the graph for which 1 is the first coordinate. From that point, look to the vertical axis to find the corresponding y-coordinate, 1. Thus, $f(1) = 1$.

   b) The domain is the set of all x-values in the graph. It is $\{x \mid -4 < x \le 5\}$.

   c) To determine which member(s) of the domain are paired with 2, locate 2 on the vertical axis. From there look left and right on the graph to find any points for which 2 is the second coordinate. All points in the set $\{x \mid 2 < x \le 5\}$ satisfy this condition. These are the x-values for which $f(x) = 2$.

   d) The range is the set of all y-values in the graph. It is $\{y \mid -1, 1, 2\}$.

39. Domain: $\mathbb{R}$; range: $\mathbb{R}$

41. Domain: $\mathbb{R}$; range: $\{4\}$

43. Domain: $\mathbb{R}$; range: $\{y \mid y \ge 1\}$, or $[1, \infty)$

45. Domain: $\{x \mid x \text{ is a real number } and \ x \ne -2\}$; range: $\{y \mid y \text{ is a real number } and \ y \ne -4\}$

47. Domain: $\{x \mid x \ge 0\}$, or $[0, \infty)$; range: $\{y \mid y \ge 0\}$, or $[0, \infty)$

49. We can use the vertical line test.

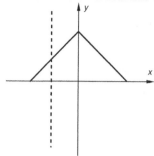

Visualize moving the vertical line across the graph. No vertical line will intersect the graph more than once. Thus the graph is the graph of a function.

51. We can use the vertical line test.

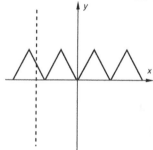

Visualize moving the vertical line across the graph. No vertical line will intersect the graph more than once. Thus the graph is the graph of a function.

53. We can use the vertical line test.

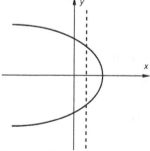

It is possible for a vertical line to intersect the graph more than once. Thus this is not the graph of a function.

55. We can use the vertical line test.

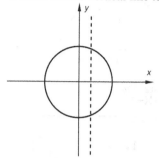

It is possible for a vertical line to intersect the graph more than once. Thus this is not the graph of a function.

57. $g(x) = 2x + 3$

a) $g(0) = 2 \cdot 0 + 3 = 0 + 3 = 3$

b) $g(-4) = 2(-4) + 3 = -8 + 3 = -5$

c) $g(-7) = 2(-7) + 3 = -14 + 3 = -11$

d) $g(8) = 2 \cdot 8 + 3 = 16 + 3 = 19$

e) $g(a+2) = 2(a+2) + 3$
$$= 2a + 4 + 3 = 2a + 7$$

f) $g(a) + 2 = (2a + 3) + 2 = 2a + 5$

59. $f(n) = 5n^2 + 4n$

a) $f(0) = 5 \cdot 0^2 + 4 \cdot 0 = 0 + 0 = 0$

b) $f(-1) = 5(-1)^2 + 4(-1) = 5 - 4 = 1$

c) $f(3) = 5 \cdot 3^2 + 4 \cdot 3 = 45 + 12 = 57$

d) $f(t) = 5t^2 + 4t$

e) $f(2a) = 5(2a)^2 + 4 \cdot 2a$
$$= 5 \cdot 4a^2 + 8a = 20a^2 + 8a$$

f) $2 \cdot f(a) = 2(5a^2 + 4a) = 10a^2 + 8a$

61. $f(x) = \dfrac{x-3}{2x-5}$

a) $f(0) = \dfrac{0-3}{2 \cdot 0 - 5} = \dfrac{-3}{0-5} = \dfrac{-3}{-5} = \dfrac{3}{5}$

b) $f(4) = \dfrac{4-3}{2 \cdot 4 - 5} = \dfrac{1}{8-5} = \dfrac{1}{3}$

c) $f(-1) = \dfrac{-1-3}{2(-1)-5} = \dfrac{-4}{-2-5} = \dfrac{-4}{-7} = \dfrac{4}{7}$

d) $f(3) = \dfrac{3-3}{2 \cdot 3 - 5} = \dfrac{0}{6-5} = \dfrac{0}{1} = 0$

e) $f(x+2) = \dfrac{x+2-3}{2(x+2)-5}$

$= \dfrac{x-1}{2x+4-5} = \dfrac{x-1}{2x-1}$

63.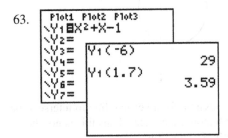

65.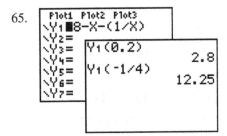

67. $A(s) = s^2 \dfrac{\sqrt{3}}{4}$

$A(4) = 4^2 \dfrac{\sqrt{3}}{4} = 4\sqrt{3} \approx 6.93$

The area is $4\sqrt{3}$ cm$^2 \approx 6.93$ cm^2.

69. $V(r) = 4\pi r^2$

$V(3) = 4\pi(3)^2 = 36\pi$

The area is 36π in$^2 \approx 113.10$ in^2.

71. $P(d) = 1 + \dfrac{d}{33}$

$P(20) = 1 + \dfrac{20}{33} = 1\dfrac{20}{33}$

The pressure at 20 ft is $1\dfrac{20}{33}$ atm.

$P(30) = 1 + \dfrac{30}{33} = 1\dfrac{10}{11}$

The pressure at 30 ft is $1\dfrac{10}{11}$ atm.

$P(100) = 1 + \dfrac{100}{33} = 1 + 3\dfrac{1}{33} = 4\dfrac{1}{33}$

The pressure at 100 ft is $4\dfrac{1}{33}$ atm.

73. $f(x) = 2x - 5$

$f(8) = 2(8) - 5$

$= 16 - 5$

$= 11$

75. $f(x) = 2x - 5$

$-5 = 2x - 5$

$0 = 2x$

$0 = x$

77. $f(x) = \frac{1}{3}x + 4$

$\frac{1}{2} = \frac{1}{3}x + 4$

$\frac{1}{2} - 4 = \frac{1}{3}x$

$\frac{1}{2} - \frac{8}{2} = \frac{1}{3}x$

$-\frac{7}{2} = \frac{1}{3}x$

$\left(-\frac{7}{2}\right)\left(\frac{3}{1}\right) = x$

$-\frac{21}{2} = x$

79. $f(x) = \frac{1}{3}x + 4$

$f\left(\frac{1}{2}\right) = \frac{1}{3}\left(\frac{1}{2}\right) + 4$

$= \frac{1}{6} + 4$

$= \frac{1}{6} + \frac{24}{6} = \frac{25}{6}$

81. $f(x) = 4 - x$

$7 = 4 - x$

$x = 4 - 7 = -3$

So, $f(-3) = 7$.

83. $f(x) = 0.1x - 0.5$

$-3 = 0.1x - 0.5$

$-3 + 0.5 = 0.1x$

$-2.5 = 0.1x$

$\dfrac{-2.5}{0.1} = x$

$-25 = x$

So, $f(-25) = -3$.

85.   The graph crosses the x-axis at only one point whose coordinate is $(-2, 0)$, so -2 is the zero of the function.

87.   The graph does not cross the x-axis at all so there are no zeros of this function.

89. The graph crosses the x-axis at two points whose coordinates are $(-2,0)$ and $(2,0)$, so the zeros of this function are –2 and 2.

91. We want to find any x-values for which $f(x) = 0$, so we substitute 0 for $f(x)$ and solve:
$$f(x) = x - 5$$
$$0 = x - 5$$
$$5 = x$$
The zero of this function is 5.

93. We want to find any x-values for which $f(x) = 0$, so we substitute 0 for $f(x)$ and solve:
$$f(x) = \tfrac{1}{2}x + 10$$
$$0 = \tfrac{1}{2}x + 10$$
$$-10 = \tfrac{1}{2}x$$
$$(-10)(2) = x$$
$$-20 = x$$
The zero of this function is –20.

95. We want to find any x-values for which $f(x) = 0$, so we substitute 0 for $f(x)$ and solve:
$$f(x) = 2.7 - x$$
$$0 = 2.7 - x$$
$$x = 2.7$$
The zero of this function is 2.7.

97. We want to find any x-values for which $f(x) = 0$, so we substitute 0 for $f(x)$ and solve:
$$f(x) = 3x + 7$$
$$0 = 3x + 7$$
$$-7 = 3x$$
$$-\tfrac{7}{3} = x$$
The zero of this function is $-\tfrac{7}{3}$.

99. $f(x) = \dfrac{5}{x-3}$

Since $\dfrac{5}{x-3}$ cannot be computed when the denominator is 0, we find the x-value that causes $x - 3$ to be 0:

$$x - 3 = 0$$
$$x = 3$$
Thus, 3 is not in the domain of f, while all other real numbers are. The domain of f is $\{x \mid x$ is a real number $and\ x \neq 3\}$.

101. $f(x) = \dfrac{x}{2x-1}$

Since $\dfrac{x}{2x-1}$ cannot be computed when the denominator is 0, we find the x-value that causes $2x - 1$ to be 0:
$$2x - 1 = 0$$
$$2x = 1$$
$$x = \frac{1}{2}$$

Thus, $\dfrac{1}{2}$ is not in the domain of f, while all other real numbers are. The domain of f is $\left\{x \mid x \text{ is a real number } and\ x \neq \dfrac{1}{2}\right\}$.

103. $f(x) = 2x + 1$

Since we can compute $2x + 1$ for any real number x, the domain is the set of all real numbers.

105. $g(x) = |5 - x|$

Since we can compute $|5 - x|$ for any real number x, the domain is the set of all real numbers.

107. $f(x) = \dfrac{5}{x-9}$

Since $\dfrac{5}{x-9}$ cannot be computed when the denominator is 0, we find the x-value that causes $x - 9$ to be 0:
$$x - 9 = 0$$
$$x = 9$$
Thus, 9 is not in the domain of f, while all other real numbers are. The domain of f is $\{x \mid x$ is a real number $and\ x \neq 9\}$.

109. $f(x) = x^2 - 9$

Since we can compute $x^2 - 9$ for any real number x, the domain is the set of all real numbers.

111. $f(x) = \dfrac{2x-7}{5}$

Since we can compute $\dfrac{2x-7}{5}$ for any real number x, the domain is the set of all real numbers.

113. $f(x) = \begin{cases} x, & \text{if } x < 0, \\ 2x+1, & \text{if } x \geq 0 \end{cases}$

a) $f(-5)$

Since $-5 < 0$, $f(x) = x$. Thus

$f(-5) = -5$.

b) $f(0)$

Since $0 \geq 0$, $f(x) = 2x+1$. Thus,

$f(0) = 2 \cdot 0 + 1 = 1$.

c) $f(10)$

Since $10 \geq 0$, $f(x) = 2x+1$. Thus,

$f(10) = 2 \cdot 10 + 1 = 20 + 1 = 21$.

115. $G(x) = \begin{cases} x-5, & \text{if } x < -1, \\ x, & \text{if } -1 \leq x \leq 2 \\ x+2, & \text{if } x > 2 \end{cases}$

a) $G(0)$

Since $-1 \leq 0 \leq 2$, $G(x) = x$. Thus,

$G(0) = 0$.

b) $G(2)$

Since $-1 \leq 2 \leq 2$, $G(x) = x$. Thus,

$G(2) = 2$.

c) $G(5)$

Since $5 > 2$, $G(x) = x+2$. Thus,

$G(5) = 5 + 2 = 7$

117. $f(x) = \begin{cases} x^2 - 10 & \text{if } x < -10 \\ x^2, & \text{if } -10 \leq x \leq 10 \\ x^2 + 10, & \text{if } x > 10 \end{cases}$

a) $f(-10)$

Since $-10 \leq -10 \leq 10$, $f(x) = x^2$.

Thus, $f(-10) = (-10)^2 = 100$.

b) $f(10)$

Since $-10 \leq 10 \leq 10$, $f(x) = x^2$. Thus,

$f(10) = 10^2 = 100$.

c) $f(11)$

Since $11 > 10$, $f(x) = x^2 + 10$. Thus,

$f(11) = 11^2 + 10 = 121 + 10 = 131$.

119. *Thinking and Writing Exercise.*

121. $\dfrac{6-3}{-2-7} = \dfrac{3}{-9} = -\dfrac{1}{3}$

123. $\dfrac{-5-(-5)}{3-(-10)} = \dfrac{0}{13} = 0$

125. $2x - y = 8$

$-y = -2x + 8$

$y = 2x - 8$

127. $2x + 3y = 6$

$3y = -2x + 6$

$y = -\dfrac{2}{3}x + 2$

129. *Thinking and Writing Exercise.*

131. To find $f(g(-4))$, we first find $g(-4)$:

$g(-4) = 2(-4) + 5 = -8 + 5 = -3$.

Then $f(g(-4)) = f(-3) = 3(-3)^2 - 1$

$= 3 \cdot 9 - 1 = 27 - 1 = 26.$

To find $g(f(-4))$, we first find $f(-4)$:

$f(-4) = 3(-4)^2 - 1 = 3 \cdot 16 - 1 = 48 - 1 = 47$.

Then $g(f(-4)) = g(47) = 2 \cdot 47 + 5$

$= 94 + 5 = 99.$

133. To find $f\big(f\big(f\big(f\,(tiger)\big)\big)\big)$, we start with the innermost function and work our way out. Since $f(tiger)=dog$, we have
$$f\big(f\big(f\big(f\,(tiger)\big)\big)\big)=f\big(f\big(f\,(dog)\big)\big).$$
Since $f(dog)=cat$, we have
$$f\big(f\,(cat)\big).$$
Since $f(cat)=fish$, we have
$$f\,(fish).$$
Finally, $f(fish)=worm$. So,
$$f\big(f\big(f\big(f\,(tiger)\big)\big)\big)=worm.$$

135. To find the time during the test when the largest contraction occurred, locate the highest point on the graph, and find the corresponding time on the x-axis, which is the first coordinate of the point. The time of the largest contraction was approximately 2 min 50 sec into the test.

137. The two largest contractions occurred at about 2 minutes, 50 seconds and 5 minute, 40 seconds. The difference in these times is 2 minutes 50 seconds, so the frequency is about 1 every 3 minutes.

139. We know that $(-1,-7)$ and $(3,8)$ are both solutions of $g(x)=mx+b$. Substituting, we have
$$-7=m(-1)+b,\ \text{or}\ -7=-m+b,$$
and $8=m(3)+b$, or $8=3m+b$.
Solve the first equation for b and substitute that expression into the second equation.

$-7=-m+b$	First equation
$m-7=b$	Solving for b
$8=3m+b$	Second equation
$8=3m+(m-7)$	Substituting
$8=3m+m-7$	
$8=4m-7$	
$15=4m$	
$\dfrac{15}{4}=m$	

We know that $m-7=b$, so $\dfrac{15}{4}-7=b$, or $-\dfrac{13}{4}=b$. We have $m=\dfrac{15}{4}$ and $b=-\dfrac{13}{4}$, so
$$g(x)=\frac{15}{4}x-\frac{13}{4}.$$

Exercise Set 2.2

1. f

3. e

5. a

7. Graph $f(x)=2x-1$.

We make a table of values. Then we plot the corresponding points and connect them.

x	$f(x)$
1	1
2	3
3	5
5	9

$f(x)=2x-1$

9. Graph $g(x)=-\tfrac{1}{3}x+2$.

We make a table of values. Then we plot the corresponding points and connect them.

x	$g(x)$
0	2
3	1
6	0
9	−1

$g(x)=-\tfrac{1}{3}x+2$

11. Graph $h(x)=\tfrac{2}{5}x-4$.

We make a table of values. Then we plot the corresponding points and connect them.

x	$h(x)$
0	−4
5	−2
10	0

$h(x)=\tfrac{2}{5}x-4$

13. $y=5x+3$

The y-intercept is $(0,3)$, or simply 3.

15. $g(x) = -x - 1$

The y-intercept is $(0,-1)$, or simply -1.

17. $y = -\frac{3}{8}x - 4.5$

The y-intercept is $(0,-4.5)$, or simply -4.5.

19. $f(x) = 1.3x - \frac{1}{4}$

The y-intercept is $\left(0,-\frac{1}{4}\right)$, or simply $-\frac{1}{4}$.

21. $y = 17x + 138$

The y-intercept is $(0,138)$, or simply 138.

23. Slope $= \dfrac{\text{Change in } y}{\text{Change in } x} = \dfrac{3-11}{8-10} = \dfrac{-8}{-2} = 4$

25. Slope $= \dfrac{\text{Change in } y}{\text{Change in } x} = \dfrac{-7-4}{-20-13} = \dfrac{-11}{-33} = \dfrac{1}{3}$

27. Slope $= \dfrac{\text{Change in } y}{\text{Change in } x} = \dfrac{\frac{1}{6}-\left(-\frac{2}{3}\right)}{\frac{1}{6}-\frac{1}{2}} = \dfrac{\frac{1}{6}+\frac{4}{6}}{\frac{1}{6}-\frac{3}{6}}$

$= \dfrac{\frac{5}{6}}{-\frac{2}{6}} = \left(\frac{5}{6}\right)\left(-\frac{6}{2}\right)$

$= -\frac{30}{12} = -\frac{5}{2}$

29. Slope $= \dfrac{\text{Change in } y}{\text{Change in } x} = \dfrac{43.6-43.6}{4.5-(-9.7)} =$

$\dfrac{0}{4.5+9.7} = \dfrac{0}{14.2} = 0$

31. a) The graph of $y = 3x - 5$ has a positive slope, 3, and the y-intercept is $(0,-5)$. Thus, graph II matches this equation.

b) The graph of $y = 0.7x + 1$ has a positive slope, 0.7, and the y-intercept is $(0,1)$. Thus graph IV matches this equation.

c) The graph of $y = -0.25x - 3$ has a negative slope, -0.25, and the y-intercept is $(0,-3)$. Thus graph III matches this equation.

d) The graph of $y = -4x + 2$ has a negative slope, -4, and the y-intercept is $(0,2)$.

33. $y = \dfrac{5}{2}x - 3$

Slope is $\dfrac{5}{2}$; y-intercept is $(0, -3)$.

From the y-intercept, we go *up* 5 units and to the *right* 2 units. This gives us the point (2, 2). We can now draw the graph.

As a check, we can rename the slope and find another point.

$$\frac{5}{2} = \frac{5}{2} \cdot \frac{-1}{-1} = \frac{-5}{-2}$$

From the y-intercept, we go *down* 5 units and to the *left* 2 units. This gives us the point $(-2, -8)$. Since $(-2, -8)$ is on the line, we have a check.

35. $f(x) = -\dfrac{5}{2}x + 2$

Slope is $-\dfrac{5}{2}$, or $\dfrac{-5}{2}$; y-intercept is $(0, 2)$.

From the y-intercept, we go *down* 5 units and to the *right* 2 units. This gives us the point $(2, -3)$. We can now draw the graph.

As a check, we can rename the slope and find another point.

$$\frac{-5}{2} = \frac{5}{-2}$$

From the y-intercept, we go *up* 5 units and to the *left* 2 units. This gives us the point $(-2, 7)$. Since $(-2, 7)$ is on the line, we have a check.

37. $F(x) = 2x + 1$

Slope is 2, or $\dfrac{2}{1}$; y-intercept is $(0, 1)$.

From the y-intercept, we go *up* 2 units and to the *right* 1 unit. This gives us the point (1, 3). We can now draw the graph.

$F(x) = 2x + 1$

As a check, we can rename the slope and find another point.

$$2 = \frac{2}{1} \cdot \frac{3}{3} = \frac{6}{3}$$

From the *y*-intercept, we go *up* 6 units and to the *right* 3 units. This gives us the point (3, 7). Since (3, 7) is on the line, we have a check.

39. Convert to a slope-intercept equation.

$$4x + y = 3$$

$$y = -4x + 3$$

Slope is –4, or $\frac{-4}{1}$; *y*-intercept is (0, 3).

From the *y*-intercept, we go *down* 4 units and to the *right* 1 unit. This gives us the point (1, –1). We can now draw the graph.

$4x + y = 3,$
or
$y = -4x + 3$

As a check, we can rename the slope and find another point.

$$\frac{-4}{1} = \frac{-4}{1} \cdot \frac{-1}{-1} = \frac{4}{-1}$$

From the *y*-intercept, we go *up* 4 units and to the *left* 1 unit. This gives us the point (–1, 7). Since (–1, 7) is on the line, we have a check.

41. Convert to a slope-intercept equation.

$$6y + x = 6$$

$$6y = -x + 6$$

$$y = -\frac{1}{6}x + 1$$

Slope is $-\frac{1}{6}$, or $\frac{-1}{6}$; *y*-intercept is (0, 1).

From the *y*-intercept, we go *down* 1 unit and to the *right* 6 units. This gives us the point (6, 0). We can now draw the graph.

$6y + x = 6,$
or
$y = -\frac{1}{6}x + 1$

As a check, we choose some other value for *x*, say –6, and determine *y*:

$$y = -\frac{1}{6}(-6) + 1 = 1 + 1 = 2$$

We plot the point (–6, 2) and see that it *is* on the line.

43. $g(x) = -0.25x$

Slope is –0.25, or $\frac{-1}{4}$; *y*-intercept is (0, 0).

From the *y*-intercept, we go *down* 1 unit and to the *right* 4 units. This gives us the point (4, – 1). We can now draw the graph.

$g(x) = -0.25x$

As a check, we can rename the slope and find another point.

$$\frac{-1}{4} = \frac{-1}{4} \cdot \frac{-1}{-1} = \frac{1}{-4}$$

From the *y*-intercept, we go *up* 1 unit and to the *left* 4 units. This gives us the point (–4, 1). Since (–4, 1) is on the line, we have a check.

45. Convert to a slope-intercept equation.

$$4x - 5y = 10$$

$$-5y = -4x + 10$$

$$y = \frac{4}{5}x - 2$$

Slope is $\frac{4}{5}$; *y*-intercept is (0, –2).

From the *y*-intercept, we go *up* 4 units and to the *right* 5 units. This gives us the point (5, 2). We can now draw the graph.

$4x - 5y = 10$, or $y = \frac{4}{5}x - 2$

As a check, we choose some other value for *x*, say –5, and determine *y*:

$$y = \frac{4}{5}(-5) - 2 = -4 - 2 = -6$$

We plot the point (–5, –6) and see that it *is* on the line.

47. Convert to a slope-intercept equation.
$$2x + 3y = 6$$
$$3y = -2x + 6$$
$$y = -\frac{2}{3}x + 2$$

Slope is $-\frac{2}{3}$; y-intercept is $(0, 2)$.

From the y-intercept, we go *down* 2 units and to the *right* 3 units. This gives us the point $(3, 0)$. We can now draw the graph.

As a check, we choose some other value for x, say -3, and determine y:
$$y = -\frac{2}{3}(-3) + 2 = 2 + 2 = 4$$

We plot the point $(-3, 4)$ and see that it *is* on the line.

49. Convert to a slope-intercept equation.
$$5 - y = 3x$$
$$-y = 3x - 5$$
$$y = -3x + 5$$

Slope is -3, or $\frac{-3}{1}$; y-intercept is $(0, 5)$.

From the y-intercept, we go *down* 3 units and to the *right* 1 unit. This gives us the point $(1, 2)$. We can now draw the graph.

As a check, we choose some other value for x, say -1, and determine y:
$$y = -3(-1) + 5 = 3 + 5 = 8$$

We plot the point $(-1, 8)$ and see that it *is* on the line.

51. $g(x) = 4.5 = 0x + 4.5$

Slope is 0; y-intercept is $(0, 4.5)$.

From the y-intercept, we go up or down 0 units and any number of nonzero units to the left or right. Any point on the graph will lie on a horizontal line 4.5 units above the x-axis. We draw the graph.

(−3, 4.5) | y (0, 4.5) $g(x) = 4.5$

53. Use the slope-intercept equation,
$$f(x) = mx + b,$$
with $m = 2$ and $b = 5$.
$$f(x) = mx + b$$
$$f(x) = 2x + 5$$

55. Use the slope-intercept equation,
$$f(x) = mx + b,$$
with $m = -\frac{2}{3}$ and $b = -2$.
$$f(x) = mx + b$$
$$f(x) = -\frac{2}{3}x - 2$$

57. Use the slope-intercept equation,
$$f(x) = mx + b,$$
with $m = -7$ and $b = \frac{1}{3}$.
$$f(x) = mx + b$$
$$f(x) = -7x + \frac{1}{3}$$

59. We can use the coordinates of any two points on the line. Let's use $(0, 5)$ and $(4, 6)$.
$$\text{Rate of change} = \frac{\text{change in } y}{\text{change in } x} = \frac{6-5}{4-0} = \frac{1}{4}$$

The distance from home is increasing at a rate of $\frac{1}{4}$ km per minute.

61. We can use the coordinates of any two points on the line. We'll use $(0, 100)$ and $(9, 40)$.
$$\text{Rate of change} = \frac{\text{change in } y}{\text{change in } x},$$
$$= \frac{40-100}{9-0} = \frac{-60}{9} = -\frac{20}{3}$$

or $-6\frac{2}{3}$

The distance from the finish line is decreasing at a rate of $6\frac{2}{3}$ m per second.

63. We can use the coordinates of any two points on the line. We'll use (3, 2.5) and (6, 4.5).

$$\text{Rate of change} = \frac{\text{change in } y}{\text{change in } x}$$

$$= \frac{2.5 - 4.5}{3 - 6} = \frac{-2}{-3} = \frac{2}{3}$$

The number of bookcases stained is increasing at a rate of $\frac{2}{3}$ bookcase per quart of stain used.

65. We can use the coordinates of any two points on the line. We'll use (35, 490) and (45, 500).

$$\text{Rate of change} = \frac{\text{change in } y}{\text{change in } x}$$

$$= \frac{500 - 490}{45 - 35} = \frac{10}{10}, \text{ or } 1$$

The average SAT math score is increasing at a rate of 1 point per thousand dollars of family income.

67. a) Graph II indicated that 200 ml of fluid was dripped in the first 3 hr, a rate of $\frac{200}{3}$ ml/hr. It also indicates that 400 ml of fluid was dripped in the next 3 hr, a rate of $\frac{400}{3}$ ml/hr, and that this rate continues until the end of the time period shown. Since the rate of $\frac{400}{3}$ ml/hr is double the rate of $\frac{200}{3}$ ml/hr, this graph is appropriate for the given situation.

b) Graph IV is the only graph that shows a slope of 0 from 7 PM to 10 PM. Thus, it is the appropriate graph for the given situation.

c) Graph I is the only graph that shows a constant rate for 5 hours, in this case from 3 PM to 8 PM. Thus, it is appropriate for the given situation.

d) Graph III indicates that 100 ml of fluid was dripped in the first 4 hr, a rate of 100/4, or 25 ml/hr. In the next 3 hr, 200 ml was dripped. This is a rate of 200/3, or $66\frac{2}{3}$ ml/hr. Then 100 ml was dripped in the next hour, a rate of 100 ml/hr. In the last hour 200 ml was dripped, a rate of 200 ml/hr. Since the rate at which the fluid was given gradually increased, this graph is appropriate for the given situation.

69. The skier's speed is given by

$$\frac{\text{change in distance}}{\text{change in time}}.$$ Note that the skier reaches the 12-km mark 45 min after the 3-km mark was reached or after $15 + 45$, or 60 min. We will express time in hours:

$15 \text{ min} = 0.25 \text{ hr}$ and $60 \text{ min} = 1 \text{ hr}$. Then

$$\frac{\text{change in distance}}{\text{change in time}} = \frac{12 - 3}{1 - 0.25} = \frac{9}{0.75} = 12.$$

The speed is 12 km/h.

71. The work rate is given by

$$\frac{\text{change in portion of house painted}}{\text{change in time}}.$$

$$\frac{\text{change in portion of house painted}}{\text{change in time}}$$

$$= \frac{\frac{2}{3} - \frac{1}{4}}{8 - 0} = \frac{\frac{5}{12}}{8} = \frac{5}{12} \cdot \frac{1}{8} = \frac{5}{96}$$

The painter's work rate is $\frac{5}{96}$ of the house per hour.

73. The rate at which the number of hits is increasing is given by

$$\frac{\text{change in number of hits}}{\text{change in time}}.$$

$$\frac{\text{change in number of hits}}{\text{change in time}} = \frac{430{,}000 - 80{,}000}{2009 - 2007}$$

$$= \frac{350{,}000}{2} = 175{,}000$$

The number of hits is increasing at a rate of 175,000 hits/yr.

75. $C(d) = 0.75d + 30$

0.75 signifies the cost per mile is \$0.75; 30 signifies that the minimum cost to rent a truck is \$30.

77. $L(t) = \frac{1}{2}t + 5$

$\frac{1}{2}$ signifies that Lauren's hair grows $\frac{1}{2}$ in. per month. 5 signifies that her hair is 5 in long immediately after she gets it cut.

79. $A(t) = \frac{1}{8}t + 75.5$

$\frac{1}{8}$ signifies that the life expectancy of

American females increases $\frac{1}{8}$ of a year, per

year, for years after 1970. 75.5 signifies that
the life expectancy for a female born in 1970
was 75.5 years.

81. $P(t) = 0.89t + 16.63$

0.89 signifies that the average price of a ticket
increases by $0.89 per year, for years after
2000. 16.63 signifies that the cost of a ticket is
$16.63 in 2000.

83. $C(t) = 849t + 5960$

849 signifies that the number of acres of
organic cotton increases by 849 acres per year,
for years after 2006. 5960 signifies that 5960
acres were planted with organic cotton in
2006.

85. $F(t) = -5000t + 90{,}000$

a) −5000 signifies that the truck's value
depreciates $5000 per year; 90,000
signifies that the original value of the
truck was $90,000.

b) We find the value of t for which
$F(t) = 0$.
$$0 = -5000t + 90{,}000$$
$$5000t = 90{,}000$$
$$t = 18$$
It will take 18 yr for the truck to
depreciate completely.

c) The truck's value goes from $90,000
when $t = 0$ to $0 when $t = 18$, so the
domain of F is $\{x \mid 0 \le t \le 18\}$.

87. $v(n) = -200n + 1800$

a) −200 signifies that the depreciation is
$200 per year; 1800 signifies that the
original value of the bike was $1800.

b) We find the value of n for which
$v(n) = 600$.
$$600 = -200n + 1800$$
$$-1200 = -200n$$
$$6 = n$$
The trade-in value is $600 after 6 yrs of
use.

c) First we find the value of n for which
$v(n) = 0$.
$$0 = -200n + 1800$$
$$-1800 = -200n$$
$$9 = n$$
The value of the bike goes from $1800
when $n = 0$, to $0 when $n = 9$, so the
domain of v is $\{n \mid 0 \le n \le 9\}$.

89. *Thinking and Writing Exercise.*

91. $\dfrac{-8 - (-8)}{6 - (-6)} = \dfrac{-8 + 8}{6 + 6} = \dfrac{0}{12} = 0$

93. $3 \cdot 0 - 2y = 9$
$$0 - 2y = 9$$
$$-2y = 9$$
$$y = -\frac{9}{2}$$

95. $f(x) = 2x - 7$
$$f(0) = 2(0) - 7 = 0 - 7 = -7$$

97. *Thinking and Writing Exercise.*

99. a) Graph III indicates that the first 2 mi and
the last 3 mi were traveled in
approximately the same length of time
and at a fairly rapid rate. The mile
following the first two miles was traveled
at a much slower rate. This could indicate
that the first two miles were driven, the
next mile was swum and the last three
miles were driven, so this graph is most
appropriate for the given situation.

b) The slope in Graph IV decreases at 2 mi
and again at 3 mi. This could indicate that
the first two miles were traveled by
bicycle, the next mile was run, and the last
3 miles were walked, so this graph is most
appropriate for the given situation.

c) The slope in Graph I decreases at 2 mi and
then increases at 3 mi. This could indicate
that the first two miles were traveled by
bicycle, the next mile was hiked, and the
last three miles were traveled by bus, so
this graph is most appropriate for the
given situation.

d) The slope in Graph II increases at 2 mi and again at 3 mi. This could indicate that the first two miles were hiked, the next mile was run, and the last three miles were traveled by bus, so this graph is most appropriate for the given situation.

101. The longest uphill climb is the widest rising line. It is the trip from Sienna to Castellina in Chianti.

103. Reading from the graph the trip from Castellina in Chianti to Ponte sul Pesa is downhill, then to Panzano is uphill and then to Creve in Chianti is downhill. All sections are about the same grade. So the trip began at Castellina in Chianti.

105.　$rx + py = s - ry$

$ry + py = -rx + s$

$y(r + p) = -rx + s$

$y = -\dfrac{r}{r + p}x + \dfrac{s}{r + p}$

The slope is $-\dfrac{r}{r + p}$, and the y-intercept

is $\left(0, \dfrac{s}{r + p}\right)$.

107. Since (x_1, y_1) and (x_2, y_2) are two points on the graph of $y = mx + b$, then $y_1 = mx_1 + b$ and $y_2 = mx_2 + b$. Using the definition of slope, we have:

$\text{Slope} = \dfrac{y_2 - y_1}{x_2 - x_1}$

$= \dfrac{(mx_2 + b) - (mx_1 + b)}{x_2 - x_1}$

$= \dfrac{m(x_2 - x_1)}{x_2 - x_1}$

$= m.$

109. Let $c = 1$ and $d = 2$. Then

$f(c + d) = f(1 + 2) = f(3) = 3m + b$, but

$f(c) + f(d) = (m + b) + (2m + b) = 3m + 2b$.

The given statement is false.

111. Let $k = 2$. Then $f(kx) = f(2x) = 2mx + b$, but

$kf(x) = 2(mx + b) = 2mx + 2b$. The given statement is false.

113.　a) $\dfrac{-c - (-6c)}{b - 5b} = \dfrac{5c}{-4b} = -\dfrac{5c}{4b}$

b) $\dfrac{(d + e) - d}{b - b} = \dfrac{e}{0}$ Since we cannot divide by 0, the slope is undefined.

c) $\dfrac{(-a - d) - (a + d)}{(c - f) - (c + f)} = \dfrac{-a - d - a - d}{c - f - c - f}$

$= \dfrac{-2a - 2d}{-2f}$

$= \dfrac{-2(a + d)}{-2f}$

$= \dfrac{a + d}{f}$

115.　$y_1 = 1.4x + 2,\; y_2 = 0.6x + 2,$
$y_3 = 1.4x + 5,\; y_4 = 0.6x + 5$

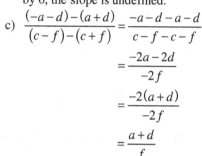

Exercise Set 2.3

1. horizontal

3. vertical

5. 0; x

7. parallel

9. linear

11.　$y - 9 = 3$

$y = 12$

The graph of $y = 12$ is a horizontal line. Since $y - 9 = 3$ is equivalent to $y = 12$, the slope of the line $y - 9 = 3$ is 0.

13. $8x = 6$

$$x = \frac{6}{8}$$

$$x = \frac{3}{4}$$

The graph of $x = \frac{3}{4}$ is a vertical line. Since

$8x = 6$ is equivalent to $x = \frac{3}{4}$, the slope of

15. $3y = 28$

$$y = \frac{28}{3}$$

The graph of $y = \frac{28}{3}$ is a horizontal line.

Since $3y = 28$ is equivalent to $y = \frac{28}{3}$, the

slope of the line $3y = 28$ is 0.

17. $9 + x = 12$

$$x = 3$$

The graph $x = 3$ is a vertical line. Since
$9 + x = 12$ is equivalent to $x = 3$, the slope of
the line $9 + x = 12$ is undefined.

19. $2x - 4 = 3$

$$2x = 7$$

$$x = \frac{7}{2}$$

The graph of $x = \frac{7}{2}$ is a vertical line. Since

$2x - 4 = 3$ is equivalent to $x = \frac{7}{2}$, the slope

of the line $2x - 4 = 3$ is undefined.

21. $5y - 4 = 35$

$$5y = 39$$

$$y = \frac{39}{5}$$

The graph of $y = \frac{39}{5}$ is a horizontal line.

Since $5y - 4 = 35$ is equivalent to $y = \frac{39}{5}$,

the slope of the line $5y - 4 = 35$ is 0.

23. $4y - 3x = 9 - 3x$

$$4y = 9$$

$$y = \frac{9}{4}$$

The graph of $y = \frac{9}{4}$ is a horizontal line.

Since $4y - 3x = 9 - 3x$ is equivalent to

$y = \frac{9}{4}$, the slope of the line is zero.

25. $5x - 2 = 2x - 7$

$$5x = 2x - 5$$

$$x = -\frac{5}{3}$$

The graph of $x = -\frac{5}{3}$ is a vertical line. Since

$5x - 2 = 2x - 7$ is equivalent to $x = -\frac{5}{3}$, the

slope of the line $5x - 2 = 2x - 7$ is undefined.

27. $y = -\frac{2}{3}x + 5$

The equation is written in slope-intercept

form. We see that the slope is $-\frac{2}{3}$.

29. Graph $y = 5$.

This is a horizontal line that crosses the y-axis
at $(0,5)$. If we find some ordered pairs, note
that, for any x-value chosen, y must be 5.

x	y
-2	5
0	5
3	5

31. Graph $x = 3$.

This is a vertical line that crosses the x-axis
at $(3,0)$. If we find some ordered pairs,
note that, for any y-value chosen, x must be
3.

x	y
3	5
3	0
3	-3

33. Graph $f(x) = -2$.

This is a horizontal line that crosses the y-axis at $(0, -2)$. If we find some ordered pairs, for any x-value chosen, y must be -2.

x	y
-3	-2
0	-2
2	-2

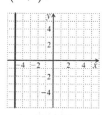

35. Graph $3x = -15$.

Since y does not appear, we solve for x.
$$3x = -15$$
$$x = -5$$

This is a vertical line that crosses the x-axis at $(-5, 0)$.

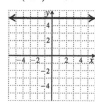

$3x = -15$

37. Graph $3 \cdot g(x) = 15$.

Solve for $g(x)$.
$$3 \cdot g(x) = 15$$
$$g(x) = 5$$

This is a horizontal line that crosses the y-axis at $(0, 5)$.

$3 \cdot g(x) = 15$

39. Graph $x + y = 4$.

To find the y-intercept, let $x = 0$ and solve for y.
$$x + y = 4$$
$$0 + y = 4$$
$$y = 4$$

The y-intercept is $(0, 4)$.

To find the x-intercept, let $y = 0$ and solve for x.
$$x + y = 4$$
$$x + 0 = 4$$
$$x = 4$$

The x-intercept is $(4, 0)$.

Plot these points and draw a line. A third point could be used as a check.

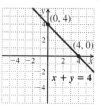

$x + y = 4$

41. Graph $f(x) = 2x - 1$.

To find the y-intercept, let $x = 0$ and solve for y.
$$f(x) = 2x - 1$$
$$f(x) = 2(0) - 1$$
$$f(x) = -1$$

The y-intercept is $(0, -1)$.

To find the x-intercept, let $f(x) = 0$ and solve for x.
$$f(x) = 2x - 1$$
$$0 = 2x - 1$$
$$2x = 1$$
$$x = \tfrac{1}{2}$$

The x-intercept is $\left(\tfrac{1}{2}, 0\right)$.

Plot these points and draw a line. A third point could be used as a check.

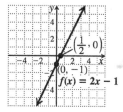

$f(x) = 2x - 1$

43. Graph $3x + 5y = -15$.

To find the *y*-intercept, let $x = 0$ and solve for *y*.

$$3x + 5y = -15$$
$$3 \cdot 0 + 5y = -15$$
$$5y = -15$$
$$y = -3$$

The *y*-intercept is $(0, -3)$.

To find the *x*-intercept, let $y = 0$ and solve for *x*.

$$3x + 5y = -15$$
$$3x + 5 \cdot 0 = -15$$
$$3x = -15$$
$$x = -5$$

The *x*-intercept is $(-5, 0)$.

Plot these points and draw a line. A third point could be used as a check.

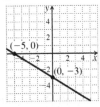

$3x + 5y = -15$

45. Graph $2x - 3y = 18$.

To find the *y*-intercept, let $x = 0$ and solve for *y*.

$$2x - 3y = 18$$
$$2 \cdot 0 - 3y = 18$$
$$-3y = 18$$
$$y = -6$$

The *y*-intercept is $(0, -6)$.

To find the *x*-intercept, let $y = 0$ and solve for *x*.

$$2x - 3y = 18$$
$$2x - 3 \cdot 0 = 18$$
$$2x = 18$$
$$x = 9$$

The *x*-intercept is $(9, 0)$.

Plot these points and draw a line. A third point could be used as a check.

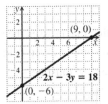

47. Graph $3y = -12x$.

To find the *y*-intercept, let $x = 0$ and solve for *y*.

$$3y = -12x$$
$$3y = -12(0)$$
$$3y = 0$$
$$y = 0$$

The *y*-intercept is $(0, 0)$.

To find the *x*-intercept, let $y = 0$ and solve for *x*.

$$3y = -12x$$
$$3(0) = -12x$$
$$0 = -12x$$
$$x = 0$$

The *x*-intercept is $(0, 0)$. Since the *x*- and *y*-intercepts are the same, we obtain one additional point. Letting $x = 2$,

$$3y = -12x$$
$$3y = -12(2)$$
$$3y = -24$$
$$y = -8$$,

We obtain the point $(2, -8)$.

Plot these points and draw a line. A third point could be used as a check.

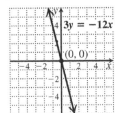

49. Graph $f(x) = 3x - 7$.

To find the y-intercept, let $x = 0$ and solve for $f(x)$.

$$f(x) = 3x - 7$$
$$f(x) = 3 \cdot 0 - 7$$
$$f(x) = 0 - 7$$
$$f(x) = -7$$

The y-intercept is $(0, -7)$.

To find the x-intercept, let $f(x) = 0$ and solve for x.

$$f(x) = 3x - 7$$
$$0 = 3x - 7$$
$$7 = 3x$$
$$\frac{7}{3} = x$$

The x-intercept is $\left(\frac{7}{3}, 0\right)$.

Plot these points and draw a line. A third point to be used as a check.

$f(x) = 3x - 7$

51. Graph $5y - x = 5$.

To find the y-intercept, let $x = 0$ and solve for y.

$$5y - x = 5$$
$$5y - 0 = 5$$
$$5y = 5$$
$$y = 1$$

The y-intercept is $(0, 1)$.

To find the x-intercept, let $y = 0$ and solve for x.

$$5y - x = 5$$
$$5 \cdot 0 - x = 5$$
$$-x = 5$$
$$x = -5$$

The x-intercept is $(-5, 0)$.

Plot these points and draw a line. A third point could be used as a check.

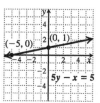

53. Graph $0.2y - 1.1x = 6.6$.

To find the y-intercept, let $x = 0$ and solve for y.

$$0.2y - 1.1x = 6.6$$
$$0.2y - 1.1 \cdot 0 = 6.6$$
$$0.2y = 6.6$$
$$y = 33$$

The y-intercept is $(0, 33)$.

To find the x-intercept, let $y = 0$ and solve for x.

$$0.2y - 1.1x = 6.6$$
$$0.2 \cdot 0 - 1.1x = 6.6$$
$$-1.1x = 6.6$$
$$x = -6$$

The x-intercept is $(-6, 0)$.

Plot these points and draw a line. A third point could be used as a check.

$0.2y - 1.1x = 6.6$

55. $f(x) = 20 - 4x$

To find the y-intercept, let $x = 0$ and solve for $f(x)$.

$$f(x) = 20 - 4x$$
$$f(x) = 20 - 4 \cdot 0$$
$$f(x) = 20 - 0$$
$$f(x) = 20$$

The y-intercept is $(0, 20)$.

To find the x-intercept, let $f(x) = 0$ and solve for x.

$$f(x) = 20 - 4x$$
$$0 = 20 - 4x$$
$$4x = 20$$
$$x = 5$$

The x-intercept is $(5, 0)$.

Choice c) with window $[-10, 10, -10, 30]$ will display both intercepts.

57. $p(x) = -35x + 7000$

To find the y-intercept, let $x = 0$ and solve for $p(x)$.

$$p(x) = -35x + 7000$$
$$p(x) = -35 \cdot 0 + 7000$$
$$p(x) = 0 + 7000$$
$$p(x) = 7000$$

The y-intercept is $(0, 7000)$.

To find the x-intercept, let $p(x) = 0$ and solve for x.

$$p(x) = -35x + 7000$$
$$0 = -35x + 7000$$
$$35x = 7000$$
$$x = 200$$

The x-intercept is $(200, 0)$.

Choice d) with window $[0, 500, 0, 10,000]$ will display both intercepts.

59. We first solve for y and determine the slope of each line.

$$x + 8 = y$$
$$y = x + 8$$

The slope of $y = x + 8$ is 1.

$$y - x = -5$$
$$y = x - 5$$

The slope of $y = x - 5$ is 1.

The slopes are the same; the lines are parallel.

61. We first solve for y and determine the slope of each line.

$$y + 9 = 3x$$
$$y = 3x - 9$$

The slope of $y + 9 = 3x$ is 3.

$$3x - y = -2$$
$$3x + 2 = y$$
$$y = 3x + 2$$

The slope of $3x - y = -2$ is 3.

The slopes are the same; the lines are parallel.

63. We determine the slope of each line.

$$f(x) = 3x + 9$$

The slope of $f(x) = 3x + 9$ is 3.

$$2y = -6x - 2$$
$$y = -3x - 1$$

The slope of $2y = -6x - 2$ is –3.

The slopes are not the same; the lines are not parallel.

65. We determine the slope of each line.

The slope of $f(x) = 4x - 3$ is 4.

$$4y = 7 - x$$
$$4y = -x + 7$$
$$y = -\frac{1}{4}x + \frac{7}{4}$$

The slope of $4y = 7 - x$ is $-\frac{1}{4}$.

The product of their slopes is $4\left(-\frac{1}{4}\right)$, or –1; the lines are perpendicular.

67. We determine the slope of each line.

$$x + 2y = 7$$
$$2y = -x + 7$$
$$y = -\frac{1}{2}x + \frac{7}{2}$$

The slope of $x + 2y = 7$ is $-\frac{1}{2}$.

$$2x + 4y = 4$$
$$4y = -2x + 4$$
$$y = -\frac{2}{4}x + \frac{4}{4}$$
$$y = -\frac{1}{2}x + 1$$

The slope of $2x + 4y = 4$ is $-\frac{1}{2}$.

The product of their slopes is $-\dfrac{1}{2}\left(-\dfrac{1}{2}\right)$, or $\dfrac{1}{4}$, so the lines are not perpendicular.

69. $y = \frac{7}{8}x - 3$

   a) The slope of this line is $\dfrac{7}{8}$, so the slope of a parallel line is also $\dfrac{7}{8}$.

   b) The reciprocal of this slope is $\dfrac{8}{7}$, and the opposite of this number is $-\dfrac{8}{7}$, so the slope of a perpendicular line is $-\dfrac{8}{7}$.

71. $y = -\frac{1}{4}x - \frac{5}{8}$

   a) The slope of this line is $-\dfrac{1}{4}$, so the slope of a parallel line is also $-\dfrac{1}{4}$.

   b) The reciprocal of this slope is -4, and the opposite of this number is 4, so the slope of a perpendicular line is 4.

73. $20x - y = 12$

   We rewrite the second equation in slope-intercept form:
   $$20x - y = 12$$
   $$y = 20x - 12$$

   a) The slope of this line is 20, so the slope of a parallel line is also 20.

   b) The reciprocal of this slope is $\dfrac{1}{20}$, and the opposite of this number is $-\dfrac{1}{20}$, so the slope of a perpendicular line is $-\dfrac{1}{20}$.

75. $x + y = 4$

   We rewrite the second equation in slope-intercept form:
   $$x + y = 4$$
   $$y = -x + 4$$

   a) The slope of this line is -1, so the slope of a parallel line is also -1.

b) The reciprocal of this slope is $-\dfrac{1}{1} = -1$, and the opposite of this number is 1, so the slope of a perpendicular line is 1.

77. The slope of the given line is 3. Therefore, the slope of a line parallel to it is also 3. The y-intercept is $(0,9)$, so the equation of the desired function is $f(x) = 3x + 9$.

79. First we find the slope of the given line.
   $$2x + y = 3$$
   $$y = -2x + 3$$
   The slope of the given line is -2. Therefore, the slope of a line parallel to it is also -2. The y-intercept is $(0,-5)$, so the equation of the desired function is $f(x) = -2x - 5$.

81. First we find the slope of the given line.
   $$2x + 5y = 8$$
   $$5y = -2x + 8$$
   $$y = -\frac{2}{5}x + \frac{8}{5}$$
   The slope of the given line is $-\dfrac{2}{5}$. Therefore, the slope of a line parallel to it is also $-\dfrac{2}{5}$. The y-intercept is $\left(0, -\dfrac{1}{3}\right)$, so the equation of the desired function is $f(x) = -\dfrac{2}{5}x - \dfrac{1}{3}$.

83. First we find the slope of the given line.
   $$3y = 12$$
   $$y = 0x + 4$$
   The slope of the given line is 0. Therefore, the slope of a line parallel to it is also 0. The y-intercept is $(0,-5)$, so the equation of the desired function is $f(x) = -5$.

85. The slope of the given line is 1. The slope of a line perpendicular to it is the opposite of the reciprocal of 1, or -1. The y-intercept is $(0,4)$, so we have $y = -x + 4$.

87. First find the slope of the given line.
$$2x + 3y = 6$$
$$3y = -2x + 6$$
$$y = -\frac{2}{3}x + 2$$

The slope of the given line is $-\frac{2}{3}$. The slope of a line perpendicular to it is the opposite of the reciprocal of $-\frac{2}{3}$, or $\frac{3}{2}$. The y-intercept is $(0, -4)$, so we have $y = \frac{3}{2}x - 4$.

89. First find the slope of the given line.
$$5x - y = 13$$
$$5x - 13 = y$$
$$y = 5x - 13$$

The slope of the given line is 5. The slope of a line perpendicular to it is the opposite of the reciprocal of 5, or $-\frac{1}{5}$. The y-intercept is $\left(0, \frac{1}{5}\right)$, so we have $y = -\frac{1}{5}x + \frac{1}{5}$.

91. This equation is in the standard form for a linear equation, $Ax + By = C$, with $A = 5$, $B = -3$, and $C = 15$. Thus, it is a linear equation.
Solve for y to find the slope.
$$5x - 3y = 15$$
$$-3y = -5x + 15$$
$$y = \frac{5}{3}x - 5$$

The slope is $\frac{5}{3}$.

93. We write $16 + 4y = 10$ in standard form for a linear equation, $Ax + By = C$, as $0x + 4y = -6$ with $A = 0$, $B = 4$, and $C = -6$. Thus, it is a linear equation.
Solve for y to find the slope.
$$4y = -6$$
$$y = 0x + -\frac{3}{2}$$

The slope is 0.

95. $xy = 10$
The equation cannot be written in standard form for a linear equation, $Ax + By = C$, since there is an xy term. Thus, the equation is not linear.

97. $3y = 7(2x - 4)$
The equation can be written in standard form for a linear equation, $Ax + By = C$
$$3y = 7(2x - 4)$$
$$3y = 14x - 28$$
$$-14x + 3y = -28$$
with $A = -14$, $B = 3$, and $C = -28$. Thus, it is a linear equation. Solve for y to find the slope.
$$-14x + 3y = -28$$
$$3y = 14x - 28$$
$$y = \frac{14}{3}x - \frac{28}{3}$$

The slope is $\frac{14}{3}$.

99. $g(x) = \frac{1}{x}$
Replacing $g(x)$ with y and attempt to write the equation in standard form.
$$y = \frac{1}{x}$$
$$xy = 1$$
The equation is not linear because it has an xy-term.

101. $\frac{f(x)}{5} = x^2$
Replace $f(x)$ with y and attempt to write the equation in standard form.
$$\frac{y}{5} = x^2$$
$$y = 5x^2$$
$$-5x^2 + y = 0$$
The equation is not linear because it has an x^2-term.

103. *Thinking and Writing Exercise.*

105. $-\frac{3}{10}\left(\frac{10}{3}\right) = -\frac{30}{30} = -1$

107. $-3\left[x-(-1)\right] = -3(x+1) = -3x-3$

109. $\dfrac{2}{3}\left[x-\left(-\dfrac{1}{2}\right)\right]-1 = \dfrac{2}{3}\left[x+\dfrac{1}{2}\right]-1$

$\qquad\qquad = \dfrac{2}{3}x+\dfrac{2}{6}-1 = \dfrac{2}{3}x-\dfrac{2}{3}$

111. *Thinking and Writing Exercise.*

113. The line contains the points $(5,0)$ and $(0,-4)$. We use the points to find the slope.

$$\text{Slope} = \frac{-4-0}{0-5} = \frac{-4}{-5} = \frac{4}{5}$$

Then the slope-intercept equation is

$y = \dfrac{4}{5}x-4$. We rewrite this equation in standard form.

$$y = \frac{4}{5}x-4$$
$$5y = 4x-20$$
$$-4x+5y = -20$$

This equation can also be written as $4x-5y = 20$.

115. $rx+3y = p^2-s$

The equation is in standard form with $A = r$, $B = 3$, and $C = p^2-s$. It is linear.

117. $r^2x = py+5$

Try to put the equation in standard form.

$r^2x = py+5$

$r^2x-py = 5$

The equation is in standard form with $A = r^2$, $B = -p$, and $C = 5$. It is linear.

119. Let equation A have intercepts $(a,0)$ and $(0,b)$. Then equation B has intercepts $(2a,0)$ and $(0,b)$.

$$\text{Slope of } A = \frac{b-0}{0-a} = -\frac{b}{a}$$

$$\text{Slope of } B = \frac{b-0}{0-2a} = -\frac{b}{2a} = \frac{1}{2}\left(-\frac{b}{a}\right)$$

The slope of equation B is $\dfrac{1}{2}$ the slope of equation A.

121. First write the equation in standard form.

$$ax+3y = 5x-by+8$$

$ax-5x+3y+by = 8$

$(a-5)x+(3+b)y = 8$

If the graph is a vertical line, then the coefficient of y is 0.

$\qquad 3+b = 0$

$\qquad b = -3$

Then we have $(a-5)x = 8$

If the line passes through $(4,0)$, we have:

$(a-5)4 = 8$ Substituting 4 for x

$\qquad a-5 = 2$

$\qquad\quad a = 7$

123. a) Solve each equation for y, enter each on the equation-editor screen, and then examine a table of values for the two functions. Since the difference between the y-values is the same for all x-values, the lines are parallel.

b) Solve each equation for y, enter each on the equation-editor screen, and then examine a table of values for the two functions. Since the difference between the y-values is not the same for all x-values, the lines are not parallel.

Exercise Set 2.4

1. True

3. False

5. True

7. True

9. False

11. The point-slope form of a line is:
$$y - y_1 = m(x - x_1).$$
Substitute $m = 3$, $x_1 = 5$, and $y_1 = 2$ to get:
$$y - 2 = 3(x - 5).$$
To graph the line, plot the point $(5, 2)$. Then
use $m = \dfrac{\text{change in } y}{\text{change in } x} = \dfrac{3}{1} = \dfrac{-3}{-1}$ to generate a
second point $(5 + (-1),\ 2 + (-3)) = (4, -1).$
Plot this second point and draw the line
passing through both points.

$y - 2 = 3(x - 5)$

13. The point-slope form of a line is:
$$y - y_1 = m(x - x_1).$$
Substitute $m = -4$, $x_1 = 1$, and $y_1 = 2$ to get:
$$y - 2 = -4(x - 1).$$
To graph the line, plot the point $(1,\ 2)$. Then
use $m = \dfrac{\text{change in } y}{\text{change in } x} = -\dfrac{4}{1} = \dfrac{-4}{1}$ to generate a
second point $(1 + 1,\ 2 + (-4)) = (2, -2).$ Plot
this second point and draw the line passing
through both points.

$y - 2 = -4(x - 1)$

15. The point-slope form of a line is:
$$y - y_1 = m(x - x_1).$$
Substitute $m = \dfrac{1}{2}$, $x_1 = -2$, and $y_1 = -4$ to get:
$$y - (-4) = \dfrac{1}{2}(x - (-2)),\ \text{or } y + 4 = \dfrac{1}{2}(x + 2).$$
To graph the line, plot the point $(-2, -4)$.
Then use $m = \dfrac{\text{change in } y}{\text{change in } x} = \dfrac{1}{2}$ to generate a
second point $(-2 + 2,\ -4 + 1) = (0, -3).$ Plot

this second point and draw the line passing
through both points.

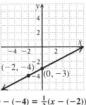

$y - (-4) = \frac{1}{2}(x - (-2)),$
or $y + 4 = \frac{1}{2}(x + 2)$

17. The point-slope form of a line is:
$$y - y_1 = m(x - x_1).$$
Substitute $m = -1$, $x_1 = 8$, and $y_1 = 0$ to get:
$$y - 0 = -1(x - 8),\ \text{or } y = -(x - 8).$$
To graph the line, plot the point $(8, 0)$. Then
use $m = \dfrac{\text{change in } y}{\text{change in } x} = -\dfrac{1}{1} = \dfrac{1}{-1}$ to generate a
second point $(8 + (-1),\ 0 + 1) = (7, 1).$ Plot this
second point and draw the line passing
through both points.

$y - 0 = -1(x - 8),$ or
$y = -(x - 8)$

19. $y - 3 = \dfrac{1}{4}(x - 5)$

$y - y_1 = m(x - x_1)$

$m = \dfrac{1}{4}$, $x_1 = 5$, $y_1 = 3$, so the slope is $\dfrac{1}{4}$ and

the point $(5, 3)$ is on the graph.

21. $y + 1 = -7(x - 2)$

$y - (-1) = -7(x - 2)$

$y - y_1 = m(x - x_1)$

$m = -7$, $x_1 = 2$, $y_1 = -1$, so the slope is -7 and

the point $(2, -1)$ is on the graph.

23. $y - 6 = -\dfrac{10}{3}(x + 4)$

$y - 6 = -\dfrac{10}{3}(x - (-4))$

$y - y_1 = m(x - x_1)$

$m = -\dfrac{10}{3}$, $x_1 = -4$, $y_1 = 6$, so the slope is $-\dfrac{10}{3}$

and the point $(-4, 6)$ is on the graph.

25. $y = 5x$

$y - 0 = 5(x - 0)$

$y - y_1 = m(x - x_1)$

$m = 5$, $x_1 = 0$, $y_1 = 0$, so the slope is 5 and the

point $(0, 0)$ is on the graph.

27. | $y - y_1 = m(x - x_1)$ | Point-slope equation |
|---|---|
| $y - (-4) = 2(x - 1)$ | Substituting 2 for m, and $(1, -4)$ for (x_1, y_1) |
| $y + 4 = 2x - 2$ | Simplifying |
| $y = 2x - 6$ | Subtracting 4 from both sides |
| $f(x) = 2x - 6$ | Using function notation |

To graph the line, plot the point $(1, -4)$ and the

y-intercept $(0, -6)$, and then draw the line

passing through both points.

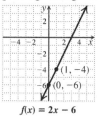

$f(x) = 2x - 6$

29. | $y - y_1 = m(x - x_1)$ | Point-slope equation |
|---|---|
| $y - 8 = -\frac{3}{5}(x - (-4))$ | Substituting $-\frac{3}{5}$ for m and $(-4, 8)$ for (x_1, y_1) |
| $y - 8 = -\frac{3}{5}x - \frac{12}{5}$ | Simplifying |
| $y = -\frac{3}{5}x + \frac{28}{5}$ | Adding 8 to both sideds |
| $f(x) = -\frac{3}{5}x + \frac{28}{5}$ | Function notation |

To graph the line, plot the point $(-4, 8)$ and the

y-intercept $\left(0, \dfrac{28}{5}\right) = \left(0, 5\dfrac{3}{5}\right)$, and then draw the

line passing through both points.

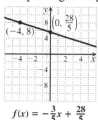

$f(x) = -\dfrac{3}{5}x + \dfrac{28}{5}$

31. | $y - y_1 = m(x - x_1)$ | Point-slope equation |
|---|---|
| $y - (-4) = -0.6(x - (-3))$ | Substituting -0.6 for m and $(-3, -4)$ for (x_1, y_1) |
| $y + 4 = -0.6x - 1.8$ | Simplifying |
| $y = -0.6x - 5.8$ | Subtracting 4 from both sides |
| $f(x) = -0.6x - 5.8$ | Function notation |

To graph the line, plot the point $(-3, -4)$ and the

y-intercept $(0, -5.8)$, and then draw the line

passing through both points.

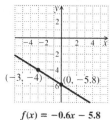

$f(x) = -0.6x - 5.8$

33. $y - y_1 = m(x - x_1)$ Point-slope equation

      $y - (-6) = \frac{2}{7}(x - 0)$ Substituting $\frac{2}{7}$ for m

                       and $(0, -6)$ for (x_1, y_1)

        $y + 6 = \frac{2}{7}x - 0$ Simplifying

          $y = \frac{2}{7}x - 6$ Subtracting 6 from

                         both sides

       $f(x) = \frac{2}{7}x - 6$ Function notation

To graph the line, plot the point $(0, -6)$. Then, since the given point is the y-intercept, use $m = \dfrac{\text{change in } y}{\text{change in } x} = \dfrac{2}{7}$ to generate a second point $(0 + 7, -6 + 2) = (7, -4)$. Plot this second point and draw the line passing through both points.

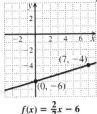

$f(x) = \frac{2}{7}x - 6$

35. $y - y_1 = m(x - x_1)$ Point-slope equation

      $y - 6 = \frac{3}{5}(x - (-4))$ Substituting $\frac{3}{5}$ for

                         m, -4 for x_1, and 6

                         for y_1

      $y - 6 = \frac{3}{5}x + \frac{12}{5}$ Simplifying

         $y = \frac{3}{5}x + \frac{42}{5}$ Simplifying

       $f(x) = \frac{3}{5}x + \frac{42}{5}$ Function notation

To graph the line, plot the point $(-4, 6)$ and the y-intercept $\left(0, \frac{42}{5}\right) = \left(0, 8\frac{2}{5}\right)$, and then draw the line passing through both points.

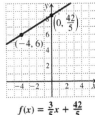

$f(x) = \frac{3}{5}x + \frac{42}{5}$

37. First find the slope of the line.

$$m = \frac{7 - 3}{3 - 2} = \frac{4}{1} = 4$$

Use the point-slope equation with $m = 4$ and $(2, 3) = (x_1, y_1)$ (or $(3, 7) = (x_1, y_1)$).

$$y - 3 = 4(x - 2)$$
$$y - 3 = 4x - 8$$
$$y = 4x - 5$$
$$f(x) = 4x - 5 \quad \text{Function notation}$$

39. First find the slope of the line.

$$m = \frac{5 - (-4)}{3.2 - 1.2} = \frac{5 + 4}{3.2 - 1.2} = \frac{9}{2} = 4.5$$

Use the point-slope equation with $m = 4.5$ and $(1.2, -4) = (x_1, y_1)$ (or $(3.2, 5) = (x_1, y_1)$).

$$y - (-4) = 4.5(x - 1.2)$$
$$y + 4 = 4.5x - 5.4$$
$$y = 4.5x - 9.4$$
$$f(x) = 4.5x - 9.4 \quad \text{Function notation}$$

41. First find the slope of the line.

$$m = \frac{-1 - (-5)}{0 - 2} = \frac{-1 + 5}{0 - 2} = \frac{4}{-2} = -2$$

Observe that the y-intercept is $(0, -1)$. Using $m = -2$ and the slope-intercept equation immediately gives:

$$y = -2x - 1$$
$$f(x) = -2x - 1 \quad \text{Function notation}$$

One could also use the point-slope equation with $m = -2$ and either point as (x_1, y_1).

43. First find the slope of the line

$$m = \frac{-5 - (-10)}{-3 - (-6)} = \frac{-5 + 10}{-3 + 6} = \frac{5}{3}$$

Use the point-slope equation with $m = \frac{5}{3}$ and $(-6, -10) = (x_1, y_1)$ (or $(-3, -5) = (x_1, y_1)$).

$$y - (-10) = \frac{5}{3}(x - (-6))$$
$$y + 10 = \frac{5}{3}(x + 6)$$
$$y + 10 = \frac{5}{3}x + 10$$
$$y = \frac{5}{3}x$$
$$f(x) = \frac{5}{3}x \quad \text{Function notation}$$

45. Plot the given data values using the incandescent wattage as the first coordinate and the corresponding CFL wattage as the second coordinate.

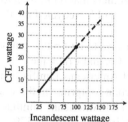

To predict the CFL wattage that creates light equivalent to a 75-watt incandescent bulb, locate the point on the line directly above 75. Then move horizontally to the vertical axis and read the CFL wattage value. The wattage is about 19 watts. To predict the CFL wattage that creates light equivalent to a 120-watt incandescent bulb, locate the point on the line directly above 120. Then move horizontally to the vertical axis and read the CFL wattage value. The wattage is about 30 watts.

47. Plot the given data values using body weight as the first coordinate and the corresponding number of drinks as the second coordinate.

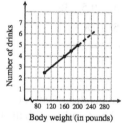

To estimate the number of drinks that a 140-lb person would have to drink to be considered intoxicated, locate the point on the line directly above 140. Then move horizontally to the vertical axis and read the corresponding number of drinks. The estimated number of drinks is 3.5. To estimate the number of drinks that a 230-lb person would have to drink to be considered intoxicated, locate the point on the line directly above 230. Then move horizontally to the vertical axis and read the corresponding number of drinks. The estimated number of drinks is 5.75.

49. a) The problem says to let t = the number of years after 2000, and $a(t)$ = the world production capacity, in millions of vehicles, for the year t. Then the production capacity of 84 million vehicles in 2008, and of 97 million in 2015 correspond to data points: $(2008-2000,84) = (8,84)$ and $(2015-2000,97) = (15,97)$.

Find the slope of the function that fits the data:

$$m = \frac{a(15)-a(8)}{15-8}$$

$$= \frac{97-84}{7} = \frac{13}{7}$$

Use the value of m, and either data point, and substitute them into the point-slope equation for a line.

$$a(t)-84 = \frac{13}{7}(t-8)$$

$$a(t) = \frac{13}{7}t - \frac{104}{7} + 84$$

$$a(t) = \frac{13}{7}t - \frac{104}{7} + \frac{588}{7}$$

$$a(t) = \frac{13}{7}t + \frac{484}{7},$$

or $a(t) = \frac{13t+484}{7}$

b) In 2013, $t = 2013 - 2000 = 13.$

$$a(13) = \frac{13(13)+484}{7} = \frac{169+484}{7}$$

$$= \frac{653}{7} = 93\frac{2}{7} \approx 93.3$$

The model predicts the production capacity in 2013 will be about 93.3 million vehicles.

c) Set $a(t) = 100$ and solve for t.

$$\frac{13t+484}{7} = 100$$

$$13t+484 = 700$$

$$13t = 700 - 484 = 216$$

$$t = \frac{216}{13} \approx 16.6$$

$2000+16.6 = 2016.6$

The model predicts that the production capacity will reach 100 million vehicles in 2016.

51. a) The problem says to let t = the number of years after 1990, and $E(t)$ = the life expectancy, in years, for females born in the year t. Then the expected value of 79.0 years in 1994, and 80.2 years in 2006 correspond to data points:

$(1994-1990, 79.0) = (4, 79.0)$ and

$(2006-1990, 80.2) = (16, 80.2)$.

Find the slope of the function that fits the data:

$$m = \frac{E(16) - E(4)}{16 - 4}$$

$$= \frac{80.2 - 79.0}{12} = \frac{1.2}{12} = 0.1$$

Use the value of m, and either data point, and substitute them into the point-slope equation for a line.

$$E(t) - 79 = 0.1(t - 4)$$

$$E(t) = 0.1t - 0.4 + 79$$

$$E(t) = 0.1t + 78.6$$

b) In 2012, $t = 2012 - 1990 = 22$.

$$E(22) = 0.1(22) + 78.6$$

$$= 2.2 + 78.6 = 80.8$$

The model predicts that the life expectancy for a female born in 2012 will be 80.8 years.

53. a) The problem says to let t = the number of years after 2000, and $N(t)$ = the amount of solid waste recycled, in millions of tons, in the year t. Then the 53 million tons recycled in 2000, and the 61 million tons recycled in 2008 correspond to data points:

$(2000-2000, 53) = (0, 53)$ and

$(2008-2000, 61) = (8, 61)$.

Find the slope of the function that fits the data:

$$m = \frac{N(8) - N(0)}{8 - 0}$$

$$= \frac{61 - 53}{8} = \frac{8}{8} = 1$$

Use the value of m, and either data point, and substitute them into the point-slope equation for a line.

$$N(t) - 53 = 1(t - 0)$$

$$N(t) = t + 53$$

b) In 2012, $t = 2012 - 2000 = 12$.

$$N(12) = 12 + 53$$

$$= 65$$

The model predicts that 65 million tons of solid waste will be recycled in 2012.

55. a) The problem says to let t = the number of years after 2006, and $C(t)$ = the percentage of Americans familiar with the term "carbon footprint" in the year t. Then the 38% of Americans familiar with the term in 2007, and the 57% familiar with the term in 2009 correspond to data points:

$(2007-2006, 38) = (1, 38)$ and

$(2009-2006, 57) = (3, 57)$.

Find the slope of the function that fits the data:

$$m = \frac{C(3) - C(1)}{3 - 1}$$

$$= \frac{57 - 38}{2} = \frac{19}{2} = 9.5$$

Use the value of m, and either data point, and substitute them into the point-slope equation for a line.

$$C(t) - 38 = 9.5(t - 1)$$

$$C(t) = 9.5t - 9.5 + 38$$

$$C(t) = 9.5t + 28.5$$

b) In 2012, $t = 2012 - 2006 = 6$.

$$C(6) = 9.5(6) + 28.5$$

$$= 57 + 28.5 = 85.5$$

The model predicts that 85.5% of Americans will be familiar with the term "carbon footprint" in 2012.

c) Set $C(t) = 100$ and solve for t.

$$9.5t + 28.5 = 100$$

$$9.5t = 100 - 28.5$$

$$9.5t = 71.5$$

$$t = \frac{71.5}{9.5} \approx 7.5$$

$2006 + 7.5 = 2013.5$

The model predicts all Americans will be familiar with the term "carbon footprint" in 2013.

57. a) The problem says to let t = the number of years after 2000, and $N(t)$ = the number of American households, in millions, that conducted some online banking in the year t. Then the 54 million households using online banking in 2009, and the 66 million households using online banking in 2014 correspond to data points:
$(2009 - 2000, 54) = (9, 54)$ and
$(2014 - 2000, 66) = (14, 66)$.
Find the slope of the function that fits the data:
$$m = \frac{N(14) - N(9)}{14 - 9}$$
$$= \frac{66 - 54}{5} = \frac{12}{5} = 2.4$$
Use the value of m, and either data point, and substitute them into the point-slope equation for a line.
$$N(t) - 54 = 2.4(t - 9)$$
$$N(t) = 2.4t - 21.6 + 54$$
$$N(t) = 2.4t + 32.4$$

b) In 2019, $t = 2019 - 2000 = 19$.
$$N(19) = 2.4(19) + 32.4$$
$$= 45.6 + 32.4 = 78$$
The model predicts that 78 million American households will use online banking in 2019.

c) Set $N(t) = 100$ and solve for t.
$$2.4t + 32.4 = 100$$
$$2.4t = 100 - 32.4 = 67.6$$
$$t = \frac{67.6}{2.4} \approx 28.2$$
$2000 + 28.2 = 2028.2$
The model predicts that the number of American households using online banking will reach 100 million in 2028.

59. a) The problem says to let t = the number of years after 1999, and $R(t)$ = the record time, in seconds, in the 100-m run in the year t. Then the record of 9.79 seconds in 1999, and 9.58 seconds in 2009, correspond to data points:
$(1999 - 1999, 9.79) = (0, 9.79)$ and
$(2009 - 1999, 9.58) = (10, 9.58)$.
Find the slope of the function that fits the data:

$$m = \frac{R(10) - R(0)}{10 - 0}$$
$$= \frac{9.58 - 9.79}{10} = -\frac{0.21}{10} = -0.021$$
Use the value of m, and either data point, and substitute them into the point-slope equation for a line.
$$R(t) - 9.79 = -0.021(t - 0)$$
$$R(t) = -0.021t + 9.79$$

b) In 2015, $t = 2015 - 1999 = 16$.
$$R(16) = -0.021(16) + 9.79$$
$$= -0.336 + 9.79 = 9.454$$
The model predicts that the record in the 100-m run will be 9.454 seconds in 2015.

In 2030, $t = 2030 - 1999 = 31$.
$$R(31) = -0.021(31) + 9.79$$
$$= -0.651 + 9.79 = 9.139$$
The model predicts that the record in the 100-m run will be 9.139 seconds in 2030.

c) Set $R(t) = 9.5$ and solve for t.
$$-0.021t + 9.79 = 9.5$$
$$-0.021t = 9.5 - 9.79 = -0.29$$
$$t = \frac{-0.29}{-0.021} \approx 13.8$$
$1999 + 13.8 = 2012.8$
The model predicts that the record in the 100-m run will be 9.5 seconds in 2012.

61. The points lie approximately in a straight line, so the graph of this data is linear.

63. The points do not lie on a straight line, so the data are not linear.

65. The points lie approximately in a straight line, so the graph of this data is linear.

67. a) The problem says to let x = the number of years since 1900, and W = the life expectancy, in years, of a female born in the year x. Enter the data with the number of years since 1900 in L_1 and life expectancy in L_2.

L1	L2	L3	1
60	73.1	------	
70	74.4		
80	77.5		
90	78.8		
100	79.7		
106	80.2		
------	------	------	

$L1(1)=60$

Use the LinReg feature of the graphing calculator to find the equation of the line.

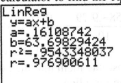

The second screen indicates that the equation is $W = 0.1611x + 63.6983$.

b) In 2012, $x = 2012 - 1900 = 112$. To predict the life expectancy of a female born in 2012, find $W(112)$.

$$W(112) = 0.1611(112) + 63.6983$$
$$= 18.0432 + 63.6983$$
$$= 81.7415 \approx 81.7$$

This estimate is $81.7 - 80.8 = 0.9$ of a year higher than the previous estimate.

69. a) The problem says to let t = the number of years since 2000, and N = the number of registered nurses, in millions, employed in the year t. Enter the data with the number of years since 2000 in L_1 and the number of registered nurses in L_2.

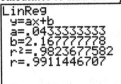

Use the LinReg feature of the graphing calculator to find the equation of the line.

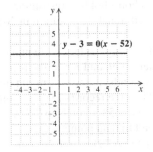

The second screen indicates that the equation is $N = 0.0433t + 2.1678$.

b) In 2012, $t = 2012 - 2000 = 12$. To estimate the number of registered nurses in 2012, find $N(12)$.

$$N(12) = 0.0433(12) + 2.1678$$
$$= 0.5196 + 2.1678$$
$$= 2.6874 \approx 2.69$$

The model predicts there will be approximately 2.69 million registered nurses in the U.S. in 2112.

71. *Thinking and Writing Exercise.*

73. $(2x^2 - x) + (3x - 5) = 2x^2 - x + 3x - 5$
$$= 2x^2 + (-1 + 3)x - 5$$
$$= 2x^2 + 2x - 5$$

75. $(2t - 1) - (t - 3) = 2t - 1 - t + 3$
$$= (2 - 1)t + (-1 + 3)$$
$$= t + 2$$

77. $f(x) = \dfrac{x}{x - 3}$

$\dfrac{x}{x - 3}$ is undefined when its denominator is 0.
Find the value(s) of x that make $x - 3 = 0$.
$$x - 3 = 0$$
$$x = 3$$
3 is not in the domain of $f(x)$. The domain of $f(x)$ is $\{x | x \text{ is a real number and } x \neq 3\}$.

79. $g(x) = |6x + 11|$

$|6x + 11|$ is well-defined for any real number x. The domain is the set of all real numbers.

81. Thinking *and Writing Exercise.*

83. First simplify the equation.
$$y - 3 = 0(x - 52)$$
$$y - 3 = 0$$
$$y = 3$$
$y = c$ is the form of a horizontal line that passes through the y-axis at $(0, c)$. In this case, $c = 3$, so the graph is:

85. Two points with coordinates $(4,-1)$ and $(3,-3)$ are shown on the graph. First determine the slope of the line.
$$m = \frac{-1-(-3)}{4-3} = \frac{-1+3}{4-3} = \frac{2}{1} = 2$$
Use the point-slope equation with $m = 2$ and $(x_1, y_1) = (4,-1)$ (or $(x_1, y_1) = (3,-3)$) to determine the required equation.
$$y - (-1) = 2(x - 4)$$
$$y + 1 = 2x - 8$$
$$y = 2x - 9$$

87. First solve the equation for y and determine the slope of the given line.
$$x + 2y = 6 \qquad \text{Given line}$$
$$2y = -x + 6$$
$$y = -\frac{1}{2}x + 3 \quad m = -\frac{1}{2}$$
The slope of the given line is $-\frac{1}{2}$. Every line parallel to the given line must also have a slope of $-\frac{1}{2}$. Find the equation of the line with a slope of $-\frac{1}{2}$, containing the point $(3,7)$.
$$y - y_1 = m(x - x_1) \quad \text{Point-slope equation}$$
$$y - 7 = -\frac{1}{2}(x - 3)$$
$$y - 7 = -\frac{1}{2}x + \frac{3}{2}$$
$$y = -\frac{1}{2}x + \frac{17}{2}$$

89. First solve the equation for y and determine the slope of the given line.
$$2x + y = -3 \qquad \text{Given line}$$
$$y = -2x - 3 \quad m = -2$$
The slope of the given line is -2. Every line perpendicular to the given line must have a slope of $-\frac{1}{-2} = \frac{1}{2}$. Find the equation of the line with a slope of $\frac{1}{2}$, containing the point

$(2,5)$.
$$y - 5 = \frac{1}{2}(x - 2) \quad \text{Substituting}$$
$$y - 5 = \frac{1}{2}x - 1$$
$$y = \frac{1}{2}x + 4$$

91. a) Use the two points $(3,-5)$ and $(7,-1)$ to determine the slope of the linear function. Then use the point-slope form of a line to find $g(x)$. :
$$m = \frac{-1-(-5)}{7-3} = \frac{-1+5}{4} = \frac{4}{4} = 1$$
$$y - (-5) = 1(x - 3)$$
$$y + 5 = x - 3$$
$$y = x - 8$$
$$g(x) = x - 8 \quad \text{Function notation}$$
   b) $g(-2) = -2 - 8 = -10$
   c) $g(a) = a - 8$
   If $g(a) = 75$, we have
$$a - 8 = 75$$
$$a = 83.$$

Mid-Chapter Review

Guided Solutions

1. To find the y-intercept of the line $y - 3x = 6$, set x to 0 in the equation and solve for y.
   y-intercept: $y - 3 \cdot 0 = 6$
   $$y = 6$$
   The y-intercept $(0,6)$.
   To find the x-intercept of the line $y - 3x = 6$, set y to 0 in the equation and solve for x.
   x-intercept: $0 - 3x = 6$
   $$-3x = 6$$
   $$x = -2$$
   The x-intercept $(-2,0)$.

2. For the line containing $(1,5)$ and $(3,-1)$, the slope is:

$$m = \frac{y_2 - y_1}{x_2 - x_1} = \frac{-1-5}{3-1}$$
$$= \frac{-6}{2}$$
$$= -3$$

Mixed Review

1. $2x + 5y = 8$ is in standard form.

2. $y = \frac{2}{3}x - \frac{11}{3}$ is in slope-intercept form.

3. $x - 13 = 5y$ is in none of these forms.

4. $y - 2 = \frac{1}{3}(x-6)$ is in point-slope form.

5. $x - y = 1$ is in standard form.

6. $y = -18x + 3.6$ is in slope-intercept form.

7. $(-5,-2)$ and $(1,8)$

$$m = \frac{8-(-2)}{1-(-5)} = \frac{8+2}{1+5} = \frac{10}{6} = \frac{5}{3}$$

8. $(0,0)$ and $(0,-2)$

$$m = \frac{0-(-2)}{0-0} = \frac{2}{0}$$

Since division by zero is undefined, the slope of this line is undefined.

9. The line $y = 4$ is a horizontal line. The slope is 0.

10. The line $x = -7$ is a vertical line. The slope is undefined.

11. $x - 3y = 1$

$$-3y = -x + 1$$
$$y = \frac{1}{3}x - \frac{1}{3}$$

Slope is $\frac{1}{3}$; y-intercept is $\left(0, -\frac{1}{3}\right)$.

12. Substitute the values of the given slope $m = -3$ and the given y-intercept $b = 7$ into the slope-intercept form of a linear function $f(x) = mx + b$.

$$f(x) = -3x + 7.$$

13. Substitute the values of the given slope $m = 5$ and the given point $(x_1, y_1) = (-3,7)$ into the point-slope form of a line $y - y_1 = m(x - x_1)$.

$$y - 7 = 5(x - (-3)), \text{ or } y - 7 = 5(x + 3).$$

14. First use the given points to determine the slope of the line:

$$m = \frac{-5-(-1)}{-2-4} = \frac{-5+1}{-6} = \frac{-4}{-6} = \frac{2}{3}$$

Then substitute the values of the slope and the given point $(x_1, y_1) = (4,-1)$ into the point-slope form of a line $y - y_1 = m(x - x_1)$.

$$y - (-1) = \frac{2}{3}(x - 4)$$
$$y + 1 = \frac{2}{3}x - \frac{8}{3}$$
$$y = \frac{2}{3}x - \frac{11}{3}$$

15. Graph $y = 2x - 1$.

16. Graph $3x + y = 6$.

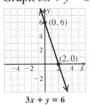

17. Graph $y - 2 = \frac{1}{2}(x - 1)$.

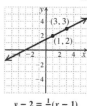

$y - 2 = \frac{1}{2}(x - 1)$

18. Graph $f(x) = 4$.

19. Graph $f(x) = -\frac{3}{4}x + 5$.

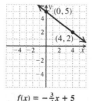

$f(x) = -\frac{3}{4}x + 5$

20. Graph $3x = 12$. First simplify the equation.

$$3x = 12$$
$$\frac{1}{3} \cdot 3x = \frac{1}{3} \cdot 12$$
$$x = 4$$

Exercise Set 2.5

1. difference

3. evaluate

5. excluding

7. $f(2) = -3 \cdot 2 + 1 = -5$
$g(2) = 2^2 + 2 = 6$
$f(2) + g(2) = -5 + 6 = 1$

9. $f(5) = -3 \cdot 5 + 1 = -15 + 1 = -14$
$g(5) = 5^2 + 2 = 25 + 2 = 27$
$f(5) - g(5) = -14 - 27 = -41$

11. $f(-1) = -3(-1) + 1 = 3 + 1 = 4$
$g(-1) = (-1)^2 + 2 = 1 + 2 = 3$
$(f \cdot g)(-1) = 4 \cdot 3 = 12$

13. $f(-4) = -3(-4) + 1 = 12 + 1 = 13$
$g(-4) = (-4)^2 + 2 = 16 + 2 = 18$
$f(-4) / g(-4) = \frac{13}{18}$

15. $g(1) = 1^2 + 2 = 1 + 2 = 3$
$f(1) = -3 \cdot 1 + 1 = -3 + 1 = -2$
$g(1) - f(1) = 3 - (-2) = 5$

17. $(f + g)(x) = f(x) + g(x)$
$= (-3x + 1) + (x^2 + 2)$
$= x^2 - 3x + 3$

19. $(f - g)(x) = f(x) - g(x)$
$= (-3x + 1) - (x^2 + 2)$
$= -3x + 1 - x^2 - 2$
$= -x^2 - 3x - 1$

21. $(F + G)(x) = F(x) + G(x)$
$= (x^2 - 2) + (5 - x)$
$= x^2 - x + 3$

23. $F(-4) = (-4)^2 - 2 = 16 - 2 = 14$
$G(-4) = 5 - (-4) = 5 + 4 = 9$
$(F + G)(-4) = F(-4) + G(-4) = 14 + 9 = 23$

25. $F(3) = 3^2 - 2 = 9 - 2 = 7$
$G(3) = 5 - 3 = 2$
$(F - G)(3) = F(3) - G(3) = 7 - 2 = 5$

27. $F(-3) = (-3)^2 - 2 = 9 - 2 = 7$
$G(-3) = 5 - (-3) = 5 + 3 = 8$
$(F \cdot G)(-3) = F(-3) \cdot G(-3) = 7 \cdot 8 = 56$

29. $(F-G)(a) = F(a) - G(a)$
$$= (a^2 - 2) - (5 - a)$$
$$= a^2 - 2 - 5 + a$$
$$= a^2 + a - 7$$

31. $(F-G)(x) = F(x) - G(x)$
$$= (x^2 - 2) - (5 - x)$$
$$= x^2 - 2 - 5 + x$$
$$= x^2 + x - 7$$

33. $F(-2) = (-2)^2 - 2 = 4 - 2 = 2$
$$G(-2) = 5 - (-2) = 5 + 2 = 7$$
$$(F/G)(-2) = \frac{F(-2)}{G(-2)} = \frac{2}{7}$$

35. Answers may vary slightly. Locate 2 on the horizontal axis then move vertically to the graph of P. $P(2) \approx 26.5$. Locate 2 on the horizontal axis then move vertically to the graph of L. $L(2) \approx 22.5$.
$(P-L)(2) = P(2) - L(2) \approx 26.5 - 22.5 = 4\%$.

37. Answers may vary slightly.
Using the graph,
$C(2004) \approx 1.2$ and $B(2004) \approx 2.9$.
$N(2004) = C(2004) + B(2004)$
$$\approx 1.2 + 2.9$$
$$= 4.1 \text{ million}$$
We estimate that 4.1 million U.S. women had children in 2004.

39. Answers may vary slightly.
Since the p and r bands are stacked on top of one another, the value of $(p+r)('05)$ can be found by subtracting the value at the bottom of the r band from the value at the top of the p band for '05. Thus:
$(p+r)('05) \approx 256 - 162 = 94$ million tons.
This value represents the amount of trash that was either composted or recycled in the U.S. in 2005.

41. Answers may vary slightly.
$F('96) \approx 215$ million tons. This value represents the total amount of trash generated in the U.S. in 1996.

43. Answers may vary slightly.
$(F-p)('04)$ is found graphically by determining the value at the bottom of the p band for 2004. Thus, $(F-p)('04) \approx 231$ million tons. This value represents the amount of trash generated in the U.S. in 2004 that was not composted.

45. $f(x) = x^2$; $g(x) = 7x - 4$
The domain of f is $\mathbb{R}$. The domain of g is $\mathbb{R}$.
The domains of $f+g$, $f-g$, and $f \cdot g$ are all equal to the set of elements common to the domains of f and g. The domain of $f+g$, $f-g$, and $f \cdot g$ is the set of elements common to the domains of f and g:
$\mathbb{R} = \{x \mid x \text{ is a real number}\}$.

47. $f(x) = \frac{1}{x-3}$; $g(x) = 4x^3$
The domain of $f(x)$ is $\{x \mid x \text{ is a real number, and } x \neq 3\}$. The domain $g(x)$ is $\mathbb{R}$. The domain of $f+g$, $f-g$, and $f \cdot g$ is the set of elements common to the domains of f and g: $\{x \mid x \text{ is a real number, and } x \neq 3\}$.

49. $f(x) = \frac{2}{x}$; $g(x) = x^2 - 4$
The domain of f is $\{x \mid x \text{ is a real number, and } x \neq 0\}$. The domain of g is $\mathbb{R}$. The domain of $f+g$, $f-g$, and $f \cdot g$ is the set of elements common to the domains of f and g: $\{x \mid x \text{ is a real number, and } x \neq 0\}$.

51. $f(x) = x + \frac{2}{x-1}$; $g(x) = 3x^3$
The domain of $f(x)$ is $\{x \mid x \text{ is a real number, and } x \neq 1\}$. The domain $g(x)$ is $\mathbb{R}$. The domain of $f+g$, $f-g$, and $f \cdot g$ is the set of elements common to the domains of f and g: $\{x \mid x \text{ is a real number, and } x \neq 1\}$.

53. $f(x) = \dfrac{x}{2x-9}$; $g(x) = \dfrac{5}{1-x}$

The domain of f is $= \{x \mid x$ in $\mathbb{R},\ x \neq \dfrac{9}{2}\}$. The

domain of g is $= \{x \mid x$ in $\mathbb{R},\ x \neq 1\}$. The

domain of $f + g$, $f - g$, and $f \cdot g$ is the set of

elements common to the domains of f and g:

$\{x \mid x$ in $\mathbb{R},\ x \neq \dfrac{9}{2},$ or $1\}$.

55. $f(x) = x^4$; $g(x) = x - 3$

The domain of f is $\mathbb{R}$. The domain of g is $\mathbb{R}$.

$g(x) = 0$ has only one solution, $x = 3$.

The domain of $f / g = \dfrac{x^4}{x-3}$ is:

$\{x \mid x$ is a real number, $x \neq 3\}$.

57. $f(x) = 3x - 2$; $g(x) = 2x - 8$

The domain of f is $\mathbb{R}$. The domain of g is $\mathbb{R}$.

$2x - 8 = 0$

$2x = 8$

$x = 4$ So, $x \neq 4$.

$g(x) = 0$ has only one solution, $x = 4$.

The domain of $f / g = \dfrac{3x-2}{2x-8}$ is:

$\{x \mid x$ is a real number, $x \neq 4\}$

59. $f(x) = \dfrac{3}{x-4}$; $g(x) = 5 - x$

The domain of f is $= \{x \mid x$ is a real number

and $x \neq 4\}$. The domain of g is $\mathbb{R}$.

$g(x) = 0$ has only one solution, $x = 5$.

The domain of

$f / g = \left(\dfrac{3}{x-4}\right) / (5-x) = \dfrac{3}{(x-4)(5-x)}$ is:

$\{x \mid x$ is a real number, $x \neq 4$ and $x \neq 5\}$.

61. $f(x) = \dfrac{2x}{x+1}$; $g(x) = 2x + 5$

The domain of f is $= \{x \mid x$ is a real number

and $x \neq -1\}$. The domain of g is $\mathbb{R}$.

$2x + 5 = 0$

$2x = -5$

$x = -\dfrac{5}{2}$

$g(x) = 0$ has only one solution, $x = -\dfrac{5}{2}$.

The domain of $f / g = \left(\dfrac{2x}{x+1}\right) / (2x+5)$

$= \dfrac{2x}{(x+1)(2x+5)}$ is:

$\{x \mid x$ is a real number, $x \neq -1$ and $x \neq -\dfrac{5}{2}\}$.

63. From the graph:

$F(5) = 1$ and $G(5) = 3$.

$(F+G)(5) = F(5) + G(5) = 1 + 3 = 4$

$F(7) = -1$ and $G(7) = 4$.

$(F+G)(7) = F(7) + G(7) = -1 + 4 = 3$

65. From the graph:

$G(7) = 4$ and $F(7) = -1$

$(G-F)(7) = G(7) - F(7)$

$= 4 - (-1)$

$= 4 + 1 = 5$

$G(3) = 1$ and $F(3) = 2$

$(G-F)(3) = G(3) - F(3)$

$= 1 - 2 = -1$

67. The x-values of f are from 0 to 9, so the

domain of $F = \{x \mid 0 \leq x \leq 9\}$.

The x-values of g are from 3 to 10, so the

domain of $G = \{x \mid 3 \leq x \leq 10\}$.

The domain of $F + G$ is the set of numbers

common to the domains of F and G, so the

domain of $F + G = \{x \mid 3 \leq x \leq 9\}$.

The domain of F / G is the set of numbers

common to the domains of F and G,

excluding any number(s) where $G = 0$.

There are no exclusions. The domain of

$F / G = \{x \mid 3 \leq x \leq 9\}$.

69. Determine the values of $F + G$ at some values in its domain and graph these points. Then join these points with a smooth curve.

	$F + G$
3	$2 + 1 = 3$
5	$1 + 3 = 4$
7	$-1 + 4 = 3$
9	$1 + 2 = 3$

71. *Thinking and Writing Exercise.*

73. $x - 6y = 3$

$$-6y = -x + 3$$

$$y = \frac{1}{6}x - \frac{1}{2}$$

75. $5x + 2y = -3$

$$2y = -5x - 3$$

$$y = -\frac{5}{2}x - \frac{3}{2}$$

77. Let n represent the number. $2n + 5 = 49$.

79. Let n represent the smaller integer.

$$n + (n + 1) = 145.$$

81. *Thinking and Writing Exercise.*

83. $f(x) = \dfrac{3x}{2x + 5}$; $g(x) = \dfrac{x^4 - 1}{3x + 9}$

$f : 2x + 5 = 0$

$$2x = -5$$

$$x = \frac{-5}{2}, \text{ So } x \neq \frac{-5}{2}$$

Domain of $f = \left\{ x \mid x \text{ is a real number, } x \neq \frac{-5}{2} \right\}$

$g : 3x + 9 = 0$

$$3x = -9$$

$$x = -3; \text{ So } x \neq -3$$

Domain of $g = \{ x \mid x \text{ is a real number, }$

$x \neq -3 \}$.

$$g(x) = 0$$

$$x^4 - 1 = 0$$

$$(x^2 + 1)(x + 1)(x - 1) = 0$$

$$x = \pm 1$$

$$f / g = \frac{3x}{2x + 5} \bigg/ \frac{x^4 - 1}{3x + 9} = \frac{3x(3x + 9)}{(2x + 5)(x^4 - 1)}$$

Therefore, the domain of $f / g = \{ x \mid x$ is a real number, $x \neq \frac{-5}{2}$, $x \neq -3$, and $x \neq \pm 1 \}$.

85. Answers may vary. The two functions must be defined over the intervals $[-2, 3]$, but must not *both* be defined anywhere else. To remove the value $x = 1$ from the domain of f / g, make $g(1) = 0$.

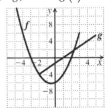

87. The domain of m is $\{ x \mid -1 < x < 5 \}$.

$n(x) = 0$ when $2x - 3 = 0$

$$2x = 3$$

$$x = \frac{3}{2}$$

We exclude this value from the domain of m / n. The domain of m / n is:

$$\left\{ x \mid -1 < x < 5 \text{ and } x \neq \frac{3}{2} \right\}.$$

89. Answers may vary. Since the domain of $f + g$ is the intersection of the individual domains of f and g, at least one of the functions needs to be undefined at $x = -2$, and at least one needs to be undefined at $x = 5$. Both functions should be well-defined everywhere else.

$$f(x) = \frac{1}{x + 2}; \quad g(x) = \frac{1}{x - 5}$$

91. $y_1 = 2.5x + 1.5$ and $y_2 = x - 3$

$$\frac{y_1}{y_2} = \frac{2.5x + 1.5}{x - 3}$$

Domain: $\{x \mid$ is any real number $and \ x \neq 3\}$

The CONNECTED mode graph contains values that cross the line $x = 3$, whereas, the DOT mode graph contains no points having 3 as the first coordinate. Thus, the DOT mode graph represents y_3 more accurately.

Chapter 2 Study Summary

1. $f(x) = 2 - 3x$

    $f(-1) = 2 - 3(-1) = 2 + 3 = 5$

2. Use the vertical line test by visualizing a vertical line moving across the graph. No vertical line will intersect the graph more than once. Thus the graph is the graph of a function.

3. The domain is the set of all real numbers, or $\mathbb{R}$. The range is the set of all y-values in the graph. It is $\{y \mid y \geq -2\}$ or $[-2, \infty)$

4. $f(x) = \dfrac{1}{4}x - 5$ is well-defined for all real numbers. The domain is, therefore, the set of all real numbers.

5. $(1, 4)$ and $(-9, 3)$

    $$m = \frac{3 - 4}{-9 - 1} = \frac{-1}{-10} = \frac{1}{10}$$

6. $y = -4x + \frac{2}{5}$

    $y = mx + b$

    The slope m is -4, and the y-intercept $(0, b)$ is $\left(0, \frac{2}{5}\right)$.

7. $y = \frac{1}{2}x + 2$

    The slope m is $\dfrac{1}{2}$ and the y-intercept $(0, b)$ is $(0, 2)$. First plot the y-intercept $(0, 2)$. Then, starting at the point $(0, 2)$, use the slope, $\dfrac{1}{2}$, to find another point by moving up

1 unit and right 2 units to $(2, 3)$. Then draw the line through the two points.

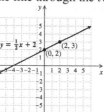

8. The line $y = -2$ is a horizontal line with y-intercept $(0, -2)$.

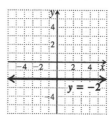

9. The line $x = 3$ is a vertical line with an x-intercept of $(3, 0)$.

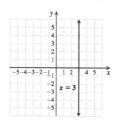

10. $10x - y = 10$

    Find the x-intercept, by letting $y = 0$ and solving for x.

    $10x - 0 = 10$

    $10x = 10$

    $x = 1$

    The x-intercept is $(1, 0)$.

    Find the y-intercept by letting $x = 0$ and solving for y.

    $10 \cdot 0 - y = 10$

    $-y = 10$

    $y = -10$

    The y-intercept is $(0, -10)$.

11. Two lines are parallel if their slopes are equal. The equation $y = 4x - 12$ is in slope-intercept form, and its slope is 4. Rewrite the second equation in slope-intercept form:
$$4y = x - 9$$
$$y = \frac{1}{4}x - \frac{9}{4}$$
Its slope is $\frac{1}{4}$. Since the lines have different slopes, their graphs are not parallel.

12. Two lines are perpendicular if the product of their slopes is -1. The equation $y = x - 7$ is in slope-intercept form. It has a slope of 1. Rewrite the second equation in slope-intercept form.
$$x + y = 3$$
$$y = -x + 3$$
Its slope is -1. Since $1(-1) = -1$, their graphs are perpendicular lines.

13. $m = \frac{1}{4}; (-1,6)$
The point-slope equation is
$$y - y_1 = m(x - x_1).$$
Substitute $\frac{1}{4}$ for m, -1 for x_1, and 6 for y_1:
$$y - 6 = \frac{1}{4}\left(x - (-1)\right).$$

14. $(f + g)(x) = f(x) + g(x)$
$$= (x - 2) + (x - 7)$$
$$= x - 2 + x - 7$$
$$= 2x - 9$$

15. $(f - g)(x) = f(x) - g(x)$
$$= (x - 2) - (x - 7)$$
$$= x - 2 - x + 7$$
$$= 5$$

16. $(f \cdot g)(5) = f(5) \cdot g(5)$
$$= (5 - 2)(5 - 7)$$
$$= 3(-2) = -6$$

17. $(f / g)(x) = f(x) / g(x) = \dfrac{x - 2}{x - 7}.$

Chapter 2 Review Exercises

1. True

2. False

3. False

4. False

5. False

6. True

7. True

8. True

9. False

10. True

11. a) To use the vertical line test visualize a vertical line moving across the graph. No vertical line will intersect the graph more than once. Thus, the graph is the graph of a function.

    b) The domain is the set of all x values: $\{x \mid -4 \le x \le 5\}$.
    The range is the set of all y-values: $\{y \mid -2 \le y \le 4\}$.

12. a) To use the vertical line test visualize a vertical line moving across the graph. No vertical line will intersect the graph more than once. Thus the graph is the graph of a function.

    b) The domain is the set of all x values: ll real numbers or $\{x \mid x \text{ is a real number}\}$.
    The range is the set of all y values: $\{y \mid y \ge 1\}$.

13. a) To use the vertical line test visualize a vertical line moving across the graph. It is possible for a vertical line to intersect the graph more than once. Thus this is not the graph of a function.

14. a) On the graph provided, find the point with an x value of 3. The point is $(3,0)$.
    Therefore $f(3) = 0$

    b) Find the point or points with a y value of -2. The y value is -2 at the point $(-4,-2)$: $f(-4) = -2$.

15. a) $g(1) = \dfrac{4}{2(1)+1} = \dfrac{4}{3}$

   b) Since $\dfrac{4}{2x+1}$ is undefined when the

   denominator is 0, find the x-value(s)
   where $2x+1=0$.
   $2x+1=0$

   $\quad 2x = -1$ Adding -1 to both sides.

   $\qquad x = -\dfrac{1}{2}$ Dividing both sides by 2

   Thus, $-\dfrac{1}{2}$ is not in the domain of g. The

   domain of g is:

   $\left\{ x \mid x \text{ is a real number } and \ x \neq -\dfrac{1}{2} \right\}$.

16.
   $f(x) = \begin{cases} x^2, & \text{if } x < 0, \\ 3x-5, & \text{if } 0 \le x \le 2 \\ x+7, & \text{if } x > 2 \end{cases}$

   a) Since $0 \le 0 \le 2$, $f(0) = 3 \cdot 0 - 5 = -5$.

   b) Since $3 > 2$, $f(3) = 3 + 7 = 10$.

17. The equation $g(x) = -4x - 9$ is in slope-
    intercept form.
    Slope is -4; y-intercept is $(0, -9)$

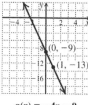

   $g(x) = -4x - 9$

18. Convert the equation to slope-intercept form.
    $-6y + 2x = 14$

    $\quad -6y = -2x + 14$

    $\qquad y = \dfrac{2}{6}x - \dfrac{14}{6} = \dfrac{1}{3}x - \dfrac{7}{3}$

    Slope is $\dfrac{1}{3}$; y-intercept is $\left(0, -\dfrac{7}{3}\right)$.

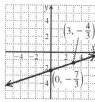

   $-6y + 2x = 14$

19. To find the rate of change, or slope, of the
    graph, select any two points, say $(2, 75)$ and
    $(8, 120)$ and use the slope formula.

    $m = \dfrac{y_2 - y_1}{x_2 - x_1} = \dfrac{120 - 75}{8 - 2} = \dfrac{45}{6} = 7.5$

    The rate of change in the apartment's value is
    $\$7500$ per year.

20. Let $(x_1, y_1) = (4, 5)$ and $(x_2, y_2) = (-3, 1)$.

    $m = \dfrac{y_2 - y_1}{x_2 - x_1} = \dfrac{1 - 5}{-3 - 4} = \dfrac{-4}{-7} = \dfrac{4}{7}$

21. Let $(x_1, y_1) = (-16.4, 2.8)$ and
    $(x_2, y_2) = (-16.4, 3.5)$.

    $m = \dfrac{y_2 - y_1}{x_2 - x_1} = \dfrac{3.5 - 2.8}{-16.4 - (-16.4)} = \dfrac{0.7}{0}$

    The slope of the line containing the points is
    undefined since division by zero is undefined.

22. $C(t) = 11t + 1542$
    The slope 11 signifies the rate of change in
    the average number of calories consumed
    each day. The average number of calories
    consumed each day has increased by 11
    calories per year since 1971. The intercept
    1542 signifies the average number of calories
    consumed each day in 1971.

23. Simplify the equation.
    $y + 3 = 7$

    $\quad y = 4$

    The graph of $y = 4$ is a horizontal line. Its
    slope is 0.

24. Simplify the equation.
    $-2x = 9$

    $\quad x = -\dfrac{9}{2}$

    The graph of $x = -\dfrac{9}{2}$ is a vertical line. Its

    slope is undefined.

25. $3x - 2y = 8$

To find the y-intercept, let $x = 0$ and solve for y.

$$3 \cdot 0 - 2y = 8$$
$$-2y = 8$$
$$y = -4$$

The y-intercept is $(0, -4)$.

To find the x-intercept, let $y = 0$ and solve for x.

$$3x - 2 \cdot 0 = 8$$
$$3x = 8$$
$$x = \frac{8}{3}$$

The x-intercept is $\left(\frac{8}{3}, 0\right)$.

26. The equation $y = -3x + 2$ is in slope-intercept form. The slope is $-3 = \frac{-3}{1}$. The y-intercept is $(0, 2)$. From the y-intercept, go down 3 units and to the right 1 unit to $(1, -1)$. Draw the line through the two points.

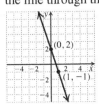

$$y = -3x + 2$$

27. Graph $-2x + 4y = 8$ using the intercepts.

To find the y-intercept, let $x = 0$ and solve for y.

$$-2 \cdot 0 + 4y = 8$$
$$y = 2$$

The y-intercept is $(0, 2)$.

To find the x-intercept, let $y = 0$ and solve for x.

$$-2x + 4 \cdot 0 = 8$$
$$-2x = 8$$
$$x = -4$$

The x-intercept is $(-4, 0)$.

Plot these points and draw the line.

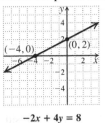

$$-2x + 4y = 8$$

28. The graph of $y = 6$ is a horizontal line that crosses the y-axis at $(0, 6)$.

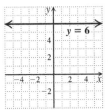

29. $y + 1 = \frac{3}{4}(x - 5)$

$$y - (-1) = \frac{3}{4}(x - 5)$$

The equation is in point-slope form. Starting at the point $(5, -1)$, use the slope to determine a second point.

$$(x, y) = (5 - 4, -1 - 3) = (1, -4)$$

Use the two points to draw the line.

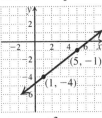

$$y + 1 = \frac{3}{4}(x - 5)$$

30. Simplify the equation.
$$8x + 32 = 0$$
$$x = -4.$$
This is a vertical line that crosses the x-axis at $(-4, 0)$.

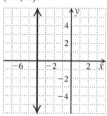

8x + 32 = 0

31. $g(x) = 15 - x = -x + 15.$.

The slope is $-1 = \dfrac{-1}{1}$. The y-intercept is $(0, 15)$. From the y-intercept, go down 1 unit and right 1 unit to the point $(1, 14)$. Use the two points to draw the graph.

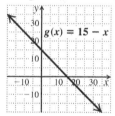

$g(x) = 15 - x$

32. The function $f(x) = \dfrac{1}{2}x - 3$ is in slope-intercept form. The slope is $\dfrac{1}{2}$. The y-intercept is $(0, -3)$. From the y-intercept, go up 1 unit and right 2 units to the point $(2, -2)$. Use the two points to draw the graph.

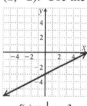

$f(x) = \frac{1}{2}x - 3$

33. The graph of $f(x) = 0$ is a horizontal line that crosses the vertical axis at $(0, 0)$.

$f(x) = 0$

34. The window should show both intercepts: $(98, 0)$ and $(0, 14)$. One possibility is $[-10, 120, -5, 15]$, with Xscl = 10.

35. First solve for y and determine the slope of each line.
$$y + 5 = -x$$
$$y = -x - 5$$
The slope of $y + 5 = -x$ is -1.
$$x - y = 2$$
$$y = x - 2$$
The slope of $x - y = 2$ is 1.
The product of their slopes is $(-1)(1) = -1$.
Therefore, the lines are perpendicular.

36. First solve for y and determine the slope of each line.
$$3x - 5 = 7y$$
$$y = \frac{3}{7}x - \frac{5}{7}$$
The slope of $3x - 5 = 7y$ is $\dfrac{3}{7}$.
$$7y - 3x = 7$$
$$y = \frac{3}{7}x + 1$$
The slope of $7y - 3x = 7$ is $\dfrac{3}{7}$.
The slopes are the same, so the lines are parallel.

37. Use the slope-intercept equation
$f(x) = mx + b$, with $m = \dfrac{2}{9}$ and b = −4.
$$f(x) = mx + b$$
$$f(x) = \frac{2}{9}x - 4$$

38. $y - y_1 = m(x - x_1)$ Point-slope equation
    $y - 10 = -5(x - 1)$ Substituting -5 for m, 1 for x_1, and 10 for y_1

39. First find the slope using the pair of given points.

$$m = \frac{5-6}{2-(-2)} = \frac{-1}{4} = -\frac{1}{4}$$

Use the point-slope equation.

$$y - 5 = -\frac{1}{4}(x - 2)$$

$$y - 5 = -\frac{1}{4}x + \frac{1}{2}$$

$$y = -\frac{1}{4}x + \frac{11}{2}$$

$$f(x) = -\frac{1}{4}x + \frac{11}{2}$$

40. The slope in the given equation is 1. The slope of any perpendicular line is, therefore,

$$-\frac{1}{1} = -1.$$ Use the slope and the given

y-intercept $(0, -3)$ to write the equation in function form. $f(x) = -x - 3$.

41. Plot and connect the points, using the years as the first coordinate and the corresponding cost per gallon as the second coordinate.

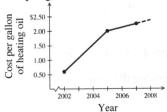

To estimate the cost in 2004, locate the point that is directly above 2004. Then move horizontally from the point to the vertical axis and read the approximate function value. The approximate cost per gallon in 2004 was $1.50.

42. To estimate the cost per gallon in 2008, extend the graph and extrapolate. The approximate cost per gallon in 2008 was $2.40.

43. a) The equation says to let t = the number of years since 1980, and $R(t)$ = the record, in seconds, for the 200-m run. Use the given data to form the pairs (3, 19.75) and (27, 19.32). Then find the slope of the function that fits the data.

$$m = \frac{19.32 - 19.75}{27 - 3} = \frac{-0.43}{24} = -\frac{43}{2400}$$

Substitute $m = -\frac{43}{2400}$, $t_1 = 3$, and

$R_1 = 19.75$ into the point-slope equation.

$$R - R_1 = m(t - t_1)$$

$$R - 19.75 = -\frac{43}{2400}(t - 3)$$

$$R = -\frac{43}{2400}t + \frac{15,843}{800}$$

$$R(t) = -\frac{43}{2400}t + \frac{15,843}{800}$$

b) In 2013, $t = 2013 - 1980 = 33$

$$R(33) = -\frac{43}{2400}(33) + \frac{15,843}{800}$$

$$\approx 19.21 \text{ sec}$$

In 2020, $t = 2020 - 1980 = 40$

$$R(40) = -\frac{43}{2400}(40) + \frac{15,843}{800}$$

$$\approx 19.09 \text{ sec}$$

44. $2x - 7 = 0$

$$x = \frac{7}{2}$$

can be written in standard form for a linear equation, $Ax + By = C$. Thus, it is a linear equation. The graph is a vertical line.

45. $3x - \dfrac{y}{8} = 7$

$24x - y = 56$

This equation can be written in standard form for a linear equation, $Ax + By = C$. Thus, it is a linear equation.

46. $2x^3 - 7y = 5$

The equation is not linear, because it contains an x^3 term.

47. $\dfrac{2}{x} = y$

$2 = xy$ Multiplying by x

The equation is not linear, because it has an xy term.

48. Enter the data with the number of years since 2000 as L_1 and the revenue, in billions of dollars, as L_2. The data appear to be linear.

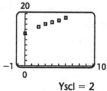

Yscl = 2

49. Enter the data with the number of years since 2000 as L_1 and the revenue, in billions of dollars, as L_2.

L1	L2	L3	1
0	12.5	------	
2	15.		
3	16.1		
4	16.8		
5	17.5		
6	18.4		
------	------		

L1(1)=0

Use the LinReg feature of the graphing calculator to find the equation of the line.

LinReg
y=ax+b
a=.96
b=12.85
r²=.9821420416
r=.9910307975

The second screen indicates that the equation is $F = 0.96t + 12.85$.

50. 2012 is 12 years after year 2000. We substitute 12 for t and solve for F.
$$F = 0.96t + 12.85$$
$$F = 0.96(12) + 12.85$$
$$F = 11.52 + 12.85$$
$$F = 24.37$$
The estimated revenue in 2012 will be about $24.4 billion.

51. $g(a+5) = 3(a+5) - 6$
$$= 3a + 15 - 6$$
$$= 3a + 9$$

52. $(g \cdot h)(4) = g(4) \cdot h(4)$
$$= (3 \cdot 4 - 6)(4^2 + 1)$$
$$= (12 - 6)(16 + 1)$$
$$= 6 \cdot 17$$
$$= 102$$

53. $(g / h)(-1) = g(-1) / h(-1)$
$$= \frac{3(-1) - 6}{(-1)^2 + 1}$$
$$= \frac{-3 - 6}{1 + 1}$$
$$= \frac{-9}{2} = -\frac{9}{2}$$

54. $(g + h)(x) = (3x - 6) + (x^2 + 1)$
$$= 3x - 6 + x^2 + 1$$
$$= x^2 + 3x - 5$$

55. To find the zeros of $g(x)$, solve the equation $g(x) = 0$ for x.
$$3x - 6 = 0$$
$$3x = 6$$
$$x = 2$$

56. $g(x)$ is well-defined for all real numbers. Its domain is $\{x \mid x \text{ is any real number}\}$.

57. The domains of g and h are both $\{x \mid x \text{ is any real number}\} = \mathbb{R}$. Thus, the domain of $g + h = \mathbb{R}$.

58. The domains of g and h are $\{x \mid x \text{ is a real number}\}$. $g(x) = 0$ when $3x - 6 = 0$, when $x = 2$. Therefore the domain of $h / g = \{x \mid x \text{ is a real number } and \ x \neq 2\}$.

59. *Thinking and Writing Exercise.*

60. *Thinking and Writing Exercise.*

61. The y-intercept is found by determining $f(0)$.
$$f(0) + 3 = 0.17 \cdot 0^2 + (5 - 2 \cdot 0)^0 - 7$$
$$f(0) + 3 = 0 + (5)^0 - 7$$
$$f(0) + 3 = 0 + 1 - 7$$
$$f(0) + 3 = -6$$
$$f(0) = -6 - 3 = -9$$
The y-intercept is $(0, -9)$.

62. Parallel lines have the same slope. Begin by writing each equation in slope-intercept form.

$3x - 4y = 12 \qquad ax + 6y = -9$

$3x - 12 = 4y \qquad ax + 9 = -6y$

$\dfrac{3}{4}x - 3 = y \qquad \dfrac{a}{-6}x + \dfrac{9}{-6} = y$

The two slopes are $\dfrac{3}{4}$ and $-\dfrac{a}{6}$. Solve the

equation $\dfrac{3}{4} = -\dfrac{a}{6}$.

$\dfrac{3}{4} = -\dfrac{a}{6}$

$4a = -18$

$a = -\dfrac{18}{4} = -\dfrac{9}{2}$

63. Each package costs \$7.99 and costs \$2.95 to ship. This makes a total charge of \$7.99 + \$2.95 = \$10.94 per package. There is also a flat fee of \$20 for overnight delivery no matter how many packages are purchased. Let x = the number of packages purchased and $f(x)$ represent the total cost. Then

$f(x) = \text{cost per pkg} \cdot \# \text{ of pkgs} + \text{flat fee}$

$= 10.94x + 20$

64. a) The slope of the line represents the distance covered per unit time, or speed. So, the steeper the line, the greater the speed. Graph III matches this situation since a slower (walking) speed is followed by a faster (train) speed, followed by a slower (walking) speed.

b) Graph IV matches this situation since biking would be faster than running, and running would be faster than walking.

c) Graph I matches this situation since sitting in one spot to fish would correspond to a speed of 0.

d) Graph II matches this situation since waiting corresponds to a speed of 0, and riding a train would be faster than running.

Chapter 2 Test

1. a) To determine $f(-2)$ locate the point directly above –2 on the x-axis. Then estimate the second coordinate by moving horizontally to the y-axis. From the graph, $f(-2) = 1$.

    b) The domain is the set of all x-values for which f is defined. From the graph the domain of f appears to be $\{x \mid -3 \le x \le 4\}$.

    c) To determine any x-value(s) for which $f(x) = \frac{1}{2}$, locate the point $\frac{1}{2}$ on the y-axis and move left or right to locate any points on the graph with a y value of $\frac{1}{2}$. Then estimate the x-coordinate by moving down from the point to the x-axis. From the graph $f(3) = \frac{1}{2}$.

    d) The range of f is the set of all outputs or second values of the function. From the graph it appears that the range of $f = \{y \mid -1 \le y \le 2\}$.

2. Use the indicated two points on the graph to determine the rate of change in y. It is the same as the slope of the line.
    For $(2005, 100)$ and $(2007, 150)$:

$$m = \frac{y_2 - y_1}{x_2 - x_1} = \frac{150 - 100}{2007 - 2005} = \frac{50}{2} = 25$$

3. Slope $= \dfrac{\text{change in } y}{\text{change in } x} = \dfrac{3 - (-2)}{6 - (-2)} = \dfrac{3 + 2}{6 + 2} = \dfrac{5}{8}$

4. Slope $= \dfrac{\text{change in } y}{\text{change in } x}$

$= \dfrac{5.2 - 5.2}{-4.4 - (-3.1)} = \dfrac{0}{-4.4 + 3.1} = 0$

5. The function $f(x) = -\dfrac{3}{5}x + 12$ is in slope-intercept form.

    The slope is $-\dfrac{3}{5}$. The y-intercept is (0, 12).

6. Convert the equation to slope-intercept form.
$$-5y - 2x = 7$$
$$-5y = 2x + 7$$
$$y = -\frac{2}{5}x - \frac{7}{5}$$
The slope is $-\frac{2}{5}$. The y-intercept is $\left(0, -\frac{7}{5}\right)$.

7. $f(x) = -3$

This is a horizontal line that crosses the vertical axis at $(0, -3)$. The slope is 0.

8. $x - 5 = 11$
$$x = 16$$
The graph of $x = 16$ is a vertical line that crosses the horizontal axis at $(16, 0)$. The slope is undefined.

9. $5x - y = 15$

To find the y-intercept, let $x = 0$ and solve for y.
$$5 \cdot 0 - y = 15$$
$$y = -15$$
The y-intercept is $(0, -15)$.
To find the x-intercept, let $y = 0$ and solve for x.
$$5x - 0 = 15$$
$$x = 3$$
The x-intercept is $(3, 0)$.

10. Graph $f(x) = -3x + 4$.

Slope is -3 or $\frac{-3}{1}$; y-intercept is $(0, 4)$.

From the y-intercept, go *down* 3 units and to the *right* 1 unit to get to $(1, 1)$. Use the two points to draw the graph.

$$y = -3x + 4$$

11. $y - 1 = -\frac{1}{2}(x + 4)$ is in point-slope form.

The slope is $-\frac{1}{2}$. The point $(-4, 1)$ is on the graph. From the point $(-4, 1)$, go down 1 unit and right 2 units to get to $(-2, 0)$. Use the two points to draw the graph.

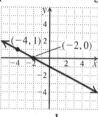

$$y - 1 = -\frac{1}{2}(x + 4)$$

12. Find the x-intercept.
$$-2x + 5 \cdot 0 = 20$$
$$x = \frac{20}{-2} = -10$$
The x-intercept is $(-10, 0)$.
Find the y-intercept.
$$-2 \cdot 0 + 5y = 20$$
$$y = \frac{20}{5} = 4$$
The y-intercept is $(0, 4)$.
Use the two points to draw the graph.

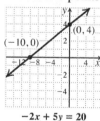

$$-2x + 5y = 20$$

13. Simplify the equation.
$$3 - x = 9$$
$$-x = 9 - 3 = 6$$
$$x = -6$$
This is a vertical line that crosses the x-axis at $(-6, 0)$.

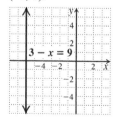

14. Yes, the standard viewing window of $[-10, 10, -10, 10]$, with Xscl = 1 and Yscl = 1, works well, because $-10 < f(0) < 10$ and

$f\left(\dfrac{-9}{2}\right) = 0$ and $-10 < -\dfrac{9}{2} < 10$ also.

15. a) $8x - 7 = 0$

$x = \dfrac{7}{8}$

The equation is linear. (Its graph is a vertical line.)

b) $4x - 9y^2 = 12$

The equation is not linear, because it has a y^2 term.

c) $2x - 5y = 3$

The equation is linear.

16. Write both equations in slope-intercept form.

$4y + 2 = 3x$ $\qquad$ $-3x + 4y = -12$

$y = \dfrac{3}{4}x - \dfrac{1}{2}$ $\qquad$ $y = \dfrac{3}{4}x - 3$

$m = \dfrac{3}{4}$ $\qquad\qquad$ $m = \dfrac{3}{4}$

The slopes are the same, so the lines are parallel.

17. Write both equations in slope-intercept form.

$y = -2x + 5$ $\quad$ $2y - x = 6$

$m = -2$

$\qquad\qquad$ $y = \dfrac{1}{2}x + 3$

$\qquad\qquad$ $m = \dfrac{1}{2}$

The product of the slopes is $(-2)\left(\dfrac{1}{2}\right) = -1$, so the lines are perpendicular.

18. Use the slope-intercept equation, $f(x) = mx + b$, with $m = -5$ and $b = -1$.

$f(x) = -5x - 1$

19. Use the point-slope equation,

$y - (-4) = 4(x - (-2))$, or $y + 4 = 4(x + 2)$

20. Determine the slope of the linear function.

$m = \dfrac{-2 - (-1)}{4 - 3} = \dfrac{-1}{1} = -1$

Use the slope and the given point in the point-slope equation.

$f(x) - (-2) = -1(x - 4)$

$f(x) + 2 = -x + 4$

$f(x) = -x + 2$

21. a) The problem says to let t = the number of years after 1982, and c = the average number of hours commuters sit in traffic in the year t. Use the given information to generate the two points $(0, 16)$ and $(25, 41)$. Determine the slope of the line containing these points.

$m = \dfrac{41 - 16}{25 - 0} = \dfrac{25}{25} = 1$

Use the point-slope equation with $m = 1$ and $(t_1, c_1) = (0, 16)$ to obtain the equation.

$c - 16 = 1(t - 0)$

$c - 16 = t$

$c = t + 16$

b) 2000 is $2000 - 1982 = 18$ years after 1982. Substitute 18 for t and solve for c.

$c = t + 16$

$c = 18 + 16$

$c = 34$

Commuters spent an average of 34 hours sitting in traffic in 2000.

c) 2012 is $2012 - 1982 = 30$ years after 1982. Substitute 30 for t and solve for c.

$c = t + 16$

$c = 30 + 16$

$c = 46$

By the year 2012, the average commuter will spend 46 hours sitting in traffic.

22. Let t = the number of years since 1980 and B = the number of twins, in thousands, born in the year t. Enter the data with the number of years since 1980 in L_1, and the number of births, in thousands, in L_2.

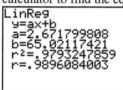

Use the LinReg feature of the graphing calculator to find the equation of the line.

The second screen indicates that the equation is $B = 2.6718t + 65.0212$.

23. 2012 is $2012 - 1980 = 32$ years after 1980. Substitute 32 for t and solve for B.

$B = 2.6718t + 65.0212$

$B = 2.6718(32) + 65.0212$

$B = 85.4976 + 65.0212$

$B = 150.5188$

The predicted number of twin births in 2012 is about 151,000.

24. $h(-5) = 2(-5) + 1 = -9$

25. $(g + h)(x) = g(x) + h(x) = \dfrac{1}{x} + 2x + 1$

26. The zeros for h are any values of x such that $h(x) = 0$.

$h(x) = 2x + 1$

$0 = 2x + 1$

$-1 = 2x$

$-\dfrac{1}{2} = x$

The only zero of h is $x = -\dfrac{1}{2}$.

27. Domain of g:

$\{x | x \text{ is a real number and } x \neq 0\}$

28. The domain of g is:

$\{x | x \text{ is a real number and } x \neq 0\}$.

The domain of h is: $\{x | x \text{ is a real number}\}$.

The domain of $g + h$ is:

$\{x | x \text{ is a real number and } x \neq 0\}$.

29. The domain of g is:

$\{x | x \text{ is a real number and } x \neq 0\}$.

The domain of h is:

$\{x | x \text{ is a real number}\}$.

$h(x) = 2x + 1 = 0$, if $x = -\dfrac{1}{2}$

The domain of g / h is:

$\left\{ x \middle| x \text{ is a real number and } x \neq 0 \text{ and } x \neq -\dfrac{1}{2} \right\}$

30. a) 1 h and 40 min is equal to

$1\dfrac{40}{60} = 1\dfrac{2}{3} = \dfrac{5}{3}$ h.

Find $f\left(\dfrac{5}{3}\right)$.

$f\left(\dfrac{5}{3}\right) = 5 + 15 \cdot \dfrac{5}{3} = 5 + 25 = 30$

The cyclist will be 30 mi from the starting point 1 h and 40 min after passing the 5-mile mark.

b) In the equation $f(t) = 5 + 15t$, 15 is the slope. The cyclist is traveling 15 mi every hour = 15 mph.

31. Determine the slope of the given line.

$2x - 5y = 8$

$y = \dfrac{2}{5}x - \dfrac{8}{5}$

The slope is $\dfrac{2}{5}$.

Any parallel line must also have a slope of $\dfrac{2}{5}$. The parallel line containing the point $(-3, 2)$ is:

$y - 2 = \dfrac{2}{5}(x + 3)$

$y - 2 = \dfrac{2}{5}x + \dfrac{6}{5}$

$y = \dfrac{2}{5}x + \dfrac{16}{5}$

32. Determine the slope of the given line.

$2x - 5y = 8$

$$y = \frac{2}{5}x - \frac{8}{5}$$

The slope is $\frac{2}{5}$.

Any perpendicular line must have a slope

given by the opposite of the reciprocal of $\frac{2}{5}$,

$-\frac{5}{2}$. The perpendicular line containing the

point $(-3, 2)$ is:

$$y - 2 = -\frac{5}{2}\left(x - (-3)\right)$$

$$y - 2 = -\frac{5}{2}x - \frac{15}{2}$$

$$y = -\frac{5}{2}x - \frac{11}{2}$$

33. Determine the slope of the given line.

$3x - 2y = 7$

$$-2y = -3x + 7$$

$$y = \frac{3}{2}x - \frac{7}{2} \qquad m = \frac{3}{2}$$

Since parallel lines have the same slope, the

slope of the function must also be $\frac{3}{2}$. An

expression for the slope of the function can

also be generated using the two points $(r, 3)$

and $(7, s)$, and the formula.

$$m = \frac{y_2 - y_1}{x_2 - x_1}.$$

Therefore, set the expression for the slope in r

and s equal to $\frac{3}{2}$ and solve for s in terms of r.

$$\frac{s - 3}{7 - r} = \frac{3}{2}$$

$$2(s - 3) = 3(7 - r)$$

$$2s - 6 = 21 - 3r$$

$$2s = 21 - 3r + 6 = -3r + 27$$

$$s = \frac{-3r + 27}{2} \text{ or } s = -\frac{3}{2}r + \frac{27}{2}$$

Chapter 3

Systems of Linear Equations and Problem Solving

Exercise Set 3.1

1. True

3. True

5. True

7. False

9. Use alphabetical order for the variables.
Substitute 1 for x and 2 for y.
$4x - y = 2$

$$\frac{4(1) - (2) \mid 2}{\begin{array}{r} 4 - 2 \mid 2 \\ 2 \mid 2 \quad \text{TRUE} \end{array}}$$

$10x - 3y = 4$

$$\frac{10(1) - 3(2) \mid 4}{\begin{array}{r} 10 - 6 \mid 4 \\ 4 \mid 4 \quad \text{TRUE} \end{array}}$$

The ordered pair $(1, 2)$ is a solution of the system of equations.

11. Use alphabetical order for the variables.
Substitute -5 for x and 1 for y.
$x + 5y = 0$

$$\frac{(-5) + 5(1) \mid 0}{\begin{array}{r} -5 + 5 \mid 0 \\ 0 \mid 0 \quad \text{TRUE} \end{array}}$$

$y = 2x + 9$

$$\frac{(1) \mid 2(-5) + 9}{\begin{array}{r} 1 \mid -10 + 9 \\ 1 \mid \quad -1 \quad \text{FALSE} \end{array}}$$

The ordered pair $(-5, 1)$ is not a solution of the system of equations.

13. Use alphabetical order for the variables.
Substitute 0 for x and -5 for y.

$x - y = 5$

$$\frac{(0) - (-5) \mid 5}{\begin{array}{r} 0 - (-5) \mid 5 \\ 5 \mid 5 \quad \text{TRUE} \end{array}}$$

$y = 3x - 5$

$$\frac{(-5) \mid 3(0) - 5}{\begin{array}{r} -5 \mid 0 - 5 \\ -5 \mid \quad -5 \quad \text{TRUE} \end{array}}$$

The ordered pair $(0, -5)$ is a solution of the system of equations.

15. The second equation is just 2 times the first equation. Thus, if the given point makes one equation true, it will make the other equation true also. Substitute 3 for x and 1 for y.
$3x + 4y = 13$

$$\frac{3(3) + 4(1) \mid 13}{\begin{array}{r} 9 + 4 \mid 13 \\ 13 \mid 13 \quad \text{TRUE} \end{array}}$$

The ordered pair $(3, 1)$ is a solution of the system of equations.

17. Graph both equations.

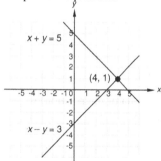

The solution (point of intersection) is apparently $(4, 1)$.
Check:

$x - y = 3 \qquad x + y = 5$

$$\frac{4 - 1 \mid 3}{3 \mid 3 \quad \text{TRUE}} \qquad \frac{4 + 1 \mid 5}{5 \mid 5 \quad \text{TRUE}}$$

The solution is $(4, 1)$.

19. Graph the equations.

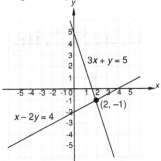

The solution (point of intersection) is apparently $(2, -1)$.

Check:

$$3x + y = 5$$

$$\frac{3 \cdot 2 + (-1)}{6 - 1} \bigg| \begin{array}{c} 5 \\ 5 \end{array}$$

$$5 \bigg| 5 \text{ TRUE}$$

$$x - 2y = 4$$

$$\frac{2 - 2(-1)}{2 + 2} \bigg| \begin{array}{c} 4 \\ 4 \end{array}$$

$$4 \bigg| 4 \text{ TRUE}$$

The solution is $(2, -1)$.

21. Graph both equations.

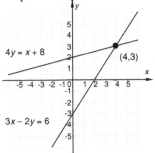

The solution (point of intersection) is apparently $(4, 3)$.

Check:

$$4y = x + 8$$

$$\frac{4 \cdot 3}{12} \bigg| \begin{array}{c} 4 + 8 \\ 12 \end{array} \text{ TRUE}$$

$$3x - 2y = 6$$

$$\frac{3 \cdot 4 - 2 \cdot 3}{12 - 6} \bigg| 6$$

$$6 \bigg| 6 \text{ TRUE}$$

The solution is $(4, 3)$.

23. Graph both equations.

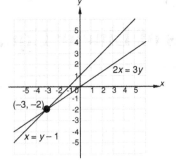

The solution (point of intersection) is apparently $(-3, -2)$.

Check:

$$x = y - 1$$

$$\frac{-3}{-3} \bigg| \begin{array}{c} -2 - 1 \\ -3 \end{array} \text{ TRUE}$$

$$2x = 3y$$

$$\frac{2 \cdot (-3)}{-6} \bigg| \begin{array}{c} 3 \cdot (-2) \\ -6 \end{array} \text{ TRUE}$$

The solution is $(-3, -2)$.

25. Graph both equations.

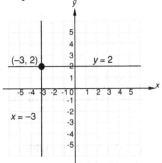

The solution (point of intersection) is $(-3, 2)$.

27. Graph both equations.

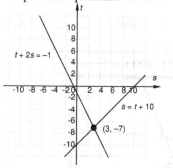

The solution (point of intersection) is apparently $(3,-7)$.

Check:

$$
\begin{array}{c|c}
t + 2s = -1 \\
\hline
-7 + 2\cdot 3 & -1 \\
-7 + 6 & \\
& -1 \mid -1 \ \text{TRUE}
\end{array}
\qquad
\begin{array}{c|c}
s = t + 10 \\
\hline
3 & -7 + 10 \\
3 & 3 \quad \text{TRUE}
\end{array}
$$

The solution is $(3,-7)$.

29. Graph both equations.

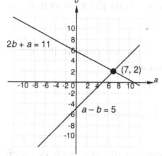

The solution (point of intersection) is apparently $(7,2)$.

Check:

$$
\begin{array}{c|c}
2b + a = 11 \\
\hline
2\cdot 2 + 7 & 11 \\
4 + 7 & \\
& 11 \mid 11 \ \text{TRUE}
\end{array}
\qquad
\begin{array}{c|c}
a - b = 5 \\
\hline
7 - 2 & 5 \\
& 5 \mid 5 \ \text{TRUE}
\end{array}
$$

The solution is $(7,2)$.

31. Graph both equations.

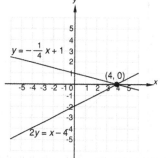

The solution (point of intersection) is apparently $(4,0)$.

Check:

$$
\begin{array}{c|c}
y = -\dfrac{1}{4}x + 1 \\
\hline
0 & -\dfrac{1}{4}\cdot(4) + 1 \\
& -1 + 1 \\
0 & 0 \qquad \text{TRUE}
\end{array}
\qquad
\begin{array}{c|c}
2y = x - 4 \\
\hline
2\cdot 0 & 4 - 4 \\
0 & 0 \quad \text{TRUE}
\end{array}
$$

The solution is $(4,0)$.

33. Graph both equations.

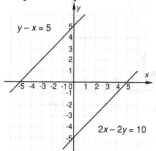

The lines are parallel. The system has no solution.

35. Graph both equations.

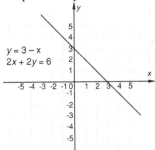

The graphs are the same. Any solution of one equation is a solution of the other. Each equation has infinitely many solutions. The solution set is the set of all pairs (x, y) for which $y = 3 - x$, or $\{(x, y) \mid y = 3 - x\}$. (In place of $y = 3 - x$, we could have used $2x + 2y = 6$ since the two equations are equivalent.)

37. Enter $y_1 = -5.43x + 10.89$ and $y_2 = 6.29x - 7.04$ on a graphing calculator and use the INTERSECT feature.

The solution is about $(1.53, 2.58)$.

39. Solve each equation for y. We get $y = \dfrac{-2.6x + 4}{-1.1}$ and $y = \dfrac{3.12x - 5.04}{1.32}$. Graph these equations on a graphing calculator, using a window that shows both graphs clearly. One choice is $[-1, 1, -5, 0]$. The graphs appear to be parallel.

This is confirmed by the error message "NO SIGN CHNG" that is returned when we use the INTERSECT feature. The system of equations has no solution.

41. Solve each equation for y. We get $y = 0.2x - 17.5$ and $y = \dfrac{10.6x + 30}{2}$. Graph these equations on a graphing calculator and use the INTERSECT feature.

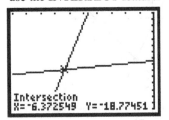

The solution is about $(-6.37, -18.77)$.

43. A system of equations is consistent if it has at least one solution. Of the odd exercises 17-41, only the systems in exercises 33 and 39 have no solution. Therefore, all except the systems in Exercise 33 and 39 are consistent.

45. A linear system of two equations in two variables is dependent only if it has infinitely many solutions. Only the system in exercise 35 is dependent.

47. The two equations described in the problem can be written as
$$x + y = 10 \quad (1)$$
$$x = \frac{2}{3}y \quad (2)$$

49. Let $c =$ the number of calls made or received each month, and $t =$ the number of text messages sent or received each month. Then:
$$c + t = 561 \qquad (1)$$
$$t = c + 153 \qquad (2)$$

51. Let $x =$ the degree measure of the first angle, and $y =$ the degree measure of the second angle. Then:
$$x + y = 180 \qquad (1)$$
$$x = 2y - 3 \qquad (2)$$

53. Let $x =$ the number of one-point shots made, and $y =$ the number of two-point shots. Then:
$$x + y = 64 \quad (1)$$
$$x + 2y = 100 \quad (2)$$

55. Let h = the number of hats sold, and s = the number of tee-shirts sold. Then:
$$h + s = 45 \qquad (1)$$
$$14.50h + 19.50s = 697.50 \quad (2)$$

57. Let x = the number of vials of Humalog insulin sold, and y = the number of vials of Lantus insulin sold. Then:
$$x + y = 50 \qquad (1)$$
$$121.88x + 105.19y = 5526.54 \quad (2)$$

59. Let l = the length, in yards, of the lacrosse field, and w = the width, in yards, of the lacrosse field. Then:
$$2l + 2w = 340 \qquad (1)$$
$$l = w + 50 \qquad (2)$$

61. Let x represent the number of years since 1980, and y represent the number of full-time faculty, in thousands, for the first regression line, and the number of part-time faculty, in thousands, for the second regression line. To determine the regression equation for each type of employment, enter the following data in lists L1, L2, and L3:

L1	L2	L3
0	450	236
5	459	256
11	536	291
15	551	381
19	591	437
25	676	615

Press **STAT** and move the cursor to **CALC**. Under **CALC** select **LinReg(ax + b)** and select lists L1 and L2 to display the equation: $y \approx 9.0524x + 430.6778$.

Press **STAT** again and move the cursor to **CALC**. Select **LinReg(ax + b)** and select lists L1 and L3 to display the equation: $y \approx 14.7175x + 185.3643$.

Graph these equations on your calculator and use the INTERSECT feature. According to these equations, the number of full-time and part-time faculty will be the same in 2023 $(x \approx 43.30)$.

63. Let x represent the number of years since 2004 and y represent the number of independent financial advisers, in thousands,

for the first regression line, and the number of financial advisers with national firms, in thousands, for the second regression line. To determine the regression equation for each type of adviser, enter the following data in lists L1, L2, and L3:

L1	L2	L3
0	21	60
1	23	62
2	25	59
3	28	57
4	32	55

Press **STAT** and move the cursor to **CALC**. Under **CALC** select **LinReg(ax + b)** and select lists L1 and L2 to display the equation: $y = 2.7x + 20.4$.

Press **STAT** again and move the cursor to **CALC**. Select **LinReg(ax + b)** and select lists L1 and L3 to display the equation: $y = -1.5x + 61.6$.

Graph these equations on your calculator and use the INTERSECT feature. According to these equations, the number of independent advisers will equal the number of financial advisers with national firms in about 2014 $(x \approx 9.8)$.

65. *Thinking and Writing Exercise.*

67. $2(4x - 3) - 7x = 9$
$$8x - 6 - 7x = 9 \quad \text{Removing parentheses}$$
$$x - 6 = 9 \quad \text{Collecting like terms}$$
$$x = 15 \quad \text{Adding 6 to both sides}$$

69. $4x - 5x = 8x - 9 + 11x$
$$-x = 19x - 9 \quad \text{Collecting like terms}$$
$$-20x = -9 \quad \text{Adding } -19x \text{ to both sides}$$
$$x = \frac{9}{20} \quad \text{Mult. both sides by } -\frac{1}{20}$$

71. $3x + 4y = 7$
$$4y = -3x + 7 \quad \text{Add } -3x \text{ both sides}$$
$$y = \frac{1}{4}(-3x + 7) \quad \text{Mult. both sides by } \frac{1}{4}$$
$$y = -\frac{3}{4}x + \frac{7}{4}$$

73. *Thinking and Writing Exercise.*

75. a) There are many correct answers. One system can be created by expressing the sum and difference of the two numbers:

$$x + y = 6$$
$$x - y = 4$$

   b) There are many correct answers. For example, write an equation in two variables not equal to 0. Then write a second equation by multiplying the left side of the first equation by one nonzero number and multiplying the right side by a different nonzero number.

$$x + y = 1$$
$$2x + 2y = 3$$

   c) There are many correct answers. One can be found by writing an equation in two variables and then writing a nonzero multiple of that equation:

$$x + y = 1$$
$$2x + 2y = 2$$

77. Substitute 4 for x and -5 for y in the first equation:

$$A(4) - 6(-5) = 13$$
$$4A + 30 = 13$$
$$4A = -17$$
$$A = -\frac{17}{4}$$

   Substitute 4 for x and -5 for y in the second equation:

$$4 - B(-5) = -8$$
$$4 + 5B = -8$$
$$5B = -12$$
$$B = -\frac{12}{5}$$

   We have $A = -\dfrac{17}{4}$ and $B = -\dfrac{12}{5}$.

79.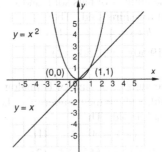

The solutions appear to be $(0,0)$ and $(1,1)$.

Both pairs check.

81. Put the equations in slope-intercept form.

$$2x - 8 = 4y$$
$$y = \frac{2}{4}x - \frac{8}{4}$$
$$y = \frac{1}{2}x - 2$$

   and

$$x - 2y = 4$$
$$-2y = -x + 4$$
$$y = \frac{-1}{-2}x + \frac{4}{-2}$$
$$y = \frac{1}{2}x - 2$$

Both lines have the same slope and the same y-intercept. Their graphs are identical. Only graph c) matches this system.

83. Put the equations in slope-intercept form.

$$x = 3y - 4$$
$$x + 4 = 3y$$
$$y = \frac{1}{3}x + \frac{4}{3}$$

   and

$$2x + 1 = 6y$$
$$y = \frac{2}{6}x + \frac{1}{6}$$
$$y = \frac{1}{3}x + \frac{1}{6}$$

The lines have the same slope, but different y-intercepts. Their graphs are parallel lines. Only graph b) matches this system.

85. a) ***Familiarize.*** The number of notebook PCs shipped in 2004 was 46 million and growing at a rate of $22\frac{2}{3} = \frac{68}{3}$ million per year. The number of desktop PCs shipped in 2004 was 140 million and growing at a rate of 4 million per year. Let $t =$ the number of years since 2004, and $n =$ the number of PCs shipped, in millions. ***Translate.*** For a linear equation, the slope of the line represents the rate of growth, and the value at $t = 0$, in this case the amount shipped in 2004, is the y-intercept. Thus for notebooks, the amount in millions shipped worldwide satisfies the equation:

$$n = \frac{68}{3}t + 46.$$

And for desktops the amount in millions is: $n = 4t + 140$

b) Graph $y_1 = \frac{68}{3}t + 46$ and $y_2 = 4t + 140$ on a graphing calculator and use the INTERSECT feature. The x value (or t value) is approximately 5. According to these equations, approximately 5 years after 2004, in 2009, the numbers of notebook PCs shipped worldwide will equal the number of desktop PCs shipped worldwide.

Exercise Set 3.2

1. d

3. a

5. c

7. $y = 3 - 2x$ (1)
   $3x + y = 5$ (2)

Substitute $3 - 2x$ for y in the second equation and solve for x.

$$\begin{aligned}
3x + y &= 5 \qquad (2) \\
3x + (3 - 2x) &= 5 \qquad \text{Substituting} \\
3x + 3 - 2x &= 5 \\
x + 3 &= 5 \\
x &= 5 - 3 \\
x &= 2
\end{aligned}$$

Next substitute 2 for x in either equation of the original system and solve for y.

$$\begin{aligned}
y &= 3 - 2x \qquad (1) \\
y &= 3 - 2 \cdot 2 \qquad \text{Substituting} \\
y &= 3 - 4 \\
y &= -1
\end{aligned}$$

Check the ordered pair $(2, -1)$.

$$\frac{y = 3 - 2x}{-1 \;\bigg|\; \begin{array}{l} 3 - 2 \cdot 2 \\[4pt] 3 - 4 \end{array}}$$
$$-1 \;\big|\; -1 \qquad \text{TRUE}$$

$$\frac{3x + y = 5}{3 \cdot 2 + (-1) \;\bigg|\; \begin{array}{l} 5 \\[4pt] 6 - 1 \end{array}}$$
$$5 \;\big|\; 5 \qquad \text{TRUE}$$

Since $(2, -1)$ checks, it is the solution.

9. $3x + 5y = 3$ (1)
   $x = 8 - 4y$ (2)

Substitute $8 - 4y$ for x in the first equation and solve for x.

$$\begin{aligned}
3(8 - 4y) + 5y &= 3 \qquad \text{Sub. for } x \text{ in } (1) \\
24 - 12y + 5y &= 3 \\
24 - 7y &= 3 \\
-7y &= -21 \\
y &= 3
\end{aligned}$$

Next substitute 3 for y in either equation of the original system and solve for x.

$$\begin{aligned}
x &= 8 - 4y \qquad (2) \\
x &= 8 - 4 \cdot 3 \qquad \text{Substituting} \\
x &= 8 - 12 \\
x &= -4
\end{aligned}$$

Check the ordered pair $(-4, 3)$.

$$\frac{3x + 5y = 3}{3 \cdot (-4) + 5 \cdot 3 \;\bigg|\; \begin{array}{l} 3 \\[4pt] -12 + 15 \end{array}}$$
$$3 \;\big|\; 3 \qquad \text{TRUE}$$

$$\frac{x = 8 - 4y}{-4 \;\bigg|\; \begin{array}{l} 8 - 4 \cdot 3 \\[4pt] 8 - 12 \end{array}}$$
$$-4 \;\big|\; -4 \qquad \text{TRUE}$$

Since $(-4, 3)$ checks, it is the solution.

11. $3s - 4t = 14$ (1)

  $5s + t = 8$ (2)

Solve the second equation for t.

$5s + t = 8$ (2)

$t = 8 - 5s$ (3)

Substitute $8 - 5s$ for t in the first equation and solve for s.

   $3s - 4t = 14$ (1)

 $3s - 4(8 - 5s) = 14$ Substituting

 $3s - 32 + 20s = 14$

  $23s - 32 = 14$

   $23s = 46$

    $s = 2$

Next substitute 2 for s in equation (1), (2), or (3). It is easiest to use equation (3) since it is already solved for t.

$t = 8 - 5 \cdot 2 = 8 - 10 = -2$

Check the ordered pair $(2, -2)$.

$$\frac{3s - 4t = 14}{\begin{array}{c|c} 3 \cdot 2 - 4 \cdot (-2) & 14 \\ 6 + 8 & \\ 14 & 14 \quad \text{TRUE} \end{array}}$$

$$\frac{5s + t = 8}{\begin{array}{c|c} 5 \cdot 2 + (-2) & 8 \\ 10 - 2 & \\ 8 & 8 \quad \text{TRUE} \end{array}}$$

Since $(2, -2)$ checks, it is the solution.

13. $4x - 2y = 6$ (1)

  $2x - 3 = y$ (2)

Substitute $2x - 3$ for y in the first equation and solve for x.

  $4x - 2y = 6$ (1)

 $4x - 2(2x - 3) = 6$

  $4x - 4x + 6 = 6$

    $6 = 6$

We have an identity, an equation that is always true. The equations are dependent, and the solution set is infinite:

$$\{(x, y) \mid 2x - 3 = y\}.$$

15. $-5s + t = 11$ (1)

  $4s + 12t = 4$ (2)

We solve the first equation for t.

 $-5s + t = 11$ (1)

   $t = 5s + 11$ (3)

We substitute $5s + 11$ for t in the second equation and solve for s.

  $4s + 12t = 4$ (2)

 $4s + 12(5s + 11) = 4$

 $4s + 60s + 132 = 4$

  $64s + 132 = 4$

   $64s = -128$

    $s = -2$

Next substitute –2 for s in equation (3).

$t = 5s + 11 = 5(-2) + 11 = -10 + 11 = 1$

Check the ordered pair $(-2, 1)$.

$$\frac{-5s + t = 11}{\begin{array}{c|c} -5(-2) + 1 & 11 \\ 10 + 1 & \\ 11 & 11 \quad \text{TRUE} \end{array}}$$

$$\frac{4s + 12t = 4}{\begin{array}{c|c} 4(-2) + 12 \cdot 1 & 4 \\ -8 + 12 & \\ 4 & 4 \quad \text{TRUE} \end{array}}$$

Since $(-2, 1)$ checks, it is the solution.

17. $2x + 2y = 2$ (1)

  $3x - y = 1$ (2)

Solve the second equation for y.

$3x - y = 1$ (2)

 $-y = -3x + 1$

  $y = 3x - 1$ (3)

Substitute $3x - 1$ for y in the first equation and solve for x.

  $2x + 2y = 2$ (1)

 $2x + 2(3x - 1) = 2$

  $2x + 6x - 2 = 2$

   $8x - 2 = 2$

    $8x = 4$

    $x = \dfrac{1}{2}$

Next substitute $\frac{1}{2}$ for x in equation (3).

$$y = 3x - 1 = 3 \cdot \frac{1}{2} - 1 = \frac{3}{2} - 1 = \frac{1}{2}$$

The ordered pair $\left(\frac{1}{2}, \frac{1}{2}\right)$ checks in both equations. It is the solution.

19. $2a + 6b = 4$ (1)
    $3a - b = 6$ (2)

Solve the second equation for b.
$3a - b = 6$ (2)
  $-b = -3a + 6$
   $b = 3a - 6$ (3)

Substitute $3a - 6$ for b in the first equation and solve for a.
$$2a + 6b = 4 \quad (1)$$
$$2a + 6(3a - 6) = 4$$
$$2a + 18a - 36 = 4$$
$$20a - 36 = 4$$
$$20a = 40$$
$$a = 2$$

Substitute 2 for a in equation (3).
$b = 3a - 6 = 3 \cdot 2 - 6 = 6 - 6 = 0$
The ordered pair $(2, 0)$ checks in both equations. It is the solution.

21. $2x - 3 = y$ (1)
    $y - 2x = 1$ (2)

Substitute $2x - 3$ for y in the second equation and solve for x.
$y - 2x = 1$ (2)
$2x - 3 - 2x = 1$
$-3 = 1$

We get a contradiction, an equation that is always false. The system has no solution.

23. $x + 3y = 7$ (1)
    $-x + 4y = 7$ (2)
    $\overline{0 + 7y = 14}$ Adding
       $7y = 14$
        $y = 2$

Substitute 2 for y in one of the original equations and solve for x.

$x + 3y = 7$ (1)
$x + 3 \cdot 2 = 7$ Substituting
$x + 6 = 7$
$x = 1$

Check:

$x + 3y = 7$		$-x + 4y = 7$	
$1 + 3 \cdot 2$	7	$-1 + 4 \cdot 2$	7
$1 + 6$		$-1 + 8$	
7	7 TRUE	7	7 TRUE

Since $(1, 2)$ checks, it is the solution.

25. $x - 2y = 11$ (1)
    $3x + 2y = 17$ (2)
    $\overline{4x + 0 = 28}$ Adding
       $x = 7$

Substitute 7 for x in equation (1) and solve for y.
$x - 2y = 11$
$7 - 2y = 11$ Substituting
  $-2y = 4$
   $y = -2$

We obtain $(7, -2)$ which checks, so it is the solution.

27. $9x + 3y = -3$ (1)
    $2x - 3y = -8$ (2)
    $\overline{11x + 0 = -11}$ Adding
       $11x = -11$
        $x = -1$

Substitute -1 for x in equation (1) and solve for y.
   $9x + 3y = -3$
$9(-1) + 3y = -3$ Substituting
   $-9 + 3y = -3$
       $3y = 6$
        $y = 2$

We obtain $(-1, 2)$ which checks, so it is the solution.

29. $5x + 3y = 19$ (1)

    $x - 6y = 11$ (2)

    Add 2 times equation (1) to equation (2) to eliminate y.

    $10x + 6y = 38$ $2 \cdot (1)$

    $\underline{x - 6y = 11}$

    $11x + 0 = 49$ Adding

    $x = \dfrac{49}{11}$

    Substitute $\dfrac{49}{11}$ for x in equation (1) and solve for y.

    $5x + 3y = 19$

    $5 \cdot \dfrac{49}{11} + 3y = 19$ Substituting

    $\dfrac{245}{11} + 3y = \dfrac{209}{11}$

    $3y = -\dfrac{36}{11}$

    $\dfrac{1}{3} \cdot 3y = \dfrac{1}{3} \cdot \left(-\dfrac{36}{11} \right)$

    $y = -\dfrac{12}{11}$

    We obtain $\left(\dfrac{49}{11}, -\dfrac{12}{11} \right)$. This checks, so it is the solution.

31. $5r = 3s + 24$

    $5r - 3s = 24$ (1)

    $3r + 5s = 28$ (2)

    $3s$ was subtracted from both sides of the first equation to align like terms. Now add 5 times equation (1) to 3 times equation (2) to eliminate s.

    $25r - 15s = 120$ $5 \cdot (1)$

    $\underline{9r + 15s = 84}$ $3 \cdot (2)$

    $34r + 0 = 204$ Adding

    $r = 6$

    Substitute 6 for r in equation (2) and solve for s.

    $3r + 5s = 28$

    $3 \cdot 6 + 5s = 28$ Substituting

    $18 + 5s = 28$

    $5s = 10$

    $s = 2$

We obtain $(6, 2)$. This checks, so it is the solution.

33. $6s + 9t = 12$ (1)

    $4s + 6t = 5$ (2)

    Add 2 times equation (1) to -3 times equation (2) to eliminate t.

    $12s + 18t = 24$ $2 \cdot (1)$

    $\underline{-12s - 18t = -15}$ $-3 \cdot (2)$

    $0 = 9$ Adding

    We get a contradiction. The system has no solution.

35. $\dfrac{1}{2}x - \dfrac{1}{6}y = 10$ (1)

    $\dfrac{2}{5}x + \dfrac{1}{2}y = 8$ (2)

    Multiply each equation by the LCM of its denominators to clear fractions.

    $3x - y = 60$ (3) $= 6 \cdot (1)$

    $4x + 5y = 80$ (4) $= 10 \cdot (2)$

    Add 5 times equation (3) to equation (4) to eliminate y.

    $15x - 5y = 380$ $5 \cdot (3)$

    $\underline{4x + 5y = 80}$ (4)

    $19x + 0 = 380$ Adding

    $x = \dfrac{380}{19}$

    $x = 20$

    Substitute 20 for x in one of the cleared equations and solve for y.

    $3x - y = 60$ (3)

    $3(20) - y = 60$ Substituting

    $60 - y = 60$

    $-y = 0$

    $y = 0$

We obtain $(20, 0)$. This checks, so it is the solution.

37. $\dfrac{x}{2} + \dfrac{y}{3} = \dfrac{7}{6}$ (1)

     $\dfrac{2x}{3} + \dfrac{3y}{4} = \dfrac{5}{4}$ (2)

Multiply each equation by the LCM of its denominators to clear fractions.

$3x + 2y = 7$ (3) $= 6 \cdot (1)$

$8x + 9y = 15$ (4) $= 12 \cdot (2)$

Add 9 times equation (3) to 2 times equation (4) to eliminate y.

$27x + 18y = 63 \quad 9 \cdot (3)$

$\underline{-16x - 18y = -30 \quad 2 \cdot (4)}$

$11x = 33 \quad \text{Adding}$

$x = 3$

Substitute 3 for x in one of the cleared equations and solve for y.

$3x + 2y = 7$ (3)

$3 \cdot 3 + 2y = 7$ Substituting

$9 + 2y = 7$

$2y = -2$

$y = -1$

We obtain $(3, -1)$. This checks, so it is the

39. $12x - 6y = -15$ (1)

     $-4x + 2y = 5$ (2)

Observe that, if we multiply equation (1) by $-\dfrac{1}{3}$, we obtain equation (2). Thus, any pair that is a solution of equation (1) is also a solution of equation (2). The equations are dependent and the solution set is infinite: $\{(x, y) \,|\, -4x + 2y = 5\}$.

41. $0.3x + 0.2y = 0.3$

     $0.5x + 0.4y = 0.4$

First multiply each equation by 10 to clear decimals.

$3x + 2y = 3$ (1)

$5x + 4y = 4$ (2)

Add -2 times equation (1) to equation (2) to eliminate y.

$-6x - 4y = -6 \quad -2 \cdot (1)$

$\underline{5x + 4y = 4 \qquad (2)}$

$-x + 0 = -2 \quad \text{Adding}$

$x = 2$

Substitute 2 for x in equation (1) and solve for y.

$3x + 2y = 3$ (1)

$3 \cdot 2 + 2y = 3$ Substituting

$6 + 2y = 3$

$2y = -3$

$y = -\dfrac{3}{2}$

We obtain $\left(2, -\dfrac{3}{2}\right)$. This checks, so it is the solution.

43. $a - 2b = 16$ (1)

     $b + 3 = 3a$ (2)

We will use the substitution method. Solve equation (1) for a.

$a - 2b = 16$

$a = 2b + 16$ (3)

Substitute $2b + 16$ for a in equation (2) and solve for b.

$b + 3 = 3a$ (2)

$b + 3 = 3(2b + 16)$ Substituting

$b + 3 = 6b + 48$

$-45 = 5b$

$-9 = b$

Substitute -9 for b in equation (3) to solve for a.

$a = 2(-9) + 16 = -2$

We obtain $(-2, -9)$. This checks, so it is the solution.

45. $10x + y = 306$ (1)

$10y + x = 90$ (2)

We will use the substitution method. First solve equation (1) for y.

$10x + y = 306$

$y = -10x + 306$ (3)

Now substitute $-10x + 306$ for y in equation (2) and solve for y.

$$10y + x = 90 \qquad (2)$$

$$10(-10x + 306) + x = 90 \qquad \text{Substituting}$$

$$-100x + 3060 + x = 90$$

$$-99x + 3060 = 90$$

$$-99x = -2970$$

$$x = 30$$

Substitute 30 for x in equation (3).

$y = -10 \cdot 30 + 306 = 6$

We obtain $(30, 6)$. This checks, so it is the solution.

47. $6x - 3y = 3$ (1)

$4x - 2y = 2$ (2)

Observe that equation (2) is $\dfrac{2}{3}$ times equation (1). Therefore, the equations are dependent and the solution set is infinite: any solution of equation (1) is also a solution of equation (2). $\{(x, y) \mid 4x - 2y = 2\}$, or $\{(x, y) \mid 2x - y = 1\}$.

49. $3s - 7t = 5$

$7t - 3s = 8$

Rewrite the second equation to align like terms with first equation. Then use the elimination method.

$$\begin{array}{ll} 3s - 7t = 5 & (1) \\ -3s + 7t = 8 & (2) \\ \hline 0 = 13 \end{array}$$

We get a contradiction. The system has no solution.

51. $0.05x + 0.25y = 22$

$0.15x + 0.05y = 24$

We first multiply each equation by 100 to clear decimals.

$5x + 25y = 2200$

$x + 5y = 440$ (1)

$15x + 5y = 2400$

$3x + y = 480$ (2)

Note that the system was simplified by dividing each equation by its common factor 5. Add equation (1) to -5 times equation (2) to eliminate y.

$$\begin{array}{ll} x + 5y = 440 & (1) \\ -15x - 5y = -2400 & -5 \cdot (2) \\ \hline -14x = -1960 & \text{Adding} \end{array}$$

$$x = \frac{-1960}{-14}$$

$$x = 140$$

Substitute 140 for x in one of the cleared equations and solve for y.

$3x + y = 480$ (1)

$3(140) + y = 480$ Substituting

$420 + y = 480$

$y = 60$

We obtain $(140, 60)$. This checks, so it is the solution.

53. $13a - 7b = 9$ (1)

$2a - 8b = 6$ (2)

Add 8 times equation (1) to -7 times equation (2) to eliminate b.

$$\begin{array}{ll} 104a - 56b = 72 & 8 \cdot (1) \\ -14a + 56b = -42 & -7 \cdot (2) \\ \hline 90a = 30 & \text{Adding} \end{array}$$

$$a = \frac{1}{3}$$

Substitute $\dfrac{1}{3}$ for a in one of the equations and solve for b.

$2a - 8b = 6$ (2)

$2 \cdot \dfrac{1}{3} - 8b = 6$

$\dfrac{2}{3} - 8b = \dfrac{18}{3}$

$-8b = \dfrac{16}{3}$

$b = -\dfrac{2}{3}$

We obtain $\left(\dfrac{1}{3}, -\dfrac{2}{3}\right)$. This checks, so it is the solution.

55. The point of intersection is $(140, 60)$, so window (d) is the correct answer.

57. The point of intersection is $(30, 6)$, so window (b) is the correct answer.

59. *Thinking and Writing Exercise.*

61. *Familiarize.* Let t = the average kWh of electricity used in a month by a toaster oven. Then $4t$ = the average kWh of electricity used in a month by a convection oven. The sum of these quantities is 15 kWh.
Translate.

Electricity used by toaster oven	plus	Electricity used by convection oven	is	15 kWh.
↓	↓	↓	↓	↓
t	$+$	$4t$	$=$	15

Carry out. Solve the equation.
$$t + 4t = 15$$
$$5t = 15$$
$$t = 3$$
If the toaster oven uses 3 kWh of electricity, then the convection oven uses $4 \cdot 3 = 12$ kWh.
Check. 12 kWh is four times as great as 3 kWh, and 3 kWh + 12 kWh = 15 kWh, which is the total electricity used in a month. The values check.
State. A toaster uses an average of 3 kWh of electricity per month, and a convection oven uses an average of 12 kWh of electricity per month.

63. *Familiarize.* Let p = the initial asking price of Gina's house. Then $\dfrac{9}{10}p$ = the amount for which the house sold. The actual sale price was $94,500.

Translate.

Nine-tenths of initial asking price	is $94,500.
↓ ↓ ↓	
$\dfrac{9}{10}p$	$= 94,500$

Carry out. Solve the equation.
$$\frac{9}{10}p = 94,500$$
$$p = \frac{10}{9} \cdot 94,500 = \frac{945,000}{9}.$$
$$p = 105,000$$
Check. $\dfrac{9}{10}$ of 105,000 is
$$\frac{9}{10} \cdot 105,000 = 94,500.$$ The answer checks.
State. The original asking price for Gina's house was $105,000.

65. *Familiarize.* Let x = the length of the first piece of wood trim, in inches. Then $2x$ = the length of the second piece and $\dfrac{1}{10} \cdot 2x = \dfrac{x}{5}$ the length of the third piece.
Translate.

Length of the three pieces	is 96.
↓	↓ ↓
$x + 2x + \dfrac{x}{5}$	$= 96$

Carry out. Solve the equation.
$$x + 2x + \frac{x}{5} = 96$$
$$3.2x = 96$$
$$x = 30$$
If $x = 30$, then $2x = 60$ and
$$\frac{1}{10}x = \frac{1}{10} \cdot 60 = 6.$$
Check. The sum of the lengths of the three pieces is $30 + 60 + 6 = 96$, which checks.
State. The lengths of the three pieces of wood trim are 30 in, 60 in, and 6 in.

67. *Thinking and Writing Exercise.*

69. First write $f(x) = mx + b$ as $y = mx + b$. Then use the points $(x, y) = (1, 2)$ and $(x, y) = (-3, 4)$ to generate two equations in m and b.

$$2 = m \cdot 1 + b$$
$$4 = m(-3) + b$$

Solve the resulting system of equations.

$$2 = m + b$$
$$4 = -3m + b$$

Multiply the second equation by -1 and add.

$$2 = m + b$$
$$\underline{-4 = 3m - b}$$
$$-2 = 4m$$
$$-\frac{1}{2} = m$$

Substitute $-\frac{1}{2}$ for m in the first equation and solve for b.

$$2 = -\frac{1}{2} + b$$
$$\frac{5}{2} = b$$

Thus, $m = -\frac{1}{2}$ and $b = \frac{5}{2}$.

71. Use the point $(x, y) = (-4, -3)$ to generate two equations in a and b.

$$a(-4) + b(-3) = -26$$
$$-4a - 3b = -26 \quad (1)$$
$$b(-4) - a(-3) = 7$$
$$3a - 4b = 7 \quad (2)$$

To use the elimination method, add 3 times equation (1) to 4 times equation (2) to eliminate a.

$$-12a - 9b = -78 \quad 3 \cdot (1)$$
$$\underline{12a - 16b = 28 \quad 4 \cdot (2)}$$
$$-25b = -50 \quad \text{Adding}$$
$$b = 2$$

Substitute 2 for b into either equation to solve for a.

$$-4 \cdot 2 + 3a = 7 \quad (1)$$
$$3a = 15$$
$$a = 5$$

Thus, $a = 5$ and $b = 2$.

73.
$$\frac{x+y}{2} - \frac{x-y}{5} = 1$$
$$\frac{x-y}{2} + \frac{x+y}{6} = -2$$

Clear fractions and combine like terms:

$$3x + 7y = 10 \quad (1)$$
$$4x - 2y = -12 \quad (2)$$

Add 2 times equation (1) to 7 times equation (2) to eliminate y.

$$6x + 14y = 20 \quad 2 \cdot (1)$$
$$\underline{28x - 14y = -84 \quad 7 \cdot (2)}$$
$$34x = -64 \quad \text{Adding}$$
$$x = -\frac{32}{17}$$

Substitute $-\frac{32}{17}$ for x in equation (1).

$$3\left(-\frac{32}{17}\right) + 7y = 10$$
$$7y = \frac{266}{17}$$
$$y = \frac{38}{17}$$

The solution is $\left(-\frac{32}{17}, \frac{38}{17}\right)$.

75.
$$\begin{array}{cc} \dfrac{2}{x} + \dfrac{1}{y} = 0 & 2 \cdot \dfrac{1}{x} + \dfrac{1}{y} = 0 \\ & \text{or} \\ \dfrac{5}{x} + \dfrac{2}{y} = -5 & 5 \cdot \dfrac{1}{x} + 2 \cdot \dfrac{1}{y} = -5 \end{array}$$

Substitute u for $\frac{1}{x}$ and v for $\frac{1}{y}$.

$$2u + v = 0 \quad (1)$$
$$5u + 2v = -5 \quad (2)$$

Add -2 times equation (1) to equation (2) to eliminate v.

$$-4u - 2v = 0 \quad -2 \cdot (1)$$
$$\underline{5u + 2v = -5 \quad (2)}$$
$$u = -5$$

Substitute -5 for u in equation (1).

$$2(-5) + v = 0$$
$$-10 + v = 0$$
$$v = 10$$

If $u = -5$, then $\dfrac{1}{x} = -5$. Thus $x = -\dfrac{1}{5}$.

If $v = 10$, then $\dfrac{1}{y} = 10$. Thus, $y = \dfrac{1}{10}$.

The solution is $\left(-\dfrac{1}{5}, \dfrac{1}{10}\right)$.

77. *Thinking and Writing Exercise.*

Exercise Set 3.3

1. The two equations are:

$$x + y = 10 \quad (1)$$
$$x = \frac{2}{3}y \quad (2)$$

Substitute $x = \dfrac{2}{3}y$ for x in the first equation and solve for y.

$$\frac{2}{3}y + y = 10 \quad \text{Substitute (2) into (1)}$$
$$\frac{2}{3}y + \frac{3}{3}y = 10 \quad \text{Create common denominator}$$
$$\frac{5}{3}y = 10 \quad \text{Addition}$$
$$5y = 30 \quad \text{Multiply by 3}$$
$$y = \frac{30}{5} \quad \text{Divide by 5}$$
$$y = 6$$

Next substitute 6 for y in either equation of the original system and solve for x.

$$x = \frac{2}{3}y \quad\quad (2)$$
$$x = \frac{2}{3} \cdot 6 = \frac{12}{3} = 4$$

Check the ordered pair $(4, 6)$.

$$x + y = 10$$

$4 + 6$	10
10	10 TRUE

$$x = \frac{2}{3}y$$

4	$\dfrac{2}{3} \cdot (6)$
4	4 TRUE

Since $(4, 6)$ checks, it is the solution.

3. Let c = the number of calls made or received each month, and t = the number of text messages sent or received each month. Then:

$$c + t = 561 \quad (1)$$
$$t = c + 153 \quad (2)$$

The system can be solved using substitution since t is solved for in equation (2).

$$c + t = 561 \quad (1)$$
$$t = c + 153 \quad (2)$$
$$c + (c + 153) = 561$$
$$c + c + 153 = 561$$
$$2c = 561 - 153$$
$$2c = 408$$
$$c = \frac{408}{2} = 204$$

Next, solve for t in either equation by using 204 for c.

$$t = c + 153 \quad (2)$$
$$t = 204 + 153$$
$$t = 357$$

In 2008, the average wireless subscriber made or received 204 phone calls and sent or received 357 text messages each month.

5. Let x = the degree measure of the first angle, and y = the degree measure of the second angle. Then:

$$x + y = 180 \quad (1)$$
$$x = 2y - 3 \quad (2)$$

The system can be solved using substitution. Substitute $2y - 3$ for x in equation (1) and solve for y.

$$x + y = 180 \quad (1)$$
$$(2y - 3) + y = 180$$
$$2y - 3 + y = 180$$
$$3y - 3 = 180$$
$$3y = 180 + 3 = 183$$
$$y = \frac{183}{3} = 61$$
$$x = 2y - 3 \quad (2)$$
$$x = 2(61) - 3 = 122 - 3 = 119$$

The two angle measures are $119°$ and $61°$.

7. Let x = the number of one-point shots made, and y = the number of two-point shots made. Then:

$$x + y = 64 \quad (1)$$

$$x + 2y = 100 \quad (2)$$

Solve equation (1) for x and substitute the result into equation (2) to find y. Then use equation (1) to find x.

$$x + y = 64$$

$$x = 64 - y$$

$$x + 2y = 100$$

$$(64 - y) + 2y = 100$$

$$64 - y + 2y = 100$$

$$y = 100 - 64 = 36$$

Find x:

$$x = 64 - y = 64 - (36) = 28$$

Check.

$$28 + 36 = 64, \text{ and}$$

$$1(28) + 2(36) = 100$$

The numbers check.
Wilt Chamberlain made 28 one-point foul shots and 36 two-point shots.

9. Let h = the number of hats sold, and s = the number of tee-shirts sold. Then:

$$h + s = 45 \quad (1)$$

$$14.50h + 19.50s = 697.50 \quad (2)$$

Solve equation (1) for x and substitute the result into equation (2) to find s. Then use equation (1) to find h.

$$h + s = 45$$

$$h = 45 - s$$

$$14.50h + 19.50s = 697.50$$

$$14.50(45 - s) + 19.50s = 697.50$$

$$652.50 - 14.50s + 19.50s = 697.50$$

$$5.00s = 697.50 - 652.50 = 45.00$$

$$s = \frac{45.00}{5.00} = \frac{4500}{500} = 9$$

Find h:

$$h = 45 - s = 45 - (9) = 36$$

Check.

$$36 + 9 = 45, \text{ and}$$

$$14.50(36) + 19.50(9) = 697.50$$

The numbers check. Simply Souvenirs sold 36 hats and 9 tee-shirts.

11. Let x = the number of vials of Humalog insulin sold, and y = the number of vials of Lantus insulin sold. Then:

$$x + y = 50 \quad (1)$$

$$121.88x + 105.19y = 5526.54 \quad (2)$$

Solve equation (1) for x and substitute the result into equation (2) to find y. Then use equation (1) to find x.

$$x + y = 50$$

$$x = 50 - y$$

$$121.88x + 105.19y = 5526.54$$

$$121.88(50 - y) + 105.19y = 5526.54$$

$$6094.00 - 121.88y + 105.19y = 5526.54$$

$$-16.69y = 5526.54 - 6094.00$$

$$-16.69y = -567.46$$

$$y = \frac{-567.46}{-16.69} = \frac{-567.46}{-16.69}$$

$$y = \frac{56,746}{1669} = 34$$

Find x:

$$x = 50 - y = 50 - (34) = 16$$

Check.

$$16 + 34 = 50, \text{ and}$$

$$121.88(16) + 105.19(34) = 5526.54$$

The numbers check.
Diabetic Express sold 16 vials of Humalog insulin and 34 vials of Lantus insulin.

13. Let l = the length, in yards, of the lacrosse field, and let w = the width, in yards, of the lacrosse field. Then:

$$2l + 2w = 340 \quad (1)$$

$$l = w + 50 \quad (2)$$

The system can be solved using substitution since l is solved for in equation (2).

$$2l + 2w = 340 \quad (1)$$

$$l = w + 50 \quad (2)$$

$$2l + 2w = 340$$

$$2(w + 50) + 2w = 340$$

$$2w + 100 + 2w = 340$$

$$4w = 340 - 100$$

$$4w = 240$$

$$w = \frac{240}{4} = 60$$

Find l:

$$l = w + 50 = (60) + 50 = 110$$

Check.

$2(110) + 2(60) = 340$

$(110) = (60) + 50$

The numbers check. An NCAA men's lacrosse field is 110 yd long and 60 yd wide.

15. Let f = the number of Facebook users, in millions, and m = the number of MySpace users, in millions. Then:

$f + m = 160 \qquad (1)$

$f = 2m - 8 \qquad (2)$

The system can be solved using substitution since f is solved for in equation (2).

$f + m = 160 \qquad (1)$

$f = 2m - 8 \qquad (2)$

$(2m - 8) + m = 160$

$2m - 8 + m = 160$

$3m - 8 = 160$

$3m = 160 + 8 = 168$

$m = 56$

Find f:

$f = 2m - 8 \qquad (2)$

$f = 2(56) - 8 = 112 - 8 = 104$

In 2009 there were 104 million Facebook users and 56 million MySpace users.

17. Let x = the number of 3-credit courses being take, and y = the number of 4-credit courses being taken. Then:

$\underbrace{\text{Total number of courses}}$ is $\quad$ 48.

$\qquad\qquad \downarrow \qquad\qquad\qquad \downarrow \quad \downarrow$

$\qquad\qquad x + y \qquad\qquad = \quad 48$

$\underbrace{\text{Total number of credits}}$ is $\quad$ 155.

$\qquad\qquad \downarrow \qquad\qquad\qquad \downarrow \quad \downarrow$

$\qquad\qquad 3x + 4y \qquad\quad = \quad 155$

Solve the system using elimination:

$x + \ y = 48 \qquad (1)$

$3x + 4y = 155 \qquad (2)$

Multiply equation (1) by –3 and add the result to equation (2).

$\begin{array}{rcr} -3x - 3y &=& -144 \\ 3x + 4y &=& 155 \\ \hline y &=& 11 \end{array}$

Find x:

$x + y = 48$

$x + 11 = 48$

$x = 48 - 11 = 37$

Check.

$37 + 11 = 48, \text{ and}$

$37 \cdot 3 + 11 \cdot 4 = 111 + 44 = 155.$

The numbers check.

The members of the swim team are taking 37 3-credit courses and 11 4-credit courses.

19. Let x = the number of cars and y = the number of motorcycles. Then:

$x + y = 5950 \qquad (1)$

$25x + 20y = 137,625 \quad (2)$

Since only the value of y, the number of motorcycles, is asked for, eliminate x by multiplying equation (1) by –25 and add the result to equation (2).

$\begin{array}{rcl} -25x - 25y &=& -25 \cdot 5950 \\ 25x + 20y &=& 137,625 \\ \hline -5y &=& 137,625 - 25 \cdot 5950 \\ &=& 137,625 - 148,750 \\ &=& -11,125 \end{array}$

$y = \dfrac{-11125}{-5} = 2225$

2225 motorcycles entered Yellowstone National Park.

21. Let x = the number of sheets of regular papers used, and y = the number of sheets of recycled papers used. Then:

$x + y = 150 \qquad (1)$

$1.9x + 2.4y = 341 \qquad (2)$

Note that the price per page was given in cents, and, therefore, the cost in equation (2) is set to 341¢. The system can be solved using elimination. Multiply equation (1) by (-1.9) and add the result to equation (2) to find y, then find x.

$-1.9 \cdot (x + y) = -1.9 \cdot 150$

$-1.9x - 1.9y = -285$

$\begin{array}{rcr} -1.9x - 1.9y &=& -285 \\ 1.9x + 2.4y &=& 341 \\ \hline 0.5y &=& 56 \end{array}$

$y = \dfrac{56}{0.5} = 112$

Now find x using equation (1).

$$x + y = 150 \quad (1)$$

$$x + 112 = 150$$

$$x = 150 - 112 = 38$$

38 regular sheets and 112 recycled paper sheets were used.

23. Let x = the number of Epson® cartridges purchased, and y = the number of HP cartridges purchased. Then:

$$x + y = 50 \quad\quad (1)$$

$$1699x + 2599y = 98,450 \quad (2)$$

Note that all prices in equation (2) are shown in cents. The system can be solved using elimination by multiplying equation (1) by (-1699) and adding the result to equation (2) to find y, then finding x.

$$-1699 \cdot (x + y) = -1699 \cdot 50$$

$$-1699x - 1699y = -84,950$$

$$\begin{array}{r} -1699x - 1699y = -84,950 \\ 1699x + 2599y = 98,450 \\ \hline 900y = 13,500 \end{array}$$

$$y = \frac{13,500}{900} = 15$$

Now find x using equation (1).

$$x + y = 50 \quad (1)$$

$$x + 15 = 50$$

$$x = 50 - 15 = 35$$

Office Depot® sold 35 Epson® cartridges and 15 HP cartridges.

25. Let x = the number of pounds of Mexican coffee used in the blend, and y = the number of pounds of Peruvian coffee used. Then:

$$x + y = 28 \quad\quad (1)$$

$$13x + 11y = 12 \cdot 28 = 336 \quad (2)$$

The system can be solved using elimination by multiplying equation (1) by (-13) and adding the result to equation (2) to find y, then finding x.

$$-13 \cdot (x + y) = -13 \cdot 28$$

$$-13x - 13y = -364$$

$$\begin{array}{r} -13x - 13y = -364 \\ 13x + 11y = 336 \\ \hline -2y = -28 \end{array}$$

$$y = \frac{-28}{-2} = 14$$

Now find x using equation (1).

$$x + y = 28 \quad (1)$$

$$x + 14 = 14$$

$$x = 28 - 14 = 14$$

The coffee blend should be made using 14 pounds of Mexican coffee and 14 pounds of Peruvian coffee.

27. Let x = the number of ounces of sumac in the blend, and y = the number of ounces of thyme used. Then:

$$x + y = 20 \quad\quad\quad (1)$$

$$1.35x + 1.85y = 1.65 \cdot 20 = 33 \quad (2)$$

The system can be solved using elimination by multiplying equation (1) by (-1.35) and adding the result to equation (2) to find y, then finding x.

$$-1.35 \cdot (x + y) = -1.35 \cdot 20$$

$$-1.35x - 1.35y = -27$$

$$\begin{array}{r} -1.35x - 1.35y = -27 \\ 1.35x + 1.85y = 33 \\ \hline 0.5y = 6 \end{array}$$

$$y = \frac{6}{0.5} = \frac{60}{5} = 12$$

Now find x using equation (1).

$$x + y = 20 \quad (1)$$

$$x + 12 = 20$$

$$x = 20 - 12 = 8$$

The Zahtar seasoning should be made using 8 ounces of sumac and 12 ounces of thyme.

29. Let x = the number of mL of 50%-acid solution used, and y = the number of mL of the 80% acid used, in the solution. Complete the table with the given information.

Note: Completing the table is part of the exercise.

Type of Solution	50%-Acid	80%-Acid	68%-Acid Mix
Amount of Solution	x	y	200
Percent Acid	50%	80%	68%
Amount of Acid in Solution	$0.5x$	$0.8y$	136

From the table we have an equation for the total number of mL: $x + y = 200$.

We also have an equation for the total amount of number of mL of acid:

$0.5x + 0.8y = 136$.

Solve the system.

$$x + y = 200 \quad (1)$$
$$0.5x + 0.8y = 136 \quad (2)$$

Multiply equation (1) by -0.5 and add the result to equation (2).

$$
\begin{aligned}
-0.5x - 0.5y &= 100 \\
\underline{0.5x + 0.8y} &= \underline{136} \\
0.3y &= 36 \\
y &= 120
\end{aligned}
$$

$$x + y = 200$$
$$x + 120 = 200$$
$$x = 200 - 120 = 80$$

Check.

$80 + 120 = 200$, and

$0.5(80) + 0.8(120) = 40 + 96 = 136$.

The numbers check.

State. Jerome should mix 80 mL of the 50%-acid solution and 120 mL of the 80%-acid solution.

31. Let x = the number of pounds of the 50% chocolate mix used, and y = the number of pounds of the 10% chocolate mix used. Then:

$$x + y = 20 \quad (1)$$
$$50x + 10y = 25 \cdot 20 = 500 \quad (2)$$

The system can be solved using elimination by multiplying equation (1) by (-50) and adding the result to equation (2) to find y, then finding x.

$$-50 \cdot (x + y) = -50 \cdot 20$$
$$-50x - 50y = -1000$$

$$
\begin{aligned}
-50x - 50y &= -1000 \\
\underline{50x + 10y} &= \underline{500} \\
-40y &= -500
\end{aligned}
$$

$$y = \frac{-500}{-40} = 12\frac{1}{2}$$

Now find x using equation (1).

$$x + y = 20 \quad (1)$$
$$x + 12\frac{1}{2} = 20$$
$$x = 20 - 12\frac{1}{2} = 7\frac{1}{2}$$

The mix should be made using 7.5 pounds of the 50% chocolate mix and 12.5 pounds of the 10% chocolate mix.

33. Let x = the amount borrowed at 6.5%, and y = the amount borrowed at 7.2%. Then:

$$x + y = 12,000 \quad (1)$$
$$0.065x + 0.072y = 811.50 \quad (2)$$

Note that since the interest on the loans (the right side of equation (2)) is not represented as a percent of the total amount borrowed, the percentage rates shown on the left side of equation (2) need to be converted to decimals (or the interest could be multiplied by 100: 81,150). The system can be solved using elimination by multiplying equation (1) by (-0.065) and adding the result to equation (2) to find y, then finding x.

$$-0.065 \cdot (x + y) = -0.065 \cdot 12,000$$
$$-0.065x - 0.065y = -780$$

$$
\begin{aligned}
-0.065x - 0.065y &= -780 \\
\underline{0.065x + 0.072y} &= \underline{811.50} \\
0.007y &= 31.50
\end{aligned}
$$

$$y = \frac{31.50}{0.007} = \frac{31,500}{7} = 4500$$

Now find x using equation (1).

$$x + y = 12,000 \quad (1)$$
$$x + 4500 = 12,000$$
$$x = 12,000 - 4500 = 7500$$

Asel borrowed $7500 at 6.5% interest and $4500 at 7.2%.

35. Let x = the number of liters of the 18% alcohol antifreeze used, and y = the number of liters of the 10% alcohol antifreeze used. Then:

$$x + y = 20 \qquad (1)$$
$$18x + 10y = 15 \cdot 20 = 300 \qquad (2)$$

The system can be solved using elimination by multiplying equation (1) by (-18) and adding the result to equation (2) to find y, then finding x.

$$-18 \cdot (x + y) = -18 \cdot 20$$
$$-18x - 18y = -360$$
$$\begin{array}{r} -18x - 18y = -360 \\ 18x + 10y = 300 \\ \hline -8y = -60 \end{array}$$
$$y = \frac{-60}{-8} = 7\frac{1}{2}$$

Now find x using equation (1).

$$x + y = 20 \qquad (1)$$
$$x + 7\frac{1}{2} = 20$$
$$x = 20 - 7\frac{1}{2} = 12\frac{1}{2}$$

The mix should be made using 12.5 L of the 18%-alcohol antifreeze and 7.5 L of the 10%-alcohol antifreeze.

37. Let x = the number of gallons of the 87-octane gasoline used, and y = the number of gallons of the 95-octane gasoline used. Then:

$$x + y = 10 \qquad (1)$$
$$87x + 95y = 93 \cdot 10 = 930 \qquad (2)$$

The system can be solved using elimination by multiplying equation (1) by (-87) and adding the result to equation (2) to find y, then finding x.

$$-87 \cdot (x + y) = -87 \cdot 10$$
$$-87x - 87y = -870$$
$$\begin{array}{r} -87x - 87y = -870 \\ 87x + 95y = 930 \\ \hline 8y = 60 \end{array}$$
$$y = \frac{60}{8} = 7\frac{1}{2}$$

Now find x using equation (1).

$$x + y = 10 \qquad (1)$$
$$x + 7\frac{1}{2} = 10$$
$$x = 10 - 7\frac{1}{2} = 2\frac{1}{2}$$

The new 93-octane gasoline should be made using 2.5 gallons of the 87-octane gasoline and 7.5 gallons of the 95-octane gasoline.

39. Let x = the number of pounds of whole milk (4% milk fat) used, and y = the number of pounds of cream (30% milk fat) used. Then:

$$x + y = 200 \qquad (1)$$
$$4x + 30y = 8 \cdot 200 = 1600 \qquad (2)$$

The system can be solved using elimination by multiplying equation (1) by (-4) and adding the result to equation (2) to find y, then finding x.

$$-4 \cdot (x + y) = -4 \cdot 200$$
$$-4x - 4y = -800$$
$$\begin{array}{r} -4x - 4y = -800 \\ 4x + 30y = 1600 \\ \hline 26y = 800 \end{array}$$
$$y = \frac{800}{26} = 30\frac{20}{26} = 30\frac{10}{13}$$

Now find x using equation (1).

$$x + y = 200 \qquad (1)$$
$$x + 30\frac{10}{13} = 200$$
$$x = 200 - 30\frac{10}{13} = 169\frac{3}{13}$$

The milk for cream cheese should be made using $169\frac{3}{13}$ pounds of whole milk and $30\frac{10}{13}$ pounds of cream.

41. *Familiarize.* First make a drawing.

Slow train
d kilometers   75 km/h $(t+2)$ hr

Fast train
d kilometers   125 km/h t hr

From the drawing, we see that the distances are the same. Now complete the chart.

$$d \;=\; r \;\cdot\; t$$

	Distance	Rate	Time
Slow train	d	75	$t+2$
Fast train	d	125	t

Translate. Using $d=rt$ for each row of the table, we get a system of equations.

$$d = 75(t+2)$$
$$d = 125t$$

Carry out. Solve the system of equations.

$$125t = 75(t+2) \quad \text{Using substitution}$$
$$125t = 75t + 150$$
$$50t = 150$$
$$t = 3$$

Check. At 125 km/h, in 3 hr the fast train will travel $125 \cdot 3 = 375$ km. At 75 km/h, in $3+2 = 5$ hr, the slow train will travel $75 \cdot 5 = 375$ km. The numbers check.

State. The trains will meet 375 km from the station.

43. *Familiarize.* We first make a drawing. Let $d =$ the distance and $r =$ the speed of the canoe in still water. Then the canoe's speed downstream is $r+6$, and its speed upstream is $r-6$. From the drawing, we see that the distances are the same.

Downstream, 6 km/h current

d km, $r+6$, 4 hr

Upstream, 6 km/h current

d km, $r-6$, 10 hr

Organize the information in a table.

$$d \;=\; r \;\cdot\; t$$

	Distance	Rate	Time
Down-stream	d	$r+6$	4
Up-stream	d	$r-6$	10

Translate. Using $d=rt$ for each row of the table, we get a system of equations.

$$d = 4(r+6)$$
$$d = 10(r-6)$$

Carry out. Solve the system of equations.

$$4(r+6) = 10(r-6) \quad \text{Substitution}$$
$$4r + 24 = 10r - 60$$
$$84 = 6r$$
$$14 = r$$

Check. Going downstream, the speed of the canoe would be $r+6 = 14+6 = 20$ km/h, and in 4 hours, it would travel $4 \cdot 20 = 80$ km. Going upstream, the speed of the canoe would be $r-6 = 14-6 = 8$ km/h, and in 10 hours it would travel $10 \cdot 8 = 80$ km. The numbers check.

State. The speed of the canoe in still water is 14 km/h.

45. *Familiarize.* Make a drawing. Note that the plane's speed traveling toward London is $360 + 50 = 410$ mph, and the speed traveling toward New York City is $360 - 50 = 310$ mph. Also, when the plane is d mi from New York City, it is $3458 - d$ mi from London.

New York City   London
310 mph t hours t hours   410 mph

|——————— 3458 mi ———————|

|——— d ———|——— 3458 mi $-d$ ———|

Organize this information in a table.

	Distance	Rate	Time
Toward NYC	d	310	t
Toward London	$3458-d$	410	t

Translate. Using $d=rt$ for each row of the table, we get a system of equations.

$$d = 310t \quad (1)$$
$$3458 - d = 410t \quad (2)$$

Carry out. Solve the system of equations.

$3458 - 310t = 410t$ Using substitution

$$3458 = 720t$$

$$4.8028 \approx t$$

Substitute 4.8028 for t in (1).

$$d \approx 310(4.8028) \approx 1489$$

Check. If the plane is 1489 mi from New York City, it can return to New York City, flying at 310 mph, in 1489/310 ≈ 4.8 hr. If the plane is 3458 − 1489 = 1969 mi from London, it can fly to London, traveling at 410 mph, in 1969/410 ≈ 4.8 hr. Since the times are the same, the answer checks.

State. The point of no return is about 1489 mi from New York City.

47. Let x = the number of minutes used for landline calls, and y = the number of minutes used for wireless calls. Then:

$$x + y = 400 \qquad (1)$$

$$9x + 15y + 399 = 5889$$

$$9x + 15y = 5889 - 399 = 5490 \qquad (2)$$

Note that the price per minute for each type of call was given in cents. Therefore, the monthly charge and bill amount were shown in cents in equation (2). The system can be solved using elimination by multiplying equation (1) by (-9) and adding the result to equation (2) to find y, then finding x.

$$-9 \cdot (x + y) = -9 \cdot 400$$

$$-9x - 9y = -3600$$

$$\begin{array}{r} -9x - 9y = -3600 \\ 9x + 15y = 5490 \\ \hline 6y = 1890 \end{array}$$

$$y = \frac{1890}{6} = 315$$

Now find x using equation (1).

$$x + y = 400 \qquad (1)$$

$$x + 315 = 400$$

$$x = 400 - 315 = 85$$

Kim made 85 minutes of landline calls and 315 minutes of wireless calls.

49. **Familiarize.** Monica tendered $20 to pay for a purchase amounting to $9.25, so she should receive $10.75 in change. Let x = the number of quarters she receives, and y = the number of fifty-cent pieces she receives.

Translate. Organize the information in a table.

	Quarters	Fifty-cent pieces	Total
Number of coins	x	y	30
Value of the coin	$0.25	$0.50	
Total change	$0.25x$	$0.5y$	$10.75

From the "Number of coins" line in the table, we have the equation:

$x + y = 30$.

From the "Total change" line in the table, we have the equation:

$0.25x + 0.50y = 10.75$

After clearing decimals, and dividing the second equation by the common factor of 25, we have a system of equations:

$$x + y = 30 \qquad (1)$$

$$25x + 50y = 1075$$

$$x + 2y = 43 \qquad (2)$$

Carry out. To use elimination to solve the system, add −1 times equation (1) to equation (2) to eliminate x.

$$\begin{array}{rl} -x - y = -30 & -1 \cdot (1) \\ x + 2y = 43 & (2) \\ \hline y = 13 \end{array}$$

Now substitute 13 for y in equation (1) to solve for x.

$$x + y = 30$$

$$x + 13 = 30$$

$$x = 17$$

Check. There are a total of 17 + 13 = 30 coins. The total value of the coins is 17($0.25) + 13($0.50) = $4.25 + $6.50 = $10.75 The answers check.

State. Monica will receive 17 quarters and 13 fifty-cent pieces.

51. *Thinking and Writing Exercise.*

53. For $x = 5$, $y = 2$, $z = 3$,

$$2x - 3y - z = 2(5) - 3(2) - (3)$$

$$= 10 - 6 - 3 = 4 - 3 = 1$$

55. For $x = 1$, $y = -4$, $z = -5$,

$$x + y + 2z = (1) + (-4) + 2(-5)$$
$$= 1 - 4 - 10 = -3 - 10 = -13$$

57. For $a = -2$, $b = 3$, $c = -5$,

$$a - 2b - 3c = (-2) - 2(3) - 3(-5)$$
$$= -2 - 6 + 15 = -8 + 15 = 7$$

59. *Thinking and Writing Exercise.*

61. ***Familiarize*** In this problem, there is only one unknown amount, the amount of pure silver that must be added to the amount of metal in the coin. Let x = the number of ounces of pure silver that must be added to the metal in the coin. Then the total amount of metal after the pure silver is added will be: $x + 32$ ounces.
Translate. Organize the information in a table.

	coin silver	pure silver	sterling silver
%	90%	100%	92.5%
amount	32 oz	x	$x + 32$
Total silver	$0.90 \cdot 32$	$1.00x$	$0.925(x + 32)$

Therefore, we get:
$$0.90 \cdot 32 + 1.00x = 0.925(x + 32)$$

Carry out. Solve the equation for x.
$$0.90 \cdot 32 + 1.00x = 0.925(x + 32)$$
$$28.8 + x = 0.925x + 29.6$$
$$(1 - 0.925)x = 29.6 - 28.8$$
$$0.075x = 0.8$$
$$x = \frac{0.8}{0.075} = \frac{800}{75} = 10\frac{50}{75} = 10\frac{2}{3}$$

Check. There are a total of 32 ounces + $10\frac{2}{3}$

ounces, or $42\frac{2}{3}$ ounces.

$$0.925\left(42\frac{2}{3}\right) \approx 39.4667 \text{ and}$$

$$0.90 \cdot 32 + 10\frac{2}{3} \approx 28.8 + 10.6667 = 39.4667$$
The answer checks.

State. $10\frac{2}{3}$ ounces of pure silver must be

added to the metal in the coin to get a mixture that is sterling silver.

63. ***Familiarize.*** Let x = the amount of the original solution that remains after some of the original solution is drained and replaced with pure antifreeze. Let y = the amount of the original solution that is drained and replaced with pure antifreeze. We organize the information in a table. Keep in mind that the table contains information regarding the solution *after* some of the original solution is drained and replaced with pure antifreeze.

Translate. Organize the information in a table.

	Original solution	Pure Anti-freeze	New Mixture
Amount of solution	x	y	6.3 L
Percent of antifreeze	30%	100%	50%
Amount of antifreeze in solution	$0.3x$	$1 \cdot y = y$	$0.5(6.3)$ $= 3.15$

One equation comes from the "Amount of solution" row of the table:
$$x + y = 6.3$$
The last row of the table gives a second equation:
$$0.3x + y = 3.15$$
After clearing decimals the problem is translated into the system of equations:
$$10x + 10y = 63 \quad (1)$$
$$30x + 100y = 315 \quad (2)$$

Carry out. Solve the system of equations using the elimination method.
$$-30x - 30y = -189 \quad -3 \cdot (1)$$
$$\underline{30x + 100y = 315 \quad (2)}$$
$$70y = 126$$
$$y = 1.8$$

Now substitute 1.8 for y in equation (1) to solve for x.
$$x + y = 6.3$$
$$x + 1.8 = 6.3$$
$$x = 4.5$$

Check. The total amount in the mixture is 4.5 L + 1.8 L, or 6.3 L. The amount of antifreeze in the mixture is 30%(4.5) + 100%(1.8) = 1.35 L + 1.8 L = 3.15 L. The answer checks.

State. Michelle should drain 1.8 L of her radiator fluid and replace it with 1.8 L of pure antifreeze.

65. ***Familiarize.*** Let x = the number of individual volumes purchased, and y = the number of 3-volume sets purchased. Then:

Translate. $x + 3y = 51$ (1)

$\qquad\qquad 39x + 88y = 1641$ (2)

Carry out. Multiply equation (1) by (-39) and add the result to equation (2) to eliminate x.

$$-39 \cdot (x + 3y) = -39 \cdot 51$$
$$-39x - 117y = -1989$$

$$\begin{array}{r} -39x - 117y = -1989 \\ 39x + 88y = 1641 \\ \hline -29y = -348 \end{array}$$

$$y = \frac{-348}{-29} = 12$$

Check. If 12 three-volume sets were purchased, then $51 - 3 \cdot 12 = 51 - 36 = 15$ individual volumes were purchased.

$$1641 = 15 \cdot 39 + 12 \cdot 88$$
$$1641 = 585 + 1056$$
$$1641 = 1641$$

The answer checks.

State. 12 three-volume sets were purchased.

67. ***Familiarize.*** Let x = the number of gallons of pure brown and y = the number of gallons of neutral stain that should be added to the original 0.5 gal. Note that the total of 1 gal of stain needs to be added to bring the amount of stain up to 1.5 gal. The original 0.5 gal of stain contains $20\%(0.5 \text{ gal})$, or $0.2(0.5 \text{ gal}) = 0.1$ gal of brown stain. The final solution contains $60\%(1.5 \text{ gal})$, or $0.6(1.5 \text{ gal}) = 0.9$ gal and the x gal that are added.

Translate.

$\underbrace{\text{The amount of stain added}}$ was 1 gal.

$\qquad\quad \downarrow \qquad\qquad\quad \downarrow \quad \downarrow$

$\qquad x + y \qquad\qquad\quad = \quad 1$

$\underbrace{\text{The amount of brown stain}}_{\text{in the final solution}}$ is 0.9 gal.

$\qquad\quad \downarrow \qquad\qquad\quad \downarrow \quad \downarrow$

$\qquad 0.1 + x \qquad\qquad = \quad 0.9$

We have a system of equations.

$x + y = 1$ (1)

$0.1 + x = 0.9$ (2)

Carry out. First Solve (2) for x.

$$0.1 + x = 0.9$$
$$x = 0.8$$

Then substitute 0.8 for x in (1) and solve for y.

$$0.8 + y = 1$$
$$y = 0.2$$

Check.

Total amount of stain: $0.5 + 0.8 + 0.2 = 1.5$ gal.

Total amount of brown stain: $0.1 + 0.8 = 0.9$ gal.

Total amount of neutral stain: $0.8(0.5) + 0.2 = 0.4 + 0.2 = 0.6$ gal $= 0.4(1.5$ gal$)$.

The answer checks.

State. 0.8 gal of pure brown and 0.2 gal of neutral stain should be added.

69. ***Familiarize.*** Let x = the number of miles driven in the city, and let y = the number of miles driven on the highway.

Translate. Organize this information in a table.

	City driving	Highway driving	
Number of miles	x	y	465
MPG	18	24	
Gallons used	$\dfrac{x}{18}$	$\dfrac{y}{24}$	23

One equation comes from the "Number of miles" row in the table.

$x + y = 465$

A second equation comes from the "Gallons used" row in the table.

$$\frac{x}{18} + \frac{y}{24} = 23$$
$$72 \cdot \frac{x}{18} + 72 \cdot \frac{y}{24} = 72 \cdot 23$$
$$4x + 3y = 1656$$

We have a system of equations:

$x + y = 465$ (1)

$4x + 3y = 1656$ (2)

Carry out. We use substitution to solve the system of equations. Begin by solving equation (1) for x.

$$x + y = 465$$
$$x = 465 - y$$

Substitute $465 - y$ for x in equation (2) and solve for y.

$$4x + 3y = 1656$$
$$4(465 - y) + 3y = 1656 \quad \text{Substituting}$$
$$4(465 - y) + 3y = 72(23) \quad \text{Mult. by 72}$$
$$1860 - 4y + 3y = 1656$$
$$1860 - y = 1656$$
$$-y = 1656 - 1860 = -204$$
$$y = 204$$

Substitute 204 for y in (1) and solve for x.

$$x + y = 465$$
$$x + 204 = 465$$
$$x = 261$$

Check. The total number of miles driven is 261 mi + 204 mi, or 465 mi. If the car is driven 261 miles in the city at 18 miles per gallon, the car would use 261/18 = 14.5 gal of fuel. If the car is driven 204 miles on the highway at 24 miles per gallon, it would use 204/24 = 8.5 gal of fuel. The total amount of fuel used would be 14.5 gal + 8.5 gal = 23 gal. The values check.

State. The car was driven 261 miles in the city and 204 miles on the highway.

71. **Familiarize.** The problem gives the unit costs for 2-count packs of Round Stic Grip pencils and 12-count packs of Matic Grip pencils. The total number of packs purchased, and the total purchase price, are given. Use the formula:

total cost = unit cost · number of units.

Translate. Let x = the number of 2-count packs purchased, and y = the number of 12-count packs purchased. Then the total number of pencils purchased is:

$$2x + 12y$$

And the total dollar amount paid is:

$$5.99x + 7.49y$$

The corresponding system of equations is:

$$2x + 12y = 138$$
$$x + 6y = 69 \qquad (1)$$
$$5.99x + 7.49y = 157.26 \qquad (2)$$

Note that a common factor of two was divided out to generate equation (1).

Carry out. The system can be solved by substitution. Solve equation (1) for x, and

substitute the result into equation (2).

$$x + 6y = 69 \qquad (1)$$
$$x = 69 - 6y$$
$$5.99x + 7.49y = 157.26 \quad (2)$$
$$5.99(69 - 6y) + 7.49y = 157.26$$
$$413.31 - 35.94y + 7.49y = 157.26$$
$$413.31 - 28.45y = 157.26$$
$$-28.45y = 157.26 - 413.31 = -256.05$$
$$y = \frac{-256.05}{-28.45} = 9$$

Use the value of y to find x:
$$x = 69 - 6y = 69 - 6 \cdot 9 = 69 - 54 = 15$$

Check.
$$138 = 15 \cdot 2 + 9 \cdot 12$$
$$138 = 30 + 108$$
$$138 = 138$$
The first condition is satisfied.
$$157.26 = 5.99 \cdot 15 + 7.49 \cdot 9$$
$$157.26 = 89.85 + 67.41$$
$$157.26 = 157.26$$
The second condition is satisfied.
The values check.

State. Wiese Accounting purchased 15 two-count packs of Round Stic Grip pencils and 9 twelve-count packs of Matic Grip pencils.

Mid-Chapter Review

Guided Solutions

1. $2x - 3(x - 1) = 5$ Substituting $x - 1$ for y

   $2x - 3x + 3 = 5$ Using the distributive law

      $-x + 3 = 5$ Combining like terms

         $-x = 2$ Subtracting 3 from both sides

          $x = -2$ Dividing both sides by -1

   $y = x - 1$

   $y = -2 - 1$ Substituting

   $y = -3$

The solution is $(-2, -3)$.

2. $2x - 5y = 1$

   $\underline{x + 5y = 8}$

   $\quad\quad 3x = 9$

   $\quad\quad x = 3$

   $x + 5y = 8$

   $3 + 5y = 8$ Substituting

   $\quad\quad 5y = 5$

   $\quad\quad y = 1$

   The solution is $(3, 1)$.

Mixed Review

1. $\quad x = y \quad (1)$

   $x + y = 2 \quad (2)$

   This system is easily solved using substitution because equation (1) already has y solved for in terms of x. Substitute y for x in equation (2) and solve for y. Since $y = x$ finding the value of x is trivial.

   $x + y = 2 \quad (2)$

   $y + y = 2$

   $\quad 2y = 2$

   $\quad y = 1$

   $x = y = 1$

   The ordered pair (1, 1) is the solution.

2. $x + y = 10 \quad (1)$

   $x - y = 8 \quad (2)$

   This system is easily solved using elimination because the coefficients of y in equations (1) and (2) are opposites. Add equations (1) and (2) and solve for x.

   $\quad x + y = 10 \quad (1)$

   $\quad \underline{x - y = 8 \quad (2)}$

   $2x \quad\quad = 18$

   $\quad\quad x = 9$

   $x + y = 10 \quad (1)$

   $9 + y = 10$

   $\quad\quad y = 1$

   The ordered pair (9, 1) is the solution.

3. $y = \dfrac{1}{2}x + 1 \quad (1)$

   $y = 2x - 5 \quad (2)$

This system is easily solved using substitution because both equations already have y solved for in terms of x. Substitute the expression for y, from equation (1), into equation (2) and solve for x.

$\quad y = 2x - 5 \quad (2)$

$\dfrac{1}{2}x + 1 = 2x - 5$

Multiply both sides of the equation by 2 to clear denominators and find x.

$2 \cdot \left[\dfrac{1}{2}x + 1\right] = 2 \cdot [2x - 5]$

$\quad\quad x + 2 = 4x - 10$

$\quad 2 + 10 = 4x - x$

$\quad\quad 12 = 3x$

$\quad\quad x = 4$

Use the value of x to find y.

$y = 2x - 5 \quad (2)$

$y = 2(4) - 5 = 8 - 5 = 3$

The ordered pair (4, 3) is the solution.

4. $\quad y = 2x - 3 \quad (1)$

   $x + y = 12 \quad\quad (2)$

This system is easily solved using substitution because equation (1) already has y solved for in terms of x. Substitute $2x - 3$ for y into equation (2) and solve for x.

$\quad\quad x + y = 12 \quad (2)$

$\quad x + (2x - 3) = 12$

$\quad\quad x + 2x - 3 = 12$

$\quad\quad\quad 3x - 3 = 12$

$\quad\quad\quad\quad 3x = 15$

$\quad\quad\quad\quad x = 5$

Use the value of x to find y.

$y = 2x - 3 \quad (1)$

$\quad = 2(5) - 3 = 10 - 3 = 7$

The ordered pair (5, 7) is the solution.

5. $x = 5$

   $y = 10$

The system requires no work to solve. The values of x and y are already given. The ordered pair (5, 10) is the solution. Note, however, that graphically the point (5, 10) is the intersection of the vertical line $x = 5$, and the horizontal line $y = 10$.

6. $3x + 5y = 8$ (1)

   $3x - 5y = 4$ (2)

This system is easily solved using elimination because the coefficients of y in equations (1) and (2) are opposites. Add equations (1) and (2) and solve for x.

$3x + 5y = 8$ (1)

$\underline{3x - 5y = 4 \quad (2)}$

$6x \qquad = 12$

$\qquad\quad x = 2$

Use the value of x to find y.

$3x + 5y = 8$ (1)

$3(2) + 5y = 8$

$6 + 5y = 8$

$5y = 2$

$y = \dfrac{2}{5}$

The ordered pair $\left(2, \dfrac{2}{5}\right)$ is the solution.

7. $2x - y = 1$ (1)

   $2y - 4x = 3$ (2)

This system is easily solved using substitution because equation (1) is easily solved for y in terms of x. Solve equation (1) for y, substitute the expression for y into equation (2), and solve for x.

$2x - y = 1$ (1)

$-y = -2x + 1$

$y = 2x - 1$

$2y - 4x = 3$ (2)

$2(2x - 1) - 4x = 3$

$4x - 2 - 4x = 3$

$-2 = 3$

The process led to a contradiction. The system has no solution.

8. $x = 2 - y$ (1)

   $3x + 3y = 6$ (2)

This system is easily solved using substitution because equation (1) already has x solved for in terms of y. Substitute the expression for x from equation (1) into equation (2) and solve for y.

$3x + 3y = 6$ (2)

$3(2 - y) + 3y = 6$

$6 - 3y + 3y = 6$

$6 = 6$

The process led to an identity. The system has an infinite number of solutions: any solution to the first equation will also be a solution to the second. The solution set is given by the ordered pairs $\{(x, y) \mid x = 2 - y\}$, or $(2 - y, y)$.

9. $x + 2y = 3$ (1)

   $3x = 4 - y$ (2)

This system is easily solved using substitution because equation (1) is easily solved for x in terms of y. Solve equation (1) for x. Then substitute the result for x into equation (2) and solve for y.

$x + 2y = 3$ (1)

$x = 3 - 2y$

$3x = 4 - y$ (2)

$3(3 - 2y) = 4 - y$

$9 - 6y = 4 - y$

$9 - 4 = -y + 6y$

$5 = 5y$

$y = 1$

Use the value of y to find x.

$x = 3 - 2y$

$= 3 - 2(1) = 3 - 2 = 1$

The ordered pair $(1, 1)$ is the solution.

10. $9x + 8y = 0$ (1)

$11x - 7y = 0$ (2)

This system is more easily solved using elimination. Solving either equation for x or y would mean introducing fractions into the system. Add 7 times equation (1) to 8 times equation (2) to eliminate y and find x.

$$\begin{array}{ll} 63x + 56y = 0 & 7 \cdot (1) \\ 88x - 56y = 0 & 8 \cdot (2) \\ \hline \quad 151x = 0 & \\ \quad\quad x = 0 & \end{array}$$

Use the value of x to find y.

$9x + 8y = 0$ (1)

$9(0) + 8y = 0$

$0 + 8y = 0$

$8y = 0$

$y = 0$

The ordered pair (0, 0) is the solution.

11. $10x + 20y = 40$ (1)

$x - y = 7$ (2)

Every term in equation (1) is divisible by 10. Use this fact to replace the original system with the equivalent system:

$x + 2y = 4$ (3)

$x - y = 7$ (2)

where equation (3) is the result of dividing each term in equation (1) by 10. To solve the system using substitution, solve equation (3) for x, substitute the result into equation (2) to find y.

$x + 2y = 4$ (3)

$x = 4 - 2y$

$x - y = 7$ (2)

$(4 - 2y) - y = 7$

$4 - 2y - y = 7$

$4 - 3y = 7$

$-3y = 7 - 4$

$-3y = 3$

$y = -1$

Use the value of y to find x.

$x = 4 - 2y = 4 - 2(-1)$

$= 4 - (-2) = 4 + 2 = 6$

The ordered pair (6, −1) is the solution.

12. To solve this system, note that the two equations are in slope-intercept form. The slopes of the two equations are equal, but the y-intercepts are different. This system corresponds to two parallel lines, and, therefore, has no solution.

13. $2x - 5y = 1$ (1)

$3x + 2y = 11$ (2)

This system is more easily solved using elimination. Solving either equation for x or y would mean introducing fractions into the system. Add 3 times equation (1) to −2 times equation (2) to eliminate x and find y.

$$\begin{array}{ll} 6x - 15y = 3 & 3 \cdot (1) \\ -6x - 4y = -22 & -2 \cdot (2) \\ \hline \quad -19y = -19 & \end{array}$$

$$y = \frac{-19}{-19} = 1$$

Use the value of y to find x.

$3x + 2y = 11$ (2)

$3x + 2(1) = 11$

$3x + 2 = 11$

$3x = 9$

$x = 3$

The ordered pair (3, 1) is the solution.

14. Begin by clearing denominators in both equations to create an equivalent system of equations: the LCM of the first equation is 6, the LCM of the second equation is 20.

$$6 \cdot \left(\frac{x}{2} + \frac{y}{3} \right) = 6 \cdot \left(\frac{2}{3} \right)$$

$3x + 2y = 4$ (1)

$$20 \cdot \left(\frac{x}{5} + \frac{5y}{2} \right) = 20 \cdot \left(\frac{1}{4} \right)$$

$4x + 50y = 5$ (2)

The resulting system can then be solved using elimination. Add 4 times equation (1) to −3 times equation (2) to eliminate x and solve for y.

$$\begin{array}{ll} 12x + 8y = 16 & 4 \cdot (1) \\ -12x - 150y = -15 & -3 \cdot (2) \\ \hline \quad -142y = 1 & \end{array}$$

$$y = -\frac{1}{142}$$

Use the value of y to find x.

$$3x + 2y = 4 \qquad (1)$$

$$3x + 2\left(-\frac{1}{142}\right) = 4$$

$$3x - \frac{1}{71} = 4$$

$$3x = 4 + \frac{1}{71} = \frac{4 \cdot 71 + 1}{71} = \frac{285}{71}$$

$$x = \frac{1}{3} \cdot \frac{285}{71} = \frac{95}{71}$$

(Alternatively, one could have gone back to the original system and eliminated y to find x.)

The ordered pair $\left(\dfrac{95}{71}, -\dfrac{1}{142}\right)$ is the solution.

15. One can begin by clearing decimals to create an equivalent system of equations. Multiply each equation by 10.

$$10 \cdot (1.1x - 0.3y) = 10 \cdot (0.8)$$

$$10 \cdot (2.3x + 0.3y) = 10 \cdot (2.6)$$

$$11x - 3y = 8 \qquad (1)$$

$$23x + 3y = 26 \qquad (2)$$

Since the coefficients of y are opposites, one can easily find the solution using elimination.

$$11x - 3y = 8 \qquad (1)$$

$$\underline{23x + 3y = 26 \qquad (2)}$$

$$34x = 34$$

$$x = 1$$

Use the value of x to find y.

$$11x - 3y = 8$$

$$11(1) - 3y = 8$$

$$11 - 3y = 8$$

$$-3y = -3$$

$$y = 1$$

The ordered pair $(1,1)$ is the solution.

16. Begin by clearing denominators in both equations to create an equivalent system of equations: the LCM of the first equation is 12, the LCM of the second equation is 30.

$$12 \cdot \left(\frac{1}{4}x\right) = 12 \cdot \left(\frac{1}{3}y\right)$$

$$3x = 4y$$

$$3x - 4y = 0 \qquad (1)$$

$$30 \cdot \left(\frac{1}{2}x - \frac{1}{15}y\right) = 30 \cdot 2$$

$$15x - 2y = 60 \qquad (2)$$

The resulting system can then be solved using elimination. Add equation (1) to -2 times equation (2) to eliminate y and find x.

$$3x - 4y = 0 \qquad (1)$$

$$\underline{-30x + 4y = -120 \qquad -2 \cdot (2)}$$

$$-27x = -120$$

$$x = \frac{-120}{-27} = \frac{40}{9}$$

Use the value of x to find y.

$$3\left(\frac{40}{9}\right) = 4y$$

$$\frac{1}{4}\left(\frac{40}{3}\right) = \frac{1}{4}(4y)$$

$$y = \frac{10}{3}$$

The ordered pair $\left(\dfrac{40}{9}, \dfrac{10}{3}\right)$ is the solution to the system.

17. *Familiarize.* Let p = the number of personal e-mails sent per week, and b = the number of business e-mails sent per week.
Translate.

$$p + b = 578 \qquad (1)$$

$$p = b - 30 \qquad (2)$$

Carry out. The system of equations can be solved using substitution since p is solved for in terms of b in equation (2).

$$p + b = 578 \qquad (1)$$

$$p = b - 30 \qquad (2)$$

$$(b - 30) + b = 578$$

$$b + b - 30 = 578$$

$$2b = 578 + 30$$

$$2b = 608$$

$$b = \frac{608}{2} = 304$$

Use the value of b to solve for p.

$$p = b - 30 \qquad (2)$$

$$p = 304 - 30 = 274$$

Check.

$$578 = 304 + 274$$

$$578 = 578$$

The first condition is satisfied.

$$30 = 304 - 274$$

$$30 = 30$$

The second condition is satisfied.
The values check.

State. In 2007, the average e-mail user sent 274 personal e-mail messages and 304 business e-mail messages each week.

18. *Familiarize.* Let x = the number of 5-cent bottles and cans recycled, and y = the number of 10-cent bottles and cans recycled.
Translate.
$$x + y = 430 \quad (1)$$
$$5x + 10y = 2620 \quad (2)$$
Note all prices in equation (2) are shown in cents.
Carry out. This system can be solved using elimination by adding -5 times equation (1) to equation (2) to eliminate x and solve for y.

$$
\begin{array}{ll}
-5x - 5y = -2150 & 5 \cdot (1) \\
5x + 10y = 2620 & (2) \\
\hline
5y = 470 &
\end{array}
$$

$$y = \frac{470}{5} = 94$$

Use the value of y to find x.
$$x + y = 430$$
$$x = 430 - y = 430 - 94 = 336$$
Check.
$$430 = 336 + 94$$
$$430 = 430$$
The first condition is satisfied.
$$2620 = 5 \cdot 336 + 10 \cdot 94$$
$$2620 = 1680 + 940$$
$$2620 = 2620$$
The second condition is satisfied.
The values check.
State. 336 five-cent bottles and cans, and 94 ten-cent bottles and cans were recycled.

19. *Familiarize.* Let x = the number of pounds of Pecan Morning granola used, and y = the number of pounds of Oat Dream granola used.
Translate.
$$x + y = 20 \quad (1)$$
$$25x + 10y = 19 \cdot 20 = 380 \quad (2)$$
Carry out. This system can be solved using elimination by adding -10 times equation (1) to equation (2) to eliminate y and solving for x.

$$
\begin{array}{ll}
-10x - 10y = -200 & -10 \cdot (1) \\
25x + 10y = 380 & (2) \\
\hline
15x = 180 &
\end{array}
$$

$$x = \frac{180}{15} = 12$$

Use the value of x to find y.
$$x + y = 20$$
$$y = 20 - x = 20 - 12 = 8$$
Check.
$$12 + 8 = 20, \text{ and}$$
$$25(12) + 10(8) = 19 \cdot 20$$
$$300 + 80 = 380.$$
The numbers check.
State. 12 pounds of Pecan Morning granola, and 8 pounds of Oat Dream granola, should be used to make 20 pounds of granola that is 19% nuts and dried fruit.

20. *Familiarize.* Let r = the speed of the ship, in mph, in still water. Since the current has a speed of 6 mph, the ship's speed going downstream is $r + 6$, and the ship's speed going upstream is $r - 6$. Summarize the information in a table.

$$d = r \cdot t$$

	Distance	Rate	Time
Down-stream	d	$r+6$	1.5
Up-stream	d	$r-6$	3.0

Translate. Using $d = rt$ for each row of the table, we get a system of equations.
$$d = 1.5(r + 6)$$
$$d = 3.0(r - 6)$$
Carry out. Solve the system of equations.
$$1.5(r + 6) = 3.0(r - 6) \quad \text{Using substitution}$$
$$1.5r + 9 = 3r - 18$$
$$27 = 1.5r = \frac{3}{2}r$$
$$27\left(\frac{2}{3}\right) = r$$
$$r = 18$$

Check. Going downstream, the ship's speed would be $r + 6 = 18 + 6 = 24$ mph, and in 1.5 hours, it would travel $1.5 \cdot 24 = 36$ miles. Going upstream, the ship's speed would be $r - 6 = 18 - 6 = 12$ mph, and in 3 hours it would travel $3 \cdot 12 = 36$ miles. The value checks.

State. The ship's speed in still water is 18 mph.

Exercise Set 3.4

---

1. True

3. False

5. True

7. Substitute $(2, -1, -2)$ into the three equations, using alphabetical order.

$$x + y - 2z = 5$$
$$\overline{2 + (-1) - 2(-2)\,\big|\,5}$$
$$2 - 1 + 4$$
$$5 \,\big|\, 5 \quad \text{True}$$

$$2x - y - z = 7$$
$$\overline{2 \cdot 2 - (-1) - (-2)\,\big|\,7}$$
$$4 + 1 + 2$$
$$7 \,\big|\, 7 \quad \text{True}$$

$$-x - 2y - 3z = 6$$
$$\overline{-2 - 2(-1) - 3(-2)\,\big|\,6}$$
$$-2 + 2 + 6$$
$$6 \,\big|\, 6 \quad \text{True}$$

The triple $(2, -1, -2)$ is a solution to the system.

9. $$x - y - z = 0 \quad (1)$$
    $$2x - 3y + 2z = 7 \quad (2)$$
    $$-x + 2y + z = 1 \quad (3)$$

1., 2. The equations are already in standard form with no fractions or decimals.

3. Use Equations (1) and (3) to eliminate x:

$$x - y - z = 0 \quad (1)$$
$$\underline{-x + 2y + z = 1} \quad (3)$$
$$y = 1 \quad\quad (4)$$

4. Use a different pair of equations and eliminate x:

$$2x - 3y + 2z = 7 \quad (2)$$
$$\underline{-2x + 4y + 2z = 2} \quad \text{Mult. (3) by 2}$$
$$y + 4z = 9 \quad\quad (5)$$

5. Now solve the system of Equations (4) and (5).

$$y = 1 \quad (4)$$
$$y + 4z = 9 \quad (5)$$

Substitute $y = 1$ into (5):

$$1 + 4z = 9$$
$$4z = 8$$
$$z = 2$$

6. Substitute in one of the original equations to find x.

$$x - 1 - 2 = 0 \quad \text{Substituting in (1)}$$
$$x - 3 = 0$$
$$x = 3$$

We obtain $(3, 1, 2)$. This checks, so it is the solution.

11. $$x - y - z = 1 \quad (1)$$
     $$2x + y + 2z = 4 \quad (2)$$
     $$x + y + 3z = 5 \quad (3)$$

1., 2. The equations are already in standard form with no fractions or decimals.

3. We eliminate y from two different pairs of equations.

$$x - y - z = 1 \quad (1)$$
$$\underline{2x + y + 2z = 4} \quad (2)$$
$$3x \quad\;\; + z = 5 \quad (4) \quad \text{Adding}$$

4. Use a different pair of equations and eliminate y:

$$x - y - z = 1 \quad (1)$$
$$\underline{x + y + 3z = 5} \quad (3)$$
$$2x \quad\;\; + 2z = 6 \quad (5)$$

5. Now solve the system of Equations (4) and (5).

$$3x + z = 5 \quad (4)$$
$$2x + 2z = 6 \quad (5)$$

$$-6x - 2z = -10 \quad \text{Mult. } (4) \text{ by } -2$$
$$\underline{2x + 2z = 6 \quad (5)}$$
$$-4x = -4$$
$$x = 1$$

$$3(1) + z = 5 \quad \text{Substituting in } (4)$$
$$3 + z = 5$$
$$z = 2$$

6.   Substitute in one of the original equations to find y.
$$1 - y - 2 = 1 \quad \text{Substituting in } (1)$$
$$-y - 1 = 1$$
$$-y = 2$$
$$y = -2$$

We obtain $(1, -2, 2)$. This checks, so it is the solution.

13. $3x + 4y - 3z = 4 \quad (1)$
       $5x - y + 2z = 3 \quad (2)$
       $x + 2y - z = -2 \quad (3)$

1., 2. The equations are already in standard form with no fractions or decimals.
3., 4. We eliminate y from two different pairs of equations.

$$3x + 4y - 3z = 4 \quad (1)$$
$$\underline{-2x - 4y + 2z = 4 \quad \text{Mult. } (3) \text{ by } -2}$$
$$x - \qquad z = 8 \quad (4) \text{ Adding}$$

Use a different pair of equations and eliminate y:
$$10x - 2y + 4z = 6 \quad \text{Mult. } (2) \text{ by } 2$$
$$\underline{x + 2y - z = -2 \quad (3)}$$
$$11x \qquad + 3z = 4 \quad (5) \text{ Adding}$$

5.   Now solve the system of Equations (4) and (5).
$$x - z = 8 \quad (4)$$
$$11x + 3z = 4 \quad (5)$$

$$3x - 3z = 24 \quad \text{M. } (4) \text{ by } 3$$
$$\underline{11x + 3z = 4 \qquad (5)}$$
$$14x = 28$$
$$x = 2$$

$$2 - z = 8 \quad \text{Substituting in } (4)$$
$$-z = 6$$
$$z = -6$$

6.   Substitute in one of the original equations to find y.
$$2 + 2y - (-6) = -2 \quad \text{Substituting in } (3)$$
$$2y + 8 = -2$$
$$2y = -10$$
$$y = -5$$

We obtain $(2, -5, -6)$. This checks, so it is the solution.

15. $x + y + z = 0 \quad (1)$
       $2x + 3y + 2z = -3 \quad (2)$
       $-x - 2y - z = 1 \quad (3)$

1., 2. The equations are already in standard form with no fractions or decimals.
3., 4. We eliminate z from two different pairs of equations.

$$-2x - 2y - 2z = 0 \quad \text{Mult. } (1) \text{ by } -2$$
$$\underline{2x + 3y + 2z = -3 \quad (2)}$$
$$y = -3 \quad (4) \text{ Adding}$$

Use a different pair of equations and eliminate x:
$$x + y + z = 0 \quad (1)$$
$$\underline{-x - 2y - z = 1 \quad (3)}$$
$$-y = 1 \quad \text{Adding}$$
$$y = -1 \quad (5)$$

5.   Now solve the system of Equations (4) and (5).
$$y = -3 \quad (4)$$
$$y = -1 \quad (5)$$

$$y = -3 \quad (4)$$
$$\underline{-y = 1 \quad \text{Mult. } (5) \text{ by } -1}$$
$$0 = -2$$

6.   We get a false equation in step 5, or contradiction. There is no solution.

17. $2x - 3y - z = -9$ (1)

$2x + 5y + z = 1$ (2)

$x - y + z = 3$ (3)

1., 2. The equations are already in standard form with no fractions or decimals.

3., 4. We eliminate z from two different pairs of equations.

$2x - 3y - z = -9$ (1)

$\underline{2x + 5y + z = 1}$ (2)

$4x + 2y = -8$ (4) Adding

Use a different pair of equations and eliminate z:

$2x - 3y - z = -9$ (1)

$\underline{x - y + z = 3}$ (3)

$3x - 4y \quad = -6$ (5) Adding

5. Now solve the system of Equations (4) and (5).

$4x + 2y = -8$ (4)

$3x - 4y = -6$ (5)

$8x + 4y = -16$ M. (4) by 2

$\underline{3x - 4y = -6}$ (5)

$11x \quad = -22$

$x = -2$

$3(-2) - 4y = -6$ Subst. in (5)

$-6 - 4y = -6$

$-4y = 0$

$y = 0$

6. Substitute in one of the original equations to find z.

$x - y + z = 3$ Sub. (3)

$-2 - 0 + z = 3$

$z = 5$

We obtain $(-2, 0, 5)$. This checks, so it is the solution.

19. $a + b + c = 5$ (1)

$2a + 3b - c = 2$ (2)

$2a + 3b - 2c = 4$ (3)

1., 2. The equations are already in standard form with no fractions or decimals.

3., 4. We eliminate a from two different pairs of equations.

$-2a - 2b - 2c = -10$ Mult. (1) by -2

$\underline{2a + 3b - 2c = 4}$ (2)

$b - 4c = -6$ (4) Adding

Use a different pair of equations and eliminate a:

$2a + 3b - c = 2$ (2)

$\underline{-2a - 3b + 2c = -4}$ M. (3) by -1

$c = -2$ (5) Adding

5. Now solve the system of Equations (4) and (5).

$b - 4c = -6$ (4)

$c = -2$ (5)

Substitute $c = -2$ into (4):

$b - 4(-2) = -6$

$b = -14$

6. Substitute in one of the original equations to find a.

$a + (-14) + (-2) = 5$ Sub. in (1)

$a - 16 = 5$

$a = 21$

We obtain $(21, -14, -2)$. This checks, so it is the solution.

21. $-2x + 8y + 2z = 4$ (1)

$x + 6y + 3z = 4$ (2)

$3x - 2y + z = 0$ (3)

1., 2. The equations are already in standard form with no fractions or decimals.

3., 4. We eliminate z from two different pairs of equations.

$-2x + 8y + 2z = 4$ (1)

$\underline{-6x + 4y - 2z = 0}$ Mult. (3) by -2

$-8x + 12y \quad = 4$ (4) Adding

Use a different pair of equations and eliminate z:

$$x + 6y + 3z = 4 \quad (2)$$
$$\underline{-9x + 6y - 3z = 0} \quad \text{M. (3) by } -3$$
$$-8x + 12y \quad\quad = 4 \quad (5) \text{ Adding}$$

5. Now solve the system of Equations (4) and (5).

$$-8x + 12y = 4 \quad (4)$$
$$-8x + 12y = 4 \quad (5)$$

$$-8x + 12y = 4 \quad (4)$$
$$\underline{8x - 12y = -4} \quad \text{M. (5) by } -1$$
$$0 = 0 \quad (6)$$

6. Equation (6) indicates that the original system of equations is a dependent system. (Note that if Equation (1) is subtracted from Equation (2), the result is Equation (3).) We could also have concluded that the equations are dependent by observing that Equations (4) and (5) are identical.

23. $2u - 4v - w = 8 \quad (1)$
     $3u + 2v + w = 6 \quad (2)$
     $5u - 2v + 3w = 2 \quad (3)$

1., 2. The equations are already in standard form with no fractions or decimals.

3., 4. We eliminate w from two different pairs of equations.

$$2u - 4v - w = 8 \quad (1)$$
$$\underline{3u + 2v + w = 6} \quad (2)$$
$$5u - 2v = 14 \quad (4) \text{ Adding}$$

Use a different pair of equations and eliminate w:

$$6u - 12v - 3w = 24 \quad \text{M. (1) by 3}$$
$$\underline{5u - 2v + 3w = 2} \quad (3)$$
$$11u - 14v = 26 \quad (5) \text{ Adding}$$

5. Now solve the system of Equations (4) and (5).

$$5u - 2v = 14 \quad (4)$$
$$11u - 14v = 26 \quad (5)$$

$$-35u + 14v = -98 \quad \text{M. (4) by } -7$$
$$\underline{11u - 14v = 26} \quad (5)$$
$$-24u \quad\quad = -72$$
$$u = 3$$

$$5(3) - 2v = 14 \quad \text{Subst. in (4)}$$
$$15 - 2v = 14$$
$$-2v = -1$$
$$v = \frac{1}{2}$$

6. Substitute in one of the original equations to find w.

$$2(3) - 4\left(\frac{1}{2}\right) - w = 8 \quad \text{Sub. in (1)}$$
$$-w = 4$$
$$w = -4$$

We obtain $\left(3, \frac{1}{2}, -4\right)$. This checks, so it is the solution.

25. $r + \dfrac{3}{2}s + 6t = 2 \quad (1)$
     $2r - 3s + 3t = 0.5 \quad (2)$
     $r + s + t = 1 \quad (3)$

1. The equations are already in standard form.

2. Multiply the first equation by 2 to clear the fraction. Also, multiply the second equation by 10 to clear the decimal.

$$2r + 3s + 12t = 4 \quad (1)$$
$$20r - 30s + 30t = 5 \quad (2)$$
$$r + s + t = 1 \quad (3)$$

3., 4. We eliminate s from two different pairs of equations.

$$20r + 30s + 120t = 40 \quad \text{Mult. (1) by 10}$$
$$\underline{20r - 30s + 30t = 5} \quad (2)$$
$$40r \quad\quad + 150t = 45 \quad (4) \text{ Adding}$$

Use a different pair of equations and eliminate s:

$$20r - 30s + 30t = 5 \quad (2)$$
$$\underline{30r + 30s + 30t = 30} \quad \text{M. (3) by 30}$$
$$50r \quad\quad + 60t = 35 \quad (5) \text{ Adding}$$

5. Now solve the system of Equations (4) and (5).

$$40r + 150t = 45 \quad (4)$$
$$50r + 60t = 35 \quad (5)$$

$$200r + 750t = 225 \quad \text{M. (4) by 5}$$
$$\underline{-200r - 240t = -140 \quad \text{M. (5) by } -4}$$
$$510t = 85$$

$$t = \frac{85}{510}$$

$$t = \frac{1}{6}$$

$$40r + 150\left(\frac{1}{6}\right) = 45 \quad \text{Subst. in (4)}$$
$$40r + 25 = 45$$
$$40r = 20$$
$$r = \frac{1}{2}$$

6. Substitute in one of the original
    equations to find s.

$$\frac{1}{2} + s + \frac{1}{6} = 1 \quad \text{Sub. in (3)}$$

$$s + \frac{2}{3} = 1$$

$$s = \frac{1}{3}$$

We obtain $\left(\frac{1}{2}, \frac{1}{3}, \frac{1}{6}\right)$. This checks, so

it is the solution.

27. $4a + 9b = 8 \quad (1)$
$8a + 6c = -1 \quad (2)$
$6b + 6c = -1 \quad (3)$

1., 2. The equations are already in standard
form with no fractions or decimals.

3., 4. Note that there is no c in Equation
(1). We will use Equations (2) and
(3) to obtain another equation with no
c –term.

$$8a + 6c = -1 \quad (2)$$
$$\underline{-6b - 6c = 1 \quad \text{Mult. (3) by } -1}$$
$$8a - 6b = 0 \quad (4) \text{ Adding}$$

5. Now solve the system of Equations (1)
and (4).

$$-8a - 18b = -16 \quad \text{Mult. (1) by } -2$$
$$\underline{8a - 6b = 0}$$
$$-24b = -16$$
$$b = \frac{2}{3}$$

$$8a - 6\left(\frac{2}{3}\right) = 0 \quad \text{Subst. in (4)}$$
$$8a - 4 = 0$$
$$8a = 4$$
$$a = \frac{1}{2}$$

6. Substitute in Equations (2) or (3) to
find c.

$$8\left(\frac{1}{2}\right) + 6c = -1 \quad \text{Sub. in (2)}$$
$$4 + 6c = -1$$
$$6c = -5$$
$$c = -\frac{5}{6}$$

We obtain $\left(\frac{1}{2}, \frac{2}{3}, -\frac{5}{6}\right)$. This checks,

so it is the solution.

29. $x + y + z = 57 \quad (1)$
$-2x + y = 3 \quad (2)$
$x - z = 6 \quad (3)$

1., 2. The equations are already in standard
form with no fractions or decimals.

3., 4. Note that there is no z in Equation
(2). We will use Equations (1) and
(3) to obtain another equation with no
z –term.

$$x + y + z = 57 \quad (1)$$
$$\underline{x - z = 6 \quad (3)}$$
$$2x + y = 63 \quad (4) \text{ Adding}$$

5. Now solve the system of Equations (2)
and (4).

$$-2x + y = 3 \quad (2)$$
$$2x + y = 63 \quad (4)$$

$$-2x + y = 3 \quad (2)$$
$$\underline{2x + y = 63 \quad (4)}$$
$$2y = 66$$
$$y = 33$$

$$2x + 33 = 63 \quad \text{Subst. in (4)}$$
$$2x = 30$$
$$x = 15$$

6. Substitute in Equations (1) or (3) to find z.

$15 - z = 6$ Sub. 15 for x in (3)

$9 = z$

We obtain $(15, 33, 9)$. This checks, so it is the solution.

31. $a \qquad - 3c = 6 \quad (1)$

$b + 2c = 2 \quad (2)$

$7a - 3b - 5c = 14 \quad (3)$

1., 2. The equations are already in standard form with no fractions or decimals.

3., 4. Note that there is no b in Equation (1). We will use Equations (2) and (3) to obtain another equation with no b –term.

$3b + 6c = 6$ Mult. (2) by 3

$\underline{7a - 3b - 5c = 14 \quad (3)}$

$7a \qquad + c = 20 \quad (4)$ Adding

5. Now solve the system of Equations (1) and (4).

$a - 3c = 6 \qquad (1)$

$7a + c = 20 \qquad (4)$

$a - 3c = 6 \quad (1)$

$\underline{21a + 3c = 60}$ Mult. (4) by 3

$22a = 66$

$a = 3$

$3 - 3c = 6$ Subst. in (1)

$-3c = 3$

$c = -1$

6. Substitute in Equations (2) or (3) to find b.

$b + 2(-1) = 2$ Sub. in (2)

$b - 2 = 2$

$b = 4$

We obtain $(3, 4, -1)$. This checks, so it is the solution.

33. $x + y + z = 83 \quad (1)$

$y = 2x + 3 \qquad (2)$

$z = 40 + x \qquad (3)$

1., 2. Equations (2) and (3) are not in standard form. However, both equations (2) and (3) have variables

y and z written in terms of the variable x. Substitute $2x + 3$ for y and $40 + x$ for z in Equation (1).

$x + y + z = 83$

$x + 2x + 3 + 40 + x = 83$

$4x + 43 = 83$

$4x = 40$

$x = 10$

3. Substitute 10 for x in Equation (2) to solve for y.

$y = 2x + 3$

$y = 2 \cdot 10 + 3$

$y = 23$

Substitute 10 for x in Equation (3) to solve for z.

$z = 40 + x$

$z = 40 + 10$

$z = 50$

We obtain $(10, 23, 50)$. This checks, so it is the solution.

35. $x \qquad + z = 0 \quad (1)$

$x + y + 2z = 3 \quad (2)$

$y \ + z = 2 \quad (3)$

1., 2. The equations are already in standard form with no fractions or decimals.

3., 4. Note that there is no y in Equation (1). We will use Equations (2) and (3) to obtain another equation with no y –term.

$x + y + 2z = 3 \quad (2)$

$\underline{-y \ - z = -2}$ Mult. (3) by -1

$x \qquad + z = 1 \quad (4)$ Adding

5. Now solve the system of Equations (1) and (4).

$x + z = 0 \quad (1)$

$x + z = 1 \quad (4)$

$x + z = 0 \quad (1)$

$\underline{-x - z = -1}$ Mult. (4) by -1

$0 = -1$ Adding

6. We get a false equation in step 5, or contradiction. There is no solution.

37. $x + y + z = 1$ (1)

    $-x + 2y + z = 2$ (2)

    $2x - y = -1$ (3)

1., 2. The equations are already in standard form with no fractions or decimals.

3. Note that there is no z in Equation (3). We will use Equations (1) and (2) to eliminate z:

$x + y + z = 1$ (1)

$\underline{x - 2y - z = -2}$ Mult. (2) by -1

$2x - y = -1$ (4) Adding

Equations (3) and (4) are identical, so Equations (1), (2), and (3) are dependent. (We have seen that if Equation (2) is multiplied by -1 and added to Equation (1), the result is Equation (3).)

39. *Thinking and Writing Exercise.*

41. One number is half another.
Let x and y represent the two numbers, then translate.

$x = \dfrac{1}{2}y$

43. The sum of three consecutive numbers is 100.
Let x represent the first number, so $(x + 1)$ and $(x + 2)$ represent the other numbers, then translate.

$x + (x + 1) + (x + 2) = 100$

45. The product of two numbers is five times a third number.
Let x, y, and z represent the three numbers, then translate.

$xy = 5z$

47. *Thinking and Writing Exercise.*

49. $\dfrac{x+2}{3} - \dfrac{y+4}{2} + \dfrac{z+1}{6} = 0$

$\dfrac{x-4}{3} + \dfrac{y+1}{4} - \dfrac{z-2}{2} = -1$

$\dfrac{x+1}{2} + \dfrac{y}{2} + \dfrac{z-1}{4} = \dfrac{3}{4}$

1., 2. We clear fractions and write each equation in standard form.

To clear fractions we multiply both sides of each equation by the LCM of its

denominators. The LCMs are 6, 12, and 4, respectively.

$6\left(\dfrac{x+2}{3} - \dfrac{y+4}{2} + \dfrac{z+1}{6}\right) = 6 \cdot 0$

$2(x+2) - 3(y+4) + (z+1) = 0$

$2x + 4 - 3y - 12 + z + 1 = 0$

$2x - 3y + z = 7$

$12\left(\dfrac{x-4}{3} + \dfrac{y+1}{4} - \dfrac{z-2}{2}\right) = 12 \cdot (-1)$

$4(x-4) + 3(y+1) - 6(z-2) = -12$

$4x - 16 + 3y + 3 - 6z + 12 = -12$

$4x + 3y - 6z = -11$

$4\left(\dfrac{x+1}{2} + \dfrac{y}{2} + \dfrac{z-1}{4}\right) = 4 \cdot \dfrac{3}{4}$

$2(x+1) + 2(y) + (z-1) = 3$

$2x + 2 + 2y + z - 1 = 3$

$2x + 2y + z = 2$

The resulting system is

$2x - 3y + z = 7$ (1)

$4x + 3y - 6z = -11$ (2)

$2x + 2y + z = 2$ (3)

3., 4. We eliminate z from two different pairs of equations.

$12x - 18y + 6z = 42$ Mult. (1) by 6

$\underline{4x + 3y - 6z = -11}$ (2)

$16x - 15y = 31$ (4) Adding

$2x - 3y + z = 7$ (1)

$\underline{-2x - 2y - z = -2}$ Mult. (3) by -1

$-5y = 5$ (5) Adding

5. Solve (5) for y: $-5y = 5$

$y = -1$

Substitute -1 for y in (4):

$16x - 15(-1) = 31$

$16x + 15 = 31$

$16x = 16$

$x = 1$

6. Substitute 1 for x and -1 for y in (1):

$2 \cdot 1 - 3(-1) + z = 7$

$5 + z = 7$

$z = 2$

We obtain $(1,-1,2)$. This checks, so it is the solution.

51. $\begin{array}{rl} w+ \ x- \ y+ \ z = \ 0 & (1) \\ w-2x-2y- \ z = -5 & (2) \\ w-3x- \ y+ \ z = \ 4 & (3) \\ 2w- \ x- \ y+3z = \ 7 & (4) \end{array}$

The equations are already in standard form with no fractions or decimals.
Start by eliminating z from three different pairs of equations.

$\begin{array}{rl} w+ \ x- \ y+z = \ 0 & (1) \\ w-2x-2y-z = -5 & (2) \\ \hline 2w \ -x-3y \ = -5 & (5) \text{ Adding} \end{array}$

$\begin{array}{rl} w-2x-2y-z = -5 & (2) \\ w-3x- \ y+z = \ 4 & (3) \\ \hline 2w-5x-3y \ = -1 & (6) \text{ Adding} \end{array}$

$\begin{array}{rl} 3w-6x-6y-3z = -15 & \text{Mult. (2) by 3} \\ 2w- \ x- \ y+3z = \ 7 & (4) \\ \hline 5w-7x-7y \ = -8 & (7) \text{ Adding} \end{array}$

Now solve the system of equations (5), (6), and (7).

$\begin{array}{rl} 2w-x-3y = -5 & (5) \\ 2w-5x-3y = -1 & (6) \\ 5w-7x-7y = -8 & (7) \end{array}$

$\begin{array}{rl} 2w-x-3y = -5 & (5) \\ -2w+5x+3y = \ 1 & \text{Mult. (6) by } -1 \\ \hline 4x \ = -4 & \\ x \ = -1 & \end{array}$

Substituting -1 for x in (5) and (7) and simplifying, we have
$\begin{array}{rl} 2w-3y = -6 & (8) \\ 5w-7y = -15 & (9) \end{array}$

Now solve the system of Equations (8) and (9).
$\begin{array}{rl} 10w-15y = -30 & \text{Mult. (8) by 5} \\ -10w+14y = \ 30 & \text{Mult. (9) by } -2 \\ \hline -y = 0 & \\ y = 0 & \end{array}$

Substitute 0 for y in Equation (8) or (9) and solve for w.

$\begin{array}{rl} 2w-3\cdot 0 = -6 & \text{Subst. in (8)} \\ 2w = -6 & \\ w = -3 & \end{array}$

Substitute in one of the original equations to find z.

$\begin{array}{rl} -3-1-0+z = 0 & \text{Subst. in (1)} \\ -4+z = 0 & \\ z = 4 & \end{array}$

We obtain $(-3,-1,0,4)$. This checks, so it is the solution.

53. $\begin{array}{rl} \dfrac{2}{x}+\dfrac{2}{y}-\dfrac{3}{z} = \ 3 & (1) \\[2mm] \dfrac{1}{x}-\dfrac{2}{y}-\dfrac{3}{z} = \ 9 & (2) \\[2mm] \dfrac{7}{x}-\dfrac{2}{y}+\dfrac{9}{z} = -39 & (3) \end{array}$

Let u represent $\dfrac{1}{x}$, v represent $\dfrac{1}{y}$, and w represent $\dfrac{1}{z}$. Substituting, we have

$\begin{array}{rl} 2u+2v-3w = \ 3 & (1) \\ u-2v-3w = \ 9 & (2) \\ 7u-2v+9w = -39 & (3) \end{array}$

1., 2. The equations in u, v, and w are in standard form with no fractions or decimals.

3., 4. We eliminate v from two different pairs of equations.

$\begin{array}{rl} 2u+2v-3w = 3 & (1) \\ u-2v-3w = 9 & (2) \\ \hline 3u \ -6w = 12 & (4) \text{ Adding} \end{array}$

$\begin{array}{rl} 2u+2v-3w = \ 3 & (1) \\ 7u-2v+9w = -39 & (3) \\ \hline 9u \ +6w = -36 & (5) \text{ Adding} \end{array}$

5.   Now solve the system of Equations (4) and (5).

$\begin{array}{rl} 3u-6w = \ 12 & (4) \\ 9u+6w = -36 & (5) \\ \hline 12u \ = -24 & \\ u = -2 & \end{array}$

$3(-2) - 6w = 12$ Subst. in (4)

$-6 - 6w = 12$

$-6w = 18$

$w = -3$

6. Substitute in Equation (1), (2), or (3) to find v.

$2(-2) + 2v - 3(-3) = 3$ Subst. in (1)

$2v + 5 = 3$

$2v = -2$

$v = -1$

Solve for x, y, and z. We substitute -2 for u, -1 for v and -3 for w.

$$u = \frac{1}{x} \qquad v = \frac{1}{y} \qquad w = \frac{1}{z}$$

$$-2 = \frac{1}{x} \qquad -1 = \frac{1}{y} \qquad -3 = \frac{1}{z}$$

$$x = -\frac{1}{2} \qquad y = -1 \qquad z = -\frac{1}{3}$$

We obtain $\left(-\frac{1}{2}, -1, -\frac{1}{3}\right)$. This checks, so it is the solution.

55. $5x - 6y + kz = -5$ (1)

$x + 3y - 2z = 2$ (2)

$2x - y + 4z = -1$ (3)

1., 2. The equations in x, y, and z are in standard form with no fractions or decimals.

3., 4. We eliminate y from two different pairs of equations.

$5x - 6y \qquad + kz = -5$ (1)

$2x + 6y \qquad - 4z = 4$ Mult. (2) by 2

$\overline{7x \qquad + (k-4)z = -1}$ (4) Adding

$x + 3y - 2z = 2$ (2)

$6x - 3y + 12z = -3$ Mult. (3) by 3

$\overline{7x \qquad + 10z = -1}$ (5)

5. Now solve the system of equations (4) and (5).

$7x + (k-4)z = -1$ (4)

$7x \qquad + 10z = -1$ (5)

$-7x \quad + (k-4)z = -1$ M. (4) by -1

$7x \qquad + 10z = -1$ (5)

$\overline{(-k + 14)z = 0}$ (6)

The system is dependent for the value of k that makes Equation (6) true. This occurs when $-k + 14$ is 0. We solve for k:

$-k + 14 = 0$

$14 = k$

57. $z = b - mx - ny$

Three solutions are $(1, 1, 2)$, $(3, 2, -6)$, and $\left(\frac{3}{2}, 1, 1\right)$. We substitute for x, y, and z and then solve for b, m, and n.

$2 = b - m - n$

$-6 = b - 3m - 2n$

$1 = b - \frac{3}{2}m - n$

1., 2. Write the equations in standard form. Also clear the fractions in the last equation.

$b - m - n = 2$ (1)

$b - 3m - 2n = -6$ (2)

$2b - 3m - 2n = 2$ (3)

3., 4. Eliminate b from two different pairs of equations.

$b - m - n = 2$ (1)

$-b + 3m + 2n = 6$ Mult. (2) by -1

$\overline{2m + n = 8}$ (4)

$-2b + 2m + 2n = -4$ Mult. (1)

by -2

$$\underline{2b - 3m - 2n = 2 \quad (3)}$$

$-m \quad\quad = -2$ (5) Adding

5.   Solve Equation (5) for m:

$-m = -2$

$m = 2$

Substitute in Equation (4) and solve for n.

$2 \cdot 2 + n = 8$

$4 + n = 8$

$n = 4$

6.   Substitute in one of the original equations to find b.

$b - 2 - 4 = 2$ Subst. 2 for m and

4 for n in (1)

$b - 6 = 2$

$b = 8$

The solution is $(8, 2, 4)$, so the equation is $z = 8 - 2x - 4y$.

Exercise Set 3.5

---

1. **Familiarize.** Let x = the first number, y = the second number, and z = the third number.
   Translate.

The sum of three numbers   is   57.

$\downarrow$ $\downarrow$ $\downarrow$

$x + y + z$ $=$ 57

The second   is   3 more than the first.

$\downarrow$ $\downarrow$ $\downarrow$ $\downarrow$ $\downarrow$

y $=$ 3 $+$ x

The third   is   6 more than the first.

$\downarrow$ $\downarrow$ $\downarrow$ $\downarrow$ $\downarrow$

z $=$ 6 $+$ x

We now have a system of equations.

$x + y + z = 57$ or $x + y + z = 57$

$y = 3 + x$ $-x + y = 3$

$z = 6 + x$ $-x + z = 6$

Carry out. Solving the system we get $(16, 19, 22)$.

Check. The sum of the three numbers is $16 + 19 + 22$, or 57. The second number, 19, is three more than the first number, 16. The third number, 22, is 6 more than the first

number, 16. The numbers check.
State. The numbers are 16, 19, and 22.

3. **Familiarize.** Let x = the first number, y = the second number, and z = the third number.
   Translate.

The sum of three numbers   is   26.

$\downarrow$ $\downarrow$ $\downarrow$

$x + y + z$ $=$ 26

Twice   minus   the   is   the   less 2.
the first   second   third

$\downarrow$ $\downarrow$ $\downarrow$ $\downarrow$ $\downarrow$ $\downarrow$ $\downarrow$

$2x$ $-$ y $=$ z $-$ 2

The   is   the   minus   3 times
third   second   the first.

$\downarrow$ $\downarrow$ $\downarrow$ $\downarrow$ $\downarrow$

z $=$ y $-$ $3x$

We now have a system of equations.

$x + y + z = 26$ or $x + y + z = 26$

$2x - y = z - 2$ $2x - y - z = -2$

$z = y - 3x$ $3x - y + z = 0$

Carry out. Solving the system we get $(8, 21, -3)$.

Check. The sum of the three numbers is $8 + 21 + (-3)$, or 26. Twice the first minus the second is $2 \cdot 8 - 21$, or -5, which is 2 less than the third. The second minus three times the first is $21 - 3 \cdot 8$, or -3, which is the third. The numbers check.
State. The numbers are 8, 21, and -3.

5. **Familiarize.** We first make a drawing.

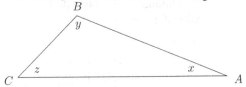

We let x, y, and z represent the measures of angles A, B, and C, respectively The measures of the angles of a triangle add up to $180°$.
   Translate.

The sum of the measures is $180°$.

$\downarrow$ $\downarrow$ $\downarrow$

$x + y + z$ $=$ 180

The measure of angle B   is   three times the measure of angle A.

$$\downarrow \qquad \downarrow \qquad \downarrow$$
$$y \quad = \quad 3x$$

The measure of angle C   is   20° more than the measure of angle A.

$$\downarrow \qquad \downarrow \qquad \downarrow$$
$$z \quad = \quad x+20$$

We now have a system of equations.

$$x+y+z=180$$
$$y=3x$$
$$z=x+20$$

Carry out. Solving the system we get $(32, 96, 52)$.

Check. The sum of the measures is $32° + 96° + 52°$, or $180°$. Three times the measure of angle A is $3 \cdot 32°$, or $96°$, the measure of angle B. 20° more than the measure of angle A is $32° + 20°$, or $52°$, the measure of angle C. The numbers check.

State. The measures of the angles A, B, and C, are $32°$, $96°$, and $52°$, respectively.

7. **Familiarize.** Let $x =$ the average score in math, $y =$ the average score in reading, and $z =$ the average score in writing. Then

Translate.

$$x+y+z=1509 \quad (1)$$
$$x=y+14 \quad (2)$$
$$z=y-8 \quad (3)$$

Carry out. Equations (2) and (3) respectively define x and z in terms of y. Through substitution, we rewrite equation (1) entirely in terms of y and solve for y.

$$x+y+z=1509 \quad (1)$$
$$(y+14)+y+(y-8)=1509$$
$$y+14+y+y-8=1509$$
$$3y+6=1509$$
$$3y=1503$$
$$y=501$$

Substitute the result into equations (2) and (3) to find x and z:

$$x=y+14 \quad (2)$$
$$x=501+14=515$$
$$z=y-8 \quad (3)$$
$$z=501-8=493$$

Check.

$$515+501+493=1509 \quad (1) \checkmark$$
$$515=501+14 \quad (2) \checkmark$$
$$493=501-8 \quad (3) \checkmark$$

State. On the 2009 SAT, the average math score was 515, the average reading score was 501, and the average writing score was 493.

9. **Familiarize.** Let $x =$ the number of grams of fiber in a bran muffin. Let $y =$ the number of grams of fiber in a banana. Let $z =$ the number of grams of fiber in 1-cup of serving of Wheaties®.

Translate. We can summarize this information in a table.

Bran Muffin	Banana	Wheaties 1-cup	Fiber
$2x$	y	z	9
x	$2y$	z	10.5
$2x$	0	z	6

We now have a system of equations.

$$2x+y+z=9$$
$$x+2y+z=10.5$$
$$2x+z=6$$

Carry out. Solving the system we get $(1.5, 3, 3)$.

Check. A breakfast of 2 bran muffins (1.5 grams of fiber each), one banana (3 grams of fiber), and a 1-cup serving of Wheaties (3 grams of fiber) provides $2(1.5)+3+3$, or 9 grams of fiber. A breakfast of 1 bran muffin, 2 bananas, and a 1-cup serving of Wheaties provides $1.5+2\cdot3+3$, or 10.5 grams of fiber. A breakfast of 2 bran muffins and a 1-cup serving of Wheaties provides $2(1.5)+3$, or 6 grams of fiber. These numbers check.

State. Each bran muffin contains 1.5 g of fiber, each banana contains 3 g of fiber, and each 1-cup serving of Wheaties contains 3 g of fiber.

11. **_Familiarize._** Let x = the price of the basic model in 2010, y = the price of the car cover, and z = the price of satellite radio. Then:
Translate.

$$x + y = 24,030 \quad (1)$$
$$x + y + z = 24,340 \quad (2)$$
$$x + z = 24,110 \quad (3)$$

Carry out. Subtracting equation (1) from equation (2) will allow us to find the value of z.

$$x + y + z = 24,340 \quad (2)$$
$$x + y \quad\quad = 24,030 \quad (1)$$
$$z = 24,340 - 24,030 = 310$$

Use the value of z and equation (3) to find x:

$$x + z = 24,110 \quad (3)$$
$$x = 24,110 - z = 24,110 - 310 = 23,800$$

Use the value of x and equation (1) to find y:

$$x + y = 24,030 \quad (1)$$
$$y = 24,030 - x = 24,030 - 23,800 = 230$$

Check.

$$23,800 + 230 = 24,030 \quad (1)\,\checkmark$$
$$23,800 + 230 + 310 = 24,340 \quad (2)\,\checkmark$$
$$23,800 + 310 = 24,110 \quad (3)\,\checkmark$$

State. In 2010, the basic price of a Honda Civic Hybrid was \$23,800. The car cover cost \$230, and satellite radio for the car cost \$310.

13. **_Familiarize._** Let x = the number of 12 oz cups sold. Let y = the number of 16 oz cups sold, and let z = the number of 20 oz cups sold. 12, 16, and 20 oz cups sell for \$1.65, \$1.85, and \$1.95, respectively. If Reba empties six 144-oz "brewers," then a total of $6 \cdot 144 = 864$ oz of coffee was sold.
Translate.

	12 oz	16 oz	20 oz	Total
Cups sold	x	y	z	55
Oz sold	$12x$	$16y$	$20z$	864
Sales	$165x$	$185y$	$195z$	9965

Now we have a system of equations.

$$x + y + z = 55 \quad (1)$$
$$12x + 16y + 20z = 864 \quad (2)$$
$$165x + 185y + 195z = 9965 \quad (3)$$

Note that equation (3) is in cents. We can reduce the coefficients in equations (2) and (3) if we note that all the coefficients in

equation (2) are divisible by 4, and all the coefficients in equation (3) are divisible by 5.

$$12x + 16y + 20z = 864$$
$$\frac{1}{4} \cdot [12x + 16y + 20z] = \frac{1}{4} \cdot [864]$$
$$3x + 4y + 5z = 216 \quad\quad (5)$$
$$165x + 185y + 195z = 9965$$
$$\frac{1}{5} \cdot [165x + 185y + 195z] = \frac{1}{5} \cdot [9965]$$
$$33x + 37y + 39z = 1993 \quad (6)$$

We can now solve the new system:

$$x + y + z = 55 \quad\quad (1)$$
$$3x + 4y + 5z = 216 \quad\quad (5)$$
$$33x + 37y + 39z = 1993 \quad (6)$$

Carry out.
Adding $-3 \cdot (1)$ to (5), and $-33(1)$ to (6) will give two equations in y and z only.

$$-3x - 3y - 3z = -165 \quad -3 \cdot (1)$$
$$\underline{3x + 4y + 5z = 216 \quad\quad (5)}$$
$$y + 2z = 51 \quad\quad\quad (7)$$

$$-33x - 33y - 33z = -1815 \quad -33 \cdot (1)$$
$$\underline{33x + 37y + 39z = 1993 \quad\quad (6)}$$
$$4y + 6z = 178$$
$$\frac{1}{2} \cdot [4y + 6z] = \frac{1}{2} \cdot [178]$$
$$2y + 3z = 89 \quad\quad (8)$$

Adding $-2 \cdot (7)$ to (8) will give an equation in z only:

$$-2y - 4z = -102 \quad -2 \cdot (7)$$
$$\underline{2y + 3z = 89 \quad\quad (8)}$$
$$-z = -13 \Rightarrow z = 13$$

Knowing z, use equation (7) to find y and then equation (1) to find x.

$$y + 2z = 51 \quad (7)$$
$$y = 51 - 2z = 51 - 2(13) = 51 - 26 = 25$$
$$x + y + z = 55 \quad (1)$$
$$x = 55 - y - z = 55 - 25 - 13 = 17$$

Check.

$$17 + 13 + 25 = 55 \quad\quad (1)\,\checkmark$$
$$12(17) + 16(25) + 20(13) = 864 \quad\quad (2)\,\checkmark$$
$$165(17) + 185(25) + 195(13) = 9965 \quad (3)\,\checkmark$$

State. Reba sold 17 twelve-oz coffees, 25 sixteen-oz coffees, and 13 twenty-oz coffees.

15. **Familiarize.** Let x = amount borrowed at 8%, y = amount borrowed at 5%, and z = amount borrowed at 4%.
Translate.

$$x + y + z = 120,000 \qquad (1)$$
$$0.08x + 0.05y + 0.04z = 5750$$
$$\text{or } 8x + 5y + 4z = 575,000 \qquad (2)$$
$$0.04z = 0.08x + 1600$$
$$\text{or } 8x - 4z = -160,000$$
$$\text{or } 2x - z = -40,000 \qquad (3)$$

Carry out. Using $(1) + (3)$ and $(2) + 4 \cdot (3)$ will generate a two-variable system in x and y.

$$\begin{array}{l} x + y + z = 120,000 \quad (1) \\ \underline{2x \quad - z = -40,000 \quad (2)} \\ 3x + y \quad = 80,000 \quad (4) \end{array}$$

and

$$\begin{array}{l} 8x + 5y + 4z = 575,000 \quad (2) \\ \underline{8x \quad - 4z = -160,000 \quad -4 \cdot (3)} \\ 16x + 5y \quad = 415,000 \quad (5) \end{array}$$

Using $-5 \cdot (4) + (5)$ we can find x and then back-substitute into equations (4) and (1) to find y and z.

$$\begin{array}{l} -15x - 5y \quad = -400,000 \quad -5 \cdot (4) \\ \underline{16x + 5y \quad = 415,000 \quad (5)} \\ x = 15,000 \end{array}$$

So: $3x + y \quad = 80,000 \quad (4)$
$y = 80,000 - 3x = 80,000 - 3(15,000)$
$y = 80,000 - 45,000 = 35,000$
and: $x + y + z = 120,000 \quad (1)$
$z = 120,000 - x - y$
$z = 120,000 - 15,000 - 35,000 = 70,000$

Check.

$$15,000 + 35,000 + 70,000 = 120,000 \quad (1) \checkmark$$
$$8(15,000) + 5(35,000) + 4(70,000) = 575,000 \quad (2) \checkmark$$
$$2(15,000) - 70,000 = -40,000 \quad (3) \checkmark$$

State. Chelsea borrowed $15,000 at 8%, $35,000 at 5%, and $70,000 at 4% interest.

17. **Familiarize.** Let x = cost per gram of gold, y = cost per gram of silver, and z = cost per gram of copper. Note that $x\%$ of 100 grams equals x.
Translate.

$$75x + 5y + 20z = 2265.40$$
$$\text{or } 15x + y + 4z = 453.08 \qquad (1)$$
$$75x + 12.5y + 12.5z = 2287.75$$
$$\text{or } 6x + y + z = 183.02 \qquad (2)$$
$$37.5x + 62.5y = 1312.50$$
$$\text{or } 3x + 5y = 105 \qquad (3)$$

Carry out.
Using $4 \cdot (2) - (1)$ and (3) will generate a two-variable system in x and y.

$$\begin{array}{l} 24x + 4y + 4z = 732.08 \quad 4 \cdot (2) \\ \underline{-15x \; - y - 4z = -453.08 \quad -1 \cdot (1)} \\ 9x + 3y \quad = 279 \\ \text{or } 3x + y = 93 \quad (5) \end{array}$$

Using $(3) - 1 \cdot (5)$ we can find y and then back-substitute into equations (5) and (2) to find x and z.

$$\begin{array}{l} 3x + 5y = 105 \quad (3) \\ \underline{-3x - y = -93 \quad -1 \cdot (5)} \\ 4y = 12 \end{array}$$

$$y = \frac{12}{4} = 3$$

So: $3x + y \quad = 93 \quad (5)$
$3x = 93 - y = 93 - 3 = 90$

$$x = \frac{90}{3} = 30$$

and: $6x + y + z = 183.02 \quad (2)$
$z = 183.02 - 6x - y = 183.02 - 6(30) - 3$
$z = 183.02 - 180 - 3 = 0.02$

Check.

$$75(30) + 5(3) + 20(0.02) = 2265.40 \quad (1) \checkmark$$
$$75(30) + 12.5(3) + 12.5(0.02) = 2287.75 \quad (2) \checkmark$$
$$37.5(30) + 62.5(3) = 1312.50 \quad (3) \checkmark$$

State. Gold is $30 a gram, silver is $3 a gram and copper is $0.02 a gram.

19. **Familiarize.** Let r = the number of servings of roast beef, p = the number of baked potatoes, and b = the number of servings of broccoli. Then r servings of roast beef contain $300r$ Calories, $20r$ g of protein, and no vitamin C. In p baked potatoes there are $100p$ Calories, $5p$ g of protein, and $20p$ mg of vitamin C. And b servings of broccoli contain $50b$ Calories, $5b$ g of protein, and $100b$ mg of vitamin C. The patient requires 800 Calories, 55 g of protein, and 220 mg of

vitamin C.

Translate. Write equations for the total number of calories, the total amount of protein, and the total amount of vitamin C.

$$300r + 100p + 50b = 800 \quad \text{(Calories)}$$
$$20r + 5p + 5b = 55 \quad \text{(Protein)}$$
$$20p + 100b = 220 \quad \text{(Vitamin C)}$$

We now have a system of equations.

Carry out. Solving the system we get $(2,1,2)$.

Check. Two servings of roast beef provide 600 Calories, 40 g of protein, and no vitamin C. One baked potato provides 100 Calories, 5 g of protein, and 20 gm of vitamin C. And 2 servings of broccoli provide 100 Calories, 10 g of protein, and 200 mg of vitamin C. Together, then, they provide 800 Calories, 55 g of protein, and 220 mg of vitamin C. The values check.

State. The dietician should prepare 2 servings of roast beef, 1 baked potato, and 2 servings of broccoli.

21. ***Familiarize.*** Let x = number of main floor tickets sold, y = number of first mezzanine tickets sold, and z = number of second mezzanine tickets sold.

Translate.
$$x + y + z = 40 \quad (1)$$
$$x + y = z$$
$$\text{or } x + y - z = 0 \quad (2)$$
$$38x + 52y + 28z = 1432$$
$$\text{or } 19x + 26y + 14z = 716 \quad (3)$$

Carry out.

Using $(1)+(2)$ and $(3)-14\cdot(1)$ will generate a two-variable system in x and y.

$$\begin{array}{l} x + y + z = 40 \quad (1) \\ \underline{x + y - z = 0 \quad (2)} \\ 2x + 2y \quad = 40 \end{array}$$

$$\text{or } x + y = 20 \quad (5)$$

This result actually allows us to find z right away using equation (2)

$$x + y - z = 0 \quad (2)$$
$$z = x + y = 20$$

Back to $(3)-14\cdot(1)$ to find x and y.

$$\begin{array}{l} 19x + 26y + 14z = 716 \quad (3) \\ \underline{-14x - 14y - 14z = -560 \quad -14\cdot(1)} \\ 5x + 12y = 156 \quad (6) \end{array}$$

Using $(6)-5\cdot(5)$ we can find y and then back-substitute into equation (5) to find x.

$$\begin{array}{l} 5x + 12y = 156 \quad (6) \\ \underline{-5x - 5y = -100 \quad -5\cdot(5)} \\ 7y = 56 \end{array}$$

$$y = \frac{56}{7} = 8$$

So: $x + y = 20 \quad (5)$

$$x = 20 - y = 20 - 8 = 12$$

Check.

$$12 + 8 + 20 = 40 \quad (1) \checkmark$$
$$12 + 8 = 20 \quad (2) \checkmark$$
$$38(12) + 52(8) + 28(20) = 1432 \quad (3) \checkmark$$

State. 12 floor seats were sold, 8 first mezzanine seats were sold, and 20 second mezzanine seats were sold.

23. ***Familiarize.*** Let x = the number of two-point field goals, y = the number of 3-point field goals, and z = the number of 1-point foul shots.

Translate.

Total number of baskets	is	50.
↓	↓	↓
$x + y + z$	$=$	50

Total number of points	is	92.
↓	↓	↓
$2x + 3y + z$	$=$	92

Number of 2-pointers	is 19	more than	the number of foul shots.
↓	↓ ↓	↓	↓
x	$= 19$	$+$	z

Now we have a system of equations.

$$x + y + z = 50$$
$$2x + 3y + z = 92$$
$$x = 19 + z$$

Carry out. Solving the system, we get $(32,5,13)$.

Check. The total number of baskets made was $32 + 5 + 13$, or 50. The total number of points made was $32\cdot 2 + 5\cdot 3 + 13\cdot 1$, or $64 +$

15 + 13 = 92. The number of 2-pointers, 32, was 19 more than the number of foul shots, 13. These numbers check.

State. The number of two-point field goals, three-point field goals, and the number of foul shots was 32, 5, and 13, respectively.

25. *Thinking and Writing Exercise.*

27. $-2(2x - 3y) = -4x + 6y$

29. $-6(x - 2y) + (6x - 5y)$
$= -6x + 12y + 6x - 5y = 7y$

31. $-(2a - b - 6c) = -2a + b + 6c$

33. $-2(3x - y + z) + 3(-2x + y - 2z)$
$= -6x + 2y - 2z - 6x + 3y - 6z$
$= -12x + 5y - 8z$

35. *Thinking and Writing Exercise.*

37. **Familiarize.** Let x = the cost for the applicant, y = the cost for the spouse, z = the cost for the first child, and w = the cost for the second child.
Translate.

Monthly cost for an applicant and spouse is 135.
$$x + y = 135$$

Monthly cost for an applicant and spouse and one child is 154.
$$x + y + z = 154$$

Monthly cost for an applicant and spouse and two children is 173.
$$x + y + z + w = 173$$

Monthly cost for an applicant, one child is 102.
$$x + z = 102$$

We now have a system of equations.

$$x + y = 135$$
$$x + y + z = 154$$
$$x + y + z + w = 173$$
$$x + z = 102$$

Carry out. Solving the system we get $(83, 52, 19, 19)$.

Check. The cost for an applicant and his or her spouse is \$83 + \$52 = \$135. The cost when just one child is added is: \$83 + \$52 + \$19 = \$154. When an additional child is added, the cost is \$83 + \$52 + \$19 + \$19 = \$173. Finally, the cost for an applicant and just one child is \$83 + \$19 = \$102. These numbers check.

State. The monthly costs for an applicant, spouse, first child, and second child are \$83, \$52, \$19, and \$19, respectively.

39. **Familiarize.** Let t = Tammy's age, let c = Carmen's age, let d = Dennis's age, and let m = Mark's age.
Translate.

Tammy's age is the sum of Carmen and Dennis's ages.
$$t = c + d$$

Carmen's age is 2 more than the sum of Dennis and Mark's ages.
$$c = 2 + d + m$$

Dennis's age is four times Mark's age.
$$d = 4m$$

Sum of all four ages is 42.
$$t + c + d + m = 42$$

Now we have a system of equations.
$$t = c + d$$
$$c = 2 + d + m$$
$$d = 4m$$
$$t + c + d + m = 42$$

Carry out. We are only asked to determine Tammy's age, but we will solve the whole system in order to check our work. Solving the system we get $(20, 12, 8, 2)$.

Check. The sum of Carmen's and Dennis's

ages are 12 + 8, or 20, which is Tammy's age. Carmen's age, 12, is 2 more than the sum of Dennis and Mark's age, 8 + 2 = 10. Dennis's age is 8, which is 4 times Mark's age. The sum of all four ages is 20 + 12 + 8 + 2, or 42. These numbers check.

State. Tammy is 20 years old. In addition, Carmen is 12, Dennis is 8, and Mark is 2.

41. **Familiarize.** Let T, G, and H represent the number of tickets Tom, Gary, and Hal begin with respectively.

Translate. After Hal gives tickets to Tom and Gary, each has the following number of tickets:

Tom: $T + T$, or $2T$

Gary: $G + G$, or $2G$

Hal: $H - T - G$

After Tom gives tickets to Gary and Hal, each has the following number of tickets:

Gary:

$2G + 2G$, or $4G$

Hal:

$(H - T - G) + (H - T - G)$, or

$2(H - T - G)$

Tom:

$2T - 2G - (H - T - G)$, or

$3T - H - G$

After Gary gives tickets to Hal and Tom, each has the following number of tickets:

Hal:

$2(H - T - G) + 2(H - T - G)$, or

$4(H - T - G)$

Tom:

$(3T - H - G) + (3T - H - G)$, or

$2(3T - H - G)$

Gary:

$4G - 2(H - T - G) - (3T - H - G)$, or

$7G - H - T$

Since Hal, Tom, and Gary each finish with 40 tickets, we write the following system of equations:

$4(H - T - G) = 40$

$2(3T - H - G) = 40$

$7G - H - T = 40$

Carry out. Solving the system we get

$(35, 20, 65)$.

Check. Hal has 65 tickets to start with and gives Tom 35 tickets and gives Gary 20 tickets. As a result, Tom has 70 tickets and Gary has 40 tickets. Hal's supply of tickets has been reduced by 35 + 20, or 55, leaving him with 10 tickets. So, Tom, Gary, and Hal now have 70, 40, and 10 tickets, respectively. Tom then gives Hal 10 tickets and Gary 40 tickets. As a result, Hal has 20 tickets and Gary has 80 tickets. Tom's supply of tickets has been reduced by 10 + 40, or 50, leaving him with 20 tickets. So, Tom, Gary, and Hal now have 20, 80, 20 tickets, respectively. Finally, Gary gives Hal 20 tickets and gives Tom 20 tickets, so they now each have 40 tickets. Gary had 80 and gave away 20 + 20, or 40, so he has 40 left. So, all three ended up with 40 tickets apiece.

State. Tom had 35 tickets to start with. Additionally, Gary had 20, and Hal had 65.

Exercise Set 3.6

1. matrix

3. entry

5. rows

7. $x + 2y = 11$

$3x - y = 5$

Write a matrix using only the constants.

$$\begin{bmatrix} 1 & 2 & | & 11 \\ 3 & -1 & | & 5 \end{bmatrix}$$

Multiply row 1 by −3 and add it to row 2.

$$\begin{bmatrix} 1 & 2 & | & 11 \\ 0 & -7 & | & -28 \end{bmatrix} \text{ New R2} = -3(\text{R1}) + \text{R2}$$

Reinserting the variables, we have

$x + 2y = 11$ (1)

$\quad -7y = -28$ (2)

Solve Equation (2) for y.

$-7y = -28$

$\quad y = 4$

Substitute 4 for y in Equation (1) and solve for x.

$x + 2y = 11 \Rightarrow x = 11 - 2y = 11 - 2(4) = 3$

The solution is $(3, 4)$.

9. $x + 4y = 8$

$3x + 5y = 3$

Write a matrix using only the constants.

$$\begin{bmatrix} 1 & 4 & | & 8 \\ 3 & 5 & | & 3 \end{bmatrix}$$

Multiply the first row by –3 and add it to the second row.

$$\begin{bmatrix} 1 & 4 & | & 8 \\ 0 & -7 & | & -21 \end{bmatrix}$$ New Row 2 $= -3($Row 1$)$
                                                   $+$ Row 2

Reinserting the variables, we have

$x + 4y = 8$ (1)

$-7y = -21$ (2)

Solve Equation (2) for y.

$-7y = -21$

$y = 3$

Substitute 3 for y in Equation (1) and solve for x.

$x + 4 \cdot 3 = 8$

$x + 12 = 8$

$x = -4$

The solution is $(-4, 3)$.

11. $6x - 2y = 4$

$7x + y = 13$

Write a matrix using only the constants.

$$\begin{bmatrix} 6 & -2 & | & 4 \\ 7 & 1 & | & 13 \end{bmatrix}$$

Multiply the second row by 6 to make the first number in row 2 a multiple of 6.

$$\begin{bmatrix} 6 & -2 & | & 4 \\ 42 & 6 & | & 78 \end{bmatrix}$$ New Row 2 $= 6($Row 2$)$

Multiply the first row by –7 and add it to the second row.

$$\begin{bmatrix} 6 & -2 & | & 4 \\ 0 & 20 & | & 50 \end{bmatrix}$$
New Row 2 $= -7($Row 1$) +$Row 2

Reinserting the variables, we have

$6x - 2y = 4$ (1)

$20y = 50$ (2)

Solve Equation (2) for y.

$20y = 50$

$y = \dfrac{5}{2}$

Substitute $\dfrac{5}{2}$ for y in Equation (1) and solve for x.

$6x - 2y = 4$

$6x - 2\left(\dfrac{5}{2}\right) = 4$

$6x - 5 = 4$

$6x = 9$

$x = \dfrac{3}{2}$

The solution is $\left(\dfrac{3}{2}, \dfrac{5}{2}\right)$.

13. $3x + 2y + 2z = 3$

$x + 2y - z = 5$

$2x - 4y + z = 0$

Write a matrix using only the constants.

$$\begin{bmatrix} 3 & 2 & 2 & | & 3 \\ 1 & 2 & -1 & | & 5 \\ 2 & -4 & 1 & | & 0 \end{bmatrix}$$

First interchange rows 1 and 2 so that each number below the first number in the first row is a multiple of that number.

$$\begin{bmatrix} 1 & 2 & -1 & | & 5 \\ 3 & 2 & 2 & | & 3 \\ 2 & -4 & 1 & | & 0 \end{bmatrix}$$

Multiply row 1 by –3 and add it to row 2. Multiply row 1 by –2 and add it to row 3.

$$\begin{bmatrix} 1 & 2 & -1 & | & 5 \\ 0 & -4 & 5 & | & -12 \\ 0 & -8 & 3 & | & -10 \end{bmatrix}$$

Multiply row 2 by –2 and add it to row 3.

$$\begin{bmatrix} 1 & 2 & -1 & | & 5 \\ 0 & -4 & 5 & | & -12 \\ 0 & 0 & -7 & | & 14 \end{bmatrix}$$

Reinserting the variables, we have

$x + 2y - z = 5$ (1)

$-4y + 5z = -12$ (2)

$-7z = 14$ (3)

Solve Equation (3) for z.

$-7z = 14$

$z = -2$

Substitute –2 for z in Equation (2) and solve for y.

$$-4y + 5(-2) = -12$$
$$-4y - 10 = -12$$
$$-4y = -2$$
$$y = \frac{1}{2}$$

Substitute $\frac{1}{2}$ for y and -2 for z in (1) and solve for x.

$$x + 2 \cdot \frac{1}{2} - (-2) = 5$$
$$x + 1 + 2 = 5$$
$$x + 3 = 5$$
$$x = 2$$

The solution is $\left(2, \frac{1}{2}, -2\right)$.

15. $a - 2b - 3c = 3$
$2a - b - 2c = 4$
$4a + 5b + 6c = 4$
Write a matrix using only the constants.

$$\begin{bmatrix} 1 & -2 & -3 & | & 3 \\ 2 & -1 & -2 & | & 4 \\ 4 & 5 & 6 & | & 4 \end{bmatrix}$$

Multiply row 1 by -2 and add it to row 2.
Multiply row 1 by -4 and add it to row 3.

$$\begin{bmatrix} 1 & -2 & -3 & | & 3 \\ 0 & 3 & 4 & | & -2 \\ 0 & 13 & 18 & | & -8 \end{bmatrix} \quad \begin{array}{l} \text{New } R2 = -2R1 + R2 \\ \\ \text{New } R3 = -4R1 + R3 \end{array}$$

Multiply row 3 by 3.

$$\begin{bmatrix} 1 & -2 & -3 & | & 3 \\ 0 & 3 & 4 & | & -2 \\ 0 & 39 & 54 & | & -24 \end{bmatrix} \quad \text{New } R3 = 3R3$$

Multiply row 2 by -13 and add it to row 3.

$$\begin{bmatrix} 1 & -2 & -3 & | & 3 \\ 0 & 3 & 4 & | & -2 \\ 0 & 0 & 2 & | & 2 \end{bmatrix} \quad \text{New } R3 = -13R2 + R3$$

Reinserting the variables, we have

$a - 2b - 3c = 3 \quad (1)$
$3b + 4c = -2 \quad (2)$
$2c = 2 \quad (3)$

Solve Equation (3) for c.
$2c = 2 \Rightarrow c = 1$
Substitute $c = 1$ into equation (2) and solve

for b.
$$3b + 4 \cdot 1 = -2$$
$$3b + 4 = -2$$
$$3b = -6$$
$$b = -2$$

Substitute $b = -2$ and $c = 1$ into (1) and solve for a.

$$a - 2(-2) - 3 \cdot 1 = 3$$
$$a + 4 - 3 = 3$$
$$a + 1 = 3$$
$$a = 2$$

The solution is $(2, -2, 1)$.

17. $3u + 2w = 11$
$v - 7w = 4$
$u - 6v = 1$
Write a matrix using only the constants.

$$\begin{bmatrix} 3 & 0 & 2 & | & 11 \\ 0 & 1 & -7 & | & 4 \\ 1 & -6 & 0 & | & 1 \end{bmatrix}$$

Interchange row 1 and row 3.

$$\begin{bmatrix} 1 & -6 & 0 & | & 1 \\ 0 & 1 & -7 & | & 4 \\ 3 & 0 & 2 & | & 11 \end{bmatrix} \quad \begin{array}{l}\text{Interchange R1 and R3}\end{array}$$

Multiply row 1 by -3 and add it to row 3.

$$\begin{bmatrix} 1 & -6 & 0 & | & 1 \\ 0 & 1 & -7 & | & 4 \\ 0 & 18 & 2 & | & 8 \end{bmatrix} \quad \text{New } R3 = -3R1 + R3$$

Multiply row 2 by -18 and add it to row 3.

$$\begin{bmatrix} 1 & -6 & 0 & | & 1 \\ 0 & 1 & -7 & | & 4 \\ 0 & 0 & 128 & | & -64 \end{bmatrix} \quad \text{New } R3 = -18R2 + R3$$

Reinserting the variables, we have

$u - 6v = 1 \quad (1)$
$v - 7w = 4 \quad (2)$
$128w = -64 \quad (3)$

Solve Equation (3) for w.
$128w = -64$

$$w = -\frac{1}{2}$$

Substitute $w = -\dfrac{1}{2}$ into equation (2) and

solve for v.

$$v - 7w = 4$$
$$v - 7\left(-\dfrac{1}{2}\right) = 4$$
$$v + \dfrac{7}{2} = 4$$
$$v = \dfrac{1}{2}$$

Substitute $v = \dfrac{1}{2}$ in (1) and solve for u.

$$u - 6 \cdot \dfrac{1}{2} = 1$$
$$u - 3 = 1$$
$$u = 4$$

The solution is $\left(4, \dfrac{1}{2}, -\dfrac{1}{2}\right)$.

19. We will rewrite the equations with the variables in alphabetical order:
$$-2w + 2x + 2y - 2z = -10$$
$$w + x + y + z = -5$$
$$3w + x - y + 4z = -2$$
$$w + 3x - 2y + 2z = -6$$

Write a matrix using only the constants.

$$\begin{bmatrix} -2 & 2 & 2 & -2 & | & -10 \\ 1 & 1 & 1 & 1 & | & -5 \\ 3 & 1 & -1 & 4 & | & -2 \\ 1 & 3 & -2 & 2 & | & -6 \end{bmatrix}$$

Multiply row 1 by $\dfrac{1}{2}$.

$$\begin{bmatrix} -1 & 1 & 1 & -1 & | & -5 \\ 1 & 1 & 1 & 1 & | & -5 \\ 3 & 1 & -1 & 4 & | & -2 \\ 1 & 3 & -2 & 2 & | & -6 \end{bmatrix}$$ New Row 1 = $\dfrac{1}{2}$(Row 1)

Add row 1 to row 2.
Add 3 times row 1 to row 3.
Add row 1 to row 4.

$$\begin{bmatrix} -1 & 1 & 1 & -1 & | & -5 \\ 0 & 2 & 2 & 0 & | & -10 \\ 0 & 4 & 2 & 1 & | & -17 \\ 0 & 4 & -1 & 1 & | & -11 \end{bmatrix}$$

New Row 2 = Row 1 + Row 2
New Row 3 = 3(Row 1) + Row 3
New Row 4 = Row 1 + Row 4

Multiply row 2 by –2 and add it to row 3.
Multiply row 2 by –2 and add it to row 4.

$$\begin{bmatrix} -1 & 1 & 1 & -1 & | & -5 \\ 0 & 2 & 2 & 0 & | & -10 \\ 0 & 0 & -2 & 1 & | & 3 \\ 0 & 0 & -5 & 1 & | & 9 \end{bmatrix}$$

New Row 3 = –2(Row 2) + Row 3
New Row 4 = –2(Row 2) + Row 4

Multiply row 4 by 2.

$$\begin{bmatrix} -1 & 1 & 1 & -1 & | & -5 \\ 0 & 2 & 2 & 0 & | & -10 \\ 0 & 0 & -2 & 1 & | & 3 \\ 0 & 0 & -10 & 2 & | & 18 \end{bmatrix}$$

New Row 4 = 2(Row 4)

Multiply row 3 by –5 and add it to row 4.

$$\begin{bmatrix} -1 & 1 & 1 & -1 & | & -5 \\ 0 & 2 & 2 & 0 & | & -10 \\ 0 & 0 & -2 & 1 & | & 3 \\ 0 & 0 & 0 & -3 & | & 3 \end{bmatrix}$$

New Row 4 = –5(Row 3) + Row 4

Reinserting the variables, we have
$$-w + x + y - z = -5 \quad (1)$$
$$2x + 2y = -10 \quad (2)$$
$$-2y + z = 3 \quad (3)$$
$$-3z = 3 \quad (4)$$

Solve (4) for z.
$$-3z = 3$$
$$z = -1$$

Substitute –1 for z in (3) and solve for y.
$$-2y + (-1) = 3$$
$$-2y = 4$$
$$y = -2$$

Substitute –2 for y in (2) and solve for x.

$$2x + 2(-2) = -10$$
$$2x - 4 = -10$$
$$2x = -6$$
$$x = -3$$

Substitute -3 for x, -2 for y, and -1 for z and solve for w.

$$-w + (-3) + (-2) - (-1) = -5$$
$$-w - 3 - 2 + 1 = -5$$
$$-w - 4 = -5$$
$$-w = -1$$
$$w = 1$$

The solution is $(1, -3, -2, -1)$.

21. **Familiarize.** Let d = the number of dimes and n = the number of nickels. The value of the dimes is \$0.10d. The value of the nickels is \$0.05n.

Translate.

$\underbrace{\text{Total number of coins}}$ is 42.

$\downarrow\downarrow\downarrow$
$d + n= 42$

$\underbrace{\text{Total value of the coins}}$ is \$3.00.

$\downarrow\downarrow\downarrow$
$0.10d + 0.05n= 3.00$

After clearing decimals, we have this system.

$$d + n = 42$$
$$10d + 5n = 300$$

Carry out. Solve using matrices.

$$\begin{bmatrix} 1 & 1 & | & 42 \\ 10 & 5 & | & 300 \end{bmatrix}$$

$$\begin{bmatrix} 1 & 1 & | & 42 \\ 0 & -5 & | & -120 \end{bmatrix} \text{New Row 2}$$
$$= -10(\text{Row 1}) + \text{Row 2}$$

Reinserting the variables, we have

$$d + n = 42 \quad (1)$$
$$-5n = -120 \quad (2)$$

Solve (2) for n.

$$-5n = -120$$
$$n = 24$$

$d + n = 42$ Substituting in (2)

$$d + 24 = 42$$
$$d = 18$$

Check. The sum of the two numbers is 42. The total value is \$0.10(18) + \$0.05(24) = \$1.80 + \$1.20 = \$3.00. The answer checks.
State. There are 18 dimes and 24 nickels.

23. **Familiarize.** Let x = pounds of dried-fruit used, and y = pounds of macadamia nuts used. Then:
Translate.

$$x + y = 15 \qquad (1)$$
$$580x + 1475y = 938 \cdot 15 = 14,070$$
$$\text{or } 116x + 295y = 2814 \qquad (2)$$

or

$$\begin{bmatrix} 1 & 1 & | & 15 \\ 116 & 295 & | & 2814 \end{bmatrix}$$

Carry out.
Multiply row 1 by -116 and add to row 2.

$$\begin{bmatrix} 1 & 1 & | & 15 \\ 0 & 179 & | & 1074 \end{bmatrix}$$

Reinsert variables and solve for x and y.

$$x + y = 15 \quad (1)$$
$$179y = 1074 \quad (2)$$
$$y = \frac{1074}{179} = 6$$
$$x + 6 = 15 \Rightarrow x = 9$$

Check.

$$9 + 6 = 15 \qquad (1)\sqrt{}$$
$$580(9) + 1475(6) = 14,070 \quad (2)\sqrt{}$$

State. The snack mix should be made with 9 pounds of dried fruit and 6 pounds of macadamia nuts.

25. **Familiarize.** We let x, y, and z represent the amounts invested at 7%, 8%, and 9%, respectively. Recall the formula for simple interest.

Interest = Principal × Rate × Time

Translate. We organize the information in a table.

	First Invest-ment	Second Invest-ment	Third Invest-ment	Total
P	x	y	z	\$2500
R	7%	8%	9%	
T	1 yr	1 yr	1 yr	
I	$0.07x$	$0.08y$	$0.09z$	\$212

The first row give us one equation:
$x + y + z = 2500$

The last row gives a second equation:
$0.07x + 0.08y + 0.09z = 212$

Amount invested is at 9%		$1100	more than	amount invested at 8%.
↓	↓	↓	↓	↓
z	=	$1100	+	y

After clearing decimals, we have this system:
$$x + y + z = 2500$$
$$7x + 8y + 9z = 21,200$$
$$-y + z = 1100$$

Carry out. Solve using matrices.

$$\begin{bmatrix} 1 & 1 & 1 & | & 2500 \\ 7 & 8 & 9 & | & 21,200 \\ 0 & -1 & 1 & | & 1100 \end{bmatrix}$$

$$\begin{bmatrix} 1 & 1 & 1 & | & 2500 \\ 0 & 1 & 2 & | & 3700 \\ 0 & -1 & 1 & | & 1100 \end{bmatrix}$$ New Row 2 = -7(Row 1) + Row 2

$$\begin{bmatrix} 1 & 1 & 1 & | & 2500 \\ 0 & 1 & 2 & | & 3700 \\ 0 & 0 & 3 & | & 4800 \end{bmatrix}$$ New Row 3 = Row 2 + Row 3

Reinserting the variables, we have
$$x + y + z = 2500 \quad (1)$$
$$y + 2z = 3700 \quad (2)$$
$$3z = 4800 \quad (3)$$

Solve (3) for z.
$$3z = 4800$$
$$z = 1600$$

Substitute 1600 for z in (2) and solve for y.
$$y + 2 \cdot 1600 = 3700$$
$$y + 3200 = 3700$$
$$y = 500$$

Substitute 500 for y and 1600 for z in (1) and solve for x.
$$x + 500 + 1600 = 2500$$
$$x + 2100 = 2500$$
$$x = 400$$

Check. The total investment is $400 + $500 + $1600, or $2500. The total interest is $0.07(\$400) + 0.08(\$500) + 0.09(\$1600) = \$28 + \$40 + \$144 = \$212$. The amount invested at 9%, $1600, is $1100 more than the amount invested at 8%, $500. The numbers check.
State. The amounts invested at 7%, 8%, and 9%, are $400, $500, and $1600, respectively.

27. *Thinking and Writing Exercise.*

29. $5(-3) - (-7)4 = -15 - (-28)$
$$= -15 + 28 = 13$$

31. $-2(5 \cdot 3 - 4 \cdot 6) - 3(2 \cdot 7 - 15) + 4(3 \cdot 8 - 5 \cdot 4)$
$$= -2(15 - 24) - 3(14 - 15) + 4(24 - 20)$$
$$= -2(-9) - 3(-1) + 4(4)$$
$$= 18 + 3 + 16$$
$$= 21 + 16$$
$$= 37$$

33. *Thinking and Writing Exercise.*

35. **Familiarize.** Let w, x, y, and z represent the thousand's, hundred's, ten's, and one's digits, respectively.

Translate.

Total number of the digits is 10.
↓	↓	↓
$w + x + y + z$		$= 10$

Twice the sum of the thousand's and ten's digits	is	the sum of the hundred's and one's digits	less one.
↓	↓	↓	↓
$2(w + y)$	=	$x + z$	-1

The ten's digit	is twice	the thousand's digit.
↓	↓ ↓	↓
y	$= 2 \cdot$	w

The one's digit	equals	the sum of the thousand's and hundred's digits.
↓	↓	↓
z	=	$w + x$

We have a system of equations which can be written as

$$w + x + y + z = 10$$
$$2w - x + 2y - z = -1$$
$$-2w + y = 0$$
$$w + x - z = 0$$

Carry out. We can use matrices to solve the system. We get $(1, 3, 2, 4)$.

Check. The sum of the digits is 10. Twice the sum of 1 and 2 is 6. This is one less than the sum of 3 and 4. The ten's digit, 2, is twice the thousand's digit, 1. The one's digit, 4, equals 1 + 3. The numbers check.
State. The number is 1324.

Exercise Set 3.7

---

1. True

3. True

5. False

7. $\begin{vmatrix} 5 & 1 \\ 2 & 4 \end{vmatrix}$

$= 5 \cdot 4 - 2 \cdot 1$

$= 20 - 2 = 18$

9. $\begin{vmatrix} 10 & 8 \\ -5 & -9 \end{vmatrix}$

$= 10 \cdot (-9) - (-5) \cdot 8$

$= -90 + 40 = -50$

11. $\begin{vmatrix} 1 & 4 & 0 \\ 0 & -1 & 2 \\ 3 & -2 & 1 \end{vmatrix}$

$= 1 \begin{vmatrix} -1 & 2 \\ -2 & 1 \end{vmatrix} - 0 \begin{vmatrix} 4 & 0 \\ -2 & 1 \end{vmatrix} + 3 \begin{vmatrix} 4 & 0 \\ -1 & 2 \end{vmatrix}$

$= 1(-1 + 4) - 0(4 + 0) + 3(8 + 0)$

$= 1(3) - 0(4) + 3(8)$

$= 3 - 0 + 24 = 27$

13. $\begin{vmatrix} -4 & -2 & 3 \\ -3 & 1 & 2 \\ 3 & 4 & -2 \end{vmatrix}$

$= -4 \begin{vmatrix} 1 & 2 \\ 4 & -2 \end{vmatrix} - (-3) \begin{vmatrix} -2 & 3 \\ 4 & -2 \end{vmatrix} + 3 \begin{vmatrix} -2 & 3 \\ 1 & 2 \end{vmatrix}$

$= -4(-2 - 8) + 3(4 - 12) + 3(-4 - 3)$

$= -4(-10) + 3(-8) + 3(-7)$

$= 40 - 24 - 21$

$= -5$

15. $5x + 8y = 1$

$3x + 7y = 5$

We have

$x = \dfrac{\begin{vmatrix} 1 & 8 \\ 5 & 7 \end{vmatrix}}{\begin{vmatrix} 5 & 8 \\ 3 & 7 \end{vmatrix}}$

$= \dfrac{1 \cdot 7 - 5 \cdot 8}{5 \cdot 7 - 3 \cdot 8} = \dfrac{7 - 40}{35 - 24} = \dfrac{-33}{11} = -3$

and

$y = \dfrac{\begin{vmatrix} 5 & 1 \\ 3 & 5 \end{vmatrix}}{\begin{vmatrix} 5 & 8 \\ 3 & 7 \end{vmatrix}} = \dfrac{5 \cdot 5 - 3 \cdot 1}{11} = \dfrac{25 - 3}{11} = \dfrac{22}{11} = 2$

The solution is $(-3, 2)$ which checks.

17. $5x - 4y = -3$

$7x + 2y = 6$

We have

$x = \dfrac{\begin{vmatrix} -3 & -4 \\ 6 & 2 \end{vmatrix}}{\begin{vmatrix} 5 & -4 \\ 7 & 2 \end{vmatrix}} = \dfrac{(-3) \cdot 2 - 6(-4)}{5 \cdot 2 - 7 \cdot (-4)}$

$= \dfrac{-6 + 24}{10 + 28} = \dfrac{18}{38} = \dfrac{9}{19}$

and

$y = \dfrac{\begin{vmatrix} 5 & -3 \\ 7 & 6 \end{vmatrix}}{\begin{vmatrix} 5 & -4 \\ 7 & 2 \end{vmatrix}} = \dfrac{5 \cdot 6 - 7 \cdot (-3)}{38} = \dfrac{30 + 21}{38} = \dfrac{51}{38}$

The solution is $\left(\dfrac{9}{19}, \dfrac{51}{38}\right)$ which checks.

19. $3x - y + 2z = 1$

$x - y + 2z = 3$

$-2x + 3y + z = 1$

We compute D, D_x, D_y, and D_z.

$D = \begin{vmatrix} 3 & -1 & 2 \\ 1 & -1 & 2 \\ -2 & 3 & 1 \end{vmatrix}$

$= 3\begin{vmatrix} -1 & 2 \\ 3 & 1 \end{vmatrix} - 1\begin{vmatrix} -1 & 2 \\ 3 & 1 \end{vmatrix} - 2\begin{vmatrix} -1 & 2 \\ -1 & 2 \end{vmatrix}$

$= 3(-1-6) - 1(-1-6) - 2(-2+2)$

$= 3(-7) - 1(-7) - 2 \cdot 0 = -21 + 7 - 0 = -14$

$D_x = \begin{vmatrix} 1 & -1 & 2 \\ 3 & -1 & 2 \\ 1 & 3 & 1 \end{vmatrix}$

$= 1\begin{vmatrix} -1 & 2 \\ 3 & 1 \end{vmatrix} - 3\begin{vmatrix} -1 & 2 \\ 3 & 1 \end{vmatrix} + 1\begin{vmatrix} -1 & 2 \\ -1 & 2 \end{vmatrix}$

$= 1(-1-6) - 3(-1-6) + 1(-2+2)$

$= 1(-7) - 3(-7) + 1(0)$

$= -7 + 21 + 0 = 14$

$D_y = \begin{vmatrix} 3 & 1 & 2 \\ 1 & 3 & 2 \\ -2 & 1 & 1 \end{vmatrix}$

$= 3\begin{vmatrix} 3 & 2 \\ 1 & 1 \end{vmatrix} - 1\begin{vmatrix} 1 & 2 \\ 1 & 1 \end{vmatrix} - 2\begin{vmatrix} 1 & 2 \\ 3 & 2 \end{vmatrix}$

$= 3(3-2) - 1(1-2) - 2(2-6)$

$= 3 \cdot 1 - 1(-1) - 2(-4)$

$= 3 + 1 + 8 = 12$

$D_z = \begin{vmatrix} 3 & -1 & 1 \\ 1 & -1 & 3 \\ -2 & 3 & 1 \end{vmatrix}$

$= 3\begin{vmatrix} -1 & 3 \\ 3 & 1 \end{vmatrix} - 1\begin{vmatrix} -1 & 1 \\ 3 & 1 \end{vmatrix} - 2\begin{vmatrix} -1 & 1 \\ -1 & 3 \end{vmatrix}$

$= 3(-1-9) - 1(-1-3) - 2(-3+1)$

$= 3(-10) - 1(-4) - 2(-2)$

$= -30 + 4 + 4 = -22$

$x = \dfrac{D_x}{D} = \dfrac{14}{-14} = -1$

$y = \dfrac{D_y}{D} = \dfrac{12}{-14} = -\dfrac{6}{7}$

$z = \dfrac{D_z}{D} = \dfrac{-22}{-14} = \dfrac{11}{7}$

The solution is $\left(-1, -\dfrac{6}{7}, \dfrac{11}{7}\right)$ which checks.

21. $2x - 3y + 5z = 27$

$x + 2y - z = -4$

$5x - y + 4z = 27$

We compute D, D_x, D_y, and D_z.

$D = \begin{vmatrix} 2 & -3 & 5 \\ 1 & 2 & -1 \\ 5 & -1 & 4 \end{vmatrix}$

$= 2\begin{vmatrix} 2 & -1 \\ -1 & 4 \end{vmatrix} - 1\begin{vmatrix} -3 & 5 \\ -1 & 4 \end{vmatrix} + 5\begin{vmatrix} -3 & 5 \\ 2 & -1 \end{vmatrix}$

$= 2(8-1) - 1(-12+5) + 5(3-10)$

$= 2 \cdot 7 - 1(-7) + 5(-7)$

$= 14 + 7 - 35$

$= -14$

$D_x = \begin{vmatrix} 27 & -3 & 5 \\ -4 & 2 & -1 \\ 27 & -1 & 4 \end{vmatrix}$

$= 27\begin{vmatrix} 2 & -1 \\ -1 & 4 \end{vmatrix} - (-4)\begin{vmatrix} -3 & 5 \\ -1 & 4 \end{vmatrix}$

$\quad + 27\begin{vmatrix} -3 & 5 \\ 2 & -1 \end{vmatrix}$

$= 27(8-1) + 4(-12+5) + 27(3-10)$

$= 27(7) + 4(-7) + 27(-7)$

$= 189 - 28 - 189$

$= -28$

$D_y = \begin{vmatrix} 2 & 27 & 5 \\ 1 & -4 & -1 \\ 5 & 27 & 4 \end{vmatrix}$

$= 2\begin{vmatrix} -4 & -1 \\ 27 & 4 \end{vmatrix} - 1\begin{vmatrix} 27 & 5 \\ 27 & 4 \end{vmatrix} + 5\begin{vmatrix} 27 & 5 \\ -4 & -1 \end{vmatrix}$

$= 2(-16+27) - 1(108-135)$

$\qquad\qquad + 5(-27+20)$

$= 2(11) - 1(-27) + 5(-7)$

$= 22 + 27 - 35$

$= 14$

$$D_z = \begin{vmatrix} 2 & -3 & 27 \\ 1 & 2 & -4 \\ 5 & -1 & 27 \end{vmatrix}$$

$$= 2\begin{vmatrix} 2 & -4 \\ -1 & 27 \end{vmatrix} - 1\begin{vmatrix} -3 & 27 \\ -1 & 27 \end{vmatrix} + 5\begin{vmatrix} -3 & 27 \\ 2 & -4 \end{vmatrix}$$

$$= 2(54-4) - 1(-81+27) + 5(12-54)$$

$$= 2(50) - 1(-54) + 5(-42)$$

$$= 100 + 54 - 210$$

$$= -56$$

$$x = \frac{D_x}{D} = \frac{-28}{-14} = 2$$

$$y = \frac{D_y}{D} = \frac{39}{-14} = \frac{14}{-14} = -1$$

$$z = \frac{D_z}{D} = \frac{-56}{-14} = 4$$

The solution is $(2, -1, 4)$ which checks.

23. $r - 2s + 3t = 6$

 $2r - s - t = -3$

 $r + s + t = 6$

We compute D, D_r, D_s, and D_t.

$$D = \begin{vmatrix} 1 & -2 & 3 \\ 2 & -1 & -1 \\ 1 & 1 & 1 \end{vmatrix}$$

$$= 1\begin{vmatrix} -1 & -1 \\ 1 & 1 \end{vmatrix} - 2\begin{vmatrix} -2 & 3 \\ 1 & 1 \end{vmatrix} + 1\begin{vmatrix} -2 & 3 \\ -1 & -1 \end{vmatrix}$$

$$= 1(-1+1) - 2(-2-3) + 1(2+3)$$

$$= 1(0) - 2(-5) + 1(5)$$

$$= 0 + 10 + 5$$

$$= 15$$

$$D_r = \begin{vmatrix} 6 & -2 & 3 \\ -3 & -1 & -1 \\ 6 & 1 & 1 \end{vmatrix}$$

$$= 6\begin{vmatrix} -1 & -1 \\ 1 & 1 \end{vmatrix} - (-3)\begin{vmatrix} -2 & 3 \\ 1 & 1 \end{vmatrix} + 6\begin{vmatrix} -2 & 3 \\ -1 & -1 \end{vmatrix}$$

$$= 6(-1+1) + 3(-2-3) + 6(2+3)$$

$$= 6(0) + 3(-5) + 6(5)$$

$$= 0 - 15 + 30$$

$$= 15$$

$$D_s = \begin{vmatrix} 1 & 6 & 3 \\ 2 & -3 & -1 \\ 1 & 6 & 1 \end{vmatrix}$$

$$= 1\begin{vmatrix} -3 & -1 \\ 6 & 1 \end{vmatrix} - 2\begin{vmatrix} 6 & 3 \\ 6 & 1 \end{vmatrix} + 1\begin{vmatrix} 6 & 3 \\ -3 & -1 \end{vmatrix}$$

$$= 1(-3+6) - 2(6-18) + 1(-6+9)$$

$$= 1(3) - 2(-12) + 1(3)$$

$$= 3 + 24 + 3$$

$$= 30$$

$$D_t = \begin{vmatrix} 1 & -2 & 6 \\ 2 & -1 & -3 \\ 1 & 1 & 6 \end{vmatrix}$$

$$= 1\begin{vmatrix} -1 & -3 \\ 1 & 6 \end{vmatrix} - 2\begin{vmatrix} -2 & 6 \\ 1 & 6 \end{vmatrix} + 1\begin{vmatrix} -2 & 6 \\ -1 & -3 \end{vmatrix}$$

$$= 1(-6+3) - 2(-12-6) + 1(6+6)$$

$$= 1(-3) - 2(-18) + 1(12)$$

$$= -3 + 36 + 12$$

$$= 45$$

$$r = \frac{D_r}{D} = \frac{15}{15} = 1$$

$$s = \frac{D_s}{D} = \frac{30}{15} = 2$$

$$t = \frac{D_t}{D} = \frac{45}{15} = 3$$

The solution is $(1, 2, 3)$ which checks.

25. *Thinking and Writing Exercise.*

27. $f(90) = 80(90) + 2500 = 7200 + 2500 = 9700$

29. $(g - f)(10) = 70(10) - 2500 = 700 - 2500$

 $= -1800$

31. $g(x) = f(x)$

 $50x = 80x + 2500$

 $70x = 2500$

 $x = \dfrac{250\cancel{0}}{7\cancel{0}} = \dfrac{250}{7} = 35\dfrac{5}{7}$

33. *Thinking and Writing Exercise.*

35. $\begin{vmatrix} y & -2 \\ 4 & 3 \end{vmatrix} = 44$

Evaluating the determinant, we have
$3y + 8 = 44$
$\qquad 3y = 36$
$\qquad y = 12$

37. $\begin{vmatrix} m+1 & -2 \\ m-2 & 1 \end{vmatrix} = 27$

First evaluate the determinant.
$\begin{vmatrix} m+1 & -2 \\ m-2 & 1 \end{vmatrix} = (m+1)(1) - (m-2)(-2)$

$\qquad\qquad = m + 1 + 2m - 4$

$\qquad\qquad = 3m - 3$

Next set $3m - 3$ equal to 27 and solve the resulting equation.
$3m - 3 = 27$
$\qquad 3m = 30$
$\qquad m = 10$
The solution is 10.

Exercise Set 3.8

1. b

3. e

5. h

7. g

9. $C(x) = 45x + 300,000 \quad R(x) = 65x$

a) $P(x) = R(x) - C(x)$
$\qquad = 65x - (45x + 300,000)$
$\qquad = 65x - 45x - 300,000$
$\qquad = 20x - 300,000$

b) To find the break-even point we solve the system:
$R(x) = 65x$
$C(x) = 45x + 300,000$

Since $R(x) = C(x)$ at the break-even point, we can rewrite the system:
$R(x) = 65x \qquad\qquad (1)$
$R(x) = 45x + 300,000 \quad (2)$

We solve using substitution.
$65x = 45x + 300,000 \quad$ Substituting $65x$ for $R(x)$ in (2)
$20x = 300,000$
$\quad x = 15,000$

Thus, 15,000 units must be produced and sold in order to break even. Also,
$R(15,000) = C(15,000) = 975,000$, so the breakeven point is
(15,000 units, \$975,000).

11. $C(x) = 15x + 3100 \quad R(x) = 40x$

a) $P(x) = R(x) - C(x)$
$\qquad = 40x - (15x + 3100)$
$\qquad = 25x - 3100$

b) To find the break-even point we solve the system, set $P(x) = 0$.
$P(x) = 0$
$25x - 3100 = 0$
$\qquad x = \dfrac{3100}{25} = 124$

Thus, 124 units must be produced and sold in order to break even.
$C(124) = R(124) = 40(124) = \$4960$
So the break-even point is
(124 units, \$4960).

13. $C(x) = 40x + 22,500 \quad R(x) = 85x$

a) $P(x) = R(x) - C(x)$
$\qquad = 85x - (40x + 22,500)$
$\qquad = 85x - 40x - 22,500$
$\qquad = 45x - 22,500$

b) To find the break-even point we solve the system:
$R(x) = 85x$
$C(x) = 40x + 22,500$

Since $R(x) = C(x)$ at the break-even point, we can rewrite the system:
$R(x) = 85x \qquad\qquad (1)$
$R(x) = 40x + 22,500 \quad (2)$

We solve using substitution.
$85x = 40x + 22,500 \quad$ Substituting $85x$ T for $R(x)$ in (2)

$45x = 22,500$
$\quad x = 500$

hus, 500 units must be produced and sold in order to break even. Also, $R(500) = C(500) = 42,500$, so the breakeven point is (500 units, \$42,500)

15. $C(x) = 24x + 50,000 \quad R(x) = 40x$

 a) $P(x) = R(x) - C(x)$

   $= 40x - (24x + 50,000)$

   $= 16x - 50,000$

 b) To find the break-even point we solve the system, set $P(x) = 0$.

   $P(x) = 0$

   $16x - 50,000 = 0$

   $x = \dfrac{50,000}{16} = 3125$

   3125 units must be produced and sold in order to break even.
   $C(3125) = R(3125) = 40(3125) = \$125,000$

17. $C(x) = 75x + 100,000 \quad R(x) = 125x$

 a) $P(x) = R(x) - C(x)$

   $= 125x - (75x + 100,000)$

   $= 125x - 75x - 100,000$

   $= 50x - 100,000$

 b) To find the break-even point we solve the system:

   $R(x) = 125x$

   $C(x) = 75x + 100,000$

   Since $R(x) = C(x)$ at the break-even point, we can rewrite the system:

   $R(x) = 125x \qquad\qquad (1)$

   $R(x) = 75x + 100,000 \quad (2)$

   We solve using substitution.
   $125x = 75x + 100,000 \quad$ Substituting $125x$ for $R(x)$ in (2)

   $50x = 100,000$

   $x = 2000$

   Thus, 2000 units must be produced and sold in order to break even.
   Also, $R(2000) = C(2000) = 250,000$, so the breakeven point is (2000 units, \$250,000).

19. $D(p) = 1000 - 10p$

   $S(p) = 230 + p$

   Since both demand and supply are quantities, the system can be written:

   $q = 1000 - 10p \quad (1)$

   $q = 230 + p \qquad (2)$

   Substitute $1000 - 10p$ for q in (2) and solve.

   $1000 - 10p = 230 + p$

   $770 = 11p$

   $70 = p$

   The equilibrium price is \$70 per unit. To find the equilibrium quantity we substitute \$70 into either $D(p)$ or $S(p)$.

   $D(70) = 1000 - 10 \cdot 70 = 1000 - 700 = 300$

   The equilibrium quantity is 300 units.
   The equilibrium point is $(\$70, 300 \text{ units})$.

21. $D(p) = 760 - 13p$

   $S(p) = 430 + 2p$

   Since both demand and supply are quantities, the system can be written:

   $q = 760 - 13p \quad (1)$

   $q = 430 + 2p \quad (2)$

   Substitute $760 - 13p$ for q in (2) and solve.

   $760 - 13p = 430 + 2p$

   $330 = 15p$

   $22 = p$

   The equilibrium price is \$22 per unit. To find the equilibrium quantity we substitute \$22 into either $D(p)$ or $S(p)$.

   $S(22) = 430 + 2(22) = 430 + 44 = 474$

   The equilibrium quantity is 474 units.
   The equilibrium point is $(\$22, 474 \text{ units})$.

23. $D(p) = 7500 - 25p$

   $S(p) = 6000 + 5p$

   Since both demand and supply are quantities, the system can be written:

   $q = 7500 - 25p \quad (1)$

   $q = 6000 + 5p \quad (2)$

   Substitute $7500 - 25p$ for q in (2) and solve.

$7500 - 25p = 6000 + 5p$

$1500 = 30p$

$50 = p$

The equilibrium price is $50 per unit. To find the equilibrium quantity we substitute $50 into either $D(p)$ or $S(p)$.

$D(50) = 7500 - 25(50)$

$= 7500 - 1250 = 6250$

The equilibrium quantity is 6250 units.

The equilibrium point is $(\$50, 6250 \text{ units})$.

25. $D(p) = 1600 - 53p$

$S(p) = 320 + 75p$

Since both demand and supply are quantities, the system can be written:

$q = 1600 - 53p$ (1)

$q = 320 + 75p$ (2)

Substitute $1600 - 53p$ for q in (2) and solve.

$1600 - 53p = 320 + 75p$

$1280 = 128p$

$10 = p$

The equilibrium price is $10 per unit. To find the equilibrium quantity we substitute $10 into either $D(p)$ or $S(p)$.

$S(10) = 320 + 75(10) = 320 + 750 = 1070$

The equilibrium quantity is 1070 units.

The equilibrium point is $(\$10, \ 1070 \text{ units})$.

27. a) $C(x) = $ Fixed costs + Variable Costs

$C(x) = 45,000 + 40x$

where x is the number of MP3 phones produced

b) Each MP3 phone sells for $130. The total revenue is 130 times the number of MP3 phones sold. We assume that all MP3 phones produced are sold.

$R(x) = 130x$

c) $P(x) = R(x) - C(x)$

$P(x) = 130x - (45,000 + 40x)$

$= 90x - 45,000$

d) $P(3000) = 90(3000) - 45,000$

$= 225,000$

The company will realize a $225,000 profit when 3000 MP3 phones are

produced and sold.

$P(400) = 90(400) - 45,000 = -9000$

The company will lose $9000 if only 400 MP3 phones are produced and sold.

e) Set

$P(x) = 0$

$90x - 45,000 = 0$

$x = \dfrac{45,00\cancel{0}}{9\cancel{0}} = 500$

The break-even point is 500 MP3 phones.

$C(500) = R(500) = 130(500) = \$65,000$

29. a) $C(x) = $ Fixed costs + Variable Costs

$C(x) = 10,000 + 30x$

where x is the number of car seats produced

b) Each car seat sells for $80. The total revenue is 80 times the number of car seats sold. We assume that all car seats are sold.

$R(x) = 80x$

c) $P(x) = R(x) - C(x)$

$P(x) = 80x - (10,000 + 30x)$

$= 50x - 10,000$

d) $P(2000) = 50(2000) - 10,000 = 90,000$

The company will make a $90,000 profit if 2000 car seats are produced and sold.

$P(50) = 50(50) - 10,000 = -7500$

The company will lose $7500 if only 50 car seats are produced and sold.

e) Set

$P(x) = 0$

$50x - 10,000 = 0$

$x = \dfrac{10,00\cancel{0}}{5\cancel{0}} = 200$

The break-even point is 200 car seats.

$C(200) = R(200) = 80(200) = \$16,000$

31. a) $D(p) = -14.97p + 987.35$

$S(p) = 98.55p - 5.13$

Rewrite the system:

$q = -14.97p + 987.35$ (1)

$q = 98.55p - 5.13$ (2)

Substitute $-14.97p + 987.35$ for q in (2) and solve.

$-14.97p + 987.35 = 98.55p - 5.13$

$992.48 = 113.52p$

$8.74 \approx p$

The equilibrium price is \$8.74 per unit. A price of \$8.74 per unit should be charged in order to have equilibrium between supply and demand.

b) $R(x) = 8.74x$

$C(x) = 5.15x + 87,985$

Rewrite the system:

$d = 8.74x$ (1)

$d = 5.15x + 87,985$ (2)

We solve using substitution.

$8.74x = 5.15x + 87,985$ Substituting $8.74x$ for d in (2)

$3.59x = 87,985$

$x = \dfrac{87,985}{3.59} = 24,509$ Rounding up

Thus 24,509 units must be sold in order to break even.

33. *Thinking and Writing Exercise.*

35. $4x - 3 = 21$

$4x = 24$

$x = 6$

37. $3x - 5 = 12x + 6$

$-5 = 9x + 6$

$-11 = 9x$

$x = \dfrac{-11}{9} = -\dfrac{11}{9}$

39. $3 - (x + 2) = 7$

$3 - x - 2 = 7$

$1 - x = 7$

$-x = 6$

$x = -6$

41. *Thinking and Writing Exercise.*

43. The supply function contains the points $(\$2, 100)$ and $(\$8, 500)$. We find its equation:

$m = \dfrac{500 - 100}{8 - 2} = \dfrac{400}{6} = \dfrac{200}{3}$

$y - y_1 = m(x - x_1)$ Point-slope form

$y - 100 = \dfrac{200}{3}(x - 2)$

$y - 100 = \dfrac{200}{3}x - \dfrac{400}{3}$

$y = \dfrac{200}{3}x - \dfrac{100}{3}$

We can equivalently express supply S as a function of price p:

$S(p) = \dfrac{200}{3}p - \dfrac{100}{3}$

The demand function contains the points $(\$1, 500)$ and $(\$9, 100)$. We find its equation:

$m = \dfrac{100 - 500}{9 - 1} = \dfrac{-400}{8} = -50$

$y - y_1 = m(x - x_1)$

$y - 500 = -50(x - 1)$

$y - 500 = -50x + 50$

$y = -50x + 550$

We can equivalently express demand D as a function of price p:

$D(p) = -50p + 550$

We have a system of equations

$S(p) = \dfrac{200}{3}p - \dfrac{100}{3}$

$D(p) = -50p + 550$

Rewrite the system:

$q = \dfrac{200}{3}p - \dfrac{100}{3}$ (1)

$q = -50p + 550$ (2)

Substitute $\dfrac{200}{3}p - \dfrac{100}{3}$ for q in (2) and solve.

$\dfrac{200}{3}p - \dfrac{100}{3} = -50p + 550$

$200p - 100 = -150p + 1650$ Multiplying by 3 to clear fractions

$350p - 100 = 1650$

$350p = 1750$

$p = 5$

The equilibrium price is $5 per unit. To find the equilibrium quantity, we substitute $5 into either $S(p)$ or $D(p)$.

$$D(5) = -50(5) + 550 = -250 + 550 = 300$$

The equilibrium quantity is 300 units. The equilibrium point is $\left(\$5, 300 \text{ yo-yo's}\right)$.

45. a) Enter the data and use the linear regression feature to get
$$S(p) = 15.97p - 1.05 .$$

   b) Enter the data and use the linear regression feature to get
$$D(p) = -11.26p + 41.16 .$$

   c) Find the point of intersection of the graphs of the functions found in parts (a) and (b).

Intersection
X=1.5501285 Y=23.705553

We see that the equilibrium point is $\left(\$1.55, 23.7 \text{ million jars}\right)$.

Chapter 3 Study Summary

---

1. $x - y = 3$ (1)
    $y = 2x - 5$ (2)

The graph of a line can be made by determining two points that lie on the line using an x-y table. For the first line, defined by $x - y = 3$, substituting $x = 0$ into the equation and solving for y gives the point $(0, -3)$. And substituting $y = 0$ into the equation gives the point $(3, \ 0)$.

For the second line, defined by $y = 2x - 5$, substituting $x = 0$ into the equation and solving for y gives the point $(0, -5)$. Substituting $x = 3$ into the equation gives the point $(3, 1)$.

Using these pairs of points to graph the lines, the intersection appears to be the point

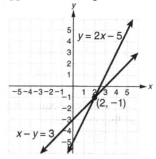

$(2, -1)$.

We can verify this by testing the point in each equation. For the first equation, we have:

$x - y$	3
$(2) - (-1)$	3
$2 + 1$	3
3	3

For the second equation, we have:

y	$2x - 5$
-1	$2(2) - 5$
-1	$4 - 5$
-1	-1

The ordered pair $(2, -1)$ is the solution to the system.

2. $x = 3y - 2$ (1)
      $y - x = 1$ (2)

Equation (1) is already solved for x. Substituting the result into equation (2) gives:
$$y - x = 1 \quad (2)$$
$$y - (3y - 2) = 1$$
$$y - 3y + 2 = 1$$
$$-2y + 2 = 1$$
$$-2y = 1 - 2 = -1$$
$$y = \frac{-1}{-2} = \frac{1}{2}$$

Substituting the value of y back into equation (1) gives the value of x,
$$x = 3y - 2 \quad (1)$$
$$x = 3\left(\frac{1}{2}\right) - 2$$
$$= \frac{3}{2} - 2$$
$$= \frac{3}{2} - \frac{4}{2} = -\frac{1}{2}$$

The ordered pair $\left(-\dfrac{1}{2}, \dfrac{1}{2}\right)$ is the solution to

the system.

3. $2x - y = 5 \qquad (1)$

    $x + 3y = 1 \qquad (2)$

    The system can be solved using elimination
    by multiplying equation (2) by (-2) and
    adding the result to equation (1) to find y,
    then finding x.

    $-2[x + 3y] = -2[1] \rightarrow -2x - 6y = -2$

    $\quad 2x - y = 5$

    $\underline{-2x - 6y = -2}$

    $\qquad -7y = 3$

    $y = \dfrac{3}{-7} = -\dfrac{3}{7}$

    Now find x using equation (2).

    $x + 3y = 1 \qquad (2)$

    $x + 3\left(-\dfrac{3}{7}\right) = 1$

    $x - \dfrac{9}{7} = 1$

    $x = 1 + \dfrac{9}{7} = 1\dfrac{9}{7} = \dfrac{16}{7}$

    The ordered pair $\left(\dfrac{16}{7}, -\dfrac{3}{7}\right)$ is the solution to

    the system.

4. Let x = the number of boxes of Roller Grip™
    pens purchased, and y = the number of boxes
    of eGEL™ pens purchased. Then:

    $x + y = 120 \qquad (1)$

    $1749x + 1649y = 201{,}080 \qquad (2)$

    Note that all prices shown in equation (2), and
    the total spent, are in cents.
    The system can be solved using elimination
    by multiplying equation (1) by (-1749) and
    adding the result to equation (2) to find y,
    then using equation (1) to find x.

    $-1749[x + y] = -1749[120]$

    $\rightarrow -1749x - 1749y = -209{,}880$

    $-1749x - 1749y = -209{,}880$

    $\underline{1749x + 1649y = 201{,}080}$

    $\qquad -100y = -8800$

    $y = \dfrac{-8800}{-100} = 88$

    Now find x using equation (1).

$x + y = 120 \qquad (1)$

$x + 88 = 120$

$x = 120 - 88 = 32$

Barlow's Office Supply purchased 32 boxes
of Roller Grip™ pens and 88 boxes of
eGEL™ pens.

5. Let x = the amount, in liters, of 40% nitric
    acid solution used, and y = the amount, in
    liters, of 15% nitric acid solution used. Then:

    $x + y = 2 \qquad\qquad (1)$

    $40x + 15y = 25 \cdot 2 = 50 \qquad (2)$

    The system can be solved using elimination
    by multiplying equation (1) by (-40) and
    adding the result to equation (2) to find y,
    then finding x.

    $-40[x + y] = -40[2] \rightarrow -40x - 40y = -80$

    $-40x - 40y = -80$

    $\underline{40x + 15y = 50}$

    $\qquad -25y = -30$

    $y = \dfrac{-30}{-25} = 1\dfrac{5}{25} = 1\dfrac{1}{5} = 1.2$

    Now find x using equation (1).

    $x + y = 2 \qquad (1)$

    $x + 1\dfrac{1}{5} = 2$

    $x = 2 - 1\dfrac{1}{5} = \dfrac{4}{5} = 0.8$

    To produce 2 L of a 25% mixture, 0.8 L of
    40% solution should be added to 1.2 L of
    15% solution.

6. Let x = the speed of Ruth paddles in still
    water. Then, using the equation $d = rt$ for the
    time traveling with the current gives:

    $d_{\text{with current}} = (x + 2) \cdot 1.5$

    and for the time traveling against the current
    gives:

    $d_{\text{against current}} = (x - 2) \cdot 2.5$

    Since the distance traveled with the current is
    the same as the distance traveled against it,
    we have:

    $d_{\text{with current}} = d_{\text{against current}}$

    $(x + 2) \cdot 1.5 = (x - 2) \cdot 2.5$

    $1.5x + 3 = 2.5x - 5$

    $5 + 3 = 2.5x - 1.5x$

    $8 = x$

Ruth paddles at a speed of 8 mph in still water.

7. $x - 2y - z = 8$ (1)
     $2x + 2y - z = 8$ (2)
     $x - 8y + z = 1$ (3)

Add (1) and (3) to generate an equation in x and y. Then add (2) and (3) to generate a second equation in x and y.

$$
\begin{array}{l}
x -\ 2y - z = 8 \quad (1) \\
\underline{x -\ 8y + z = 1 \quad (3)} \\
2x - 10y \quad\ \ = 9 \quad (4)
\end{array}
$$

$$
\begin{array}{l}
2x + 2y - z = 8 \quad (2) \\
\underline{x - 8y + z = 1 \quad (3)} \\
3x - 6y \quad\ \ = 9
\end{array}
$$

or $x - 2y = 3$ (5)

Add $-1 \cdot (4)$ to $2(5)$ to eliminate x and find y. Then back-substitute to find x and z.

$$
\begin{array}{l}
-2x + 10y = -9 \quad -1 \cdot (4) \\
\underline{2x - 4y = 6 \quad\ \ (5)} \\
6y = -3
\end{array}
$$

$$y = \frac{-3}{6} = -\frac{1}{2}$$

Use (5) and the value of y to find x.
$x - 2y = 3$ (5)

$$x = 3 + 2y = 3 + 2\left(-\frac{1}{2}\right) = 2$$

Use (3) and the values of x and y to find z.
$x - 8y + z = 1$ (3)
$z = 1 - x + 8y$

$$z = 1 - 2 + 8\left(-\frac{1}{2}\right) = -5$$

$$(x, y, z) = \left(2, -\frac{1}{2}, -5\right)$$

8. Let x = the first number, y = the second number, and z = the third number. Then
$x + y + z = 9$ (1)

$$z = \frac{1}{2}(x + y) \quad (2)$$

$y = x + z - 2$ (3)

Note that equation (2) implies that $x + y = 2z$
Substituting this result into (1) allows us to find z.

$x + y + z = 9$
$(x + y) + z = 9$
$(2z) + z = 9$
$3z = 9$
$z = 3$

Use this result to simplify equation (1) and (3).

$x + y + z = 9$ (1)
$x + y + 3 = 9$
$x + y = 6$ (4)
$y = x + z - 2$ (3)
$y = x + 3 - 2$
$y = x + 1$
$x - y = -1$ (5)

Add (4) to (5) to eliminate y and find x. Then back-substitute to find y.

$$
\begin{array}{l}
x + y = 6 \quad (4) \\
\underline{x - y = -1 \quad (5)} \\
2x \quad\ \ = 5
\end{array}
$$

$$x = \frac{5}{2}$$

Use (4) and the value of x to find y.
$x + y = 6$ (4)
$y = 6 - x$

$$y = 6 - \frac{5}{2} = \frac{12}{2} - \frac{5}{2} = \frac{7}{2}$$

$$(x, y, z) = \left(\frac{5}{2}, \frac{7}{2}, 3\right)$$

9. $3x - 2y = 10$ (1)
   $x + y = 5$ (2)

Write the augmented coefficient matrix,

$$\begin{bmatrix} 3 & -2 & | & 10 \\ 1 & 1 & | & 5 \end{bmatrix}$$

Since R2 is easy to work with, switch R1 and R2.

$$\begin{bmatrix} 1 & 1 & | & 5 \\ 3 & -2 & | & 10 \end{bmatrix}$$

Now add 3R1 to (-1)R2 to get a new R2

$$
\begin{array}{llll}
3 & 3 & 15 & 3R1 \\
\underline{-3} & \underline{2} & \underline{-10} & \underline{-R2} \\
0 & 5 & 5 & \text{New R2}
\end{array}
$$

$$\begin{bmatrix} 1 & 1 & | & 5 \\ 0 & 5 & | & 5 \end{bmatrix}$$

Reinsert the variables to find x and y.

R2 $\Rightarrow 5y = 5$

$\quad\quad\quad y = 1$

R1 $\Rightarrow x + y = 5$

$\quad\quad\quad x = 5 - y$

$\quad\quad\quad x = 5 - 1 = 4$

$(x, y) = (4, 1)$

10. $\begin{vmatrix} 3 & -5 \\ 2 & 6 \end{vmatrix} = 3 \cdot 6 - (-5)(2) = 18 + 10 = 28$

11. $\begin{vmatrix} 1 & 2 & -1 \\ 2 & 0 & 3 \\ 0 & 1 & 5 \end{vmatrix} = 1\begin{vmatrix} 0 & 3 \\ 1 & 5 \end{vmatrix} - 2\begin{vmatrix} 2 & -1 \\ 1 & 5 \end{vmatrix} + 0\begin{vmatrix} 2 & -1 \\ 0 & 3 \end{vmatrix}$

$\quad\quad\quad\quad = (0 - 3) - 2(10 - (-1)) + 0$

$\quad\quad\quad\quad = -3 - 22 = -25$

12. $3x - 5y = 12$ (1)

$\quad 2x + 6y = 1$ (2)

$D = \begin{vmatrix} 3 & -5 \\ 2 & 6 \end{vmatrix} = 18 + 10 = 28$

$D_x = \begin{vmatrix} 12 & -5 \\ 1 & 6 \end{vmatrix} = 72 + 5 = 77$

$D_y = \begin{vmatrix} 3 & 12 \\ 2 & 1 \end{vmatrix} = 3 - 24 = -21$

$x = \dfrac{D_x}{D} = \dfrac{77}{28} = \dfrac{11}{4}$

$y = \dfrac{D_y}{D} = \dfrac{-21}{28} = -\dfrac{3}{4}$

13. $C(x) = 15x + 9000$

$\quad R(x) = 90x$

a) $P(x) = R(x) - C(x)$

$\quad\quad\quad = 90x - (15x + 9000)$

$\quad\quad\quad = 75x - 9000$

b) $P(x) = 0$

$\quad 75x - 9000 = 0$

$\quad\quad\quad x = \dfrac{9000}{75} = 120$

$\quad C(120) = R(120) = 90(120) = \$10,800$

14. $S(p) = 60 + 9p$

$\quad D(p) = 195 - 6p$

$\quad S(p) = D(p)$

$\quad 60 + 9p = 195 - 6p$

$\quad\quad\quad 15p = 135$

$\quad\quad\quad p = \dfrac{135}{15} = 9$

$\quad S(9) = D(9) = 60 + 9(9) = 60 + 81 = 141$

Chapter 3 Review Exercises

1. substitution

2. elimination

3. dependent

4. consistent

5. parallel

6. alphabetical

7. determinant

8. total profit

9. equilibrium point

10. zero

11. Graph each line and determine the point of intersection.

$y = x - 3$ $y = \frac{1}{4}x$

x	y
0	-3
3	0

$m = \frac{1}{4}$

y-intercept is $(0, 0)$

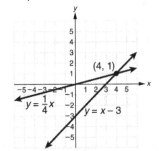

The point of intersection is $(4, 1)$.

12. We first solve each equation for y.

$$16x - 7y = 25 \qquad\qquad 8x + 3y = 19$$

$$-7y = -16x + 25 \qquad\qquad 3y = -8x + 19$$

$$y = \frac{16}{7}x - \frac{25}{7} \qquad\qquad y = -\frac{8}{3}x + \frac{19}{3}$$

Then we enter and graph $y_1 = \dfrac{16}{7}x - \dfrac{25}{7}$

and $y_2 = -\dfrac{8}{3}x + \dfrac{19}{3}$

Using the INTERSECT option, we determine the point of intersection.

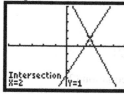

[–5, 5, Xscl = 1, –5, 5, Yscl = 1]

The point of intersection is (2, 1).

13. $x - y = 8 \qquad (1)$

$\quad y = 3x + 2 \quad (2)$

Substitute $3x + 2$ for y in Equation (1).

$$x - (3x + 2) = 8$$

$$x - 3x - 2 = 8$$

$$-2x = 8 + 2 = 10$$

$$x = \frac{10}{-2} = -5$$

Substitute -5 for x in Equation (1).

$$x - y = 8 \quad (1)$$

$$(-5) - y = 8$$

$$-5 - y = 8$$

$$-y = 8 + 5 = 13$$

$$y = -13$$

The ordered pair $(-5, -13)$ is the solution to the system.

14. $\quad y = x + 2 \quad (1)$

$y - x = 8 \qquad (2)$

Substitute $x + 2$ for y in equation (2).

$$(x + 2) - x = 8$$

$$2 = 8$$

This is a contradiction. The system has no solution.

15. $x - 3y = -2 \quad (1)$

$\quad 7y - 4x = 6 \quad (2)$

Solve equation (1) for x.

$$x - 3y = -2$$

$$x = 3y - 2 \quad (3)$$

Substitute $3y - 2$ for x in Equation (2).

$$7y - 4(3y - 2) = 6$$

$$7y - 12y + 8 = 6$$

$$-5y = -2$$

$$y = \frac{2}{5}$$

Substitute $\dfrac{2}{5}$ for y in Equation (3).

$$x = 3\left(\frac{2}{5}\right) - 2$$

$$x = \frac{6}{5} - \frac{10}{5}$$

$$x = -\frac{4}{5}$$

The solution is $\left(-\frac{4}{5}, \frac{2}{5}\right)$.

16. First, rearrange the second equation to align the x's and y's.

$$2x - 5y = 11 \qquad (1)$$

$$y - 2x = 5$$

$$-2x + y = 5 \qquad (2)$$

The coefficients of x are already opposites, so add the equations to solve for y. Then solve for x.

$$2x - 5y = 11$$

$$\underline{-2x + y = 5}$$

$$-4y = 16$$

$$y = \frac{16}{-4} = -4$$

Now find x using equation (1).

$$2x - 5y = 11$$

$$2x - 5(-4) = 11$$

$$2x + 20 = 11$$

$$2x = 11 - 20 = -9$$

$$x = \frac{-9}{2} = -\frac{9}{2}$$

The ordered pair $\left(-\dfrac{9}{2}, -4\right)$ is the solution to the system.

17. $4x - 7y = 18$ (1)

    $9x + 14y = 40$ (2)

    Multiply equation (1) by 2 and add the result to Equation (2).

    $$\begin{array}{rcr} 8x \; - \; 14y & = & 36 \\ 9x \; + \; 14y & = & 40 \\ \hline 17x & = & 76 \\ x & = & \dfrac{76}{17} \end{array}$$

    Substitute $\frac{76}{17}$ for x in Equation (1).

    $$4\left(\frac{76}{17}\right) - 7y = 18$$

    $$\frac{304}{17} - 7y = \frac{306}{17}$$

    $$\frac{-1}{7} \cdot (-7y) = \frac{2}{17} \cdot \frac{-1}{7}$$

    $$y = \frac{-2}{119}$$

    The solution is $\left(\frac{76}{17}, -\frac{2}{119}\right)$.

18. $3x - 5y = 9$ (1)

    $5x - 3y = -1$ (2)

    Multiply Equation (1) by -5, Equation (2) by 3, and add the resulting equations.

    $$\begin{array}{rcr} -15x \; + \; 25y & = & -45 \\ 15x \; - \; 9y & = & -3 \\ \hline 16y & = & -48 \\ y & = & -3 \end{array}$$

    Substitute -3 for y in Equation (2).

    $$5x - 3(-3) = -1$$

    $$5x + 9 = -1$$

    $$5x = -1 - 9 = -10$$

    $$x = \frac{-10}{5} = -2$$

    The ordered pair $(-2,\ -3)$ is the solution to the system.

19. $1.5x - 3 = -2y \rightarrow 1.5x + 2y = 3$ (1)

    $ 3x + 4y = 6$ (2)

    Multiply Equation (1) by -2 and add the result to Equation (2).

    $$\begin{array}{rcr} -3x \; - \; 4y & = & -6 \\ 3x \; + \; 4y & = & 6 \\ \hline 0 & = & 0 \end{array}$$

This is an identity.

$\{(x, y) \mid 3x + 4y = 6\}$

20. Let $x =$ the number of Nintendo Wii game machines sold, in millions, and $y =$ the number of PlayStation 3 consoles sold, in millions. Then:

    $x + y = 4.84$ (1)

    $x = 3y$ (2)

    Substitute $3y$ for x in equation (1) to find y, then use equation (2) to find x.

    $x + y = 4.84$ (1)

    $3y + y = 4.84$

    $4y = 4.84$

    $y = \dfrac{4.84}{4} = 1.21$

    Now find x using equation (2).

    $x = 3y = 3(1.21) = 3.63$

    There were 3.63 million Nintendo Wii game machines and 1.21 million PlayStation 3 consoles sold.

21. **Familiarize.** Let $x =$ the number of students taking private lessons and $y =$ the number of students taking group lessons.

    Translate.

    Total # of students is 12 $\rightarrow$ $x + y = 12$

    Total \$ is \$265 $\rightarrow$ \$25x + \$18y = \$265

    Carry out. Solve the system.

    $x \; + \; y = 12$ (1)

    $25x + 18y = 265$ (2)

    Multiply equation (1) by -18 and add the result to equation (2)

    $$\begin{array}{rcr} -18x \; - \; 18y & = & -216 \\ 25x \; + \; 18y & = & 265 \\ \hline 7x & = & 49 \\ x & = & 7 \end{array}$$

    $x + y = 12 \rightarrow 7 + y = 12$

    $ y = 5$

    Check.

    $7 + 5 = 12$, and

    $25 \cdot 7 + 18 \cdot 5 = 175 + 90 = 265$.

    The numbers check.

    State. Jillian had 7 students who took private lessons and 5 students who took group lessons.

22. *Familiarize.* We will summarize the information given in a chart and determine the equation using $D = R \times T$.
Translate.

	Distance	Rate	Time
Freight Train	d	44 mph	t
Passenger Train	d	55 mph	$t-1$

We have two equations:

$d = 44t$

$d = 55(t-1)$

Carry out. Solve the system.

$d = 44t$ (1)

$d = 55(t-1) \rightarrow d = 55t - 55$ (2)

Substitute $44t$ for d in Equation (2).

$44t = 55t - 55$

$-11t = -55$

$t = 5$

$t - 1 = 5 - 1 = 4$

Check. The freight train's distance $= 5 \cdot 44 = 220$ m., and the passenger train's distance $= 4 \cdot 55 = 220$ m. The distances are equal; the numbers check.

State. The passenger train will travel 4 hr before it overtakes the freight train.

23. *Familiarize.* Let $x =$ the number of liters of the 15% juice punch and $y =$ the number of liters of the 8% juice punch.
Translate.

Total liters is $14 \rightarrow x + y = 14$

Total amount of juice	is	10% of total liters
$\downarrow$	$\downarrow$	
$0.15x + 0.08y$	$=$	$0.10 \cdot 14$
$0.15x + 0.08y$	$=$	1.4

Carry out. Solve the system.

$x \ + \ y \ = \ 14$ (1)

$0.15x \ + \ 0.08y \ = \ 1.4$ (2)

Multiply Equation (1) by -0.08 and add the result to Equation (2).

$$
\begin{array}{rcr}
-0.08x \ - \ 0.08y &=& -1.12 \\
0.15x \ + \ 0.08y &=& 1.40 \\
\hline
0.07x \quad\quad\quad &=& 0.28 \\
x \quad\quad\quad &=& 4
\end{array}
$$

$x + y = 14 \rightarrow 4 + y = 14$

$y = 10$

Check.

$4 + 10 = 14$, and

$0.15 \cdot 4 + 0.08 \cdot 10 = 0.6 + 0.8$

$= 1.4 = 0.10 \cdot 14$

The numbers check.

State. D'Andre should purchase 4 L of the 15% juice punch and 10 L of the 8% juice punch.

24. $x + 4y + 3z = 2$ (1)

    $2x + y + z = 10$ (2)

   $-x + y + 2z = 8$ (3)

1., 2.   The equations are already in standard form with no fractions or decimals.

3., 4.   We eliminate x from two different pairs of equations.

$$
\begin{array}{rcl}
x + 4y + 3z &=& 2 \quad (1) \\
-x + \ y + 2z &=& 8 \quad (3) \\
\hline
5y + 5z &=& 10 \quad (4)
\end{array}
$$

$$
\begin{array}{rcll}
2x \ + \ y \ + z &=& 10 & (2) \\
-2x + 2y + 4z &=& 16 & \text{Mult. (3) by 2} \\
\hline
3y + 5z &=& 26 & (5)
\end{array}
$$

5.   Now solve the system of equations (4) and (5)

$5y + 5z = 10$ (4)

$3y + 5z = 26$ (5)

$$
\begin{array}{rcll}
5y + 5z &=& 10 & (4) \\
-3y - 5z &=& -26 & \text{Mult. (5) by } -1 \\
\hline
2y &=& -16 & \\
y &=& -8 &
\end{array}
$$

$5y + 5z = 10$ Substituting in (4)

$5(-8) + 5z = 10$

$-40 + 5z = 10$

$5z = 50$

$z = 10$

6. Substitute in one of the original equations
    to find x.

$$x + 4y + 3z = 2$$
$$x + 4(-8) + 3(10) = 2$$
$$x - 32 + 30 = 2$$
$$x - 2 = 2$$
$$x = 4$$

We obtain $(4, -8, 10)$. This checks, so it
is the solution.

25. $4x + 2y - 6z = 34$ (1)
       $2x + y + 3z = 3$ (2)
       $6x + 3y - 3z = 37$ (3)

1., 2. The equations are already in standard
        form with no fractions or decimals.
3., 4. We eliminate z from two different
        pairs of equations.

$$
\begin{array}{ll}
4x + 2y - 6z = 34 & (1) \\
\underline{4x + 2y + 6z = 6} & \text{Mult. } (2) \text{ by } 2 \\
8x + 4y = 40 & (4)
\end{array}
$$

$$
\begin{array}{ll}
2x + y + 3z = 3 & (2) \\
\underline{6x + 3y - 3z = 37} & (3) \\
8x + 4y = 40 & (5)
\end{array}
$$

5. Now solve the system of equations (4)
    and (5)

$$
\begin{array}{ll}
8x + 4y = 40 & (4) \\
\underline{-8x - 4y = -40} & \text{Mult } (5) \text{ by } -1 \\
0 = 0 & (6)
\end{array}
$$

6. Equation (6) indicates that equations (1),
    (2) and (3) are dependent.

26. $2x - 5y - 2z = -4$ (1)
       $7x + 2y - 5z = -6$ (2)
     $-2x + 3y + 2z = 4$ (3)

1., 2. The equations are already in standard
        form with no fractions or decimals.
3., 4. We eliminate x from two different
        pairs of equations.

$$
\begin{array}{ll}
2x - 5y - 2z = -4 & (1) \\
\underline{-2x + 3y + 2z = 4} & (3) \\
-2y = 0 & (4)
\end{array}
$$

$$
\begin{array}{ll}
-14x + 35y + 14z = 28 & \text{M. } (1) \text{ by } -7 \\
\underline{14x + 4y - 10z = -12} & \text{M. } (2) \text{ by } 2 \\
39y + 4z = 16 & (5)
\end{array}
$$

5. From equation (4) we have

$$-2y = 0$$
$$y = 0$$

Substituting 0 for y in equation (5) we
have

$$39y + 4z = 16$$
$$39(0) + 4z = 16$$
$$0 + 4z = 16$$
$$4z = 16$$
$$z = 4$$

6. Substitute in one of the original equations
    to find x.

$$2x - 5y - 2z = -4$$
$$2x - 5(0) - 2(4) = -4$$
$$2x - 0 - 8 = -4$$
$$2x - 8 = -4$$
$$2x = 4$$
$$x = 2$$

We obtain $(2, 0, 4)$. This checks, so it is
the solution.

27. $2x - 3y + z = 1$ (1)
       $x - y + 2z = 5$ (2)
      $3x - 4y + 3z = -2$ (3)

Multiply Equation (1) by -2 and add the
result to Equation (2) to Eliminate z.

$$-2(2x - 3y + z = 1)$$
$$-4x + 6y - 2z = -c)$$

$$
\begin{array}{rcrcrcrcr}
\rightarrow -4x & + & 6y & - & 2z & = & -2 \\
x & - & y & + & 2z & = & 5 \\
\hline
-3x & + & 5y & & & = & 3 & (4)
\end{array}
$$

Multiply Equation (1) by -3 and add the
result to Equation (3) to Eliminate z.

$$-3(2x - 3y + z = 1)$$
$$-6x + 9y - 3z = -3$$

$$
\begin{array}{rcrcrcrcr}
-6x & + & 9y & - & 3z & = & -3 \\
3x & - & 4y & + & 3z & = & -2 \\
\hline
-3x & + & 5y & & & = & -5 & (5)
\end{array}
$$

We solve the resulting system:

$$
\begin{array}{rcrcrcr}
-3x & + & 5y & = & 3 & (4) \\
-(-3x & + & 5y & = & -5) & (5) \\
\hline
& & 0 & = & 8 &
\end{array}
$$

We have a contradiction. The system has no
solution.

28. $3x + y = 2$ (1)

   $x + 3y + z = 0$ (2)

      $x + z = 2$ (3)

   1., 2.   The equations are already in standard form with no fractions or decimals.

   3., 4.   We eliminate x from equations (2) and (3).

$$\begin{array}{l} x + 3y + z = 0 \quad (2) \\ \underline{-x - z = -2} \quad \text{Mult. (3) by } -1 \\ 3y = -2 \quad (4) \end{array}$$

   Solve equation (4), we have
   $3y = -2$.

   $$y = -\frac{2}{3}$$

   Now substitute $-\dfrac{2}{3}$ for y in equation (1) and solve for x.

   $$3x + y = 2$$

   $$3x + \left(-\frac{2}{3}\right) = 2$$

   $$3x = 2 + \frac{2}{3}$$

   $$3x = \frac{8}{3}$$

   $$x = \frac{8}{9}$$

   Now substitute in equation (3) to solve for z.

   $$x + z = 2$$

   $$\frac{8}{9} + z = 2$$

   $$z = 2 - \frac{8}{9} = \frac{10}{9}$$

   We obtain $\left(\dfrac{8}{9}, -\dfrac{2}{3}, \dfrac{10}{9}\right)$. This

   checks, so it is the solution.

29. *Familiarize.* Let x = the measure of angle A, y = the measure of angle B, and z = the measure of angle C.

   Translate.

   The sum of the measures of the angles is 180°.

   $$x + y + z = 180$$

Measure of angle A is 4 times the measure of angle C.

   $$x = 4 \cdot z$$

Measure of angle B is 45° more than the measure of angle C

   $$y = 45 + z$$

We now have a system of equations.

   $x + y + z = 180$ (1)

   $x = 4z$ (2)

   $y = 45 + z$ (3)

Carry out. Substituting $4z$ for x and $45 + z$ for y in equation (1), we solve for z.

   $$x + y + z = 180$$

   $$4z + (45 + z) + z = 180$$

   $$6z + 45 = 180$$

   $$6z + 45 = 180$$

   $$6z = 135$$

   $$z = 22.5$$

Substituting 22.5 for z in equation (2), we can solve for x.

   $x = 4z$

   $x = 4(22.5)$

   $x = 90$

Substituting in equation (1), we can solve for y.

   $$x + y + z = 180$$

   $$90 + y + 22.5 = 180$$

   $$y + 112.5 = 180$$

   $$y = 67.5$$

Check. The sum of the measures of the three angles is 90° + 67.5° + 22.5°, or 180°. The measure of angle A, which is 90°, is 4 times the measure of angle C, which is 22.5. The measure of angle B is equal to the measure of angle C, plus 45°, or 22.5° + 45° = 67.5°. These numbers check.

State. Angles A, B, and C measure 90°, 67.5°, and 22.5°, respectively.

30. *Familiarize.* Let x = the average times a man cries each month, y = the average times a woman cries each month, and z = the average times a one-year-old cries each month. Then:

Translate.

$$x+y+z = 56.7 \qquad (1)$$
$$y = 3.9x \qquad (2)$$
$$z = x+y+43.3 \quad (3)$$

Carry out. Use (2) to rewrite (1) and (3) in terms of x and z.

$$x+y+z = 56.7 \qquad (1)$$
$$x+3.9x+z = 56.7$$
$$4.9x+z = 56.7 \quad (4)$$
$$z = x+y+43.3 \qquad (3)$$
$$z = x+3.9x+43.3$$
$$z = 4.9x+43.3 \quad (5)$$

Use (5) to rewrite (4) in terms of x and solve for x.

$$4.9x+z = 56.7 \quad (4)$$
$$4.9x+(4.9x+43.3) = 56.7$$
$$9.8x = 56.7-43.3 = 13.4$$
$$x = \frac{13.4}{9.8} \approx 1.367$$

Use (2) and the value of x to find y.

$$y = 3.9x \quad (5)$$
$$y \approx 3.9(1.367) \approx 5.33$$

Use (5) and the value of x to find z.

$$z = 4.9x+43.3 \quad (5)$$
$$z \approx 4.9(1.367)+43.3 \approx 50$$

State. Rounding to the nearest tenth gives:

$$(x,y,z) = (1.4, 5.3, 50)$$

31. $3x+4y = -13$

$5x+6y = 8$

Write a matrix using only the constants.

$$\begin{bmatrix} 3 & 4 & \vline & -13 \\ 5 & 6 & \vline & 8 \end{bmatrix}$$

Multiply the second row by 3 to make the first number in row 2 a multiply of 3.

$$\begin{bmatrix} 3 & 4 & \vline & -13 \\ 15 & 18 & \vline & 24 \end{bmatrix} \text{New Row } 2 = 3(\text{Row } 2)$$

Now multiply the first row by -5 and add it to the second row.

$$\begin{bmatrix} 3 & 4 & \vline & -13 \\ 0 & -2 & \vline & 89 \end{bmatrix} \text{Row } 2 = -5(\text{Row } 1)+\text{Row } 2$$

Reinserting the variables, we have

$$3x+4y = -13 \ (1)$$
$$-2y = 89 \ (2)$$

Solve Equation (2) for y.

$$-2y = 89$$
$$y = -\frac{89}{2}$$

Substituting $-\dfrac{89}{2}$ for y in Equation (1) and solve for x.

$$3x+4y = -13$$
$$3x+4\left(-\frac{89}{2}\right) = -13$$
$$3x-178 = -13$$
$$3x = 165$$
$$x = 55$$

The solution is $\left(55, -\dfrac{89}{2}\right)$.

32. $3x-y+z = -1$

$2x+3y+z = 4$

$5x+4y+2z = 5$

We first write a matrix using only the constants.

$$\begin{bmatrix} 3 & -1 & 1 & \vline & -1 \\ 2 & 3 & 1 & \vline & 4 \\ 5 & 4 & 2 & \vline & 5 \end{bmatrix}$$

Multiply the first row by 10.
Multiply the second row by -15.
Multiply the third row by -6.

$$\begin{bmatrix} 30 & -10 & 10 & \vline & -10 \\ -30 & -45 & -15 & \vline & -60 \\ -30 & -24 & -12 & \vline & -30 \end{bmatrix} \begin{matrix} \text{Multiply by 10.} \\ \text{Multiply by } -15. \\ \text{Multiply by } -6 \end{matrix}$$

Now add row 1 to row 2 and add row 1 to row 3.

$$\begin{bmatrix} 30 & -10 & 10 & \vline & -10 \\ 0 & -55 & -5 & \vline & -70 \\ 0 & -34 & -2 & \vline & -40 \end{bmatrix} \begin{matrix} \\ R2 = R1 + R2. \\ R3 = R1 + R3 \end{matrix}$$

Multiply row 2 by -34 and multiply row 3 by 55.

$$\begin{bmatrix} 30 & -10 & 10 & \vline & -10 \\ 0 & 1870 & 170 & \vline & 2380 \\ 0 & -1870 & -110 & \vline & -2200 \end{bmatrix} \begin{matrix} \\ \text{M. by } -34. \\ \text{M. by 55.} \end{matrix}$$

Add row 2 to row 3.

$$\begin{bmatrix} 30 & -10 & 10 & \vline & -10 \\ 0 & 1870 & 170 & \vline & 2380 \\ 0 & 0 & 60 & \vline & 180 \end{bmatrix} \begin{matrix} \\ \\ R3 = R2 + R3 \end{matrix}$$

Reinserting the variables, we have

$30x - 10y + 10z = -10$ (1)

$1870y + 170z = 2380$ (2)

$60z = 180$ (3)

Solve (3) for z.

$60z = 180$

$z = 3$

Substitute 3 for z in (2) and solve for y.

$1870y + 170z = 2380$

$1870y + 170(3) = 2380$

$1870y + 510 = 2380$

$1870y = 1870$

$y = 1$

Substitute 1 for y and 3 for z in (1) and solve for x.

$30x - 10y + 10z = -10$

$30x - 10(1) + 10(3) = -10$

$30x - 10 + 30 = -10$

$30x + 20 = -10$

$30x = -30$

$x = -1$

The solution is $(-1, 1, 3)$.

33. $\begin{vmatrix} -2 & -5 \\ 3 & 10 \end{vmatrix} = -2(10) - 3(-5) = -20 + 15 = -5$

34. $\begin{vmatrix} 2 & 3 & 0 \\ 1 & 4 & -2 \\ 2 & -1 & 5 \end{vmatrix}$

$= 2\begin{vmatrix} 4 & -2 \\ -1 & 5 \end{vmatrix} - 1\begin{vmatrix} 3 & 0 \\ -1 & 5 \end{vmatrix} + 2\begin{vmatrix} 3 & 0 \\ 4 & -2 \end{vmatrix}$

$= 2(20 - 2) - 1(15 + 0) + 2(-6 - 0)$

$= 2(18) - 1(15) + 2(-6)$

$= 36 - 15 - 12$

$= 9$

35. $2x + 3y = 6$

$x - 4y = 14$

$x = \dfrac{\begin{vmatrix} 6 & 3 \\ 14 & -4 \end{vmatrix}}{\begin{vmatrix} 2 & 3 \\ 1 & -4 \end{vmatrix}} = \dfrac{-24 - 42}{-8 - 3} = \dfrac{-66}{-11} = 6$

$y = \dfrac{\begin{vmatrix} 2 & 6 \\ 1 & 14 \end{vmatrix}}{\begin{vmatrix} 2 & 3 \\ 1 & -4 \end{vmatrix}} = \dfrac{28 - 6}{-8 - 3} = \dfrac{22}{-11} = -2$

The solution is $(6, -2)$.

36. $2x + y + z = -2$

$2x - y + 3z = 6$

$3x - 5y + 4z = 7$

First find D, D_x, D_y and D_z.

$D = \begin{vmatrix} 2 & 1 & 1 \\ 2 & -1 & 3 \\ 3 & -5 & 4 \end{vmatrix}$

$= 2\begin{vmatrix} -1 & 3 \\ -5 & 4 \end{vmatrix} - 2\begin{vmatrix} 1 & 1 \\ -5 & 4 \end{vmatrix} + 3\begin{vmatrix} 1 & 1 \\ -1 & 3 \end{vmatrix}$

$= 2(-4 + 15) - 2(4 + 5) + 3(3 + 1)$

$= 2(11) - 2(9) + 3(4)$

$= 22 - 18 + 12$

$= 16$

$D_x = \begin{vmatrix} -2 & 1 & 1 \\ 6 & -1 & 3 \\ 7 & -5 & 4 \end{vmatrix}$

$= -2\begin{vmatrix} -1 & 3 \\ -5 & 4 \end{vmatrix} - 6\begin{vmatrix} 1 & 1 \\ -5 & 4 \end{vmatrix} + 7\begin{vmatrix} 1 & 1 \\ -1 & 3 \end{vmatrix}$

$= -2(-4 + 15) - 6(4 + 5) + 7(3 + 1)$

$= -2(11) - 6(9) + 7(4)$

$= -22 - 54 + 28$

$= -48$

$D_y = \begin{vmatrix} 2 & -2 & 1 \\ 2 & 6 & 3 \\ 3 & 7 & 4 \end{vmatrix}$

$= 2\begin{vmatrix} 6 & 3 \\ 7 & 4 \end{vmatrix} - 2\begin{vmatrix} -2 & 1 \\ 7 & 4 \end{vmatrix} + 3\begin{vmatrix} -2 & 1 \\ 6 & 3 \end{vmatrix}$

$= 2(24 - 21) - 2(-8 - 7) + 3(-6 - 6)$

$= 2(3) - 2(-15) + 3(-12)$

$= 6 + 30 - 36$

$= 0$

$$D_z = \begin{vmatrix} 2 & 1 & -2 \\ 2 & -1 & 6 \\ 3 & -5 & 7 \end{vmatrix}$$

$$= 2\begin{vmatrix} -1 & 6 \\ -5 & 7 \end{vmatrix} - 2\begin{vmatrix} 1 & -2 \\ -5 & 7 \end{vmatrix} + 3\begin{vmatrix} 1 & -2 \\ -1 & 6 \end{vmatrix}$$

$$= 2(-7+30) - 2(7-10) + 3(6-2)$$

$$= 2(23) - 2(-3) + 3(4)$$

$$= 46 + 6 + 12$$

$$= 64$$

$$x = \frac{D_x}{D} = \frac{-48}{16} = -3$$

$$y = \frac{D_y}{D} = \frac{0}{16} = 0$$

$$z = \frac{D_z}{D} = \frac{64}{16} = 4$$

The solution is $(-3,0,4)$.

37. a) $P(x) = R(x) - C(x)$

$$= (50x) - (30x + 15,800)$$

$$= 50x - 30x - 15,800$$

$$P(x) = 20x - 15,800$$

   b) $R(x) = C(x)$

$$50x = 30x + 15,800$$

$$20x = 15,800$$

$$x = 790$$

$$R(x) = 50x$$

$$R(790) = 50 \cdot 790 = \$39,500$$

The break-even point is
(790, \$39,500) which represents 790
units and revenue of \$39,500.

38. $D(p) = 120 - 13p$

$$S(p) = 60 + 7p$$

Rewrite the system:

$$q = 120 - 13p \quad (1)$$

$$q = 60 + 7p \quad (2)$$

Substitute $120 - 13p$ for q in (2) and solve.

$$120 - 13p = 60 + 7p$$

$$60 = 20p$$

$$3 = p$$

The equilibrium price is \$3 per unit.

To find the equilibrium quantity we substitute
\$3 into either $D(p)$ or $S(p)$.

$$S(3) = 60 + 7(3)$$

$$= 60 + 21$$

$$= 81$$

The equilibrium quantity is 81 units.
The equilibrium point is $(\$3, 81)$.

39. Let $x=$ the number of pints of honey produced
and sold.

   a) $C(x) = 4.75x + 54,000$

   b) $R(x) = 9.25x$

   c) $P(x) = R(x) - C(x)$

$$= 9.25x - (4.75x + 54,000)$$

$$= 4.50x - 54,000$$

   d) If 5000 pints are produced and sold:

$$P(5000) = 4.5(5000) - 54,000$$

$$= -31,500$$

there is a loss of \$31,500.
If 15,000 pints are produced and sold:

$$P(15,000) = 4.5(15,000) - 54,000$$

$$= 13,500$$

there is a profit of \$13,500.

   e) $P(x) = 0 \Rightarrow 4.5x - 54,000 = 0$

$$\Rightarrow x = \frac{54,000}{4.5} = 12,000$$

$$C(12,000) = R(12,000) = 9.25(12,000)$$

$$= \$111,000$$

40. *Thinking and Writing Exercise.*

41. *Thinking and Writing Exercise.*

42. Danae must make a profit producing and
selling honey equal to the pay of her previous
job, or \$36,000, so $P(x) = 36,000$. We must
solve for x to determine the number of units
of honey she needs to produce and sell.

$$P(x) = 4.5x - 54,000 = 36,000$$

$$4.5x = 90,000$$

$$x = \frac{90,000}{4.5} = 20,000$$

Danae must produce and sell 20,000 pints of
honey.

43. Enter and graph $y_1 = x + 2$ and $y_2 = x^2 + 2$.
    Using the INTERSECT option, determine the
    points of intersection.

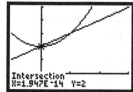

[−1, 3, Xscl=1, −1, 5, Yscl=1]
Note: 1.947E-14 is extremely close to 0.

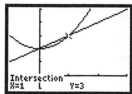

[−1, 3, Xscl=1, −1, 5, Yscl=1]
The points of intersection are $(0, 2)$ and $(1, 3)$.

Chapter 3 Test

1. Graph each line and determine the points of
   intersection.
   $$2x + y = 8 \qquad y - x = 2$$

x	y
0	8
4	0

x	y
−2	0
0	2

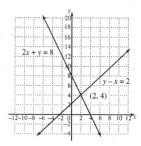

   The solution is $(2, 4)$.

2. $x + 3y = -8$ (1)
   $4x - 3y = 23$ (2)
   Solve equation (1) for x.
   $$x + 3y = -8$$
   $$x = -3y - 8$$
   Substitute $-3y - 8$ for x in equation (2) and
   solve for y.

$$4x - 3y = 23$$
$$4(-3y - 8) - 3y = 23$$
$$-12y - 32 - 3y = 23$$
$$-15y - 32 = 23$$
$$-15y = 55$$
$$y = -\frac{55}{15}$$
$$y = -\frac{11}{3}$$

Finally substitute $-\dfrac{11}{3}$ for y in either of the
original equation and solve for x. We
choose equation (1).
$$x + 3y = -8$$
$$x + 3\left(-\frac{11}{3}\right) = -8$$
$$x - 11 = -8$$
$$x = 3$$
The solution is $\left(3, -\dfrac{11}{3}\right)$ which checks.

3. $2x - 4y = -6$ (1)
   $x = 2y - 3$ (2)
   Substitute $2y - 3$ for x in equation (1) and
   solve for x.
   $$2(2y - 3) - 4y = -6$$
   $$4y - 6 - 4y = -6$$
   $$-6 = -6$$
   The equations are dependent. The solutions
   are $\{(x, y) \mid x = 2y - 3\}$.

4. $3x - y = 7$ (1)
   $x + y = 1$ (2)
   Add equation (1) to (2).
   $$\begin{array}{ll} 3x - y = 7 & (1) \\ \underline{x + y = 1} & (2) \\ 4x = 8 & \end{array}$$
   $$x = \frac{8}{4} = 2$$

   Substitute $x = 2$ in either of the original
   equations and solve for y.
   $$x + y = 1 \quad (2)$$
   $$2 + y = 1$$
   $$y = 1 - 2 = -1$$

The solution is $(2,-1)$ which checks.

5. $4y + 2x = 18$
    $3x + 6y = 26$
    Rearrange terms in form of $A + By = C$.
    $2x + 4y = 18$ (1)
    $3x + 6y = 26$ (2)

    Multiply equation (1) by $-\dfrac{3}{2}$ and add it to equation (2)

    $-3x - 6y = -27$ Multiply by $-\frac{3}{2}$
    $\underline{3x + 6y = 26}$
    $0 = -1$ Adding

    Because $0 = -1$ is a false statement, this system does not have a solution.

6. $4x - 6y = 3$ (1)
    $6x - 4y = -3$ (2)
    Multiply Equation (1) by 3, Equation (2) by -2, and add the resulting equations.
    $12x - 18y = 9$
    $\underline{-\ 12x + 8y = 6}$
    $-\ 10y = 15$

    $y = \dfrac{-15}{10} = \dfrac{-3}{2}$

    $4x - 6y = 3 \rightarrow 4x - 6\left(\dfrac{-3}{2}\right) = 3$
    $4x + 9 = 3$
    $4x = -6$
    $x = \dfrac{-6}{4} = -\dfrac{3}{2}$

    The solution is $\left(-\dfrac{3}{2}, -\dfrac{3}{2}\right)$.

7. **Familiarize.** Let $w =$ the width of the court and let $l =$ the length of the court. Recall that the perimeter of a rectangle is given by the formula $P = 2w + 2l$.
    Translate.
    The perimeter of the court is 288.
    $2w + 2l = 288$

The length is 44 ft longer than the width.
$l = w + 44$
We now have a system of equations.
$2w + 2l = 288$ (1)
$l = w + 44$ (2)
Carry out.
Substitute $w + 44$ for l in the first equation and solve for w.
$2w + 2l = 288$
$2w + 2(w + 44) = 288$
$2w + 2w + 88 = 288$
$4w + 88 = 288$
$4w = 200$
$w = 50$
Finally substitute 50 for w in equation (2) and solve for l.
$l = w + 44$
$l = 50 + 44$
$l = 94$
Check.
The perimeter of the rectangle is 2(50 ft) + 2(94 ft) = 100 ft + 188 ft = 288 ft. The length, 94 ft is 44 ft longer than the width, or 50 + 44 = 94 ft.
State. The standard basketball court is 94 ft long and 50 ft wide.

8. **Familiarize.** Let $x =$ the number of grams of Pepperidge Farm Goldfish and $y =$ the number of grams of Rold Gold Pretzels. The total number of grams of fat in the mixture is 15%(620 g), or 93 g.
    Translate. We can represent the information in a table.

	Peppe-ridge Farm Goldfish	Rold Gold Pretzels	Mix
No. of grams	x	y	620
% cal. from fat	40%	9%	15%
Amt. of fat	$0.40x$	$0.09y$	$0.15(620)$ $= 93$

From the "Number of grams" row in the table, we get one equation.
$x + y = 620$

From the last row, "Amount of fat," we get a second equation.

$0.40x + 0.09y = 93$

We now have a system of equations:

$$x + y = 620 \quad (1)$$
$$0.40x + 0.09y = 93 \quad (2)$$

After clearing decimals we have:

$$x + y = 620 \quad (3)$$
$$40x + 9y = 9300 \quad (4)$$

Carry out. We will use substitution to solve the system. First solve equation (3) for the variable x.

$$x + y = 620$$
$$x = -y + 620$$

Substitute $-y + 620$ for x in the second equation and solve for y.

$$40x + 9y = 9300$$
$$40(-y + 620) + 9y = 9300$$
$$-40y + 24{,}800 + 9y = 9300$$
$$-31y + 24{,}800 = 9300$$
$$-31y = -15{,}500$$
$$y = 500$$

Finally substitute 500 for y in equation (1) and solve for x.

$$x + y = 620$$
$$x + 500 = 620$$
$$x = 120$$

Check. The total number of grams in the mixture is $120 + 500$, or 620. The amount of fat in the mixture is $40\%(120 \text{ g}) + 9\%(500 \text{ g}) = 48 + 45 = 93$. These numbers check.

State. The mixture should contain 120 g of Pepperidge Farm Goldfish and 500 g of Rold Gold Pretzels.

9. **Familiarize.** Let s = speed of boat, in mph.
Translate. We make a table

	Distance	Rate	Time
With current	d	$s+5$	3
Against current	d	$s-5$	5

We have two equations:

$d = 3(s + 5) \rightarrow d = 3s + 15 \quad (1)$

$d = 5(s - 5) \rightarrow d = 5s - 25 \quad (2)$

Carry out. Solve the system.
Substitute $3s + 15$ for d in Equation (2).

$$3s + 15 = 5s - 25$$
$$3s + 40 = 5s$$
$$40 = 2s$$
$$20 = s$$

Check. Downstream distance is $3(20 + 5) = 3 \cdot 25 = 75$; upstream distance is $5(20 - 5) = 5 \cdot 15 = 75$. The distances are the same; our solution checks.
State. The speed of the boat is 20 mph.

10. $-3x + y - 2z = 8 \quad (1)$
    $-x + 2y - z = 5 \quad (2)$
    $2x + y + z = -3 \quad (3)$

1., 2. The equations are already in standard form with no fractions or decimals.

3., 4. We eliminate x from equations (1) and (2).

$$-3x + y - 2z = 8$$
$$\underline{3x - 6y + 3z = -15} \quad \text{M. eq. 2 by } -3$$
$$-5y + z = -7 \quad (4)$$

We eliminate x from equations (2) and (3).

$$-2x + 4y - 2z = 10 \quad \text{M. eq. 2 by 2}$$
$$\underline{2x + y + z = -3}$$
$$5y - z = 7 \quad (5)$$

Solve equations (4) and (5):

$$-5y + z = -7 \quad (4)$$
$$\underline{5y - z = 7 \quad (5)}$$
$$0 = 0 \quad \text{Adding}$$

Because $0 = 0$ is a true statement, the system is dependent.

11. $6x + 2y - 4z = 15 \quad (1)$
    $-3x - 4y + 2z = -6 \quad (2)$
    $4x - 6y + 3z = 8 \quad (3)$

1., 2. The equations are already in standard form with no fractions or decimals.

3., 4. We eliminate x from equations (1) and (2).

$$6x + 2y - 4z = 15$$
$$\underline{-6x - 8y + 4z = -12} \quad \text{M. eq. 2 by 2}$$
$$-6y = 3 \quad (4)$$
$$y = -\frac{1}{2}$$

We eliminate x from equations (2)

and (3).

$$-12x - 16y + 8z = -24 \quad \text{M. eq. 2 by 4}$$
$$\underline{12x - 18y + 9z = 24} \quad \text{M. eq. 3 by 3}$$
$$-34y + 17z = 0 \quad (5)$$

Substituting $-\dfrac{1}{2}$ for y in equation (5)

and solving for z, we have:

$$-34y + 17z = 0$$
$$-34\left(-\dfrac{1}{2}\right) + 17z = 0$$
$$17 + 17z = 0$$
$$17z = -17$$
$$z = -1$$

Substituting $-\dfrac{1}{2}$ for y and -1 for z in

any of the original equations, we can solve for x. We choose equation (1).

$$6x + 2y - 4z = 15$$
$$6x + 2\left(-\dfrac{1}{2}\right) - 4(-1) = 15$$
$$6x - 1 + 4 = 15$$
$$6x + 3 = 15$$
$$6x = 12$$
$$x = 2$$

The solution is $\left(2, -\dfrac{1}{2}, -1\right)$ which

checks.

12. $2x + 2y = 0 \ \ (1)$
      $4x + 4z = 4 \ \ (2)$
      $2x + y + z = 2 \ \ (3)$

1., 2.   The equations are already in standard form with no fractions or decimals.

3., 4.   We can simplify equations (1) and (2) by multiplying by $\dfrac{1}{2}$ and $\dfrac{1}{4}$, respectively.

$$x + y = 0 \ \ (4)$$
$$x + z = 1 \ \ (5)$$
$$2x + y + z = 2 \ \ (3)$$

We can eliminate the variable x from equations (4) and (5).

$$x + y \quad\quad = 0$$
$$\underline{-x \quad\quad - z = -1} \quad \text{M. eq. 5 by } -1$$
$$y - z = -1 \ \ (6)$$

We can eliminate the variable x from equations (5) and (3).

$$-2x \quad\quad - 2z = -2 \quad \text{M. eq. 5 by } -2$$
$$\underline{2x + y + z = \ \ 2}$$
$$y - z = 0 \ \ (7)$$

Solving equations (6) and (7), we have:

$$-y + z = 1 \quad \text{M. eq. 6 by } -1$$
$$\underline{y - z = 0}$$
$$0 = 1$$

Because the statement $0 = 1$ is false, the system has no solution.

13. $3x + 3z = 0 \ \ (1)$
      $2x + 2y = 2 \ \ (2)$
      $3y + 3z = 3 \ \ (3)$

1., 2.   The equations are already in standard form with no fractions or decimals.

3., 4.   We can simplify all three equations by multiplying by $\dfrac{1}{3}$, $\dfrac{1}{2}$, and $\dfrac{1}{3}$, respectively.

$$x + z = 0 \ \ (4)$$
$$x + y = 1 \ \ (5)$$
$$y + z = 1 \ \ (6)$$

We can eliminate the variable x from equations (4) and (5).

$$x + z = 0$$
$$\underline{-x - y = -1} \quad \text{M. eq. 5 by } -1$$
$$-y + z = -1 \ \ (7)$$

Now we can solve equations (6) and (7).

$$y + z = 1 \ \ (6)$$
$$\underline{-y + z = -1} \ \ (7)$$
$$2z = 0$$
$$z = 0$$

We can substitute 0 for z in equation (6) and solve for y.

$$y + z = 1$$
$$y + 0 = 1$$
$$y = 1$$

We can substitute 0 for z in equation (4) and solve for x.

$x + z = 0$

$x + 0 = 0$

$x = 0$

The solution is $(0,1,0)$, which

checks.

14. $4x + y = 12$ (1)

$3x + 2y = 2$ (2)

Write a matrix using only the coefficients,

$$\begin{bmatrix} 4 & 1 & | & 12 \\ 3 & 2 & | & 2 \end{bmatrix}$$

Add (-3) R1 to 4R2 to get a new R2

-12	-3	-36	-3R1
12	8	8	4R2
0	5	-28	New R2

$$\begin{bmatrix} 4 & 1 & | & 12 \\ 0 & 5 & | & -28 \end{bmatrix}$$

Reinsert the variables to find x and y.

NEW R2 : $5y = -28$

$$y = -\frac{28}{5}$$

R1 : $4x + y = 12$

$$4x = 12 - \left(-\frac{28}{5}\right)$$

$$x = \frac{12}{4} + \left(\frac{1}{4}\right)\left(\frac{28}{5}\right)$$

$$x = 3 + \frac{7}{5} = \frac{22}{5}$$

The solution is $\left(\frac{22}{5}, -\frac{28}{5}\right)$, which checks.

15. $x + 3y - 3z = 12$

$3x - y + 4z = 0$

$-x + 2y - z = 1$

We first write a matrix with only constants.

$$\begin{bmatrix} 1 & 3 & -3 & | & 12 \\ 3 & -1 & 4 & | & 0 \\ -1 & 2 & -1 & | & 1 \end{bmatrix}$$

Multiply row 1 by -3 and add it to row 2.

Add row 1 to row 3.

$$\begin{bmatrix} 1 & 3 & -3 & | & 12 \\ 0 & -10 & 13 & | & -36 \\ 0 & 5 & -4 & | & 13 \end{bmatrix}$$

Reverse rows 2 and 3.

$$\begin{bmatrix} 1 & 3 & -3 & | & 12 \\ 0 & 5 & -4 & | & 13 \\ 0 & -10 & 13 & | & -36 \end{bmatrix}$$

Multiply row 2 by 2 and add it to row 3.

$$\begin{bmatrix} 1 & 3 & -3 & | & 12 \\ 0 & 5 & -4 & | & 13 \\ 0 & 0 & 5 & | & -10 \end{bmatrix}$$

Reinserting the variables, we have

$x + 3y - 3z = 12$ (1)

$5y - 4z = 13$ (2)

$5z = -10$ (3)

Solve (3) for z.

$5z = -10$

$z = -2$

Substitute -2 for z in (2) and solve for y.

$5y - 4z = 13$

$5y - 4(-2) = 13$

$5y + 8 = 13$

$5y = 5$

$y = 1$

Substitute -2 for z and 1 for y in equation (1)

and solve for x.

$x + 3y - 3z = 12$

$x + 3(1) - 3(-2) = 12$

$x + 3 + 6 = 12$

$x + 9 = 12$

$x = 3$

The solution is $(3, 1, -2)$ which checks.

16. $\begin{vmatrix} 4 & -2 \\ 3 & -5 \end{vmatrix} = 4(-5) - 3(-2) = -20 + 6 = -14$

17. $\begin{vmatrix} 3 & 4 & 2 \\ -2 & -5 & 4 \\ 0 & 5 & -3 \end{vmatrix}$

$= 3\begin{vmatrix} -5 & 4 \\ 5 & -3 \end{vmatrix} + 2\begin{vmatrix} 4 & 2 \\ 5 & -3 \end{vmatrix} + 0\begin{vmatrix} 4 & 2 \\ -5 & 4 \end{vmatrix}$

$= 3(15 - 20) + 2(-12 - 10) + 0$

$= -59$

18. $3x + 4y = -1$

$5x - 2y = 4$

$D = \begin{vmatrix} 3 & 4 \\ 5 & -2 \end{vmatrix} = -6 - 20 = -26$

$D_x = \begin{vmatrix} -1 & 4 \\ 4 & -2 \end{vmatrix} = 2 - 16 = -14$

$D_Y = \begin{vmatrix} 3 & -1 \\ 5 & 4 \end{vmatrix} = 12 + 5 = 17$

$x = \dfrac{D_X}{D} = \dfrac{-14}{-26} = \dfrac{7}{13}$

$y = \dfrac{D_Y}{D} = \dfrac{17}{-26} = -\dfrac{17}{26}$

19. **Familiarize.** Let x = the number of hours the electrician worked, y = the number of hours the carpenter worked, and z = the number of hours the plumber worked.

Translate.

$\underbrace{\text{The total number of hours worked}}$ is 21.5.

$\downarrow \qquad\qquad\qquad \downarrow \ \downarrow$

$x + y + z \qquad\qquad = 21.5$

$\underbrace{\text{The total pay for all three}}$ was \$469.50.

$\downarrow \qquad\qquad\quad \downarrow \quad \downarrow$

$21x + 19.50y + 24z \quad = \quad 469.50$

$\underbrace{\begin{matrix}\text{Hours the} \\ \text{plumber} \\ \text{worked}\end{matrix}}$ is $\underbrace{\text{2 more than}}$ $\underbrace{\begin{matrix}\text{the hours} \\ \text{worked by} \\ \text{carpenter.}\end{matrix}}$

$\downarrow \quad \downarrow \quad \downarrow \qquad \downarrow$

$z \quad = \quad 2 + \qquad y$

We now have a system of equations.

$x + y + z = 21.5$

$21x + 19.50y + 24z = 469.50$

$z = 2 + y$

Carry out. Clearing fractions we have the system:

$10x + 10y + 10z = 215 \quad (1)$

$210x + 195y + 240z = 4695 \quad (2)$

$z = 2 + y \quad (3)$

Substitute $2 + y$ for z in both equations (1) and (2), we have:

$10x + 10y + 10(2 + y) = 215$

$10x + 10y + 20 + 10y = 215$

$10x + 20y + 20 = 215$

$10x + 20y = 195 \quad (3)$

$210x + 195y + 240(2 + y) = 4695$

$210x + 195y + 480 + 240y = 4695$

$210x + 435y + 480 = 4695$

$210x + 435y = 4215 \quad (4)$

We now solve the system:

$10x + 20y = 195 \quad (3)$

$210x + 435y = 4215 \quad (4)$

Multiply Equation (3) by -21 and add it to equation (4).

$-210x - 420y = -4095$

$\underline{210x + 435y = 4215}$

$15y = 120$

$y = 8$

Substituting 8 for y in equation (3), solve for x.

$10x + 20y = 195$

$10x + 20(8) = 195$

$10x + 160 = 195$

$10x = 35$

$x = 3.5$

Substituting 8 for y in equation (3), we solve for z.

$z = 2 + y$

$z = 2 + 8$

$z = 10$

Check. The three workers put in 3.5 + 8 + 10, or 21.5 hours. The plumber worked two more hours than the carpenter. Finally, the total cost for their work was 3.5(\$21) + 8(\$19.50) + 10(\$24) = \$73.50 + \$156 + \$240 = \$469.50. The numbers check.

State. The electrician worked 3.5 hours; the carpenter worked 8 hours, and the plumber worked 10 hours.

20. $D(p) = 79 - 8p$

$S(p) = 37 + 6p$

Rewrite the system.

$q = 79 - 8p \quad (1)$

$q = 37 + 6p \quad (2)$

Substitute $79 - 8p$ for q in (2) and solve.

$79 - 8p = 37 + 6p$

$79 - 37 = 6p + 8p$

$42 = 14p$

$3 = p$

The equilibrium price is \$3 per unit.
To find the equilibrium quantity we substitute
\$3 into either $D(p)$ or $S(p)$.

$D(3) = 79 - 8(3)$

$= 79 - 24$

$= 55$

The equilibrium quantity is 55 units.
The equilibrium point is $(\$3, 55 \text{ units})$.

21. a) $C(x) = 25x + 44{,}000$

b) $R(x) = 80x$

c) $P(x) = R(x) - C(x)$

$= 80x - (25x + 44{,}000)$

$= 80x - 25x - 44{,}000$

$= 55x - 44{,}000$

d)   For 300 hammocks,

$P(300) = 55(300) - 44{,}000$

$= 16{,}500 - 44{,}000$

$= -27{,}500$

the company will have a loss of \$27,500.
For 900 hammocks,

$P(900) = 55(900) - 44{,}000$

$= 49{,}500 - 44{,}000$

$= 5000$

the company will realize a \$5000 profit.

e)   Since both $R(x)$ and $C(x)$ are in dollars
and they are equal at the break-even
point, we can rewrite the system:

$R(x) = 80x$

$C(x) = 25x + 44{,}000$

as

$q = 80x$ \hspace{2em} (1)

$q = 25x + 44{,}000$ \hspace{1em} (2)

We solve using substitution.

$80x = 25x + 44{,}000$

$55x = 44{,}000$

$x = \dfrac{44{,}000}{55} = 800 \text{ units}$

Thus, 800 hammocks must be produced
in order to break even.

$R(800) = 80 \cdot 800 = \$64{,}000$

So the break-even point is
(800 hammocks, \$64,000).

22. Since $f(x)$ contains $(-1, 3)$ and $(-2, -4)$,
those points must satisfy the equation.
For $(-1, 3)$:

$f(-1) = 3 = m(-1) + b$

$= -m + b$

which gives us $-m + b = 3$
For $(-2, -4)$:

$f(-2) = -4 = m(-2) + b$

$= -2m + b$

which gives us $-2m + b = -4$
Hence we have the system of equations:

$-m + b = 3$ \hspace{1em} (1)

$-2m + b = -4$ \hspace{0.5em} (2)

Solving equation (1) for b, we have

$-m + b = 3$

$b = m + 3$

Substituting $m + 3$ for b in equation (2) we
can solve for m.

$-2m + b = -4$

$-2m + (m + 3) = -4$

$-m + 3 = -4$

$-m = -7$

$m = 7$

Substituting 7 for m in equation (1) we can
solve for b.

$-m + b = 3$

$-(7) + b = 3$

$b = 10$

So $m = 7$ and $b = 10$. Thus the function is
$f(x) = 7x + 10$.

23. ***Familiarize.*** Let k = the number of pounds of Kona coffee that must be added to the Mexican coffee, and m = the number of pounds of coffee in the mixture.

Translate. We organize the information in a table.

	Mexican	Kona	Mixture
No. of pounds	40	k	m
% of Kona	0%	100%	30%
Amount of Kona	0	k	$0.3m$

We get one equation from the "Number of pounds" row of the table:

$40 + k = m$

The last row of the table gives us a second equation:

$k = 0.3m$

After clearing the decimal we have the problem translated to a system of equations:

$40 + k = m$ (1)

$10k = 3m$ (2)

Carry out. We use substitution to solve the system of equations. First we substitute $40 + k$ for m in (2).

$10k = 3m$ (2)

$10k = 3(40 + k)$ Substituting

$10k = 120 + 3k$

$7k = 120$

$k = \dfrac{120}{7}$

Although the problem asks for only k, the amount of Kona coffee that should be used, we will also find m in order to check the answer.

$40 + k = m$ (1)

$40 + \dfrac{120}{7} = m$ Substituting $\dfrac{120}{7}$ for k

$\dfrac{280}{7} + \dfrac{120}{7} = m$

$\dfrac{400}{7} = m$

Check. If $\dfrac{400}{7}$ lb of coffee contains $\dfrac{120}{7}$ lb of Kona coffee, then the percent of Kona beans in the mixture is

$\dfrac{120/7}{400/7} = \dfrac{120}{7} \cdot \dfrac{7}{400} = \dfrac{3}{10}$, or 30%. The answer checks.

State. $\dfrac{120}{7}$ lb of Kona coffee should be added to the Mexican coffee.

Chapters 1 – 3

Cumulative Review

1. $x^4 \cdot x^{-6} \cdot x^{13} = x^{4+(-6)+13} = x^{11}$

2. $\left(6x^2 y^3\right)^2 \left(-2x^0 y^4\right)^{-3}$

$= 6^2 x^{2 \cdot 2} y^{2 \cdot 3} \left(-2y^4\right)^{-3}$

$= 36x^4 y^6 \left(\dfrac{-1}{2y^4}\right)^3$

$= 36x^4 y^6 \left(\dfrac{(-1)^3}{2^3 y^{4 \cdot 3}}\right)$

$= 36x^4 y^6 \left(\dfrac{-1}{8y^{12}}\right)$

$= -\dfrac{36x^4 y^6}{8y^{12}} = -\dfrac{9x^4}{2y^6}$

3. $\dfrac{-10a^7 b^{-11}}{25a^{-4} b^{22}} = -\dfrac{10}{25} \cdot a^{7-(-4)} b^{-11-22}$

$= -\dfrac{2}{5} a^{11} b^{-33} = -\dfrac{2a^{11}}{5b^{33}}$

4. $\left(\dfrac{3x^4 y^{-2}}{4x^{-5}}\right)^4 = \dfrac{3^4 x^{16} y^{-8}}{4^4 x^{-20}}$

$= \dfrac{81x^{16-(-20)}}{256y^8} = \dfrac{81x^{36}}{256y^8}$

5. $\left(1.95 \times 10^{-3}\right)\left(5.73 \times 10^8\right)$

$= 11.1735 \times 10^{-3+8}$

$= 1.12 \times 10^1 \times 10^5 = 1.12 \times 10^6$

6. $\dfrac{2.42 \times 10^5}{6.05 \times 10^{-2}}$

$= \dfrac{2.42}{6.05} \times 10^{5-(-2)} = 0.40 \times 10^{5+2}$

$= 4.00 \times 10^{-1} \times 10^7 = 4.00 \times 10^6$

7. $A = \frac{1}{2} h(b+t)$

$2 \cdot A = \cancel{2} \cdot \frac{1}{\cancel{2}} h(b+t)$

$2A = h(b+t)$

$2A = hb + ht$

$2A - ht = hb$

$\dfrac{2A - ht}{h} = b$

$b = \dfrac{2A - ht}{h}$, or $b = \dfrac{2A}{h} - \dfrac{ht}{h} = \dfrac{2A}{h} - t$

8. $5a - 2b = -23$

$5(-3) - 2(4)$	-23
$-15 - 8$	-23
-23	-23 True

Since $-23 = -23$ is a true statement, $(-3, 4)$ is a solution of the equation.

9. $x + 9.4 = -12.6$

$x = -9.4 - 12.6$

$x = -22$

10. $-2.4x = -48$

$x = \dfrac{-48}{-2.4} = 20$

11. $\frac{3}{8} x + 7 = -14$

$\frac{3}{8} x = -21$

$x = \frac{8}{3}(-21) = -56$

12. $-3 + 5x = 2x + 15$

$5x - 2x = 3 + 15$

$3x = 18$

$x = 6$

13. $3n - (4n - 2) = 7$

$3n - 4n + 2 = 7$

$-n = 5$

$n = -5$

14. $6y - 5(3y - 4) = 10$

$$6y - 15y + 20 = 10$$
$$-9y + 20 = 10$$
$$-9y = -10$$
$$y = \frac{10}{9}$$

15. $9c - \left[3 - 4(2 - c)\right] = 10$

$$9c - \left[3 - 8 + 4c\right] = 10$$
$$9c - \left[-5 + 4c\right] = 10$$
$$9c + 5 - 4c = 10$$
$$5c + 5 = 10$$
$$5c = 5$$
$$c = \frac{5}{5} = 1$$

16. $3x + y = 4 \quad (1)$

$6x - y = 5 \quad (2)$

Add equation (1) to equation (2) to eliminate the y-variable.

$$3x + y = 4 \quad (1)$$
$$\underline{6x - y = 5 \quad (2)}$$
$$9x = 9 \quad \text{Adding}$$
$$x = 1$$

Substitute 1 for x in either equation and solve for y. We choose equation (1).

$$3x + y = 4$$
$$3(1) + y = 4$$
$$3 + y = 4$$
$$y = 1$$

The solution is $(1, 1)$, which checks.

17. $4x + 4y = 4 \quad (1)$

$5x + 4y = 2 \quad (2)$

We can simplify equation (1) by multiplying each term by $\frac{1}{4}$. This gives the equivalent system:

$$x + y = 1 \quad (1)$$
$$5x + 4y = 2 \quad (2)$$

We can solve this system by substitution. Solve equation (1) for y.

$$x + y = 1 \qquad (1)$$
$$y = 1 - x \quad (3)$$

Substitute $1 - x$ for y into equation (2) and solve for x.

$$5x + 4y = 2 \quad (2)$$
$$5x + 4(1 - x) = 2$$
$$5x + 4 - 4x = 2$$
$$x + 4 = 2$$
$$x = -2$$

Substitute $x = -2$ into equation (3) to find y.

$$y = 1 - x \qquad (3)$$
$$y = 1 - (-2)$$
$$= 1 + 2$$
$$= 3$$

The solution is $(-2, 3)$, which checks.

18. $6x - 10y = -22 \quad (1)$

$-11x - 15y = 27 \quad (2)$

We can use elimination to solve this system. Multiply equation (1) by 11 and equation (2) by 6 and add the equations to eliminate the x-variable.

$$66x - 110y = -242 \quad \text{Multiply by 11}$$
$$\underline{-66x - 90y = 162 \quad \text{Multiply by 6}}$$
$$-200y = -80 \quad \text{Adding}$$
$$y = \frac{2}{5}$$

Substitute $\frac{2}{5}$ for y in either equation and solve for x. We choose equation (1).

$$6x - 10y = -22$$
$$6x - 10\left(\frac{2}{5}\right) = -22$$
$$6x - 4 = -22$$
$$6x = -18$$
$$x = -3$$

The solution is $\left(-3, \frac{2}{5}\right)$.

19. $x + y + z = -5 \quad (1)$

$2x + 3y - 2z = 8 \quad (2)$

$x - y + 4z = -21 \quad (3)$

We can use elimination to solve this system. Add -3 times equation (1) to equation (2) to eliminate the y variable.

$-3x-3y-3z=15$ Multiply by -3

$\underline{2x+3y-2z=8}$

$-x-5z=23$ (4)Adding

Adding equation (1) and (3) will also eliminate the y variable.

$x+y+z=-5$ (1)

$\underline{x-y+4z=-21}$ (3)

$2x+5z=-26$ (5)Adding

Next, add equation (4) to equation (5) to eliminate the z variable.

$-x-5z=23$

$\underline{2x+5z=-26}$

$x=-3$

Substitute -3 for x in equation (4) and solve for z.

$-x-5z=23$

$-(-3)-5z=23$

$3-5z=23$

$-5z=20$

$z=-4$

Finally substitute -3 for x and -4 for z in any of the original equations and solve for y. We choose equation (1).

$x+y+z=-5$

$-3+y+(-4)=-5$

$-7+y=-5$

$y=2$

The solution is $(-3,2,-4)$, which checks.

20. $2x+5y-3z=-11$ (1)

$-5x+3y-2z=-7$ (2)

$3x-2y+5z=12$ (3)

We use elimination to solve the system. Multiply equation (1) by 5 *and* multiply equation (2) by 2 and add them together to eliminate variable x.

$10x+25y-15z=-55$ Multiply by 5

$\underline{-10x+6y-4z=-14}$ Multiply by 2

$31y-19z=-69$ (4)Adding

Multiply equation (2) by 3 *and* multiply equation (3) by 5 and add them together to eliminate variable x.

$-15x+9y-6z=-21$ Multiply by 3

$\underline{15x-10y+25z=60}$ Multiply by 5

$-y+19z=39$ (5)Adding

Add equations (4) and (5) to eliminate variable z.

$31y-19z=-69$

$\underline{-y+19z=39}$

$30y=-30$ Adding

$y=-1$

Substitute -1 for y in equation (5) and solve for z.

$-y+19z=39$

$-(-1)+19z=39$

$1+19z=39$

$19z=38$

$z=2$

Substitute -1 for y and 2 for z in any of the original equations and solve for x. We choose equation (1).

$2x+5y-3z=-11$

$2x+5(-1)-3(2)=-11$

$2x-5-6=-11$

$2x-11=-11$

$2x=0$

$x=0$

The solution is $(0,-1,2)$, which checks.

21. Graph $f(x)=-2x+8$.

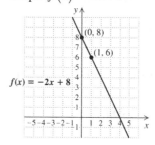

22. Graph $y=x^2-1$.

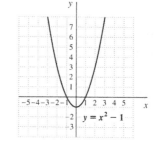

23. Graph $4x+16=0$.

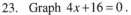

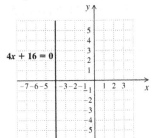

24. Graph $-3x+2y=6$.

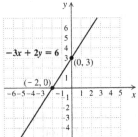

25. $-4y+9x=12$

Put the equation in slope-intercept form.
$-4y+9x=12$

$$-4y=-9x+12$$

$$y=\frac{9}{4}x-3$$

The slope is $\frac{9}{4}$ and the y-intercept is $(0,-3)$.

26. Points: $(2,7)$ and $(-1,3)$.

$$\text{Slope} = m = \frac{y_2-y_1}{x_2-x_1} = \frac{3-7}{-1-2} = \frac{-4}{-3} = \frac{4}{3}.$$

27. The line has slope 4 and contains the point $(2,-11)$.

Using the point-slope form, we have:
$$y-(-11)=4(x-2)$$
$$y+11=4x-8$$
$$y=4x-19$$

28. Points: $(-6,3)$ and $(4,2)$

First determine the slope.
$$m=\frac{y_2-y_1}{x_2-x_1}=\frac{2-3}{4-(-6)}=\frac{-1}{4+6}=-\frac{1}{10}$$

Using the formula for the slope-intercept form, we determine b. We could use either point; we choose $(4,2)$.

$$y=mx+b$$

$$2=-\frac{1}{10}(4)+b$$

$$2=-\frac{2}{5}+b$$

$$2\frac{2}{5}=\frac{12}{5}=b$$

The equation is $y=-\frac{1}{10}x+\frac{12}{5}$.

29. $\qquad 2x=4y+7$

$\qquad x-2y=5$

Determine the slopes of both lines.
For $2x=4y+7$:
$$2x-7=4y$$
$$\frac{1}{2}x-\frac{7}{4}=y$$

The slope of this line is $\frac{1}{2}$.

For $x-2y=5$
$$-2y=-x+5$$
$$y=\frac{1}{2}x-\frac{5}{2}$$

The slope of this line is $\frac{1}{2}$.

Since both lines have the same slope, $\frac{1}{2}$, they are parallel.

30. The line has a y-intercept of $(0,5)$ and is perpendicular to the line $x-2y=5$.
First, determine the slope of the line $x-2y=5$.
$$x-2y=5$$
$$-2y=-x+5$$
$$y=\frac{1}{2}x-\frac{5}{2}$$

The slope is $\frac{1}{2}$. Therefore, the line perpendicular to this line must have slope of $-\frac{1}{\left(\frac{1}{2}\right)}=-2$. The line has the equation
$$y=-2x+5.$$

31. The domain $= \{-5, -3, -1, 1, 3\}$ and the range
is $\{-3, -2, 1, 4, 5\}$. From the graph,
$f(-3) = -2$ and $f(x) = 5$ when $x = 3$.

32. $f(x) = \dfrac{7}{x+10}$

The rational function $f(x)$ is defined for all
values of x except those that make its
denominator 0. Find the value(s) of x that
make $x + 10 = 0$.

$x + 10 = 0$

$\qquad x = -10$

Thus, -10 is not in the domain of f.
The domain of f is
$\{x | x \text{ is a real number, } x \neq -10\}$.

33. $h(4) = -2(4)^2 + 1$
$\qquad = -2(16) + 1$
$\qquad = -32 + 1$
$\qquad = -31$

34. $-g(0) = -(4 \cdot 0 - 3) = -(-3) = 3$

35. $(g \cdot h)(-1) = g(-1) \cdot h(-1)$
$\qquad = \left[4(-1) - 3\right]\left[-2(-1)^2 + 1\right]$
$\qquad = \left[-4 - 3\right]\left[-2(1) + 1\right]$
$\qquad = \left[-7\right]\left[-1\right] = 7$

36. $(g - h)(a) = g(a) - h(a)$
$\qquad = (4a - 3) - \left[-2(a)^2 + 1\right]$
$\qquad = 4a - 3 + 2a^2 - 1$
$\qquad = 2a^2 + 4a - 4$

37. $\begin{vmatrix} 2 & -3 \\ 4 & 1 \end{vmatrix} = 2(1) - 4(-3) = 2 + 12 = 14$

38. $\begin{vmatrix} 1 & 0 & 1 \\ -1 & 2 & 1 \\ 2 & 1 & 3 \end{vmatrix}$

$= 1\begin{vmatrix} 2 & 1 \\ 1 & 3 \end{vmatrix} - (-1)\begin{vmatrix} 0 & 1 \\ 1 & 3 \end{vmatrix} + 2\begin{vmatrix} 0 & 1 \\ 2 & 1 \end{vmatrix}$

$= 1(6 - 1) + 1(0 - 1) + 2(0 - 2)$

$= 1(5) + 1(-1) + 2(-2)$

$= 5 - 1 - 4$

$= 0$

39. a) *Familiarize.* The problem gives the
formula for calculating the average out-
of-network ATM fee $f(t)$ charged by
banks based on the year since 2003, t.
Determine the correct value of t to use in
the formula to determine the average out-
of-network ATM fee in 2008.
Translate. $t = 2008 - 2003 = 5$. The
average out-of-network ATM fee in 2008
will be given by $f(5)$.
Carry out. $f(5) = 0.15(5) + 1.03$
$\qquad\qquad\qquad = 0.75 + 1.03 = 1.78$

Check.

1.78	$f(5) = 0.15(5) + 1.03$
1.78	$0.75 + 1.03$
1.78	1.78

The answer checks.
State. The average out-of-network ATM
fee in 2008 was $1.78.

b) The function is given in the slope-
intercept form of a line, where 0.15 is the
slope and 1.03 is the y-intercept. The
slope here represents the amount by
which the fee changes per year. That is,
the average fee increases by 15 cents per
year. 1.03 represents the fee when $t = 0$,
that is, the fee in 2003.

40. a) *Familiarize.* The problem gives the
number of ATMs in the United States in
the years 1996 and 2008 and says to let
$t =$ the number of years after 1992 and
$A(t) =$ the number of ATMs, in
thousands, in the United States in year t.
Translate. Assuming the number of
ATMs increases linearly, then the values
of 139,000 ATMs in 1996 and 400,000
ATMs in 2008 correspond to points
$(t, A(t))$, $(1996 - 1992, 139) = (4, 139)$
and $(2008 - 1992, 400) = (16, 400)$.

Carry out. Use the points $(4, 139)$ and $(16, 400)$ to determine the slope of the linear function.

$$m = \frac{A(t_2) - A(t_1)}{t_2 - t_1} = \frac{400 - 139}{16 - 4} = \frac{261}{12} = \frac{87}{4}.$$

Use one of the points to determine the y-intercept b.

$$139 = \frac{87}{4}(4) + b$$

$$139 = 87 + b$$

$$b = 139 - 87 = 52$$

Therefore $A(t) = \frac{87}{4}t + 52$.

Check.

$$A(4) = \frac{87}{4} \cdot 4 + 52 = 139$$

$$A(16) = \frac{87}{4} \cdot 16 + 52 = 87 \cdot 4 + 52 = 400$$

The linear function $A(t)$ passes through the points $(4, 139)$ and $(16, 400)$ as required.

State. The number of ATMs in the United States can be modeled by the linear function $A(t) = \frac{87}{4}t + 139$, where $t =$ the number of years since 1992 and $A(t)$ is in thousands.

b) In 2014, $t = 2014 - 1992 = 22$.

$$A(22) = \frac{87}{4}(22) + 52 = 530.5. \text{ Therefore,}$$

the expected number of ATMs in the U.S. in 2014 is 530,500.

c) To determine the year when the number of ATMs will equal 500,000, set $A(t) = 500$ and solve for t.

$$A(t) = 500$$

$$\frac{87}{4}t + 52 = 500$$

$$\frac{87}{4}t = 448$$

$$t = \frac{4}{87} \cdot 448 \approx 20.6$$

$1992 + 20.6 = 2012.6$

The number of ATMs in the U.S. is predicted to reach 500,000 in the year 2012, or by 2013.

41. ***Familiarize.*** The problem tells us there are two types of one-year IRE memberships, professional and student, and the cost of each type. It then tells us the total amount of money collected for both types and the total number of memberships and asks us to find how many of each kind there are.

Translate. Let $p =$ the number of one-year professional memberships, and $s =$ the number of one-year student memberships. Then based on the information given, we can build the following system of equations:

$$p + s = 150 \quad (1)$$

$$60p + 25s = 6130 \quad (2)$$

Carry out. We can solve this system by substitution. Solve equation (1) for p.

$$p + s = 150 \quad (1)$$

$$p = 150 - s \quad (3)$$

Substitute $150 - s$ for p into equation (2) and solve for s.

$$60p + 25s = 6130 \quad\quad (2)$$

$$60(150 - s) + 25s = 6130$$

$$9000 - 60s + 25s = 6130$$

$$9000 - 35s = 6130$$

$$-35s = -2870$$

$$s = \frac{-2870}{-35} = 82$$

Substitute $s = 82$ into equation (3) to find p.

$$p = 150 - s \quad (3)$$

$$p = 150 - 82$$

$$= 68$$

Check. $68 + 82 = 150$

$$68 \cdot 60 + 82 \cdot 25 = 4080 + 2050$$

$$= 6130$$

The values for p and s are solutions to both equations (1) and (2).

State. 68 professional and 82 student one-year IRE memberships were sold.

42. **Familiarize.** Let x = the number of ounces of "Sea Spray" needed, and y = the number of ounces of "Ocean Mist" needed, to create the 50% mixture.

Translate.

We can organize this information in a table.

	"Sea Spray"	"Ocean Mist"	Total
No. of ounces	x	y	120
% salt	25%	65%	50%
Amt. of salt	$0.25x$	$0.65y$	$0.50 \cdot 120$ oz $= 60$ oz

From the second row of the table, we obtain the equation

$x + y = 120$.

From the last row of the table, we obtain the equation

$0.25x + 0.65y = 60$.

After clearing decimals, we obtain the system:

$$x + y = 120 \quad (1)$$
$$25x + 65y = 6000 \quad (2)$$

Dividing each term in equation (2) by the common factor of 5, we obtain the new system:

$$x + y = 120 \quad (1)$$
$$5x + 13y = 1200 \quad (2)$$

Carry out.

We can use substitution to solve the system. Solve equation (1) for x.

$x + y = 120$

$x = 120 - y$

Substitute $120 - y$ for x in equation (2) and solve for y.

$$5x + 13y = 1200$$
$$5(120 - y) + 13y = 1200$$
$$600 - 5y + 13y = 1200$$
$$600 + 8y = 1200$$
$$8y = 600$$
$$y = 75$$

Substitute 75 for y into equation (1) and solve for x.

$x + y = 120$

$x + 75 = 120$

$x = 45$

Check. The total number of ounces in the mixture is 45 oz $+ 75$ oz $= 120$ oz. The total amount of salt in the mixture is

$0.25(45$ oz$) + 0.65(75$ oz$)$

$= 11.25$ oz $+ 48.75$ oz $= 60$ oz.

The values check.

State. The mixture should contain 45 oz of "Sea Spray" and 75 oz of "Ocean Mist".

43. **Familiarize.** The problem states that we are to find three consecutive odd numbers that satisfy the condition stated. Consecutive odd numbers are 2 units away from each other on a number line.

Translate. Let x = the smallest number. Then $x + 2$ = the middle number, and $x + 2 + 2$ $= x + 4$ = the largest number. Therefore, "five times the third" $= 5(x + 4) = 5x + 20$, and "three times the first" $= 3x$. Their difference is $5x + 20 - 3x = 2x + 20$. Therefore,

$2x + 20 = 54$.

Carry out. $2x + 20 = 54$

$$2x = 34$$
$$x = \frac{34}{2} = 17,$$
$$x + 2 = 17 + 2 = 19$$
$$x + 4 = 17 + 4 = 21$$

Check. $5 \cdot 21 - 3 \cdot 17 = 105 - 51 = 54$

The values check

State. The three numbers are 17, 19, and 21.

44. **Familiarize.** The perimeter of a rectangular field, and a numerical relationship between the field's length and width, are given. Use the given information and the formula for the perimeter of a rectangle to find the dimensions of the field.

Translate. Let w = the width of the field in yds. Then $l = w + 40$ = the length of the field in yds. The perimeter of a rectangle is given by $P = 2w + 2l$. Therefore,

$$320 = 2w + 2l$$
$$= 2w + 2(w + 40)$$
$$= 2w + 2w + 80$$
$$= 4w + 80$$

Carry out.

$$320 = 4w + 80$$
$$240 = 4w$$
$$w = \frac{240}{4} = 60$$
$$l = w + 40 = 60 + 40 = 100$$

Check.

$$100 - 60 = 40$$

$$2 \cdot 100 + 2 \cdot 60 = 200 + 120 = 320$$

These values check.

State. The rectangular field is 100 yd long and 60 yd wide.

45. *Familiarize.* The problem states there are three types of tickets (child, adult, and senior) and the cost of each type. It then states a relationship between the number of adult, senior and child tickets sold, and a second relationship between the number of adult and senior tickets sold. Finally, the problem states the total amount of money collected for all three types of tickets.

Translate. Let c = the number of child tickets sold, a = the number of adult tickets sold, and s = the number of senior tickets sold. Then based on the information given, we can build the following system of equations:

$$a + s = c + 30 \qquad (1)$$

$$a = 4s + 6 \qquad (2)$$

$$5.50a + 4.00s + 1.50c = 11,219.50 \quad (3)$$

Carry out. This system can be solved in several ways, but note that equation (2) defines a in terms of s only. Therefore, equations (1) and (2) can be used to determine c in terms of s only as well. With the result, equation (3) can be rewritten in terms of s only and solved for s. a and c can then be determined using back-substitution. Use equations (1) and (2) to solve for c in terms of s:

$$a + s = c + 30 \quad (1)$$

$$(4s + 6) + s = c + 30 \quad \text{Use equation (2)}$$

$$5s + 6 = c + 30$$

$$5s - 24 = c$$

$$c = 5s - 24 \quad (4)$$

Use equations (2) and (4) to rewrite (3) in terms of s, and solve for s.

$$5.50a + 4.00s + 1.50c = 11,219.50$$

$$5.50(4s + 6) + 4.00s + 1.50(5s - 24) = 11,219.50$$

$$22s + 33 + 4s + 7.5s - 36 = 11,219.5$$

$$33.5s - 3 = 11,219.5$$

$$33.5s = 11,222.5$$

$$s = \frac{11,222.5}{33.5}$$

$$= 335$$

Use equation (2) to find a:

$$a = 4s + 6 = 4 \cdot 335 + 6 = 1346$$

Use equation (4) to find c:

$$c = 5s - 24 = 5 \cdot 335 - 24 = 1651$$

Check.

$$1346 + 335 = 1651 + 30 \quad (1)$$

$$1681 = 1681$$

$$1346 = 4 \cdot 335 + 6 \quad (2)$$

$$1346 = 1340 + 6$$

$$1346 = 1346$$

$$5.50 \cdot 1346 + 4.00 \cdot 335 + 1.50 \cdot 1651 = 11,219.50 \quad (3)$$

$$7403 + 1340 + 2476.5 = 11,219.5$$

$$11,219.5 = 11,219.5$$

All three equations check.

State. 1651 child tickets, 1346 adult tickets and 335 senior tickets were sold.

46. *Familiarize.* Let x = the score on Franco's fifth test.

Translate.

The average of Franco's exams is 90.

$$\frac{93 + 85 + 100 + 86 + x}{5} = 90$$

Carry out.

$$\frac{93 + 85 + 100 + 86 + x}{5} = 90$$

$$93 + 85 + 100 + 86 + x = 5 \cdot 90$$

$$364 + x = 450$$

$$x = 450 - 364 = 86$$

Check.

$$90 = \frac{93 + 85 + 100 + 86 + 86}{5}$$

$$90 = \frac{450}{5}$$

$$90 = 90$$

The answer checks.

State. Franco will need to get an 86 on his fifth exam to have an exam average of 90.

47. $\left(6x^{a+2}y^{b+2}\right)\left(-2x^{a-2}y^{y+1}\right)$

$$= 6(-2)x^{(a+2)+(a-2)}y^{(b+2)+(y+1)}$$

$$= -12x^{2a}y^{b+y+3}$$

48. Let r = the amount spent on radio advertising and S = the amount of sales increase when r dollars are spent. We have two points of the form (r, S), namely $(\$1000, \$101,000)$ and $(\$1250, \$126,000)$.

    First determine the rate of change of sales per amount spent on radio advertising.

    $$\text{Slope} = m = \frac{S_2 - S_1}{r_2 - r_1}$$

    $$= \frac{\$126,000 - \$101,000}{\$1250 - \$1000}$$

    $$= \frac{\$25,000}{\$250} = 100$$

    Using the point-slope form we have

    $$S - 101,000 = m(r - 1000)$$

    $$S - 101,000 = 100(r - 1000)$$

    $$S - 101,000 = 100r - 100,000$$

    $$S = 100r + 1000$$

    Using function notation, we have

    $$S(r) = 100r + 1000.$$

    When $\$1500$ is spent on radio advertising, the predicted sales, assuming a linear relationship between sales and advertising, would be

    $$S(1500) = 100(1500) + 1000$$

    $$= 150000 + 1000$$

    $$= \$151,000$$

49. Given $f(x) = mx + b$, we know that

    $$f(5) = m(5) + b = 5m + b$$

    and

    $$f(-4) = m(-4) + b = -4m + b.$$

    Now we have a system of equations.

    $$5m + b = -3 \quad (1)$$

    $$-4m + b = 2 \quad (2)$$

    We solve this system by elimination. Add -1 times equation (1) to equation (2) to eliminate the b variable.

    $$\begin{array}{ll} -5m - b = 3 & \text{Multiply by } -1 \\ -4m + b = 2 & \\ \hline -9m = 5 & \text{Adding} \end{array}$$

    $$m = -\frac{5}{9}$$

Substituting $-\frac{5}{9}$ for m in either equation (1) or equation (2), we solve for b. We choose equation (1).

$$5m + b = -3$$

$$5\left(-\frac{5}{9}\right) + b = -3$$

$$-\frac{25}{9} + b = -3$$

$$b = -3 + \frac{25}{9}$$

$$b = -\frac{27}{9} + \frac{25}{9}$$

$$b = -\frac{2}{9}$$

We find that $m = -\frac{5}{9}$ and $b = -\frac{2}{9}$.

Chapter 4

Inequalities

1. $\geq$

3. $<$

5. $<$

7. Equivalent

9. Equivalent

11. $x - 4 \geq 1$
    -4: We substitute and get $-4 - 4 \geq 1$, or $-8 \geq 1$, a false statement. Therefore, -4 is not a solution.
    4: We substitute and get $4 - 4 \geq 1$, or $0 \geq 1$, a false statement. Therefore, 0 is not a solution.
    5: We substitute and get $5 - 4 \geq 1$, or $1 \geq 1$, a true statement. Therefore, 5 is a solution.
    8: We substitute and get $8 - 4 \geq 1$, or $4 \geq 1$, a true statement. Therefore 8 is a solution.

13. $2y + 3 < 6 - y$
    0: We substitute and get $2(0) + 3 < 6 - 0$, or $3 > 6$, a true statement. Therefore, 0 is a solution.
    1: We substitute and get $2(1) + 3 < 6 - 1$, or $5 < 5$, a false statement Therefore, 1 is not a solution.
    -1: We substitute and get $2(-1) + 3 < 6 - (-1)$, or $1 < 7$, a true statement. Therefore, -1 is a solution.
    4: We substitute and get $2(4) + 3 < 6 - 4$, or $11 < 2$, a false statement. Therefore, 4 is not a solution.

15. $y < 6$
    Graph: The solutions consist of all real numbers less than 6, so we shade all numbers to the left of 6 and use a parenthesis at 6 to indicate that it is not a solution.

    Set builder notation: $\{y \mid y < 6\}$
    Interval notation: $(-\infty, 6)$

17. $x \geq -4$
    Graph: We shade all numbers to the right of -4 and use a bracket at -4 to indicate that it is also a solution.

    Set builder notation: $\{x \mid x \geq -4\}$
    Interval notation: $[-4, \infty)$

19. $t > -3$
    Graph: We shade all numbers to the right of -3 and use a parenthesis at -3 to indicate that it is not a solution.

    Set builder notation: $\{t \mid t > -3\}$
    Interval notation: $(-3, \infty)$

21. $x \leq -7$
    Graph: We shade all numbers to the left of -7 and use a bracket at -7 to indicate that it is also a solution.

    Set builder notation: $\{x \mid x \leq -7\}$
    Interval notation: $(-\infty, -7]$

23. $x + 2 > 1$
    $x + 2 + (-2) > 1 + (-2)$ Adding -2
    $x > -1$
    The solution set is $\{x \mid x > -1\}$, or $(-1, \infty)$.

25. $t - 6 \leq 4$
    $t - 6 + 6 \leq 4 + 6$ Adding 6
    $t \leq 10$
    The solution set is $\{t \mid t \leq 10\}$, or $(-\infty, 10]$.

27. $x - 12 \geq -11$

   $x - 12 + 12 \geq -11 + 12$ Adding 12

            $x \geq 1$

   The solution set is $\{x \mid x \geq 1\}$, or $[1, \infty)$.

29. $9t < -81$

   $\frac{1}{9} \cdot 9t < \frac{1}{9}(-81)$ Multiplying by $\frac{1}{9}$

        $t < -9$

   The solution set is $\{t \mid t < -9\}$, or $(-\infty, -9)$.

31. $-0.3x > -15$

   $-\frac{1}{0.3}(-0.3x) < -\frac{1}{0.3}(-15)$ Multiplying by $-\frac{1}{0.3}$

                        and reversing the

                        inequality symbol

            $x < 50$

   The solution set is $\{x \mid x < 50\}$, or $(-\infty, 50)$.

33. $-9x \geq 8.1$

   $-\frac{1}{9}(-9x) \leq -\frac{1}{9}(8.1)$ Multiplying by $-\frac{1}{9}$ and

                        reversing the inequality

                        symbol

            $x \leq -0.9$

   The solution set is $\{x \mid x \leq -0.9\}$, or

   $(-\infty, -0.9]$.

35. $\frac{3}{4}y \geq -\frac{5}{8}$

   $\frac{4}{3}\left(\frac{3}{4}y\right) \geq \frac{4}{3}\left(-\frac{5}{8}\right)$ Multiplying by $\frac{4}{3}$

            $y \geq -\frac{5}{6}$

   The solution set is

   $\left\{y \,\middle|\, y \geq -\frac{5}{6}\right\}$, or $\left[-\frac{5}{6}, \infty\right)$.

37. $3x + 1 < 7$

      $3x < 6$ Adding -1

       $x < 2$ Dividing by 3

   The solution set is $\{x \mid x < 2\}$, or $(-\infty, 2)$.

39. $3 - x \geq 12$

      $-x \geq 9$ Adding -3

       $x \leq -9$ Dividing by -1 and reversing

                   the inequality symbol

   The solution set is $\{x \mid x \leq -9\}$, or $(-\infty, -9]$.

41. $\frac{2x + 7}{5} < -9$

   $5 \cdot \frac{2x + 7}{5} < 5(-9)$ Multiplying by 5

      $2x + 7 < -45$

       $2x < -52$ Adding -7

        $x < -26$ Dividing by 2

   The solution set is $\left\{x \mid x < -26\right\}$, or

   $(-\infty, -26)$.

43. $\frac{3t - 7}{-4} \leq 5$

   $-4 \cdot \frac{3t - 7}{-4} \geq -4 \cdot 5$ Multiplying by -4 and

                        reversing the inequality

                        symbol

      $3t - 7 \geq -20$

       $3t \geq -13$ Adding 7

        $t \geq -\frac{13}{3}$ Dividing by 3

   The solution set is $\left\{t \mid t \geq -\frac{13}{3}\right\}$, or $\left[-\frac{13}{3}, \infty\right)$.

45. $\dfrac{9-x}{-2} \geq -6$

    $9-x \leq 12$ Multipling by -2 and reversing
                     the inequality symbol

     $-x \leq 3$ Adding -9

      $x \geq -3$ Multiplying by -1 and reversing
                  the inequality symbol

The solution set is $\{x \mid x \geq -3\}$, or $[-3, \infty)$.

47. $f(x) = 7 - 3x$, $g(x) = 2x - 3$

       $f(x) \leq g(x)$

     $7 - 3x \leq 2x - 3$

   $7 - 5x \leq -3$ Adding $-2x$

    $-5x \leq -10$ Adding -7

         $x \geq 2$ Multiplying by $-\dfrac{1}{5}$

                   and reversing the
                   inequality symbol

The solution set is $\{x \mid x \geq 2\}$, or $[2, \infty)$.

49. $f(x) = 2x - 7$, $g(x) = 5x - 9$

       $f(x) < g(x)$

     $2x - 7 < 5x - 9$

   $-3x - 7 < -9$ Adding $-5x$

    $-3x < -2$ Adding 7

       $x > \dfrac{2}{3}$ Dividing by -3

The solution set is $\left\{x \mid x > \dfrac{2}{3}\right\}$, or $\left(\dfrac{2}{3}, \infty\right)$.

51. $y_1 = \dfrac{3}{8} + 2x$, $y_2 = 3x - \dfrac{1}{8}$

        $y_2 \geq y_1$

   $3x - \dfrac{1}{8} \geq \dfrac{3}{8} + 2x$

    $x - \dfrac{1}{8} \geq \dfrac{3}{8}$ Adding $-2x$

      $x \geq \dfrac{1}{2}$ Adding $\dfrac{1}{8}$

The solution set is $\left\{x \mid x \geq \dfrac{1}{2}\right\}$, or $\left[\dfrac{1}{2}, \infty\right)$.

53. $3 - 8y \geq 9 - 4y$

    $-4y + 3 \geq 9$

      $-4y \geq 6$

        $y \leq -\dfrac{3}{2}$

The solution set is

$\left\{y \,\middle|\, y \leq -\dfrac{3}{2}\right\}$, or $\left(-\infty, -\dfrac{3}{2}\right]$.

55. $5(t - 3) + 4t < 2(7 + 2t)$

   $5t - 15 + 4t < 14 + 4t$

       $9t - 15 < 14 + 4t$

       $5t - 15 < 14$

          $5t < 29$

           $t < \dfrac{29}{5}$

The solution set is $\left\{t \,\middle|\, t < \dfrac{29}{5}\right\}$, or $\left(-\infty, \dfrac{29}{5}\right)$.

57. $5[3m - (m + 4)] > -2(m - 4)$

   $5(3m - m - 4) > -2(m - 4)$

     $5(2m - 4) > -2(m - 4)$

     $10m - 20 > -2m + 8$

      $12m - 20 > 8$

        $12m > 28$

          $m > \dfrac{28}{12}$

          $m > \dfrac{7}{3}$

The solution set is $\left\{m \,\middle|\, m > \dfrac{7}{3}\right\}$, or $\left(\dfrac{7}{3}, \infty\right)$.

59. $19 - (2x + 3) \leq 2(x + 3) + x$

   $19 - 2x - 3 \leq 2x + 6 + x$

      $16 - 2x \leq 3x + 6$

      $16 - 5x \leq 6$

        $-5x \leq -10$

          $x \geq 2$

The solution set is $\{x \mid x \geq 2\}$, or $[2, \infty)$.

61. $\frac{1}{4}(8y+4)-17 < -\frac{1}{2}(4y-8)$

$2y+1-17 < -2y+4$

$2y-16 < -2y+4$

$4y-16 < 4$

$4y < 20$

$y < 5$

The solution set is $\{y \mid y < 5\}$, or $(-\infty, 5)$.

63. $2\left[8-4(3-x)\right]-2 \geq 8\left[2(4x-3)+7\right]-50$

$2[8-12+4x]-2 \geq 8[8x-6+7]-50$

$2[-4+4x]-2 \geq 8[8x+1]-50$

$-8+8x-2 \geq 64x+8-50$

$8x-10 \geq 64x-42$

$-56x-10 \geq -42$

$-56x \geq -32$

$x \leq \frac{32}{56}$

$x \leq \frac{4}{7}$

The solution set is $\left\{x \mid x \leq \frac{4}{7}\right\}$, or $\left(-\infty, \frac{4}{7}\right]$.

65. Let n represent the number. Then we have $n < 10$.

67. Let t represent the temperature. Then we have $t \leq -3$.

69. Let a represent the age of the Mayan altar. Then we have $a > 1200$.

71. Let d represent the distance to Normandale Community College. Then we have $d \leq 15$.

73. Let d represent the number of years of driving experience. Then we have $d \geq 5$.

75. Let c represent the cost of production. Then we have $c \leq 12,500$.

77. ***Familiarize***. Let n = the number of hours. Then the total fee using the hourly plan is $120n$.

Translate. We write an inequality stating that the hourly plan costs less than the flat fee.

$120n < 900$

Carry out.

$120n < 900$

$n < \frac{15}{2}$, or $7\frac{1}{2}$

Check. We can do a partial check by substituting a value for n less than $\frac{15}{2}$. When $n = 7$, the hourly plan costs $120(7)$, or \$840, so the hourly plan is less than the flat fee of \$900. When $n = 8$, the hourly plan costs $120(8)$, or \$960, so the hourly plan is more expensive than the flat fee of \$900.

State. The hourly rate is less expensive for lengths of time less than $7\frac{1}{2}$ hours.

79. ***Familiarize***. Let n = Chloe's grade point average. An unconditional acceptance is given to students whose GMAT score plus 200 times the undergraduate grade point average is at least 950

Translate.

GMAT score	plus	200 times the undergrad GPA		at least 950	
↓	↓	↓		↓	↓
500	+	200n		≥	950

Carry out. We solve the inequality.

$500+200n \geq 950$

$200n \geq 450$

$n \geq 2.25$

Check. As a partial check, we can determine the score for a grade point average of 2.

$500+200(2) = 500+400 = 900$. 900 is less than the 950 score required, so it appears that $n \geq 2.25$ is correct.

State. Chloe must earn at least a 2.25 grade points average to an unconditional acceptance into the Master of Business Administration (MBA) program at Arkansas State University.

81. **Familiarize.** Let n = the number of correct answers. Then the points earned are $2n$, and the points deducted are $\frac{1}{2}$ of the rest of the questions, $80 - n$, or $\frac{1}{2}(80 - n)$.

Translate. We write an inequality stating the score is at least 100.

$$2n - \frac{1}{2}(80 - n) \geq 100.$$

Carry out.

$$2n - \frac{1}{2}(80 - n) \geq 100$$

$$2n - 40 + \frac{1}{2}n \geq 100$$

$$\frac{5}{2}n - 40 \geq 100$$

$$\frac{5}{2}n \geq 140$$

$$n \geq 56$$

Check. When $n = 56$, the score earned is $2(56) - \frac{1}{2}(80 - 56)$, or $112 - \frac{1}{2}(24)$, or $112 - 12$, or 100. When $n = 58$, the score earned is $2(58) - \frac{1}{2}(80 - 58)$, or $116 - \frac{1}{2}(22)$, or $116 - 11$, or 105. Since the score is exactly 100 when 56 questions are answered correctly and more than 100 when 58 questions are correct, we have performed a partial check.

State. At least 56 questions are correct for a score of at least 100.

83. **Familiarize.** Let d = the depth of the well. Under the "pay-as-you-go" plan, the charge is $\$500 + \$8d$. Under the "guaranteed-water" plan, the charge is $\$4000$.

Translate.

Pay-as-you-go plan	is less than	the guaranteed-water plan.
↓	↓	↓
$500 + 8d$	$<$	4000

Carry out. We solve the inequality.

$$500 + 8d < 4000$$

$$8d < 3500$$

$$d < 437.50$$

Check. We compute the cost of the well under the "pay-as-you-go" plan for various depths. If $d = 437$ ft, the cost is

$\$500 + \$8 \cdot 437 = \$3996$. If $d = 438$ ft, the cost is $\$500 + \$8 \cdot 438 = \$4004$.

85. **Familiarize.** We list the given information in a table.

Plan A: Monthly Income	Plan B: Monthly Income
$400 salary	$610
8% of sales	5% of sales
Total: 400+8% of sales	Total: 610+5% of sales

Suppose Toni had gross sales of $5000 one month. Then under plan A she would earn $\$400 + 0.08(\$5000)$, or $\$800$.

Under plan B she would earn $\$610 + 0.05(\$5000)$, or $\$860$.

This shows that, for gross sales of $5000, plan B is better. If Toni had gross sales of $10,000 one month, then under plan A she would earn $\$400 + 0.08(\$10,000)$, or $\$1200$.

Under plan B she would earn $\$610 + 0.05(\$10,000)$, or $\$1110$.

This shows that, for gross sales of $10,000, plan A is better. To determine all values for which plan A is better we solve an inequality.

Translate.

Income from plan A	is greater than	Income from plan B.
↓	↓	↓
$400 + 0.08s$	$>$	$610 + 0.05s$

Carry out.

$$400 + 0.08s > 610 + 0.05s$$

$$400 + 0.03s > 610$$

$$0.03s > 210$$

$$s > 7000$$

Check. For $s = \$7000$, the income from plan A is $\$400 + 0.08(\$7000)$, or $\$960$ and the income from plan B is $\$610 + 0.05(\$7000)$, or $\$960$.

This shows that for sales of $7000 Toni's income is the same from each plan. In the Familiarize step we show that, for a value less than $7000, plan B is better and, for a value greater than $7000, plan A is better. Since we cannot check all possible values, we stop here.

State. Toni should select plan A for gross sales greater than $7000.

87. *Familiarize*. Let m = the medical bill. Then the "Green Badge" medical insurance plan will cost $2000 + 0.30(\$m - \$2000)$ and the "Blue Seal" plan will cost $2500 + 0.20(\$m - \$2500)$.

Translate. We write an inequality stating that the "Blue Seal" plan costs less than the "Green Badge" plan.

$2000 + 0.30(m - 2000) > 2500 + 0.20(m - 2500)$

Carry out.

$$2000 + 0.30(m - 2000) > 2500 + 0.20(m - 2500)$$
$$2000 + 0.3m - 600 > 2500 + 0.2m - 500$$
$$1400 + 0.3m > 2000 + 0.2m$$
$$0.1m > 600$$
$$m > 6000$$

Check. When $m = 5000$, the "Green Badge" plan cost is $2000 + 0.30(\$5000 - \$2000)$, or $2000 + 0.30(\$3000)$, or $2900 and the "Blue Seal" plan cost is $2500 + 0.20(\$5000 - \$2500)$, or $2500 + 0.20(\$2500)$, or $3000. So the "Green Badge" plan is less expensive. When $m = 7000$, the "Green Badge" plan cost is $2000 + 0.30(\$7000 - \$2000)$, or $2000 + 0.30(\$5000)$, or $3500 and the "Blue Seal" plan is $2500 + 0.20(\$7000 - \$2500)$, or $2500 + 0.20(\$4500)$, or $3400. So the "Blue Seal" plan is less expensive.

State. For medical bills of more than $6000, the "Blue Seal" plan is less expensive.

89. *Familiarize*. Find the values of t for which $C(t) < 1750$.

Translate. $-40.5t + 2159 < 1750$

Carry out.

$$-40.5t + 2159 < 1750$$
$$-40.5t < -409$$
$$t > \frac{818}{81}, \text{ or } 10\frac{8}{81}$$

Check. $C\left(\dfrac{818}{81}\right) = 1750$.

When $t = 10$,

$C(10) = -40.5(10) + 2159$, or 1754.

When $t = 11$,

$C(11) = -40.5(11) + 2159$, or 1713.5.

State. The domestic production will drop below 1750 million barrels later than 10 years after 2000, or the years after 2010.

91. a) *Familiarize*. Find the values of d for which $F(d) > 25$.

Translate.

$$\left(\frac{4.95}{d} - 4.50\right) \times 100 > 25$$

Carry out.

$$\left(\frac{4.95}{d} - 4.50\right) \times 100 > 25$$
$$\frac{495}{d} - 450 > 25$$
$$\frac{495}{d} > 475$$
$$495 > 475d$$
$$\frac{495}{475} > d$$
$$\frac{99}{95} > d, \text{ or } d < 1.04$$

Check. When $d = 1$,

$$F(1) = \left(\frac{4.95}{1} - 4.50\right) \times 100, \text{ or } 45 \text{ percent}.$$

When $d = 1.05$,

$$F(1.05) = \left(\frac{4.95}{1.05} - 4.50\right) \times 100, \text{ or } 21$$

percent.

State. A man is considered obese for body density less than $\dfrac{99}{95}$ kg/L, or about 1.04 kg/L.

b) *Familiarize*. Find the values of d for which $F(d) > 32$.

Translate.

$$\left(\frac{4.95}{d} - 4.50\right) \times 100 > 32$$

Carry out.

$$\left(\frac{4.95}{d} - 4.50\right) \times 100 > 32$$
$$\frac{495}{d} - 450 > 32$$
$$\frac{495}{d} > 482$$
$$495 > 482d$$
$$\frac{495}{482} > d, \text{ or } d < 1.03$$

Check. Our check from part (a) leads to the result that $F(d) > 32$ when $d < 1.03$.

State. A woman is considered obese for body density less than $\frac{495}{482}$ kg/L, or about 1.03 kg/L.

93. a) *Familiarize*. Find the values of x for which $R(x) < C(x)$.

    Translate.
    $48x < 90{,}000 + 25x$

    Carry out.
    $23x < 90{,}000$
    $x < 3913\frac{1}{23}$

    Check.
    $R\left(3913\frac{1}{23}\right) = \$187{,}826.09 = C\left(3913\frac{1}{23}\right)$.

    Calculate $R(x)$ and $C(x)$ for some x greater than $3913\frac{1}{23}$ and for some x less than $3913\frac{1}{23}$.

    Suppose $x = 4000$:
    $\quad R(x) = 48(4000) = 192{,}000$ and
    $\quad C(x) = 90{,}000 + 25(4000) = 190{,}000$.

    In this case $R(x) > C(x)$.

    Suppose $x = 3900$:
    $R(x) = 48(3900) = 187{,}200$ and
    $C(x) = 90{,}000 + 25(3900) = 187{,}500$.

    In this case $R(x) < C(x)$.

    Then for $x < 3913\frac{1}{23}$, $R(x) < C(x)$.

    State. We will state the result in terms of integers, since the company cannot sell a fraction of a lamp. For 3913 or fewer lamps the company loses money.

    b) Our check in part a) shows that for $x > 3913\frac{1}{23}$, $R(x) > C(x)$ and the company makes a profit. Again, we will state the result in terms of an integer. For more than 3913 lamps the company makes money.

95. *Thinking and Writing Exercise.*

97. $y = 2x - 3$

    Slope: 2; y-intercept: $(0, -3)$

    We begin by plotting the y-intercept $(0, -3)$ and thinking of the slope as $\frac{2}{1}$, we move up 2 units and right 1 unit to the point $(1, -1)$.

    Thinking of the slope as $\frac{2}{1}$ we can begin at $(1, -1)$ and move up 2 units and right 1 unit to

the point $(2, 1)$. We finish by connecting these points to form a line.

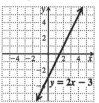

99. $y = 2$

    We can write this equation as $0 \cdot x + y = 2$. No matter what number we choose for x, we find that y must be 2.

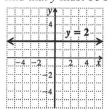

101. $f(x) = -\frac{2}{3}x + 1 \rightarrow y = -\frac{2}{3}x + 1$

    Slope: $-\frac{2}{3}$; y-intercept: $(0, 1)$

    We begin by plotting the y-intercept $(0, 1)$ and thinking of the slope as $\frac{-2}{3}$, we move down 2 units and right 3 units to the point $(3, -1)$. Thinking of the slope as $\frac{2}{-3}$ we can begin at $(0, 1)$ and move up 2 units and left 3 units to the point $(-3, 3)$. We finish by connecting these points to form a line.

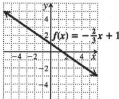

103. *Thinking and Writing Exercise.*

105. $3ax + 2x \ge 5ax - 4$
    $2x - 2ax \ge -4$
    $2x(1 - a) \ge -4$
    $x(1 - a) \ge -2$
    $x \le -\frac{2}{1 - a}, \text{ or } \frac{2}{a - 1}$

    We reversed the inequality symbol when we

divided because when $a > 1$, then $1 - a < 0$.

The solution set is $\left\{ x \,\middle|\, x \leq \dfrac{2}{a-1} \right\}$.

107.　$a(by - 2) \geq b(2y + 5)$

$aby - 2a \geq 2by + 5b$

$aby - 2by \geq 2a + 5b$

$y(ab - 2b) \geq 2a + 5b$

$\qquad y \geq \dfrac{2a + 5b}{ab - 2b}, \text{ or } \dfrac{2a + 5b}{b(a-2)}$

The inequality symbol remained unchanged when we divided because when $a > 2$ and $b > 0$, then
$$ab - 2b > 0.$$

The solution set is $\left\{ y \,\middle|\, y \geq \dfrac{2a + 5b}{b(a-2)} \right\}$.

109.　$c(2 - 5x) + dx > m(4 + 2x)$

$2c - 5cx + dx > 4m + 2mx$

$-5cx + dx - 2mx > 4m - 2c$

$x(-5c + d - 2m) > 4m - 2c$

$x[d - (5c + 2m)] > 4m - 2c$

$\qquad x > \dfrac{4m - 2c}{d - (5c + 2m)}$

The inequality symbol remained unchanged when we divided because when $5c + 2m < d$, then $d - (5c + 2m) > 0$.

The solution set is $\left\{ x \,\middle|\, x > \dfrac{4m - 2c}{d - (5c + 2m)} \right\}$.

111. False. If $a = 2$, $b = 3$, $c = 4$, and $d = 5$, then $2 < 3$ and $4 < 5$ but $2 - 4 = 3 - 5$.

113. *Thinking and Writing Exercise.*

115. $x + 5 \leq 5 + x$

$\qquad 5 \leq 5$

We get an inequality that is true for all real numbers x. Thus the solution set is all real numbers.

117. $0^2 = 0$, $x^2 > 0$ for $x \neq 0$

The solution is

$\{ x \mid x \text{ is a real number } and \ x \neq 0 \}$.

119. From Exercise 118, the first option, prepayment, was more economical if more than 6.8 gal of gas was used. If the car gets 30 mph, then Abriana would need to drive 6.8 mi × 30 mpg, or about 204 miles to make the first option more economical.

Exercise Set 4.2

1.　e

3.　f

5.　a

7.　g

9.　$2x - 1 = -5$

The solution is the x-coordinate of the point of intersection of the graph $f(x) = 2x - 1$ and $g(x) = -5$. Inspecting the graph suggests that –2 is the solution.

Check:　$\begin{array}{c|c} 2x - 1 = -5 \\ \hline 2(-2) - 1 & -5 \\ -4 - 1 & -5 \\ -5 = -5 & \text{TRUE} \end{array}$

The solution is –2.

11.　$2x + 3 = x - 1$

The solution is the x-coordinate of the point of intersection of the graph $f(x) = 2x + 3$ and $g(x) = x - 1$. Inspecting the graph suggests that –4 is the solution.

Check:　$\begin{array}{c|c} 2x + 3 = x - 1 \\ \hline 2(-4) + 3 & -4 - 1 \\ -8 + 3 & -5 \\ -5 = -5 & \text{TRUE} \end{array}$

The solution is –4.

13.　$\frac{1}{2}x + 3 = x - 1$

The solution is the x-coordinate of the point of intersection of the graph $f(x) = \frac{1}{2}x + 3$ and $g(x) = x - 1$. Inspecting the graph suggests that 8 is the solution.

Check:　$\begin{array}{c|c} \frac{1}{2}x + 3 = x - 1 \\ \hline \frac{1}{2}(8) + 3 & 8 - 1 \\ 4 + 3 & 7 \\ 7 = 7 & \text{TRUE} \end{array}$

The solution is 8.

15. $f(x) = g(x)$

The solution is the x-coordinate of the point of intersection of the graph $f(x)$ and $g(x)$.

Inspecting the graph suggests that 0 is the solution.

17. Inspecting the graph suggests that $y_1 = y_2$ when $x = 5$.

19. Solve $x - 3 = 4$.

We graph $f(x) = x - 3$ and $g(x) = 4$ on the same axes.

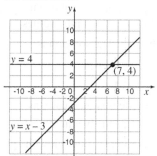

It appears that the lines intersect at $(7, 4)$.

$$x - 3 = 4$$
$$\overline{7 - 3}\,|\,4$$
$$\quad 4\,|\,4 \text{ TRUE}$$

The solution is 7.

21. Solve $2x + 1 = 7$.

We graph $f(x) = 2x + 1$ and $g(x) = 7$ on the same axes.

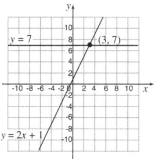

It appears that the lines intersect at $(3, 7)$.

$$2x + 1 = 7$$
$$\overline{2(3) + 1}\,|\,7$$
$$\qquad 7\,|\,7 \text{ TRUE}$$

The solution is 3.

23. Solve $\frac{1}{3}x - 2 = 1$.

We graph $f(x) = \frac{1}{3}x - 2$ and $g(x) = 1$ on the same axes.

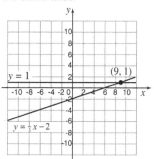

It appears that the lines intersect at $(9, 1)$.

$$\frac{1}{3}x - 2 = 1$$
$$\overline{\frac{1}{3}(9) - 2}\,|\,1$$
$$\quad 3 - 2\,|\,1 \text{ TRUE}$$

The solution is 9.

25. Solve $x + 3 = 5 - x$.

We graph $f(x) = x + 3$ and $g(x) = 5 - x$ on the same axes.

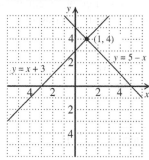

It appears that the lines intersect at $(1, 4)$.

$$x + 3 = 5 - x$$
$$\overline{1 + 3}\,|\,\overline{5 - 1}$$
$$\quad 4\,|\,4 \quad \text{TRUE}$$

The solution is 1.

27. Solve $5 - \frac{1}{2}x = x - 4$.

We graph $f(x) = 5 - \frac{1}{2}x$ and $g(x) = x - 4$
on the same axes.

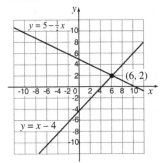

It appears that the lines intersect at $(6, 2)$.

$$
\begin{array}{c|c}
\multicolumn{2}{c}{5 - \frac{1}{2}x = x - 4} \\
\hline
5 - \frac{1}{2}(6) & 6 - 4 \\
5 - 3 & 2 \\
2 = 2 & \text{TRUE}
\end{array}
$$

The solution is 6.

29. Solve $2x - 1 = -x + 3$.

We graph $y_1 = 2x - 1$ and $y_2 = -x + 3$

on the same axes using a graphing calculator.
We then use the INTERSECT option from the
CALC menu.

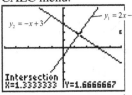

It the lines intersect at $\left(1\frac{1}{3}, 1\frac{2}{3}\right)$.

31. $f(x) \geq g(x)$ when the graph of f lies above,
or is at the same height as the graph of g.
From the graph, for all values greater than or
equal to 2, $f(x) \geq g(x)$. The solution is
$\{x \mid x \geq 2\}$, or $[2, \infty)$.

33. $y_1 < y_2$ when the graph of y_1 lies below
the graph of y_2. From the graph, for all
values less than 3, $y_1 < y_2$. The solution is
$\{x \mid x < 3\}$, or $(-\infty, 3)$.

35. a) $2x + 1 \leq x - 1$
We can rewrite this inequality as
$f(x) \leq h(x)$. This inequality is true
when the graph of f lies at the same
height or below the graph of $h(x)$. This
occurs for all x values less than or equal
to -2. The solution is $\{x \mid x \leq -2\}$, or
$(-\infty, -2]$.

b) $x - 1 > -\frac{1}{2}x + 3$
We can rewrite this inequality as
$h(x) > g(x)$. This inequality is true
when the graph of h lies above the graph
of g. This occurs for all x values
greater than $\frac{8}{3}$. The solution is
$\{x \mid x > \frac{8}{3}\}$, or $\left(\frac{8}{3}, \infty\right)$.

c) $-\frac{1}{2}x + 3 < 2x + 1$
We can rewrite this inequality as
$g(x) < f(x)$. This inequality is true
when the graph of g lies below the graph
of $f(x)$. This occurs for all x values
greater than $\frac{4}{5}$. The solution is
$\{x \mid x > \frac{4}{5}\}$, or $\left(\frac{4}{5}, \infty\right)$.

37. $x - 3 < 4$
We graph $f(x) = x - 3$ and $g(x) = 4$ on
the same axes.

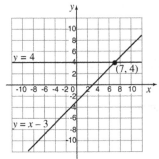

$f(x) < g(x)$ when the graph of f lies below
the graph of g. From the graph, for all values
less than 7, $f(x) < g(x)$. The solution is
$\{x \mid x < 7\}$, or $(-\infty, 7)$.

39. $2x - 3 \geq 1$

We graph $f(x) = 2x - 3$ and $g(x) = 1$ on the same axes.

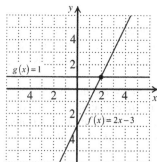

$f(x) \geq g(x)$ when the graph of f lies above, or is at the same height as the graph of g. From the graph, for all values greater than or equal to 2, $f(x) \geq g(x)$. The solution is $\{x \mid x \geq 2\}$, or $[2, \infty)$.

41. $x + 3 > 2x - 5$

We graph $f(x) = x + 3$ and $g(x) = 2x - 5$ on the same axes.

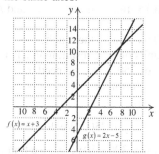

$f(x) > g(x)$ when the graph of f lies above, the graph of g. From the graph, for all values less than 8, $f(x) > g(x)$. The solution is $\{x \mid x < 8\}$, or $(-\infty, 8)$.

43. $\frac{1}{2}x - 2 \leq 1 - x$

We graph $f(x) = \frac{1}{2}x - 2$ and $g(x) = 1 - x$ on the same axes.

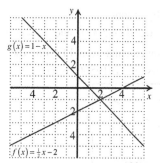

$f(x) \leq g(x)$ when the graph of f lies below, or is at the same height as the graph of g. From the graph, for all values less than or equal to 2, $f(x) \leq g(x)$. The solution is $\{x \mid x \leq 2\}$, or $(-\infty, 2]$.

45. $4x + 7 \leq 3 - 5x$

We graph $y_1 = 4x + 7$ and $y_2 = 3 - 5x$ on a graphing calculator in the same window. We can also graph $y_3 = y_1 \leq y_2$.

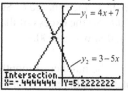

$y_1 \leq y_2$ when the graph of y_1 lies below, or is at the same height as the graph of y_2. From the graph, for all values less than or equal to $-\frac{4}{9}$, $y_1 \leq y_2$. The solution is $\{x \mid x \leq -\frac{4}{9}\}$, or $(-\infty, -\frac{4}{9}]$.

47. ***Familiarize.*** Let C = the cost to the patient and let b = the hospital bill. The problem asks that we determine how much the patient's hospital bill exceeded \$5000 if the patient's cost was \$6350.

Translate.

$$\underbrace{\text{Patient cost}}_{\downarrow} \;\;\underset{\downarrow}{\text{is}}\;\; \underset{\downarrow}{5000}\;\; \underset{\downarrow}{\text{plus}}\;\; \underset{\downarrow}{\underline{30\%}}\;\; \underset{\downarrow}{\text{of}}\;\; \underbrace{\text{amount over \$5000.}}_{\downarrow}$$

$$C \;\; = \;\; 5000 \;\; + \;\; 0.30 \;\; \cdot \;\; (b - 5000)$$

where $b \geq 5000$ since the bill cannot be negative.

Carry out. To estimate the amount of the hospital bill that would have resulted in the stated cost to the patient, we need to estimate the solution of

$$6350 = 5000 + 0.30(b - 5000),$$

We do this by graphing

$$y_1 = 5000 + 0.30(x - 5000) \text{ and } y_2 = 6350$$

on a graphing calculator, and finding the point of intersection: $(9500,\ 6350)$.

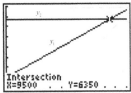

Thus, we estimate that a hospital bill of $9500 would result in a patient's after-insurance bill to be $6350.

Check. We evaluate:

$$C(9500) = 5000 + 0.30(9500 - 5000)$$
$$= 5000 + 0.3(4500)$$
$$= 5000 + 1350$$
$$= 6350$$

Our estimate turns out to be precise.

State. The amount of the hospital bill that resulted in an after-insurance charge of $6350 was a total of $9500. So the excess of that bill over the first $5000 was $4500.

49. **Familiarize.** Let t = the number of months and let C = the total cost of the telephone bill. We are to determine the number of months that would result in cumulative telephone bill of $275.

Translate.

$\underbrace{\text{Phone bill}}$ is $100 plus $35 · months.

$$\begin{array}{ccccccc} \downarrow & & \downarrow & \downarrow & \downarrow & \downarrow\downarrow & \downarrow \\ C & = & 100 & + & 35 & \cdot & t \end{array}$$

where $t \geq 0$ since months cannot be negative.

Carry out. To estimate the number of months resulting in a cumulative phone bill of $275, we need to estimate the solution of $275 = 100 + 35t$,

We do this by graphing $y_1 = 100 + 35x$ and $y_2 = 275$ on a graphing calculator, and we find the point of intersection at $(5,\ 275)$.

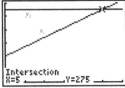

Thus, we estimate that after 5 months, the phone bill under the described plan will be $275.

Check. We evaluate:

$$C(5) = 100 + 35(5)$$
$$= 100 + 175$$
$$= 275$$

Our estimate turns out to be precise.

State. The time required for the cumulative phone bill to equal $275 is 5 months.

51. **Familiarize.** Let x = the number of 15-minute units of time a person is parked. Let F = the parking fee. We are to determine the time a person was parked resulting in a fee of $7.50.

Translate.

$\underbrace{\begin{array}{c}\text{Parking}\\ \text{fee}\end{array}}$ is $3.00 plus $0.50 $\underbrace{\begin{array}{c}\text{per 15-min}\\ \text{unit of time.}\end{array}}$

$$\begin{array}{cccccc} \downarrow & \downarrow & \downarrow & \downarrow & \downarrow & \downarrow \\ F & = & 3.00 & + & 0.50 & \cdot x \end{array}$$

where $x \geq 0$ since time units cannot be negative.

Rewriting the equation to eliminate decimals we have:

$$F = 300 + 50x$$

where x is in 15-minute units of time and F is parking fee in cents.

Carry out. To estimate the number of 15-minute units of time resulting in a parking fee of $7.50, or 750 cents, we need to estimate the solution of

$$F(x) = 300 + 50x,$$

replacing $F(x)$ with 750. We do this by graphing $F(x) = 300 + 50x$, and we find the point of intersection at $(9, 750)$.

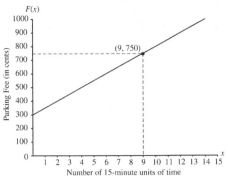

Thus, we estimate that after 9 15-minute units of time, the parking fee will be 750 cents, or $7.50.

Check. We evaluate:

$$F(9) = 300 + 50(9)$$
$$= 300 + 450$$
$$= 750$$

Our estimate turns out to be precise.
State. A parking fee of 750 cents, or $7.50, would result from parking a total of 9 15-minute units of time, or 2 hr and 15 minutes.

53. **Familiarize.** Let p = the weight of the package, in pounds. Let C = the cost to ship the package. We are to determine the weight of a package that costs $325 to ship.
Translate.

$$\underbrace{\text{Shipping charge}} \text{ is } \$130 \text{ plus } \underbrace{\$1.30 \text{ for each pound over 100.}}$$
$$\downarrow \quad \downarrow \quad \downarrow \quad \downarrow \quad\quad \downarrow$$
$$C \ = \ 130 \ + \ 1.30(p-100)$$

where $p \geq 100$ since weight cannot be negative.
Carry out. To estimate the weight of a package that cost $325 to ship, we need to estimate the solution of

$$325 = 130 + 1.30(p - 100),$$

We do this by graphing

$y_1 = 130 + 1.30(x - 100)$, $y_2 = 325$ and graphing their intersection on a graphing calculator. We find the point of intersection at $(250, 325)$

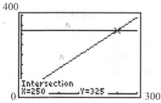

Thus, we estimate that a package weighing 250 lb would cost $325 to ship.
Check.

$$C(250) = 130 + 1.30(250 - 100)$$
$$= 130 + 1.30(150)$$
$$= 130 + 195$$
$$= 325$$

Our estimate turns out to be precise.
State. The weight of a package that cost $325 to ship is 250 lb.

55. **Familiarize.** Let n = the number of people who must attend and let R = the amount the band will receive. We are to determine the number of people who must attend in order for the band to be paid at least $1200.
Translate.

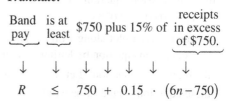

$$\underbrace{\text{Band pay}} \quad \underbrace{\text{is at least}} \quad \$750 \text{ plus } 15\% \text{ of } \underbrace{\begin{array}{c}\text{receipts in excess of \$750.}\end{array}}$$
$$\downarrow \quad \downarrow \quad \downarrow \ \downarrow \ \downarrow \ \downarrow \quad \downarrow$$
$$R \quad \leq \quad 750 \ + \ 0.15 \ \cdot \ (6n - 750)$$

where $n \geq 125$ since number of people attending cannot be negative.
Carry out. To estimate the number of people attending the band's concert in order for the band to receive at least $1200, we need to estimate the solution of

$$R(n) \leq 750 + 0.15(6n - 750)$$

replacing $R(n)$ with 1200. We do this by letting $y_1 = 1200$,

$y_2 = 750 + 0.15(6x - 750)$ and graphing their intersection on a graphing calculator. We find the point of intersection at $(625, 1200)$

and note that y_2 is greater than or equal to y_1 for all x-values greater than or equal to 625.

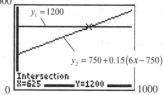

Thus, we estimate that at least 625 people will have to attend the show for the band to receive at least $1200.
Check. If $n = 624$, then the band would receive $750 + 0.15[$6(624) − $750], or $1199.10. However, when $n = 626$, the band would receive $750 + 0.15[$6(626) − $750], or $1200.90. We cannot check all possible numbers so we stop here.
State. In order for the band to receive at least $1200, at least 625 people must attend the show.

57. Let x = the number of years after 2000 and let
r = the estimated annual ridership at
amusement parks in billions. Enter the data in
the table on a graphing calculator and use the
linear regression feature to determine the
linear regression equation. The equation is

$r(x) = -0.04x + 2.0233$.

Create a scatter plot along with the graph of
the regression equation by letting y_1 equal

the regression equation. Let $y_2 = 1.5$ and plot
these equations in the same window.
Determine the intersection of y_1 and y_2.

Note   that $y_1 < y_2$ when $x > 13$, or 13 years
after   2000. After 2013 there will be fewer than
1.5   billion riders.

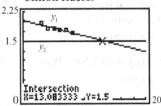

59. Let x = the number of years after 2006 and let
n = the advertising revenue for newspapers in
billions of dollars. Let t = the advertising
revenue for the Internet in billions of dollars.
Enter the data in two tables on a graphing
calculator and use the linear regression feature
to determine the linear regression equations for
both. The equations are:

$n(x) = -6.4x + 56.5$

$t(x) = 3.7x + 12$

Create a scatter plot along with the graph of
the regression equations by letting y_1 equal

the regression equation for $n(x)$. Let y_2

equal the regression equation for $t(x)$ and
plot both equations in the same window.
Determine the intersection of y_1 and y_2. To
determine those years for which advertising
revenue for the Internet will *exceed* that for
newspapers use $y_2 > y_1$. The solution is any
year after 2010 (2006 + 4.4).

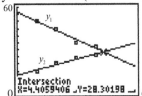

61. Let x = the number of years after 1900 and let f
= the world record for the women's 100 meter
freestyle, in seconds. Enter the data in a table
on a graphing calculator and use the linear
regression feature to determine the linear
regression equation. The equation is:

$f(x) = -0.08259x + 54.541$.

Create a scatter plot along with the graph of
the regression equation by letting y_1 equal

the regression equation for $f(x)$. Let

$y_2 = 50$ and plot both equations in the same
window. Determine the intersection of y_1

and y_2. To determine the years in which the
world record will be less than 50 seconds, we
note those x-values for which $y_1 < y_2$. The
solution is any year after 2045 (1990 + 55).

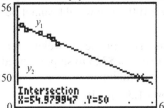

63. *Thinking and Writing Exercise.*

65. $f(x) = \dfrac{5}{x}$

Since $\dfrac{5}{x}$ cannot be computed when

the denominator is 0, we find the value
$x = 0$ is not in the domain of f, while
all other real numbers are. The domain
of f is $\{x \mid x$ is a real number *and* $x \neq 0\}$

67. $f(x) = \dfrac{x-2}{2x+1}$

Since $\dfrac{x-2}{2x+1}$ cannot be computed when

the denominator is 0, we find the x-value
that causes $2x + 1$ to be 0:

$2x + 1 = 0$

$2x = -1$

$x = \dfrac{-1}{2}$

Thus, $\frac{-1}{2}$ is not in the domain of f, while all other real numbers are. The domain of f is $\left\{ x \mid x \text{ is a real number } and \ x \neq \frac{-1}{2} \right\}$.

69. $f(x) = \dfrac{x+10}{8}$

Since there are no values of x that make the denominator equal 0, the domain of f is all real numbers.

71. *Thinking and Writing Exercise.*

73. $f(x) = g(x)$ is true at those points where the graphs of f and g intersect. The solutions are the first coordinate of each of the points of intersection, or –2, and 2. The solution set is $\{-2, 2\}$.

75. Solve $2x = |x+1|$. Let $y_1 = 2x$ and $y_2 = |x+1|$. Graph both in the same window on a graphing calculator and locate the point of intersection.

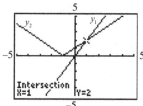

From the graph the solution is 1.

Check.
$$\begin{array}{c|c} 2x = |x+1| \\ \hline 2 \cdot 1 & |1+1| \\ 2 & 2 \\ 2 = 2 & \text{is TRUE} \end{array}$$

77. Solve $\frac{1}{2}x = 3 - |x|$. Let $y_1 = \frac{1}{2}x$ and $y_2 = 3 - |x|$. Graph both in the same window on a graphing calculator and locate the points of intersection.

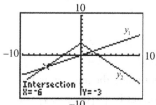

From the graph, the first solution is –6.

Check.
$$\begin{array}{c|c} \frac{1}{2}x = 3 - |x| \\ \hline \frac{1}{2}(-6) & 3 - |-6| \\ -3 & 3 - 6 \\ -3 & -3 \\ -3 = -3 & \text{is TRUE} \end{array}$$

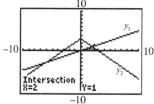

From the graph, the second solution is 2.

Check.
$$\begin{array}{c|c} \frac{1}{2}x = 3 - |x| \\ \hline \frac{1}{2}(2) & 3 - |2| \\ 1 & 3 - 2 \\ 1 & 1 \\ 1 = 1 & \text{is TRUE} \end{array}$$

The two solutions are –6 and 2.

79. Solve $x^2 = x + 2$. Let $y_1 = x^2$ and $y_2 = x + 2$. Graph both in the same window on a graphing calculator and locate the point of intersection.

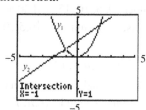

From the graph, the first solution is –1.

Check.
$$\begin{array}{c|c} x^2 = x + 2 \\ \hline (-1)^2 & -1 + 2 \\ 1 & 1 \\ 1 = 1 & \text{is TRUE} \end{array}$$

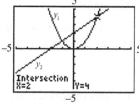

From the graph, the second solution is 2.

Check.
$$\begin{array}{c|c} x^2 = x + 2 \\ \hline 2^2 & 2 + 2 \\ 4 & 4 \\ 4 = 4 & \text{is TRUE} \end{array}$$

The two solutions are –1 and 2.

81.

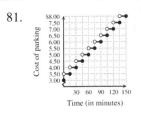

Exercise Set 4.3

1. h

3. f

5. e

7. b

9. c

11. $\{5,9,11\}\cap\{9,11,18\}$

The numbers 9 and 11 are common to both sets, so they intersection is $\{9,11\}$.

13. $\{0,5,10,15\}\cup\{5,15,20\}$

The numbers 0, 5, 10, 15, and 20 are in either or both sets, so the union is $\{0,5,10,15,20\}$.

15. $\{a,b,c,d,e,f\}\cap\{b,d,f\}$

The letters b, d, and f are common to both sets, so the intersection is $\{b,d,f\}$.

17. $\{r,s,t\}\cup\{r,u,t,s,v\}$

The letters r, s, t, u, and v, are in either or both sets, so the union is $\{r,s,t,u,v\}$.

19. $\{3,6,9,12\}\cap\{5,10,15\}$

There are no numbers common to both sets, so the solution has no members. It is $\varnothing$.

21. $\{3,5,7\}\cup\varnothing$

The members in either or both sets are 3, 5, and 7, so the union is $\{3,5,7\}$.

23. $3<x<7$

$(3,7)$

25. $-6\le y\le 0$

$[-6,0]$

27. $x<-1$ or $x>4$

$(-\infty,-1)\cup(4,\infty)$

29. $x\le-2$ or $x>1$

$(-\infty,-2]\cup(1,\infty)$

31. $x>-2$ and $x<4$

$(-2,4)$

33. $-4\le-x<2$

$(-2,4]$

35. $5>a$ or $a>7$

$(-\infty,5)\cup(7,\infty)$

37. $x\ge5$ or $-x\ge4$

$(-\infty,-4]\cup[5,\infty)$

39. $7>y$ and $y\ge-3$

$[-3,7)$

41. $x<7$ and $x\ge3$

$[3,7)$

43. $t<2$ or $t<5$

$(-\infty,5)$

45. $-2<t+1<8$

$-3<t<7$

The solution set is $\{t\,|-3<t<7\}$, or $(-3,7)$.

47. $4<x+4$ and $x-1<3$

$0<x$ and $x<4$

We can abbreviate the answer as $0<x<4$.

The solution set is $\{x\,|\,0<x<4\}$, or $(0,4)$.

49. $-7 \ge 2a - 3$ *or* $3a + 1 > 7$

     $-4 \ge 2a$ *or* $3a > 6$

     $-2 \ge a$ *or* $a > 2$

     The solution set is $\{a \mid a \le -2 \ or \ a > 2\}$, or

     $(-\infty, -2] \cup (2, \infty)$.

51. $x + 7 \le -2$ *or* $x + 7 \ge -3$

     Observe that any real number is either less than or equal to –2 or greater than or equal to –3. We can abbreviate the answer as $\mathbb{R}$. The solution set is

     $\{x \mid x \text{ is any real number}\}$, or $(-\infty, \infty)$.

53. $-7 \le 4x + 5 \le 13$

     $-12 \le 4x \le 8$

     $-3 \le x \le 2$

     The solution set is $\{x \mid -3 \le x \le 2\}$, or

     $[-3, 2]$.

55. $5 > \dfrac{x-3}{4} > 1$

     $20 > x - 3 > 4$

     $23 > x > 7$

     The solution set is $\{x \mid 7 < x < 23\}$, or

     $(7, 23)$.

57. $-2 \le \dfrac{x+2}{-5} \le 6$

     $10 \ge x + 2 \ge -30$

     $8 \ge x \ge -32$

     The solution set is $\{x \mid -32 \le x \le 8\}$, or

     $[-32, 8]$.

59. $2 \le 3x - 1 \le 8$

     $3 \le 3x \le 9$

     $1 \le x \le 3$

     The solution set is $\{x \mid 1 \le x \le 3\}$, or $[1, 3]$.

61. $-21 \le -2x - 7 < 0$

     $-14 \le -2x < 7$

     $7 \ge x > -\dfrac{7}{2}$, or

     $-\dfrac{7}{2} < x \le 7$

     The solution set is $\left\{x \middle| -\dfrac{7}{2} < x \le 7\right\}$, or

     $\left(-\dfrac{7}{2}, 7\right]$.

63. $5t + 3 < 3$ *or* $5t + 3 > 8$

     $5t < 0$ *or* $5t > 5$

     $t < 0$ *or* $t > 1$

     The solution set is $\{t \mid t < 0 \ or \ t > 1\}$, or

     $(-\infty, 0) \cup (1, \infty)$.

65. $6 > 2a - 1$ *or* $-4 \le -3a + 2$

     $7 > 2a$ *or* $-6 \le -3a$

     $\dfrac{7}{2} > a$ *or* $2 \ge a$

     The solution set is $\left\{a \middle| \dfrac{7}{2} > a\right\} \cup \{a \mid 2 \ge a\} =$

     $\left\{a \middle| \dfrac{7}{2} > a\right\}$, or $\left\{a \middle| a < \dfrac{7}{2}\right\}$, or $\left(-\infty, \dfrac{7}{2}\right)$.

67. $a + 3 < -2$ *and* $3a - 4 < 8$

     $a < -5$ *and* $3a < 12$

     $a < -5$ *and* $a < 4$

     The solution set is

     $\{a \mid a < -5\} \cap \{a \mid a < 4\} = \{a < -5\}$, or

     $(-\infty, -5)$.

69. $3x + 2 < 2$ *and* $3 - x < 1$

     $3x < 0$ *and* $-x < -2$

     $x < 0$ *and* $x > 2$

     Since no number is both greater than 2 and less than 0, the solution set is the empty set, $\varnothing$.

71. $2t - 7 \leq 5$ *or* $5 - 2t > 3$

$2t \leq 12$ *or* $-2t > -2$

$t \leq 6$ *or* $t < 1$

The solution set is

$\{t \mid t \leq 6\} \cup \{t \mid t < 1\} = \{t \mid t \leq 6\}$, or

$(-\infty, 6]$.

73. From the graph, we observe that the values of x for which $2x - 5 > -7$ *and* $2x - 5 < 7$ are $\{x \mid -1 < x < 6\}$, or $(-1, 6)$.

75. $f(x) = \dfrac{9}{x + 8}$

$f(x)$ cannot be computed when $x + 8 = 0$. Since $x + 8 = 0$ is equivalent to $x = -8$, we have a domain of $f = (-\infty, -8) \cup (-8, \infty)$.

77. $f(x) = \dfrac{-8}{x}$

$f(x)$ cannot be computed when $\dfrac{-8}{x} = 0$.

Since $\dfrac{-8}{x} = 0$ is equivalent to $x = 0$, we have a domain of $f = (-\infty, 0) \cup (0, \infty)$.

79. $f(x) = \sqrt{x - 6}$

The expression $\sqrt{x - 6}$ is not a real number when $x - 6$ is negative. Thus, the domain of f is the set of all x-values for which $x - 6 \geq 0$. Since $x - 6 \geq 0$ is equivalent to $x \geq 6$, we have a domain of $f = [6, \infty)$.

81. $f(x) = \sqrt{2x + 7}$

The expression $\sqrt{2x + 7}$ is not a real number when $2x + 7$ is negative. Thus, the domain of f is the set of all x-values for which $2x + 7 \geq 0$. Since $2x + 7 \geq 0$ is equivalent to $x \geq -\dfrac{7}{2}$, we have a domain of $f = \left[-\dfrac{7}{2}, \infty\right)$.

83. $f(x) = \sqrt{8 - 2x}$

The expression $\sqrt{8 - 2x}$ is not a real number when $8 - 2x$ is negative. Thus, the domain of f is the set of all x-values for which

$8 - 2x \geq 0$. Since $8 - 2x \geq 0$ is equivalent to $x \leq 4$, we have a domain of $f = (-\infty, 4]$.

85. $f(x) = \sqrt{x - 5}$, $g(x) = \sqrt{\frac{1}{2}x + 1}$

The domain of f is the set of all x-values for which $x - 5 \geq 0$, or $[5, \infty)$. The domain of g is the set of all x-values for which $\frac{1}{2}x + 1 \geq 0$, or $[-2, \infty)$. The intersection of the domains is $[5, \infty)$.

87. $f(x) = \sqrt{3 - x}$, $g(x) = \sqrt{3x - 2}$

The domain of f is the set of all x-values for which $3 - x \geq 0$, or $(-\infty, 3]$. The domain of g is the set of all x-values for which $3x - 2 \geq 0$, or $\left[\frac{2}{3}, \infty\right)$. The intersection of the domains is $\left[\frac{2}{3}, 3\right]$.

89. *Thinking and Writing Exercise.*

91. $|-5 - 2| = |-7| = 7$

93. Graph: $g(x) = 2x$

We make a table of values, plot points, and draw the graph.

x	$g(x)$
-2	-4
-1	-2
0	0
1	2
2	4

95. Graph: $g(x) = -3$

The graph of any constant function $y = c$ is a horizontal line that crosses the vertical axis at $(0, c)$. Thus, the graph of $g(x) = -3$ is a horizontal line that crosses the vertical axis at $(0, -3)$.

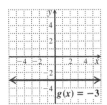

97. *Thinking and Writing Exercise.*

99. Solve $18,000 \le s(t) \le 21,000$, or
$18,000 \le 500t + 16,500 \le 21,000$.
$$18,000 \le 500t + 16,500 \le 21,000$$
$$1500 \le 500t \le 4500$$
$$3 \le t \le 9$$
Thus, from 3 through 9 years after 2000, from 2003 to 2009 the number of student visits to the counseling center is between 18,000 and 21,000.

Exercise Set 4.4

1. True

3. True

5. True

7. False

9. g

11. d

13. a

15. The solutions of $|x+2| = 3$ are the first coordinates of the points of intersection of $y_1 = abs(x+2)$ and $y_2 = 3$. They are –5 and 1, so the solution set is $\{-5,1\}$.

17. The graph of $y_1 = abs(x+2)$ lies below the graph of $y_2 = 3$ for $\{x \,|\, -5 < x < -1\}$, or on $(-5,1)$.

19. The graph of $y_1 = abs(x+2)$ lies on or above the graph of $y_2 = 3$ for $\{x \,|\, x \le -5 \text{ or } x \ge 1\}$, or on $(\infty, -5] \cup [1, \infty)$.

21. $|x| = 7$
$x = -7 \text{ or } x = 7$
The solution set is $\{-7, 7\}$.

23. $|x| = -6$
The absolute value of a number is always nonnegative. Therefore, the solution is $\varnothing$.

25. $|p| = 0$
The only number whose absolute value is 0 is 0. The solution set is $\{0\}$.

27. $|2x - 3| = 4$
$2x - 3 = -4 \text{ or } 2x - 3 = 4$
$2x = -1 \text{ or } \quad 2x = 7$
$x = -\dfrac{1}{2} \text{ or } \quad x = \dfrac{7}{2}$
The solution set is $\left\{-\dfrac{1}{2}, \dfrac{7}{2}\right\}$.

29. $|3x - 5| = -8$
The absolute value of a number is always nonnegative. Therefore, the solution is $\varnothing$.

31. $|x - 2| = 6$
$x - 2 = -6 \text{ or } x - 2 = 6$
$x = -4 \text{ or } \quad x = 8$
The solution set is $\{-4, 8\}$.

33. $|x - 5| = 3$
$x - 5 = -3 \text{ or } x - 5 = 3$
$x = 2 \text{ or } \quad x = 8$
The solution set is $\{2, 8\}$.

35. $|t| + 1.1 = 6.6 \Rightarrow |t| = 5.5$
$t = -5.5 \text{ or } t = 5.5$
The solution set is $\{-5.5, 5.5\}$.

37. $|5x| - 3 = 37$
$|5x| = 40$
$5x = -40 \text{ or } 5x = 40$
$x = -8 \text{ or } \quad x = 8$
The solution set is $\{-8, 8\}$.

39. $7|q| - 2 = 9$

$7|q| = 11$

$|q| = \dfrac{11}{7}$

$q = -\dfrac{11}{7} \ or \ \dfrac{11}{7}$

The solution set is $\left\{ -\dfrac{11}{7}, \dfrac{11}{7} \right\}$.

41. $\left| \dfrac{2x-1}{3} \right| = 4$

$\dfrac{2x-1}{3} = -4 \quad or \quad \dfrac{2x-1}{3} = 4$

$2x - 1 = -12 \ or \ 2x - 1 = 12$

$2x = -11 \ or \ 2x = 13$

$x = -\dfrac{11}{2} \quad or \ x = \dfrac{13}{2}$

The solution set is $\left\{ -\dfrac{11}{2}, \dfrac{13}{2} \right\}$

43. $|5 - m| + 9 = 16$

$|5 - m| = 7$

$5 - m = -7 \quad or \ 5 - m = 7$

$-m = -12 \ or \quad -m = 2$

$m = 12 \quad or \quad m = -2$

The solution set is $\{-2, 12\}$.

45. $5 - 2|3x - 4| = -5$

$-2|3x - 4| = -10$

$|3x - 4| = 5$

$3x - 4 = -5 \ or \ 3x - 4 = 5$

$3x = -1 \ or \ 3x = 9$

$x = -\dfrac{1}{3} \ or \ x = 3$

The solution set is $\left\{ -\dfrac{1}{3}, 3 \right\}$.

47. $f(x) = |2x + 6|; \ f(x) = 8$

$|2x + 6| = 8$

$2x + 6 = -8 \ or \ 2x + 6 = 8$

$2x = -14 \ or \quad 2x = 2$

$x = -7 \quad or \qquad x = 1$

The solution set is $\{-7, 1\}$.

49. $f(x) = |x| - 3; \ f(x) = 5.7$

$|x| - 3 = 5.7$

$|x| = 8.7$

$x = -8.7 \ or \ x = 8.7$

The solution set is $\{-8.7, 8.7\}$.

51. $f(x) = \left| \dfrac{3x - 2}{5} \right|; \ f(x) = 2$

$\left| \dfrac{3x - 2}{5} \right| = 2$

$\dfrac{3x - 2}{5} = -2 \quad or \quad \dfrac{3x - 2}{5} = 2$

$3x - 2 = -10 \ or \ 3x - 2 = 10$

$3x = -8 \quad or \qquad 3x = 12$

$x = -\dfrac{8}{3} \quad or \qquad x = 4$

The solution set is $\left\{ -\dfrac{8}{3}, 4 \right\}$.

53. $|x + 4| = |2x - 7|$

$x + 4 = 2x - 7 \ or \ x + 4 = -(2x - 7)$

$4 = x - 7 \quad or \ x + 4 = -2x + 7$

$11 = x \qquad or \ 3x + 4 = 7$

$3x = 3$

$x = 1$

The solution set is $\{1, 11\}$.

55. $|x + 4| = |x - 3|$

$x + 4 = x - 3 \ or \ x + 4 = -(x - 3)$

$4 = -3 \quad or \ x + 4 = -x + 3$

$\text{False} \qquad 2x = -1$

$x = -\dfrac{1}{2}$

The solution set is $\left\{ -\dfrac{1}{2} \right\}$.

57. $\left|3a-1\right|=\left|2a+4\right|$

$3a-1=2a+4$ $\;or\;$ $3a-1=-(2a+4)$

$\quad a-1=4$ $\quad or\;$ $3a-1=-2a-4$

$\qquad a=5$ $\quad or\;$ $5a-1=-4$

$\qquad\qquad\qquad\qquad\qquad 5a=-3$

$\qquad\qquad\qquad\qquad\qquad a=-\dfrac{3}{5}$

The solution set is $\left\{-\dfrac{3}{5},5\right\}$.

59. $\left|n-3\right|=\left|3-n\right|$

$n-3=3-n$ $\;or\;$ $n-3=-(3-n)$

$2n-3=3$ $\;or\;$ $n-3=-3+n$

$\quad 2n=6$ $\;or\;$ $-3=-3$

$\qquad n=3$ $\qquad$ True for all real

$\qquad\qquad\qquad\qquad$ values of n

The solution set is the set of all real numbers.

61. $\left|7-4a\right|=\left|4a+5\right|$

$7-4a=4a+5$ $\;or\;$ $7-4a=-(4a+5)$

$\quad 7=8a+5$ $\;or\;$ $7-4a=-4a-5$

$\quad 2=8a$ $\qquad or\;$ $\qquad 7=-5$

$\dfrac{1}{4}=a$ $\qquad\qquad$ False

The solution set is $\left\{\dfrac{1}{4}\right\}$.

63. $\left|a\right|\le 9$

$-9\le a\le 9$

The solution set is $\left\{a\,|-9\le a\le 9\right\}$, or

$\left[-9,9\right]$.

65. $\left|t\right|>0$

$t<0$ $\;or\;$ $0<t$

The solution set is $\left\{t\,|\,t<0\;or\;t>0\right\}$, or

$\left\{t\,|\,t\ne 0\right\}$, or $\left(-\infty,0\right)\cup\left(0,\infty\right)$.

67. $\left|x-1\right|<4$

$-4<x-1<4$

$-3<x<5$

The solution set is $\left\{x\,|-3<x<5\right\}$, or

$\left(-3,5\right)$.

69. $\left|x+2\right|\le 6$

$-6\le x+2\le 6$

$-8\le x\le 4$

The solution set is $\left\{x\,|-8\le x\le 4\right\}$, or

$\left[-8,4\right]$.

71. $\left|x-3\right|+2>7$

$\left|x-3\right|>5$

$x-3<-5$ $\;or\;$ $5<x-3$

$\quad x<-2$ $\;or\;$ $8<x$

The solution set is $\left\{x\,|\,x<-2\;or\;x>8\right\}$, or

$\left(-\infty,-2\right)\cup\left(8,\infty\right)$.

73. $\left|2y-9\right|>-5$

Since the absolute value is never negative, any value of $2y-9$, and hence any value of y, will satisfy the inequality. The solution set is the set of all real numbers, or $\left(-\infty,\infty\right)$.

75. $\left|3a-4\right|+2\ge 8$

$\left|3a-4\right|\ge 6$

$3a-4\le -6$ $\;or\;$ $6\le 3a-4$

$\quad 3a\le -2$ $\;or\;$ $10\le 3a$

$\quad a\le -\dfrac{2}{3}$ $\;or\;$ $\dfrac{10}{3}\le a$

The solution set is $\left\{a\,\Big|\,a\le -\dfrac{2}{3}\;or\;a\ge\dfrac{10}{3}\right\}$,

or $\left(-\infty,-\dfrac{2}{3}\right]\cup\left[\dfrac{10}{3},\infty\right)$.

77. $\left|y-3\right|<12$

$-12<y-3<12$

$-9<y<15$

The solution set is $\left\{y\,|-9<y<15\right\}$, or

$\left(-9,15\right)$.

79. $9-\left|x+4\right|\le 5$

$-\left|x+4\right|\le -4$

$\left|x+4\right|\ge 4$

$x+4\le -4$ $\;or\;$ $4\le x+4$

$\quad x\le -8$ $\;or\;$ $0\le x$

The solution set is $\{x \mid x \le -8 \ or \ x \ge 0\}$, or

$(-\infty, -8] \cup [0, \infty)$.

81. $6 + |3 - 2x| > 10$

$|3 - 2x| > 4$

$3 - 2x < -4 \ or \ 4 < 3 - 2x$

$-2x < -7 \ or \ 1 < -2x$

$x > \dfrac{7}{2} \ or \ -\dfrac{1}{2} > x$

The solution set is $\left\{x \mid x < -\dfrac{1}{2} \ or \ x > \dfrac{7}{2}\right\}$, or

$\left(-\infty, -\dfrac{1}{2}\right) \cup \left(\dfrac{7}{2}, \infty\right)$.

83. $|5 - 4x| < -6$

Absolute value is always nonnegative, so the inequality has no solution. The solution set is $\varnothing$.

85. $\left|\dfrac{2 - 5x}{4}\right| \ge \dfrac{2}{3}$

$\dfrac{2 - 5x}{4} \le -\dfrac{2}{3} \ or \ \dfrac{2}{3} \le \dfrac{2 - 5x}{4}$

$2 - 5x \le -\dfrac{8}{3} \ or \ \dfrac{8}{3} \le 2 - 5x$

$-5x \le -\dfrac{14}{3} \ or \ \dfrac{2}{3} \le -5x$

$x \ge \dfrac{14}{15} \ or \ -\dfrac{2}{15} \ge x$

The solution set is $\left\{x \mid x \le -\dfrac{2}{15} \ or \ x \ge \dfrac{14}{15}\right\}$,

or $\left(-\infty, -\dfrac{2}{15}\right] \cup \left[\dfrac{14}{15}, \infty\right)$.

87. $|m + 3| + 8 \le 14$

$|m + 3| \le 6$

$-6 \le m + 3 \le 6$

$-9 \le m \le 3$

The solution set is $\{m \mid -9 \le m \le 3\}$, or

$[-9, 3]$.

89. $25 - 2|a + 3| > 19$

$-2|a + 3| > -6$

$|a + 3| < 3$

$-3 < a + 3 < 3$

$-6 < a < 0$

The solution set is $\{a \mid -6 < a < 0\}$, or

$(-6, 0)$.

91. $|2x - 3| \le 4$

$-4 \le 2x - 3 \le 4$

$-1 \le 2x \le 7$

$-\dfrac{1}{2} \le x \le \dfrac{7}{2}$

The solution set is $\left\{x \mid -\dfrac{1}{2} \le x \le \dfrac{7}{2}\right\}$, or

$\left[-\dfrac{1}{2}, \dfrac{7}{2}\right]$.

93. $5 + |3x - 4| \ge 16$

$|3x - 4| \ge 11$

$3x - 4 \le -11 \ or \ 11 \le 3x - 4$

$3x \le -7 \ or \ 3x \ge 15$

$x \le -\dfrac{7}{3} \ or \ x \ge 5$

The solution set is $\left\{x \mid x \le -\dfrac{7}{3} \ or \ x \ge 5\right\}$, or

$\left(-\infty, -\dfrac{7}{3}\right] \cup [5, \infty)$.

95. $7 + |2x - 1| < 16$

$|2x - 1| < 9$

$-9 < 2x - 1 < 9$

$-8 < 2x < 10$

$-4 < x < 5$

The solution set is $\{x \mid -4 < x < 5\}$, or

$(-4, 5)$.

97. *Thinking and Writing Exercise.*

99. $3x - y = 6$

To find the y-intercept, let $x = 0$ and solve for y:

$3 \cdot 0 - y = 6$

$y = -6$

The y-intercept is $(0, -6)$.

To find the x-intercept, let $y = 0$ and solve for x:

$3x - 0 = 6$

$3x = 6$

$x = 2$

The x-intercept is $(2, 0)$.

Before drawing the line, plot a third point as a check. Substitute any convenient value for x and solve for y. For $x = 1$,

$3 \cdot 1 - y = 6$

$3 - y = 6$

$y = -3$

The point $(1, -3)$ appears to line up with the intercepts. To finish, draw and label the line.

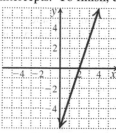

$3x - y = 6$

101. $x = -2$

We can write this equation as $x + 0 \cdot y = -2$.

No matter what number is chosen for y, x must be -2. Consider the following table.

x	y	(x, y)
-2	-4	$(-2, -4)$
-2	0	$(-2, 0)$
-2	4	$(-2, 4)$

Plot the ordered pairs and connect the points.

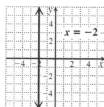

103. $x - 3y = 8$ (1)

$2x + 3y = 4$ (2)

Solve the first equation for x.

$x - 3y = 8$

$x = 3y + 8$ (3)

Substitute $3y + 8$ for x in the second equation and solve for y.

$2x + 3y = 4$ (2)

$2(3y + 8) + 3y = 4$

$6y + 16 + 3y = 4$

$9y + 16 = 4$

$9y = -12$

$y = -\dfrac{4}{3}$

Next substitute $-\dfrac{4}{3}$ for y in Equation (3).

$x = 3y + 8$ (3)

$x = 3 \cdot \left(-\dfrac{4}{3}\right) + 8$

$x = -4 + 8$

$x = 4$

The ordered pair $\left(4, -\dfrac{4}{3}\right)$ checks in both equations; it is the solution.

105. $y = 1 - 5x$ (1)

$2x - y = 4$ (2)

Substitute $1 - 5x$ for y in the second equation and solve for y.

$2x - y = 4$ (2)

$2x - (1 - 5x) = 4$

$2x - 1 + 5x = 4$

$7x - 1 = 4$

$7x = 5$

$x = \dfrac{5}{7}$

Next substitute $\dfrac{5}{7}$ for x in Equation (1).

$y = 1 - 5x$ (1)

$y = 1 - 5\left(\dfrac{5}{7}\right)$

$y = \dfrac{7}{7} - \dfrac{25}{7}$

$y = -\dfrac{18}{7}$

The ordered pair $\left(\dfrac{5}{7}, -\dfrac{18}{7}\right)$ checks in both

equations; it is the solution.

107. *Thinking and Writing Exercise.*

109. From the definition of absolute value,
$|3t - 5| = 3t - 5$ only when $3t - 5 \geq 0$. Solve
$$3t - 5 \geq 0$$
$$3t \geq 5$$
$$t \geq \frac{5}{3}$$
The solution set is $\left\{t \,\middle|\, t \geq \dfrac{5}{3}\right\}$, or $\left[\dfrac{5}{3}, \infty\right)$.

111. $|x + 2| > x$

The inequality is true for all $x < 0$ (because
absolute value must be nonnegative). The
solution set in this case is $\{x \mid x < 0\}$. If
$x = 0$, we have $|0 + 2| > 0$, which is true.
The solution set in this case is $\{0\}$. If $x > 0$,
we have the following:
$$x + 2 < -x \ or \ x < x + 2$$
$$2x < -2 \ or \ 0 < 2$$
$$x < -1$$
Although $x > 0$ *and* $x < -1$ yields no
solutions, $x > 0$ and $2 > 0$ (true for all x)
yields the solution set $\{x \mid x > 0\}$ in this case.
The solution set for the inequality is
$\{x \mid x < 0\} \cup \{0\} \cup \{x \mid x > 0\}$, or
$\{x \mid x$ is a real number$\}$, or $(-\infty, \infty)$.

113. $|5t - 3| = 2t + 4$

From the definition of absolute value, we
know that $2t + 4 \geq 0$, or $t \geq -2$. So we have
$t \geq -2$ *and*
$$5t - 3 = -(2t + 4) \ or \ 5t - 3 = 2t + 4$$
$$5t - 3 = -2t - 4 \quad or \qquad 3t = 7$$
$$7t = -1 \qquad\quad or \qquad t = \frac{7}{3}$$
$$t = -\frac{1}{7} \qquad or \qquad t = \frac{7}{3}$$

Since $-\dfrac{1}{7} \geq -2$ and $\dfrac{7}{3} \geq -2$, the solution set

is $\left\{-\dfrac{1}{7}, \dfrac{7}{3}\right\}$.

115. Using part (b) we find that $-3 < x < 3$ is
equivalent to $|x| < 3$.

117. $x < -8 \ or \ 2 < x$
$$x + 3 < -5 \ or \ 5 < x + 3 \ \text{Adding 3}$$
$$|x + 3| > 5 \qquad\qquad\quad \text{Using part (c)}$$

119. The distance from x to 7 is $|x - 7|$ or $|7 - x|$,
so we have $|x - 7| < 2$, or $|7 - x| < 2$.

121. The length of the segment from -1 to 7 is
$|-1 - 7| = |-8| = 8$ units. The midpoint of the

segment is $\dfrac{-1 + 7}{2} = \dfrac{6}{2} = 3$. Thus, the interval

extends 8/2, or 4 units, on each side of 3. An
inequality for which the closed interval is the
solution set is then $|x - 3| \leq 4$.

123. The length of the segment from -7 to -1 is
$|-7 - (-1)| = |-6| = 6$ units. The midpoint of

the segment is $\dfrac{-7 + (-1)}{2} = \dfrac{-8}{2} = -4$. Thus,

the interval extends 6/2, or 3 units on each
side of -4. An inequality for which the open
interval is the solution set is $|x - (-4)| < 3$, or
$|x + 4| < 3$.

125. Let $d =$ the distance above the river. This
distance must satisfy the inequality
$|d - 60| \leq 10$. First solve for d.
$$|d - 60| \leq 10$$
$$-10 \leq d - 60 \leq 10$$
$$50 \leq d \leq 70$$
Since the bridge is 150 ft from the river, the
bungee jumper will, at any point in time, be
$150 - d$ feet from the bridge, so the jumper
will be a maximum of 150 ft $-$ 50 ft, or 100 ft
from the bridge *and* a minimum of 150 ft $-$ 70
ft, or 80 ft from the bridge.
The solution is between 80 ft and 100 ft.

Mid-Chapter Review

Guided Solutions

1. $-3 < x - 5 < 6$
    $2 < x < 11$
    The solution is $(2,11)$.

2. $|x-1| > 9$
    $x - 1 < -9 \quad or \quad 9 < x - 1$
    $x < -8 \quad or \quad 10 < x$
    The solution is $(-\infty, -8) \cup (10, \infty)$.

Mixed Review

1. $|x| = 15$
    $x = -15 \quad or \quad x = 15$
    The solution set is $\{-15, 15\}$.

2. $|t| < 10$
    $-10 < t < 10$
    The solution set is $\{t \mid -10 < t < 10\}$, or
    $(-10, 10)$.

3. $|p| > 15$
    $p < -15 \quad or \quad p > 15$
    The solution set is $\{p \mid p < -15 \ or \ p > 15\}$,
    or $(-\infty, -15) \cup (15, \infty)$.

4. $|2x + 1| = 7$
    $2x + 1 = -7 \quad or \quad 2x + 1 = 7$
    $2x = -8 \quad or \quad 2x = 6$
    $x = -4 \quad or \quad x = 3$
    The solution set is $\{-4, 3\}$.

5. $-1 < 10 - x < 8$
    $-11 < -x < -2$
    $2 < x < 11$
    The solution set is $\{x \mid 2 < x < 11\}$, or $(2, 11)$.

6. $5|t| < 20$
    $|t| < 4$
    $-4 < t < 4$

The solution set is $\{t \mid -4 < t < 4\}$, or $(-4, 4)$.

7. $x + 8 < 2 \quad or \quad x - 4 > 9$
    $x < -6 \quad or \quad \quad x > 13$
    The solution set is $\{x \mid x < -6 \ or \ x > 13\}$, or
    $(-\infty, -6) \cup (13, \infty)$.

8. $|x + 2| \le 5$
    $-5 \le x + 2 \le 5$
    $-7 \le x \le 3$
    The solution set is $\{x \mid -7 \le x \le 3\}$, or
    $[-7, 3]$.

9. $2 + |3x| = 10$
    $|3x| = 8 \quad$ Subtracting 2
    $3x = -8 \quad or \quad 3x = 8$
    $x = -\dfrac{8}{3} \quad or \quad x = \dfrac{8}{3}$
    The solution set is $\left\{-\dfrac{8}{3}, \dfrac{8}{3}\right\}$.

10. $2(x - 7) - 5x > 4 - (x + 5)$
    $2x - 14 - 5x > 4 - x - 5$
    $-3x - 14 > -x - 1$
    $-13 > 2x$
    $x < -\dfrac{13}{2}$
    The solution set is $\left\{x \mid x < -\dfrac{13}{2}\right\}$,
    or $\left(-\infty, -\dfrac{13}{2}\right)$.

11. $-12 < 2n + 6 \quad and \quad 3n - 1 \le 7$
    $-18 < 2n \quad \quad and \quad \quad 3n \le 8$
    $-9 < n \quad \quad and \quad \quad n \le \dfrac{8}{3}$
    The solution set is $\left\{n \mid -9 < n \le \dfrac{8}{3}\right\}$, or
    $\left(-9, \dfrac{8}{3}\right]$.

12. $|2x+5|+1 \geq 13$

$|2x+5| \geq 12$

$2x+5 \leq -12$ *or* $12 \leq 2x+5$

$2x \leq -17$ *or* $7 \leq 2x$

$x \leq -\dfrac{17}{2}$ *or* $\dfrac{7}{2} \leq x$

The solution set is $\left\{ x \,\middle|\, x \leq -\dfrac{17}{2} \text{ or } x \geq \dfrac{7}{2} \right\}$, or

$\left(-\infty, -\dfrac{17}{2} \right] \cup \left[\dfrac{7}{2}, \infty \right)$.

13. $\dfrac{1}{2}(2x-6) \leq \dfrac{1}{3}(9x+3)$

$x-3 \leq 3x+1$

$-4 \leq 2x$

$-2 \leq x$

The solution set is $\{x \,|\, x \geq -2\}$, or $[-2, \infty)$.

14. $\left| \dfrac{x+2}{5} \right| = 8$

$\dfrac{x+2}{5} = -8$ *or* $\dfrac{x+2}{5} = 8$

$x+2 = -40$ *or* $x+2 = 40$

$x = -42$ *or* $x = 38$

The solution set is $\{-42, 38\}$.

15. $|8x-11|+6 < 2$

$|8x-11| < -4$

Absolute values are always nonnegative, so there is no value for x that can make the expression in the absolute value symbol negative. There is no solution, $\varnothing$.

16. $8-5|a+6| > 3$

$-5|a+6| > -5$

$|a+6| < 1$

$-1 < a+6 < 1$

$-7 < a < -5$

The solution set is $\{a \,|\, -7 < a < -5\}$, or $(-7, -5)$.

17. $|5x+7|+9 \geq 4$

$|5x+7| \geq -5$

Since absolute values are never negative, any value of x will make $|5x+7| \geq 0$. The solution set is the set of all real numbers, or $(-\infty, \infty)$.

18. We graph $f(x) = \dfrac{1}{3}x+1$ and $g(x) = x+5$ on the same axes.

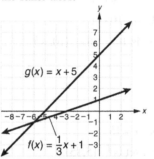

It appears that the lines intersect at $(-6, -1)$.

$\dfrac{\frac{1}{3}x+1 = -1}{\dfrac{1}{3}(-6)+1 \,|\, -1}$

$-2+1 \,|\, -1$ TRUE

The solution is -6.

19. $2x+3 > \dfrac{1}{3}x-2$

We graph $f(x) = 2x+3$ and $g(x) = \dfrac{1}{3}x-2$ on the same axes.

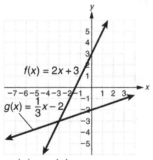

$f(x) > g(x)$ when the graph of f lies above the graph of g. From the graph, for all values greater than -3, $f(x) > g(x)$. The solution is $\{x \,|\, x > -3\}$, or $(-3, \infty)$.

20. $5x - 6 \le \dfrac{1}{2}x - 15$

We graph $y_1 = 5x - 6$ and $y_2 = \dfrac{1}{2}x - 15$ on a

graphing calculator in the same window. We

can also graph $y_3 = y_1 \le y_2$.

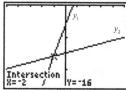

$y_1 \le y_2$ when the graph of y_1 lies below, or is

at the same height as the graph of y_2. From

the graph, for all values less than or equal to

-2, $y_1 \le y_2$. The solution is $\{x \mid x \le -2\}$, or

$(-\infty, -2]$.

___

Exercise Set 4.5
___

1. e

3. d

5. b

7. We replace x with -4 and y with 2.

$2x + 3y < -1$

$$\begin{array}{c|c} 2(-4) + 3 \cdot 2 & -1 \\ \hline -8 + 6 & -1 \\ \hline -2 & -1 \text{ TRUE} \end{array}$$

Since $-2 < -1$ is true, $(-4, 2)$ is a solution.

9. We replace x with 8 and y with 14.

$2y - 3x \ge 9$

$$\begin{array}{c|c} 2 \cdot 14 - 3 \cdot 8 & 9 \\ \hline 28 - 24 & 9 \\ \hline 4 & 9 \text{ FALSE} \end{array}$$

Since $4 \ge 9$ is false, $(8, 14)$ is not a solution.

11. Graph: $y \ge \dfrac{1}{2}x$

We first graph the boundary line $y = \dfrac{1}{2}x$.

We draw the line solid since the inequality
symbol is $\ge$. To determine which half-plane

to shade, test a point not on the line. We try

$(0, 1)$:

$y \ge \dfrac{1}{2}x$

$$\begin{array}{c|c} 1 & \dfrac{1}{2} \cdot 0 \\ \hline 1 & 0 \text{ TRUE} \end{array}$$

Since $1 \ge 0$ is true, $(0, 1)$ is a solution as are

all of the points in the half-plane containing

$(0, 1)$. We shade that half-plane and obtain

the graph.

13. Graph: $y > x - 3$

We first graph the boundary line $y = x - 3$.

We draw the line dashed since the inequality
symbol is $>$. To determine which half-plane
to shade, test a point not on the line. We try

$(0, 0)$:

$y > x - 3$

$$\begin{array}{c|c} 0 & 0 - 3 \\ \hline 0 & -3 \quad \text{TRUE} \end{array}$$

Since $0 > -3$ is true, $(0, 0)$ is a solution as

are all of the points in the half-plane

containing $(0, 0)$. We shade that half-plane

and obtain the graph.

15. Graph: $y \le x + 5$

We first graph the boundary line $y = x + 5$.

We draw the line solid since the inequality
symbol is $\le$. To determine which half-plane
to shade, test a point not on the line. We try

$(0, 0)$:

$y \le x + 5$

$$\begin{array}{c|c} 0 & 0 + 5 \\ \hline 0 & 5 \quad \text{TRUE} \end{array}$$

Since $0 \le 5$ is true, $(0, 0)$ is a solution as are

all of the points in the half-plane containing

$(0,0)$. We shade that half-plane and obtain the graph.

17. Graph: $x - y \leq 4$

We first graph the boundary line $x - y = 4$.

We draw the line solid since the inequality symbol is $\leq$. To determine which half-plane to shade, test a point not on the line. We try $(0,0)$:

$$x - y \leq 4$$

$$\overline{0 - 0 \quad | \quad 4}$$

$$0 \quad | \quad 4 \text{ TRUE}$$

Since $0 \leq 4$ is true, $(0,0)$ is a solution as are all of the points in the half-plane containing $(0,0)$. We shade that half-plane and obtain the graph.

19. Graph: $2x + 3y > 6$

We first graph the boundary line $2x + 3y = 6$.

We draw the line dashed since the inequality symbol is $>$. To determine which half-plane to shade, test a point not on the line. We try $(0,0)$:

$$2x + 3y > 6$$

$$\overline{2 \cdot 0 + 3 \cdot 0 \quad | \quad 6}$$

$$0 \quad | \quad 6 \text{ FALSE}$$

Since $0 > 6$ is false, $(0,0)$ is not a solution nor are any of the points in the half-plane containing $(0,0)$. The points in the other half-plane are the solutions so we shade that half-plane and obtain the graph.

21. Graph: $2y - x \leq 4$

We first graph the boundary line $2y - x \leq 4$.

We draw the line solid since the inequality symbol is $\leq$. To determine which half-plane to shade, test a point not on the line. We try $(0,0)$:

$$2y - x \leq 4$$

$$\overline{2 \cdot 0 - 0 \quad | \quad 4}$$

$$0 \quad | \quad 4 \text{ TRUE}$$

Since $0 \leq 4$ is true, $(0,0)$ is a solution as are all of the points in the half-plane containing $(0,0)$. We shade that half-plane and obtain the graph.

23. Graph: $2x - 2y \geq 8 + 2y$

$$2x - 4y \geq 8$$

We first graph the boundary line $2x - 4y = 8$.

We draw the line solid since the inequality symbol is $\geq$. To determine which half-plane to shade, test a point not on the line. We try $(0,0)$:

$$2x - 4y \geq 8$$

$$\overline{2 \cdot 0 - 4 \cdot 0 \quad | \quad 8}$$

$$0 \quad | \quad 8 \text{ FALSE}$$

Since $0 \geq 8$ is false, $(0,0)$ is not a solution nor are any of the points in the half-plane containing $(0,0)$. The points in the other half-plane are the solutions so we shade that half-plane and obtain the graph.

$2x - 2y \geq 8 + 2y$

25. Graph: $x > -2$

We first graph the boundary line $x = -2$. We draw the line dashed since the inequality symbol is >. To determine which half-plane to shade, test a point not on the line. We try $(0, 0)$:

$x > -2$

$\overline{0 \,\vert\, -2}$

$0 \,\vert\, -2$ TRUE

Since $0 > -2$ is true, $(0, 0)$ is a solution as are all of the points in the half-plane containing $(0, 0)$. We shade that half-plane and obtain the graph.

27. Graph: $y \le 6$

We first graph the boundary line $y = 6$. We draw the line solid since the inequality symbol is $\le$. To determine which half-plane to shade, test a point not on the line. We try $(0, 0)$:

$y \le 6$

$\overline{0 \,\vert\, 6}$

$0 \,\vert\, 6$ TRUE

Since $0 < 6$ is true, $(0, 0)$ is a solution as are all of the points in the half-plane containing $(0, 0)$. We shade that half-plane and obtain the graph.

29. Graph: $-2 < y < 7$

This is a system of inequalities:
$-2 < y$

$y < 7$

We graph the boundary line $-2 = y$ and see that the graph of $-2 < y$ is the half-plane above the line $-2 = y$. We also graph the boundary line $y = 7$ and see that the graph of $y < 7$ is the half-plane below the line $y = 7$.

Finally, we shade the intersection of these graphs.

31. Graph $-4 \le x \le 2$

This is a system of inequalities.
$-4 \le x$

$x \le 2$

We graph the equations $-4 = x$ and see that the graph of $-4 \le x$ is the half-plane right of the line $-4 = x$. We also graph $x = 2$ and see that the graph of $x \le 2$ is the half-plane left of the line $x = 2$.

Finally, we shade the intersection of these graphs.

33. Graph: $0 \le y \le 3$

This is a system of inequalities:
$0 \le y$

$y \le 3$

We graph the equations $0 = y$ and see that the graph of $0 \le y$ is the half-plane above the line $0 = y$. We also graph $y = 3$ and see that the graph of $y \le 3$ is the half-plane below the line $y = 3$. Finally, we shade the intersection of these graphs.

35. $y > x + 3.5$

37.   First get y alone on one side of the inequality.

$$8x - 2y < 11$$
$$-2y < -8x + 11$$
$$y > \frac{-8x + 11}{-2}$$

39.   Graph: $y > x$

$$y < -x + 3$$

We graph the boundary lines $y = x$ and
$y = -x + 3$, using dashed lines. Note where
the regions overlap and shade the region of
solutions.

41.   Graph: $y \leq x$

$$y \leq 2x - 5$$

We graph the boundary lines $y = x$ and
$y = 2x - 5$, using solid lines. Note where the
regions overlap and shade the region of
solutions.

43.   Graph: $y \leq -3$

$$x \geq -1$$

We graph the boundary lines $y = -3$ and
$x = -1$, using solid lines. We indicate the
region for each inequality by the arrows at the
ends of the lines. Note where the regions
overlap and shade the region of solutions.

45.   Graph: $x > -4$

$$y < -2x + 3$$

We graph the lines $x = -4$ and
$y = -2x + 3$ using dashed lines. We indicate
the region for each inequality by the arrows at
the ends of the lines. Note where the regions
overlap and shade the region of solutions.

47.   Graph: $y \leq 5$

$$y \geq -x + 4$$

We graph the lines $y = 5$ and $y = -x + 4$,
using solid lines. We indicate the region for
each inequality by the arrows at the ends of
the lines. Note where the regions overlap and
shade the region of solutions.

49.   Graph: $x + y \leq 6$

$$x - y \leq 4$$

We graph the lines $x + y = 6$ and $x - y = 4$,
using solid lines. We indicate the region for
each inequality by the arrows at the ends of
the lines. Note where the regions overlap and
shade the region of solutions.

51.   Graph: $y + 3x > 0$

$$y + 3x < 2$$

We graph the lines $y + 3x = 0$ and
$y + 3x = 2$, using dashed lines. We indicate
the region for each inequality by the arrows at
the ends of the lines. Note where the regions
overlap and shade the region of solutions.

53. Graph: $y \le 2x - 3$ (1)

          $y \ge -2x + 1$ (2)

          $x \le 5$ (3)

Graph the lines $y = 2x - 3$, $y = -2x + 1$, and $x = 5$ using solid lines. Indicate the region for each inequality by arrows, and shade the region where they overlap.

To find the vertices we solve three different systems of related equations.

From (1) and (2) we have $y = 2x - 3$

                         $y = -2x + 1$.

Solving, we obtain the vertex $(1, -1)$.

From (1) and (3) we have $y = 2x - 3$

                         $x = 5$.

Solving, we obtain the vertex $(5, 7)$.

From (2) and (3) we have $y = -2x + 1$

                         $x = 5$.

Solving, we obtain the vertex $(5, -9)$.

55. Graph: $x + 2y \le 12$ (1)

          $2x + y \le 12$ (2)

          $x \ge 0$ (3)

          $y \ge 0$ (4)

Graph the lines $x + 2y = 12$, $2x + y = 12$, $x = 0$, and $y = 0$ using solid lines. Indicate the region for each inequality by arrows, and shade the region where they overlap.

To find the vertices we solve four different systems of related equations.

From (1) and (2) we have $x + 2y = 12$

                         $2x + y = 12$.

Solving, we obtain the vertex $(4, 4)$.

From (1) and (3) we have $x + 2y = 12$

                         $x = 0$.

Solving, we obtain the vertex $(0, 6)$.

From (2) and (4) we have $2x + y = 12$

                         $y = 0$.

Solving, we obtain the vertex $(6, 0)$.

From (3) and (4) we have $x = 0$

                         $y = 0$.

Solving, we obtain the vertex $(0, 0)$.

57. Graph: $8x + 5y \le 40$ (1)

          $x + 2y \le 8$ (2)

          $x \ge 0$ (3)

          $y \ge 0$ (4)

Graph the lines $8x + 5y = 40$, $x + 2y = 8$, $x = 0$, and $y = 0$ using solid lines. Indicate the region for each inequality by arrows, and shade the region where they overlap.

To find the vertices we solve four different systems of related equations.

From (1) and (2) we have $8x + 5y = 40$

                         $x + 2y = 8$.

Solving, we obtain the vertex $\left(\dfrac{40}{11}, \dfrac{24}{11} \right)$.

From (1) and (4) we have $8x + 5y = 40$

                         $y = 0$.

Solving, we obtain the vertex $(5, 0)$.

From (2) and (3) we have $x + 2y = 8$

                         $x = 0$.

Solving, we obtain the vertex $(0, 4)$.

From (3) and (4) we have $x = 0$

                         $y = 0$.

Solving, we obtain the vertex $(0, 0)$.

59. Graph: $y - x \ge 2$ (1)

          $y - x \le 4$ (2)

          $2 \le x \le 5$ (3)

Think of (3) as two inequalities:

          $2 \le x$ (4)

          $x \le 5$ (5)

Graph the lines $y - x = 2$, $y - x = 4$, $2 = x$, and $x = 5$ using solid lines. Indicate the region for each inequality by arrows, and shade the region where they overlap.

To find the vertices we solve three different systems of related equations.

From (1) and (4) we have $y - x = 2$

$$2 = x.$$

Solving, we obtain the vertex $(2, 4)$.

From (1) and (5) we have $y - x = 2$

$$x = 5.$$

Solving, we obtain the vertex $(5, 7)$.

From (2) and (5) we have $y - x = 4$

$$x = 5.$$

Solving, we obtain the vertex $(5, 9)$.

From (2) and (4) we have $y - x = 4$

$$x = 2.$$

Solving, we obtain the vertex $(2, 6)$.

61. *Thinking and Writing Exercise.*

63. For $3x^3 - 5x^2 - 8x + 7$, when $x = -1$,

$$3(-1)^3 - 5(-1)^2 - 8(-1) + 7$$

$$= -3 - 5 + 8 + 7 = 7$$

65. $3(2t - 7) + 5(3t + 1)$

$$= 6t - 21 + 15t + 5$$

$$= 21t - 16$$

67. $(8t + 6) - (7t + 6)$

$$= 8t + 6 - 7t - 6$$

$$= t$$

69. $(2a - 3) - 4(a + 6)$

$$= 2a - 3 - 4a - 24$$

$$= -2a - 27$$

71. *Thinking and Writing Exercise.*

73. Graph: $x + y > 8$

$$x + y \leq -2$$

Graph the line $x + y = 8$ using a dashed line and graph $x + y = -2$ using a solid line. Indicate the region for each inequality by arrows. The regions do not overlap (the solution set is $\varnothing$), so we do not shade any portion of the graph.

75. Graph: $x - 2y \leq 0$

$$-2x + y \leq 2$$

$$x \leq 2$$

$$y \leq 2$$

$$x + y \leq 4$$

Graph the lines $x - 2y = 0$, $-2x + y = 2$, $x = 2$, $y = 2$, and $x + y = 4$ using solid lines. Indicate the regions for each inequality by arrows. Note where the regions overlap and shade the region of solutions.

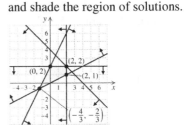

77. $w > 0$

$$h > 0$$

$$w + h + 30 \leq 62, \text{ or}$$

$$w + h \leq 32$$

$$2w + 2h + 30 \leq 130, \text{ or}$$

$$w + h \leq 50$$

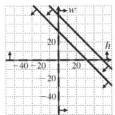

79. Graph: $h \leq 2w$

$w \leq 1.5h$

$h \leq 3200$

$h \geq 0$

$w \geq 0$

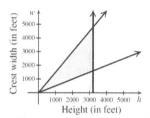

81. Graph: $q + v \geq 1150$

$q \geq 700$

$q \leq 800$

$v \geq 400$

$v \leq 800$

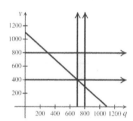

83. The shaded region lies below the graphs of
$y = x$ and $y = 2$, and both lines are solid.
Thus, we have

$y \leq x$

$y \leq 2$

85. The shaded region lies below the graphs of
$y = x + 2$ and $y = -x + 4$, and above $y = 0$.
All of the lines are solid. Thus, we have

$y \leq x + 2$

$y \leq -x + 4$

$y \geq 0$

___

Chapter 4 Study Summary

1. $\{x \mid x \leq 0\}$ in interval notation is $(-\infty, 0]$

2. $x - 11 > -4$

$x - 11 + 11 > -4 + 11$ Adding 11

$x > 7$ Simplifying

The solution set is $\{x \mid x > 7\}$, or $(7, \infty)$.

3. $-8x \leq 2$

$\dfrac{-8x}{-8} \geq \dfrac{2}{-8}$ Dividing by -8

$x \geq -\dfrac{1}{4}$ Simplifying

The solution set is $\left\{ x \mid x \geq -\frac{1}{4} \right\}$, or $\left[-\frac{1}{4}, \infty \right)$.

4. Let d represent the distance Luke runs, in
miles. If Luke runs no less than 3 mi per day,
then $d \geq 3$.

5. Let $y_1 = x - 3$ and $y_2 = 5x + 1$

The graph of a line can be made by
determining two points that lie on the line
using an x-y table. For the first line, defined
by $y = x - 3$, substituting $x = 0$ into the
equation and solving for y gives the point
$(0, -3)$. And substituting $y = 0$ into the
equation gives the point $(3, \ 0)$.

For the second line, defined
by $y = 5x + 1$, substituting $x = 0$ into the
equation and solving for y gives the
point $(0, 1)$. Substituting $x = 2$ into the
equation gives the point $(2, \ 11)$

Using these pairs of points to graph the lines,
the intersection appears to be the point
$(-1, -4)$.

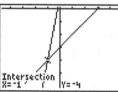

We can verify this by testing the point in each
equation. For the first equation, we have:

$$\begin{array}{r|l} x - 3 & -4 \\ \hline (-1) - 3 & -4 \\ -1 - 3 & -4 \\ -4 & -4 \end{array}$$

For the second equation, we have:

$$\begin{array}{r|l} 5x + 1 & -4 \\ \hline 5(-1) + 1 & -4 \\ -5 + 1 & -4 \\ -4 & -4 \end{array}$$

The value $x = -1$ is the solution to the
equation.

6. $2x - 1 < x$

On a graphing calculator, let $y_1 = 2x - 1$ and $y_2 = x$. We graph the equations in the window $[-5, 5, -5, 5]$.

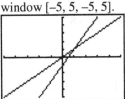

We can see from the graph that the point of intersection is $(1, 1)$, and that $y_1 < y_2$ when $x < 1$. The solution set is $\{x \mid x < 1\}$, or $(-\infty, 1)$.

7. $-5 < 4x + 3 \le 0$

$-8 < 4x \le -3$

$-2 < x \le -\dfrac{3}{4}$

The solution set is $\left\{ x \mid -2 < x \le -\dfrac{3}{4} \right\}$, or $\left(-2, -\frac{3}{4} \right]$.

8. $x - 3 \le 10 \ \ or \ \ 25 - x < 3$

$\quad x \le 13 \ \ or \quad -x < -22$

$\quad x \le 13 \ \ or \quad\quad x > 22$

The solution set is $\{ x \mid x \le 13 \ \ or \ \ x > 22 \}$, or $(-\infty, 13] \cup (22, \infty)$.

9. $|4x - 7| = 11$

$4x - 7 = -11 \ \ or \ \ 4x - 7 = 11$

$\quad 4x = -4 \ \ or \quad\quad 4x = 18$

$\quad\quad x = -1 \ \ or \quad\quad\quad x = \dfrac{9}{2}$

The solution set is $\left\{ -1, \dfrac{9}{2} \right\}$.

10. $|x - 12| \le 1$

$-1 \le x - 12 \le 1$

$11 \le x \le 13$

The solution set is $\{ x \mid 11 \le x \le 13 \}$, or $[11, 13]$.

11. $|2x + 3| > 7$

$2x + 3 < -7 \ \ or \ \ 7 < 2x + 3$

$\quad 2x < -10 \ or \ \ 4 < 2x$

$\quad\quad x < -5 \ \ or \ \ 2 < x$

The solution set is $\{ x \mid x < -5 \ \ or \ \ x > 2 \}$, or $(-\infty, -5) \cup (2, \infty)$.

12. $2x - y < 5$

First graph the boundary line $2x - y = 5$ using a dashed line. Since the test point $(0, 0)$ is a solution and $(0, 0)$ is above the line, we shade the half-plane above the line.
The solution is the region which is shaded.

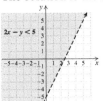

___

Chapter 4 Review Exercises

1. True

2. False

3. True

4. True

5. False

6. True

7. True

8. True

9. False

10. False

11. $x \le -2$

Graph: The solution consists of all real numbers to the left of –2, so we shade all numbers to the left of –2 and use a bracket at –2 to indicate that it is also a solution.

$\{x \mid x \le -2\}$, or $(-\infty, -2]$

12. $a + 7 \le -14$

$a \le -21$

Graph: The solution consists of all real numbers to the left of –21, so we shade all numbers to the left of –21 and use a bracket at –21 to indicate that it is also a solution.

Set builder notation: $\{a \mid a \le -21\}$

Interval notation: $(-\infty, -21]$

13. $4y > -15$

$y > -\dfrac{15}{4}$

Graph: The solution consists of all real numbers to the right of $-\dfrac{15}{4}$, so we shade all numbers to the right of $-\dfrac{15}{4}$ and use a parenthesis at $-\dfrac{15}{4}$ to indicate that it is not a solution.

Set builder notation: $\left\{ y \mid y > -\dfrac{15}{4} \right\}$

Interval notation: $\left(-\dfrac{15}{4}, \infty \right)$

14. $-0.3y < 9$

$y > -30$

Graph: The solution consists of all real numbers to the right of –30, so we shade all numbers to the right of –30 and use a parenthesis at –30 to indicate that it is not a solution.

Set builder notation: $\{y \mid y > -30\}$

Interval notation: $(-30, \infty)$

15. $-6x - 5 < 4$

$-6x < 9$

$x > -\dfrac{3}{2}$

Graph: The solution consists of all real numbers to the right of $-\dfrac{3}{2}$, so we shade all numbers to the right of $-\dfrac{3}{2}$ and use a parenthesis at $-\dfrac{3}{2}$ - to indicate that it is not a solution.

Set builder notation: $\left\{ x \mid x > -\dfrac{3}{2} \right\}$

Interval notation: $\left(-\dfrac{3}{2}, \infty \right)$

16. $-\frac{1}{2}x - \frac{1}{4} > \frac{1}{2} - \frac{1}{4}x$

$-\frac{1}{2}x + \frac{1}{4}x > \frac{1}{2} + \frac{1}{4}$

$-\frac{1}{4}x > \frac{3}{4}$

$x < -3$

Graph: The solution consists of all real numbers to the left of –3, so we shade all numbers to the left of –3 and use a parenthesis at –3 to indicate that it is not a solution.

Set builder notation: $\{x \mid x < -3\}$

Interval notation: $(-\infty, -3)$

17. $0.3y - 7 < 2.6y + 15$

$-22 < 2.3y$

$-\dfrac{220}{23} < y$

Graph: The solution consists of all real numbers to the right of $-\dfrac{220}{23}$, so we shade all numbers to the left of $-\dfrac{220}{23}$ and use a parenthesis at $-\dfrac{220}{23}$ to indicate that it is not a solution.

Set builder notation: $\left\{ y \middle| y > -\dfrac{220}{23} \right\}$

Interval notation: $\left(-\dfrac{220}{23}, \infty \right)$

18. $-2(x-5) \geq 6(x+7) - 12$

$-2x + 10 \geq 6x + 42 - 12$

$-2x + 10 \geq 6x + 30$

$-8x \geq 20$

$x \leq -\dfrac{5}{2}$

Graph: The solution consists of all real numbers to the left of $-\dfrac{5}{2}$, so we shade all numbers to the left of $-\dfrac{5}{2}$ and use a bracket at $-\dfrac{5}{2}$ to indicate that it is also a solution.

Set builder notation: $\left\{ x \middle| x \leq -\dfrac{5}{2} \right\}$

Interval notation: $\left(-\infty, -\dfrac{5}{2} \right]$

19. $f(x) = 3x - 5;\ g(x) = 11 - x$

$f(x) \leq g(x)$

$3x - 5 \leq 11 - x$

$4x \leq 16$

$x \leq 4$

Set builder notation: $\{ x \mid x \leq 4 \}$

Interval notation: $(-\infty, 4]$

20. **Familiarize.** Let x = the number of hours Mariah will work at either job. Then Mariah will make $8.40x$ at the sandwich shop, and she will make $16.00x - \$950$ if she does carpentry work.

Translate. We write an inequality stating that the earnings from carpentry work will exceed earnings from the sandwich shop.

$16.00x - 950 > 8.40x$

Carry out.

$16.00x - 950 > 8.40x$

$7.60x > 950$

$x > 125$

Check. We can do a partial check by substituting a value for x that is less than 125 and a value that is greater than 125. If $x =$ 124, Mariah will make $\$16.00(124) - \$950$, or $\$1984 - \$950 = \$1032$ doing carpentry work, and she will make $\$8.40(124)$, or $\$1041.60$ at the sandwich shop. However, if $x = 126$, Rose will make $\$16.00(126) - \$950$, or $\$2016 - \$950$, or $\$1066$ doing carpentry work, and she will make $\$8.40(126)$, or $\$1058.40$ at the sandwich shop. We cannot check all possible values for x, so we stop here.

State. As long as Mariah works more than 125 hr, she will make more money doing carpentry work.

21. **Familiarize.** Let x = the amount Clay should invest at 3%. Then $\$9000 - x$ is the amount he will invest at 3.5%. He will make $0.03x$ on the money invested at 3%, and he will make $0.035(\$9000 - x)$ on the money he invests at 3.5%.

Translate. We write an inequality such that his interest income will be at least $300 each year.

$0.03x + 0.035(9000 - x) \geq 300$

Carry out.

$0.03x + 0.035(9000 - x) \geq 300$

$0.03x + 315 - 0.035x \geq 300$

$-0.005x \geq -15$

$x \leq 3000$

Check. If Clay invests $2990 at 3%, then he would invest $9000 - 2990$, or $6010 at 3.5%, and he would earn $0.03(\$2990) +$ $0.035(\$6010)$, or $\$89.70 + \$210.35$, or $\$300.05$. However, if Clay invests $3010 at 3%, then he would invest $5990 at 3.5%, and he would earn $0.03(\$3010) + 0.035(\$5990)$, or $\$90.30 + \$209.65$, or $\$299.95$. We cannot check all possible values for x, so we stop here.

State. Clay must invest no more than $3000 at 3% in order to be guaranteed $300 in interest each year.

22. Solve $x - 3 = 3x + 5$.

We graph $f(x) = x - 3$ and $g(x) = 3x + 5$ on the same axes.

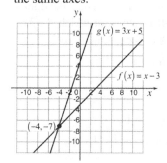

It appears that the lines intersect at $(-4, -7)$.

$$\frac{x - 3 = 3x + 5}{-4 - 3 \mid 3(-4) + 5}$$
$$\quad -7 \mid \quad\quad -7 \text{ TRUE}$$

The solution is –4.

23. Solve $x + 1 \geq \frac{1}{2}x - 2$.

We graph $f(x) = x + 1$ and $g(x) = \frac{1}{2}x - 2$ on the same axes.

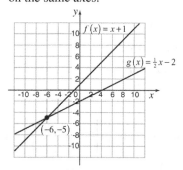

It appears that the lines intersect at $(-6, -5)$.

Note from the graphs that $f(x) \geq g(x)$ when the graph of f lies above the graph of g. In fact, for all values of x greater than or equal to –6, $f(x) \geq g(x)$. So the solution set is $\{x \mid x \geq -6\}$, or $[-6, \infty)$.

24. $\{a, b, c, d\} \cap \{a, c, e, f, g\}$

The letters a and c are common to both sets, so the intersection is $\{a, c\}$.

25. $\{a, b, c, d\} \cup \{a, c, e, f, g\}$

The numbers a, b, c, d, e, f, and g are in either or both sets, so the union is $\{a, b, c, d, e, f, g\}$.

26. $x \leq 3 \ and \ x > -5$

This conjunction can be abbreviated as
$$-5 < x \leq 3.$$

Interval notation: $(-5, 3]$

27. $x \leq 3 \ or \ x > -5$

Interval notation: $(-\infty, \infty)$

28. $-4 < x + 8 \leq 5$

$-12 < x \leq -3$

Set builder notation: $\{x \mid -12 < x \leq -3\}$

Interval notation: $(-12, -3]$

29. $-15 < -4x - 5 < 0$

$-10 < -4x < 5$

$$\frac{5}{2} > x > -\frac{5}{4}$$

Set builder notation: $\left\{ x \mid -\frac{5}{4} < x < \frac{5}{2} \right\}$

Interval notation: $\left(-\frac{5}{4}, \frac{5}{2} \right)$

30. $3x < -9 \ or \ -5x < -5$

$x < -3 \ or \ x > 1$

Set builder notation: $\{x \mid x < -3 \ or \ x > 1\}$

Interval notation: $(-\infty, -3) \cup (1, \infty)$

31. $2x+5<-17$ or $-4x+10\le 34$

$\quad\quad 2x<-22$ or $\quad\quad -4x\le 24$

$\quad\quad\quad x<-11$ or $\quad\quad\quad x\ge -6$

Set builder notation: $\{x\,|\,x<-11\ or\ x\ge -6\}$

Interval notation: $(-\infty,-11)\cup[-6,\infty)$

32. $2x+7\le -5$ or $x+7\ge 15$

$\quad\quad 2x\le -12$ or $\quad\quad x\ge 8$

$\quad\quad\quad x\le -6$ or $\quad\quad x\ge 8$

Set builder notation: $\{x\,|\,x\le -6\ or\ x\ge 8\}$

Interval notation: $(-\infty,-6]\cup[8,\infty)$

33. $f(x)<-5$ or $f(x)>5$

$\quad 3-5x<-5$ or $3-5x>5$

$\quad\quad 8<5x$ or $-2>5x$

$\quad\quad \dfrac{8}{5}<x$ or $-\dfrac{2}{5}>x$

Set builder notation: $\left\{x\,\middle|\,x<-\dfrac{2}{5}\ or\ x>\dfrac{8}{5}\right\}$

Interval notation: $\left(-\infty,-\dfrac{2}{5}\right)\cup\left(\dfrac{8}{5},\infty\right)$

34. $f(x)=\dfrac{2x}{x-8}$

$f(x)$ cannot be computed when the denominator is 0. Since $x-8=0$ is equivalent for $x=8$, we have Domain of $f=\{x\,|\,x\ is\ any\ real\ number\ and\ x\ne 8\}=$ $(-\infty,8)\cup(8,\infty)$.

35. $f(x)=\sqrt{x+5}$

The expression $\sqrt{x+5}$ is not a real number when $x+5$ is negative. Thus, the domain of f is the set of all x-values for which $x+5\ge 0$. Since $x+5\ge 0$ is equivalent to $x\ge -5$, we have the Domain of $f=[-5,\infty)$.

36. $f(x)=\sqrt{8-3x}$

The expression $\sqrt{8-3x}$ is not a real number when $8-3x$ is negative. Thus, the domain of f is the set of all x-values for which $8-3x\ge 0$. Since $8-3x\ge 0$ is equivalent to $\frac{8}{3}\ge x$, we have the Domain of $f=\left(-\infty,\frac{8}{3}\right]$.

37. $|x|=11$

$\quad x=-11$ or $x=11$

The solution set is $\{-11,11\}$.

38. $|t|\ge 21$

$\quad t\le -21$ or $t\ge 21$

The solution set is $\{t\,|\,t\le -21\ or\ t\ge 21\}$, or $(-\infty,-21]\cup[21,\infty)$.

39. $|x-3|=7$

$\quad x-3=-7$ or $x-3=7$

$\quad\quad x=-4$ or $\quad\quad x=10$

The solution set is $\{-4,10\}$.

40. $|2x+5|<12$

$\quad -12<2x+5<12$

$\quad -17<2x<7$

$\quad -\dfrac{17}{2}<x<\dfrac{7}{2}$

The solution set is $\left\{x\,\middle|-\dfrac{17}{2}<x<\dfrac{7}{2}\right\}$, or $\left(-\dfrac{17}{2},\dfrac{7}{2}\right)$.

41. $|3x-4|\ge 15$

$\quad -15\ge 3x-4$ or $15\le 3x-4$

$\quad -11\ge 3x$ or $19\le 3x$

$\quad -\dfrac{11}{3}\ge x$ or $\dfrac{19}{3}\le x$

The solution set is $\left\{x\,\middle|\,x\le -\dfrac{11}{3}\ or\ x\ge \dfrac{19}{3}\right\}$, or $\left(-\infty,-\dfrac{11}{3}\right]\cup\left[\dfrac{19}{3},\infty\right)$.

42. $|2x+5|=|x-9|$

$2x+5=x-9 \ or \ 2x+5=-(x-9)$

$x=-14 \ or \ 2x+5=-x+9$

$x=-14 \ or \ 3x=4$

$x=-14 \ or \ x=\dfrac{4}{3}$

The solution set is $\left\{-14,\dfrac{4}{3}\right\}$.

43. $|5n+6|=-8$

The absolute value is always nonnegative, so the equation has not solution. The solution set is $\varnothing$.

44. $\left|\dfrac{x+4}{6}\right|\le 2$

$-2\le\dfrac{x+4}{6}\le 2$

$-12\le x+4\le 12$

$-16\le x\le 8$

The solution set is $\{x\,|-16\le x\le 8\}$, or $[-16,8]$.

45. $2|x-5|-7>3$

$2|x-5|>10$

$|x-5|>5$

$x-5<-5 \ or \ x-5>5$

$x<0 \ or \ x>10$

The solution set is $\{x\,|\,x<0 \ or \ x>10\}$, or $(-\infty,0)\cup(10,\infty)$.

46. $19-3|x+1|\ge 4$

$-3|x+1|\ge -15$

$|x+1|\le 5$

$-5\le x+1\le 5$

$-6\le x\le 4$

The solution set is $\{x\,|-6\le x\le 4\}$, or $[-6,4]$.

47. $f(x)=|3x-5|$

$f(x)<0$

$|3x-5|<0$

The absolute value is always nonnegative, so the equation has not solution. The solution set is $\varnothing$.

48. Graph: $x-2y\ge 6$

First graph the line $x-2y=6$. Draw it solid since the inequality symbol is $\ge$. Test the point $(0,0)$ to determine if it is a solution.

$$\dfrac{x-2y\ge 6}{0-2\cdot 0 \ \bigl|\ 6}$$

$$0 \ \bigl|\ 6 \quad\text{False}$$

Since $0\ge 6$ is false, $(0,0)$ is not a solution, nor are any points in the half-plane containing $(0,0)$. The points in the other half-plane are solutions, so we shade that half-plane and obtain the graph.

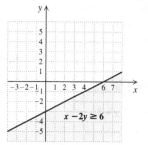

49. Graph: $x+3y>-1$

$x+3y<4$

We graph the lines $x+3y=-1$ and $x+3y=4$, using dashed lines. We indicate the region for each inequality by arrows at the ends of the lines. Note where the regions overlap and shade the region of solutions.

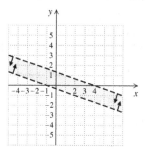

50. Graph: $x-3y\le 3$

$x+3y\ge 9$

$y\le 6$

We graph the lines $x-3y=3$, $x+3y=9$, and $y=6$ using solid lines. We indicate the region for each inequality by arrows at the ends of the lines. Note where the regions overlap and shade the region of solutions.

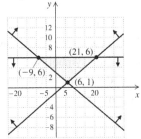

51. *Thinking and Writing Exercise*

52. *Thinking and Writing Exercise.*

53. $|2x+5| \le |x+3|$

This inequality says that the distance from $x+3$ to zero is greater than or equal than the distance from $2x+5$ to zero.

Case 1: $2x+5 \ge 0$ and $x+3 \ge 0$, which

imply that $x \ge -\dfrac{5}{2}$ and $x \ge -3$.

$2x+5 \le x+3$

$x \le -2$

The intersection of these three

intervals is $\left[-\dfrac{5}{2}, -2 \right]$.

Case 2: $2x+5 < 0$ and $x+3 < 0$, which

imply that $x < -\dfrac{5}{2}$ and $x < -3$.

$-(2x+5) \le -(x+3)$

$-2x-5 \le -x-3$

$-2 \le x$

This is impossible so we rule it out.

Case 3: $2x+5 < 0$ and $x+3 \ge 0$, which

imply that $x < -\dfrac{5}{2}$ and $x \ge -3$.

$-(2x+5) \le x+3$

$-2x-5 \le x+3$

$-8 \le 3x$

$-\dfrac{8}{3} \le x$

The intersection of these three

intervals is $\left[-\dfrac{8}{3}, -\dfrac{5}{2} \right]$.

Case 4: $2x+5 \ge 0$ and $x+3 < 0$, which

imply that $x \ge -\dfrac{5}{2}$ and $x < -3$.

$2x+5 \le -(x+3)$

$2x+5 \le -x-3$

$3x \le -8$

$x \le -\dfrac{8}{3}$

This case is impossible so we rule it out.

The union of the intervals for Cases 1 and 3 is

$\left[-\dfrac{5}{2}, -2 \right] \cup \left[-\dfrac{8}{3}, -\dfrac{5}{2} \right] = \left[-\dfrac{8}{3}, -2 \right]$, or

$\left\{ x \left| -\dfrac{8}{3} \le x \le -2 \right. \right\}$.

54. The statement: *If $x < 3$, then $x^2 < 9$*, is false. Assume $x = -4$. Then it is true that $-4 < 3$, however $(-4)^2 = 16$ which is not less than 9.

55. Let d represent the diameter of a doorknob. The tolerance, or allowable variation in diameter is then $|d-2.5| \le 0.003$.

Chapter 4 Test

1. $x-3 < 8$

$x < 11$

The solution set is $\{x \mid x < 11\}$, or $(-\infty, 11)$.

2. $-\dfrac{1}{2}t < 12$

$t > -24$

The solution set is $\{t \mid t > -24\}$, or $(-24, \infty)$.

3. $-4y-3 \ge 5$

$-4y \ge 8$

$y \le -2$

The solution set is $\{y \mid y \le -2\}$, or $(-\infty, -2]$.

4. $3a - 5 \le -2a + 6$

    $5a \le 11$

    $a \le \dfrac{11}{5}$

    The solution set is $\left\{ a \middle| a \le \dfrac{11}{5} \right\}$, or $\left(-\infty, \dfrac{11}{5} \right]$.

    ← —————|———|——→
         0   $\frac{11}{5}$

5. $3(7 - x) < 2x + 5$

    $21 - 3x < 2x + 5$

    $-5x < -16$

    $x > \dfrac{16}{5}$

    The solution set is $\left\{ x \middle| x > \dfrac{16}{5} \right\}$, or $\left(\dfrac{16}{5}, \infty \right)$.

    ← —————|———|——→
         0   $\frac{16}{5}$

6. $-2(3x - 1) - 5 \ge 6x - 4(3 - x)$

    $-6x + 2 - 5 \ge 6x - 12 + 4x$

    $-6x - 3 \ge 10x - 12$

    $9 \ge 16x$

    $x \le \dfrac{9}{16}$

    The solution set is $\left\{ x \middle| x \le \dfrac{9}{16} \right\}$, or $\left(-\infty, \dfrac{9}{16} \right]$.

    ← —————|———|——→
         0   $\frac{9}{16}$

7. $f(x) = -5x - 1;\ g(x) = -9x + 3$

    Find $f(x) > g(x)$:

    $-5x - 1 > -9x + 3$

    $4x > 4$

    $x > 1$

    The solution set is $\{ x \mid x > 1 \}$, or $(1, \infty)$.

8. **Familiarize.** Let x = the number of miles Dani will drive the van. The cost for the unlimited mileage plan is \$80, and the cost of the per-mile plan is \$45 + \$0.40(x – 100).

    Translate. We write an inequality stating that \$80 is less than \$45 + \$0.40(x – 100).

    $80 < 45 + 0.40(x - 100)$

    Carry out.

    $80 < 45 + 0.40(x - 100)$

    $80 < 45 + 0.40x - 40$

    $80 < 5 + 0.40x$

    $75 < 0.40x$

    $\dfrac{75}{0.40} < x$

    $187\tfrac{1}{2} < x$

    Check. We can do a partial check by substituting a value for x that is less than $187\tfrac{1}{2}$. When $x = 187$, the unlimited plan costs \$80 and the per-mile plan costs \$45 + \$0.40(187 – 100) = \$45 + \$0.40(87), or \$45 + \$34.80, or \$79.80. This is less than the unlimited plan. When $x = 188$, the per-mile plan costs \$45 + \$0.40(188 – 100), or \$45 + \$0.40(88), or \$45 + \$35.20, or \$80.20. So, as long as the miles driven exceed $187\tfrac{1}{2}$, the unlimited plan is cheaper. We cannot check all possible values for x, so we stop here.

    State. The unlimited plan is cheaper if the miles driven exceed $166\tfrac{2}{3}$ miles.

9. **Familiarize.** Let x = the time, in hours, required for the service call after the first half-hour. The refrigeration company charges \$80 for the first half-hour and \$60 for each additional hour, so they would charge \$80 + \$60x, assuming the service call takes more than one half-hour.

    Translate. Write an inequality stating that the service call will cost no more than \$200.

    $80 + 60x \le 200$

    Carry out.

    $80 + 60x \le 200$

    $60x \le 120$

    $x \le 2$

    Check. Recall, x means time for service, in hours, *after* the first half-hour. If the service call takes more than 2.5 hours, say 2.6 hours, the cost will be \$80 + \$60(2.1), or \$80 + \$126, or \$206, which exceeds the camp's budget. We cannot check all possible values for x, so we stop here.

    State. As long as the service takes less than or equal to 2.5 hours, the camp can afford to repair the cooler.

10. $2x+1=3x-2$

We graph $f(x)=2x+1$ and $g(x)=3x-2$.

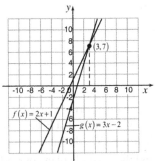

It appears that the lines intersect at the point $(3,7)$. We check

$$2x+1=3x-2$$

$2 \cdot 3 + 1$	$3 \cdot 3 - 2$
$6+1$	$9-2$
7	7 True

Our check shows that 3 is the solution.

11. $x+2>\frac{1}{3}x$

On a graphing calculator, let $y_1=x+2$ and $y_2=\frac{1}{3}x$. We let $y_3=y_1>y_2$. We graph all three equations in the window $[-5, 5, -5, 5]$.

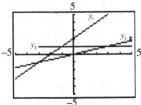

We can see from the graph that $y_1>y_2$ when $x>-3$. The solution set is $\{x\,|\,x>-3\}$, or $(-3,\infty)$.

12. $\{a, e, i, o, u\} \cap \{a, b, c, d, e\}$

The letters a and e are common to both sets, so the intersection is $\{a, e\}$.

13. $\{a, e, i, o, u\} \cup \{a, b, c, d, e\}$

The letters $a, b, c, d, e, i, o,$ and u are in either or both sets, so the union is $\{a, b, c, d, e, i, o, u\}$.

14. $f(x)=\sqrt{6-3x}$

The expression $\sqrt{6-3x}$ is not a real number when $6-3x$ is negative. Thus, the domain of f is the set of all x-values for which $6-3x\geq 0$. Since $6-3x\geq 0$ is equivalent to $x\leq 2$, we have Domain of $f=(-\infty,2]$.

15. $f(x)=\dfrac{x}{x-7}$

$f(x)$ cannot be computed when the denominator is 0. Since $x-7=0$ is equivalent for $x=7$, we have Domain of $f\{x\,|\,x$ is any real number $and\ \ x\neq 7\}=(-\infty,7)\cup(7,\infty)$.

16. $-5<4x+1\leq 3$

$-6<4x\leq 2$

$\dfrac{-3}{2}<x\leq\dfrac{1}{2},$

The solution set is $\left\{x\,\middle|-\dfrac{3}{2}<x\leq\dfrac{1}{2}\right\}$, or $\left(-\dfrac{3}{2},\dfrac{1}{2}\right]$.

17. $3x-2<7\ \ or\ \ x-2>4$

$3x<9\ \ or\ \ x>6$

$x<3\ \ or\ \ x>6$

The solution set is $\{x\,|\,x<3\ \ or\ \ x>6\}$, or $(-\infty,3)\cup(6,\infty)$.

18. $-3x>12\ \ or\ \ 4x>-10$

$x<-4\ \ or\ \ x>-\frac{5}{2}$

The solution set is $\left\{x\,\middle|\,x<-4\ \ or\ \ x>-\dfrac{5}{2}\right\}$, or $(-\infty,-4)\cup\left(\dfrac{5}{2},\infty\right)$.

19. $1\leq 3-2x\leq 9$

$-2\leq -2x\leq 6$

$-3\leq x\leq 1$

The solution set is $\{x\,|\,-3\leq x\leq 1\}$, or $[-3,1]$.

20. $|n| = 15$

$n = -15 \ or \ n = 15$

The solution set is $\{-15, 15\}$.

21. $|a| > 5$

$a < -5 \ or \ a > 5$

The solution set is $\{a \mid a < -5 \ or \ a > 5\}$, or

$(-\infty, -5) \cup (5, \infty)$.

22. $|3x - 1| < 7$

$-7 < 3x - 1 < 7$

$-6 < 3x < 8$

$-2 < x < \dfrac{8}{3}$

The solution set is $\left\{ x \mid -2 < x < \dfrac{8}{3} \right\}$, or

$\left(-2, \dfrac{8}{3} \right)$.

23. $|-5t - 3| \geq 10$

$-5t - 3 \leq -10 \ or \ 10 \leq -5t - 3$

$-5t \leq -7 \ or \ 13 \leq -5t$

$t \geq \dfrac{7}{5} \ or \ -\dfrac{13}{5} \geq t$

The solution set is $\left\{ t \mid t \leq -\dfrac{13}{5} \ or \ t \geq \dfrac{7}{5} \right\}$, or

$\left(-\infty, -\dfrac{13}{5} \right] \cup \left[\dfrac{7}{5}, \infty \right)$.

24. $|2 - 5x| = -12$

Absolute value is always nonnegative, so the question has no solution. The solution set is $\varnothing$.

25. $g(x) < -3 \ or \ g(x) > 3; \ g(x) = 4 - 2x$

$4 - 2x < -3 \ or \ 4 - 2x > 3$

$-2x < -7 \ or \ -2x > -1$

$x > \dfrac{7}{2} \ or \ x < \dfrac{1}{2}$

The solution set is $\left\{ x \mid x < \dfrac{1}{2} \ or \ x > \dfrac{7}{2} \right\}$, or

$\left(-\infty, \dfrac{1}{2} \right) \cup \left(\dfrac{7}{2}, \infty \right)$.

26. $f(x) = |2x - 1|$ and $g(x) = |2x + 7|$

$f(x) = g(x)$

$|2x - 1| = |2x + 7|$

$2x - 1 = 2x + 7 \ or \ 2x - 1 = -(2x + 7)$

$0 = 8 \quad or \ 2x - 1 = -2x - 7$

False - $\qquad\qquad 4x = -6$

yields no $\qquad\qquad x = -\dfrac{6}{4}$

solution

The solution set is $\left\{ -\dfrac{3}{2} \right\}$.

27. $y \leq 2x + 1$

28. $x + y \geq 3 \quad (1)$

$x - y \geq 5 \quad (2)$

First sketch the graph of $x + y = 3$ using a solid line. Since the test point $(0, 0)$ is not a solution and $(0, 0)$ is below the line, we shade the half-plane above the line (note the arrows pointing up-and-right).

Next, sketch the graph of $x - y = 5$ using a solid line. Since the test point $(0, 0)$ is not a solution and $(0, 0)$ is above the line, we shade the half-plane below the line (note the arrows pointing down-and-right).

The solution is the region which is shaded.

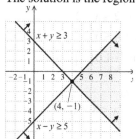

29. $2y - x \geq -7$

$2y + 3x \leq 15$

$y \leq 0$

$x \leq 0$

First sketch the graph of $2y - x = -7$ using a solid line. Since the test point $(0,0)$ is a solution and $(0,0)$ is above the line, we shade the half-plane above the line (note the arrows pointing up-and-left).

Next, sketch the graph of $2y + 3x = 15$ using a solid line. Since the test point $(0,0)$ is a solution and $(0,0)$ is below the line, we shade the half-plane below the line (note the arrows pointing down-and-left).

We graph the equation $y = 0$ and note that the graph of $y \leq 0$ is the half-plane below the x-axis.

We graph the equation $x = 0$ and note that the graph of $x \leq 0$ is the half-plane left of the y-axis.

The solution is the region which is shaded which is where all the regions overlap.

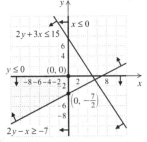

30. $|2x - 5| \leq 7 \ \ and \ \ |x - 2| \geq 2$

For the inequality $|2x - 5| \leq 7$:

$-7 \leq 2x - 5 \leq 7$

$-2 \leq 2x \leq 12$

$-1 \leq x \leq 6$

For the inequality $|x - 2| \geq 2$:

$x - 2 \leq -2 \ \ or \ \ 2 \leq x - 2$

$x \leq 0 \ \ \ or \ \ 4 \leq x$

The solution set is $[-1, 0] \cup [4, 6]$.

31. $7x < 8 - 3x < 6 + 7x$

For the inequality $7x < 8 - 3x$:

$7x < 8 - 3x$

$10x < 8$

$x < \dfrac{4}{5}$

For the inequality $8 - 3x < 6 + 7x$:

$-10x < -2$

$x > \dfrac{1}{5}$

The solution is $\left(\dfrac{1}{5}, \dfrac{4}{5} \right)$.

32. The length of the segment from -8 to 2 is $|-8 - 2| = |-10| = 10$ units. The midpoint of the segment is $\dfrac{-8 + 2}{2} = \dfrac{-6}{2} = -3$. Thus, the interval extends 10/2, or 5, units on each side of -3. An inequality for which the closed interval is the solution set is $|x - (-3)| \leq 5$, or $|x + 3| \leq 5$.

Chapter 5

Polynomials and Polynomial Functions

1. b

3. h

5. g

7. a

9. $3x - 7$ can be written as a sum of monomials, so it is a polynomial.

11. $\dfrac{x^2 + x + 1}{x^3 - 7}$ cannot be written as a sum of monomials, so it is not a polynomial.

13. $\dfrac{1}{4}x^{10} - 8.6$ can be written as a sum of monomials, so it is a polynomial.

15. $7x^4 + x^3 - 5x + 8 = 7x^4 + x^3 + (-5x) + 8$. The terms are $7x^4$, x^3, $-5x$, and 8.

17. $-t^6 + 7t^3 - 3t^2 + 6 = -t^4 + 7t^3 + (-3t^2) + 6$. The terms are $-t^6$, $7t^3$, $-3t^2$, and 6.

19. Three monomials are added, so $x^2 - 23x + 17$ is a trinomial.

21. The polynomial $x^3 - 7x^2 + 2x - 4$ is a polynomial with no special name.

23. Two monomials are added, so $y + 5$ is a binomial.

25. The polynomial 17 is a monomial because it is the product of a constant and a variable raised to a whole number power. (In this case the variable is raised to the power 0.)

27. $3x^2 - 5x$

Term	$3x^2$	$-5x$
Degree	2	1

29. $2t^5 - t^2 + 1$

Term	$2t^5$	$-t^2$	1
Degree	5	2	0

31. $8x^2y - 3x^4y^3 + y^4$

Term	$8x^2y$	$-3x^4y^3$	y^4
Degree	3	7	4

33. $4x^5 + 7x - 3$

Term	$4x^5$	$7x$	-3
Coefficient	4	7	-3

35. $x^4 - x^3 + 4x$

Term	x^4	$-x^3$	$4x$
Coefficient	1	-1	$4x$

37. $a^2b^3 - 5ab + 7b^2 + 1$

Term	a^2b^3	$-5ab$	$7b^2$	1
Coefficient	3	-5	7	1

39. $-5x^6 + x^4 + 7x^3 - 2x - 10$
   a) Number of terms: 5
   b)

Term	$-5x^6$	x^4	$7x^3$	$-2x$	-10
Degree	6	4	3	1	0

   c) Degree of polynomial: 6
   d) Leading term: $-5x^6$
   e) Leading coefficient: -5

41. $7a^4 + a^3b^2 - 5a^2b + 3$
   a) Number of terms: 4
   b)

Term	$7a^4$	a^3b^2	$-5a^2b$	3
Degree	4	5	3	0

   c) Degree of polynomial: 5
   d) Leading term: a^3b^2
   e) Leading coefficient: 1

43. $8y^2 + y^5 - 9 - 2y + 3y^4$

Term	$8y^2$	y^5	-9	$-2y$	$3y^4$
Degree	2	5	0	1	4
Degree of polynomial			5		

45. $3p^4 - 5pq + 2p^3q^3 + 8pq^2 - 7$

Term	$3p^4$	$-5pq$	$2p^3q^3$	$8pq^2$	-7
Degree	4	2	6	3	0
Degree of polynomial			6		

47. $-15t^4 + 2t^3 + 5t^2 - 8t + 4$; $-15t^4$; -15

49. $-x^6 + 6x^5 + 7x^2 + 3x - 5$; $-x^6$; -1

51. $-9 + 4x + 5x^3 - x^6$

53. $8y + 5xy^3 + 2x^2y - x^3$

55. $g(x) = x - 5x^2 + 4$
$g(3) = 3 - 5(3)^2 + 4 = -38$

57. $f(x) = -3x^4 + 5x^3 + 6x - 2$
$f(x) = -3(-1)^4 + 5(-1)^3 + 6(-1) - 2$
$\quad = -3 - 5 - 6 - 2$
$\quad = -16$

59. $F(x) = 2x^2 - 6x - 9$
$F(2) = 2 \cdot 2^2 - 6 \cdot 2 - 9 = 8 - 12 - 9 = -13$
$F(5) = 2 \cdot 5^2 - 6 \cdot 5 - 9 = 50 - 30 - 9 = 11$

61. $Q(y) = -8y^3 + 7y^2 - 4y - 9$
$Q(-3) = -8(-3)^3 + 7(-3)^2 - 4(-3) - 9$
$\quad = 216 + 63 + 12 - 9 = 282$
$Q(0) = -8 \cdot 0^3 + 7 \cdot 0^2 - 4 \cdot 0 - 9$
$\quad = 0 + 0 + 0 - 9 = -9$

63. Use $t = 2$ because 2006 is 2 years after 2004.
$a(t) = 0.4x + 1.13$
$a(2) = 0.4(2) + 1.13$
$\quad = 0.8 + 1.13$
$\quad = 1.93$
The amount spent on shoes for college in 2006 is $1.93 billion.

65. $n(t) = 11.12t^2$
$n(10) = 11.12 \cdot 10^2$
$\quad = 11.12 \cdot 100$
$\quad = 1112$
The skydivers had fallen approximately 1112 ft. 10 sec. after jumping from a plane.

67. $A = \pi r^2 \quad \pi \approx 3.14$
$\quad \approx 3.14 \cdot (7\,\text{m})^2$
$\quad \approx 3.14 \cdot 49\,\text{m}^2$
$\quad \approx 153.86\,\text{m}^2$

69. $s(t) = 16t^2$
$s(2.9) = 16 \cdot (2.9)^2 = 16 \cdot 8.41 = 134.56$
Approximately 135 feet

71. From the graph, there appears to be about 55 million Web sites in 2004. Using the polynomial, the number is about 54 million:
$w(t) = 4.03t^2 + 6.78t + 42.86$
$w(2004 - 2003) = w(1)$
$\quad = 4.03(1)^2 + 6.78(1) + 42.86 = 53.67$

73. From the graph, 5 years after 2003 appears to have about 175 million Web sites. Using the polynomial, the number is about 178 million:
$w(t) = 4.03t^2 + 6.78t + 42.86$
$w(5) = 4.03(5)^2 + 6.78(5) + 42.86$
$\quad = 4.03(25) + 6.78(5) + 42.86 = 177.51$

75. $N(x) = \frac{1}{3}x^3 + \frac{1}{2}x^2 + \frac{1}{6}x$

$N(3) = \frac{1}{3} \cdot 3^3 + \frac{1}{2} \cdot 3^2 + \frac{1}{6} \cdot 3$

$= \frac{1}{3} \cdot 27 + \frac{1}{2} \cdot 9 + \frac{1}{6} \cdot 3$

$= 9 + \frac{9}{2} + \frac{1}{2}$

$= 14$

From the diagram we see that the bottom layer contains 9 balls, the second layer contains 4, and the top layer contains 1 for a total of $9 + 4 + 1$, or 14.

$N(5) = \frac{1}{3} \cdot 5^3 + \frac{1}{2} \cdot 5^2 + \frac{1}{6} \cdot 5$

$= \frac{125}{3} + \frac{25}{2} + \frac{5}{6}$ LCD is 6.

$= \frac{125}{3} \cdot \frac{2}{2} + \frac{25}{2} \cdot \frac{3}{3} + \frac{5}{6}$

$= \frac{250 + 75 + 5}{6}$

$= \frac{330}{6} = 55$ oranges

77. Locate 2 on the horizontal axis. From there move vertically to the graph and then horizontally to the $C(t)$-axis. This locates a value of about 2.3 mcg/mL.

79. $C(t)$ has a minimum value of 0 and a maximum value of 10, so the range is [0, 10].

81. The function has a maximum value of 3 and no minimum is indicated, so the range is $(-\infty, 3]$.

83. There is no maximum or minimum value indicated by the graph, so the range is $(-\infty, \infty)$.

85. The function has a minimum value of –4 and no maximum is indicated, so the range is $[-4, \infty)$.

87. The function has a minimum value of –65 and no maximum is indicated, so the range is $[-65, \infty)$.

89. We graph $f(x) = x^2 + 2x + 1$ in the standard viewing window.

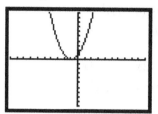

The range appears to be [0, ∞).

91. We graph $q(x) = -2x^2 + 5$ in the standard viewing window.

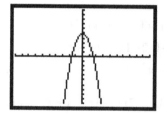

The range appears to be (−∞, 5].

93. We graph $p(x) = -2x^3 + x + 5$ in the standard viewing window.

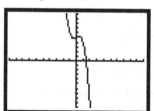

The range appears to be (−∞, ∞).

95. We graph $g(x) = x^4 + 2x^3 - 5$ in the window $[-5, 5, -10, 10]$.

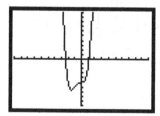

We estimate that the range is [−6.7, ∞).

97. $8x + 2 - 5x + 3x^3 - 4x - 1$

$= 3x^3 + (8 - 5 - 4)x + 2 - 1$

$= 3x^3 - x + 1$

99. $3a^2b + 4b^2 - 9a^2b - 7b^2$

$= (3-9)a^2b + (4-7)b^2$

$= -6a^2b - 3b^2$

101. $6x^2 + 2x^4 - 2x^2 - x^4 - 4x^2$

$= (6-2-4)x^2 + (2-1)x^4$

$= x^4$

103. $9x^2 - 3xy + 12y^2 + x^2 - y^2 + 5xy + 4y^2$

$= (9+1)x^2 + (-3+5)xy + (12-1+4)y^2$

$= 10x^2 + 2xy + 15y^2$

105. $\left(5t^4 - 2t^3 + t\right) + \left(-t^4 - t^3 + 6t^2\right)$

$= (5-1)t^4 + (-2-1)t^3 + 6t^2 + t$

$= 4t^4 - 3t^3 + 6t^2 + t$

107. $\left(x^2 + 2x - 3xy - 7\right) + \left(-3x^2 - x + 2y^2 + 6\right)$

$= (1-3)x^2 + (2-1)x - 3xy + 2y^2 + (-7+6)$

$= -2x^2 + x - 3xy + 2y^2 - 1$

109. $\left(8x^2y - 3xy^2 + 4xy\right) + \left(-2x^2y - xy^2 + xy\right)$

$= (8-2)x^2y + (-3-1)xy^2 + (4+1)xy$

$= 6x^2y - 4xy^2 + 5xy$

111. $\left(2r^2 + 12r - 11\right) + \left(6r^2 - 2r + 4\right) + \left(r^2 - r - 2\right)$

$= (2+6+1)r^2 + (12-2-1)r + (-11+4-2)$

$= 9r^2 + 9r - 9$

113. $\left(\frac{1}{8}xy - \frac{3}{5}x^3y^2 + 4.3y^3\right) + \left(-\frac{1}{3}xy - \frac{3}{4}x^3y^2 - 2.9y^3\right)$

$= \left(\frac{1}{8} - \frac{1}{3}\right)xy + \left(-\frac{3}{5} - \frac{3}{4}\right)x^3y^2 + (4.3 - 2.9)y^3$

$= \left(\frac{3}{24} - \frac{8}{24}\right)xy + \left(-\frac{12}{20} - \frac{15}{20}\right)x^3y^2 + 1.4y^3$

$= -\frac{5}{24}xy - \frac{27}{20}x^3y^2 + 1.4y^3$

115. $3t^4 + 8t^2 - 7t - 1$

$-\left(3t^4 + 8t^2 - 7t - 1\right),$

$-3t^4 - 8t^2 + 7t + 1$

117. $-12y^5 + 4ay^4 - 7by^2$

$-\left(-12y^5 + 4ay^4 - 7by^2\right),$

$12y^5 - 4ay^4 + 7by^2$

119. $(4x - 6) - (-3x + 2)$

$= (4x - 6) + (3x - 2)$

$= 7x - 8$

121. $\left(-3x^2 + 2x + 9\right) - \left(x^2 + 5x - 4\right)$

$= \left(-3x^2 + 2x + 9\right) + \left(-x^2 - 5x + 4\right)$

$= -4x^2 - 3x + 13$

123. $(8a - 3b + c) - (2a + 3b - 4c)$

$= (8a - 3b + c) + (-2a - 3b + 4c)$

$= 6a - 6b + 5c$

125. $\left(6a^2 + 5ab - 4b^2\right) - \left(8a^2 - 7ab + 3b^2\right)$

$= \left(6a^2 + 5ab - 4b^2\right) + \left(-8a^2 + 7ab - 3b^2\right)$

$= -2a^2 + 12ab - 7b^2$

127. $\left(6ab - 4a^2b + 6ab^2\right) - \left(3ab^2 - 10ab - 12a^2b\right)$

$= \left(6ab - 4a^2b + 6ab^2\right) + \left(-3ab^2 + 10ab + 12a^2b\right)$

$= 8a^2b + 16ab + 3ab^2$

129. $\left(\frac{5}{8}x^4 - \frac{1}{4}x^2 - \frac{1}{2}\right) - \left(-\frac{3}{8}x^4 + \frac{3}{4}x^2 + \frac{1}{2}\right)$

$= \left(\frac{5}{8}x^4 - \frac{1}{4}x^2 - \frac{1}{2}\right) + \left(\frac{3}{8}x^4 - \frac{3}{4}x^2 - \frac{1}{2}\right)$

$= x^4 - x^2 - 1$

131. $\left(6t^2 + 7\right) - \left(2t^2 + 3\right) + \left(t^2 + t\right)$

$= \left(6t^2 + 7\right) + \left(-2t^2 - 3\right) + t^2 + t$

$= (6 - 2 + 1)t^2 + t + 7 - 3$

$= 5t^2 + t + 4$

133. $\left(8r^2 - 6r\right) - (2r - 6) + \left(5r^2 - 7\right)$

$= \left(8r^2 - 6r\right) + (-2r + 6) + \left(5r^2 - 7\right)$

$= (8 + 5)r^2 + (-6 - 2)r + (6 - 7)$

$= 13r^2 - 8r - 1$

135. $\left(x^2 - 4x + 7\right) + \left(3x^2 - 9\right) - \left(x^2 - 4x + 7\right)$

Note that $x^2 - 4x + 7$ and $-(x^2 - 4x + 7)$ are opposites so their sum is 0. Then the result is $3x^2 - 9$.

137. Comparing the graphs, they are clearly not identical. So, the addition is not correct.

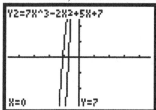

139. The graphs appear to be the same. By subtracting the expressions, the result appears to be 0 showing that the two expressions are equal.

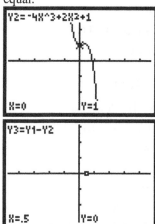

141. Comparing the graphs, they are clearly not identical. So, it is not correct.

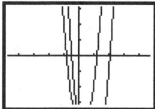

143. *Thinking and Writing Exercise.*

145. $2(x^2 - x + 3) = 2 \cdot x^2 - 2 \cdot x + 2 \cdot 3$
$$= 2x^2 - 2x + 6$$

147. $t^2 t^{11} = t^{2+11} = t^{13}$

149. $2n \cdot n^6 = 2n^1 \cdot n^6 = 2n^{1+6} = 2n^7$

151. *Thinking and Writing Exercise.*

153. Answer may vary. Use an ax^5-term, where a is an integer, and 3 other terms with different degrees, each less than degree 5, and consecutive even integer coefficients. Three answers are

$6x^5 + 8x^4 + 10x^2 + 12,$

$-8x^5 - 6x^4 - 4x^2 - 2,$ and

$2x^5 + 4x^4 + 6x^2 + 8.$

155. $(5m^5)^2 = 5^2 m^{5 \cdot 2} = 25m^{10}$

The degree is 10.

157. $2[P(x)]$
$= 2(13x^5 - 22x^4 - 36x^3 + 40x^2 - 16x + 75)$
$= 26x^5 - 44x^4 - 72x^3 + 80x^2 - 32x + 150$
Use columns to add:

$$
\begin{array}{r}
26x^5 - 44x^4 - 72x^3 + 80x^2 - 32x + 150 \\
42x^5 - 37x^4 + 50x^3 - 28x^2 + 34x + 100 \\
\hline
68x^5 - 81x^4 - 22x^3 + 52x^2 + 2x + 250
\end{array}
$$

159. $2[Q(x)]$
$= 2(42x^5 - 37x^4 + 50x^3 - 28x^2 + 34x + 100)$
$= 84x^5 - 74x^4 + 100x^3 - 56x^2 + 68x + 200$
$3[P(x)]$
$= 3(13x^5 - 22x^4 - 36x^3 + 40x^2 - 16x + 75)$
$= 39x^5 - 66x^4 - 108x^3 + 120x^2 - 48x + 225$
Use columns to subtract, adding the opposite of $3[P(x)]$:

$$
\begin{array}{r}
84x^5 - 74x^4 + 100x^3 - 56x^2 + 68x + 200 \\
-39x^5 + 66x^4 + 108x^3 - 120x^2 + 48x - 225 \\
\hline
45x^5 - 8x^4 + 208x^3 - 176x^2 + 116x - 25
\end{array}
$$

161. First we find the number of truffles in the display.

$$N(x) = \frac{1}{6}x^3 + \frac{1}{2}x^2 + \frac{1}{3}x$$
$$N(5) = \frac{1}{6} \cdot 5^3 + \frac{1}{2} \cdot 5^2 + \frac{1}{3} \cdot 5$$
$$= \frac{1}{6} \cdot 125 + \frac{1}{2} \cdot 25 + \frac{5}{3}$$
$$= \frac{125}{6} + \frac{25}{2} + \frac{5}{3}$$
$$= \frac{125}{6} + \frac{75}{6} + \frac{10}{6}$$
$$= \frac{210}{6} = 35$$

There are 35 truffles in the display. Now find the volume of one truffle. Each truffle's diameter is 3 cm, so the radius is $\frac{3}{2}$, or 1.5 cm.

$$V(r) = \frac{4}{3}\pi r^3$$

$$V(1.5) \approx \frac{4}{3}(3.14)(1.5)^3 \approx 14.13 \text{ cm}^3$$

Finally, multiply the number of truffles and the volume of a truffle to find the total volume of chocolate.

$$35(14.13 \text{ cm}^3) = 494.55 \text{ cm}^3$$

The display contains about 494.55 cm^3 of chocolate.

163. The area of the base is $x \cdot x$, or x^2.

The area of each side is $x \cdot (x-2)$.

The total area of all four sides is $4x(x-2)$.

The surface area of this box can be expressed as a polynomial function.

$$\begin{aligned} S(x) &= x^2 + 4x(x-2) \\ &= x^2 + 4x^2 - 8x \\ &= 5x^2 - 8x \end{aligned}$$

165. $(3x^{6a} - 5x^{5a} + 4x^{3a} + 8) -$
$$(2x^{6a} + 4x^{4a} + 3x^{3a} + 2x^{2a})$$
$$= (3-2)x^{6a} - 5x^{5a} - 4x^{4a} + (4-3)x^{3a} - 2x^{2a} + 8$$
$$= x^{6a} - 5x^{5a} - 4x^{4a} + x^{3a} - 2x^{2a} + 8$$

Exercise Set 5.2

1. False

3. True

5. False

7. True

9. $3x^4 \cdot 5x = (3 \cdot 5)(x^4 \cdot x) = 15x^5$

11. $6a^2(-8ab^2) = 6(-8)(a^2 \cdot a)b^2 = -48a^3b^2$

13. $\left(-4x^3y^2\right)\left(-9x^2y^4\right)$
$$= (-4)(-9)(x^3 \cdot x^2)(y^2 \cdot y^4)$$
$$= 36x^5y^6$$

15. $7x(3-x) = 7x \cdot 3 - 7x \cdot x$
$$= 21x - 7x^2$$

17. $5cd(4c^2d - 5cd^2)$
$$= 5cd \cdot 4c^2d - 5cd \cdot 5cd^2$$
$$= 20c^3d^2 - 25c^2d^3$$

19. $(x+3)(x+5)$
$$= x^2 + 5x + 3x + 15 \quad \text{FOIL}$$
$$= x^2 + 8x + 15$$

21. $(2a+3)(4a-1)$
$$= 8a^2 - 2a + 12a - 3 \quad \text{FOIL}$$
$$= 8a^2 + 10a - 3$$

23. $(x+2)(x^2 - 3x + 1)$
$$= x(x^2 - 3x + 1) + 2(x^2 - 3x + 1)$$
$$= x^3 - 3x^2 + x + 2x^2 - 6x + 2$$
$$= x^3 - x^2 - 5x + 2$$

25. $(t-5)(t^2 + 2t - 3)$
$$= t(t^2 + 2t - 3) - 5(t^2 + 2t - 3)$$
$$= t^3 + 2t^2 - 3t - 5t^2 - 10t + 15$$
$$= t^3 - 3t^2 - 13t + 15$$

27.
$$\begin{array}{rl}
a^2 + a - 1 & \\
a^2 + 4a - 5 & \\
\hline
-5a^2 - 5a + 5 & \text{Multiplying by } -5 \\
4a^3 + 4a^2 - 4a & \text{Multiplying by } 4a \\
a^4 + a^3 - a^2 & \text{Multiplying by } a^2 \\
\hline
a^4 + 5a^3 - 2a^2 - 9a + 5 & \text{Adding}
\end{array}$$

29. $(x+3)(x^2 - 3x + 9)$
$$= x(x^2 - 3x + 9) + 3(x^2 - 3x + 9)$$
$$= x^3 - 3x^2 + 9x + 3x^2 - 9x + 27$$
$$= x^3 + 27$$

31. $(a-b)(a^2 + ab + b^2)$
$$= a(a^2 + ab + b^2) - b(a^2 + ab + b^2)$$
$$= a^3 + a^2b + ab^2 - a^2b - ab^2 - b^3$$
$$= a^3 - b^3$$

33. $(t-3)(t+2)$

$= t^2 + 2t - 3t - 6$

$= t^2 - t - 6$

35. $(5x+2y)(4x+y)$

$= 20x^2 + 5xy + 8xy + 2y^2$

$= 20x^2 + 13xy + 2y^2$

37. $\left(t - \frac{1}{3}\right)\left(t - \frac{1}{4}\right)$

$= t^2 - \frac{1}{4}t - \frac{1}{3}t + \frac{1}{12}$

$= t^2 - \frac{3}{12}t - \frac{4}{12}t + \frac{1}{12}$

$= t^2 - \frac{7}{12}t + \frac{1}{12}$

39. $(1.2t+3s)(2.5t-5s)$

$= 3t^2 - 6st + 7.5st - 15s^2$

$= 3t^2 + 1.5st - 15s^2$

41. $(r+3)(r+2)(r-1)$

$= (r^2 + 2r + 3r + 6)(r-1)$

$= (r^2 + 5r + 6)(r-1)$

$= (r^2 + 5r + 6)(r) + (r^2 + 5r + 6)(-1)$

$= r^3 + 5r^2 + 6r - r^2 - 5r - 6$

$= r^3 + 4r^2 + r - 6$

43. $(x+5)^2$

$= x^2 + 2 \cdot x \cdot 5 + 5^2$ $(A+B)^2 = A^2 + 2AB + B^2$

$= x^2 + 10x + 25$

45. $(2y-7)^2$

$= (2y)^2 - 2 \cdot 2y \cdot 7 + 7^2$ $(A-B)^2$

$= A^2 - 2AB + B^2$

$= 4y^2 - 28y + 49$

47. $(5c-2d)^2$

$= (5c)^2 - 2 \cdot 5c \cdot 2d + (2d)^2$

$(A-B)^2 = A^2 - 2AB + B^2$

$= 25c^2 - 20cd + 4d^2$

49. $\left(3a^3 - 10b^2\right)^2$

$= \left(3a^3\right)^2 - 2 \cdot 3a^3 \cdot 10b^2 + \left(10b^2\right)^2$

$(A-B)^2 = A^2 - 2AB + B^2$

$= 9a^6 - 60a^3b^2 + 100b^4$

51. $(x^3 y^4 + 5)^2$

$= (x^3 y^4)^2 + 2 \cdot x^3 y^4 \cdot 5 + 5^2$

$(A+B)^2 = A^2 + 2AB + B^2$

$= x^6 y^8 + 10x^3 y^4 + 25$

53. $P(x) \cdot Q(x)$

$= (3x^2 - 5)(4x^2 - 7x + 1)$

$= 3x^2(4x^2 - 7x + 1) - 5(4x^2 - 7x + 1)$

$= 12x^4 - 21x^3 + 3x^2 - 20x^2 + 35x - 5$

$= 12x^4 - 21x^3 - 17x^2 + 35x - 5$

55. $P(x) \cdot P(x)$

$= (5x-2)(5x-2)$

$= (5x-2)^2$

$= (5x)^2 - 2 \cdot 5x \cdot 2 + 2^2$

$(A-B)^2 = A^2 - 2AB + B^2$

$= 25x^2 - 20x + 4$

57. $\left[F(x)\right]^2 = \left(2x - \frac{1}{3}\right)^2$

$= (2x)^2 - 2 \cdot 2x \cdot \frac{1}{3} + \left(\frac{1}{3}\right)^2$

$(A-B)^2 = A^2 - 2AB + B^2$

$= 4x^2 - \frac{4}{3}x + \frac{1}{9}$

59. $(c+7)(c-7)$

$= c^2 - 7^2$ $(A+B)(A-B) = A^2 - B^2$

$= c^2 - 49$

61. $(1-4x)(1+4x)$

$= 1^2 - (4x)^2$ $(A+B)(A-B) = A^2 - B^2$

$= 1 - 16x^2$

63. $\left(3m - \dfrac{1}{2}n\right)\left(3m + \dfrac{1}{2}n\right)$

$= (3m)^2 - \left(\dfrac{1}{2}n\right)^2 \quad (A+B)(A-B) = A^2 - B^2$

$= 9m^2 - \dfrac{1}{4}n^2$

65. $(x^3 + yz)(x^3 - yz)$

$= \left(x^3\right)^2 - (yz)^2 \quad (A+B)(A-B) = A^2 - B^2$

$= x^6 - y^2 z^2$

67. $\left(-mn + 3m^2\right)\left(mn + 3m^2\right)$

$= (3m^2 - mn)(3m^2 + mn)$

$= \left(3m^2\right)^2 - (mn)^2 \quad (A+B)(A-B) = A^2 - B^2$

$= 9m^4 - m^2 n^2, \text{ or } -m^2 n^2 + 9m^4$

69. $(x+7)^2 - (x+3)(x-3)$

$= x^2 + 2 \cdot x \cdot 7 + 7^2 - (x^2 - 3^2)$

$= x^2 + 14x + 49 - (x^2 - 9)$

$= x^2 + 14x + 49 - x^2 + 9$

$= 14x + 58$

71. $(2m - n)(2m + n) - (m - 2n)^2$

$= \left[(2m)^2 - n^2\right] - \left[m^2 - 2 \cdot m \cdot 2n + (2n)^2\right]$

$= 4m^2 - n^2 - (m^2 - 4mn + 4n^2)$

$= 4m^2 - n^2 - m^2 + 4mn - 4n^2$

$= 3m^2 + 4mn - 5n^2$

73. $(a + b + 1)(a + b - 1)$

$= [(a+b) + 1][(a+b) - 1]$

$= (a+b)^2 - 1^2$

$= a^2 + 2ab + b^2 - 1$

75. $(2x + 3y + 4)(2x + 3y - 4)$

$= [(2x+3y) + 4][(2x+3y) - 4]$

$= (2x + 3y)^2 - 4^2$

$= 4x^2 + 12xy + 9y^2 - 16$

77. $A = P(1+r)^2$

$A = P\left(1 + 2r + r^2\right)$

$A = P + 2Pr + Pr^2$

79. a) $f(t-1) = (t-1)^2 + 5$

$= t^2 - 2t + 1 + 5$

$= t^2 - 2t + 6$

b) $f(a+h) - f(a)$

$= \left[(a+h)^2 + 5\right] - (a^2 + 5)$

$= a^2 + 2ah + h^2 + 5 - a^2 - 5$

$= 2ah + h^2$

c) $f(a) - f(a-h)$

$= (a^2 + 5) - \left[(a-h)^2 + 5\right]$

$= a^2 + 5 - (a^2 - 2ah + h^2 + 5)$

$= a^2 + 5 - a^2 + 2ah - h^2 - 5$

$= 2ah - h^2$

81. a) $f(a) + f(-a)$

$= (a^2 + a) + \left[(-a)^2 + (-a)\right]$

$= a^2 + a + a^2 - a$

$= 2a^2$

b) $f(a+h)$

$= (a+h)^2 + (a+h)$

$= a^2 + 2ah + h^2 + a + h$

c) $f(a+h) - f(a)$

$= (a+h)^2 + (a+h) - (a^2 + a)$

$= a^2 + 2ah + h^2 + a + h - a^2 - a$

$= 2ah + h^2 + h$

83. *Thinking and Writing Exercise.*

85. $5x + 15y - 5 = 5(x + 3y - 1)$

87. $14t - 49 = 7(2t - 7)$

89. $ax + bx - cx = x(a + b - c)$

91. *Thinking and Writing Exercise.*

93. $\left(x^2 + y^n\right)\left(x^2 - y^n\right) = \left(x^2\right)^2 - \left(y^n\right)^2 = x^4 - y^{2n}$

95. $x^2 y^3 (5x^n + 4y^n) = x^2 y^3 \cdot 5x^n + x^2 y^3 \cdot 4y^n$

$= 5x^{n+2} y^3 + 4x^2 y^{n+3}$

97. $(x^n - 4)(x^{2n} + 3x^n - 2)$

$= x^n(x^{2n} + 3x^n - 2) - 4(x^{2n} + 3x^n - 2)$

$= x^{3n} + 3x^{2n} - 2x^n - 4x^{2n} - 12x^n + 8$

$= x^{3n} - x^{2n} - 14x^n + 8$

99. $(a - b + c - d)(a + b + c + d)$

$= [(a + c) - (b + d)][(a + c) + (b + d)]$

$= (a + c)^2 - (b + d)^2$

$= (a^2 + 2ac + c^2) - (b^2 + 2bd + d^2)$

$= a^2 + 2ac + c^2 - b^2 - 2bd - d^2$

101. $(x^2 - 3x + 5)(x^2 + 3x + 5)$

$= [(x^2 + 5) - 3x][(x^2 + 5) + 3x]$

$= (x^2 + 5)^2 - (3x)^2$

$= x^4 + 10x^2 + 25 - 9x^2$

$= x^4 + x^2 + 25$

103. $(x - 1)(x^2 + x + 1)(x^3 + 1)$

$= (x^3 + x^2 + x - x^2 - x - 1)(x^3 + 1)$

$= (x^3 - 1)(x^3 + 1)$

$= x^6 - 1$

105. $\left(x^{a-b}\right)^{a+b} = x^{(a-b)(a+b)} = x^{a^2 - b^2}$

107. $(x - a)(x - b)(x - c) \cdots (x - z)$

$= (x - a)(x - b) \cdots (x - x)(x - y)(x - z)$

$= (x - a)(x - b) \cdots (0)(x - y)(x - z)$

$= 0$

109. $\dfrac{g(a + h) - g(a)}{h}$

$= \dfrac{(a + h)^2 - 9 - (a^2 - 9)}{h}$

$= \dfrac{a^2 + 2ah + h^2 - 9 - a^2 + 9}{h}$

$= \dfrac{2ah + h^2}{h}$

$= \dfrac{h(2a + h)}{h}$

$= 2a + h$

111. $(A - B)^2 = A^2 - 2AB + B^2$

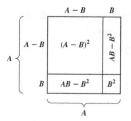

Exercise Set 5.3

1. False

3. True

5. True

7. True

9. $x^2 + 6x + 9$ has no equals sign, so it is an expression.

11. $3x^2 = 3x$ has an equals sign, so it is an equation.

13. $2x^3 + x^2 = 0$ has an equals sign, so it is an equation.

15. From the graph we see that $f(x) = 0$ when $x = -3$ or $x = 5$.

17. From the graph we see that $f(x) = 0$ when $x = -2$ or $x = 0$. -2 and 0 are the zeros of the function.

19. From the graph we see that $f(x) = 3$ when $x = -3$ or $x = 1$.

21. From the graph we see that $f(x) = 0$ when $x = -4$ or $x = 2$.

23. We can graph $y_1 = x^2$ and $y_2 = 5x$ and use the Intersect feature to find the first coordinates of the points of intersection, or we can begin by rewriting the equation so that one side is 0:

$$x^2 = 5x$$

$x^2 - 5x = 0$ Subtracting $5x$ on both sides

Then graph $y = x^2 - 5x$ and use the Zero feature to find the roots of the equation. In either case, we find that the solutions are 0 and 5.

25. We can graph $y_1 = 4x$ and $y_2 = x^2 + 3$ and use the Intersect feature to find the first coordinates of the points of intersection, or we can begin by rewriting the equation so that one side is 0:

$$4x = x^2 + 3$$
$$0 = x^2 - 4x + 3$$

Then graph $y = x^2 - 4x + 3$ and use the Zero feature to find the roots of the equation. In either case, we find that the solutions are 1 and 3.

27. We can graph $y_1 = x^2 + 150$ and $y_2 = 25x$ and use the Intersect feature to find the first coordinates of the points of intersection, or we can begin by rewriting the equation so that one side is 0:

$$x^2 + 150 = 25x$$
$$x^2 - 25x + 150 = 0$$

Then graph $y = x^2 - 25x + 150$ and use the Zero feature to find the roots of the equation. In either case, we find that the solutions are 10 and 15.

29. Graph $y = x^3 - 3x^2 + 2x$ and use the Zero feature to find the roots of the equation. The solutions are 0, 1, and 2.

31. Graph $y = x^3 - 3x^2 - 198x + 1080$ and use the Zero feature to find the roots of the equation. The solutions are -15, 6, and 12.

33. Graph $y = 21x^2 + 2x - 3$ and use the Zero feature to find the roots of the equation. The solutions are approximately -0.42857 and 0.33333.

35. Graph $y = x^2 - 4x - 45$ and use the Zero feature to find the zeros of the function. They are -5 and 9.

37. Graph $y = 2x^2 - 13x - 7$ and use the Zero feature to find the zeros of the function. They are -0.5 and 7.

39. Graph $y = x^3 - 2x^2 - 3x$ and use the Zero feature to find the zeros of the function. They are -1, 0, and 3.

41. We see that $2x - 1 = 0$ when $x = 0.5$ and $3x + 1 = 0$ when $x = -0.\overline{3}$, so Graph III corresponds to the given function.

43. We see that $4 - x = 0$ when $x = 4$ and $2x - 11 = 0$ when $x = 5.5$, so Graph I corresponds to the given function.

45. $2t^2 + 8t = 2t \cdot t + 2t \cdot 4$
$$= 2t(t + 4)$$

47. $9y^3 - y^2 = y^2 \cdot 9y - y^2 \cdot 1$
$$= y^2(9y - 1)$$

49. $15x^2 - 5x^4 + 5x = 5x \cdot 3x - 5x \cdot x^3 + 5x \cdot 1$
$$= 5x(3x - x^3 + 1)$$

51. $4x^2y - 12xy^2 = 4xy \cdot x - 4xy \cdot 3y$
$$= 4xy(x - 3y)$$

53. $3y^2 - 3y - 9 = 3 \cdot y^2 - 3 \cdot y - 3 \cdot 3$
$$= 3(y^2 - y - 3)$$

55. $6ab - 4ad + 12ac = 2a \cdot 3b - 2a \cdot 2d + 2a \cdot 6c$
$$= 2a(3b - 2d + 6c)$$

57. $72x^3 - 36x^2 + 24x$
$$= 12x \cdot 6x^2 - 12x \cdot 3x + 12x \cdot 2$$
$$= 12x(6x^2 - 3x + 2)$$

59. $x^5y^5 + x^4y^3 + x^3y^3 - xy^2$
$= xy^2 \cdot x^4y^3 + xy^2 \cdot x^3y + xy^2 \cdot x^2y - xy^2 \cdot 1$
$= xy^2(x^4y^3 + x^3y + x^2y - 1)$

61. $9x^3y^6z^2 - 12x^4y^4z^4 + 15x^2y^5z^3$
$= 3x^2y^4z^2 \cdot 3xy^2 - 3x^2y^4z^2 \cdot 4x^2z^2 +$
$\qquad\qquad 3x^2y^4z^2 \cdot 5yz$
$= 3x^2y^4z^2(3xy^2 - 4x^2z^2 + 5yz)$

63. $-5x + 35 = -5(x - 7)$

65. $-2x^2 + 4x - 12 = -2(x^2 - 2x + 6)$

67. $3y - 24x = -3(-y + 8x)$, or $-3(8x - y)$

69. $-x^2 + 5x - 9 = -(x^2 - 5x + 9)$

71. $-a^4 + 2a^3 - 13a = -a(a^3 - 2a^2 + 13)$

73. $a(b-5) + c(b-5) = (b-5)(a+c)$

75. $(x+7)(x-1) + (x+7)(x-2)$
$= (x+7)(x-1+x-2)$
$= (x+7)(2x-3)$

77. $a^2(x-y) + 5(y-x)$
$= a^2(x-y) + 5(-1)(x-y)$ Factoring out -1
$= a^2(x-y) - 5(x-y)$ to reverse the
$= (x-y)(a^2 - 5)$ second
$\qquad\qquad$ subtraction

79. $ac + ad + bc + bd = a(c+d) + b(c+d)$
$\qquad\qquad\qquad\qquad = (c+d)(a+b)$

81. $b^3 - b^2 + 2b - 2 = b^2(b-1) + 2(b-1)$
$\qquad\qquad\qquad = (b-1)(b^2 + 2)$

83. $x^3 - x^2 - 2x + 5 = x^2(x-1) - 1(2x-5)$
There is no common factor to the polynomial, so the polynomial is not factorable by grouping.

85. $a^3 - 3a^2 + 6 - 2a$
$= a^2(a-3) + 2(3-a)$ Factoring out -1
$= a^2(a-3) + 2(-1)(a-3)$ to reverse the
$= a^2(a-3) - 2(a-3)$ second
$= (a-3)(a^2 - 2)$ subtraction

87. $x^6 - x^5 - x^3 + x^4 = x^3(x^3 - x^2 - 1 + x)$
$\qquad\qquad\qquad = x^3[x^2(x-1) - 1 + x]$
$\qquad\qquad\qquad = x^3[x^2(x-1) + 1(x-1)]$
$\qquad\qquad\qquad = x^3(x-1)(x^2 + 1)$

89. $2y^4 + 6y^2 + 5y^2 + 15$
$= 2y^2(y^2 + 3) + 5(y^2 + 3)$
$= (y^2 + 3)(2y^2 + 5)$

91. a) $h(t) = -16t^2 + 72t$
$\qquad h(t) = -8t(2t - 9)$

b) Using $h(t) = -16t^2 + 72t$:
$\qquad h(1) = -16 \cdot 1^2 + 72 \cdot 1 = -16 \cdot 1 + 72$
$\qquad\qquad = -16 + 72 = 56$ ft
Using $h(t) = -8t(2t-9)$:
$\qquad h(1) = -8(1)(2 \cdot 1 - 9) = -8(1)(-7) = 56$ ft
The expressions have the same value for $t = 1$, so the factorization is probably correct.

93. $R(n) = n^2 - n$
$\qquad R(n) = n(n-1)$

95. $P(x) = x^2 - 3x$
$\qquad P(x) = x(x-3)$

97. $R(x) = 280x - 0.4x^2$
$\qquad R(x) = 0.4x(700 - x)$

99. $N(x) = \frac{1}{6}x^3 + \frac{1}{2}x^2 + \frac{1}{3}x$
$\qquad = \frac{1}{6} \cdot x^3 + \frac{1}{6} \cdot 3x^2 + \frac{1}{6} \cdot 2x$
$\qquad = \frac{1}{6}\left(x^3 + 3x^2 + 2x\right)$

101. $H(n) = \frac{1}{2}n^2 - \frac{1}{2}n$
$\qquad = \frac{1}{2}n(n-1)$

103. $(x+3)(x-4)=0$

$x+3=0$ or $x-4=0$

$x=-3$ or $x=4$

The solutions are –3 and 4.

105. $x(x+1)=0$

$x=0$ or $x+1=0$

$x=0$ or $x=-1$

The solutions are 0 and –1.

107. $x^2-3x=0$

$x(x-3)=0$

$x=0$ or $x-3=0$

$x=0$ or $x=3$

The solutions are 0 and 3.

109. To use the principle of *zero* products, we set the equation equal to *zero*.

$-5x^2=15x$

$0=5x^2+15x$

$0=5x(x+3)$

$5x=0$ or $x+3=0$

$x=0$ or $x=-3$

The solutions are –3 and 0.

111. $12x^4+4x^3=0$

$4x^3(3x+1)=0$

$4x^3=0$ or $3x+1=0$

$x^3=0$ or $3x=-1$

$x=0$ or $x=-\dfrac{1}{3}$

The solutions are $-\frac{1}{3}$ and 0.

113. $f(x)=(x-3)(x+7)$

If $f(a)=0,$ then

$0=(a-3)(a+7)$

$a-3=0$ or $a+7=0$

$a=3$ or $a=-7$

The solutions are 3 and –7.

115. $f(x)=2x(5x+9)$

If $f(a)=0,$ then

$0=2a(5a+9)$

$5a+9=0$ or $2a=0$

$5a=-9$ or $a=0$

$a=-\dfrac{9}{5}$

The solutions are 0 and $-\frac{9}{5}$.

117. $f(x)=x^3-3x^2$

If $f(a)=0,$ then

$0=a^3-3a^2=a^2(a-3)$

$a^2=0$ or $a-3=0$

$a=0$ or $a=3$

The solutions are 0 and 3.

119. *Thinking and Writing Exercise.*

121. $(x+2)(x+7)=x\cdot x+x\cdot 7+2\cdot x+2\cdot 7$

$=x^2+7x+2x+14$

$=x^2+9x+14$

123. $(x+2)(x-7)=x\cdot x+x\cdot(-7)+2\cdot x+2\cdot(-7)$

$=x^2-7x+2x-14$

$=x^2-5x-14$

125. $(a-1)(a-3)=a\cdot a+a\cdot(-3)$

$+(-1)\cdot a+(-1)\cdot(-3)$

$=a^2-3a-a+3$

$=a^2-4a+3$

127. $(t-5)(t+10)=t\cdot t+t\cdot 10+(-5)\cdot t+(-5)\cdot 10$

$=t^2+10t-5t-50$

$=t^2+5t-50$

129. *Thinking and Writing Exercise.*

131. We use the principle of zero products in reverse. Since the zeros of $f(x)=x^2+2x-8$ are –4 and 2, we have

$x=-4$ or $x=2$

$x+4=0$ or $x-2=0,$

so $x^2+2x-8=(x+4)(x-2).$

133. $x^5 y^4 + \underline{\quad} = x^4 y^4 \left(\underline{\quad} + y^2 \right)$

The term that goes in the first blank is the product of $x^4 y^4$ and y^2, or $x^4 y^6$.

The term that goes in the second blank is the expression that is multiplied with $x^4 y^4$ to obtain $x^5 y^4$, or x. Thus, we have

$x^5 y^4 + x^4 y^6 = x^4 y^4 (x + y^2)$.

135. $x^{-6} + x^{-9} + x^{-3} = x^{-9}(x^3 + 1 + x^6)$

137. $x^{1/3} - 5x^{1/2} + 3x^{3/4}$

$= x^{4/12} - 5x^{6/12} + 3x^{9/12}$

$= x^{4/12}(1 - 5x^{2/12} + 3x^{5/12})$

$= x^{1/3}(1 - 5x^{1/6} + 3x^{5/12})$

139. $5x^5 - 5x^4 + x^3 - x^2 + 3x - 3$

$= 5x^4(x-1) + x^2(x-1) + 3(x-1)$

$= (5x^4 + x^2 + 3)(x-1)$

141. $2x^{3a} + 8x^a + 4x^{2a} = 2x^{a+2a} + 8x^a + 4x^{a+a}$

$= 2x^a \cdot x^{2a} + 2x^a \cdot 4$

$\qquad + 2x^a \cdot 2x^a$

$= 2x^a(x^{2a} + 4 + 2x^a)$

143. The shaded area is equal to the difference between the area of the rectangle and the area of the circles and partial circles. Because the diameter $(2x)$ of each circle is equal to the width of the rectangle, and two diameters $(4x)$ are equal to the length of the rectangle, the area of the rectangle (A_r) is

$A_r = l \cdot w = 4x \cdot 2x = 8x^2$.

The area of the full circle and the two half-circles equals 2 times the area of one circle, or $A_c = 2(\pi r^2) = 2\pi x^2$.

The area of the shaded region is therefore

$A_s = A_r - A_c = 8x^2 - 2\pi x^2$

$\qquad = 2x^2(4 - \pi)$

1. True

3. False

5. True

7. True

9. $x^2 + 8x + 12$

We look for two numbers whose product is 12 and whose sum is 8. Since 12 and 8 are both positive, we need only consider positive factors.

Pairs of Factors	Sum of Factors
1, 12	13
2, 6	8

The numbers we need are 2 and 6. The factorization is $(x+2)(x+6)$.

11. $t^2 + 8t + 15$

Since the constant term is positive and the coefficient of the middle term is also positive, we look for a factorization of 15 in which both factors are positive. Their sum must be 8.

Pair of Factors	Sum of Factors
1, 15	16
3, 5	8

The numbers we need are 3 and 5. The factorization is $(t+3)(t+5)$.

13. $a^2 - 7a + 12$

Since the constant term is positive and the coefficient of the middle term is negative, we look for two negative terms whose product is 12 and whose sum must be -7.

Pair of Factors	Sum of Factors
$-12, -1$	-13
$-6, -2$	-8
$-4, -3$	-7

The numbers we want are -4 and -3. The factorization is $(a-4)(a-3)$.

15. $x^2 - 2x - 15$

Since the constant term is negative, we look for factorization of -15 in which one factor is positive and one factor is negative. Their sum must be -2, so the negative factor must have the larger absolute value. Thus we consider only pairs of factors in which the negative factor has the larger absolute value.

Pair of Factors	Sum of Factors
1, -15	-14
3, -5	-2

The numbers we need are 3 and -5. The factorization is $(x+3)(x-5)$.

17. $x^2 + 2x - 15$

Since the constant term is negative, we look for factorization of -15 in which one factor is positive and one factor is negative. Their sum must be 2, so the positive factor must have the larger absolute value. Thus we consider only pairs of factors in which the positive factor has the larger absolute value.

Pair of Factors	Sum of Factors
-1, 15	14
-3, 5	2

The numbers we need are 5 and -3. The factorization is $(x+5)(x-3)$.

19. $2n^2 - 20n + 50$

$= 2(n^2 - 10n + 25)$ Removing the common factor

We now factor $n^2 - 10n + 25$. We look for two numbers whose product is 25 and whose sum is -10. Since the constant term is positive and the coefficient of the middle term is negative, we look for factorization of 25 in which both factors are negative.

Pair of Factors	Sum of Factors
-1, -25	-26
-5, -5	-10

The numbers we need are -5 and -5.

$n^2 - 10n + 25 = (n-5)(n-5)$

We must not forget to include the common factor 2.

$2n^2 - 20n + 50 = 2(n-5)(n-5)$, or

$2(n-5)^2$

21. $a^3 - a^2 - 72a$

$= a(a^2 - a - 72)$ Removing the common factor

We now factor $a^2 - a - 72$. Since the constant term is negative, we look for a factorization of -72 in which one factor is positive and one factor is negative. We consider only pairs of factors in which the negative factor has the larger absolute value, since the sum of the factors, -1, is negative.

Pair of Factors	Sum of Factors
-72, 1	-71
-36, 2	-34
-18, 4	-14
-9, 8	-1

The numbers we need are -9 and 8.

$a^2 - a - 72 = (a-9)(a+8)$

We must not forget to include the common factor a.

$a^3 - a^2 - 72a = a(a-9)(a+8)$

23. $14x + x^2 + 45 = x^2 + 14x + 45$

Since the constant term and the middle term are both positive, we look for a factorization of 45 in which both factors are positive. Their sum must be 14.

Pair of Factors	Sum of Factors
45, 1	46
15, 3	18
9, 5	14

The numbers we need are 9 and 5. The factorization is $(x+9)(x+5)$.

25. $3x + x^2 - 10 = x^2 + 3x - 10$

Since the constant term is negative, we look for a factorization of –10 in which one factor is positive and one factor is negative. We consider only pairs of factors in which the positive factor has the larger absolute value, since the sum of the factors, 3, is positive.

Pair of Factors	Sum of Factors
10, –1	9
5, –2	3

The numbers we need are 5 and –2. The factorization is $(x + 5)(x - 2)$.

27. $3x^2 - 15x + 18$

$= 3(x^2 - 5x + 6)$ Removing the common factor

We now factor $x^2 - 5x + 6$. Since the constant term is positive and the coefficient of the middle term is negative, we look for two negative terms whose product is 6 and whose sum is –5.

Pair of Factors	Sum of Factors
–1, –6	–7
–2, –3	–5

The numbers we need are –2 and –3.

$x^2 - 5x + 6 = (x - 2)(x - 3)$

We must not forget to include the common factor 3.

$3x^2 - 15x + 18 = 3(x - 2)(x - 3)$

29. $56 + x - x^2 = -x^2 + x + 56 = -(x^2 - x - 56)$

We now factor $x^2 - x - 56$. Since the constant term is negative, we look for a factorization of –56 in which one factor is positive and one factor is negative. We consider only pairs of factors in which the negative factor has the larger absolute value, since the sum of the factors, –1, is negative.

Pair of Factors	Sum of Factors
–56, 1	–55
–28, 2	–26
–14, 4	–10
–8, 7	–1

The numbers we need are –8 and 7. Thus, $x^2 - x - 56 = (x - 8)(x + 7)$. We must not forget to include the common factors –1.

$56 + x - x^2 = -(x - 8)(x + 7)$, or $(-x + 8)(x + 7)$, or $(8 - x)(7 + x)$

31. $32y + 4y^2 - y^3$

There is a common factor, y. We also factor out –1 in order to make the leading coefficient positive.

$32y + 4y^2 - y^3 = -y(-32 - 4y + y^2)$

$= -y(y^2 - 4y - 32)$

Now we factor $y^2 - 4y - 32$. Since the constant term is negative, we look for factorization of –32 in which one factor is positive and one factor is negative. We consider only pairs of factors in which the negative factor has the larger absolute value, since the sum of the factors, –4, is negative.

Pair of Factors	Sum of Factors
–32, 1	–31
–16, 2	–14
–8, 4	–4

The numbers we need are –8 and 4. Thus, $y^2 - 4y - 32 = (y - 8)(y + 4)$. We must not forget to include the common factor $-y$.

$32y + 4y^2 - y^3 = -y(y - 8)(y + 4)$, or $y(-y + 8)(y + 4)$, or $y(8 - y)(4 + y)$

33. $x^4 + 11x^3 - 80x^2$

$= x^2(x^2 + 11x - 80)$ Removing the
common factor

We now factor $x^2 + 11x - 80$. We look for pairs of factors of -80, one positive and one negative, such that the positive factor has the larger absolute value and the sum of the factors is 11.

Pair of Factors	Sum of Factors
80, −1	79
40, −2	38
20, −4	16
16, −5	11
10, −8	2

The numbers we need are 16 and −5.
Then $x^2 + 11x - 80 = (x + 16)(x - 5)$. We must not forget to include the common factor.
$x^4 + 11x^3 - 80x^2 = x^2(x + 16)(x - 5)$

35. $x^2 + 12x + 13$

There are no factors of 13 whose sum is 12. This trinomial is not factorable into binomials with integer coefficients. The polynomial is prime.

37. $p^2 - 5pq - 24q^2$

We look for numbers r and s such that
$p^2 - 5pq - 24q^2 = (p + rq)(p + sq)$.

Our thinking is much the same as if we were factoring $p^2 - 5p - 24$. We look for factors of -24 whose sum is -5, one positive and one negative, such that the negative factor has the larger absolute value.

Pair of Factors	Sum of Factors
−24, 1	−23
−12, 2	−10
−8, 3	−5
−6, 4	−2

The numbers we need are −8 and 3. The factorization is $(p - 8q)(p + 3q)$.

39. $y^2 + 8yz + 16z^2$

We look for numbers p and q such that $y^2 + 8yz + 16z^2 = (y + pz)(y + qz)$. Our thinking is much the same as if we factor $y^2 + 8y + 16$. Since the constant term is positive and the coefficient of the middle term is positive, we look for a factorization of 16 in which both factors are positive. Their sum must be 8.

Pair of Factors	Sum of Factors
1, 16	17
2, 8	10
4, 4	8

The numbers we need are 4 and 4. The factorization is
$(y + 4z)(y + 4z)$ or $(y + 4z)^2$.

41. $p^4 - 80p^3 + 79p^2$

$= p^2(p^2 - 80p + 79)$ Factor out p^2

We now factor $p^2 - 80p + 79$. We look for a pair of factors of 79 whose sum is -80. The only negative pair of factors is -1 and -79. These are the numbers we need, so
$p^2 - 80p + 79 = (p - 1)(p - 79)$.

We must not forget to include the common factor.
$p^4 - 80p^3 + 79p^2 = p^2(p - 1)(p - 79)$

43. $x^2 + 8x + 12 = 0$

$(x + 2)(x + 6) = 0$ From Exercise 9
$x + 2 = 0$ or $x + 6 = 0$ Using the principle of zero products:
$x = -2$ or $x = -6$
The solutions are −2 and −6.

45. $2n^2 + 50 = 20n$

$2n^2 - 20n + 50 = 0$ Subtracting $20n$
from both sides

$2(n - 5)(n - 5) = 0$ From Exercise 19

$n - 5 = 0$ or $n - 5 = 0$
Using the principle of zero products:
$n = 5$ or $n = 5$
The solution is 5.

47. The x-intercepts are $(-5, 0)$ and $(1, 0)$, so the solutions are -5 and 1.

Check: For -5:

$$\begin{array}{c|c} x^2 + 4x - 5 = 0 & \\ \hline (-5)^2 + 4(-5) - 5 & 0 \\ 25 - 20 - 5 & \\ \hline 0 & 0 \quad \text{TRUE} \end{array}$$

For 1:

$$\begin{array}{c|c} x^2 + 4x - 5 = 0 & \\ \hline 1^2 + 4 \cdot 1 - 5 & 0 \\ 1 + 4 - 5 & \\ \hline 0 & 0 \quad \text{TRUE} \end{array}$$

Both numbers check, so they are the solutions.

49. The intercepts are $(-3, 0)$ and $(2, 0)$, so the solutions are -3 and 2.

Check: For -3:

$$\begin{array}{c|c} x^2 + x - 6 = 0 & \\ \hline (-3)^2 + (-3) - 6 & 0 \\ 9 - 3 - 6 & \\ \hline 0 & 0 \quad \text{TRUE} \end{array}$$

For 2:

$$\begin{array}{c|c} x^2 + x - 6 = 0 & \\ \hline 2^2 + 2 - 6 & 0 \\ 4 + 2 - 6 & \\ \hline 0 & 0 \quad \text{TRUE} \end{array}$$

Both numbers check, so they are the solutions.

51. The zeros of $f(x) = x^2 - 4x - 45$ are the solutions of the equation $x^2 - 4x - 45 = 0$. We factor and use the principle of zero products.

$$x^2 - 4x - 45 = 0$$
$$(x - 9)(x + 5) = 0$$
$$x - 9 = 0 \ or \ x + 5 = 0$$
$$x = 9 \ or \quad x = -5$$

The zeros are 9 and -5.

53. The zeros of $r(x) = x^3 + 4x^2 + 3x$ are the solutions of the equation $x^3 + 4x^2 + 3x = 0$. We factor and use the principle of zero products.

$$x^3 + 4x^2 + 3x = 0$$
$$x(x^2 + 4x + 3) = 0$$
$$x(x + 1)(x + 3) = 0$$
$$x = 0 \ or \ x + 1 = 0 \ \ or \ x + 3 = 0$$
$$x = 0 \ or \quad x = -1 \ or \quad x = -3$$

The zeros are 0, -1, and -3.

55.
$$x^2 + 4x = 45$$
$$x^2 + 4x - 45 = 0$$
$$(x + 9)(x - 5) = 0$$
$$x + 9 = 0 \quad or \ x - 5 = 0$$
$$x = -9 \ or \quad x = 5$$

The solutions are -9 and 5.

57.
$$x^2 - 9x = 0$$
$$x(x - 9) = 0$$
$$x = 0 \ \ or \ x - 9 = 0$$
$$x = 0 \ \ or \quad x = 9$$

The solutions are 0 and 9.

59.
$$a^3 + 40a = 13a^2$$
$$a^3 - 13a^2 + 40a = 0$$
$$a(a^2 - 13a + 40) = 0$$
$$a(a - 5)(a - 8) = 0$$
$$a = 0 \ or \ a - 5 = 0 \ or \ a - 8 = 0$$
$$a = 0 \ or \quad a = 5 \ or \quad a = 8$$

The solutions are 0, 5, and 8.

61.
$$(x - 3)(x + 2) = 14$$
$$x^2 - x - 6 = 14$$
$$x^2 - x - 20 = 0$$
$$(x - 5)(x + 4) = 0$$
$$x - 5 = 0 \ or \ x + 4 = 0$$
$$x = 5 \ or \quad x = -4$$

The solutions are 5 and -4.

63.
$$35 - x^2 = 2x$$
$$35 - 2x - x^2 = 0$$
$$(7 + x)(5 - x) = 0$$
$$7 + x = 0 \quad or \ 5 - x = 0$$
$$x = -7 \ or \quad 5 = x$$

The solutions are -7 and 5.

65. From the graph we see that the zeros of $f(x) = x^2 + 10x - 264$ are –22 and 12. We also know that –22 is a zero of $g(x) = x + 22$ and 12 is a zero of $h(x) = x - 12$. Using the principle of zero products in reverse, we have $x^2 + 10x - 264 = (x + 22)(x - 12)$.

67. Graph $y = x^2 + 40x + 384$ and find the zeros. They are –24 and –16. We know that –24 is a zero of $g(x) = x + 24$ and –16 is a zero of $h(x) = x + 16$. Using the principle of zero products in reverse, we have $x^2 + 40x + 384 = (x + 24)(x + 16)$.

69. Graph $y = x^2 + 26x - 2432$ and find the zeros. They are –64 and 38. We know that –64 is a zero of $g(x) = x + 64$ and 38 is a zero of $h(x) = x - 38$. Using the principle of zero products in reverse, we have $x^2 + 26x - 2432 = (x + 64)(x - 38)$.

71. We write a linear function for each zero: –1 is a zero of $g(x) = x + 1$; 2 is a zero of $h(x) = x - 2$.
Then $f(x) = (x + 1)(x - 2)$, or $f(x) = x^2 - x - 2$.

73. We write a linear function for each zero: –7 is a zero of $g(x) = x + 7$; –10 is a zero of $h(x) = x + 10$.
Then $f(x) = (x + 7)(x + 10)$, or $f(x) = x^2 + 17x + 70$.

75. We write a linear function for each zero:
0 is a zero of $g(x) = x$;
1 is a zero of $h(x) = x - 1$;
2 is a zero of $k(x) = x - 2$.
Then $f(x) = x(x - 1)(x - 2)$, or $f(x) = x^3 - 3x^2 + 2x$.

77. *Thinking and Writing Exercise.*

79. $(2x + 3)(3x + 4) = 2x \cdot 3x + 2x \cdot 4 + 3 \cdot 3x + 3 \cdot 4$
$= 6x^2 + 8x + 9x + 12$
$= 6x^2 + 17x + 12$

81. $(2x - 3)(3x + 4) = 2x \cdot 3x + 2x \cdot 4$
$\qquad\qquad\qquad + (-3) \cdot 3x + (-3) \cdot 4$
$= 6x^2 + 8x - 9x - 12$
$= 6x^2 - x - 12$

83. $(5x - 1)(x - 7) = 5x \cdot x + 5x \cdot (-7) + (-1) \cdot x$
$\qquad\qquad\qquad\qquad + (-1) \cdot (-7)$
$= 5x^2 - 35x - x + 7$
$= 5x^2 - 36x + 7$

85. *Thinking and Writing Exercise.*

87. The x-coordinates of the x-intercepts are –1 and 3. These are the solutions of $x^2 - 2x - 3 = 0$. From the graph we see that the x-values for which $f(x) < 5$ are in the interval (–2, 4). We could also express the solution set as $\{x | -2 < x < 4\}$.

89. Answers may vary. A polynomial function of lowest degree that meets the given criteria is of the form $f(x) = ax^3 + bx^2 + cx + d$.
Substituting we have
$a \cdot 2^3 + b \cdot 2^2 + c \cdot 2 + d = 0,$
$a(-1)^3 + b(-1)^2 + c(-1) + d = 0,$
$a \cdot 3^3 + b \cdot 3^2 + c \cdot 3 + d = 0,$
$a \cdot 0^3 + b \cdot 0^2 + c \cdot 0 + d = 30,$ or
$8a + 4b + 2c + d = 0,$
$-a + b - c + d = 0,$
$27a + 9b + 3c + d = 0,$
$d = 30$
Solving the system of equations, we get $(5, -20, 5, 30)$, so the corresponding function is $f(x) = 5x^3 - 20x^2 + 5x + 30$.

91. Graph $y_1 = -x^2 + 13.80x$ and $y_2 = 47.61$ and use the Intersect feature to find the first coordinate of the point of intersection. The solution is 6.90

93. Graph $y_1 = x^3 - 3.48x^2 + x$ and $y_2 = 3.48$
and use the Intersect feature to find the first
coordinate of the point of intersection. The
solution is 3.48.

95. $x^2 + \frac{1}{2}x - \frac{3}{16}$

We look for factors of $-\frac{3}{16}$ whose sum is $\frac{1}{2}$.

The factors are $\frac{3}{4}$ and $-\frac{1}{4}$. The factorization

is $x^2 + \frac{1}{2}x - \frac{3}{16} = \left(x + \frac{3}{4}\right)\left(x - \frac{1}{4}\right)$.

97. $x^{2a} + 5x^a - 24$

Substitute u for x^a $\left(\text{and } u^2 \text{ for } x^{2a}\right)$. We

factor $u^2 + 5u - 24$. We look for factors of $-$
24 whose sum is 5. The factors are 8 and
-3. We have $u^2 + 5u - 24 = (u+8)(u-3)$.

Replace u by x^a : $x^{2a} + 5x^a - 24$

$= \left(x^a + 8\right)\left(x^a - 3\right)$.

99. $(a+1)x^2 + (a+1)3x + (a+1)2$

$= (a+1)\left(x^2 + 3x + 2\right)$

We factor $x^2 + 3x + 2$ by looking for factors
of 2 whose sum is 3. These factors are 2 and
1, so we have $x^2 + 3x + 2 = (x+2)(x+1)$.

$(a+1)x^2 + (a+1)3x + (a+1)2$

$= (a+1)(x+2)(x+1)$.

101. $(x+3)^2 - 2(x+3) - 35$

We want two factors of -35 whose sum is -2.
They are -7 and 5.

$[(x+3)-7][(x+3)+5] = (x-4)(x+8)$

103. $x^2 + qx - 32$

All such q are the sums of the factors of -32.

Pair of Factors	Sum of Factors
$32, -1$	31
$-32, 1$	-31
$16, -2$	14
$-16, 2$	-14
$8, -4$	4
$-8, 4$	-4

q can be 31, -31, 14, -14, 4, or -4.

105. The area can be described as the sum of the
individual areas:

$x^2 + 5x + 4x + 20 = x^2 + 9x + 20$.

We factor $x^2 + 9x + 20$ by looking for factors
of 20 whose sum is 9. These factors are 4 and
5, so we have

$x^2 + 9x + 20 = (x+4)(x+5)$.

Exercise Set 5.5

1. f

3. e

5. g

7. h

9. $2x^2 + 7x - 4$

We will use the FOIL method.
1. There is no common factor (other than 1
   or -1).
2. Factor the first term, $2x^2$. The factors
   are $2x, x$. The only possibility is
   $(2x+\ )(x+\ )$.
3. Factor the last term, -4. The
   possibilities are $4(-1)$, $-1 \cdot 4$, $-4 \cdot 1$,
   $1(-4)$, $-2 \cdot 2$, and $2(-2)$.
4. We need factors for which the sum of the
   products (the "outer" and "inner" parts of
   FOIL) is the middle term, $7x$. Try some
   possibilities and check by multiplying.
   $(2x+4)(x-1) = 2x^2 + 2x - 4$
   We try again.
   $(2x-1)(x+4) = 2x^2 + 7x - 4$
   The factorization is $(2x-1)(x+4)$.

11. $3x^2 - 17x - 6$

We will use the FOIL method.
1. There is no common factor (other than 1
   or -1).
2. Factor the first term, $3x^2$. The only
   factors are $3x, x$. The only possibility is
   $(3x+\ )(x+\ )$.

3. Factor the last term, −6. The possibilities are $6(-1)$, $-1 \cdot 6$, $-6 \cdot 1$, $1(-6)$, $-2 \cdot 3$, $3(-2)$, $-3 \cdot 2$, and $2(-3)$.

4. We need factors for which the sum of the products (the "outer" and "inner" parts of FOIL) is the middle term, $-17x$. Try various possibilities, and check by multiplying.
    The factorization is $(3x+1)(x-6)$.

13. $15a^2 - 14a + 3$
    We will use the FOIL method.
    1. There is no common factor (other than 1 or −1).
    2. Factor the first term, $15a^2$. The factors are $15a, a$ and $5a, 3a$. We have these possibilities:
        $(15a+ \ )(a+ \ ), (5a+ \ )(3a+ \ )$.
    3. Factor the last term, 3. The possibilities are $1 \cdot 3$, $3 \cdot 1$, $(-1)(-3)$, and $(-3)(-1)$.
    4. Look for factors such that the sum of the products is the middle term, $-14a$. Trial and error leads us to the correct factorization
        $15a^2 - 14a + 3 = (5a-3)(3a-1)$.

15. $6t^2 + 17t + 7$
    We will use the FOIL method.
    1. There is no common factor (other than 1 or −1).
    2. Factor the first term, $6t^2$. The factors are $6t, t$ and $3t, 2t$. We have these possibilities:
        $(6t+ \ )(t+ \ ), (3t+ \ )(2t+ \ )$.
    3. Factor the last term, 7. The possibilities are $1 \cdot 7$, $7 \cdot 1$, $(-1)(-7)$, and $(-7)(-1)$.
    4. Look for factors such that the sum of the products is the middle term, $17t$. Trial and error leads us to the correct factorization
        $6t^2 + 17t + 7 = (3t+7)(2t+1)$.

17. $6x^2 - 10x - 4$
    We will use the FOIL method.
    1. Factor out the common factor, 2:
        $2(3x^2 - 5x - 2)$.

2. Now we factor the trinomial $3x^2 - 5x - 2$. Factor the first term, $3x^2$. The factors are $3x$ and x. We have one possibility: $(3x+ \ )(x+ \ )$.

3. Factor the last term, −2. The possibilities are $2(-1)$, $-1 \cdot 2$, $-2 \cdot 1$, and $1(-2)$.

4. Look for factors such that the sum of the products is the middle term, $-10x$. Trial and error leads us to the correct factorization:
    $3x^2 - 5x - 2 = (3x+1)(x-2)$
    We must include the common factor to get a factorization of the original trinomial:
    $6x^2 - 10x - 4 = 2(3x+1)(x-2)$

19. $8x^2 - 16 - 28x = 8x^2 - 28x - 16$
    We will use the grouping method.
    1. Factor out the common factor, 4:
        $4(2x^2 - 7x - 4)$
    2. Now we factor the trinomial $2x^2 - 7x - 4$. Multiply the leading coefficient, 2, and the constant, −4:
        $2(-4) = -8$
    3. Factor −8 so the sum of the factors is −7. We need only consider pairs of factors in which the negative factor has the larger absolute value, since their sum is negative.

Pair of Factors	Sum of Factors
−4, 2	−2
−8, 1	−7

    4. Split $-7x$ using the results of step (3):
        $-7x = -8x + x$
    5. Factor by grouping:
        $2x^2 - 7x - 4 = 2x^2 - 8x + x - 4$
        $\qquad = 2x(x-4) + 1(x-4)$
        $\qquad = (x-4)(2x+1)$
    We must include the common factor to get a factorization of the original trinomial:
    $8x^2 - 16 - 28x = 4(x-4)(2x+1)$

21. $14x^4 - 19x^3 - 3x^2$

We will use the grouping method.

1. Factor out the common factor, x^2.

   $x^2\left(14x^2 - 19x - 3\right)$

2. Now we factor the trinomial $14x^2 - 19x - 3$. Multiply the leading coefficient, 14, and the constant, –3:

   $14(-3) = -42$

3. Factor –42 so the sum of the factors is –19. We need only consider pairs of factors in which the negative factor has the larger absolute value, since the sum is negative.

Pair of Factors	Sum of Factors
-42, 1	-41
-21, 2	-19
-14, 3	-11
-7, 6	-1

4. Split $-19x$ using the results of step (3):

   $-19x = -21x + 2x$

5. Factor by grouping:

   $14x^2 - 19x - 3 = 14x^2 - 21x + 2x - 3$

   $\qquad = 7x(2x - 3) + 2x - 3$

   $\qquad = (2x - 3)(7x + 1)$

   We must include the common factor to get a factorization of the original trinomial:

   $14x^4 - 19x^3 - 3x^2 = x^2(2x - 3)(7x + 1)$

23. $10 - 23x + 12x^2 = 12x^2 - 23x + 10$

We will use the grouping method.

1. There is no common factor (other than 1 or –1).

2. We now factor the trinomial $12x^2 - 23x + 10$. Multiply the leading coefficient and the constant $12 \cdot 10 = 120$.

3. We want factors of 120 whose sum equals –23. Notice that $-15(-8) = 120$ and $-15 - 8 = -23$.

4. Split $-23x$ into $-15x$ and $-8x$.

5. Factor by grouping:

   $12x^2 - 23x + 10 = 12x^2 - 8x - 15x + 10$

   $\qquad = 4x(3x - 2) - 5(3x - 2)$

   $\qquad = (4x - 5)(3x - 2)$

25. $9x^2 + 15x + 4$

We will use the grouping method.

1. There is no common factor (other than 1 or –1).

2. Multiply the leading coefficient and constant: $9(4) = 36$

3. Factor 36 so the sum of the factors is 15. We need only consider pairs of positive factors since 36 and 15 are both positive.

Pair of Factors	Sum of Factors
36, 1	37
18, 2	20
12, 3	15
9, 4	13
6, 6	12

4. Split $15x$ using the results of step (3):

   $15x = 12x + 3x$

5. Factor by grouping:

   $9x^2 + 15x + 4 = 9x^2 + 12x + 3x + 4$

   $\qquad = 3x(3x + 4) + 3x + 4$

   $\qquad = (3x + 4)(3x + 1)$

27. $4x^2 + 15x + 9$

We will use the FOIL method.

1. There is no common factor other than 1 or –1)

2. Factor the first term, $4x^2$. The possibilities are $(4x + \quad)(x + \quad)$ and $(2x + \quad)(2x + \quad)$.

3. Factor the last term, 9. We consider only positive factors since the middle term and the last term are positive. The possibilities are $9 \cdot 1$ and $3 \cdot 3$.

4. We need factors for which the sum of products is the middle term, $15x$. Trial and error leads us to the correct factorization: $(4x + 3)(x + 3)$

29. $4 + 6t^2 - 13t = 6t^2 - 13t + 4$
    We will use the FOIL method.
    1. There is no common factor (other than 1 or -1).
    2. Factor the first term, $6t^2$, to get the possibilities $(6t +\ )(t +\ )$ and $(3t +\ )(2t +\ )$.
    3. Factor 4. The possibilities are $1 \cdot 4,\ 4 \cdot 1$, $(-4)(-1),\ (-1)(-4),\ 2 \cdot 2$ and $(-2)(-2)$.
    4. Look for factors such that the sum of the products is the middle term, $-13t$. Trial and error indicates that no possible combination of factors will produce the polynomial $6t^2 - 13t + 4$. Therefore, the polynomial $4 + 6t^2 - 13t$ is prime.

31. $-8t^2 - 8t + 30$
    We will use the grouping method.
    1. Factor out -2: $-2(4t^2 + 4t - 15)$
    2. Now we factor the trinomial $4t^2 + 4t - 15$. Multiply the leading coefficient and the constant: $4(-15) = -60$
    3. Factor -60 so the sum of the factors is 4. The desired factorization is $10(-6)$.
    4. Split $4t$ using the results of step (3): $4t = 10t - 6t$
    5. Factor by grouping:
       $4t^2 + 4t - 15 = 4t^2 + 10t - 6t - 15$
       $$= 2t(2t + 5) - 3(2t + 5)$$
       $$= (2t + 5)(2t - 3)$$
       We must include the common factor to get a factorization of the original trinomial:
       $-8t^2 - 8t + 30 = -2(2t + 5)(2t - 3)$

33. $8 - 6z - 9z^2$
    We will use the FOIL method.
    1. There is no common factor (other than 1 or -1).
    2. Factor the first term, 8. The possibilities are $(8 +\ )(1 +\ )$ and $(4 +\ )(2 +\ )$.

3. Factor the last term, $-9z^2$. The possibilities are $-9z \cdot z, -3z \cdot 3z,$ and $9z(-z)$.
4. We need factors for which the sum of products is the middle term, $-6z$. Trial and error leads us to the correct factorization: $(4 + 3z)(2 - 3z)$

35. $18xy^3 + 3xy^2 - 10xy$
    We will use the FOIL method.
    1. Factor out the common factor, xy.
       $xy(18y^2 + 3y - 10)$
    2. We now factor the trinomial $18y^2 + 3y - 10$. Factor the first term, $18y^2$. The possibilities are $(18y +\ )(y +\ ), (9y +\ )(2y +\ ),$ and $(6y +\ )(3y +\ )$.
    3. Factor the last term, -10. The possibilities are $-10 \cdot 1, -5 \cdot 2, 10(-1)$ and $5(-2)$.
    4. We need factors for which the sum of the products is the middle term, $3y$. Trial and error leads us to the correct factorization.
       $18y^2 + 3y - 10 = (6y + 5)(3y - 2)$
       We must include the common factor to get a factorization of the original trinomial:
       $18xy^3 + 3xy^2 - 10xy$
       $$= xy(6y + 5)(3y - 2)$$

37. $24x^2 - 2 - 47x = 24x^2 - 47x - 2$
    We will use the grouping method.
    1. There is no common factor (other than 1 or -1).
    2. Multiply the leading coefficient and the constant: $24(-2) = -48$
    3. Factor -48 so the sum of the factors is -47. The desired factorization is $-48 \cdot 1$.
    4. Split $-47x$ using the results of step (3): $-47x = -48x + x$
    5. Factor by grouping:
       $24x^2 - 47x - 2 = 24x^2 - 48x + x - 2$
       $$= 24x(x - 2) + (x - 2)$$
       $$= (x - 2)(24x + 1)$$

39. $63x^3 + 111x^2 + 36x$

We will use the FOIL method.

1. Factor out the common factor, $3x$.

   $3x(21x^2 + 37x + 12)$

2. Now we will factor the trinomial $21x^2 + 37x + 12$. Factor the first term, $21x^2$. The factors are $21x, x$ and $7x, 3x$. We have these possibilities:

   $(21x +)(x +)$ and $(7x +)(3x +)$.

3. Factor the last term, 12. The possibilities are $12 \cdot 1, (-12)(-1), 6 \cdot 2, (-6)(-2), 4 \cdot 3$, and $(-4)(-3)$ as well as $1 \cdot 12$, $(-1)(-12)$, $2 \cdot 6$, $(-2)(-6)$, $3 \cdot 4$, and $(-3)(-4)$.

4. Look for factors such that the sum of the products is the middle term, $37x$. Trial and error leads us to the correct factorization: $(7x + 3)(3x + 4)$

   We must include the common factor to get a factorization of the original trinomial:

   $63x^3 + 111x^2 + 36x = 3x(7x + 3)(3x + 4)$

41. $48x^4 + 4x^3 - 30x^2$

We will use the grouping method.

1. We factor out the common factor, $2x^2$.

   $2x^2(24x^2 + 2x - 15)$

2. We now factor $24x^2 + 2x - 15$. Multiply the leading coefficient and the constant: $24(-15) = -360$

3. Factor -360 so the sum of the factors is 2. The desired factorization is $-18 \cdot 20$.

4. Split $2x$ using the results of step (3): $2x = -18x + 20x$

5. Factor by grouping:

   $24x^2 + 2x - 15 = 24x^2 - 18x + 20x - 15$

   $\qquad = 6x(4x - 3) + 5(4x - 3)$

   $\qquad = (4x - 3)(6x + 5)$

   We must not forget to include the common factor:

   $48x^4 + 4x^3 - 30x^2 = 2x^2(4x - 3)(6x + 5)$

43. $12a^2 - 17ab + 6b^2$

We will use the FOIL method. (Our thinking is much the same as if we were factoring $12a^2 - 17a + 6$.)

1. There is no common factor (other than 1 or -1).

2. Factor the first term, $12a^2$. The factors are $12a, a$ and $6a, 2a$ and $4a, 3a$. We have these possibilities:

   $(12a +)(a +), (6a +)(2a +)$ and $(4a +)(3a +)$.

3. Factor the last term, $6b^2$. The possibilities are $6b \cdot b, (-6b)(-b), 3b \cdot 2b$, and $(-3b)(-2b)$ as well as $b \cdot 6b, (-b)(-6b), 2b \cdot 3b$, and $(-2b)(-3b)$.

4. Look for factors such that the sum of the products is the middle term, $-17ab$. Trial and error leads us to the correct factorization: $(4a - 3b)(3a - 2b)$

45. $2x^2 + xy - 6y^2$

We will use the grouping method.

1. There is no common factor (other than 1 or -1).

2. Multiply the coefficients of the first and last terms: $2(-6) = -12$

3. Factor -12 so the sum of the factors is 1. The desired factorization is $4(-3)$.

4. Split xy using the results of step (3): $xy = 4xy - 3xy$

5. Factor by grouping:

   $2x^2 + xy - 6y^2$

   $= 2x^2 + 4xy - 3xy - 6y^2$

   $= 2x(x + 2y) - 3y(x + 2y)$

   $= (x + 2y)(2x - 3y)$

47. $8s^2 + 22st + 14t^2$

We will use the FOIL method.

1. We factor out the common factor, 2.
    $2(4s^2 + 11st + 7t^2)$

2. Now we will factor the trinomial
    $4s^2 + 11st + 7t^2$. Factor the first term,
    $4s^2$. The factors are $4s, s$ and $2s, 2s$.
    We have these possibilities:
    $(4s + \quad)(s + \quad)$ and $(2s + \quad)(2s + \quad)$.

3. Factor the last term, $7t^2$. the
    possibilities are $7t \cdot t$, $(-7t)(-t)$,
    $t \cdot 7t$, and $(-t)(-7t)$.

4. Look for factors such that the sum of the
    products is the middle term, $11st$. Trial
    and error leads us to the correct
    factorization: $(s + t)(4s + 7t)$

    We must include the common factor to
    get a factorization of the original
    trinomial:
    $8s^2 + 22st + 14t^2 = 2(s + t)(4s + 7t)$

49. $9x^2 - 30xy + 25y^2$

We will use the grouping method.

1. There is no common factor (other than 1
    or -1).

2. Multiply the coefficient of the first and
    last terms: $9(25) = 225$

3. Factor 225 so the sum of the factors is
    -30. The desired factorization is
    $-15(-15)$.

4. Split $-30xy$ using the results of step (3):
    $-30xy = -15xy - 15xy$

5. Factor by grouping:
    $9x^2 - 30xy + 25y^2$
    $= 9x^2 - 15xy - 15xy + 25y^2$
    $= 3x(3x - 5y) - 5y(3x - 5y)$
    $= (3x - 5y)(3x - 5y)$ or $(3x - 5y)^2$

51. $9x^2y^2 + 5xy - 4$

Let $u = xy$ and $u^2 = x^2y^2$. Factor

$9u^2 + 5u - 4$. We will use the FOIL method.

1. There is no common factor (other than 1
    or -1).

2. Factor the first term, $9u^2$. The factors
    are $9u, u$ and $3u, 3u$. We have these
    possibilities: $(9u + \quad)(u + \quad)$ and
    $(3u + \quad)(3u + \quad)$.

3. Factor the last term, -4. The possibilities
    are: $-4 \cdot 1, -2 \cdot 2, 2 \cdot -2$ and $-1 \cdot 4$.

4. We need factors for which the sum of the
    products is the middle term, $5u$. Trial
    and error leads us to the factorization:
    $(9u - 4)(u + 1)$. Replace u by xy. We
    have $9x^2y^2 + 5xy - 4 = (9xy - 4)(xy + 1)$.

53. $9z^2 + 6z = 8$
    $0 = 8 - 6z - 9z^2$
    $0 = (4 + 3z)(2 - 3z)$ From Ex. 33
    $4 + 3z = 0$ or $2 - 3z = 0$ Using the principle
                                      of zero products
    $3z = -4$ or $2 = 3z$
    $z = -\dfrac{4}{3}$ or $\dfrac{2}{3} = z$

    The solutions are $-\dfrac{4}{3}$ and $\dfrac{2}{3}$.

55. $63x^3 + 111x^2 + 36x = 0$
    $3x(7x + 3)(3x + 4) = 0$ From Ex. 39
    $3x = 0$ or $7x + 3 = 0$ or $3x + 4 = 0$
    $x = 0$ or $\quad 7x = -3$ or $\quad 3x = -4$
    $x = 0$ or $\quad x = -\dfrac{3}{7}$ or $\quad x = -\dfrac{4}{3}$

    The solutions are $0, -\dfrac{3}{7}, -\dfrac{4}{3}$.

57. $3x^2 - 8x + 4 = 0$
    $(3x - 2)(x - 2) = 0$
    $3x - 2 = 0$ or $x - 2 = 0$
    $3x = 2$ or $\quad x = 2$
    $x = \dfrac{2}{3}$ or $\quad x = 2$

    The solutions are $\dfrac{2}{3}$ and 2.

59. $4t^3 + 11t^2 + 6t = 0$

$t(4t^2 + 11t + 6) = 0$

$t(4t + 3)(t + 2) = 0$

$t = 0 \; or \; 4t + 3 = 0 \quad or \; t + 2 = 0$

$t = 0 \; or \quad 4t = -3 \; or \quad t = -2$

$t = 0 \; or \quad t = -\dfrac{3}{4} \; or \quad t = -2$

The solutions are 0, $-\dfrac{3}{4}$, and -2.

61. $\qquad\qquad 6x^2 = 13x + 5$

$6x^2 - 13x - 5 = 0$

$(2x - 5)(3x + 1) = 0$

$2x - 5 = 0 \; or \; 3x + 1 = 0$

$2x = 5 \; or \quad 3x = -1$

$x = \dfrac{5}{2} \; or \quad x = -\dfrac{1}{3}$

The solutions are $\dfrac{5}{2}$ and $-\dfrac{1}{3}$.

63. $\qquad x(5 + 12x) = 28$

$5x + 12x^2 = 28$

$5x + 12x^2 - 28 = 0$

$12x^2 + 5x - 28 = 0$

$(4x + 7)(3x - 4) = 0$

$4x + 7 = 0 \; or \; 3x - 4 = 0$

$4x = -7 \; or \quad 3x = 4$

$x = -\dfrac{7}{4} \; or \quad x = \dfrac{4}{3}$

The solutions are $-\dfrac{7}{4}$ and $\dfrac{4}{3}$.

65. The zeros of $f(x) = 2x^2 - 13x - 7$ are the roots, or solutions, of the equation $2x^2 - 13x - 7 = 0$.

$2x^2 - 13x - 7 = 0$

$(2x + 1)(x - 7) = 0$

$2x + 1 = 0 \quad or \; x - 7 = 0$

$2x = -1 \; or \quad x = 7$

$x = -\dfrac{1}{2} \; or \quad x = 7$

The zeros are $-\dfrac{1}{2}$ and 7.

67. $f(x) = x^2 + 12x + 40$

We set $f(a) = 8$.

$a^2 + 12a + 40 = 8$

$a^2 + 12a + 32 = 0$

$(a + 8)(a + 4) = 0$

$a + 8 = 0 \; or \; a + 4 = 0$

$a = -8 \; or \quad a = -4$

The values of a for which $f(a) = 8$ are -8 and -4.

69. $g(x) = 2x^2 + 5x$

We set $g(a) = 12$.

$\qquad 2a^2 + 5a = 12$

$2a^2 + 5a - 12 = 0$

$(2a - 3)(a + 4) = 0$

$2a - 3 = 0 \; or \; a + 4 = 0$

$2a = 3 \; or \quad a = -4$

$a = \dfrac{3}{2} \; or \quad a = -4$

The values for a for which $g(a) = 12$ are $\dfrac{3}{2}$ and -4.

71. $f(x) = \dfrac{3}{x^2 - 4x - 5}$

$f(x)$ cannot be calculated for any x-value for which the denominator, $x^2 - 4x - 5$, is 0. To find the excluded values, we solve:

$x^2 - 4x - 5 = 0$

$(x - 5)(x + 1) = 0$

$x - 5 = 0 \; or \; x + 1 = 0$

$x = 5 \; or \quad x = -1$

The domain of f is { $x | x$ is a real number *and* $x \neq 5$ *and* $x \neq -1$ }.

73. $f(x) = \dfrac{x-5}{9x-18x^2}$

$f(x)$ cannot be calculated for any x-value for which the denominator, $9x-18x^2$, is 0. To find the excluded values, we solve:

$9x-18x^2 = 0$

$9x(1-2x) = 0$

$9x = 0 \ or \ 1-2x = 0$

$x = 0 \ or \ -2x = -1$

$x = 0 \ or \ \quad x = \dfrac{1}{2}$

The domain of f is { $x|x$ is a real number and

$x \neq 0 \ and \ x \neq \dfrac{1}{2}$ }.

75. $f(x) = \dfrac{3x}{2x^2-9x+4}$

$f(x)$ cannot be calculated for any x-value for which the denominator, $2x^2-9x+4$, is 0. To find the excluded values, we solve:

$2x^2-9x+4 = 0$

$(2x-1)(x-4) = 0$

$2x-1 = 0 \ or \ x-4 = 0$

$2x = 1$

$x = \dfrac{1}{2} \ or \quad x = 4$

The domain of f is { $x|x$ is a real number and

$x \neq \dfrac{1}{2} \ and \ x \neq 4$ }.

77. $f(x) = \dfrac{7}{5x^3-35x^2+50x}$

$f(x)$ cannot be calculated for any x-value for which the denominator, $5x^3-35x^2+50x$, is 0. To find the excluded values, we solve:

$5x^3-35x^2+50x = 0$

$5x(x^2-7x+10) = 0$

$5x(x-2)(x-5) = 0$

$5x = 0 \ or \ x-2 = 0 \ or \ x-5 = 0$

$x = 0 \ or \quad x = 2 \ or \quad x = 5$

The domain of f is { $x|x$ is a real number and

$x \neq 0 \ and \ x \neq 2 \ and \ x \neq 5$ }.

79. *Thinking and Writing Exercise.*

81. $(x-2)^2 = x^2 - 2 \cdot x \cdot 2 + 2^2$

$= x^2 - 4x + 4$

83. $(x+2)(x-2) = x^2 - 2^2$

$= x^2 - 4$

85. $(4a+1)^2 = (4a)^2 + 2 \cdot (4a) \cdot 1 + 1^2$

$= 16a^2 + 8a + 1$

87. $(3c-10)^2 = (3c)^2 - 2 \cdot (3c) \cdot 10 + 10^2$

$= 9c^2 - 60c + 100$

89. $(8n+3)(8n-3) = (8n)^2 - 3^2$

$= 64n^2 - 9$

91. *Thinking and Writing Exercise.*

93. Graph $y = 4x^2 + 120x + 675$ and find the zeros. They are –7.5 and –22.5, or $-\dfrac{15}{2}$ and $-\dfrac{45}{2}$. We know that $-\dfrac{15}{2}$ is a zero of $g(x) = 2x+15$ and $-\dfrac{45}{2}$ is a zero of $h(x) = 2x+45$. We have

$y = 4x^2 + 120x + 675 = (2x+15)(2x+45)$.

95. First factor out the largest common factor.

$3x^3 + 150x^2 - 3672x = 3x(x^2 + 50x - 1224)$

Now graph $y = x^2 + 50x - 1224$ and find the zeros. They are –68 and 18. We know that –68 is a zero of $g(x) = x+68$, and 18 is a zero of $h(x) = x-18$.

We have $x^2 + 50x - 1224 = (x+68)(x-18)$, so $3x^3 + 150x^2 - 3672x = 3x(x+68)(x-18)$.

97. $(8x+11)(12x^2-5x-2)=0$

$(8x+11)(3x-2)(4x+1)=0$

$8x+11=0 \quad or \quad 3x-2=0 \quad or \quad 4x+1=0$

$8x=-11 \quad or \quad 3x=2 \quad or \quad 4x=-1$

$x=-\dfrac{11}{8} \quad or \quad x=\dfrac{2}{3} \quad or \quad x=-\dfrac{1}{4}$

The solutions are $-\dfrac{11}{8},\dfrac{2}{3}$, and $-\dfrac{1}{4}$.

99. $(x-2)^3=x^3-2$

$x^3-6x^2+12x-8=x^3-2$

$0=6x^2-12x+6$

$0=6(x^2-2x+1)$

$0=6(x-1)(x-1)$

$x-1=0 \; or \; x-1=0$

$x=1 \; or \quad x=1$

The solution is 1.

101. $18a^2b^2-3ab-10$

Let $u=ab$ (and $u^2=a^2b^2$). Factor $18u^2-3u-10$.

We will use the FOIL method.

1. There is no common factor (other than 1 or –1).
2. Factor the first term, $18u^2$. The factors are $18u,u$ and $9u,2u$ and $6u,3u$. We have these possibilities: $(18u+\;\;)(u+\;\;)$ and $(9u+\;\;)(2u+\;\;)$ and $(6u+\;\;)(3u+\;\;)$.
3. Factor the last term, –10. The possibilities are $-10\cdot 1, -5\cdot 2$, $-2\cdot 5, -1\cdot 10$.
4. We need factors for which the sum of the products is the middle term, $-3u$. Trial and error leads us to the factorization: $(6u-5)(3u+2)$. Replace u by ab. The factorization is $(6ab-5)(3ab+2)$.

103. $16a^2b^3+25ab^2+9$

There is no common factor (other than 1 or –1).

The trinomial is not of the form ax^2+bx+c, and also is not of the form au^2+bu+c, using substitution. Therefore, $16a^2b^3+25ab^2+9$ is prime.

105. $25t^{10}-10t^5+1$

Let $u=t^5$ $(\text{and } u^2=t^{10})$. Factor $25u^2-10u+1$. We will use the grouping method. Multiply the leading coefficient and the constant: $25(1)=25$. Factor 25 so the sum of the factors is –10. The desired factorization is –5, –5. Split the middle term and factor by grouping.

$25u^2-10u+1=25u^2-5u-5u+1$

$=5u(5u-1)-1(5u-1)$

$=(5u-1)(5u-1)$

or $(5u-1)^2$

Replace u by t^5. The factorization is $(5t^5-1)^2$.

107. $20x^{2n}+16x^n+3$

Let $u=x^n$ $(\text{and } u^2=x^{2n})$. Factor $20u^2+16u+3$. We will use the FOIL method.

1. There is no common factor (other than 1 or –1).
2. Factor the first term, $20u^2$. The factors are $20u,u$ and $10u,2u$ and $5u,4u$. We have these possibilities: $(20u+\;\;)(u+\;\;)$ and $(10u+\;\;)(2u+\;\;)$ and $(5u+\;\;)(4u+\;\;)$.
3. Factor the last term, 3. The possibilities are $3\cdot 1$ and $-3\cdot -1$.
4. We need factors for which the sum of the products is the middle term, $16u$. Trial and error leads us to the factorization $(10u+3)(2u+1)$. Replace u by x^n.

The factorization is $(10x^n+3)(2x^n+1)$.

109. $7(t-3)^{2n} + 5(t-3)^n - 2$

Let $u = (t-3)^n$ $\left(\text{and } u^2 = (t-3)^{2n}\right)$. Factor

$7u^2 + 5u - 2$. We will use the FOIL method.

1. There is no common factor (other than 1 or -1).
2. Factor the first term, $7u^2$. The factors are $7u, u$. The possibility is

   $(7u + \quad)(u + \quad)$.
3. Factor the last term, -2. The possibilities are $-2 \cdot 1$ and $-1, 2$.
4. We need factors for which the sum of the product is the middle term, $5u$. Trial and error leads us to the factorization

   $(7u - 2)(u + 1)$. Replace u by $(t-3)^n$.

   The factorization is

   $\left[7(t-3)^n - 2\right]\left[(t-3)^n + 1\right]$

111. $2a^4 b^6 - 3a^2 b^3 - 20$

Let $u = a^2 b^3$ $\left(\text{and } u^2 = a^4 b^6\right)$. Factor

$2u^2 - 3u - 20$. We will use the FOIL method.

1. There is no common factor (other than 1 or -1).
2. Factor the first term, $2u^2$. The factors are $2u, u$. The possibility is

   $(2u + \quad)(u + \quad)$.
3. Factor the last term, -20. The possibilities are $-20 \cdot 1, -10 \cdot 2$, $-5 \cdot 4, -4 \cdot 5, -2 \cdot 10$, and $-1 \cdot 20$.
4. We need factors for which the sum of the products is the middle term, $-3u$. Trial and error leads us to the factorization:

   $(2u + 5)(u - 4)$. Replace u by $a^2 b^3$.

   We have $(2a^2 b^3 + 5)(a^2 b^3 - 4)$.

113. $ax^2 + bx + c = (mx + r)(nx + s)$

Let $P = ms$ and $Q = rn$. Then

$$ax^2 + bx + c = (mx + r)(nx + s)$$
$$= mnx^2 + msx + rnx + rs$$
$$= mnx^2 + Px + Qx + rs$$
$$= mnx^2 + (P + Q)x + rs$$

Comparing the terms on the right and left sides of the equation,

$ax^2 = mnx^2$, or $a = mn$

$bx = (P + Q)x$, or $b = P + Q$

$c = rs$

and

$ac = (mn)(rs) = (ms)(rn) = PQ$.

Exercise Set 5.6

1. Differences of two squares

3. Perfect-square trinomial

5. None of these

7. Prime polynomial

9. $x^2 + 18x + 81 = x^2 + 2 \cdot 9 \cdot x + 9^2$

   This is of the form $A^2 + 2BA + B^2$, which is a perfect-square trinomial.

11. $x^2 - 10x - 25$

    The sign of the third term must be "+" for the polynomial to be a perfect-square trinomial. Therefore, $x^2 - 10x - 25$ is not a perfect-square trinomial.

13. $x^2 - 3x + 9$

    The second term must equal $-6x$ for $x^2 - 3x + 9$ to be a perfect-square trinomial. Therefore, $x^2 - 3x + 9$ is not a perfect-square trinomial.

15. $9x^2 + 25 - 30x = 9x^2 - 30x + 25$

    $$= (3x)^2 - 2 \cdot 5 \cdot 3x + 5^2$$

    This is of the form $A^2 - 2BA + B^2$, which is a perfect-square trinomial.

17. $t^2 + 6t + 9 = (t + 3)^2$

    Find the square terms and write the quantities that were squared with a plus sign between them.

19. $a^2 - 14a + 49 = (a - 7)^2$

    Find the square terms and write the quantities that were squared with a minus sign between them.

21. $4a^2 - 16a + 16$

$= 4(a^2 - 4a + 4)$

    Factoring out the common factor

$= 4(a - 2)^2$

    Factoring the perfect-square trinomial

23. $1 - 2t + t^2$

$= t^2 - 2t + 1$

    Changing order

$= (t - 1)^2$, or $(1 - t)^2$

    Factoring the perfect-square trinomial

25. $24a^2 + a^3 + 144a$

$= a^3 + 24a^2 + 144a$

    Changing order

$= a(a^2 + 24a + 144)$

    Factoring out the common factor

$= a(a + 12)^2$

    Factoring the perfect-square trinomial

27. $20x^2 + 100x + 125$

$= 5(4x^2 + 20x + 25)$

    Factoring out the common factor

$= 5(2x + 5)^2$

    Factoring the perfect-square trinomial

29. $1 + 8d^3 + 16d^6 = (1 + 4d^3)^2$ or $(4d^3 + 1)^2$

Find the square terms and write the quantities that were squared with an addition sign between them.

31. $-y^3 + 8y^2 - 16y$

$= -y(y^2 - 8y + 16)$

    Factoring out the common factor

$= -y(y - 4)^2$

    Factoring the perfect-square trinomial

33. $0.25x^2 + 0.30x + 0.09 = (0.5x + 0.3)^2$

Find the square terms and write the quantities that were squared with a plus sign between them. Square this binomial.

35. $x^2 - 2xy + y^2 = (x - y)^2$

Find the square terms and write the quantities that were squared with a minus sign between them. Square this binomial.

37. $25a^6 + 30a^3b^3 + 9b^6 = (5a^3 + 3b^3)^2$

Find the square terms and write the quantities that were squared with a plus sign between them. Square this binomial.

39. $5a^2 - 10ab + 5b^2$

$= 5(a^2 - 2ab + b^2)$

    Factoring out the common factor

$= 5(a - b)^2$

    Factoring the perfect-square trinomial

41. $x^2 - 100$ is a difference of squares, because $x^2 = (x)^2$ and $100 = 10^2$, and the terms have different signs.

43. $n^4 + 1$ is not a difference of squares, because the terms have the same sign,.

45. $-1 + 64t^2$ is a difference of squares, because $1 = (1)^2$ and $64t^2 = (8t)^2$, and the terms have different signs.

47. $y^2 - 100 = y^2 - 10^2 = (y + 10)(y - 10)$

49. $m^2 - 64 = m^2 - 8^2 = (m + 8)(m - 8)$

51. $-49 + t^2 = t^2 - 7^2 = (t + 7)(t - 7)$

53. $8x^2 - 8y^2$

$= 8(x^2 - y^2)$

    Factoring out the common factor

$= 8(x + y)(x - y)$

    Factoring the difference of squares

55. $-80a^6 + 45 = -5(16a^6 - 9)$

$= -5\left[(4a^3)^2 - 3^2\right]$

$= -5(4a^3 + 3)(4a^3 - 3)$

57. $49a^4 + 100$

This expression is the sum of two squares, and therefore cannot be factored any further. It is therefore prime.

59. $t^4 - 1 = \left(t^2\right)^2 - 1^2$

$= \left(t^2 + 1\right)\left(t^2 - 1\right)$

$= \left(t^2 + 1\right)\left(t^2 - 1^2\right)$

$= \left(t^2 + 1\right)(t + 1)(t - 1)$

61. $9a^4 - 25a^2b^4 = a^2\left(9a^2 - 25b^4\right)$

$= a^2\left[(3a)^2 - \left(5b^2\right)^2\right]$

$= a^2\left(3a + 5b^2\right)\left(3a - 5b^2\right)$

63. $16x^4 - y^4 = \left(4x^2\right)^2 - \left(y^2\right)^2$

$= \left(4x^2 + y^2\right)\left(4x^2 - y^2\right)$

$= \left(4x^2 + y^2\right)\left((2x)^2 - y^2\right)$

$= \left(4x^2 + y^2\right)(2x + y)(2x - y)$

65. $\dfrac{1}{49} - x^2 = \left(\dfrac{1}{7}\right)^2 - x^2 = \left(\dfrac{1}{7} + x\right)\left(\dfrac{1}{7} - x\right)$

67. $(a + b)^2 - 9 = (a + b)^2 - 3^2$

$= \left[(a + b) + 3\right]\left[(a + b) - 3\right]$

$= (a + b + 3)(a + b - 3)$

69. $x^2 - 6x + 9 - y^2$

$= \left(x^2 - 6x + 9\right) - y^2$

Grouping as a difference of squares

$= (x - 3)^2 - y^2$

$= (x - 3 + y)(x - 3 - y)$

71. $t^3 + 8t^2 - t - 8$

$= t^2(t + 8) - 1(t + 8)$

Factoring by grouping

$= (t + 8)\left(t^2 - 1\right)$

$= (t + 8)(t + 1)(t - 1)$

Factoring the difference of squares

73. $r^3 - 3r^2 - 9r + 27$

$= r^2(r - 3) - 9(r - 3)$

Factoring by grouping

$= (r - 3)\left(r^2 - 9\right)$

$= (r - 3)(r + 3)(r - 3)$ or $(r - 3)^2(r + 3)$

Factoring the difference of squares

75. $m^2 - 2mn + n^2 - 25$

$= \left(m^2 - 2mn + n^2\right) - 25$

Grouping as a difference of squares

$= (m - n)^2 - 5^2$

$= (m - n + 5)(m - n - 5)$

77. $36 - (x + y)^2 = 6^2 - (x + y)^2$

$= \left[6 + (x + y)\right]\left[6 - (x + y)\right]$

$= (6 + x + y)(6 - x - y)$

79. $16 - a^2 - 2ab - b^2$

$= 16 - \left(a^2 + 2ab + b^2\right)$

Grouping as a difference of squares

$= 4^2 - (a + b)^2$

$= \left[4 + (a + b)\right]\left[4 - (a + b)\right]$

$= (4 + a + b)(4 - a - b)$

81. $a^3 - ab^2 - 2a^2 + 2b^2$

$= a\left(a^2 - b^2\right) - 2\left(a^2 - b^2\right)$

$= \left(a^2 - b^2\right)(a - 2)$

$= (a + b)(a - b)(a - 2)$

83. $\qquad a^2 + 1 = 2a$

$a^2 - 2a + 1 = 0$

$(a - 1)(a - 1) = 0$

$a - 1 = 0 \ or \ a - 1 = 0$

$\quad a = 1 \ or \quad a = 1$

The solution is 1.

85. $2x^2 - 24x + 72 = 0$

$2(x^2 - 12x + 36) = 0$

$2(x-6)(x-6) = 0$

$x - 6 = 0 \; or \; x - 6 = 0$

$x = 6 \; or \quad x = 6$

The solution is 6.

87. $x^2 - 9 = 0$

$(x+3)(x-3) = 0$

$x + 3 = 0 \quad or \; x - 3 = 0$

$x = -3 \; or \quad x = 3$

The solutions are –3 and 3.

89. $a^2 = \dfrac{1}{25}$

$a^2 - \dfrac{1}{25} = 0$

$\left(a + \dfrac{1}{5}\right)\left(a - \dfrac{1}{5}\right) = 0$

$a + \dfrac{1}{5} = 0 \quad or \; a - \dfrac{1}{5} = 0$

$a = -\dfrac{1}{5} \; or \quad a = \dfrac{1}{5}$

The solutions are $-\dfrac{1}{5}$ and $\dfrac{1}{5}$.

91. $8x^3 + 1 = 4x^2 + 2x$

$8x^3 - 4x^2 - 2x + 1 = 0$

$4x^2(2x-1) - (2x-1) = 0$

$(2x-1)(4x^2 - 1) = 0$

$(2x-1)(2x+1)(2x-1) = 0$

$2x - 1 = 0 \; or \; 2x + 1 = 0 \quad or \; 2x - 1 = 0$

$2x = 1 \; or \quad 2x = -1 \; or \quad 2x = 1$

$x = \dfrac{1}{2} \; or \quad x = -\dfrac{1}{2} \; or \quad x = \dfrac{1}{2}$

The solutions are $\dfrac{1}{2}$ and $-\dfrac{1}{2}$.

93. $x^3 + 3 = 3x^2 + x$

$x^3 - 3x^2 - x + 3 = 0$

$x^2(x-3) - (x-3) = 0$

$(x-3)(x^2 - 1) = 0$

$(x-3)(x+1)(x-1) = 0$

$x - 3 = 0 \; or \; x + 1 = 0 \quad or \; x - 1 = 0$

$x = 3 \; or \quad x = -1 \; or \quad x = 1$

The solutions are 3, –1, and 1.

95. The polynomial $x^2 - 3x - 7$ is prime. We solve the equation by graphing $y = x^2 - 3x - 7$ and finding the zeros. They are approximately –1.541 and 4.541. These are the solutions.

97. The polynomial $2x^2 + 8x + 1$ is prime. We solve the equation by graphing $y = 2x^2 + 8x + 1$ and finding the zeros. They are approximately –3.871 and –0.129. These are the solutions.

99. The polynomial $x^3 + 3x^2 + x - 1$ is prime. We solve the equation by graphing $y = x^3 + 3x^2 + x - 1$ and finding the zeros. They are approximately –2.414, –1, and approximately 0.414. These are the solutions.

101. $f(x) = x^2 - 12x$

We set $f(a)$ equal to –36.

$a^2 - 12a = -36$

$a^2 - 12a + 36 = 0$

$(a-6)(a-6) = 0$

$a - 6 = 0 \; or \; a - 6 = 0$

$a = 6 \; or \quad a = 6$

The value of a for which $f(a) = -36$ is 6.

103. To find the zeros of $f(x) = x^2 - 16$, we find the roots of the equation $x^2 - 16 = 0$.

$x^2 - 16 = 0$

$(x+4)(x-4) = 0$

$x + 4 = 0 \quad or \; x - 4 = 0$

$x = -4 \; or \quad x = 4$

The zeros are –4 and 4.

105. To find the zeros of $f(x) = 2x^2 + 4x + 2$, we find the roots of the equation
$$2x^2 + 4x + 2 = 0$$
$$2(x^2 + 2x + 1) = 0$$
$$2(x+1)(x+1) = 0$$
$$x + 1 = 0 \quad or \quad x + 1 = 0$$
$$x = -1 \, or \quad x = -1$$
The zero is -1.

107. To find the zeros of $f(x) = x^3 - 2x^2 - x + 2$, we find the roots of the equation $x^3 - 2x^2 - x + 2 = 0$.
$$x^3 - 2x^2 - x + 2 = 0$$
$$x^2(x - 2) - (x - 2) = 0$$
$$(x - 2)(x^2 - 1) = 0$$
$$(x - 2)(x + 1)(x - 1) = 0$$
$$x - 2 = 0 \quad or \quad x + 1 = 0 \quad or \quad x - 1 = 0$$
$$x = 2 \, or \quad x = -1 \, or \quad x = 1$$
The zeros are 2, -1, and 1.

109. *Thinking and Writing Exercise.*

111. $(2x^2 y^4)^3 = 2^3 (x^2)^3 (y^4)^3$
$$= 8 \cdot x^6 \cdot y^{12}$$
$$= 8x^6 y^{12}$$

113. $(x+1)(x+1)(x+1)$
$$= (x+1)(x^2 + 2 \cdot x \cdot 1 + 1)$$
$$= (x+1)(x^2 + 2x + 1)$$
$$= x(x^2 + 2x + 1) + (x^2 + 2x + 1)$$
$$= x^3 + 2x^2 + x + x^2 + 2x + 1$$
$$= x^3 + 2x^2 + x^2 + 2x + x + 1$$
$$= x^3 + 3x^2 + 3x + 1$$

115. $(m+n)^3$
$$= (m+n)(m+n)(m+n)$$
$$= (m+n)(m^2 + 2 \cdot m \cdot n + n^2)$$
$$= (m+n)(m^2 + 2mn + n^2)$$
$$= m(m^2 + 2mn + n^2) + n(m^2 + 2mn + n^2)$$
$$= m^3 + 2m^2 n + mn^2 + nm^2 + 2mn^2 + n^3$$
$$= m^3 + 2m^2 n + nm^2 + 2mn^2 + mn^2 + n^3$$
$$= m^3 + 3m^2 n + 3mn^2 + n^3$$

117. *Thinking and Writing Exercise.*

119. $x^8 - 2^8 = (x^4)^2 - (2^4)^2$
$$= (x^4 + 2^4)(x^4 - 2^4)$$
$$= (x^4 + 2^4)((x^2)^2 - (2^2)^2)$$
$$= (x^4 + 2^4)(x^2 + 2^2)(x^2 - 2^2)$$
$$= (x^4 + 2^4)(x^2 + 2^2)(x+2)(x-2)$$
or $(x^4 + 16)(x^2 + 4)(x+2)(x-2)$

121. $3x^2 - \dfrac{1}{3} = 3\left(x^2 - \dfrac{1}{9}\right)$
$$= 3\left(x^2 - \left(\dfrac{1}{3}\right)^2\right)$$
$$= 3\left(x + \dfrac{1}{3}\right)\left(x - \dfrac{1}{3}\right)$$

123. $0.09x^8 + 0.48x^4 + 0.64 = (0.3x^4 + 0.8)^2$,
or $\dfrac{1}{100}(3x^4 + 8)^2$

125. $r^2 - 8r - 25 - s^2 - 10s + 16$
$$= (r^2 - 8r + 16) - (s^2 + 10s + 25)$$
$$= (r - 4)^2 - (s + 5)^2$$
$$= [(r-4) + (s+5)][(r-4) - (s+5)]$$
$$= (r - 4 + s + 5)(r - 4 - s - 5)$$
$$= (r + s + 1)(r - s - 9)$$

127. $x^{4a} - 49y^{2a} = \left(x^{2a}\right)^2 - \left(7y^a\right)^2$

$= \left(x^{2a} + 7y^a\right)\left(x^{2a} - 7y^a\right)$

129. $3(x+1)^2 + 12(x+1) + 12$

$= 3\left[(x+1)^2 + 4(x+1) + 4\right]$

$= 3\left[(x+1) + 2\right]^2$

$= 3(x+3)^2$

131. $9x^{2n} - 6x^n + 1 = \left(3x^n\right)^2 - 6x^n + 1 = \left(3x^n - 1\right)^2$

133. $s^2 - 4st + 4t^2 + 4s - 8t + 4$

$= \left(s^2 - 4st + 4t^2\right) + \left(4s - 8t\right) + 4$

$= (s - 2t)^2 + 4(s - 2t) + 2^2$

$= (s - 2t + 2)^2$

135. If $P(x) = x^4$, then

$P(a + h) - P(a)$

$= (a + h)^4 - a^4$

$= \left[(a+h)^2 + a^2\right]\left[(a+h)^2 - a^2\right]$

$= \left[(a+h)^2 + a^2\right]\left[(a+h) + a\right]\left[(a+h) - a\right]$

$= \left(a^2 + 2ah + h^2 + a^2\right)(2a + h)(h)$

$= h(2a + h)\left(2a^2 + 2ah + h^2\right)$

Exercise Set 5.7

1. Difference of cubes

3. Difference of squares

5. Sum of cubes

7. None of these

9. Difference of cubes

11. $x^3 + 64 = x^3 + 4^3$

$= (x + 4)\left(x^2 - 4x + 16\right)$

$A^3 + B^3 = (A + B)\left(A^2 - AB + B^2\right)$

13. $z^3 - 1 = z^3 - 1^3$

$= (z - 1)\left(z^2 + z + 1\right)$

$A^3 - B^3 = (A - B)\left(A^2 + AB + B^2\right)$

15. $t^3 - 1000 = t^3 - 10^3$

$= (t - 10)\left(t^2 + 10t + 100\right)$

$A^3 - B^3 = (A - B)\left(A^2 + AB + B^2\right)$

17. $27x^3 + 1 = (3x)^3 + 1^3$

$= (3x + 1)\left(9x^2 - 3x + 1\right)$

19. $64 - 125x^3 = (4)^3 - (5x)^3$

$= (4 - 5x)\left(16 + 20x + 25x^2\right)$

21. $8y^3 + 64 = 8\left(y^3 + 8\right)$

$= 8\left(y^3 + 2^3\right)$

$= 8(y + 2)\left(y^2 - 2y + 4\right)$

23. $x^3 - y^3 = (x - y)\left(x^2 + xy + y^2\right)$

25. $a^3 + \dfrac{1}{8} = a^3 + \left(\dfrac{1}{2}\right)^3$

$= \left(a + \dfrac{1}{2}\right)\left(a^2 - \dfrac{1}{2}a + \dfrac{1}{4}\right)$

27. $8t^3 - 8 = 8\left(t^3 - 1\right)$

$= 8\left(t^3 - 1^3\right)$

$= 8(t - 1)\left(t^2 + t + 1\right)$

29. $y^3 - \dfrac{1}{1000} = y^3 - \left(\dfrac{1}{10}\right)^3$

$= \left(y - \dfrac{1}{10}\right)\left(y^2 + \dfrac{1}{10}y + \dfrac{1}{100}\right)$

31. $ab^3 + 125a = a\left(b^3 + 125\right)$

$= a\left(b^3 + 5^3\right)$

$= a(b + 5)\left(b^2 - 5b + 25\right)$

33. $5x^3 - 40z^3 = 5(x^3 - 8z^3)$
$$= 5\left[x^3 - (2z)^3\right]$$
$$= 5(x - 2z)(x^2 + 2xz + 4z^2)$$

35. $x^3 + 0.001 = x^3 + 0.1^3$
$$= (x + 0.1)(x^2 - 0.1x + 0.01)$$

37. $64x^6 - 8t^6 = 8(8x^6 - t^6)$
$$= 8\left[(2x^2)^3 - (t^2)^3\right]$$
$$= 8(2x^2 - t^2)(4x^4 + 2x^2t^2 + t^4)$$

39. $2y^4 - 128y = 2y(y^3 - 64)$
$$= 2y(y^3 - 4^3)$$
$$= 2y(y - 4)(y^2 + 4y + 16)$$

41. $z^6 - 1 = (z^3)^2 - 1^2$
$$= (z^3 + 1)(z^3 - 1)$$

43. $t^6 + 64y^6 = (t^2)^3 + (4y^2)^3$
$$= (t^2 + 4y^2)(t^4 - 4t^2y^2 + 16y^4)$$

45. $x^{12} - y^3z^{12} = (x^4)^3 - (yz^4)^3$
$$= (x^4 - yz^4)(x^8 + x^4yz^4 + y^2z^8)$$

47. $x^3 + 1 = 0$
$$(x + 1)(x^2 - x + 1) = 0$$
$$x + 1 = 0 \quad \text{or} \quad x^2 - x + 1 = 0$$
$$x = -1$$
We cannot factor $x^2 - x + 1$. The only real-number solution is -1.

49. $8x^3 = 27$
$$8x^3 - 27 = 0$$
$$(2x - 3)(4x^2 + 6x + 9) = 0$$
$$2x - 3 = 0 \quad \text{or} \quad 4x^2 + 6x + 9 = 0$$
$$2x = 3$$
$$x = \frac{3}{2}$$

We cannot factor $4x^2 + 6x + 9$. The only real-number solution is $\frac{3}{2}$.

51. $2t^3 - 2000 = 0$
$$2(t^3 - 1000) = 0$$
$$2(t - 10)(t^2 + 10t + 100) = 0$$
$$t - 10 = 0 \quad \text{or} \quad t^2 + 10t + 100 = 0$$
$$t = 10$$
We cannot factor $t^2 + 10t + 100$. The only real-number solution is 10.

53. *Thinking and Writing Exercise.*

55. Let m and n represent the numbers. The square of the sum of two numbers is $(m + n)^2$.

57. Let x represent the first integer. Then $x + 1$ represents the second integer. The product of the two consecutive integers is therefore $x(x + 1)$.

59. *Familiarize.* Let x = the measure of the second angle, in degrees. Then the measure of the first angle is $4x$ degrees, and the third angle measures $x - 30$ degrees. Recall that the sum of the measures of the angles of any triangle is $180°$.
Translate.

The sum of the measures of the three angles is 180 degrees.
$$x + 4x + (x - 30) = 180$$
Carry out. We solve the equation.
$$x + 4x + (x - 30) = 180$$
$$6x - 30 = 180 \quad \text{Combining like terms}$$
$$6x = 210 \quad \text{Adding 30}$$
$$x = 35 \quad \text{Dividing by 6}$$
If the second angle measures $35°$, then the first angle measures $4 \cdot 35°$, or $140°$, and the third angle measures $35° - 30°$, or $5°$.
Check. The sum of the measures of the three angles is $140° + 35° + 5°$, or $180°$. These results check.
State. The angles measure $140°$, $35°$, and $5°$, respectively.

61. *Thinking and Writing Exercise.*

63. $x^{6a} - y^{3b} = \left(x^{2a}\right)^3 - \left(y^b\right)^3$
$$= \left(x^{2a} - y^b\right)\left(x^{4a} + x^{2a}y^b + y^{2b}\right)$$

65. $(x+5)^3 + (x-5)^3$
$$= \left[(x+5) + (x-5)\right] \cdot$$
$$\left[(x+5)^2 - (x+5)(x-5) + (x-5)^2\right]$$
$$= 2x\left[\left(x^2 + 10x + 25\right) - \left(x^2 - 25\right)\right.$$
$$\left. + \left(x^2 - 10x + 25\right)\right]$$
$$= 2x\left(x^2 + 10x + 25 - x^2 + 25 + x^2 - 10x + 25\right)$$
$$= 2x\left(x^2 + 75\right)$$

67. $5x^3y^6 - \dfrac{5}{8} = 5\left(\left(xy^2\right)^3 - \left(\dfrac{1}{2}\right)^3\right)$
$$= 5\left(x^3y^6 - \dfrac{1}{8}\right)$$
$$= 5\left(xy^2 - \dfrac{1}{2}\right)\left(x^2y^4 + \dfrac{1}{2}xy^2 + \dfrac{1}{4}\right)$$

69. $x^{6a} - \left(x^{2a} + 1\right)^3$
$$= \left(x^{2a}\right)^3 - \left(x^{2a} + 1\right)^3$$
$$= \left[x^{2a} - \left(x^{2a} + 1\right)\right]\left[x^{4a} + x^{2a}\left(x^{2a} + 1\right) + \left(x^{2a} + 1\right)^2\right]$$
$$= \left(x^{2a} - x^{2a} - 1\right)\left(x^{4a} + x^{4a} + x^{2a} + x^{4a} + 2x^{2a} + 1\right)$$
$$= -\left(3x^{4a} + 3x^{2a} + 1\right)$$

71. $t^4 - 8t^3 - t + 8$
$$= t^3(t-8) - (t-8)$$
$$= (t-8)\left(t^3 - 1\right)$$
$$= (t-8)(t-1)\left(t^2 + t + 1\right)$$

73. If $Q(x) = x^6$, then

$Q(a+h) - Q(a)$
$$= (a+h)^6 - a^6$$
$$= \left[(a+h)^3 + a^3\right]\left[(a+h)^3 - a^3\right]$$
$$= \left[(a+h) + a\right]\left[(a+h)^2 - (a+h)a + a^2\right] \cdot$$
$$\left[(a+h) - a\right]\left[(a+h)^2 + (a+h)a + a^2\right]$$
$$= (2a+h)\left(a^2 + 2ah + h^2 - a^2 - ah + a^2\right) \cdot$$
$$(h)\left(a^2 + 2ah + h^2 + a^2 + ah + a^2\right)$$
$$= h(2a+h)\left(a^2 + ah + h^2\right)\left(3a^2 + 3ah + h^2\right)$$

Mid-Chapter Review

Guided Solutions

1. $12x^2 - 3y^2 = 3\left(4x^2 - y^2\right)$
$$= 3(2x+y)(2x-y)$$

2. $x^2y^2 - 5xy^2 - 6y = y\left(x^2y - 5xy - 6\right)$
$$= y(xy-6)(xy+1)$$

Mixed Review

1. $t^2 - 2t + 1 = (t-1)^2$ Perfect-square trinomial

2. $2x^2 - 16x + 30 = 2\left(x^2 - 8x + 15\right)$
$$= 2(x-3)(x-5)$$

3. $x^3 - 64x$
$$= x\left(x^2 - 64\right)$$
$$= x\left(x^2 - 8^2\right) \quad \text{Difference of squares}$$
$$= x(x+8)(x-8)$$

4. $6a^2 - a - 1 = (3a+1)(2a-1)$

5. $5t^3 + 500t = 5t\left(t^2 + 100\right)$

6. $x^3 - 64$
$$= (x-4)\left(x^2 + 4x + 16\right) \quad \text{Difference of cubes}$$

7. $4x^3 + 100x + 40x^2$

    $= 4x^3 + 40x^2 + 100x$

    $= 4x(x^2 + 10x + 25)$ Perfect square trinomial

    $= 4x(x+5)^2$

8. $2x^3 - 3 - 6x^2 + x$

    $= 2x^3 - 6x^2 + x - 3$

    $= 2x^2(x-3) + 1(x-3)$ Factoring by grouping

    $= (x-3)(2x^2 + 1)$

9. $12y^3 + y^2 - 6y$

    $= y(12y^2 + y - 6)$

    $= y(4y+3)(3y-2)$

10. $24 + n^6 - 10n^3 = n^6 - 10n^3 + 24$

    $\qquad\qquad\qquad = (n^3 - 6)(n^3 - 4)$

11. $7t^3 + 7$

    $= 7(t^3 + 1)$ Sum of cubes

    $= 7(t+1)(t^2 - t + 1)$

12. $6m^4 + 96m^3 + 384m^2$

    $= 6m^2(m^2 + 16m + 64)$ Perfect square trinomial

    $= 6m^2(m+8)^2$

13. $x^3 + 3x^2 - x - 3$

    $= x^2(x+3) - 1(x+3)$ Factoring by grouping

    $= (x+3)(x^2 - 1)$ Difference of squares

    $= (x+3)(x+1)(x-1)$

14. $a^2 - \dfrac{1}{9}$

    $= \left(a + \dfrac{1}{3}\right)\left(a - \dfrac{1}{3}\right)$ Difference of squares

15. $0.25 - y^2$

    $= (0.5 + y)(0.5 - y)$ Difference of squares

16. $3n^2 - 21n = 3n(n-7)$

17. $x^4 + 4 - 5x^2$

    $= x^4 - 5x^2 + 4$

    $= (x^2 - 4)(x^2 - 1)$ Difference of squares

    $= (x+2)(x-2)(x+1)(x-1)$

18. $-10c^3 + 25c^2 = -5c^2(2c - 5)$

19. $1 - 64t^6$

    $= 1 - (8t^3)^2$ Difference of squares

    $= (1 + 8t^3)(1 - 8t^3)$ Sum and difference of cubes

    $= \left[1 + (2t)^3\right]\left[1 - (2t)^3\right]$

    $= (1 + 2t)(1 - 2t + 4t^2)(1 - 2t)(1 + 2t + 4t^2)$

20. $6x^5 - 15x^4 + 18x^2 + 9x$

    $= 3x(2x^4 - 5x^3 + 6x + 3)$

21. $2x^5 + 6x^4 + 3x^2 + 9x$

    $= x(2x^4 + 6x^3 + 3x + 9)$

    $= x\left[2x^3(x+3) + 3(x+3)\right]$ Factoring by grouping

    $= x(x+3)(2x^3 + 3)$

22. $yz - 2tx - ty + 2xz$

    $= yz - ty + 2xz - 2tx$

    $= y(z-t) + 2x(z-t)$ Factoring by grouping

    $= (z-t)(y+2x)$

23. $50a^2b^2 - 32c^4$

    $= 2(25a^2b^2 - 16c^4)$

    $= 2\left[(5ab)^2 - (4c^2)^2\right]$ Difference of squares

    $= 2(5ab + 4c^2)(5ab - 4c^2)$

24. $x^2 + 10x + 25 - y^2$

    $= (x+5)^2 - y^2$ Difference of squares

    $= (x+5+y)(x+5-y)$

25. $2x^2 - 12xy - 32y^2$

    $= 2(x^2 - 6xy - 16y^2)$

    $= 2(x - 8y)(x + 2y)$

26. $4x^2y^6 + 20xy^3 + 25$

$= \left(2xy^3 + 5\right)^2$ Perfect square trinomial

27. $m^2 - n^2 + 12n - 36$

$= m^2 - \left(n^2 - 12n + 36\right)$ Perfect-square trinomial

$= m^2 - (n-6)^2$ Difference of squares

$= \left[m + (n-6)\right]\left[m - (n-6)\right]$

$= (m + n - 6)(m - n + 6)$

28. $6a^2b - 9ab - 60b = 3b\left(2a^2 - 3a - 20\right)$

$= 3b(2a + 5)(a - 4)$

29. $p^2 + 121q^2 - 22pq$

$= p^2 - 22pq + 121q^2$ Perfect-square trinomial

$= \left(p - 11q\right)^2$

30. $8a^2b^3c - 40ab^2c^3 + 4ab^2c$

$= 4ab^2c\left(2ab - 10c^2 + 1\right)$

Exercise Set 5.8

1. ***Familiarize.*** Let x represent the number.

Translate.

Square of

$\underbrace{\text{number}}_{\downarrow}$ $\underbrace{\text{plus}}_{\downarrow}$ $\underbrace{\text{number}}_{\downarrow}$ $\underset{\downarrow}{\text{is}}$ $\underset{\downarrow}{132}$

$\quad x^2 \quad\quad + \quad\quad x \quad\quad = \quad 132$

Carry out. We solve the equation:

$x^2 + x = 132$

$x^2 + x - 132 = 0$

$(x + 12)(x - 11) = 0$

$x + 12 = 0 \quad or \ x - 11 = 0$

$x = -12 \ or \quad\quad x = 11$

Check. The square of –12, which is 144, plus –12 is 132. The square of 11, which is 121, plus 11 is 132. Both numbers check.
State. The number is –12 or 11.

3. ***Familiarize.*** Let x represent the number of the first parking space, and $x + 1$ the number of the next parking space.
Translate.

First space   second space

$\underbrace{\text{number}}_{\downarrow}$ $\underbrace{\text{times}}_{\downarrow}$ $\underbrace{\text{number}}_{\downarrow}$ $\underset{\downarrow}{\text{is}}$ $\underset{\downarrow}{110}$

$\quad x \quad\quad \times \quad\quad (x+1) \quad = \quad 110$

Carry out. We solve the equation:

$x(x + 1) = 110$

$x^2 + x - 110 = 0$

$(x + 11)(x - 10) = 0$

$x + 11 = 0 \quad or \ x - 10 = 0$

$x = -11 \ or \quad\quad x = 10$

Check. The number –11 is not a solution, because consecutive numbering of objects uses natural numbers. The product of 10 and $10 + 1 = 11$ is 110. The answer checks.
State. The parking spaces are numbered 10 and 11.

5. ***Familiarize.*** We let w represent the width and $5w$ represent the length. Recall that the formula for the area of a rectangle is $A =$ length $\times$ width.
Translate.

$\underbrace{\text{Area}}_{\downarrow}$ $\underset{\downarrow}{\text{is}}$ $\underset{\downarrow}{180 \text{ ft}^2}$

$\quad w(5w) \quad = \quad 180$

Carry out. We solve the equation:

$w(5w) = 180$

$5w^2 - 180 = 0$

$5\left(w^2 - 36\right) = 0$

$5(w + 6)(w - 6) = 0$

$w = -6 \quad or \quad w = 6$

Check. The number –6 is not a solution, because width cannot be negative. If the width is 6 ft, and the length is $5 \cdot 6$, or 30 ft, then the area is $6 \cdot 30 = 180 \text{ ft}^2$. The answer checks.
State. The length is 30 ft, and the width is 6 ft.

7. *Familiarize.* We let w represent the width and $w + 5$ represent the length. We make a drawing and label it.

$$\begin{array}{|c|}\hline \text{Area} \\ 84 \text{ cm}^2 \\ \hline \end{array} \; w$$
$$w + 5$$

Recall that the formula for the area of a rectangle is $A = \text{length} \times \text{width}$.
Translate.

$$\underbrace{\text{Area}}_{\downarrow} \;\; \underbrace{\text{is}}_{\downarrow} \;\; \underbrace{84 \text{ cm}^2}_{\downarrow}$$
$$w(w + 5) = \;\;\; 84$$

Carry out. We solve the equation:

$$w(w + 5) = 84$$
$$w^2 + 5w = 84$$
$$w^2 + 5w - 84 = 0$$
$$(w + 12)(w - 7) = 0$$
$$w + 12 = 0 \;\;\; or \;\;\; w - 7 = 0$$
$$w = -12 \; or \;\;\;\;\; w = 7$$

Check. The number -12 is not a solution, because width cannot be negative. If the width is 7 cm and the length is 5 cm more, or 12 cm, then the area is $12 \cdot 7$, or 84 cm^2. This is a solution.
State. The length is 12 cm, and the width is 7 cm.

9. *Familiarize.* Let x represent the length of the foot of the sail, and $x + 5$ represents the height. Recall that the formula for the area of a triangle is $A = \dfrac{1}{2} \times \text{base} \times \text{height}$.
Translate. The area is 42 ft^2.

$$\underbrace{\text{The area}}_{\downarrow} \;\; \underbrace{\text{is}}_{\downarrow} \;\; \underbrace{42 \text{ ft}^2.}_{\downarrow}$$
$$\frac{1}{2}x(x + 5) = \;\;\; 42$$

Carry out. We solve the equation:

$$\frac{1}{2}x(x + 5) = 42$$
$$x(x + 5) = 84$$
$$x^2 + 5x = 84$$
$$x^2 + 5x - 84 = 0$$
$$(x + 12)(x - 7) = 0$$

$$x + 12 = 0 \;\;\; or \; x - 7 = 0$$
$$x = -12 \; or \;\;\;\;\; x = 7$$

Check. We check only 7, since the length of the foot cannot be negative. If the base is 7 ft, the height is $7 + 5$, or 12 ft, and the area is $\dfrac{1}{2} \cdot 7 \cdot 12$, or 42 ft^2. The answer checks.
State. The foot is 7 ft, and the height is 12 ft.

11. *Familiarize.* Let h represent the height of the triangle, and $\dfrac{1}{2}h$ represents the width of the base. Recall that the formula for the area of a triangle is $A = \dfrac{1}{2} \times \text{base} \times \text{height}$.
Translate. The area is 64 ft^2.

$$\underbrace{\text{the area}}_{\downarrow} \;\; \underbrace{\text{is}}_{\downarrow} \;\; \underbrace{64 \text{ ft}^2.}_{\downarrow}$$
$$\frac{1}{2}\left(\frac{h}{2}\right)h = \;\;\; 64$$

Carry out. We solve the equation:

$$\frac{1}{2}\left(\frac{h}{2}\right)h = 64$$
$$\frac{h^2}{4} = 64$$
$$h^2 = 256$$
$$h^2 - 256 = 0$$
$$(h + 16)(h - 16) = 0$$
$$h = -16 \;\;\; or \;\;\; h = 16$$

Check. We check only 16, since the height cannot be negative. If the height is 16 ft, the width of the base is $\dfrac{16}{2}$, or 8 ft, and the area is $\dfrac{1}{2} \cdot 8 \cdot 16$, or 64 ft^2. The answer checks.
State. The base is 8 ft, and the height is 16 ft.

13. *Familiarize.* We use the function $x^2 - x = N$ and set N equal to 240 games.
Translate.

$$\underbrace{\text{Number of games}} \;\; \underbrace{\text{equals}} \;\; \underbrace{240}$$
$$\;\;\;\;\;\;\;\;\;\;\;\; \downarrow \;\;\;\;\;\;\;\;\;\;\;\;\;\;\; \downarrow \;\;\;\;\;\; \downarrow$$
$$x^2 - x \;\;\;\;\;\;\;\;\;\; = \;\;\;\; 240$$

Carry out. We solve the equation:

$$x^2 - x = 240$$

$$x^2 - x - 240 = 0$$

$$(x+15)(x-16) = 0$$

$$x+15 = 0 \quad or \quad x-16 = 0$$

$$x = -15 \quad or \qquad x = 16$$

Check. We check only 16 since the number of teams cannot be negative. If 16 teams are in the league, the number of games played is $16^2 - 16 = 256 - 16 = 240$. The answer checks.

State. The number of teams in the league is 16.

15. **Familiarize.** We use the function $A = -50t^2 + 200t$ and set A equal to 150 micrograms.

Translate.

$$\underbrace{\text{Size of dose}}_{\downarrow} \underbrace{\text{equals}}_{\downarrow} \underbrace{150\ \mu g}_{\downarrow}$$

$$-50t^2 + 200t \quad = \qquad 150$$

Carry out. We solve the equation:

$$-50t^2 + 200t = 150$$

$$-50t^2 + 200t - 150 = 0$$

$$-50(t^2 - 4t + 3) = 0$$

$$t^2 - 4t + 3 = 0$$

$$(t-3)(t-1) = 0$$

$$t-3 = 0 \quad or \quad t-1 = 0$$

$$t = 3 \quad or \qquad t = 1$$

Check. If the time is 3 minutes, the amount of Albuterol in the bloodstream is

$$-50(3)^2 + 200 \cdot 3 = -50 \cdot 9 + 600$$

$$= -450 + 600$$

$$= 150$$

If the time is 1 minute, the amount of Albuterol in the bloodstream is

$$-50(1)^2 + 200 \cdot 1 = -50 + 200$$

$$= 150$$

Both answers check.

State. The time at which 150 micrograms of Albuterol is present in the bloodstream is either 1 minute or 3 minutes.

17. **Familiarize.** We use the function $H = 0.006x^2 + 0.6x$ and set H equal to 6.6 ft.

Translate.

$$\underbrace{\text{Height of wave}}_{\downarrow} \underbrace{\text{equals}}_{\downarrow} \underbrace{6.6\ \text{ft}}_{\downarrow}$$

$$0.006x^2 + 0.6x \quad = \qquad 6.6$$

Carry out. We solve the equation:

$$0.006x^2 + 0.6x = 6.6$$

$$0.006x^2 + 0.6x - 6.6 = 0$$

$$0.006(x^2 + 100x - 1100) = 0$$

$$x^2 + 100x - 1100 = 0$$

$$(x+110)(x-10) = 0$$

$$x+110 = 0 \quad or \quad x-10 = 0$$

$$x = -110 \quad or \qquad x = 10$$

Check. We check only 10 since the wind speed cannot be negative. If the wind speed is 10 knots, the height of the wave is

$$0.006(10)^2 + 0.6 \cdot 10 = 0.006 \cdot 100 + 6$$

$$= 0.6 + 6$$

$$= 6.6$$

The answer checks.

State. The speed of the wind required to produce a wave 6.6 ft high is 10 knots.

19. **Familiarize.** We use the function $h(t) = -15t^2 + 75t + 10$. Note that t cannot be negative since it represents time after launch.

Translate. We need to determine the value of t for which $h(t) = 70$ ft.

$$-15t^2 + 75t + 10 = 70$$

Carry out. We solve the equation.

$$-15t^2 + 75t + 10 = 70$$

$$-15t^2 + 75t = 60$$

$$-15(t^2 - 5t) = 60$$

$$t^2 - 5t = -4$$

$$t^2 - 5t + 4 = 0$$

$$(t-4)(t-1) = 0$$

$$t-4 = 0 \quad or \quad t-1 = 0$$

$$t = 4 \quad or \qquad t = 1$$

Check. We reject 1, since this represents the moment of launch when the tee shirt starts traveling upwards.

$$h(4) = -15 \cdot 4^2 + 75 \cdot 4 + 10$$

$$= -240 + 300 + 10 = 70$$

State. The tee shirt was airborne for 4 sec before it was caught.

21. ***Familiarize.*** We will use the given formula, $h(t) = -16t^2 + 64t + 80$. Note that t cannot be negative since it represents time after launch. ***Translate.*** We need to find the value of t for which $h(t) = 0$. We have:

$$-16t^2 + 64t + 80 = 0$$

Carry out. We solve the equation.

$$-16t^2 + 64t + 80 = 0$$
$$-16(t^2 - 4t - 5) = 0$$
$$-16(t - 5)(t + 1) = 0$$
$$t - 5 = 0 \ or \ t + 1 = 0$$
$$t = 5 \ or \quad t = -1$$

Check. Since t cannot be negative, we check only 5. $h(5) = -16 \cdot 5^2 + 64 \cdot 5 + 80$
$= -400 + 320 + 80 = 0$. The number 5 checks.

State. The cardboard shell will reach the ground in 5 sec after it is launched.

23. ***Familiarize.*** We make a drawing and label it. We let x represent the length of a side of the original square, in meters.

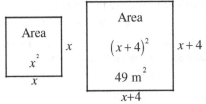

Translate.

$$\underbrace{\text{Area of}}_{} \quad \text{is} \quad 49 \text{ m}^2$$
$$\text{new square}$$
$$\downarrow \qquad \downarrow \quad \downarrow$$
$$(x + 4)^2 \quad = \quad 49$$

Carry out. We solve the equation.

$$(x + 4)^2 = 49$$
$$x^2 + 8x + 16 = 49$$
$$x^2 + 8x - 33 = 0$$
$$(x - 3)(x + 11) = 0$$
$$x - 3 = 0 \ or \ x + 11 = 0$$
$$x = 3 \ or \quad x = -11$$

Check. We check only 3 since the length of a side cannot be negative. If we increase the length by 4, the new length is $3 + 4$, or 7 m. Then the new area is $7 \cdot 7$, or 49 m^2. We have a solution.

State. The length of a side of the original square is 3 m.

25. ***Familiarize.*** We make a drawing and label it with both known and unknown information. We let x represent the width of the frame. The dimensions of the picture that show are represented by $20 - 2x$ and $12 - 2x$. The area of the picture that shows is 84 cm^2.

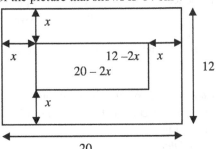

Translate. Using the formula for the area of a rectangle $A = l \cdot w$, we have

$$84 = (20 - 2x)(12 - 2x).$$

Carry out. We solve the equation:

$$84 = 240 - 64x + 4x^2$$
$$84 = 4(60 - 16x + x^2)$$
$$21 = 60 - 16x + x^2$$
$$0 = x^2 - 16x + 39$$
$$0 = (x - 3)(x - 13)$$
$$x - 3 = 0 \ or \ x - 13 = 0$$
$$x = 3 \ or \quad x = 13$$

Check. We see that 13 is not a solution because when $x = 13$, $20 - 2x = -6$ and $12 - 26 = -14$, and the length and width of the picture cannot be negative. We check 3. When $x = 3$, $20 - 2x = 14$ and $12 - 2x = 6$ and $14 \cdot 6 = 84$. The area is 84. The value checks.

State. The width of the frame is 3 cm.

27. ***Familiarize.*** We let x represent the width of the sidewalk. We make a drawing and label it with both the known and unknown information.

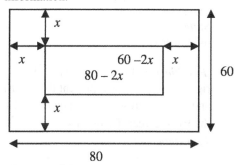

The area of the new lawn is $(80-2x)(60-2x)$.

Translate.

$$\underbrace{\text{Area of new lawn}}_{\downarrow} \quad \underset{\downarrow}{\text{is}} \quad \underset{\downarrow}{2400\,ft^2}$$

$$(80-2x)(60-2x) \quad = \quad 2400$$

Carry out. We solve the equation:

$$(80-2x)(60-2x)=2400$$

$$4800-280x+4x^2=2400$$

$$4x^2-280x+2400=0$$

$$x^2-70x+600=0$$

$$(x-10)(x-60)=0$$

$$x-10=0 \quad or \quad x-60=0$$

$$x=10 \ or \qquad x=60$$

Check. If the sidewalk is 10 ft wide, the length of the new lawn will be $80-2\cdot10,$ or 60 ft, and its width will be $60-2\cdot10,$ or 40ft. Then the area of the new lawn will be $60\cdot40,$ or 2400 ft^2. This answer checks. If the sidewalk is 60 ft wide, the length of the new lawn will be $80-2\cdot60,$ or –40 ft. Since the length cannot be negative, 60 is not a solution.

State. The sidewalk is 10 ft wide.

29. ***Familiarize.*** Let h = the height of the tower braces. We use the Pythagorean theorem $a^2+b^2=c^2$.

Translate.
$$\underset{\downarrow}{a^2} \quad + \quad \underset{\downarrow}{b^2} \quad = \quad \underset{\downarrow}{c^2}$$

$$12^2 \quad + \quad h^2 \quad = \quad 15^2$$

Carry out. We solve the equation:

$$12^2+h^2=15^2$$

$$h^2=15^2-12^2$$

$$h^2=225-144$$

$$h^2=81$$

$$h^2-81=0$$

$$(h+9)(h-9)=0$$

$$h=-9 \quad or \quad h=9$$

Check. We check only 9 since the height of a side cannot be negative. Using the Pythagorean theorem to find the length of the diagonal, $12^2+9^2=144+81=225=15^2$. The answer checks.

State. The brace is 9 ft high.

31. ***Familiarize.*** Let w = the width of Main Street. We use the Pythagorean theorem $a^2+b^2=c^2$.

Translate.
$$\underset{\downarrow}{a^2} \quad + \quad \underset{\downarrow}{b^2} \quad = \quad \underset{\downarrow}{c^2}$$

$$24^2 \quad + \quad w^2 \quad = \quad 40^2$$

Carry out. We solve the equation:

$$24^2+w^2=40^2$$

$$w^2=40^2-24^2$$

$$w^2=1600-576$$

$$w^2=1024$$

$$w^2-1024=0$$

$$(w+32)(w-32)=0$$

$$w=-32 \quad or \quad w=32$$

Check. We check only 32 since the width cannot be negative. Using the Pythagorean theorem to find the diagonal across Main and Elliot Streets, $24^2+32^2=576+1024=1600=40^2$. The answer checks.

State. Main Street is 32 ft across.

33. ***Familiarize.*** Let x represent the unknown side of the right triangle and $x+200$ represent the hypotenuse. We use the Pythagorean Theorem $a^2 + b^2 = c^2$.

Translate.
$$\begin{array}{ccc} a^2 & + \; b^2 & = \; c^2 \\ \downarrow & \downarrow & \downarrow \end{array}$$
$$400^2 + \; x^2 \; = (x+200)^2$$

Carry out. We solve the equation:
$$400^2 + x^2 = (x+200)^2$$
$$400^2 + x^2 = x^2 + 400x + 200^2$$
$$400^2 - 200^2 = 400x$$
$$160,000 - 40,000 = 400x$$
$$120,000 = 400x$$
$$300 = x$$

Check. Using the Pythagorean theorem to find the length of the garden's diagonal:
$$400^2 + 300^2 = 160,000 + 90,000$$
$$= 250,000$$
$$= 500^2$$
$$= (300 + 200)^2$$
The answer checks.

State. The sides are 300 ft and 400 ft in length, and the diagonal is 500 ft long.

35. ***Familiarize.*** Let d represent the base of the right triangle and $d+4$ represent the height. These are the legs of the right triangle and 20 is the length of the hypotenuse. Recall: Pythagorean Theorem $a^2 + b^2 = c^2$.

Translate.
$$\begin{array}{ccc} a^2 + & b^2 & = \; c^2 \\ \downarrow & \downarrow & \downarrow \end{array}$$
$$d^2 + (d+4)^2 = 20^2$$

Carry out. We solve the equation:
$$d^2 + (d+4)^2 = 20^2$$
$$d^2 + d^2 + 8d + 16 = 400$$
$$2d^2 + 8d - 384 = 0$$
$$2(d^2 + 4d - 192) = 0$$
$$d^2 + 4d - 192 = 0$$
$$(d+16)(d-12) = 0$$
$$d+16 = 0 \quad or \quad d-12 = 0$$
$$d = -16 \quad or \qquad d = 12$$

Check. Since measure cannot be negative, we know -16 is not a solution. If $d = 12$, $d + 4 = 16$ and

$$12^2 + 16^2 = 20^2$$
$$144 + 256 = 400$$
$$400 = 400$$
The answer checks.

State. The distance is 12 ft, and the height of the tower is 16 ft.

37. ***Familiarize.*** Let x represent the base of the right triangle, $x+1$ represent the other leg, and $x+2$ represent the hypotenuse. Recall the Pythagorean Theorem $a^2 + b^2 = c^2$.

Translate.
$$\begin{array}{ccc} a^2 + & b^2 & = \; c^2 \\ \downarrow & \downarrow & \downarrow \end{array}$$
$$x^2 + (x+1)^2 = (x+2)^2$$

Carry out. We solve the equation:
$$x^2 + (x+1)^2 = (x+2)^2$$
$$x^2 + x^2 + 2x + 1 = x^2 + 4x + 4$$
$$x^2 + 2x - 4x + 1 - 4 = 0$$
$$x^2 - 2x - 3 = 0$$
$$(x-3)(x+1) = 0$$
$$x - 3 = 0 \quad or \quad x + 1 = 0$$
$$x = 3 \quad or \qquad x = -1$$

Check. Since measure cannot be negative, we know -1 is not a solution. If $x = 3$, $x + 1 = 4$ and
$$3^2 + 4^2 = (3+2)^2$$
$$9 + 16 = 5^2$$
$$25 = 25$$

State. The lengths of the sides are 3, 4 and 5 units.

39. **Familiarize.** Let w represent the width and $w+10$ represent the length.

$Area = length \times width$

Translate.   Area   is $264\,\text{ft}^2$

$$w(w+10) = 264$$

Carry out. We solve the equation:

$$w(w+10) = 264$$
$$w^2 + 10w = 264$$
$$w^2 + 10w - 264 = 0$$
$$(w+22)(w-12) = 0$$
$$w+22 = 0 \quad or \quad w-12 = 0$$
$$w = -22 \quad or \quad w = 12$$

Check. -22 cannot be a solution because width cannot be negative. If $w=12$ then the length is $w+10$, which is 22. The area is $12 \cdot 22$ which is $264\,\text{ft}^2$.

State. The dimensions for the total space will be 12 ft by 22 ft. This will divide as 12 ft by 12 ft for the dining room, and 12 ft by 10 ft for the kitchen.

41. a) Enter the data in the graphing calculator, letting x represent the number of years after 1990. Then use the quartic regression feature to find the desired function $P(x) = 0.01823x^4 - 0.77199x^3 + 11.62153x^2 - 73.65807x + 179.76190$.

   b) In 2005, $x = 15$.
   $$P(15) \approx 7\%$$

   c) We solve $P(x) = 12$ graphically. Graph $y_1 = P(x)$ and $y_2 = 12$. Both $x \approx 13$ years and $x \approx 17$ years after 1990 satisfy the equation. These values correspond to the years 2003 and 2007.

43. a) Enter the data in the graphing calculator, letting x represent the number of years after 1900. Then use the cubic regression feature to find the desired function $F(x) = -0.03587x^3 + 10.35169x^2 - 871.97543x + 23,423.45189$.

   b) In 2012, $x = 112$.
   $$F(112) \approx 5220 \text{ athletes}$$

   c) We solve $F(x) = 6000$ graphically. Graph $y_1 = F(x)$ and $y_2 = 6000$. Both

$x \approx 122$ years and $x \approx 137$ after 1900 satisfy the equation. These correspond to the years 2022 and 2037; however, since the Summer Olympics occur every four years, we estimate these years as 2024 and 2036. Given that the number of female athletes was increasing from 1960 to 2008, we should only consider that the cubic model is meaningful while it is increasing. This means the only meaningful answer would be 2024.

45. **Thinking and Writing Exercise.**

47. $-\dfrac{3}{5} \cdot \dfrac{4}{7} = -\dfrac{3 \cdot 4}{5 \cdot 7} = -\dfrac{12}{35}$

49. $-\dfrac{5}{6} - \dfrac{1}{6} = \dfrac{-5-1}{6} = \dfrac{-6}{6} = -1$

51. $-\dfrac{3}{8} \cdot \left(-\dfrac{10}{15}\right) = -\left(-\dfrac{3 \cdot 10}{8 \cdot 15}\right)$

$$= \dfrac{3 \cdot 10}{8 \cdot 15}$$
$$= \dfrac{1 \cdot 3 \cdot 2 \cdot 5}{2 \cdot 4 \cdot 3 \cdot 5}$$
$$= \dfrac{1}{4}$$

53. $\dfrac{5}{24} + \dfrac{3}{28} = \dfrac{5}{4 \cdot 6} + \dfrac{3}{4 \cdot 7}$

$$= \dfrac{5 \cdot 7}{4 \cdot 6 \cdot 7} + \dfrac{3 \cdot 6}{4 \cdot 7 \cdot 6}$$
$$= \dfrac{35}{168} + \dfrac{18}{168}$$
$$= \dfrac{35+18}{168}$$
$$= \dfrac{53}{168}$$

55. **Thinking and Writing Exercise.**

57. *Familiarize.* From the drawing in the text, we note that the roof has two sides of equal area. The length of each side is 32 ft, while the width must be found using the Pythagorean Theorem. The width equals the diagonal of a right triangle whose height equals the difference between the heights of the peak of the house (25 ft) and the base of the roof (16 ft), while the base of the right triangle equals half of the house width (24 ft). The number of squares of shingles equals the total area, in square feet, divided by 100.

Translate.

$\underbrace{\text{Area}}\underbrace{\text{is}}\underbrace{\text{twice}}\underbrace{\text{length}}\underbrace{\text{times}}\qquad\underbrace{\text{width}}.$

$$A \;=\; 2\;\left[\,32\;\cdot\;\sqrt{\left(\frac{24}{2}\right)^2+(25-16)^2}\,\right]$$

$\underbrace{\text{Number of squares}}\underbrace{\text{is}}\underbrace{\text{area}}\underbrace{\text{divided by}}\underbrace{100}.$

$$N \;=\; A \;\div\; 100$$

Carry out. We solve the equations.

$$A = 2\left[32\cdot\sqrt{\left(\frac{24}{2}\right)^2+(25-16)^2}\,\right]$$

$$= 2\left[32\cdot\sqrt{12^2+9^2}\,\right]$$

$$= 2\left[32\cdot\sqrt{144+81}\,\right]$$

$$= 2\left[32\cdot\sqrt{225}\,\right]$$

$$= 2[32\cdot15]$$

$$= 2\cdot480$$

$$= 960$$

$$N = \frac{A}{100} = \frac{960}{100} = 9.6 \approx 10$$

Check. Find the height of the roof by taking the approximate area for either side,

$\dfrac{10\cdot100\ \text{ft}^2}{2}$, or $500\ \text{ft}^2$, and applying the area formula in reverse.

$$w = \frac{500}{32} = 15.625$$

$$h = \sqrt{15.625^2 - \left(\frac{24}{2}\right)^2} \approx \sqrt{244^2 - 144^2} \approx 10$$

The approximate roof height of 10 ft is close to the actual height of $25-16$, or 9 ft. The answer checks.

State. A total of 10 squares of shingles will be needed to cover the area of the roof.

59. We solve $N(t)=18$ graphically. Graph $y_1 = -0.009t(t-12)^3$ and $y_2 = 18$. Both $t=2$ hours and $t \approx 4.2$ hours satisfy the equation.

61. We solve for the time at which $N(t)$ is at a maximum by graphing $y_1 = -0.009t(t-12)^3$ and using the "maximum" feature to find the peak value $y_1 = 19.683$. This values is reached at $t=3$ hours.

63. *Familiarize.* Using the labels on the drawing in the text, we let x represent the width of the piece of tin and $2x$ represent the length. Then the width and length of the base of the box are represented by $x-4$ and $2x-4$, respectively. Recall that the formula for the volume of a rectangular solid with length l, width w, and height h is $l\cdot w\cdot h$.

Translate. $\underbrace{\text{The volume}}\quad\underbrace{\text{is}}\;\underbrace{480cm^3}.$

$$(2x-4)(x-4)(2) \;=\; 480$$

Carry out. We solve the equation.

$$(2x-4)(x-4)(2) = 480$$

$$(2x-4)(x-4) = 240$$

$$2x^2 - 12x + 16 = 240$$

$$2x^2 - 12x - 224 = 0$$

$$x^2 - 6x - 112 = 0$$

$$(x+8)(x-14) = 0$$

$$x+8 = 0\ \ or\ \ x-14 = 0$$

$$x = -8\ or\qquad x = 14$$

Check. We check only 14 since the width cannot be negative. If the width of the piece of tin is 14 cm, then its length is $2\cdot14$, or 28 cm, and the dimensions of the base of the box are $14-4$, or 10 cm by $28-4$, or 24 cm. The volume of the box is $24\cdot10\cdot2$, or 480 cm^3. The answer checks.

State. The dimensions of the piece of tin are 14 cm by 28 cm.

65. Graph $y_1 = 11.12(x+1)^2$ and $y_2 = 15.4x^2$ in a window that shows the point of intersection of the graphs. The window $[0,10,0,1000]$, Xscl = 1, Yscl = 100 is one good choice. Then find the first coordinate of the point of intersection. It is approximately 5.7, so it will take the camera about 5.7 sec to catch up to the skydiver.

___

Chapter 5 Study Summary

1. $x^2 - 10 + 5x - 8x^6$ has four terms: $x^2, -10, 5x,$ and $-8x^6$

2. $5x = 5x^1$, so the term is of degree 1.

3. $x^2 = 1x^2$, so the coefficient is 1.

4. The polynomial is not written in descending order, but the "leading term" is still the term containing the highest power of x: $-8x^6$.

5. The leading coefficient is the coefficient of the leading term: -8.

6. The degree of the polynomial is the degree of the leading term: 6.

7. Three terms: trinomial.

8. $(9x^2 - 3x) + (4x - x^2) = 9x^2 - 3x + 4x - x^2$
$$= (9-1)x^2 + (-3+4)x$$
$$= 8x^2 + x$$

9. $(9x^2 - 3x) - (4x - x^2) = 9x^2 - 3x - 4x + x^2$
$$= (9+1)x^2 + (-3-4)x$$
$$= 10x^2 + (-7)x$$
$$= 10x^2 - 7x$$

10. $(x-1)(x^2 - x - 2)$
$$= x(x^2 - x - 2) - 1(x^2 - x - 2)$$
$$= x \cdot x^2 - x \cdot x - x \cdot 2 - 1 \cdot x^2 + 1 \cdot x + 1 \cdot 2$$
$$= x^3 - x^2 - 2x - x^2 + x + 2$$
$$= x^3 - 2x^2 - x + 2$$

11. $(x - 2y)(x + 2y) = x^2 - (2y)^2$
$$= x^2 - 4y^2$$

12. $12x^4 - 18x^3 + 30x$
$$= 6x(2x^3 - 3x^2 + 5)$$

13. $2x^3 - 6x^2 - x + 3 = 2x^2(x-3) - 1(x-3)$
$$= (2x^2 - 1)(x-3)$$

14. $$8x = 6x^2$$
$$6x^2 - 8x = 0$$
$$2x(3x-4) = 0$$
$$2x = 0 \quad or \quad 3x - 4 = 0$$
$$x = 0 \quad or \quad x = \frac{4}{3}$$

15. $x^2 - 7x - 18$
Since the constant term is negative, we look for factorization of -18 in which one factor is positive and one factor is negative. Their sum must be -7, so the negative factor must have the larger absolute value. Thus we consider only pairs of factors in which the negative factor has the larger absolute value.

Pair of Factors	Sum of Factors
1, -18	-17
2, -9	-7
3, -6	-3

The numbers we need are 2 and -9. The factorization is $(x-9)(x+2)$.

16. $6x^2 + x - 2$
We will use the FOIL method.
1. There is no common factor (other than 1 or -1).
2. Factor the first term, $6x^2$. The factors are $6x$ and x, and $3x$ and $2x$. We have the possibilities $(6x +) (x +)$ and $(3x +) (2x +)$.
3. Factor the last term, -2. The possibilities are $-1 \cdot 2$, $1 \cdot (-2)$, $-2 \cdot 1$, and $2(-1)$.
4. Look for factors for which the sum of the products is x. Trial and error leads us to $(3x+2)(2x-1)$.

17. $8x^2 - 22x + 15$

We will use the grouping method.

1. There is no common factor (other than 1 or −1).

2. Multiply the leading coefficient and the constant $8(15) = 120$.

3. We want factors of 120 whose sum is −22.

Pair of Factors	Sum of Factors
−1, −120	−121
−2, −60	−62
−3, −40	−43
−4, −30	−34
−5, −24	−29
−6, −20	−26
−8, −15	−23
−10, −12	−22

Since $-10 \cdot (-12) = 120$ and $-10 - 12 = -22$, we will split $-22x$ into $-10x$ and $-12x$.

4. Factor by grouping:

$8x^2 - 22x + 15 = 8x^2 - 10x - 12x + 15$
$\qquad = 2x(4x - 5) - 3(4x - 5)$
$\qquad = (2x - 3)(4x - 5)$

18. $100n^2 + 81 + 180n = 100n^2 + 180n + 81$
$\qquad\qquad\qquad = (10n)^2 + 2 \cdot 9 \cdot 10n + 9^2$

This is of the form $A^2 + 2BA + B^2$, which is a perfect-square trinomial $(A + B)^2$, so

$100n^2 + 81 + 180n = (10n + 9)^2$

19. $144t^2 - 25$

This is of the form $A^2 - B^2$ (difference of squares) because $144t^2 = (12t)^2$ and $25 = 5^2$, and the terms have different signs.

$144t^2 - 25 = (12t)^2 - 5^2 = (12t + 5)(12t - 5)$

20. $a^3 - 1 = a^3 - 1^3$
$\qquad = (a - 1)(a^2 + a + 1)$
$A^3 - B^3 = (A - B)(A^2 + AB + B^2)$

21. $\begin{array}{ccc} a^2 & + \ b^2 & = \ c^2 \\ \downarrow & \downarrow & \downarrow \end{array}$

$5^2 \ + \ x^2 = (x + 1)^2$

$25 + x^2 = x^2 + 2x + 1$

$24 = 2x$

$x = 12$

The lengths of the sides are 5, 12 and 13 units.

Chapter 5 Review Exercises

1. g

2. b

3. a

4. d

5. e

6. j

7. h

8. c

9. i

10. f

11.

Term	$2xy^6$	$-7x^8y^3$	$2x^3$	9
Degree	7	11	3	0

Degree of polynomial: 11

12. $-5x^3 + 2x^2 + 3x + 9$; $-5x^3$; -5

13. $-3x^2 + 2x^3 + 8x^6y - 7x^8y^3$

14. $P(x) = x^3 - x^2 + 4x$

$P(0) = 0^3 - 0^2 + 4 \cdot 0 = 0$

$P(-1) = (-1)^3 - (-1)^2 + 4(-1) = -6$

15. $P(x) = x^2 + 10x$

$P(a + h) - P(a)$

$= (a + h)^2 + 10(a + h) - (a^2 + 10a)$

$= a^2 + 2ah + h^2 + 10a + 10h - a^2 - 10a$

$= 2ah + h^2 + 10h$

16. $6 - 4a + a^2 - 2a^3 - 10 + a$
    $= -2a^3 + a^2 - 3a - 4$

17. $4x^2 y - 3xy^2 - 5x^2 y + xy^2$
    $= (4 - 5)x^2 y + (-3 + 1)xy^2$
    $= -x^2 y - 2xy^2$

18. $\left(-7x^3 - 4x^2 + 3x + 2\right) + \left(5x^3 + 2x + 6x^2 + 1\right)$
    $= (-7 + 5)x^3 + (-4 + 6)x^2 + (3 + 2)x + 2 + 1$
    $= -2x^3 + 2x^2 + 5x + 3$

19. $\left(4n^3 + 2n^2 - 12n + 7\right) + \left(-6n^3 + 9n + 4 + n\right)$
    $= (4 - 6)n^3 + 2n^2 + (-12 + 9 + 1)n + 7 + 4$
    $= -2n^3 + 2n^2 - 2n + 11$

20. $\left(-9xy^2 - xy - 6x^2 y\right) + \left(-5x^2 y - xy + 4xy^2\right)$
    $= (-9 + 4)xy^2 + (-1 - 1)xy + (-6 - 5)x^2 y$
    $= -5xy^2 - 2xy - 11x^2 y$

21. $(8x - 5) - (-6x + 2)$
    $= 8x - 5 + 6x - 2$
    $= 14x - 7$

22. $(4a - b - 3c) - (6a - 7b - 3c)$
    $= 4a - b - 3c - 6a + 7b + 3c$
    $= -2a + 6b$

23. $\left(8x^2 - 4xy + y^2\right) - \left(2x^2 - 3y^2 - 9y\right)$
    $= 8x^2 - 4xy + y^2 - 2x^2 + 3y^2 + 9y$
    $= 6x^2 - 4xy + 4y^2 + 9y$

24. $\left(3x^2 y\right)\left(-6xy^3\right)$
    $= 3 \cdot (-6) \cdot x^2 \cdot x \cdot y \cdot y^3$
    $= -18x^3 y^4$

25.
$$
\begin{array}{r}
x^4 - 2x^2 + 3 \\
x^4 + x^2 - 1 \\
\hline
-x^4 + 2x^2 - 3 \\
x^6 - 2x^4 + 3x^2 \\
x^8 - 2x^6 + 3x^4 \\
\hline
x^8 - x^6 \qquad + 5x^2 - 3
\end{array}
$$

26. $(4ab + 3c)(2ab - c)$
    $= 8a^2 b^2 - 4abc + 6abc - 3c^2$
    $= 8a^2 b^2 + 2abc - 3c^2$

27. $(7t + 1)(7t - 1)$
    $= (7t)^2 - 1^2$
    $= 49t^2 - 1$

28. $(3x - 4y)^2$
    $= (3x)^2 - 2 \cdot 3x \cdot 4y + (4y)^2$
    $= 9x^2 - 24xy + 16y^2$

29. $(x + 3)(2x - 1)$
    $= 2x^2 - x + 6x - 3$
    $= 2x^2 + 5x - 3$

30. $\left(x^2 + 4y^3\right)^2$
    $= \left(x^2\right)^2 + 2 \cdot x^2 \cdot 4y^3 + \left(4y^3\right)^2$
    $= x^4 + 8x^2 y^3 + 16y^6$

31. $(3t - 5)^2 - (2t + 3)^2$
    $= 9t^2 - 30t + 25 - \left(4t^2 + 12t + 9\right)$
    $= 5t^2 - 42t + 16$

32. $\left(x - \dfrac{1}{3}\right)\left(x - \dfrac{1}{6}\right)$
    $= x^2 - \dfrac{1}{6}x - \dfrac{1}{3}x + \left(-\dfrac{1}{3}\right)\left(-\dfrac{1}{6}\right)$
    $= x^2 - \dfrac{1}{2}x + \dfrac{1}{18}$

33. $7x^2 + 6x$
    $= x \cdot 7x + x \cdot 6$
    $= x(7x + 6)$

34. $-3y^4 - 9y^2 + 12y$
    $= (-3y) \cdot y^3 + (-3y) \cdot 3y - (-3y) \cdot 4$
    $= -3y(y^3 + 3y - 4)$

35. $100t^2 - 1 = (10t)^2 - 1^2$ Difference of squares
    $= (10t + 1)(10t - 1)$ $A^2 - B^2 = (A + B)(A - B)$

36. $a^2 - 12a + 27$
We want two negative factors of 27, whose sum is -12. They are -9 and -3.
$= (a-9)(a-3)$

37. $3m^2 + 14m + 8$
Using the grouping method, multiply the leading coefficient, 3, and the constant, 8. $3 \cdot 8 = 24$. Factor 24 so the sum of factors is 14; both factors will be positive since $14m$ and 8 are both positive. They are 12 and 2. Split the middle term using $14m = 12m + 2m$ and factor by grouping.
$3m^2 + 14m + 8 = 3m^2 + 12m + 2m + 8$
$= 3m(m+4) + 2(m+4)$
$= (m+4)(3m+2)$

38. $25x^2 + 20x + 4$
$= (5x)^2 + 2(5x)2 + 2^2$ Perfect square
$= (5x+2)^2$ trinomial

39. $4y^2 - 16$
$= 4(y^2 - 4)$ Common Factor
$= 4(y^2 - 2^2)$ Difference of two squares
$= 4(y+2)(y-2)$

40. $5x^2 + x^3 - 14x = x^3 + 5x^2 - 14x$
$= x(x^2 + 5x - 14)$ Common Factor
To factor $x^2 + 5x - 14$, we want to find two factors of -14 whose sum is 5. They are 7 and -2:
$= x(x+7)(x-2)$

41. $ax + 2bx - ay - 2by$
$= x(a+2b) - y(a+2b)$ Factor by grouping
$= (a+2b)(x-y)$

42. $3y^3 + 6y^2 - 5y - 10$
$= 3y^2(y+2) - 5(y+2)$ Factor by grouping
$= (y+2)(3y^2 - 5)$

43. $81a^4 - 1 = (9a^2)^2 - 1^2$ Difference of squares
$= (9a^2+1)(9a^2-1)$
$= (9a^2+1)((3a)^2 - 1^2)$ Difference of squares
$= (9a^2+1)(3a+1)(3a-1)$

44. $48t^2 - 28t + 6$
$= 2(24t^2 - 14t + 3)$ Common Factor

45. $27x^3 + 8 = (3x)^3 + 2^3$ Sum of 2 cubes
$= (3x+2)\left[(3x)^2 - 3x \cdot 2 + 2^2\right]$
$= (3x+2)(9x^2 - 6x + 4)$

46. $-t^3 + t^2 + 42t$
$= -t(t^2 - t - 42)$ Common factor
$= -t(t-7)(t+6)$

47. $a^2b^4 - 64$
$= (ab^2)^2 - 8^2$ Difference of squares
$= (ab^2+8)(ab^2-8)$
$A^2 - B^2 = (A+B)(A-B)$

48. $3x + x^2 + 5 = x^2 + 3x + 5$
We want two positive factors of 5, whose sum is 3. This is not possible with integer coefficients, so $3x + x^2 + 5$ is prime.

49. To factor $x^2y^2 - xy - 2$, recognize that $x^2y^2 = (xy)^2$. Then find two factors of -2 whose sum is -1. They are 1 and -2.
$= (xy-2)(xy+1)$

50. $54x^6y - 2y$
$= 2y(27x^6 - 1)$ Common Factor
Factor $27x^6 - 1$.
$27x^6 - 1 = (3x^2)^3 - 1^3$ Difference of 2 cubes
$= (3x^2 - 1)(9x^4 + 3x^2 + 1)$
Therefore,
$54x^6y - 2y = 2y(3x^2 - 1)(9x^4 + 3x^2 + 1)$

51. $75 + 12x^2 - 60x$

$= 3(4x^2 - 20x + 25)$ Common Factor

$= 3(2x - 5)^2$ Perfect-square trinomial

52. $6t^2 + 17pt + 5p^2$

Factors of $6t^2$ are *6t, t* and *3t, 2t*.

Positive factors of $5p^2$ are *5p, p*.

To get $17pt$ we determine the factorization is $(3t + p)(2t + 5p)$.

$6t^2 + 17pt + 5p^2 = (3t + p)(2t + 5p)$

53. $x^3 + 2x^2 - 9x - 18$

$= x^2(x + 2) - 9(x + 2)$ Factor by grouping

$= (x + 2)(x^2 - 9)$ Difference of squares

$= (x + 2)(x + 3)(x - 3)$

54. $a^2 - 2ab + b^2 - 4t^2$

$= (a^2 - 2ab + b^2) - 4t^2$ Difference of squares

$= (a - b)^2 - (2t)^2$

$= (a - b + 2t)(a - b - 2t)$

55. From the graph we see that $p(x) = 0$ when $x = -2$, $x = 1$, or $x = 5$. $-2, 1$, and 5 are the solutions.

56. The zeros of $f(x)$ are the values of x which make $f(x) = 0$. We solve $0 = x^2 - 11x + 28$.

$0 = x^2 - 11x + 28$

$0 = (x - 7)(x - 4)$

$x - 7 = 0$ or $x - 4 = 0$

$x = 7$ or $x = 4$

The solutions are 7 and 4.

57. $(x - 9)(x + 11) = 0$

$x - 9 = 0$ or $x + 11 = 0$

$x = 9$ or $x = -11$

The solutions are 9 and –11.

58. $6b^2 - 13b + 6 = 0$

Factor and use principle of zero products

$(2b - 3)(3b - 2) = 0$

$2b - 3 = 0$ *or* $3b - 2 = 0$

$b = \dfrac{3}{2}$ *or* $b = \dfrac{2}{3}$

The solutions are $\dfrac{3}{2}$ and $\dfrac{2}{3}$.

59. $8t^2 = 14t$

Factor and use the principle of zero products.

$8t^2 = 14t$

$8t^2 - 14t = 0$

$2t(4t - 7) = 0$

$2t = 0$ *or* $4t - 7 = 0$

$t = 0$ *or* $t = \dfrac{7}{4}$

The solutions are 0 and $\dfrac{7}{4}$.

60. $x^2 - 20x = -100$

$x^2 - 20x + 100 = 0$

$(x - 10)^2 = 0$

$x - 10 = 0$ *or* $x - 10 = 0$

$x = 10$ *or* $x = 10$

The solution is 10.

61. $r^2 = 16$

Factor and use the principle of zero products.

$r^2 = 16$

$r^2 - 16 = 0$

$(r + 4)(r - 4) = 0$

$r + 4 = 0$ *or* $r - 4 = 0$

$r = -4$ *or* $r = 4$

The solutions are –4 and 4, or ±4.

62. $a^3 = 4a^2 + 21a$

$a^3 - 4a^2 - 21a = 0$

Factor and use the principle of zero products.

$a^3 - 4a^2 - 21a = a(a^2 - 4a - 21)$

$= a(a - 7)(a + 3)$

$a(a - 7)(a + 3) = 0$

$a = 0$ or $a - 7 = 0$ or $a + 3 = 0$

$a = 0$ or $a = 7$ *or* $a = -3$

The solutions are –3, 0, and 7.

63. $x(x-1) = 20$

$x^2 - x = 20$

$x^2 - x - 20 = 0$

Factor and use the principle of zero products.

$(x-5)(x+4) = 0$

$x - 5 = 0 \ or \ x + 4 = 0$

$x = 5 \ or \qquad x = -4$

The solutions are –4 and 5.

64. $x^3 - 5x^2 - 16x + 80 = 0$

$x^2(x-5) - 16(x-5) = 0$

$(x-5)(x^2-16) = 0$

$(x-5)(x+4)(x-4) = 0$

$x - 5 = 0 \ or \ x + 4 = 0 \quad or \ x - 4 = 0$

$x = 5 \ or \qquad x = -4 \ or \qquad x = 4$

The solutions are –4, 4, and 5.

65. $x^2 + 180 = 27x$

$x^2 - 27x + 180 = 0$

Factor and use the principle of zero products.

$(x-15)(x-12) = 0$

$x - 15 = 0 \ or \ x - 12 = 0$

$x = 15 \ or \qquad x = 12$

The solutions are 12 and 15.

66. Graph $y = x^2 - 2x - 6$ and find the zeros.
Using the zero feature we find $x \approx -1.646$
and $x \approx 3.646$, which are the solutions.

67. $f(x) = x^2 - 7x - 40$. Let $f(a) = 4$ and

solve the resulting equation by factoring and
using the principle of zero products.

$4 = x^2 - 7x - 40$

$0 = x^2 - 7x - 44$

$0 = (x-11)(x+4)$

$x - 11 = 0 \ or \ x + 4 = 0$

$x = 11 \ or \qquad x = -4$

The solutions are –4 and 11.

68. $f(x) = \dfrac{x-3}{3x^2 + 19x - 14}$

To find the excluded values solve
$3x^2 + 19x - 14 = 0$, by factoring and using the
principle of zero products.

$3x^2 + 19x - 14 = 0$

$(3x-2)(x+7) = 0$

$3x - 2 = 0 \ or \ x + 7 = 0$

$x = \dfrac{2}{3} \ or \qquad x = -7$

The domain is

$\left\{ x \middle| x \text{ is a real number and } x \neq -7 \text{ and } x \neq \dfrac{2}{3} \right\}$

69. ***Familiarize.*** We use the function
$x^2 - x = N$, and set N equal to 90 games.
Translate.

$$\underbrace{\text{Number of games}}_{\downarrow} \ \underbrace{\text{equals}}_{\downarrow} \ \underbrace{90}_{\downarrow}$$

$$x^2 - x \qquad = \qquad 90$$

Carry out. We solve the equation:

$x^2 - x = 90$

$x^2 - x - 90 = 0$

$(x+9)(x-10) = 0$

$x + 9 = 0 \quad or \ x - 10 = 0$

$x = -9 \ or \qquad x = 10$

Check. We check only 10 since the number
of teams cannot be negative. If 10 teams are
in the league, the number of games played is
$10^2 - 10 = 100 - 10 = 90$. The answer checks.
State. The number of teams in the league is
10.

70. ***Familiarize.*** Let w represent the width of the
gable, and $\dfrac{3}{4}w$ represents the height. Recall
that the formula for the area of a triangle is

$A = \dfrac{1}{2} \times \text{base} \times \text{height}$.

Translate. The area is 216 m^2.

$$\underbrace{\text{the area}}_{\downarrow} \ \underbrace{\text{is}}_{\downarrow} \ \underbrace{216 \text{ m}^2.}_{\downarrow}$$

$$\dfrac{1}{2}w\left(\dfrac{3w}{4}\right) \ = \quad 216$$

Carry out. We solve the equation:

$$\frac{1}{2}w\left(\frac{3w}{4}\right) = 216$$

$$\frac{3w^2}{8} = 216$$

$$w^2 = \frac{216 \cdot 8}{3}$$

$$w^2 = 576$$

$$w = -24 \quad or \quad w = 24$$

Check. We check only 24, since the width cannot be negative. If the width is 24 m, the height is $\frac{3}{4}(24)$, or 18 m, and the area is

$\frac{1}{2} \cdot 24 \cdot 18$, or 216 m². The answer checks.

State. The gable is 24 m wide and 18 m high.

71. ***Familiarize.*** Let x = the width of the photograph and $x + 3$ the length.

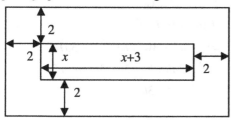

Translate.

Area of photograph
$\underbrace{\text{and border}}$ is 108 in²
$\quad\quad\downarrow \quad\quad\quad \downarrow \quad\quad \downarrow$
$(x+7)(x+4) \quad = \quad 108$

Carry out. We solve the equation.

$$(x+7)(x+4) = 108$$

$$x^2 + 11x - 80 = 0$$

$$(x+16)(x-5) = 0$$

$$x + 16 = 0 \quad or \quad x - 5 = 0$$

$$x = -16 \, or \quad x = 5$$

Check. Since length and width cannot be negative, we only check $x = 5$. The length of the frame is $(5 + 3) + 4$, or 12, and the width is $5 + 4$, or 9. So, the area of the frame is $12 \cdot 9 = 108$.

State. The width is 5 inches and the length is 8 inches.

72. ***Familiarize.*** Draw a triangle to represent the known and unknown information.

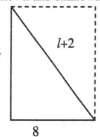

Translate. Use the Pythagorean Theorem to obtain the equation:

$$l^2 + 8^2 = (l+2)^2$$

Carry out. Solve the equation.

$$l^2 + 64 = l^2 + 4l + 4$$

$$60 = 4l$$

$$l = 15$$

Check. If $l = 15, l + 2 = 17$ and $8^2 + 15^2 = 64 + 225 = 289 = 17^2$

State. The path is 17 ft long.

73. a) Let x = the number of years after 1970. Enter the data in a graphing calculator. Use the quadratic regression feature to obtain $P(x) \approx 0.021174x^2 -$

$$0.9442x + 26.6508$$

b) In 2012, $x = 42$. $P(42) \approx 24.3\%$

c) Solve $P(x) = 25$ graphically. Graph Both $x \approx 1.8$ years (approximately 2 years after 1970, or 1972) and $x \approx 42.8$ years (approximately 43 years after 1970) or 2013.

74. *Thinking and Writing Exercise.*

75. *Thinking and Writing Exercise.*

76. $128x^6 - 2y^6$

$$= 2\left(64x^6 - y^6\right)$$

$$= 2\left[\left(8x^3\right)^2 - \left(y^3\right)^2\right]$$

$$= 2\left(8x^3 + y^3\right)\left(8x^3 - y^3\right)$$

$$= 2\left[\left(2x\right)^3 + y^3\right]\left[\left(2x\right)^3 - y^3\right]$$

$$= 2\left(2x + y\right)\left(4x^2 - 2xy + y^2\right)$$

$$\left(2x - y\right)\left(4x^2 + 2xy + y^2\right)$$

77. $\left(x-1\right)^3 - \left(x+1\right)^3$

$$\left(x-1\right)^3 = \left(x-1\right)\left(x-1\right)\left(x-1\right)$$

$$= \left(x-1\right)\left(x^2 - 2x + 1\right)$$

$$= x^3 - 2x^2 + x - x^2 + 2x - 1$$

$$= x^3 - 3x^2 + 3x - 1$$

$$\left(x+1\right)^3 = \left(x+1\right)\left(x+1\right)\left(x+1\right)$$

$$= \left(x+1\right)\left(x^2 + 2x + 1\right)$$

$$= x^3 + 2x^2 + x + x^2 + 2x + 1$$

$$= x^3 + 3x^2 + 3x + 1$$

$$\left(x-1\right)^3 - \left(x+1\right)^3 = \left(x^3 - 3x^2 + 3x - 1\right) -$$

$$\left(x^3 + 3x^2 + 3x + 1\right)$$

$$= x^3 - 3x^2 + 3x - 1 - x^3 -$$

$$3x^2 - 3x - 1$$

$$= -6x^2 - 2$$

$$= -2\left(3x^2 + 1\right)$$

78. $\left(x+1\right)^3 = x^2\left(x+1\right)$

Factor and use the principle of zero products.

$$\left(x+1\right)^3 - x^2\left(x+1\right) = 0$$

$$\left(x+1\right)\left[\left(x+1\right)^2 - x^2\right] = 0$$

$$\left(x+1\right)\left[x^2 + 2x + 1 - x^2\right] = 0$$

$$\left(x+1\right)\left(2x+1\right) = 0$$

$$x + 1 = 0 \ \ or \ 2x + 1 = 0$$

$$x = -1 \ or \ \ \ \ \ \ x = -\frac{1}{2}$$

The solutions are -1 and $-\dfrac{1}{2}$.

79. $x^2 + 100 = 0$

$$x^2 = -100$$

There is no real solution to this equation.

Chapter 5 Test

1.

Term	$8xy^3$	$-14x^2y$	$5x^5y^4$	$-9x^4y$
Degree	4	3	9	5

Degree of polynomial: 9

2. $5x^5y^4 - 9x^4y - 14x^2y + 8xy^3$

3. Leading term: $-5a^3$

4. $P(x) = 2x^3 + 3x^2 - x + 4$

$$P(0) = 2(0)^3 + 3(0)^2 - 0 + 4 = 4$$

$$P(-2) = 2(-2)^3 + 3(-2)^2 - (-2) + 4 = 2$$

5. $P(x) = x^2 - 3x$

$$P(a+h) - P(a)$$

$$= (a+h)^2 - 3(a+h) - (a^2 - 3a)$$

$$= a^2 + 2ah + h^2 - 3a - 3h - a^2 + 3a$$

$$= 2ah + h^2 - 3h$$

6. $6xy - 2xy^2 - 2xy + 5xy^2$

$$= (6-2)xy + (-2+5)xy^2$$

$$= 4xy + 3xy^2$$

7. $\left(-4y^3 + 6y^2 - y\right) + \left(3y^3 - 9y - 7\right)$

$$= (-4+3)y^3 + 6y^2 + (-1-9)y - 7$$

$$= -y^3 + 6y^2 - 10y - 7$$

8. $\left(2m^3 - 4m^2n - 5n^2\right) + \left(8m^3 - 3mn^2 + 6n^2\right)$

$$= (2+8)m^3 - 4m^2n - 3mn^2 + (-5+6)n^2$$

$$= 10m^3 - 4m^2n - 3mn^2 + n^2$$

9. $(8a - 4b) - (3a + 4b)$

$$= 8a - 4b - 3a - 4b$$

$$= 5a - 8b$$

10. $\left(9y^2 - 2y - 5y^3\right) - \left(4y^2 - 2y - 6y^3\right)$

$$= 9y^2 - 2y - 5y^3 - 4y^2 + 2y + 6y^3$$

$$= 5y^2 + y^3$$

11. $\left(-4x^2y^3\right)\left(-16xy^5\right)$

$= (-4)(-16)x^{2+1}y^{3+5}$

$= 64x^3y^8$

12. $(6a-5b)(2a+b)$

$= 12a^2 + 6ab - 10ab - 5b^2$

$= 12a^2 - 4ab - 5b^2$

13. $(x-y)\left(x^2 - xy - y^2\right)$

$= x\left(x^2 - xy - y^2\right) - y\left(x^2 - xy - y^2\right)$

$= x^3 - x^2y - xy^2 - x^2y + xy^2 + y^3$

$= x^3 - 2x^2y + y^3$

14. $(4t-3)^2$

$= (4t)^2 - 2\cdot 4t\cdot 3 + 3^2$

$= 16t^2 - 24t + 9$

15. $\left(5a^3+9\right)^2 = \left(5a^3\right)^2 + 2\cdot 5a^3\cdot 9 + 9^2$

$= 25a^6 + 90a^3 + 81$

16. $(x-2y)(x+2y)$

$= x^2 - (2y)^2$

$= x^2 - 4y^2$

17. $x^2 - 10x + 25$

$= x^2 - 5x - 5x + 25$

$= x(x-5) - 5(x-5)$ Factor by grouping

$= (x-5)^2$

18. $y^3 + 5y^2 - 4y - 20$

$= y^2(y+5) - 4(y+5)$

$= (y+5)\left(y^2 - 4\right)$

$= (y+5)(y+2)(y-2)$

19. $p^2 - 12p - 28$

We want two factors of -28 whose sum is -12. They are -14 and 2.

$p^2 - 12p - 28 = (p-14)(p+2)$

20. $t^7 - 3t^5 = t^5\left(t^2 - 3\right)$

21. $12m^2 + 20m + 3$

The factors of $12m^2$ are $12m, m$ and $6m, 2m$ and $4m, 3m$. Since $20m$ is positive, we want two positive factors of 3. They are 3 and 1. We want the sum of the products to equal $20m$. The factors are $(6m+1)(2m+3)$.

22. $9y^2 - 25 = (3y)^2 - 5^2$

$= (3y+5)(3y-5)$

23. $3r^3 - 3 = 3\left(r^3 - 1\right)$

$= 3(r-1)\left(r^2 + r + 1\right)$

24. $45x^2 + 20 + 60x = 45x^2 + 60x + 20$

$= 5\left(9x^2 + 12x + 4\right)$

$= 5\left(9x^2 + 6x + 6x + 4\right)$

$= 5\left[3x(3x+2) + 2(3x+2)\right]$

$= 5(3x+2)(3x+2)$

$= 5(3x+2)^2$

25. $3x^4 - 48y^4 = 3\left(x^4 - 16y^4\right)$

$= 3\left[\left(x^2\right)^2 - \left(4y^2\right)^2\right]$

$= 3\left(x^2 + 4y^2\right)\left(x^2 - 4y^2\right)$

$= 3\left(x^2 + 4y^2\right)\left[x^2 - (2y)^2\right]$

$= 3\left(x^2 + 4y^2\right)(x+2y)(x-2y)$

26. $y^2 + 8y + 16 - 100t^2$

$= \left(y^2 + 8y + 16\right) - 100t^2$

$= (y+4)^2 - (10t)^2$

$= \left[(y+4) + 10t\right]\left[(y+4) - 10t\right]$

$= (y+4+10t)(y+4-10t)$

27. $x^2 + 3x + 6$

We want two positive factors of 6, whose sum is 3. This is not possible with integer coefficients, so $x^2 + 3x + 6$ is prime.

28. $20a^2 - 5b^2$

$= 5\left(4a^2 - b^2\right)$

$= 5(2a + b)(2a - b)$

29. $24x^2 - 46x + 10$

$= 2\left(12x^2 - 23x + 5\right)$

Factor $12x^2 - 23x + 5$ by the grouping method. The product of the leading coefficient, 12 and constant, 5, is $12 \cdot 5 = 60$. We want two factors of 60 whose sum is –23. They are –20 and –3. Split $-23x$ into $-20x + (-3x)$, and factor by grouping.

$12x^2 - 23x + 5 = 12x^2 - 20x + (-3x) + 5$

$= 4x(3x - 5) - 1(3x - 5)$

$= (3x - 5)(4x - 1)$

The original trinomial has a factorization that includes the common factor.

$24x^2 - 46x + 10 = 2(3x - 5)(4x - 1)$

30. $3m^2 - 9mn - 30n^2$

$= 3\left(m^2 - 3mn - 10n^2\right)$

Factor $m^2 - 3mn - 10n^2$ by the grouping method. The product of the leading coefficient, 1 and constant, –10, is $1 \cdot (-10) = -10$. We want two factors of –10 whose sum is –3. They are –5 and 2. Split $-3mn$ into $-5mn + 2mn$, and factor by grouping.

$3m^2 - 9mn - 30n^2$

$= 3\left(m^2 - 3mn - 10n^2\right)$

$= 3\left(m^2 - 5mn + 2mn - 10n^2\right)$

$= 3\left[m(m - 5n) + 2n(m - 5n)\right]$

$= 3(m + 2n)(m - 5n)$

31. $16a^7b + 54ab^7$

$= 2ab\left(8a^6 + 27b^6\right)$

$= 2ab\left(\left(2a^2\right)^3 + \left(3b^2\right)^3\right)$

$= 2ab\left(2a^2 + 3b^2\right)\left(4a^4 - 6a^2b^2 + 9b^4\right)$

32. From the graph we see $p(x) = 0$ when $x = -3$, $x = -1$, $x = 2$, and $x = 4$. These are the zeros of p.

33. Let $f(x) = 0$ and solve the resulting equation using factoring and the principle of zero products.

$0 = 2x^2 - 11x - 40$

$0 = (2x + 5)(x - 8)$

$2x + 5 = 0 \quad or \quad x - 8 = 0$

$x = -\dfrac{5}{2} \ or \quad x = 8$

The zeros are $-\dfrac{5}{2}$ and 8.

34. $x^2 - 3x - 18 = 0$

$(x + 3)(x - 6) = 0$

$x + 3 = 0 \quad or \quad x - 6 = 0$

$x = -3 \ or \quad x = 6$

The solutions are –3 and 6.

35. $5t^2 = 125$

$5t^2 - 125 = 0$

$5\left(t^2 - 25\right) = 0$

$t^2 - 25 = 0$

$(t + 5)(t - 5) = 0$

$t + 5 = 0 \quad or \quad t - 5 = 0$

$t = -5 \ or \quad t = 5$

The solutions are ±5.

36. $2x^2 + 21 = -17x$

$2x^2 + 17x + 21 = 0$

$(2x + 3)(x + 7) = 0$

$2x + 3 = 0 \quad or \ x + 7 = 0$

$x = -\dfrac{3}{2} \ or \quad x = -7$

The solutions are –7 and $-\dfrac{3}{2}$.

37. $9x^2 + 3x = 0$

$3x(3x+1) = 0$

$3x = 0 \ or \ 3x+1 = 0$

$x = 0 \ or \quad x = -\dfrac{1}{3}$

The solutions are $-\dfrac{1}{3}$ and 0.

38. $x^2 + 81 = 18x$

$x^2 - 18x + 81 = 0$

$(x-9)^2 = 0$

$x - 9 = 0 \ or \ x - 9 = 0$

$x = 9 \ or \quad x = 9$

The solution is 9.

39. $x^2(x+1) = 8x$

$x^3 + x^2 - 8x = 0$

Graph $y = x^3 + x^2 - 8x$ and find the zeros. They are $x \approx -3.372$, $x = 0$, and $x \approx 2.372$, which are the solutions.

40. $f(x) = 3x^2 - 15x + 11$

Let $f(a) = 11$ and solve the equation.

$11 = 3a^2 - 15a + 11$

$0 = 3a(a-5)$

$3a = 0 \ or \ a - 5 = 0$

$a = 0 \ or \quad a = 5$

$f(0) = 11$ and $f(5) = 11$

41. $f(x) = \dfrac{3-x}{x^2 + 2x + 1}$

To find the excluded values solve $x^2 + 2x + 1 = 0$.

$x^2 + 2x + 1 = 0$

$(x+1)^2 = 0$

$(x+1)(x+1) = 0$

$x = -1$

The domain of $f(x)$ is $\{x | x$ is a real number and $x \neq -1\}$

42. **Familiarize.** Let $x =$ width of the photograph and $x + 3$ the length.

Translate. Area is $40cm^2$

$x(x+3) = \quad 40$

Carry out. Solve the equation.

$x(x+3) = 40$

$x^2 + 3x - 40 = 0$

$(x+8)(x-5) = 0$

$x + 8 = 0 \ or \ x - 5 = 0$

$x = -8 \ or \quad x = 5$

Check. Since measure cannot be negative, we check only $x = 5$. When $x = 5$, $x + 3 = 5 + 3 = 8$. The area is $5 \cdot 8 = 40$.

State. The photo is 5 cm wide and 8 cm long.

43. Let $h(t) = 0$ and solve using factoring and the principle of zero products.

$h(t) = -16t^2 + 64t + 36$

$0 = -4(4t^2 - 16t - 9)$

$0 = 4t^2 - 16t - 9$ Multiplying by $-\dfrac{1}{4}$

$0 = (2t - 9)(2t + 1)$

$2t - 9 = 0 \quad or \ 2t + 1 = 0$

$t = \dfrac{9}{2} \ or \quad t = -\dfrac{1}{2}$

Since time cannot be negative, we check $t = \dfrac{9}{2}$.

$h\left(\dfrac{9}{2}\right) = -16 \cdot \left(\dfrac{9}{2}\right)^2 + 64 \cdot \dfrac{9}{2} + 36$

$= -324 + 288 + 36 = 0$

The shell will reach the water in $4\dfrac{1}{2}$ sec.

44. ***Familiarize.*** Let h = the height the ladder reaches on the wall. The ladder is 2 ft longer, so $h+2$ represents its length.

Translate. $a^2+b^2 \;=\; c^2$

$$h^2+10^2=(h+2)^2$$

Carry out. We solve the equation:

$h^2+100 = h^2+4h+4$

$\qquad 96 = 4h$

$\qquad h = 24$

Check. If $h=24$, then $h+2=26$

$24^2+10^2=26^2$

State. The ladder reaches 24 ft up the wall.

45. a) Let x = the number of years after 2000. Enter the data in a graphing calculator and use the quadratic regression feature to obtain:

$E(x) = 0.50625x^2 + 0.245x + 32.86375$.

   b) In 2009, $x=9$. $E(9) \approx \$76.1$ billion.

   c) Solve $E(x)=100$ graphically. Graph

   $y_1 = 0.50625x^2 + 0.245x + 32.86375$ and

   $y_2 = 100$. Determine the point of intersection for which x is positive, which is $x \approx 11.3$ years. Eleven years after 2000 is 2011.

46. $(a+3)^2 - 2(a+3) - 35$

   Let $u = a+3$. Using substitution, we have $u^2 - 2u - 35$, which factors as

   $u^2 - 2u - 35 = (u-7)(u+5)$

   Substituting $u = a+3$ produces

   $(a+3)^2 - 2(a+3) - 35 = (a+3-7)(a+3+5)$

   $\qquad\qquad\qquad\qquad\quad = (a-4)(a+8)$

47. $20x(x+2)(x-1) = 5x^3 - 24x - 14x^2$

   $20x(x^2+x-2) = 5x^3 - 24x - 14x^2$

   $20x^3 + 20x^2 - 40x = 5x^3 - 24x - 14x^2$

   $20x^3 - 5x^3 + 20x^2 + 14x^2 - 40x + 24x = 0$

   $15x^3 + 34x^2 - 16x = 0$

   $x(15x^2 + 34x - 16) = 0$

   Factor $15x^2 + 34x - 16$ by the grouping method. The product of the leading

coefficient, 15 and constant, -16, is $15 \cdot (-16) = -240$. We want two factors of -240 whose sum is 34. They are -6 and 40. Split $34x$ into $-6x + 40x$, and factor by grouping into factors of the form $(5x+\ \ )(3x+\ \ )$.

$x(15x^2 + 34x - 16)$

$= x(15x^2 - 6x + 40x - 16)$

$= x[3x(5x-2) + 8(5x-2)]$

$= x(3x+8)(5x-2)$

$3x+8=0 \quad or \quad 5x-2=0 \quad or \quad x=0$

$\qquad 3x=-8 \quad or \qquad 5x=2$

$\qquad x=-\dfrac{8}{3} \quad or \qquad x=\dfrac{2}{5} \quad or \quad x=0$

The solutions are $-\dfrac{8}{3}, \dfrac{2}{5}$, and 0.

Chapter 6

Rational Expressions, Equations, and Functions

Exercise Set 6.1

1. e

3. i

5. d

7. $H(t) = \dfrac{t^2 + 3t}{2t + 3}$

$H(5) = \dfrac{5^2 + 3 \cdot 5}{2 \cdot 5 + 3} = \dfrac{25 + 15}{10 + 3}$

$= \dfrac{40}{13}$ hr, or $3\dfrac{1}{13}$ hr

9. $v(t) = \dfrac{4t^2 - 5t + 2}{t + 3}$

$v(0) = \dfrac{4 \cdot 0^2 - 5 \cdot 0 + 2}{0 + 3} = \dfrac{0 - 0 + 2}{0 + 3} = \dfrac{2}{3}$

$v(-2) = \dfrac{4(-2)^2 - 5(-2) + 2}{-2 + 3} = \dfrac{16 + 10 + 2}{-2 + 3} = 28$

$v(7) = \dfrac{4 \cdot 7^2 - 5 \cdot 7 + 2}{7 + 3} = \dfrac{196 - 35 + 2}{7 + 3} = \dfrac{163}{10}$

11. $g(x) = \dfrac{2x^3 - 9}{x^2 - 4x + 4}$

$g(0) = \dfrac{2 \cdot 0^3 - 9}{0^2 - 4 \cdot 0 + 4} = \dfrac{0 - 9}{0 - 0 + 4} = -\dfrac{9}{4}$

$g(2) = \dfrac{2 \cdot 2^3 - 9}{2^2 - 4 \cdot 2 + 4} = \dfrac{16 - 9}{4 - 8 + 4} = \dfrac{7}{0}$

Since division by zero is not defined, $g(2)$ does not exist.

$g(-1) = \dfrac{2(-1)^3 - 9}{(-1)^2 - 4(-1) + 4} = \dfrac{-2 - 9}{1 + 4 + 4} = -\dfrac{11}{9}$

13. $f(x) = \dfrac{25}{-7x}$

We find the real number(s) that make the denominator 0. To do so, we set the denominator equal to 0 and solve for x:

$-7x = 0$

$x = 0$

The expression is undefined for $x = 0$.

15. $r(t) = \dfrac{t - 3}{t + 8}$

Set the denominator equal to 0 and solve for t:

$t + 8 = 0$

$t = -8$

The expression is undefined for $t = -8$.

17. $f(x) = \dfrac{x^2 - 16}{x^2 - 3x - 28}$

Set the denominator equal to 0 and solve for x:

$x^2 - 3x - 28 = 0$

$(x - 7)(x + 4) = 0$

$x - 7 = 0$ or $x + 4 = 0$

$x = 7$ or $x = -4$

The expression is undefined for $x = 7$ and $x = -4$.

19. $g(m) = \dfrac{m^3 - 2m}{m^2 - 25}$

Set the denominator equal to 0 and solve for m:

$m^2 - 25 = 0$

$(m + 5)(m - 5) = 0$

$m + 5 = 0$ or $m - 5 = 0$

$m = -5$ or $m = 5$

The expression is undefined for $m = -5$ and $m = 5$.

21. $\dfrac{15x}{5x^2} = \dfrac{5x \cdot 3}{5x \cdot x}$ Factoring; the greatest common factor is $5x$.

$= \dfrac{5x}{5x} \cdot \dfrac{3}{x}$ Factoring the rational expression

$= 1 \cdot \dfrac{3}{x}$ $\dfrac{5x}{5x} = 1$

$= \dfrac{3}{x}$ Removing a factor equal to 1

23. $\dfrac{18t^3w^2}{27t^7w} = \dfrac{9t^3w \cdot 2w}{9t^3w \cdot 3t^4}$ Factor the numerator and denominator

$= \dfrac{9t^3w}{9t^3w} \cdot \dfrac{2w}{3t^4}$ Factor the rational expression

$= \dfrac{2w}{3t^4}$ Remove a factor of 1

25. $\dfrac{2a-10}{2} = \dfrac{2}{2} \cdot \dfrac{a-5}{1}$ Factoring the rational expression

$= \dfrac{a-5}{1} = a-5$ Removing a factor equal to 1

27. $\dfrac{5x}{25xy-30x} = \dfrac{5x}{5x(5y-6)}$

$= \dfrac{5x}{5x} \cdot \dfrac{1}{5y-6} = \dfrac{1}{5y-6}$

29. $\dfrac{20-4x}{3x-15} = \dfrac{4(5-x)}{3(x-5)} = -\dfrac{x-5}{x-5} \cdot \dfrac{4}{3} = -\dfrac{4}{3}$

31. $f(x) = \dfrac{5x+30}{x^2+6x}$

$= \dfrac{5(x+6)}{x(x+6)}$ Note that $x \ne 0, -6$

$= \dfrac{5}{x} \cdot \dfrac{x+6}{x+6}$

$= \dfrac{5}{x}, x \ne -6, 0$

33. $g(x) = \dfrac{x^2-9}{5x+15}$

$= \dfrac{(x+3)(x-3)}{5(x+3)}$ Note that $x \ne -3$

$= \dfrac{(x+3)}{(x+3)} \cdot \dfrac{x-3}{5}$

$= \dfrac{x-3}{5}, x \ne -3$

35. $h(x) = \dfrac{2-x}{7x-14}$

$= \dfrac{-1(x-2)}{7(x-2)}$ Factoring out -1 in the numerator reverses the subtraction

$= -\dfrac{1}{7} \cdot \dfrac{x-2}{x-2}$

$= -\dfrac{1}{7}, x \ne 2$

37. $f(t) = \dfrac{t^2-16}{t^2-8t+16} = \dfrac{(t+4)(t-4)}{(t-4)^2}$

$= \dfrac{t+4}{t-4} \cdot \dfrac{t-4}{t-4} = \dfrac{t+4}{t-4}, t \ne 4$

39. $g(t) = \dfrac{21-7t}{3t-9}$

$= \dfrac{-7(t-3)}{3(t-3)}$ Factoring out a -7 in the numerator reverses the subtraction

$= -\dfrac{7}{3} \cdot \dfrac{t-3}{t-3} = -\dfrac{7}{3}, t \ne 3$

41. $h(t) = \dfrac{t^2+5t+4}{t^2-8t-9}$

$= \dfrac{(t+4)(t+1)}{(t-9)(t+1)}$ Note $t \ne 9$ and $t \ne -1$

$= \dfrac{t+4}{t-9} \cdot \dfrac{t+1}{t+1} = \dfrac{t+4}{t-9}, t \ne -1, 9$

43. $f(x) = \dfrac{9x^2-4}{3x-2}$

$= \dfrac{(3x+2)(3x-2)}{3x-2}$ Note $x \ne \dfrac{2}{3}$

$= \dfrac{3x+2}{1} \cdot \dfrac{3x-2}{3x-2}$

$= 3x+2; x \ne \dfrac{2}{3}$

45. $g(t) = \dfrac{16-t^2}{t^2-8t+16} = \dfrac{16-t^2}{16-8t+t^2}$

$= \dfrac{(4+t)(4-t)}{(4-t)(4-t)} = \dfrac{4+t}{4-t} \cdot \dfrac{4-t}{4-t}$

$= \dfrac{4+t}{4-t}, t \ne 4$

47. $\dfrac{3y^3}{5z} \cdot \dfrac{10z^4}{7y^6}$

$= \dfrac{3y^3 \cdot 10z^4}{5z \cdot 7y^6}$ Multiply the numerators and the denominators

$= \dfrac{3 \cdot y^3 \cdot 5 \cdot 2 \cdot z \cdot z^3}{5 \cdot z \cdot 7 \cdot y^3 \cdot y^3}$ Factor the numerator and denominator

$= \dfrac{3 \cdot \cancel{y^3} \cdot \cancel{5} \cdot 2 \cdot \cancel{z} \cdot z^3}{\cancel{5} \cdot \cancel{z} \cdot 7 \cdot \cancel{y^3} \cdot y^3}$ Remove a factor equal to 1

$= \dfrac{6z^3}{7y^3}$

49. $\dfrac{8x-16}{5x} \cdot \dfrac{x^3}{5x-10}$

$= \dfrac{(8x-16)(x^3)}{5x(5x-10)}$ Multiplying the numerators and also the denominators

$= \dfrac{8(x-2)(x)(x^2)}{5 \cdot x \cdot 5(x-2)}$ Factoring the numerator and denominator

$= \dfrac{8\cancel{(x-2)}\cancel{(x)}(x^2)}{5 \cdot \cancel{x} \cdot 5\cancel{(x-2)}}$ Removing a factor equal to 1

$= \dfrac{8x^2}{25}$

51. $\dfrac{y^2-9}{y^2} \cdot \dfrac{y^2-3y}{y^2-y-6}$

$= \dfrac{(y^2-9)(y^2-3y)}{y^2(y^2-y-6)}$

$= \dfrac{(y+3)(y-3)\,y(y-3)}{y \cdot y(y-3)(y+2)}$

$= \dfrac{(y+3)\cancel{(y-3)}\,\cancel{y}(y-3)}{y \cdot \cancel{y}\cancel{(y-3)}(y+2)}$

$= \dfrac{(y+3)(y-3)}{y(y+2)}$

53. $\dfrac{7a-14}{4-a^2} \cdot \dfrac{5a^2+6a+1}{35a+7}$

$= \dfrac{(7a-14)(5a^2+6a+1)}{(4-a^2)(35a+7)}$

$= \dfrac{7(a-2) \cdot (5a+1)(a+1)}{(2+a)(2-a) \cdot 7(5a+1)}$

$= -\dfrac{\cancel{7}\,\cancel{(2-a)} \cdot \cancel{(5a+1)}(a+1)}{(2+a)\cancel{(2-a)} \cdot \cancel{7}\,\cancel{(5a+1)}}$

Note the negative sign needed to change $a-2$ to $2-a$ in the numerator.

$= -\dfrac{a+1}{a+2}$

55. $\dfrac{t^3-4t}{t-t^4} \cdot \dfrac{t^4-t}{4t-t^3} = \dfrac{t^3-4t}{t-t^4} \cdot \dfrac{-1(t-t^4)}{-1(t^3-4t)}$

$= \dfrac{(t^3-4t)(-1)(t-t^4)}{(t-t^4)(-1)(t^3-4t)} = 1$

57. $\dfrac{c^3+8}{c^5-4c^3} \cdot \dfrac{c^6-4c^5+4c^4}{c^2-2c+4}$

$= \dfrac{(c^3+8)(c^6-4c^5+4c)}{(c^5-4c^3)(c^2-2c+4)}$

$= \dfrac{(c+2)(c^2-2c+4)(c^4)(c-2)(c-2)}{c^3(c+2)(c-2)(c^2-2c+4)}$

$= \dfrac{c^3(c+2)(c^2-2c+4)(c-2)}{c^3(c+2)(c^2-2c+4)(c-2)} \cdot \dfrac{c(c-2)}{1}$

$= c(c-2)$

59. $\dfrac{a^3-b^3}{3a^2+9ab+6b^2} \cdot \dfrac{a^2+2ab+b^2}{a^2-b^2}$

$= \dfrac{(a^3-b^3)(a^2+2ab+b^2)}{(3a^2+9ab+6b^2)(a^2-b^2)}$

$= \dfrac{(a-b)(a^2+ab+b^2)(a+b)(a+b)}{3(a+b)(a+2b)(a+b)(a-b)}$

$= \dfrac{\cancel{(a-b)}(a^2+ab+b^2)\cancel{(a+b)}\cancel{(a+b)}}{3\cancel{(a+b)}(a+2b)\cancel{(a+b)}\cancel{(a-b)}}$

$= \dfrac{(a^2+ab+b^2)}{3(a+2b)}$

61. $\dfrac{12a^3}{5b^2} \div \dfrac{4a^2}{15b}$

$= \dfrac{12a^3}{5b^2} \cdot \dfrac{15b}{4a^2}$ Multiply by the
                                  reciprocal of the divisor

$= \dfrac{12a^3(15b)}{5b^2(4a^2)} = \dfrac{4\cdot 3\cdot a\cdot a^2\cdot 3\cdot 5\cdot b}{5\cdot b\cdot b\cdot 4\cdot a^2}$

$= \dfrac{\cancel{4}\cdot 3\cdot a\cdot \cancel{a^2}\cdot 3\cdot \cancel{5}\cdot \cancel{b}}{\cancel{5}\cdot \cancel{b}\cdot b\cdot \cancel{4}\cdot \cancel{a^2}} = \dfrac{9a}{b}$

63. $\dfrac{5x+20}{x^6} \div \dfrac{x+4}{x^2} = \dfrac{5x+20}{x^6} \cdot \dfrac{x^2}{x+4}$

$= \dfrac{(5x+20)(x^2)}{(x^6)(x+4)} = \dfrac{5(x+4)(x^2)}{x^2\cdot x^4(x+4)}$

$= \dfrac{5\cancel{(x+4)}\cancel{(x^2)}}{\cancel{x^2}\cdot x^4\cancel{(x+4)}} = \dfrac{5}{x^4}$

65. $\dfrac{25x^2-4}{x^2-9} \div \dfrac{2-5x}{x+3}$

$= \dfrac{25x^2-4}{x^2-9} \cdot \dfrac{x+3}{2-5x}$

$= \dfrac{(25x^2-4)(x+3)}{(x^2-9)(2-5x)}$

$= \dfrac{(5x+2)(5x-2)(x+3)}{(x+3)(x-3)(-1)(5x-2)}$

$= \dfrac{(5x+2)\cancel{(5x-2)}\cancel{(x+3)}}{\cancel{(x+3)}(x-3)(-1)\cancel{(5x-2)}}$

$= \dfrac{5x+2}{-x+3}$, or $-\dfrac{5x+2}{x-3}$

67. $\dfrac{5y-5x}{15y^3} \div \dfrac{x^2-y^2}{3x+3y}$

$= \dfrac{5y-5x}{15y^3} \cdot \dfrac{3x+3y}{x^2-y^2}$

$= \dfrac{(5y-5x)(3x+3y)}{(15y^3)(x^2-y^2)}$

$= \dfrac{5(y-x)(3)(x+y)}{5\cdot 3\cdot y^3(x+y)(x-y)}$

$= \dfrac{5(-1)(x-y)(3)(x+y)}{5\cdot 3\cdot y^3(x+y)(x-y)}$

$= \dfrac{\cancel{5}(-1)\cancel{(x-y)}\cancel{(3)}\cancel{(x+y)}}{\cancel{5}\cdot \cancel{3}\cdot y^3\cancel{(x+y)}\cancel{(x-y)}}$

$= \dfrac{-1}{y^3}$, or $-\dfrac{1}{y^3}$

69. $\dfrac{y^2-36}{y^2-8y+16} \div \dfrac{3y-18}{y^2-y-12}$

$= \dfrac{y^2-36}{y^2-8y+16} \cdot \dfrac{y^2-y-12}{3y-18}$

$= \dfrac{(y+6)(y-6)(y-4)(y+3)}{(y-4)(y-4)3(y-6)}$

$= \dfrac{(y+6)\cancel{(y-6)}\cancel{(y-4)}(y+3)}{\cancel{(y-4)}(y-4)(3)\cancel{(y-6)}}$

$= \dfrac{(y+6)(y+3)}{3(y-4)}$

71. $\dfrac{x^3-64}{x^3+64} \div \dfrac{x^2-16}{x^2-4x+16}$

$= \dfrac{x^3-64}{x^3+64} \cdot \dfrac{x^2-4x+16}{x^2-16}$

$= \dfrac{(x^3-64)(x^2-4x+16)}{(x^3+64)(x^2-16)}$

$= \dfrac{(x-4)(x^2+4x+16)(x^2-4x+16)}{(x+4)(x^2-4x+16)(x+4)(x-4)}$

$= \dfrac{(x-4)(x^2-4x+16)}{(x-4)(x^2-4x+16)} \cdot \dfrac{x^2+4x+16}{(x+4)(x+4)}$

$= \dfrac{x^2+4x+16}{(x+4)(x+4)}$, or $\dfrac{x^2+4x+16}{(x+4)^2}$

73. $f(t) = \dfrac{t^2 - 100}{5t + 20} \cdot \dfrac{t+4}{t-10}$

$= \dfrac{(t+10)(t-10)(t+4)}{5(t+4)(t-10)}$ Note that $t \neq -4$

$\qquad\qquad\qquad\qquad$ and $t \neq 10$

$= \dfrac{(t+10)\,\cancel{(t-10)}\,\cancel{(t+4)}}{5\,\cancel{(t+4)}\,\cancel{(t-10)}}$

$= \dfrac{t+10}{5}, t \neq -4, 10$

75. $g(x) = \dfrac{x^2 - 2x - 35}{2x^3 - 3x^2} \cdot \dfrac{4x^3 - 9x}{7x - 49}$

$= \dfrac{(x^2 - 2x - 35)(4x^3 - 9x)}{(2x^3 - 3x^2)(7x - 49)}$

$= \dfrac{(x-7)(x+5)(x)(2x+3)(2x-3)}{x^2(2x-3)(7)(x-7)}$

Note that $x \neq 0$, $x \neq \dfrac{3}{2}$, and $x \neq 7$

$= \dfrac{\cancel{(x-7)}(x+5)\,\cancel{(x)}(2x+3)\,\cancel{(2x-3)}}{\cancel{x} \cdot x\,\cancel{(2x-3)}(7)\,\cancel{(x-7)}}$

$= \dfrac{(x+5)(2x+3)}{7x}, \ x \neq 0, \dfrac{3}{2}, 7$

77. $f(x) = \dfrac{x^2 - 4}{x^3} \div \dfrac{x^5 - 2x^4}{x+4} = \dfrac{x^2 - 4}{x^3} \cdot \dfrac{x+4}{x^5 - 2x^4}$

Note that $x \neq -4$, since the divisor

cannot equal zero

$= \dfrac{(x^2 - 4)(x+4)}{x^3(x^5 - 2x^4)} = \dfrac{(x+2)(x-2)(x+4)}{x^3(x^4)(x-2)}$

Note that $x \neq 0$ and $x \neq 2$

$= \dfrac{(x+2)\,\cancel{(x-2)}(x+4)}{x^3(x^4)\,\cancel{(x-2)}}$

$= \dfrac{(x+2)(x+4)}{x^7}, x \neq -4, 0, 2$

79. $h(n) = \dfrac{n^3 + 3n}{n^2 - 9} \div \dfrac{n^2 + 5n - 14}{n^2 + 4n - 21}$

$= \dfrac{n(n^2 + 3)}{(n+3)(n-3)} \div \dfrac{(n+7)(n-2)}{(n+7)(n-3)}$

From $n^2 + 5n - 14$, we note $n \neq -7$ and

$n \neq 2$, since the divisor cannot equal zero.

From $n^2 - 9$ and $n^2 + 4n - 21$, we note

$n \neq -3, n \neq 3, n \neq -7$, and $n \neq 3$, since the

denominators cannot equal zero.

$= \dfrac{n(n^2 + 3)}{(n+3)(n-3)} \cdot \dfrac{(n+7)(n-3)}{(n+7)(n-2)}$

$= \dfrac{n(n^2 + 3)(n+7)(n-3)}{(n+3)(n-3)(n+7)(n-2)}$

$= \dfrac{n(n^2 + 3)\,\cancel{(n+7)}\,\cancel{(n-3)}}{(n+3)\,\cancel{(n-3)}\,\cancel{(n+7)}(n-2)}$

$= \dfrac{n(n^2 + 3)}{(n+3)(n-2)}, n \neq -7, -3, 2, 3$

81. $\dfrac{4x^2 - 9y^2}{8x^3 - 27y^3} \div \dfrac{4x + 6y}{3x - 9y} \cdot \dfrac{4x^2 + 6xy + 9y^2}{4x^2 - 8xy + 3y^2}$

$= \dfrac{4x^2 - 9y^2}{8x^3 - 27y^3} \cdot \dfrac{3x - 9y}{4x + 6y} \cdot \dfrac{4x^2 + 6xy + 9y^2}{4x^2 - 8xy + 3y^2}$

$= \dfrac{(2x+3y)(2x-3y)}{(2x-3y)(4x^2 + 6xy + 9y^2)} \cdot \dfrac{3(x-3y)}{2(2x+3y)}$

$\qquad \cdot \dfrac{4x^2 + 6xy + 9y^2}{(2x-3y)(2x-y)}$

$= \dfrac{\cancel{(2x+3y)}\,\cancel{(2x-3y)}\,3(x-3y)}{\cancel{(2x-3y)}\,\cancel{(4x^2 + 6xy + 9y^2)}\,2\,\cancel{(2x+3y)}}$

$\qquad \cdot \dfrac{\cancel{(4x^2 + 6xy + 9y^2)}}{(2x-3y)(2x-y)}$

$= \dfrac{3(x-3y)}{2(2x-3y)(2x-y)}$

83. $\dfrac{a^3 - ab^2}{2a^2 + 3ab + b^2} \cdot \dfrac{4a^2 - b^2}{a^2 - 2ab + b^2} \div \dfrac{a^2 + a}{a - 1}$

$= \dfrac{a^3 - ab^2}{2a^2 + 3ab + b^2} \cdot \dfrac{4a^2 - b^2}{a^2 - 2ab + b^2} \cdot \dfrac{a - 1}{a^2 + a}$

$= \dfrac{a(a+b)(a-b)}{(2a+b)(a+b)} \cdot \dfrac{(2a+b)(2a-b)}{(a-b)^2} \cdot \dfrac{a-1}{a(a+1)}$

$= \dfrac{\cancel{a}\,\cancel{(a+b)}\,\cancel{(a-b)}\,\cancel{(2a+b)}(2a-b)(a-1)}{\cancel{(2a+b)}\,\cancel{(a+b)}\,(a-b)^{\cancel{2}}\,\cancel{a}(a+1)}$

$= \dfrac{(2a-b)(a-1)}{(a-b)(a+1)}$

85. First we simplify the rational expression describing the function.

$$\frac{3x-12}{3x+15} = \frac{3(x-4)}{3(x+5)} = \frac{3}{3} \cdot \frac{x-4}{x+5} = \frac{x-4}{x+5}$$

$x+5=0$ when $x=-5$. Thus, the vertical asymptote is $x=-5$.

87. First we simplify the rational expression describing the function.

$$\frac{12-6x}{5x-10} = \frac{-6(-2+x)}{5(x-2)}$$

$$= \frac{-6(x-2)}{5(x-2)}$$

$$= \frac{-6}{5} \cdot \frac{x-2}{x-2}$$

$$= -\frac{6}{5}$$

The denominator of the simplified expression is not equal to 0 for any value of x, so there are no vertical asymptotes.

89. First we simplify the rational expression describing the function.

$$\frac{x^3+3x^2}{x^2+6x+9} = \frac{x^2(x+3)}{(x+3)(x+3)}$$

$$= \frac{x^2}{x+3} \cdot \frac{x+3}{x+3} = \frac{x^2}{x+3}$$

$x+3=0$ when $x=-3$. Thus, the vertical asymptote is $x=-3$.

91. First we simplify the rational expression describing the function.

$$\frac{x^2-x-6}{x^2-6x+8} = \frac{(x-3)(x+2)}{(x-4)(x-2)}$$

We cannot remove a factor equal to 1. Observe that $x-4=0$ when $x=4$ and $x-2=0$ when $x=2$. Thus, the vertical asymptotes are $x=4$ and $x=2$.

93. The vertical asymptote of $h(x)=\frac{1}{x}$ is $x=0$. Observe that $h(x)>0$ for $x>0$ and $h(x)<0$ for $x<0$. Thus, graph (b) corresponds to this function.

95. The vertical asymptote of $f(x)=\frac{x}{x-3}$ is $x=3$. Thus, graph (f) corresponds to this function.

97. $\dfrac{4x-2}{x^2-2x+1} = \dfrac{2(2x-1)}{(x-1)(x-1)}$

The vertical asymptote of $r(x)$ is $x=1$.

Thus, graph (a) corresponds to this function.

99. *Thinking and Writing Exercise.*

101. $-\dfrac{2}{15} \cdot \dfrac{10}{7} = -\dfrac{2}{3 \cdot \cancel{5}} \cdot \dfrac{2 \cdot \cancel{5}}{7} = -\dfrac{4}{21}$

103. $\dfrac{5}{8} \div \left(-\dfrac{1}{6}\right) = -\dfrac{5}{8} \cdot \dfrac{6}{1} = -\dfrac{5}{\cancel{2} \cdot 4} \cdot \dfrac{\cancel{2} \cdot 3}{1} = -\dfrac{15}{4}$

105. $\dfrac{7}{9} - \dfrac{2}{3} \cdot \dfrac{6}{7} = \dfrac{7}{9} - \dfrac{2}{\cancel{3}} \cdot \dfrac{\cancel{3} \cdot 2}{7} = \dfrac{7}{9} - \dfrac{4}{7}$

$$= \dfrac{7}{9} \cdot \dfrac{7}{7} - \dfrac{4}{7} \cdot \dfrac{9}{9} = \dfrac{49-36}{63} = \dfrac{13}{63}$$

107. *Thinking and Writing Exercise.*

109. $(a, f(a))$ and $(a+h, f(a+h))$

$$m = \frac{y_2-y_1}{x_2-x_1} = \frac{f(a+h)-f(a)}{a+h-a}$$

$$= \frac{f(a+h)-f(a)}{h}$$

Substituting: $f(a)=a^2+5$ and

$f(a+h)=(a+h)^2+5$

$$= a^2+2ah+h^2+5$$

we have:

$$\frac{(a^2+2ah+h^2+5)-(a^2+5)}{h}$$

$$= \frac{a^2+2ah+h^2+5-a^2-5}{h}$$

$$= \frac{2ah+h^2}{h} = \frac{h \cdot (2a+h)}{h \cdot 1}$$

$$= 2a+h$$

111. $g(x) = \dfrac{2x+3}{4x-1}$

 a) $g(x+h) = \dfrac{2(x+h)+3}{4(x+h)-1}$

  $= \dfrac{2x+2h+3}{4x+4h-1}$

 b) $g(2x-2) \cdot g(x) = \dfrac{2(2x-2)+3}{4(2x-2)-1} \cdot \dfrac{2x+3}{4x-1}$

  $= \dfrac{4x-1}{8x-9} \cdot \dfrac{2x+3}{4x-1}$

  $= \dfrac{2x+3}{8x-9}$

 c) $g(\tfrac{1}{2}x+1) \cdot g(x) = \dfrac{2(\tfrac{1}{2}x+1)+3}{4(\tfrac{1}{2}x+1)-1} \cdot \dfrac{2x+3}{4x-1}$

  $= \dfrac{x+5}{2x+3} \cdot \dfrac{2x+3}{4x-1}$

  $= \dfrac{x+5}{4x-1}$

113. $\dfrac{r^2-4s^2}{r+2s} \div (r+2s)^2 \left(\dfrac{2s}{r-2s}\right)^2$

 $= \dfrac{(r-2s)(r+2s)}{r+2s} \cdot \dfrac{1}{(r+2s)^2} \cdot \dfrac{4s^2}{(r-2s)^2}$

 $= \dfrac{4s^2}{(r+2s)^2(r-2s)}$

115. $\dfrac{6t^2-26t+30}{8t^2-15t-21} \cdot \dfrac{5t^2-9t-15}{6t^2-14t-20} \div \dfrac{5t^2-9t-15}{6t^2-14t-20}$

 $= \dfrac{6t^2-26t+30}{8t^2-15t-21} \cdot \dfrac{5t^2-9t-15}{6t^2-14t-20} \cdot \dfrac{6t^2-14t-20}{5t^2-9t-15}$

 $= \dfrac{6t^2-26t+30}{8t^2-15t-21}$

117. $\dfrac{a^3-2a^2+2a-4}{a^3-2a^2-3a+6} = \dfrac{(a^2+2)(a-2)}{(a^2-3)(a-2)}$

  $= \dfrac{a^2+2}{a^2-3}$

119. $\dfrac{u^6+v^6+2u^3v^3}{u^3-v^3+u^2v-uv^2} = \dfrac{(u^3+v^3)^2}{(u^2-v^2)(u+v)}$

  $= \dfrac{[(u+v)(u^2-uv+v^2)]^2}{(u-v)(u+v)(u+v)}$

  $= \dfrac{(u+v)^2(u^2-uv+v^2)^2}{(u-v)(u+v)^2}$

  $= \dfrac{(u^2-uv+v^2)^2}{u-v}$

121. a) $(f \cdot g)(x) = \dfrac{4}{x^2-1} \cdot \dfrac{4x^2+8x+4}{x^3-1}$

  $= \dfrac{16(x+1)^2}{(x-1)(x+1)(x-1)(x^2+x+1)}$

  $= \dfrac{16(x+1)}{(x-1)^2(x^2+x+1)}$

 b) $(f/g)(x) = \dfrac{4}{x^2-1} \div \dfrac{4x^2+8x+4}{x^3-1}$

  $= \dfrac{4}{(x-1)(x+1)} \cdot \dfrac{(x-1)(x^2+x+1)}{4(x+1)^2}$

  $= \dfrac{x^2+x+1}{(x+1)^3}$

 c) $(g/f)(x) = \dfrac{4x^2+8x+4}{x^3-1} \div \dfrac{4}{x^2-1}$

  $= \dfrac{4(x+1)^2}{(x-1)(x^2+x+1)} \cdot \dfrac{(x-1)(x+1)}{4}$

  $= \dfrac{(x+1)^3}{x^2+x+1}$

123. From the graph we see that the domain consists of all real numbers except −1, 0, and 1, so the domain is $(-\infty,-1) \cup (-1,0) \cup (0,1) \cup (1,\infty)$. We also see that the range consists of all real numbers except −3, −1, and 0, so the range is $(-\infty,-3) \cup (-3,-1) \cup (-1,0) \cup (0,\infty)$.

125. *Thinking and Writing Exercise.*

Exercise Set 6.2

1. True

3. False

5. False

7. False

9. $\dfrac{4}{3a}+\dfrac{11}{3a}$

$=\dfrac{15}{3a}$ Adding the numerators. The denominator is unchanged.

$=\dfrac{3\cdot 5}{3a}=\dfrac{\cancel{3}\cdot 5}{\cancel{3}a}=\dfrac{5}{a}$

11. $\dfrac{5}{3m^2n^2}-\dfrac{4}{3m^2n^2}$

$=\dfrac{1}{3m^2n^2}$ Subtracting the numerators. The denominator is unchanged.

13. $\dfrac{x-3y}{x+y}+\dfrac{x+5y}{x+y}=\dfrac{2x+2y}{x+y}$

$=\dfrac{2(x+y)}{x+y}$

$=\dfrac{2\cancel{(x+y)}}{1\cancel{(x+y)}}$

$=2$

15. $\dfrac{3t+2}{t-4}-\dfrac{t-2}{t-4}=\dfrac{3t+2-(t-2)}{t-4}$

$=\dfrac{3t+2-t+2}{t-4}$

$=\dfrac{2t+4}{t-4}$

17. $\dfrac{5-7x}{x^2-3x-10}+\dfrac{8x-3}{x^2-3x-10}=\dfrac{5-7x+8x-3}{x^2-3x-10}$

$=\dfrac{x+2}{x^2-3x-10}=\dfrac{x+2}{(x-5)(x+2)}=\dfrac{1}{x-5}$

19. $\dfrac{a-2}{a^2-25}-\dfrac{2a-7}{a^2-25}=\dfrac{a-2-(2a-7)}{a^2-25}$

$=\dfrac{a-2-2a+7}{a^2-25}=\dfrac{-a+5}{a^2-25}$

$=\dfrac{-(a-5)}{(a+5)(a-5)}=\dfrac{-1}{a+5}$

21. $f(x)=\dfrac{2x+1}{x^2+6x+5}+\dfrac{x-2}{x^2+6x+5}$

$=\dfrac{2x+1+x-2}{x^2+6x+5}=\dfrac{3x-1}{x^2+6x+5}$

$x^2+6x+5=(x+5)(x+1)$, so $x\neq -5,-1$

23. $f(x)=\dfrac{x-4}{x^2-1}-\dfrac{2x+1}{x^2-1}=\dfrac{x-4-(2x+1)}{x^2-1}$

$=\dfrac{x-4-2x-1}{x^2-1}=\dfrac{-x-5}{x^2-1}$, or $-\dfrac{x+5}{x^2-1}$

$x^2-1=(x+1)(x-1)$, so $x\neq -1,1$

25. $8x^2=2^3x^2,\ 12x^5=2^2\cdot 3x^5$

LCM $=2^3\cdot 3x^5=24x^5$

27. $x^2-9=(x+3)(x-3),\ x^2-6x+9=(x-3)^2$

LCM $=(x+3)(x-3)^2$

29. $\dfrac{2}{15x^2}+\dfrac{3}{5x}$ LCD is $15x^2$

$=\dfrac{2}{15x^2}+\dfrac{3}{5x}\cdot\dfrac{3x}{3x}$

$=\dfrac{2}{15x^2}+\dfrac{9x}{15x^2}$

$=\dfrac{9x+2}{15x^2}$

31. $\dfrac{y+1}{y-2} - \dfrac{y-1}{2y-4}$

$= \dfrac{y+1}{y-2} - \dfrac{y-1}{2(y-2)}$ LCD is $2(y-2)$

$= \dfrac{y+1}{y-2} \cdot \dfrac{2}{2} - \dfrac{y-1}{2(y-2)}$

$= \dfrac{2(y+1)-(y-1)}{2(y-2)}$

$= \dfrac{2y+2-y+1}{2(y-2)}$

$= \dfrac{y+3}{2(y-2)}$

33. $\dfrac{4xy}{x^2-y^2} + \dfrac{x-y}{x+y}$

$= \dfrac{4xy}{(x+y)(x-y)} + \dfrac{x-y}{x+y}$

$\left[\text{LCD is } (x+y)(x-y).\right]$

$= \dfrac{4xy}{(x+y)(x-y)} + \dfrac{x-y}{x+y} \cdot \dfrac{x-y}{x-y}$

$= \dfrac{4xy + x^2 - 2xy + y^2}{(x+y)(x-y)}$

$= \dfrac{x^2 + 2xy + y^2}{(x+y)(x-y)} = \dfrac{(x+y)(x+y)}{(x+y)(x-y)}$

$= \dfrac{\cancel{(x+y)}(x+y)}{\cancel{(x+y)}(x-y)} = \dfrac{x+y}{x-y}$

35. $\dfrac{8}{2x^2-7x+5} + \dfrac{3x+2}{2x^2-x-10}$

$= \dfrac{8}{(2x-5)(x-1)} + \dfrac{3x+2}{(2x-5)(x+2)}$

$\left[\text{LCD is } (2x-5)(x-1)(x+2).\right]$

$= \dfrac{8}{(2x-5)(x-1)} \cdot \dfrac{x+2}{x+2}$

$\quad + \dfrac{3x+2}{(2x-5)(x+2)} \cdot \dfrac{x-1}{x-1}$

$= \dfrac{8x+16+3x^2-x-2}{(2x-5)(x-1)(x+2)}$

$= \dfrac{3x^2+7x+14}{(2x-5)(x-1)(x+2)}$

37. $\dfrac{5ab}{a^2-b^2} - \dfrac{a-b}{a+b}$

$= \dfrac{5ab}{(a+b)(a-b)} - \dfrac{a-b}{a+b}$

$\left[\text{LCD is } (a+b)(a-b).\right]$

$= \dfrac{5ab}{(a+b)(a-b)} - \dfrac{a-b}{a+b} \cdot \dfrac{a-b}{a-b}$

$= \dfrac{5ab-(a^2-2ab+b^2)}{(a+b)(a-b)}$

$= \dfrac{5ab-a^2+2ab-b^2}{(a+b)(a-b)}$

$= \dfrac{-a^2+7ab-b^2}{(a+b)(a-b)}$

39. $\dfrac{x}{x^2+9x+20} - \dfrac{4}{x^2+7x+12}$

$= \dfrac{x}{(x+5)(x+4)} - \dfrac{4}{(x+3)(x+4)}$

$\left[\text{LCD is } (x+5)(x+4)(x+3).\right]$

$= \dfrac{x}{(x+5)(x+4)} \cdot \dfrac{x+3}{x+3}$

$\quad - \dfrac{4}{(x+3)(x+4)} \cdot \dfrac{x+5}{x+5}$

$= \dfrac{x^2+3x-(4x+20)}{(x+5)(x+4)(x+3)}$

$= \dfrac{x^2+3x-4x-20}{(x+5)(x+4)(x+3)}$

$= \dfrac{x^2-x-20}{(x+5)(x+4)(x+3)}$

$= \dfrac{(x-5)(x+4)}{(x+5)(x+4)(x+3)}$

$= \dfrac{(x-5)\cancel{(x+4)}}{(x+5)\cancel{(x+4)}(x+3)}$

$= \dfrac{x-5}{(x+5)(x+3)}$

41. $\dfrac{3}{t} - \dfrac{6}{-t} = \dfrac{3}{t} + \dfrac{6}{t} = \dfrac{9}{t}$

43. $\dfrac{s^2}{r-s}+\dfrac{r^2}{s-r}$

$=\dfrac{s^2}{r-s}+\dfrac{-1}{-1}\cdot\dfrac{r^2}{s-r}$

$=\dfrac{s^2}{r-s}+\dfrac{-r^2}{r-s}$

$=\dfrac{s^2-r^2}{r-s}$

$=\dfrac{(s+r)(s-r)}{r-s}$

$=\dfrac{(s+r)(-1)(r-s)}{r-s}$

$=\dfrac{(s+r)(-1)\cancel{(r-s)}}{\cancel{r-s}}$

$=-(s+r)$

45. $\dfrac{a+2}{a-4}+\dfrac{a-2}{a+3}$

$\left[\text{LCD is }(a-4)(a+3).\right]$

$=\dfrac{a+2}{a-4}\cdot\dfrac{a+3}{a+3}+\dfrac{a-2}{a+3}\cdot\dfrac{a-4}{a-4}$

$=\dfrac{\left(a^2+5a+6\right)+\left(a^2-6a+8\right)}{(a-4)(a+3)}$

$=\dfrac{2a^2-a+14}{(a-4)(a+3)}$

47. $4+\dfrac{x-3}{x+1}=\dfrac{4}{1}+\dfrac{x-3}{x+1}$

$\left[\text{LCD is }x+1.\right]$

$=\dfrac{4}{1}\cdot\dfrac{x+1}{x+1}+\dfrac{x-3}{x+1}$

$=\dfrac{(4x+4)+(x-3)}{x+1}$

$=\dfrac{5x+1}{x+1}$

49. $\dfrac{x+6}{5x+10}-\dfrac{x-2}{4x+8}$

$=\dfrac{x+6}{5(x+2)}-\dfrac{x-2}{4(x+2)}$

$\left[\text{LCD is }5\cdot4(x+2).\right]$

$=\dfrac{x+6}{5(x+2)}\cdot\dfrac{4}{4}-\dfrac{x-2}{4(x+2)}\cdot\dfrac{5}{5}$

$=\dfrac{4(x+6)-5(x-2)}{5\cdot4(x+2)}$

$=\dfrac{4x+24-5x+10}{5\cdot4(x+2)}$

$=\dfrac{-x+34}{5\cdot4(x+2)}\text{, or }\dfrac{-x+34}{20(x+2)}$

51. $\dfrac{4}{x+1}+\dfrac{x+2}{x^2-1}+\dfrac{3}{x-1}$

$=\dfrac{4}{x+1}+\dfrac{x+2}{(x+1)(x-1)}+\dfrac{3}{x-1}$

$\left[\text{LCD is }(x+1)(x-1).\right]$

$=\dfrac{4}{x+1}\cdot\dfrac{x-1}{x-1}+\dfrac{x+2}{(x+1)(x-1)}+\dfrac{3}{x-1}\cdot\dfrac{x+1}{x+1}$

$=\dfrac{4x-4+x+2+3x+3}{(x+1)(x-1)}$

$=\dfrac{8x+1}{(x+1)(x-1)}$

53. $\dfrac{y-4}{y^2-25}-\dfrac{9-2y}{25-y^2}$

$=\dfrac{y-4}{y^2-25}+(-1)\cdot\dfrac{9-2y}{25-y^2}$

$=\dfrac{y-4}{y^2-25}+\dfrac{1}{-1}\cdot\dfrac{9-2y}{25-y^2}$

$=\dfrac{y-4}{y^2-25}+\dfrac{9-2y}{y^2-25}$

$=\dfrac{-y+5}{y^2-25}$

$=\dfrac{-1(y-5)}{(y+5)(y-5)}$

$=\dfrac{-1\cancel{(y-5)}}{(y+5)\cancel{(y-5)}}$

$=\dfrac{-1}{y+5}\text{ or }-\dfrac{1}{y+5}$

55. $\dfrac{y^2-5}{y^4-81}+\dfrac{4}{81-y^4}$

$=\dfrac{y^2-5}{y^4-81}+\dfrac{-1}{-1}\cdot\dfrac{4}{81-y^4}$

$=\dfrac{y^2-5}{y^4-81}+\dfrac{-4}{y^4-81}$

$=\dfrac{y^2-9}{y^4-81}$

$=\dfrac{1\left(y^2-9\right)}{\left(y^2+9\right)\left(y^2-9\right)}$

$=\dfrac{1\cancel{\left(y^2-9\right)}}{\left(y^2+9\right)\cancel{\left(y^2-9\right)}}$

$=\dfrac{1}{y^2+9}$

57. $\dfrac{r-6s}{r^3-s^3}-\dfrac{5s}{s^3-r^3}$

$=\dfrac{r-6s}{r^3-s^3}+(-1)\dfrac{5s}{s^3-r^3}$

$=\dfrac{r-6s}{r^3-s^3}+\dfrac{1}{-1}\cdot\dfrac{5s}{s^3-r^3}$

$=\dfrac{r-6s}{r^3-s^3}+\dfrac{5s}{r^3-s^3}$

$=\dfrac{r-s}{r^3-s^3}$

$=\dfrac{1(r-s)}{(r-s)\left(r^2+rs+s^2\right)}$

$=\dfrac{1\cancel{(r-s)}}{\cancel{(r-s)}\left(r^2+rs+s^2\right)}$

$=\dfrac{1}{r^2+rs+s^2}$

59. $\dfrac{3y}{y^2-7y+10}-\dfrac{2y}{y^2-8y+15}$

$=\dfrac{3y}{(y-5)(y-2)}-\dfrac{2y}{(y-5)(y-3)}$

$\left[\text{LCD is }(y-5)(y-2)(y-3).\right]$

$=\dfrac{3y}{(y-5)(y-2)}\cdot\dfrac{y-3}{y-3}$

$\quad-\dfrac{2y}{(y-5)(y-3)}\cdot\dfrac{y-2}{y-2}$

$=\dfrac{3y^2-9y-\left(2y^2-4y\right)}{(y-5)(y-2)(y-3)}$

$=\dfrac{3y^2-9y-2y^2+4y}{(y-5)(y-2)(y-3)}$

$=\dfrac{y^2-5y}{(y-5)(y-2)(y-3)}$

$=\dfrac{y(y-5)}{(y-5)(y-2)(y-3)}$

$=\dfrac{y\cancel{(y-5)}}{\cancel{(y-5)}(y-2)(y-3)}=\dfrac{y}{(y-2)(y-3)}$

61. $\dfrac{2x+1}{x-y}+\dfrac{5x^2-5xy}{x^2-2xy+y^2}$

$=\dfrac{2x+1}{x-y}+\dfrac{5x(x-y)}{(x-y)(x-y)}$

$=\dfrac{2x+1}{x-y}+\dfrac{5x\cancel{(x-y)}}{\cancel{(x-y)}(x-y)}$

$=\dfrac{2x+1}{x-y}+\dfrac{5x}{x-y}$

$=\dfrac{7x+1}{x-y}$

63. $\dfrac{2y-6}{y^2-9} - \dfrac{y}{y-1} + \dfrac{y^2+2}{y^2+2y-3}$

$= \dfrac{2\cancel{(y-3)}}{(y+3)\cancel{(y-3)}} - \dfrac{y}{y-1} + \dfrac{y^2+2}{(y+3)(y-1)}$

$= \dfrac{2}{(y+3)} - \dfrac{y}{y-1} + \dfrac{y^2+2}{(y+3)(y-1)}$

$\left[\text{LCD is } (y+3)(y-1).\right]$

$= \dfrac{2}{(y+3)} \cdot \dfrac{y-1}{y-1} - \dfrac{y}{y-1} \cdot \dfrac{y+3}{y+3} +$

$\qquad\qquad\qquad\qquad \dfrac{y^2+2}{(y+3)(y-1)}$

$= \dfrac{2(y-1) - y(y+3) + y^2+2}{(y+3)(y-1)}$

$= \dfrac{2y-2 - y^2 - 3y + y^2 + 2}{(y+3)(y-1)}$

$= \dfrac{-y}{(y+3)(y-1)} \text{ or } -\dfrac{y}{(y+3)(y-1)}$

65. $\dfrac{5y}{1-4y^2} - \dfrac{2y}{2y+1} + \dfrac{5y}{4y^2-1}$

Observe that $\dfrac{5y}{1-4y^2}$ and $\dfrac{5y}{4y^2-1}$ are

opposites, so their sum is 0. Then the result is

the remaining expression, $-\dfrac{2y}{2y+1}$.

67. $f(x) = 2 + \dfrac{x}{x-3} - \dfrac{18}{x^2-9}$

$= \dfrac{2}{1} + \dfrac{x}{x-3} - \dfrac{18}{(x+3)(x-3)} \qquad x \neq \pm 3$

$\left[\text{LCD is } (x+3)(x-3).\right]$

$= \dfrac{2}{1} \cdot \dfrac{(x+3)(x-3)}{(x+3)(x-3)} + \dfrac{x}{x-3} \cdot \dfrac{x+3}{x+3}$

$\qquad - \dfrac{18}{(x+3)(x-3)}$

$= \dfrac{2x^2 - 18 + x^2 + 3x - 18}{(x+3)(x-3)}$

$= \dfrac{3x^2 + 3x - 36}{(x+3)(x-3)}$

$= \dfrac{3(x+4)(x-3)}{(x+3)(x-3)}$

$= \dfrac{3(x+4)\cancel{(x-3)}}{(x+3)\cancel{(x-3)}}$

$= \dfrac{3(x+4)}{(x+3)}, x \neq \pm 3$

69. $f(x) = \dfrac{3x-1}{x^2+2x-3} - \dfrac{x+4}{x^2-16}$

$= \dfrac{3x-1}{(x+3)(x-1)} - \dfrac{x+4}{(x+4)(x-4)}$

$= \dfrac{3x-1}{(x+3)(x-1)} - \dfrac{\cancel{x+4}}{\cancel{(x+4)}(x-4)}$

$\left[\text{LCD is } (x+3)(x-1)(x-4) \right]$

$= \dfrac{3x-1}{(x+3)(x-1)} \cdot \dfrac{x-4}{x-4}$

$\quad - \dfrac{1}{x-4} \cdot \dfrac{(x+3)(x-1)}{(x+3)(x-1)}$

$= \dfrac{(3x-1)(x-4)}{(x+3)(x-1)(x-4)}$

$\quad - \dfrac{(x+3)(x-1)}{(x+3)(x-1)(x-4)}$

$= \dfrac{3x^2-12x-x+4-(x^2+2x-3)}{(x+3)(x-1)(x-4)}$

$= \dfrac{3x^2-13x+4-x^2-2x+3}{(x+3)(x-1)(x-4)}$

$= \dfrac{2x^2-15x+7}{(x+3)(x-1)(x-4)},$

$\quad \text{or } \dfrac{(x-7)(2x-1)}{(x+3)(x-1)(x-4)},$

$x \neq -4, -3, 1, 4$

71. $f(x) = \dfrac{1}{x^2+5x+6} - \dfrac{2}{x^2+3x+2} - \dfrac{1}{x^2+5x+6}$

$x^2+5x+6 = (x+3)(x+2), \text{ so } x \neq -3, -2$

$x^2+3x+2 = (x+2)(x+1), \text{ so } x \neq -2, -1$

Notice $\dfrac{1}{x^2+5x+6} - \dfrac{1}{x^2+5x+6} = 0$

$f(x) = \dfrac{-2}{x^2+3x+2}, x \neq -3, -2, -1$

73. *Thinking and Writing Exercise.*

75. $2x^{-1} = 2 \cdot \dfrac{1}{x} = \dfrac{2}{x}$

77. $ab(a+b)^{-2} = ab \cdot \dfrac{1}{(a+b)^2} = \dfrac{ab}{(a+b)^2}$

79. $9x^3 \left(\dfrac{1}{x^2} - \dfrac{2}{3x^3} \right) = \dfrac{1 \cdot 9x^3}{x^2} - \dfrac{2 \cdot 9x^3}{3x^3} = 9x - 6$

81. *Thinking and Writing Exercise.*

83. We find the least common multiple of
14 (2 weeks = 14 days), 20, and 30.

$\quad 14 = 2 \cdot 7$

$\quad 20 = 2 \cdot 2 \cdot 5$

$\quad 30 = 2 \cdot 3 \cdot 5$

$\quad \text{LCM} = 2 \cdot 2 \cdot 3 \cdot 5 \cdot 7 = 420$

It will be 420 days until Jinney can refill all
three prescriptions on the same day.

85. The least number of parts possible is the least
common multiple of 6 and 4.

$\quad 6 = 2 \cdot 3$

$\quad 4 = 2 \cdot 2$

$\quad \text{LCM} = 2 \cdot 3 \cdot 2, \text{ or } 12$

A measure should be divided into 12 parts.

87. $x^8 - x^4 = x^4(x^2+1)(x+1)(x-1)$

$\quad x^5 - x^2 = x^2(x-1)(x^2+x+1)$

$\quad x^5 - x^3 = x^3(x+1)(x-1)$

$\quad x^5 + x^2 = x^2(x+1)(x^2-x+1)$

The LCM is

$\quad x^4(x^2+1)(x+1)(x-1)(x^2+x+1)(x^2-x+1).$

89. The LCM is $8a^4b^7$.

One expression is $2a^3b^7$.

Then the other expression must contain 8, a^4,
and one of the following:

no factor of b, b, b^2, b^3, b^4, b^5, b^6, or b^7.
Thus, all the possibilities for the other
expression are $8a^4$, $8a^4b$, $8a^4b^2$, $8a^4b^3$,
$8a^4b^4$, $8a^4b^5$, $8a^4b^6$, $8a^4b^7$.

91. $(f+g)(x) = \dfrac{x^3}{x^2-4} + \dfrac{x^2}{x^2+3x-10}$

$\qquad = \dfrac{x^3}{(x+2)(x-2)} + \dfrac{x^2}{(x+5)(x-2)}$

$\qquad = \dfrac{x^3(x+5) + x^2(x+2)}{(x+2)(x-2)(x+5)}$

$\qquad = \dfrac{x^4 + 5x^3 + x^3 + 2x^2}{(x+2)(x-2)(x+5)}$

$\qquad = \dfrac{x^4 + 6x^3 + 2x^2}{(x+2)(x-2)(x+5)}$

93. $(f \cdot g)(x) = \dfrac{x^3}{x^2-4} \cdot \dfrac{x^2}{x^2+3x-10}$

$\qquad = \dfrac{x^5}{(x^2-4)(x^2+3x-10)}$

95. The denominator of f is 0 when $x = 2$ or $x = -2$ and the denominator of g is 0 when $x = 2$ or $x = -5$. Thus the domain of $f + g$ is $\{x \,|\, x$ is a real number and $x \neq -5, -2, 2\}$, or $(-\infty, -5) \cup (-5, -2) \cup (-2, 2) \cup (2, \infty)$.

97. $x^{-2} + 2x^{-1} = \dfrac{1}{x^2} + \dfrac{2}{x}$

$\qquad = \dfrac{1}{x^2} + \dfrac{2}{x} \cdot \dfrac{x}{x}$

$\qquad = \dfrac{1}{x^2} + \dfrac{2x}{x^2}$

$\qquad = \dfrac{2x+1}{x^2}$

99. $5(x-3)^{-1} + 4(x+3)^{-1} - 2(x+3)^{-2}$

$\qquad = \dfrac{5}{x-3} + \dfrac{4}{x+3} - \dfrac{2}{(x+3)^2}$

$\qquad \left[\text{LCD is } (x-3)(x+3)^2 \right]$

$\qquad = \dfrac{5(x+3)^2 + 4(x-3)(x+3) - 2(x-3)}{(x-3)(x+3)^2}$

$\qquad = \dfrac{5x^2 + 30x + 45 + 4x^2 - 36 - 2x + 6}{(x-3)(x+3)^2}$

$\qquad = \dfrac{9x^2 + 28x + 15}{(x-3)(x+3)^2}$

101. $\dfrac{x+4}{6x^2-20x}\left(\dfrac{x}{x^2-x-20} + \dfrac{2}{x+4} \right)$

$\qquad = \dfrac{x+4}{2x(3x-10)}\left(\dfrac{x}{(x-5)(x+4)} + \dfrac{2}{x+4} \right)$

$\qquad = \dfrac{x+4}{2x(3x-10)}\left(\dfrac{x+2(x-5)}{(x-5)(x+4)} \right)$

$\qquad = \dfrac{x+4}{2x(3x-10)}\left(\dfrac{x+2x-10}{(x-5)(x+4)} \right)$

$\qquad = \dfrac{(x+4)(3x-10)}{2x(3x-10)(x-5)(x+4)}$

$\qquad = \dfrac{\cancel{(x+4)}\cancel{(3x-10)}(1)}{2x\cancel{(3x-10)}(x-5)\cancel{(x+4)}}$

$\qquad = \dfrac{1}{2x(x-5)}$

103. $\dfrac{8t^5}{2t^2-10t+12} \div \left(\dfrac{2t}{t^2-8t+15} - \dfrac{3t}{t^2-7t+10} \right)$

$\qquad = \dfrac{8t^5}{2t^2-10t+12}$

$\qquad \div \left(\dfrac{2t}{(t-5)(t-3)} - \dfrac{3t}{(t-5)(t-2)} \right)$

$\qquad = \dfrac{8t^5}{2t^2-10t+12} \div \left(\dfrac{2t(t-2) - 3t(t-3)}{(t-5)(t-3)(t-2)} \right)$

$\qquad = \dfrac{8t^5}{2t^2-10t+12} \div \left(\dfrac{2t^2 - 4t - 3t^2 + 9t}{(t-5)(t-3)(t-2)} \right)$

$\qquad = \dfrac{8t^5}{2t^2-10t+12} \div \dfrac{-t^2+5t}{(t-5)(t-3)(t-2)}$

$\qquad = \dfrac{8t^5}{2t^2-10t+12} \cdot \dfrac{(t-5)(t-3)(t-2)}{-t(t-5)}$

$\qquad = \dfrac{\cancel{2} \cdot 4 \cdot \cancel{t} \cdot t^4 \cancel{(t-5)}\cancel{(t-3)}\cancel{(t-2)}}{\cancel{2}\cancel{(t-3)}\cancel{(t-2)}(-1)\cancel{(t)}\cancel{(t-5)}}$

$\qquad = -4t^4$

105.

From the graph (shown in the standard window) we see that the domain of the function consists of all real numbers except -1, so the domain of f is $\{x \mid x$ is a real number $and \ x \neq -1\}$, or $(-\infty, -1) \cup (-1, \infty)$.

We also see that the range consists of all real numbers except 3, so the range of f is $\{y \mid y$ is a real number and $y \neq 3\}$, or $(-\infty, 3) \cup (3, \infty)$.

107.

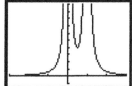

From the graph (shown in the window $|-3, 3, -2, 20|$, Yscl $= 2$), we see that the domain consists of all real numbers except 0 and 1, so the domain of r is $\{x \mid x$ is a real number $and \ x \neq 0 \ and \ x \neq 1\}$, or $(-\infty, 0) \cup (0, 1) \cup (1, \infty)$. We also see that the range consists of all real numbers greater than 0, so the range of r is $\{y \mid y > 0\}$, or $(0, \infty)$.

Exercise Set 6.3

---

1. a

3. b

5. $\dfrac{\dfrac{1}{2} + \dfrac{1}{3}}{\dfrac{1}{4} - \dfrac{1}{6}} = \dfrac{\dfrac{1}{2} + \dfrac{1}{3}}{\dfrac{1}{4} - \dfrac{1}{6}} \cdot \dfrac{12}{12} = \dfrac{12 \cdot \dfrac{1}{2} + 12 \cdot \dfrac{1}{3}}{12 \cdot \dfrac{1}{4} - 12 \cdot \dfrac{1}{6}}$

$= \dfrac{6 \cdot \cancel{2} \cdot \dfrac{1}{\cancel{2}} + 4 \cdot \cancel{3} \cdot \dfrac{1}{\cancel{3}}}{3 \cdot \cancel{4} \cdot \dfrac{1}{\cancel{4}} - 2 \cdot \cancel{6} \cdot \dfrac{1}{\cancel{6}}}$

$= \dfrac{6 \cdot 1 + 4 \cdot 1}{3 \cdot 1 - 2 \cdot 1} = \dfrac{6 + 4}{3 - 2} = \dfrac{10}{1} = 10$

7. $\dfrac{1 + \dfrac{1}{4}}{2 + \dfrac{3}{4}} = \dfrac{1 + \dfrac{1}{4}}{2 + \dfrac{3}{4}} \cdot \dfrac{4}{4} = \dfrac{4 \cdot 1 + \cancel{4} \cdot \dfrac{1}{\cancel{4}}}{4 \cdot 2 + \cancel{4} \cdot \dfrac{3}{\cancel{4}}}$

$= \dfrac{4 \cdot 1 + 1 \cdot 1}{4 \cdot 2 + 1 \cdot 3} = \dfrac{4 + 1}{8 + 3} = \dfrac{5}{11}$

9. $\dfrac{\dfrac{x}{4} + x}{\dfrac{4}{x} + x} = \dfrac{\dfrac{x}{4} + x}{\dfrac{4}{x} + x} \cdot \dfrac{4x}{4x} = \dfrac{\cancel{4}x \cdot \dfrac{x}{\cancel{4}} + 4x \cdot x}{4\cancel{x} \cdot \dfrac{4}{\cancel{x}} + 4x \cdot x}$

$= \dfrac{x \cdot x + 4x \cdot x}{4 \cdot 4 + 4x \cdot x} = \dfrac{x^2 + 4x^2}{16 + 4x^2}$

$= \dfrac{5x^2}{16 + 4x^2}$ or $\dfrac{5x^2}{4(x^2 + 4)}$

11. $\dfrac{\dfrac{x+2}{x-1}}{\dfrac{x+4}{x-3}} = \dfrac{x+2}{x-1} \cdot \dfrac{x-3}{x+4} = \dfrac{(x+2)(x-3)}{(x-1)(x+4)}$

13. $\dfrac{\dfrac{5}{a} - \dfrac{4}{b}}{\dfrac{2}{a} + \dfrac{3}{b}} = \dfrac{\dfrac{5}{a} - \dfrac{4}{b}}{\dfrac{2}{a} + \dfrac{3}{b}} \cdot \dfrac{ab}{ab} = \dfrac{\cancel{a}b \cdot \dfrac{5}{\cancel{a}} - a\cancel{b} \cdot \dfrac{4}{\cancel{b}}}{\cancel{a}b \cdot \dfrac{2}{\cancel{a}} + a\cancel{b} \cdot \dfrac{3}{\cancel{b}}}$

$= \dfrac{b \cdot 5 - a \cdot 4}{b \cdot 2 + a \cdot 3} = \dfrac{5b - 4a}{2b + 3a}$

15. $\dfrac{\dfrac{3}{z^2} + \dfrac{2}{yz}}{\dfrac{4}{zy^2} - \dfrac{1}{y}} = \dfrac{\dfrac{3}{z^2} + \dfrac{2}{yz}}{\dfrac{4}{zy^2} - \dfrac{1}{y}} \cdot \dfrac{y^2 z^2}{y^2 z^2}$

$= \dfrac{y^2 \cancel{z^2} \cdot \dfrac{3}{\cancel{z^2}} + y^{\cancel{2}1} z^{\cancel{2}1} \cdot \dfrac{2}{\cancel{yz}}}{y^{\cancel{2}1} z^{\cancel{2}1} \cdot \dfrac{4}{\cancel{zy^2}} - y^{\cancel{2}1} z^2 \cdot \dfrac{1}{\cancel{y}}}$

$= \dfrac{3y^2 + 2yz}{4z - yz^2} = \dfrac{y(3y + 2z)}{z(4 - yz)}$

17. $\dfrac{\dfrac{a^2-b^2}{ab}}{\dfrac{a-b}{b}} = \dfrac{a^2-b^2}{ab} \cdot \dfrac{b}{a-b}$

$\quad = \dfrac{(a-b)(a+b)}{ab} \cdot \dfrac{b}{a-b}$

$\quad = \dfrac{a+b}{a}$

19. $\dfrac{1-\dfrac{2}{3x}}{x-\dfrac{4}{9x}} = \dfrac{1-\dfrac{2}{3x}}{x-\dfrac{4}{9x}} \cdot \dfrac{9x}{9x}$

$\quad = \dfrac{9x\cdot 1 - 9x\cdot\dfrac{2}{3x}}{9x\cdot x - 9x\cdot\dfrac{4}{9x}}$

$\quad = \dfrac{9x\cdot 1 - 3\cdot 3x\cdot\dfrac{2}{3x}}{9x\cdot x - 9x\cdot\dfrac{4}{9x}} = \dfrac{9x\cdot 1 - 3\cdot 2}{9x\cdot x - 1\cdot 4}$

$\quad = \dfrac{9x-6}{9x^2-4} = \dfrac{3(3x-2)}{(3x+2)(3x-2)}$

$\quad = \dfrac{3}{3x+2}$

21. $\dfrac{y^{-1}-x^{-1}}{\dfrac{x^2-y^2}{xy}} = \dfrac{\dfrac{1}{y}-\dfrac{1}{x}}{\dfrac{x^2-y^2}{xy}} = \dfrac{\dfrac{1}{y}-\dfrac{1}{x}}{\dfrac{x^2-y^2}{xy}} \cdot \dfrac{xy}{xy}$

$\quad = \dfrac{xy\cdot\dfrac{1}{y}-xy\cdot\dfrac{1}{x}}{xy\cdot\dfrac{x^2-y^2}{xy}} = \dfrac{x\cdot 1 - y\cdot 1}{1\cdot(x^2-y^2)}$

$\quad = \dfrac{x-y}{(x-y)(x+y)} = \dfrac{1}{x+y}$

23. $\dfrac{\dfrac{1}{a-h}-\dfrac{1}{a}}{h}$

$\quad = \dfrac{\dfrac{1}{a-h}\cdot\dfrac{a}{a}-\dfrac{1}{a}\cdot\dfrac{a-h}{a-h}}{h}$ Combining terms in the numerator

$\quad = \dfrac{\dfrac{a-a+h}{a(a-h)}}{h} = \dfrac{\dfrac{h}{a(a-h)}}{h}$

$\quad = \dfrac{h}{a(a-h)}\cdot\dfrac{1}{h}$ Multiplying by the reciprocal of the divisor

$\quad = \dfrac{1\cdot h\cdot 1}{a(a-h)\cdot h}$

$\quad = \dfrac{1}{a(a-h)}$

25. $\dfrac{\dfrac{a^2-4}{a^2+3a+2}}{\dfrac{a^2-5a-6}{a^2-6a-7}}$

$\quad = \dfrac{a^2-4}{a^2+3a+2}\cdot\dfrac{a^2-6a-7}{a^2-5a-6}$ Multiplying by the reciprocal of the divisor

$\quad = \dfrac{(a+2)(a-2)}{(a+2)(a+1)}\cdot\dfrac{(a+1)(a-7)}{(a+1)(a-6)}$

$\quad = \dfrac{(a+2)(a-2)(a+1)(a-7)}{(a+2)(a+1)(a+1)(a-6)}$

$\quad = \dfrac{(a-2)(a-7)}{(a+1)(a-6)}$

27. $\dfrac{\dfrac{x}{x^2+3x-4}-\dfrac{1}{x^2+3x-4}}{\dfrac{x}{x^2+6x+8}+\dfrac{3}{x^2+6x+8}}$

$=\dfrac{\dfrac{x-1}{x^2+3x-4}}{\dfrac{x+3}{x^2+6x+8}}$ Combining terms in the numerator and denominator

$=\dfrac{x-1}{x^2+3x-4}\cdot\dfrac{x^2+6x+8}{x+3}$

$=\dfrac{(x-1)(x+4)(x+2)}{(x+4)(x-1)(x+3)}$

$=\dfrac{\cancel{(x-1)}\,\cancel{(x+4)}\,(x+2)}{\cancel{(x+4)}\,\cancel{(x-1)}\,(x+3)}=\dfrac{x+2}{x+3}$

29. $\dfrac{y+y^{-1}}{y-y^{-1}}=\dfrac{y+\dfrac{1}{y}}{y-\dfrac{1}{y}}$ Rewriting with positive exponents

$=\dfrac{y+\dfrac{1}{y}}{y-\dfrac{1}{y}}\cdot\dfrac{y}{y}$ Multiply by 1, using the LCD

$\dfrac{y\cdot y+\dfrac{1}{y}\cdot y}{y\cdot y-\dfrac{1}{y}\cdot y}=\dfrac{y^2+1}{y^2-1}$

(Although the denominator can be factored, doing so does not lead to further simplification.)

31. $\dfrac{\dfrac{1}{y}+2}{\dfrac{1}{y}-3}=\dfrac{\dfrac{1}{y}+2}{\dfrac{1}{y}-3}\cdot\dfrac{y}{y}$ Multiplying by 1, using the LCD

$=\dfrac{y\cdot\dfrac{1}{y}+y\cdot2}{y\cdot\dfrac{1}{y}-y\cdot3}=\dfrac{1+2y}{1-3y}$

33. $\dfrac{y+y^{-2}}{y-y^{-2}}=\dfrac{y+\dfrac{1}{y^2}}{y-\dfrac{1}{y^2}}$ Rewriting with positive exponents

$=\dfrac{y+\dfrac{1}{y^2}}{y-\dfrac{1}{y^2}}\cdot\dfrac{y^2}{y^2}$ Multiply by 1, using the LCD

$\dfrac{y^2\cdot y+\dfrac{1}{y^2}\cdot y^2}{y^2\cdot y-\dfrac{1}{y^2}\cdot y^2}=\dfrac{y^3+1}{y^3-1}$

(Although the numerator and denominator can be factored, doing so does not lead to further simplification.)

35. $\dfrac{\dfrac{x^2}{x^2-y^2}}{\dfrac{x}{x+y}}=\dfrac{x^2}{x^2-y^2}\cdot\dfrac{y+x}{x}$

$=\dfrac{x\cdot\cancel{x}}{(x+y)(x-y)}\cdot\dfrac{y+x}{\cancel{x}}$

$=\dfrac{x}{x-y}$

37. $\dfrac{\dfrac{x}{5y^3}+\dfrac{3}{10y}}{\dfrac{3}{10y}+\dfrac{x}{5y^3}}=\dfrac{\dfrac{x}{5y^3}+\dfrac{3}{10y}}{\dfrac{x}{5y^3}+\dfrac{3}{10y}}=1$

39. $\dfrac{\dfrac{3}{ab^4}+\dfrac{4}{a^3b}}{ab}=\dfrac{\dfrac{3}{ab^4}+\dfrac{4}{a^3b}}{ab}\cdot\dfrac{a^3b^4}{a^3b^4}$

$=\dfrac{a^3b^4\cdot\dfrac{3}{ab^4}+a^3b^4\cdot\dfrac{4}{a^3b}}{a^3b^4\cdot ab}$

$=\dfrac{a^2\cdot\cancel{ab^4}\cdot\dfrac{3}{\cancel{ab^4}}+b^3\cdot\cancel{a^3b}\cdot\dfrac{4}{\cancel{a^3b}}}{a^3b^4\cdot ab}$

$=\dfrac{a^2\cdot3+b^3\cdot4}{a^4b^5}=\dfrac{3a^2+4b^3}{a^4b^5}$

41. $\dfrac{x-y}{\dfrac{1}{x^3}-\dfrac{1}{y^3}} = \dfrac{x-y}{\dfrac{1}{x^3}-\dfrac{1}{y^3}} \cdot \dfrac{x^3 y^3}{x^3 y^3}$

$= \dfrac{x^3 y^3 (x-y)}{\cancel{x^3} y^3 \cdot \dfrac{1}{\cancel{x^3}} - x^3 \cancel{y^3} \cdot \dfrac{1}{\cancel{y^3}}}$

$= \dfrac{x^3 y^3 (x-y)}{y^3 - x^3} = \dfrac{x^3 y^3 (x-y)}{(y-x)(y^2+xy+x^2)}$

$= \dfrac{x^3 y^3 \cancel{(x-y)}}{-\cancel{(x-y)}(y^2+xy+x^2)}$

$= -\dfrac{x^3 y^3}{y^2+xy+x^2}$

43. $\dfrac{\dfrac{1}{x-2}+\dfrac{3}{x-1}}{\dfrac{2}{x-1}+\dfrac{5}{x-2}}$

$= \dfrac{\dfrac{1}{x-2}+\dfrac{3}{x-1}}{\dfrac{2}{x-1}+\dfrac{5}{x-2}} \cdot \dfrac{(x-2)(x-1)}{(x-2)(x-1)}$

Multiplying by 1, using the LCD

$= \dfrac{\dfrac{1}{x-2}\cdot(x-2)(x-1)+\dfrac{3}{x-1}\cdot(x-2)(x-1)}{\dfrac{2}{x-1}\cdot(x-2)(x-1)+\dfrac{5}{x-2}\cdot(x-2)(x-1)}$

$= \dfrac{x-1+3(x-2)}{2(x-2)+5(x-1)}$

$= \dfrac{x-1+3x-6}{2x-4+5x-5}$

$= \dfrac{4x-7}{7x-9}$

45. $\dfrac{a(a+3)^{-1}-2(a-1)^{-1}}{a(a+3)^{-1}-(a-1)^{-1}}$

$= \dfrac{\dfrac{a}{a+3}-\dfrac{2}{a-1}}{\dfrac{a}{a+3}-\dfrac{1}{a-1}}$

$= \dfrac{\dfrac{a}{a+3}-\dfrac{2}{a-1}}{\dfrac{a}{a+3}-\dfrac{1}{a-1}} \cdot \dfrac{(a+3)(a-1)}{(a+3)(a-1)}$

Multiplying by 1, using the LCD

$= \dfrac{\dfrac{a}{a+3}\cdot(a+3)(a-1)-\dfrac{2}{a-1}\cdot(a+3)(a-1)}{\dfrac{a}{a+3}\cdot(a+3)(a-1)-\dfrac{1}{a-1}\cdot(a+3)(a-1)}$

$= \dfrac{a(a-1)-2(a+3)}{a(a-1)-(a+3)}$

$= \dfrac{a^2-a-2a-6}{a^2-a-a-3} = \dfrac{a^2-3a-6}{a^2-2a-3}$

(The denominator can be factored, but doing so does not lead to further simplification.)

47. $\dfrac{\dfrac{2}{a^2-1}+\dfrac{1}{a+1}}{\dfrac{3}{a^2-1}+\dfrac{2}{a-1}}$

$= \dfrac{\dfrac{2}{(a+1)(a-1)}+\dfrac{1}{a+1}}{\dfrac{3}{(a+1)(a-1)}+\dfrac{2}{a-1}}$

$= \dfrac{\dfrac{2}{(a+1)(a-1)}+\dfrac{1}{a+1}}{\dfrac{3}{(a+1)(a-1)}+\dfrac{2}{a-1}} \cdot \dfrac{(a+1)(a-1)}{(a+1)(a-1)}$

Multiplying by 1, using the LCD

$= \dfrac{\dfrac{2}{(a+1)(a-1)}\cdot(a+1)(a-1)+\dfrac{1}{a+1}\cdot(a+1)(a-1)}{\dfrac{3}{(a+1)(a-1)}\cdot(a+1)(a-1)+\dfrac{2}{a-1}\cdot(a+1)(a-1)}$

$= \dfrac{2+a-1}{3+2(a+1)} = \dfrac{a+1}{3+2a+2} = \dfrac{a+1}{2a+5}$

49. $\dfrac{\dfrac{5}{x^2-4}-\dfrac{3}{x-2}}{\dfrac{4}{x^2-4}-\dfrac{2}{x+2}}=\dfrac{\dfrac{5}{(x+2)(x-2)}-\dfrac{3}{x-2}}{\dfrac{4}{(x+2)(x-2)}-\dfrac{2}{x+2}}$

$=\dfrac{\dfrac{5}{(x+2)(x-2)}-\dfrac{3}{x-2}}{\dfrac{4}{(x+2)(x-2)}-\dfrac{2}{x+2}}\cdot\dfrac{(x+2)(x-2)}{(x+2)(x-2)}$

Multiplying by 1, using the LCD

$=\dfrac{\dfrac{5}{(x+2)(x-2)}\cdot(x+2)(x-2)-\dfrac{3}{x-2}\cdot(x+2)(x-2)}{\dfrac{4}{(x+2)(x-2)}\cdot(x+2)(x-2)-\dfrac{2}{x+2}\cdot(x+2)(x-2)}$

$=\dfrac{5-3(x+2)}{4-2(x-2)}=\dfrac{5-3x-6}{4-2x+4}$

$=\dfrac{-1-3x}{8-2x}$ or $\dfrac{3x+1}{2x-8}$

51. $\dfrac{\dfrac{y^3}{y^2-4}+\dfrac{125}{4-y^2}}{\dfrac{y}{y^2-4}+\dfrac{5}{4-y^2}}=\dfrac{\dfrac{y^3}{y^2-4}-\dfrac{125}{y^2-4}}{\dfrac{y}{y^2-4}-\dfrac{5}{y^2-4}}$

$=\dfrac{\dfrac{y^3-125}{y^2-4}}{\dfrac{y-5}{y^2-4}}$ Subtracting in the numerator and the denominator

$=\dfrac{y^3-125}{y^2\!\!\!\diagup\,4}\cdot\dfrac{y^2\!\!\!\diagup\,4}{y-5}$ Multiplying by the reciprocal of the divisor

$=\dfrac{(y\!\!\!\diagup\,5)(y^2+5y+25)}{y\!\!\!\diagup\,5}$

$=y^2+5y+25$

53. $\dfrac{\dfrac{y^2}{y^2-25}-\dfrac{y}{y-5}}{\dfrac{y}{y^2-25}-\dfrac{1}{y+5}}=\dfrac{\dfrac{y^2}{(y+5)(y-5)}-\dfrac{y}{y-5}}{\dfrac{y}{(y+5)(y-5)}-\dfrac{1}{y+5}}$

$=\dfrac{\dfrac{y^2}{(y+5)(y-5)}-\dfrac{y}{y-5}}{\dfrac{y}{(y+5)(y-5)}-\dfrac{1}{y+5}}\cdot\dfrac{(y+5)(y-5)}{(y+5)(y-5)}$

Multiplying by 1, using the LCD

$=\dfrac{\dfrac{y^2}{(y+5)(y-5)}\cdot(y+5)(y-5)-\dfrac{y}{y-5}\cdot(y+5)(y-5)}{\dfrac{y}{(y+5)(y-5)}\cdot(y+5)(y-5)-\dfrac{1}{y+5}\cdot(y+5)(y-5)}$

$=\dfrac{y^2-y(y+5)}{y-(y-5)}=\dfrac{y^2-y^2-5y}{y-y+5}=\dfrac{-5y}{5}=\dfrac{-1\cdot\cancel{5}\,y}{\cancel{5}}=-y$

55. $\dfrac{\dfrac{a}{a+2}+\dfrac{5}{a}}{\dfrac{a}{2a+4}+\dfrac{1}{3a}}=\dfrac{\dfrac{a}{a+2}+\dfrac{5}{a}}{\dfrac{a}{2(a+2)}+\dfrac{1}{3a}}$

$=\dfrac{\dfrac{a}{a+2}+\dfrac{5}{a}}{\dfrac{a}{2(a+2)}+\dfrac{1}{3a}}\cdot\dfrac{6a(a+2)}{6a(a+2)}$

Multiplying by 1, using the LCD

$=\dfrac{\dfrac{a}{a+2}\cdot6a(a+2)+\dfrac{5}{a}\cdot6a(a+2)}{\dfrac{a}{2(a+2)}\cdot6a(a+2)+\dfrac{1}{3a}\cdot6a(a+2)}$

$=\dfrac{6a^2+30(a+2)}{3a^2+2(a+2)}$

$=\dfrac{6a^2+30a+60}{3a^2+2a+4}$

(Although the numerator can be factored, doing so does not lead to further simplification.)

57. $\dfrac{\dfrac{1}{x^2-3x+2}+\dfrac{1}{x^2-4}}{\dfrac{1}{x^2+4x+4}+\dfrac{1}{x^2-4}}$

$=\dfrac{\dfrac{1}{(x-1)(x-2)}+\dfrac{1}{(x+2)(x-2)}}{\dfrac{1}{(x+2)(x+2)}+\dfrac{1}{(x+2)(x-2)}}$

$=\dfrac{\dfrac{1}{(x-1)(x-2)}+\dfrac{1}{(x+2)(x-2)}}{\dfrac{1}{(x+2)(x+2)}+\dfrac{1}{(x+2)(x-2)}}\cdot$

$\dfrac{(x-1)(x-2)(x+2)(x+2)}{(x-1)(x-2)(x+2)(x+2)}$

Multiplying by 1, using the LCD

$=\dfrac{(x+2)(x+2)+(x-1)(x+2)}{(x-1)(x-2)+(x-1)(x+2)}$

$=\dfrac{x^2+4x+4+x^2+x-2}{x^2-3x+2+x^2+x-2}$

$=\dfrac{2x^2+5x+2}{2x^2-2x}$ or $\dfrac{(2x+1)(x+2)}{2x(x-1)}$

59. $\dfrac{\dfrac{3}{a^2-4a+3}+\dfrac{3}{a^2-5a+6}}{\dfrac{3}{a^2-3a+2}+\dfrac{3}{a^2+3a-10}}$

$=\dfrac{\dfrac{3}{(a-1)(a-3)}+\dfrac{3}{(a-2)(a-3)}}{\dfrac{3}{(a-1)(a-2)}+\dfrac{3}{(a+5)(a-2)}}$

$=\dfrac{\dfrac{3}{(a-1)(a-3)}+\dfrac{3}{(a-2)(a-3)}}{\dfrac{3}{(a-1)(a-2)}+\dfrac{3}{(a+5)(a-2)}}\cdot$

$\dfrac{(a-1)(a-3)(a-2)(a+5)}{(a-1)(a-3)(a-2)(a+5)}$

Multiplying by 1, using the LCD

$=\dfrac{3(a-2)(a+5)+3(a-1)(a+5)}{3(a-3)(a+5)+3(a-1)(a-3)}$

$=\dfrac{3\big[(a-2)(a+5)+(a-1)(a+5)\big]}{3\big[(a-3)(a+5)+(a-1)(a-3)\big]}$

$=\dfrac{\cancel{3}\big[(a-2)(a+5)+(a-1)(a+5)\big]}{\cancel{3}\big[(a-3)(a+5)+(a-1)(a-3)\big]}$

$=\dfrac{a^2+3a-10+a^2+4a-5}{a^2+2a-15+a^2-4a+3}$

$=\dfrac{2a^2+7a-15}{2a^2-2a-12}$ or $\dfrac{(2a-3)(a+5)}{2(a-3)(a+2)}$

61. $\dfrac{\dfrac{y}{y^2-4}-\dfrac{2y}{y^2+y-6}}{\dfrac{2y}{y^2+y-6}-\dfrac{y}{y^2-4}}$

Observe that $\dfrac{y}{y^2-4}-\dfrac{2y}{y^2+y-6}=$

$-\left(\dfrac{2y}{y^2+y-6}-\dfrac{y}{y^2-4}\right).$

Then, the numerator and denominator are opposites and thus their quotient is -1.

63. $\dfrac{t+5+\dfrac{3}{t}}{t+2+\dfrac{1}{t}}=\dfrac{t+5+\dfrac{3}{t}}{t+2+\dfrac{1}{t}}\cdot\dfrac{t}{t}=\dfrac{t\cdot t+t\cdot 5+\cancel{t}\cdot\dfrac{3}{\cancel{t}}}{t\cdot t+t\cdot 2+\cancel{t}\cdot\dfrac{1}{\cancel{t}}}$

$=\dfrac{t^2+5t+3}{t^2+2t+1}$

65. $\dfrac{x-2-\dfrac{1}{x}}{x-5-\dfrac{4}{x}} = \dfrac{x-2-\dfrac{1}{x}}{x-5-\dfrac{4}{x}} \cdot \dfrac{x}{x} = \dfrac{x\cdot x - x\cdot 2 - \cancel{x}\cdot\dfrac{1}{\cancel{x}}}{x\cdot x - x\cdot 5 - \cancel{x}\cdot\dfrac{4}{\cancel{x}}}$

$\qquad = \dfrac{x^2-2x-1}{x^2-5x-4}$

67. *Thinking and Writing Exercise.*

69. $3x-5+2(4x-1)=12x-3$

$\qquad 3x-5+8x-2=12x-3$

$\qquad 11x-7=12x-3$

$\qquad -7+3=12x-11x$

$\qquad -4=x$

71. $\dfrac{3}{4}x-\dfrac{5}{8}=\dfrac{3}{8}x+\dfrac{7}{4}$

$\qquad 8\cdot\left[\dfrac{3}{4}x-\dfrac{5}{8}\right]=8\cdot\left[\dfrac{3}{8}x+\dfrac{7}{4}\right]$

$\qquad 8\cdot\dfrac{3}{4}x-8\cdot\dfrac{5}{8}=8\cdot\dfrac{3}{8}x+8\cdot\dfrac{7}{4}$

$\qquad \cancel{4}\cdot 2\cdot\dfrac{3}{\cancel{4}}x-\cancel{8}\cdot\dfrac{5}{\cancel{8}}=\cancel{8}\cdot\dfrac{3}{\cancel{8}}x+\cancel{4}\cdot 2\cdot\dfrac{7}{\cancel{4}}$

$\qquad 2\cdot 3\cdot x-1\cdot 5=1\cdot 3\cdot x+2\cdot 7$

$\qquad 6x-5=3x+14$

$\qquad 6x-3x=14+5$

$\qquad 3x=19$

$\qquad x=\dfrac{19}{3}$

73. $x^2-7x+12=0$

$\qquad (x-3)(x-4)=0$

$\qquad \Rightarrow x-3=0 \text{ or } x-4=0$

$\qquad \Rightarrow x=3 \text{ or } x=4$

75. *Thinking and Writing Exercise.*

77. $\dfrac{\dfrac{A}{B}}{\dfrac{C}{D}} = \dfrac{\dfrac{A}{B}}{\dfrac{C}{D}} \cdot \dfrac{BD}{BD} = \dfrac{\cancel{B}D\cdot\dfrac{A}{\cancel{B}}}{B\cancel{D}\cdot\dfrac{C}{\cancel{D}}}$

$\qquad = \dfrac{D\cdot A}{B\cdot C} = \dfrac{A\cdot D}{B\cdot C} = \dfrac{A}{B}\cdot\dfrac{D}{C}$

79. $\dfrac{30{,}000\cdot\dfrac{0.075}{12}}{\left(1+\dfrac{0.075}{12}\right)^{120}-1} = \dfrac{30{,}000(0.00625)}{(1+0.00625)^{120}-1}$

$\qquad = \dfrac{187.5}{(1+0.00625)^{120}-1}$

$\qquad \approx \dfrac{187.5}{2.112064637-1}$

$\qquad \approx \dfrac{187.5}{1.112064637}$

$\qquad \approx 168.61$

Alexis' monthly investment is \$168.61.

81. $\dfrac{5x^{-2}+10x^{-1}y^{-1}+5y^{-2}}{3x^{-2}-3y^{-2}} = \dfrac{\dfrac{5}{x^2}+\dfrac{10}{xy}+\dfrac{5}{y^2}}{\dfrac{3}{x^2}-\dfrac{3}{y^2}} \cdot \dfrac{x^2y^2}{x^2y^2}$

$\qquad = \dfrac{\dfrac{5}{x^2}\cdot x^2y^2+\dfrac{10}{xy}\cdot x^2y^2+\dfrac{5}{y^2}\cdot x^2y^2}{\dfrac{3}{x^2}\cdot x^2y^2-\dfrac{3}{y^2}\cdot x^2y^2}$

$\qquad = \dfrac{5y^2+10xy+5x^2}{3y^2-3x^2}$

$\qquad = \dfrac{5(y+x)(y+x)}{3(y+x)(y-x)} = \dfrac{5(y+x)}{3(y-x)}$

83. $\dfrac{\dfrac{x-1}{x-1}-1}{\dfrac{x+1}{x-1}+1} = \dfrac{\dfrac{x-1}{x-1}-1}{\dfrac{x+1}{x-1}+1} \cdot \dfrac{x-1}{x-1}$

$\qquad = \dfrac{\cancel{(x-1)}\dfrac{x-1}{\cancel{x-1}}-(x-1)1}{\cancel{(x-1)}\dfrac{x+1}{\cancel{x-1}}+(x-1)1} = \dfrac{x-1-(x-1)}{x+1+x-1}$

$\qquad = \dfrac{x-1-x+1}{x+1+x-1} = \dfrac{0}{2x} = 0 \text{ for } x\neq 0,1$

Therefore, for $x\neq 0$ or 1,

$\qquad \left[\dfrac{\dfrac{x-1}{x-1}-1}{\dfrac{x+1}{x-1}+1}\right]^5 = 0^5 = 0$

85. $\dfrac{\dfrac{z}{1-\dfrac{z}{2+2z}}-2z}{\dfrac{2z}{5z-2}-3}=\dfrac{\dfrac{z}{1-\dfrac{z}{2+2z}}\cdot\dfrac{2+2z}{2+2z}-2z}{\dfrac{2z}{5z-2}-3}$

$=\dfrac{\dfrac{z(2+2z)}{2+2z-z}-2z}{\dfrac{2z}{5z-2}-3}=\dfrac{\dfrac{z(2+2z)}{2+z}-2z}{\dfrac{2z}{5z-2}-3}$

$=\dfrac{\dfrac{z(2+2z)}{2+z}-2z}{\dfrac{2z}{5z-2}-3}\cdot\dfrac{(2+z)(5z-2)}{(2+z)(5z-2)}$

$=\dfrac{z(2+2z)(5z-2)-2z(2+z)(5z-2)}{2z(2+z)-3(2+z)(5z-2)}$

$=\dfrac{z(5z-2)[(2+2z)-2(2+z)]}{(2+z)[2z-3(5z-2)]}$

$=\dfrac{z(5z-2)[2+2z-4-2z]}{(2+z)[2z-15z+6]}=\dfrac{-2z(5z-2)}{(2+z)(6-13z)}$

$=\dfrac{2z(5z-2)}{(z+2)(13z-6)}$

87. $f(x)=\dfrac{3}{x},\, f(x+h)=\dfrac{3}{x+h}$

$=\dfrac{f(x+h)-f(x)}{h}=\dfrac{\dfrac{3}{x+h}-\dfrac{3}{x}}{h}$

$=\dfrac{\dfrac{3x-3(x+h)}{x(x+h)}}{h}$

$=\dfrac{3x-3(x+h)}{x(x+h)}\cdot\dfrac{1}{h}$

$=\dfrac{3x-3x-3h}{xh(x+h)}$

$=\dfrac{-3h}{xh(x+h)}$

$=\dfrac{-3\cancel{h}}{x\cancel{h}(x+h)}$

$=\dfrac{-3}{x(x+h)}$

89. To avoid division by zero in $\dfrac{1}{x}$ and $\dfrac{8}{x^2}$ we must exclude 0 from the domain of F. To avoid division by zero in the complex fraction we solve:

$2-\dfrac{8}{x^2}=0$

$2x^2-8=0$

$2(x^2-4)=0$

$2(x+2)(x-2)=0$

$x+2=0\ \ or\ \ x-2=0$

$x=-2\ or\ \ \ \ x=2$

The domain of $F=\{x|x$ is a real number and $x\neq 0$ and $x\neq -2$ and $x\neq 2\}$.

91. $g(a)=\dfrac{a+3}{a-1}$

$g(g(a))=\dfrac{\dfrac{a+3}{a-1}+3}{\dfrac{a+3}{a-1}-1}=\dfrac{\dfrac{a+3}{a-1}+3}{\dfrac{a+3}{a-1}-1}\cdot\dfrac{a-1}{a-1}$

$=\dfrac{a+3+3(a-1)}{a+3-(a-1)}=\dfrac{a+3+3a-3}{a+3-a+1}$

$=\dfrac{4a}{4}=a$

for a in the domain of $g(x)$. That is, for $a\neq 1$.

Mid-Chapter Review

Guided Solutions:

1. $\dfrac{a^2}{a-10}\div\dfrac{a^2+5a}{a^2-100}=\dfrac{a^2}{a-10}\cdot\dfrac{a^2-100}{a^2+5a}$

$=\dfrac{a\cdot a\cdot(a+10)\cdot(a-10)}{(a-10)\cdot a\cdot(a+5)}$

$=\dfrac{a(a-10)}{a(a-10)}\cdot\dfrac{a(a+10)}{a+5}$

$=\dfrac{a(a+10)}{a+5}$

2. $\dfrac{2}{x}+\dfrac{1}{x^2+x}=\dfrac{2}{x}+\dfrac{1}{x(x+1)}$

$=\dfrac{2}{x}\cdot\dfrac{(x+1)}{(x+1)}+\dfrac{1}{x(x+1)}$

$=\dfrac{2x+2}{x(x+1)}+\dfrac{1}{x(x+1)}$

$=\dfrac{2x+3}{x(x+1)}$

Mixed Review:

1. $\dfrac{3}{5x}+\dfrac{2}{x^2}=\dfrac{3}{5x}\cdot\dfrac{x}{x}+\dfrac{2}{x^2}\cdot\dfrac{5}{5}$

$=\dfrac{3x}{5x^2}+\dfrac{10}{5x^2}=\dfrac{3x+10}{5x^2}$

2. $\dfrac{3}{5x}\cdot\dfrac{2}{x^2}=\dfrac{3\cdot2}{5x\cdot x^2}=\dfrac{6}{5x^3}$

3. $\dfrac{3}{5x}\div\dfrac{2}{x^2}=\dfrac{3}{5x}\cdot\dfrac{x^2}{2}$

$=\dfrac{3}{5\cdot\cancel{x}}\cdot\dfrac{x\cdot\cancel{x}}{2}$

$=\dfrac{3x}{10}$

4. $\dfrac{3}{5x}-\dfrac{2}{x^2}=\dfrac{3}{5x}\cdot\dfrac{x}{x}-\dfrac{2}{x^2}\cdot\dfrac{5}{5}$

$=\dfrac{3x}{5x^2}-\dfrac{10}{5x^2}=\dfrac{3x-10}{5x^2}$

5. $\dfrac{2x-6}{5x+10}\cdot\dfrac{x+2}{6x-12}=\dfrac{\cancel{2}\cdot(x-3)}{5\cdot(\cancel{x+2})}\cdot\dfrac{\cancel{x+2}}{\cancel{2}\cdot3\cdot(x-2)}$

$=\dfrac{x-3}{15(x-2)}$

6. $\dfrac{2}{x-5}\div\dfrac{6}{x-5}=\dfrac{\cancel{2}}{\cancel{x-5}}\cdot\dfrac{\cancel{x-5}}{\cancel{2}\cdot3}=\dfrac{1}{3}$

7. $\dfrac{x}{x+2}-\dfrac{1}{x-1}=\dfrac{x}{x+2}\cdot\dfrac{x-1}{x-1}-\dfrac{1}{x-1}\cdot\dfrac{x+2}{x+2}$

$=\dfrac{x(x-1)-(x+2)}{(x+2)(x-1)}=\dfrac{x^2-x-x-2}{(x+2)(x-1)}$

$=\dfrac{x^2-2x-2}{(x+2)(x-1)}$

8. $\dfrac{2}{x+3}+\dfrac{3}{x+4}=\dfrac{2}{x+3}\cdot\dfrac{x+4}{x+4}+\dfrac{3}{x+4}\cdot\dfrac{x+3}{x+3}$

$=\dfrac{2(x+4)+3(x+3)}{(x+3)(x+4)}=\dfrac{2x+8+3x+9}{(x+3)(x+4)}$

$=\dfrac{5x+17}{(x+3)(x+4)}$

9. $\dfrac{5}{2x-1}+\dfrac{10x}{1-2x}=\dfrac{5}{2x-1}-\dfrac{10x}{2x-1}=\dfrac{5-10x}{2x-1}$

$=\dfrac{5(1-2x)}{2x-1}=-\dfrac{5\cdot\cancel{(2x-1)}}{\cancel{(2x-1)}}=-5$

10. $\dfrac{3}{x-4}-\dfrac{2}{4-x}=\dfrac{3}{x-4}+\dfrac{2}{x-4}=\dfrac{5}{x-4}$

11. $=\dfrac{(x-2)(2x+3)}{(x+1)(x-5)}\div\dfrac{(x-2)(x+1)}{(x-5)(x+3)}$

$=\dfrac{\cancel{(x-2)}(2x+3)}{(x+1)\cancel{(x-5)}}\cdot\dfrac{\cancel{(x-5)}(x+3)}{\cancel{(x-2)}(x+1)}$

$=\dfrac{(2x+3)(x+3)}{(x+1)^2}$

12. $\dfrac{a}{6a-9b}-\dfrac{b}{4a-6b}$

$=\dfrac{a}{3(2a-3b)}\cdot\dfrac{2}{2}-\dfrac{b}{2(2a-3b)}\cdot\dfrac{3}{3}$

$=\dfrac{2a}{6(2a-3b)}-\dfrac{3b}{6(2a-3b)}=\dfrac{\cancel{2a-3b}}{6\cancel{(2a-3b)}}$

$=\dfrac{1}{6}$

13. $=\dfrac{x^2-16}{x^2-x}\cdot\dfrac{x^2}{x^2-5x+4}$

$=\dfrac{(x+4)\cancel{(x-4)}}{\cancel{x}(x-1)}\cdot\dfrac{\cancel{x}\cdot x}{\cancel{(x-4)}(x-1)}$

$=\dfrac{x(x+4)}{(x-1)^2}$

14. $\dfrac{x+1}{x^2-7x+10}+\dfrac{3}{x^2-x-2}$

$=\dfrac{x+1}{(x-5)(x-2)}\cdot\dfrac{x+1}{x+1}+\dfrac{3}{(x-2)(x+1)}\cdot\dfrac{x-5}{x-5}$

$=\dfrac{(x+1)^2+3(x-5)}{(x-5)(x-2)(x+1)}=\dfrac{x^2+2x+1+3x-15}{(x-5)(x-2)(x+1)}$

$=\dfrac{x^2+5x-14}{(x-5)(x-2)(x+1)}=\dfrac{(x+7)\cancel{(x-2)}}{(x-5)\cancel{(x-2)}(x+1)}$

$=\dfrac{x+7}{(x-5)(x+1)}$

15. $(t^2 + t - 20) \cdot \dfrac{t+5}{t-4} = \dfrac{(t+5)\,(t-4)}{1} \cdot \dfrac{t+5}{(t-4)}$

$= (t+5)^2$

16. $\dfrac{a^2 - 2a + 1}{a^2 - 4} \div (a^2 - 3a + 2)$

$= \dfrac{a^2 - 2a + 1}{a^2 - 4} \cdot \dfrac{1}{a^2 - 3a + 2}$

$= \dfrac{(a-1)(a-1)}{(a+2)(a-2)} \cdot \dfrac{1}{(a-2)(a-1)}$

$= \dfrac{a-1}{(a+2)(a-2)^2}$

17. $\dfrac{\dfrac{3}{z} + \dfrac{2}{y}}{\dfrac{4}{z} - \dfrac{1}{y}} = \dfrac{\dfrac{3}{z} + \dfrac{2}{y}}{\dfrac{4}{z} - \dfrac{1}{y}} \cdot \dfrac{yz}{yz} = \dfrac{yz \cdot \dfrac{3}{z} + yz \cdot \dfrac{2}{y}}{yz \cdot \dfrac{4}{z} - yz \cdot \dfrac{1}{y}}$

$= \dfrac{3y + 2z}{4y - z}$

18. $\dfrac{xy^{-1} + x^{-1}}{2x^{-1} + 4y^{-1}} = \dfrac{\dfrac{x}{y} + \dfrac{1}{x}}{\dfrac{2}{x} + \dfrac{4}{y}} \cdot \dfrac{xy}{xy}$

$= \dfrac{x^2 + y}{2y + 4x}$, or $\dfrac{x^2 + y}{2(2x + y)}$

19. $\dfrac{\dfrac{y}{y^2-4} + \dfrac{5}{4-y^2}}{\dfrac{y^2}{y^2-4} + \dfrac{25}{4-y^2}} = \dfrac{\dfrac{y}{y^2-4} - \dfrac{5}{y^2-4}}{\dfrac{y^2}{y^2-4} - \dfrac{25}{y^2-4}}$

$= \dfrac{\dfrac{y}{y^2-4} - \dfrac{5}{y^2-4}}{\dfrac{y^2}{y^2-4} - \dfrac{25}{y^2-4}} \cdot \dfrac{y^2-4}{y^2-4}$

$= \dfrac{y-5}{y^2-25} = \dfrac{y-5}{(y+5)(y-5)}$

$= \dfrac{1}{y+5}$

20. $\dfrac{\dfrac{1}{a} - \dfrac{1}{b}}{\dfrac{1}{a^3} - \dfrac{1}{b^3}} = \dfrac{\dfrac{1}{a} - \dfrac{1}{b}}{\dfrac{1}{a^3} - \dfrac{1}{b^3}} \cdot \dfrac{a^3 b^3}{a^3 b^3}$

$= \dfrac{a^3 b^3 \cdot \dfrac{1}{a} - a^3 b^3 \cdot \dfrac{1}{b}}{a^3 b^3 \cdot \dfrac{1}{a^3} - a^3 b^3 \cdot \dfrac{1}{b^3}}$

$= \dfrac{a^2 b^3 - a^3 b^2}{b^3 - a^3} = \dfrac{a^2 b^2 (b - a)}{(b-a)(a^2 + ab + b^2)}$

$= \dfrac{a^2 b^2}{a^2 + ab + b^2}$

Exercise Set 6.4

1. Equation

3. Expression

5. Equation

7. Equation

9. Expression

11. $\dfrac{3}{5} - \dfrac{2}{3} = \dfrac{x}{6}$, LCD is 30

$30 \cdot \dfrac{3}{5} - 30 \cdot \dfrac{2}{3} = 30 \cdot \dfrac{x}{6}$

$5 \cdot 6 \cdot \dfrac{3}{5} - 5 \cdot 10 \cdot \dfrac{2}{3} = 6 \cdot 5 \cdot \dfrac{x}{6}$

$18 - 20 = 5x$

$5x = -2$

$x = -\dfrac{2}{5}$

Check.

$$\dfrac{3}{5} - \dfrac{2}{3} = \dfrac{x}{6}$$

$\dfrac{3}{5} \cdot \dfrac{3}{3} - \dfrac{2}{3} \cdot \dfrac{5}{5}$	$-\dfrac{2}{5}$
$\dfrac{9-10}{15}$	$\dfrac{-\dfrac{2}{5}}{6}$
$-\dfrac{1}{15}$	$-\dfrac{2}{5} \cdot \dfrac{1}{6}$
	$-\dfrac{2}{5} \cdot \dfrac{1}{2 \cdot 3}$
	$-\dfrac{1}{15}$
	TRUE

13. $\dfrac{1}{8}+\dfrac{1}{12}=\dfrac{1}{t}$, LCD is $24t$

Because $\dfrac{1}{t}$ is undefined at $t=0$, note at the

onset that $t \neq 0$.

$$\dfrac{1}{8}+\dfrac{1}{12}=\dfrac{1}{t}$$

$$24t\cdot\dfrac{1}{8}+24t\cdot\dfrac{1}{12}=24t\cdot\dfrac{1}{t}$$

$$\cancel{8}\cdot3t\cdot\dfrac{1}{\cancel{8}}+\cancel{12}\cdot2t\cdot\dfrac{1}{\cancel{12}}=24\cdot\cancel{t}\cdot\dfrac{1}{\cancel{t}}$$

$$3t+2t=24$$

$$5t=24$$

$$t=\dfrac{24}{5}\neq 0$$

Check.

$$\dfrac{1}{8}+\dfrac{1}{12}=\dfrac{1}{t}$$

$\dfrac{1}{8}+\dfrac{1}{12}$	$\dfrac{1}{\frac{24}{5}}$
$\dfrac{1}{8}\cdot\dfrac{3}{3}+\dfrac{1}{12}\cdot\dfrac{2}{2}$	$1\cdot\dfrac{5}{24}$
$\dfrac{3+2}{24}$	$\dfrac{5}{24}$
$\dfrac{5}{24}$	TRUE

15. $\dfrac{x}{6}-\dfrac{6}{x}=0$, LCD is $6x$

Because $\dfrac{6}{x}$ is undefined at $x=0$, note at the

onset that $x \neq 0$.

$$\dfrac{x}{6}-\dfrac{6}{x}=0$$

$$\cancel{6}x\cdot\dfrac{x}{\cancel{6}}-6\cancel{x}\cdot\dfrac{6}{\cancel{x}}=0$$

$$x^2-36=0$$

$$(x+6)(x-6)=0$$

$$x=-6,6$$

Check

$$\dfrac{x}{6}-\dfrac{6}{x}=\dfrac{6}{6}-\dfrac{6}{6}=0 \ \sqrt{}$$

$$\dfrac{x}{6}-\dfrac{6}{x}=\dfrac{-6}{6}-\dfrac{6}{-6}=-\dfrac{6}{6}+\dfrac{6}{6}=0 \ \sqrt{}$$

17. $\dfrac{2}{3}-\dfrac{1}{t}=\dfrac{7}{3t}$

Because $\dfrac{1}{t}$ is undefined when t is 0, we note

at the outset that $t \neq 0$. Then we multiply

both sides by the LCD, $3 \cdot t = 3t$.

$$3t\left(\dfrac{2}{3}-\dfrac{1}{t}\right)=3t\cdot\dfrac{7}{3t}$$

$$3t\cdot\dfrac{2}{3}-3t\cdot\dfrac{1}{t}=3t\cdot\dfrac{7}{3t}$$

$$2t-3=7$$

$$2t=10$$

$$t=5$$

Check.

$$\dfrac{2}{3}-\dfrac{1}{t}=\dfrac{7}{3t}$$

$\dfrac{2}{3}-\dfrac{1}{5}$	$\dfrac{7}{3\cdot 5}$
$\dfrac{10}{15}-\dfrac{3}{15}$	$\dfrac{7}{15}$
$\dfrac{7}{15}$	$\dfrac{7}{15}$ TRUE

The solution is 5.

19. $\dfrac{n+2}{n-6}=\dfrac{1}{2}$, LCD is $2(n-6)$

Because $\dfrac{n+2}{n-6}$ is undefined at $n=6$, note at

the onset that $n \neq 6$.

$$\dfrac{n+2}{n-6}=\dfrac{1}{2}$$

$$2\cancel{(n-6)}\dfrac{n+2}{\cancel{n-6}}=\cancel{2}(n-6)\dfrac{1}{\cancel{2}}$$

$$2(n+2)=n-6$$

$$2n+4=n-6$$

$$2n-n=-6-4$$

$$n=-10$$

Check

$$\dfrac{(-10)+2}{(-10)-6}=\dfrac{-8}{-16}=\dfrac{1}{2} \ \sqrt{}$$

21.
$$\frac{12}{x} = \frac{x}{3}$$
$$3x \cdot \frac{12}{x} = 3x \cdot \frac{x}{3}$$
$$36 = x^2$$
$$x^2 - 36 = 0$$
$$(x+6)(x-6) = 0$$
$$x = -6, 6$$

23. $\frac{2}{6} + \frac{1}{2x} = \frac{1}{3}$

Because $\frac{1}{2x}$ is undefined when $x = 0$, we note at the outset that $x \neq 0$. We multiply both sides by the LCD, $6x$.

$$6x\left(\frac{2}{6} + \frac{1}{2x}\right) = 6x \cdot \frac{1}{3}$$
$$6x \cdot \frac{2}{6} + 6x \cdot \frac{1}{2x} = 6x \cdot \frac{1}{3}$$
$$2x + 3 = 2x$$
$$3 = 0$$

We get a false equation. The given equation has no solution.

25. $y + \frac{4}{y} = -5$

Because $\frac{4}{y}$ is undefined when y is 0, we note at the outset that $y \neq 0$. Then we multiply both sides by the LCD, y.

$$y\left(y + \frac{4}{y}\right) = y(-5)$$
$$y \cdot y + y \cdot \frac{4}{y} = -5y$$
$$y^2 + 4 = -5y$$
$$y^2 + 5y + 4 = 0$$
$$(y+1)(y+4) = 0$$
$$y + 1 = 0 \ or \ y + 4 = 0$$
$$y = -1 \ or \quad y = -4$$

Both values check. The solutions are –1 and –4.

27. $x - \frac{12}{x} = 4$

Because $\frac{12}{x}$ is undefined when $x = 0$, we note that $x \neq 0$. LCD $= x$.

$$x \cdot \left(x - \frac{12}{x}\right) = x \cdot 4$$
$$x \cdot x - x \cdot \frac{12}{x} = x \cdot 4$$
$$x^2 - 12 = 4x$$
$$x^2 - 4x - 12 = 0$$
$$(x-6)(x+2) = 0$$
$$x - 6 = 0 \ or \ x + 2 = 0$$
$$x = 6 \ or \quad x = -2$$

Both values check. The solutions are –2 and 6.

29.
$$\frac{y+3}{y-3} = \frac{6}{y-3}$$
$$(y-3)\frac{y+3}{y-3} = (y-3)\frac{6}{y-3}$$
$$y + 3 = 6$$
$$y = 3$$

But $y = 3$ is not a valid solution, because the expressions in the original equation are undefined at $y = 3$. Therefore, since the equation had no other solutions, the equation has no solution.

31. $\frac{x}{x-5} = \frac{25}{x^2 - 5x}$

To assure that neither denominator is 0, we note that $x \neq 5$ and $x \neq 0$. We then multiply both sides by the LCD, $x(x-5)$.

$$x(x-5) \cdot \frac{x}{x-5} = x(x-5) \cdot \frac{25}{x(x-5)}$$
$$x^2 = 25$$
$$x^2 - 25 = 0$$
$$(x+5)(x-5) = 0$$
$$x + 5 = 0 \ or \ x - 5 = 0$$
$$x = -5 \ or \quad x = 5$$

Recall that, because of the restriction above, 5 cannot be a solution . A check confirms this.

Check $x = 5$.

$$\frac{x}{x-5} = \frac{25}{x^2 - 5x}$$

$\frac{5}{5-5}$	$\frac{25}{5^2 - 5 \cdot 5}$
$\frac{5}{0}$	$\frac{5}{0}$ UNDEFINED

Also, Check $x = -5$.

$$\frac{x}{x-5} = \frac{25}{x^2 - 5x}$$

$\frac{-5}{-5-5}$	$\frac{25}{(-5)^2 - 5(-5)}$
$\frac{-5}{-10}$	$\frac{25}{25 + 25}$
$\frac{1}{2}$	$\frac{25}{50}$
$\frac{1}{2}$	$\frac{1}{2}$ TRUE

The solution is -5.

33.
$$\frac{n+1}{n+2} = \frac{n-3}{n+1}$$
$$(n+1)(n+2)\frac{n+1}{n+2} = (n+1)(n+2)\frac{n-3}{n+1}$$
$$(n+1)^2 = (n+2)(n-3)$$
$$n^2 + 2n + 1 = n^2 - n - 6$$
$$2n + n = -6 - 1$$
$$3n = -7$$
$$n = -\frac{7}{3}$$

35. $\dfrac{x^2 + 4}{x-1} = \dfrac{5}{x-1}$

To assure that neither denominator is 0, we note at the outset that $x \neq 1$. Then we multiply both sides by the LCD, $x - 1$.

$$(x-1)\cdot\frac{x^2+4}{x-1} = (x-1)\cdot\frac{5}{x-1}$$
$$x^2 + 4 = 5$$
$$x^2 - 1 = 0$$
$$(x+1)(x-1) = 0$$
$$x + 1 = 0 \ or \ x - 1 = 0$$
$$x = -1 \ or \quad x = 1$$

Recall that, because of the restriction above, 1 cannot be a solution. The number -1 checks and is the solution.

We might also observe that since the denominators are the same, the numerators must be equal. Solving $x^2 + 4 = 5$, we get $x = -1$ or $x = 1$ as shown above. Again because of the restriction $x \neq 1$, only -1 is a solution of the equation.

37. $\dfrac{6}{a+1} = \dfrac{a}{a-1}$

To assure that neither denominator is 0, we note at the outset that $x \neq -1$ and $x \neq 1$. Then we multiply both sides by the LCD, $(a+1)(a-1)$.

$$(a+1)(a-1)\cdot\frac{6}{a+1} = (a+1)(a-1)\cdot\frac{a}{a-1}$$
$$6(a-1) = a(a+1)$$
$$6a - 6 = a^2 + a$$
$$0 = a^2 - 5a + 6$$
$$0 = (a-2)(a-3)$$
$$a - 2 = 0 \ or \ a - 3 = 0$$
$$a = 2 \ or \quad a = 3$$

Both values check. The solutions are 2 and 3.

39. $\dfrac{60}{t-5} - \dfrac{18}{t} = \dfrac{40}{t}$

To assure that none of the denominators are 0, we note at the outset that $t \neq 5$ and $t \neq 0$. Then we multiply on both sides by the LCD, $t(t-5)$.

$$t(t-5)\left(\frac{60}{t-5} - \frac{18}{t}\right) = t(t-5)\cdot\frac{40}{t}$$
$$60t - 18(t-5) = 40(t-5)$$
$$60t - 18t + 90 = 40t - 200$$
$$2t = -290$$
$$t = -145$$

This value checks. The solution is -145.

41. $\dfrac{3}{x-3}+\dfrac{5}{x+2}=\dfrac{5x}{x^2-x-6}$

To assure that none of the denominators are 0, we note that $x \neq 3$ and $x \neq -2$. LCD is $(x-3)(x+2)$.

$(x-3)(x+2)\left(\dfrac{3}{x-3}+\dfrac{5}{x+2}\right)=$

$\qquad (x-3)(x+2)\cdot\dfrac{5x}{x^2-x-6}$

$3(x+2)+5(x-3)=5x$

$3x+6+5x-15=5x$

$3x=9$

$x=3$

Recall that, because of the restriction above, 3 cannot be a solution. The equation has no solution.

43. $\dfrac{3}{x}+\dfrac{x}{x+2}=\dfrac{4}{x^2+2x}$

$\dfrac{3}{x}+\dfrac{x}{x+2}=\dfrac{4}{x(x+2)}$

To assure that none of the denominators are 0, we note at the outset that $x \neq 0$ and $x \neq -2$. Then we multiply both sides by the LCD, $x(x+2)$.

$x(x+2)\left(\dfrac{3}{x}+\dfrac{x}{x+2}\right)=x(x+2)\cdot\dfrac{4}{x(x+2)}$

$3(x+2)+x\cdot x=4$

$3x+6+x^2=4$

$x^2+3x+2=0$

$(x+1)(x+2)=0$

$x+1=0\ \ or\ x+2=0$

$x=-1\ or\ \ \ \ x=-2$

Recall that, because of the restrictions above, -2 cannot be a solution. The number -1 checks. The solution is -1.

45. $\dfrac{5}{x+2}-\dfrac{3}{x-2}=\dfrac{2x}{4-x^2}$

$\dfrac{5}{x+2}-\dfrac{3}{x-2}=\dfrac{2x}{(2+x)(2-x)}$

$\dfrac{5}{x+2}+\dfrac{3}{2-x}=\dfrac{2x}{(2+x)(2-x)}$

$\boxed{-\dfrac{3}{x-2}=\dfrac{3}{2-x}}$

First note that $x \neq -2$ and $x \neq 2$. Then multiply on both sides by the LCD, $(2+x)(2-x)$.

$(2+x)(2-x)\left(\dfrac{5}{x+2}+\dfrac{3}{2-x}\right)=$

$\qquad (2+x)(2-x)\cdot\dfrac{2x}{(2+x)(2-x)}$

$5(2-x)+3(2+x)=2x$

$10-5x+6+3x=2x$

$16-2x=2x$

$16=4x$

$4=x$

This value checks. The solution is 4.

47. $\dfrac{3}{x^2-6x+9}+\dfrac{x-2}{3x-9}=\dfrac{x}{2x-6}$

$\dfrac{3}{(x-3)(x-3)}+\dfrac{x-2}{3(x-3)}=\dfrac{x}{2(x-3)}$

Note that $x \neq 3$.

$6(x-3)(x-3)\left(\dfrac{3}{(x-3)(x-3)}+\dfrac{x-2}{3(x-3)}\right)=$

$\qquad 6(x-3)(x-3)\cdot\dfrac{x}{2(x-3)}$

$6\cdot 3+2(x-3)(x-2)=3x(x-3)$

$18+2x^2-10x+12=3x^2-9x$

$0=x^2+x-30$

$0=(x+6)(x-5)$

$x=-6\ or\ x=5$

Both values check. The solutions are -6 and 5.

49. We find all values of a for which

$2a - \dfrac{15}{a} = 7$. First note that $a \neq 0$. Then

multiply on both sides by the LCD, a.

$$a\left(2a - \frac{15}{a}\right) = a \cdot 7$$

$$a \cdot 2a - a \cdot \frac{15}{a} = 7a$$

$$2a^2 - 15 = 7a$$

$$2a^2 - 7a - 15 = 0$$

$$(2a+3)(a-5) = 0$$

$$a = -\frac{3}{2} \ or \ a = 5$$

Both values check. The solutions are

$-\dfrac{3}{2}$ and 5.

51. We find all values of a for which $\dfrac{a-5}{a+1} = \dfrac{3}{5}$.

First note that $a \neq -1$. Then multiply on both

sides by the LCD, $5(a+1)$.

$$5(a+1) \cdot \frac{a-5}{a+1} = 5(a+1) \cdot \frac{3}{5}$$

$$5(a-5) = 3(a+1)$$

$$5a - 25 = 3a + 3$$

$$2a = 28$$

$$a = 14$$

This value checks. The solution is 14.

53. We find all values of a for which

$\dfrac{12}{a} - \dfrac{12}{2a} = 8$. First note that $a \neq 0$. then

multiply on both sides by the LCD, $2a$.

$$2a\left(\frac{12}{a} - \frac{12}{2a}\right) = 2a \cdot 8$$

$$2a \cdot \frac{12}{a} - 2a \cdot \frac{12}{2a} = 16a$$

$$24 - 12 = 16a$$

$$12 = 16a$$

$$\frac{3}{4} = a$$

This value checks. The solution is $\dfrac{3}{4}$.

55.
$$f(a) = g(a)$$

$$\frac{a+1}{3} - 1 = \frac{a-1}{2}$$

$$6 \cdot \frac{a+1}{3} - 6 \cdot 1 = 6 \cdot \frac{a-1}{2}$$

$$\cancel{6} \cdot 2 \cdot \frac{a+1}{\cancel{3}} - 6 \cdot 1 = \cancel{6} \cdot 3 \cdot \frac{a-1}{\cancel{2}}$$

$$2(a+1) - 6 = 3(a-1)$$

$$2a + 2 - 6 = 3a - 3$$

$$2a - 4 = 3a - 3$$

$$-4 + 3 = 3a - 2a$$

$$-1 = a$$

57. First note that $g(x)$ can easily be written as a

single term:

$$g(x) = \frac{4}{x-3} + \frac{2x}{x-3} = \frac{4+2x}{x-3} = \frac{2x+4}{x-3}.$$

Therefore:

$$f(a) = g(a)$$

$$\frac{12}{a^2 - 6a + 9} = \frac{2a+4}{a-3}$$

$$\frac{12}{(a-3)^2} = \frac{2a+4}{a-3}$$

$$(a-3)^2 \frac{12}{(a-3)^2} = (a-3)^2 \frac{2a+4}{a-3}$$

$$\cancel{(a-3)^2} \frac{12}{\cancel{(a-3)^2}} = (a-3)\cancel{(a-3)} \frac{2a+4}{\cancel{a-3}}$$

$$12 = (a-3)(2a+4)$$

$$(a-3)(2a+4) - 12 = 0$$

$$2a^2 + 4a - 6a - 12 - 12 = 0$$

$$2a^2 - 2a - 24 = 0$$

$$(2a+6)(a-4) = 0$$

$$2a + 6 = 0 \ \ or \ \ a - 4 = 0$$

$$2a = -6 \qquad\qquad a = 4$$

$$a = -3$$

Both -3 and 4 are in the domains of

f and g, so both are valid solutions.

59. *Thinking and Writing Exercise.*

61. Let x = the smallest number. Then $x + 2$ = the larger number. Then:

$$x + (x + 2) = 276$$
$$x + x + 2 = 276$$
$$2x + 2 = 276$$
$$2x = 274$$
$$x = \frac{274}{2} = 137.$$

The two numbers are $x = 137$, and $x + 2 = 139$.

63. The height of the triangle is defined in terms of its base. Therefore let b = the base, in cm, of the triangle. Then $b + 3$ = the height, in cm, of the triangle. Then:

$$A = \frac{1}{2}bh$$

$$\frac{1}{2}b(b+3) = 54$$

$$2 \cdot \frac{1}{2}b(b+3) = 2 \cdot 54$$

$$b(b+3) = 108$$

$$b^2 + 3b - 108 = 0$$

$$(b+12)(b-9) = 0$$

$$b = -12 \text{ or } b = 9$$

Since the base of the triangle cannot be a negative number, the base of the triangle is 9 cm, and the height is $9 + 3 = 12$ cm.

65. $\dfrac{0.9 \text{ cm}}{24 - 9 \text{ day}} = \dfrac{0.9 \text{ cm}}{15 \text{ day}} = 0.06$ cm per day

67. *Thinking and Writing Exercise.*

69. $$f(a) = g(a)$$

$$\frac{a - \frac{2}{3}}{a + \frac{1}{2}} = \frac{a + \frac{2}{3}}{a - \frac{3}{2}}$$

$$\frac{a - \frac{2}{3}}{a + \frac{1}{2}} \cdot \frac{6}{6} = \frac{a + \frac{2}{3}}{a - \frac{3}{2}} \cdot \frac{6}{6}$$

$$\frac{6a - \frac{2}{3} \cdot 6}{6a + \frac{1}{2} \cdot 6} = \frac{6a + \frac{2}{3} \cdot 6}{6a - \frac{3}{2} \cdot 6}$$

$$\frac{6a - 4}{6a + 3} = \frac{6a + 4}{6a - 9}$$

$$\frac{6a - 4}{3(2a + 1)} = \frac{6a + 4}{3(2a - 3)}$$

To assure that neither denominator is 0, we note at the outset that $a \neq -\frac{1}{2}$ and $a \neq \frac{3}{2}$. Then we multiply both sides by the LCD, $3(2a+1)(2a-3)$.

$$3(2a+1)(2a-3) \cdot \frac{6a-4}{3(2a+1)} =$$
$$3(2a+1)(2a-3) \cdot \frac{6a+4}{3(2a-3)}$$

$$(2a-3)(6a-4) = (2a+1)(6a+4)$$
$$12a^2 - 26a + 12 = 12a^2 + 14a + 4$$
$$-26a + 12 = 14a + 4$$
$$-40a + 12 = 4$$
$$-40a = -8$$
$$a = \frac{1}{5}$$

This number checks. For $a = \frac{1}{5}, f(a) = g(a).$

71.
$$f(a) = g(a)$$
$$\frac{a+3}{a+2} - \frac{a+4}{a+3} = \frac{a+5}{a+4} - \frac{a+6}{a+5}$$

Note that $a \neq -2$ and $a \neq -3$ and $a \neq -4$ and $a \neq -5$.

$$(a+2)(a+3)(a+4)(a+5)\left(\frac{a+3}{a+2} - \frac{a+4}{a+3}\right) =$$
$$(a+2)(a+3)(a+4)(a+5)\left(\frac{a+5}{a+4} - \frac{a+6}{a+5}\right)$$
$$(a+3)(a+4)(a+5)(a+3) -$$
$$(a+2)(a+4)(a+5)(a+4) =$$
$$(a+2)(a+3)(a+5)(a+5) -$$
$$(a+2)(a+3)(a+4)(a+6)$$
$$a^4 + 15a^3 + 83a^2 + 201a + 180 -$$
$$\left(a^4 + 15a^3 + 82a^2 + 192a + 160\right) =$$
$$a^4 + 15a^3 + 81a^2 + 185a + 150 -$$
$$\left(a^4 + 15a^3 + 80a^2 + 180a + 144\right)$$
$$a^2 + 9a + 20 = a^2 + 5a + 6$$
$$4a = -14$$
$$a = -\frac{7}{2}$$

This value checks. When $a = -\frac{7}{2}$, $f(a) = g(a)$.

73. Set $f(a)$ equal to $g(a)$ and solve for a.
$$\frac{0.793}{a} + 18.15 = \frac{6.034}{a} - 43.17, \ a \neq 0$$

Multiply by the LCD, a.
$$a \cdot \frac{0.793}{a} + a \cdot 18.15 = a \cdot \frac{6.034}{a} - a \cdot 43.17$$
$$0.793 + 18.15a = 6.034 - 43.17a$$
$$61.32a = 5.241$$
$$a = \frac{5.241}{61.32} \approx 0.0854696673$$

75. $\dfrac{x^2 + 6x - 16}{x - 2} = \dfrac{(x+8)(x-2)}{x-2} = x+8$, for $x \neq 2$

The expression on the left side of the equation is undefined at $x = 2$, but for all other values of x, the equation is an identity.

77.
$$1 + \frac{x-1}{x-3} = \frac{2}{x-3} - x$$
$$(x-3) \cdot 1 + (x-3) \cdot \frac{x-1}{x-3}$$
$$= (x-3) \cdot \frac{2}{x-3} - (x-3) \cdot x$$
$$x - 3 + x - 1 = 2 - x(x-3)$$
$$2x - 4 = 2 - x^2 + 3x$$
$$x^2 + 2x - 3x - 4 - 2 = 0$$
$$x^2 - x - 6 = 0$$
$$(x-3)(x+2) = 0$$
$$x = 3 \text{ or } x = -2$$

By clearing denominators we find two solutions, 3 and -2. However, 3 is not in the domain of the original equation and we must discard this solution as extraneous. Therefore, the equation has one solution: $x = -2$.

79.
$$\frac{5-3a}{a^2 + 4a + 3} - \frac{2a+2}{a+3} = \frac{3-a}{a+1}$$
$$\frac{5-3a}{(a+3)(a+1)} - \frac{2(a+1)}{a+3} = \frac{3-a}{a+1}$$
$$(a+3)(a+1) \frac{5-3a}{(a+3)(a+1)}$$
$$- (a+3)(a+1) \frac{2(a+1)}{a+3}$$
$$= (a+3)(a+1) \frac{3-a}{a+1}$$
$$5 - 3a - 2(a+1)^2 = (a+3)(3-a)$$
$$5 - 3a - 2(a^2 + 2a + 1) = 3a - a^2 + 9 - 3a$$
$$5 - 3a - 2a^2 - 4a - 2 = 9 - a^2$$
$$3 - 7a - 2a^2 = 9 - a^2$$
$$9 - a^2 = 3 - 7a - 2a^2$$
$$9 - a^2 - 3 + 7a + 2a^2 = 0$$
$$a^2 + 7a + 6 = 0$$
$$(a+6)(a+1) = 0$$
$$a = -6 \text{ or } a = -1$$

By clearing denominators we find two solutions, -6 and -1. However, -1 is not in the domain of the original equation and we must discard this solution as extraneous. Therefore, the equation has one solution: $a = -6$.

81. $\dfrac{\dfrac{1}{3}}{x} = \dfrac{1 - \dfrac{1}{x}}{x}$

$$\frac{1}{3} \cdot \frac{1}{x} = \frac{1 - \dfrac{1}{x}}{x} \cdot \frac{x}{x}$$

$$\frac{1}{3x} = \frac{x \cdot 1 - \cancel{x} \cdot \dfrac{1}{\cancel{x}}}{x \cdot x}$$

$$\frac{1}{3x} = \frac{x - 1}{x^2}$$

$$3x^2 \cdot \frac{1}{3x} = 3x^2 \cdot \frac{x - 1}{x^2}$$

$$\cancel{3x} \cdot x \cdot \frac{1}{\cancel{3x}} = 3\cancel{x} \cdot \frac{x - 1}{\cancel{x}}$$

$$x = 3(x - 1)$$

$$x = 3x - 3$$

$$3x - 3 = x$$

$$3x - x = 3$$

$$2x = 3$$

$$x = \frac{3}{2}$$

Exercise Set 6.5

1. $\dfrac{1 \text{ cake}}{2 \text{ hr}} = \dfrac{1}{2}$ cake per hour

3. $\dfrac{1}{2}$ cake per hour $+ \dfrac{1}{3}$ cake per hour

$= \left(\dfrac{1}{2} + \dfrac{1}{3}\right)$ cake per hour $= \dfrac{5}{6}$ cake per hour

5. $\dfrac{1 \text{ yard}}{3 \text{ hr}} = \dfrac{1}{3}$ lawn per hour

7. *Familiarize.* The job takes Trey 8 hours working alone and Matt 6 hours working alone. Then in 1 hour, Trey does $\dfrac{1}{8}$ of the job, and Matt does $\dfrac{1}{6}$ of the job. Working together they can do $\dfrac{1}{8} + \dfrac{1}{6}$ of the job in 1 hour. Let t represent the number of hours for Trey and Matt, working together, to do the job.

Translate. We want to find t such that

$t\left(\dfrac{1}{8}\right) + t\left(\dfrac{1}{6}\right) = 1$, or $\dfrac{t}{8} + \dfrac{t}{6} = 1$, where 1 represents one entire job.

Carry out. We solve the equation.

$$\frac{t}{8} + \frac{t}{6} = 1, \text{ LCD is } 24$$

$$24\left(\frac{t}{8} + \frac{t}{6}\right) = 24 \cdot 1$$

$$3t + 4t = 24$$

$$7t = 24$$

$$t = \frac{24}{7}$$

Check. In $\dfrac{24}{7}$ hours, Trey will do

$\dfrac{1}{8} \cdot \dfrac{24}{7}$, or $\dfrac{3}{7}$ of the job, and Matt will do

$\dfrac{1}{6} \cdot \dfrac{24}{7}$, or $\dfrac{4}{7}$ of the job. Together they do

$\dfrac{3}{7} + \dfrac{4}{7}$, or 1 entire job. The answer checks.

State. It will take $\dfrac{24}{7}$ hr, or $3\dfrac{3}{7}$ hr, for Trey and Matt, working together, to do the job.

9. *Familiarize.* The pool can be filled in 12 hours by only the pipe and in 30 hours with only the hose. Then in 1 hour the pipe fills $\dfrac{1}{12}$ of the pool, and the hose fills $\dfrac{1}{30}$.

Working together, they fill $\dfrac{1}{12} + \dfrac{1}{30}$ of the pool in an hour. Let t equal the number of hours it takes them to fill the pool together.

Translate. We want to find t such that

$t\left(\dfrac{1}{12}\right) + t\left(\dfrac{1}{30}\right) = 1$, or $\dfrac{t}{12} + \dfrac{t}{30} = 1$

Carry out. We solve the equation. LCD = 60.

$$60\left(\frac{t}{12} + \frac{t}{30}\right) = 60 \cdot 1$$

$$5t + 2t = 60$$

$$7t = 60$$

$$t = \frac{60}{7}$$

Check. The pipe fills $\dfrac{1}{12} \cdot \dfrac{60}{7}$, or $\dfrac{5}{7}$, and the hose fills $\dfrac{1}{30} \cdot \dfrac{60}{7}$, or $\dfrac{2}{7}$. Working together, they fill $\dfrac{5}{7} + \dfrac{2}{7} = 1$, or the entire pool in $\dfrac{60}{7}$ hr.

State. Working together, the pipe and hose

11. **Familiarize.** The $\dfrac{1}{4}$ HP does $\dfrac{1}{70}$ of the job in 1 minute, and the $\dfrac{1}{3}$ HP does $\dfrac{1}{30}$ of the job. Working together, they can do $\dfrac{1}{70} + \dfrac{1}{30}$ of the job in 1 minute. Let t equal the number of minutes it takes them working together.

Translate. We want to find t such that
$$t\left(\dfrac{1}{70}\right) + t\left(\dfrac{1}{30}\right) = 1, \text{ or } \dfrac{t}{70} + \dfrac{t}{30} = 1$$

Carry out. We solve the equation. LCD is 210.
$$210\left(\dfrac{t}{70} + \dfrac{t}{30}\right) = 210 \cdot 1$$
$$3t + 7t = 210$$
$$10t = 210$$
$$t = \dfrac{210}{10} = 21$$

Check. In 21 minutes the $\dfrac{1}{4}$ HP empties
$$\dfrac{1}{70} \cdot 21 = \dfrac{1}{7 \cdot 10} \cdot 7 \cdot 3 = \dfrac{3}{10} \text{ of the basement,}$$
and the $\dfrac{1}{3}$ HP empties
$$\dfrac{1}{30} \cdot 21 = \dfrac{1}{3 \cdot 10} \cdot 3 \cdot 7 = \dfrac{7}{10} \text{ of the basement.}$$
Together they remove water from
$$\dfrac{3}{10} + \dfrac{7}{10} = \dfrac{10}{10} = 1 \text{ basement. The answer}$$
checks.

State. It will take 21 minutes for the machines working together to remove the water.

13. **Familiarize.** Let t = the time, in minutes, it takes the MP C7500 to copy the proposal. Then $3t$ = the time, in minutes, it takes the MP C2500 to copy the proposal. In 1 minute, the C7500 does $\dfrac{1}{t}$ of the job, and the C2500 does $\dfrac{1}{3t}$ of the job.

Translate. Together, they can do the entire job in 1.5 min, so we want to find t such that
$$1.5 \cdot \dfrac{1}{t} + 1.5 \cdot \dfrac{1}{3t} = 1$$

Carry out. Solve the equation.
$$1.5 \cdot \dfrac{1}{t} + 1.5 \cdot \dfrac{1}{3t} = 1$$
$$10 \cdot 1.5 \cdot \dfrac{1}{t} + 10 \cdot 1.5 \cdot \dfrac{1}{3t} = 10 \cdot 1$$
$$15 \cdot \dfrac{1}{t} + 3 \cdot 5 \cdot \dfrac{1}{3t} = 10 \Rightarrow \dfrac{15}{t} + \dfrac{5}{t} = 10$$
$$t \cdot \dfrac{15}{t} + t \cdot \dfrac{5}{t} = t \cdot 10 \Rightarrow 15 + 5 = 10t$$
$$t = \dfrac{20}{10} = 2$$

Check. The C7500 does the job in 2 min; in 1.5 min, it does
$$1.5 \cdot \dfrac{1}{2} = \dfrac{1.5}{2} = \dfrac{1.5 \cdot 10}{2 \cdot 10} = \dfrac{15}{20} = \dfrac{3}{4} \text{ of the job.}$$
The C2500 does the job in $3 \cdot 2 = 6$ min; in 1.5 min. it does
$$1.5 \cdot \dfrac{1}{6} = \dfrac{1.5}{6} = \dfrac{1.5 \cdot 10}{6 \cdot 10} = \dfrac{15}{60} = \dfrac{5}{20} = \dfrac{1}{4} \text{ of the}$$
job. Together, they do $\dfrac{3}{4} + \dfrac{1}{4} = 1$ job in 1.5 min. The answer checks.

State. The C7500 can do the entire job in 2 min, and the C2500 in 6 min.

15. **Familiarize.** Let t = the time, in minutes, it takes the Airgle 750 to purify the air. Then $t + 20$ the time, in minutes, it takes the Healthmate 400 to purify the air. In 1 minute, the Airgle 750 does $\dfrac{1}{t}$ of the job, and the Healthmate 400 does $\dfrac{1}{t + 20}$ of the job.

Translate. Together, they can do the entire job in 10.5 min, so we want to find t such that
$$10.5 \cdot \dfrac{1}{t} + 10.5 \cdot \dfrac{1}{t + 20} = 1$$

Carry out. Solve the equation.

$$10.5 \cdot \frac{1}{t} + 10.5 \cdot \frac{1}{t+20} = 1$$

$$10 \cdot 10.5 \cdot \frac{1}{t} + 10 \cdot 10.5 \cdot \frac{1}{t+20} = 10 \cdot 1$$

$$\frac{105}{t} + \frac{105}{t+20} = 10$$

$$\cancel{t}(t+20) \cdot \frac{105}{\cancel{t}} + t\cancel{(t+20)} \cdot \frac{105}{\cancel{t+20}} = t(t+20) \cdot 10$$

$$105(t+20) + 105t = 10t(t+20)$$

$$105t + 2100 + 105t = 10t^2 + 200t$$

$$10t^2 + 200t - 105t - 105t - 2100 = 0$$

$$10t^2 - 10t - 2100 = 0$$

$$t^2 - t - 210 = 0$$

$$(t+14)(t-15)t = -14 \text{ or } t = 15$$

Check. Since $t > 0$ the Airgle 750 does the job in 15 min; in 10.5 min, it does

$$10.5 \cdot \frac{1}{15} = \frac{10.5}{15} = \frac{10.5 \cdot 10}{15 \cdot 10} = \frac{105}{150} = \frac{21}{30} = \frac{7}{10}$$

of the job. The Healthmate 400 does the job in $15 + 20 = 35$ min; in 10.5 min. it does

$$10.5 \cdot \frac{1}{35} = \frac{10.5}{35} = \frac{10.5 \cdot 10}{35 \cdot 10} = \frac{105}{350} = \frac{21}{70} = \frac{3}{10}$$

of the job. Together, they do $\frac{7}{10} + \frac{3}{10} = 1$ job in 10.5 min. The answer checks.

State. The Airgle 750 can do the entire job in 15 min, and the Healthmate 400 in 35 min.

17. **Familiarize.** Let t represent the number of hours that the Erickson takes, working alone to douse the fire. Then $4t$ represents the number of hours the S-58T requires. In 1 hour, the Erickson does $\frac{1}{t}$ of the job, and the S-58T does $\frac{1}{4t}$ of the job.

Translate. Together they can do the job in 8 hr, so we want to find t such that

$$8 \cdot \frac{1}{t} + 8 \cdot \frac{t}{4t} = 1, \text{ or } \frac{8}{t} + \frac{2}{t} = 1$$

Carry out. We solve the equation.

$$\frac{8}{t} + \frac{2}{t} = 1, \text{ LCD is } t$$

$$t\left(\frac{8}{t} + \frac{2}{t}\right) = t \cdot 1$$

$$8 + 2 = t$$

$$t = 10$$

Check. The Erickson takes 10 hr, working alone; it can do $8 \cdot \frac{1}{10} = \frac{4}{5}$ of the job in 8 hr.

The S-58T does $8 \cdot \frac{1}{40} = \frac{1}{5}$ of the job in 8 hr.

Together, they do $\frac{4}{5} + \frac{1}{5}$, the entire job.

State. It takes the Erickson 10 hr, and the S-58T 40 hr, working alone.

19. **Familiarize.** We will convert hours to minutes:

2 hr $= 2 \cdot 60$ min $= 120$ min

2 hr 55 min $= 120$ min $+ 55$ min $= 175$ min
Let t = the number of minutes it takes Deb to do the job alone. Then $t + 120$ = the number of minutes it takes Dawn alone. In 1 hour (60 minutes) Deb does $\frac{1}{t}$ and Dawn does $\frac{1}{t+120}$ of the job.

Translate. In 175 min Dawn and Deb will complete one entire job, so we have

$$175\left(\frac{1}{t}\right) + 175\left(\frac{1}{t+120}\right) = 1,$$

or $\dfrac{175}{t} + \dfrac{175}{t+120} = 1.$

Carry out. We solve the equation. Multiply on both sides by the LCD, $t(t+120)$.

$$t(t+120) \cdot \left(\frac{175}{t} + \frac{175}{t+120}\right) = t(t+120)(1)$$

$$175(t+120) + 175t = t^2 + 120t$$

$$175t + 21,000 + 175t = t^2 + 120t$$

$$0 = t^2 - 230t - 21,000$$

$$0 = (t-300)(t+70)$$

$t = 300 \text{ or } t = -70$

Check. Since negative time has no meaning in this problem, -70 is not a solution of the original problem. If the job takes Deb 300 min and it take Dawn $300 + 120 = 420$ min, then in 175 min they would complete

$$175\left(\frac{1}{300}\right)+175\left(\frac{1}{420}\right)=\frac{7}{12}+\frac{5}{12}=1\text{ job.}$$

The results check.

State. It would take Deb 300 min, or 5 hr, to do the job alone.

21. **Familiarize.** Let r = the speed of the AMTRAK train in km/h. Then $r-14$ = the speed of the B&M train in km/h. We complete the table.

Distance = Rate · Time

	Distance (in km)	Speed (in km/h)	Time (in hours)
B&M	330	$r-14$	$\dfrac{330}{r-14}$
AMTRAK	400	r	$\dfrac{400}{r}$

Note: "Filling In" this specific chart is part of the exercise.

Translate. Since the times are the same (equal), we have the equation:

$$\frac{330}{r-14}=\frac{400}{r}$$

Carry out. We solve the equation.

$$\frac{330}{r-14}=\frac{400}{r},\ \text{LCD is } r(r-14)$$

$$r(r-14)\cdot\frac{330}{r-14}=r(r-14)\cdot\frac{400}{r}$$

$$330r=400(r-14)$$

$$330r=400r-5600$$

$$5600=70r$$

$$80=r$$

Check. If the AMTRAK train's speed is 80 km/h, then the B&M train's speed is $80-14$, or 66 km/h. Traveling 400 km at 80 km/h takes $\frac{400}{80}=5$ hr., and traveling 330 km. at 66 km/h. takes $\frac{330}{66}=5$ hr. Since the times are equal, the answer checks.

State. The speed of the AMTRAK train is 80 km/h and the speed of the B&M train is 66 km/h.

23. **Familiarize.** We first make a drawing. Let r = the kayak's speed in still water in mph. Then $r-3$ = the speed upstream and $r+3$ = the speed downstream.

Upstream 4 miles $r-3$ mph →

← 10 miles　$r+3$ mph　Downstream

We organize the information in a table. The time is the same both upstream and downstream so we use t for each time.

	Distance	Speed	Time
Upstream	4	$r-3$	t
Downstream	10	$r+3$	t

Translate. Using the formula Time = Distance/Rate in each row of the table and the fact that the times are the same, we can write an equation.

$$\frac{4}{r-3}=\frac{10}{r+3}$$

Carry out. We solve the equation.

$$\frac{4}{r-3}=\frac{10}{r+3},\ \text{LCD is }(r-3)(r+3)$$

$$(r-3)(r+3)\cdot\frac{4}{r-3}=(r-3)(r+3)\cdot\frac{10}{r+3}$$

$$4(r+3)=10(r-3)$$

$$4r+12=10r-30$$

$$42=6r$$

$$7=r$$

Check. If $r=7$ mph, then $r-3$ is 4 mph and $r+3$ is 10 mph. the time upstream is $\frac{4}{4}$, or 1 hour. The time downstream is $\frac{10}{10}$, or 1 hour. Since the times are the same, the answer checks.

State. The speed of the kayak in still water is 7 mph.

25. Note that 38 mi is 7 mi less than 45 mi and that the local bus travels 7 mph slower than the express. Then the express travels 45 mi in one hr, or 45 mph, and the local bus travels 38 mi in one hr, or 38 mph.

27. *Familiarize.* We first make a drawing. Let r = Kaitlyn's speed on a non-moving sidewalk in ft/sec. Then his speed moving forward on the moving sidewalk is $r + 1.7$ and his speed in the opposite direction is $r - 1.7$.

Forward	$r+1.7$	120 ft →

← 52 ft. $r-1.7$ Opposite Direction		

We organize the information in a table. The time is the same both forward and in the opposite direction, so we use t for each time.

	Distance	Speed	Time
Forward	120	$r+1.7$	t
Opposite Direction	52	$r-1.7$	t

Translate. Using the formula $T = D/R$ in each row of the table and the fact the times are the same (equal), we have:

$$\frac{120}{r+1.7} = \frac{52}{r-1.7}$$

Carry out. We solve the equation.

$$\frac{120}{r+1.7} = \frac{52}{r-1.7}; \text{ LCD is } (r+1.7)(r-1.7)$$

$$(r+1.7)(r-1.7) \cdot \frac{120}{r+1.7} =$$

$$(r+1.7)(r-1.7) \cdot \frac{52}{r-1.7}$$

$$120(r-1.7) = 52(r+1.7)$$

$$120r - 204 = 52r + 88.4$$

$$68r = 292.4$$

$$r = 4.3$$

Check. If Kaitlyn's speed on a non-moving sidewalk is 4.3 ft/sec, then his speed moving forward is 4.3 + 1.7, or 6 ft/sec, and his speed moving in the opposite direction on the sidewalk is 4.3 – 1.7, or 2.6 ft/sec. Moving 120 ft at 6 ft/sec takes $\frac{120}{6} = 20$ sec. Moving 52 ft at 2.6 ft/sec takes $\frac{52}{2.6} = 20$ sec. Since the times are the same, the answer checks.

State. Kaitlyn would be walking 4.3 ft/sec on a non-moving sidewalk.

29. Let t = the time, in hours, it takes Caledonia to drive to town. Then $t + 1$ = the time, in hours, it takes Manley to drive to town. Then:

$$r_M = r_C$$

$$\frac{d_M}{t_M} = \frac{d_C}{t_C}$$

$$\frac{20}{t+1} = \frac{15}{t}$$

$$20t = 15(t+1)$$

$$20t = 15t + 15$$

$$20t - 15t = 15$$

$$5t = 15$$

$$t = 3$$

It takes Caledonia 3 hours to drive to town.

31. *Familiarize.* Let r = speed of the river in km/h. Then $15 + r$ = the speed downstream, and $15 - r$ = the speed upstream. Using a table to organize the information, we have:

	Distance	Speed	Time
Downstream	140	$15+r$	t
Upstream	35	$15-r$	t

Translate. Using $T = D/R$ in each row of the table and the fact that the times are the same, we can write the equation.

$$\frac{140}{15+r} = \frac{35}{15-r}$$

Carry out. We solve the equation.

$$\frac{140}{15+r} = \frac{35}{15-r}, \text{ LCD is } (15+r)(15-r)$$

$$(15+r)(15-r) \cdot \frac{140}{15+r} =$$

$$(15+r)(15-r) \cdot \frac{35}{15-r}$$

$$140(15-r) = 35(15+r)$$

$$2100 - 140r = 525 + 35r$$

$$1575 = 175r$$

$$9 = r$$

Check. The speed downstream is 15 + 9, or 24 km/h and the speed upstream is 15 – 9, or 6 km/h. Traveling 140 km at 24 km/h takes $\frac{140}{24} = 5\frac{5}{6}$ hr and traveling 35 km at 6 km/h takes $\frac{35}{6} = 5\frac{5}{6}$ hr. Since the times are the same, the number checks.

State. The speed of the river is 9 km/h.

33. *Familiarize.* Let c = the speed of the current, in km/h. Then $7 + c$ = the speed downriver and $7 - c$ = the speed upriver. We organize the information in a table.

	Distance	Speed	Time
Downriver	45	$7 + c$	t_1
Upriver	45	$7 - c$	t_2

Translate. Using the formula Time = Distance/Rate we see that $t_1 = \dfrac{45}{7+c}$ and $t_2 = \dfrac{45}{7-c}$. The total time upriver and back is 14 hr, so $t_1 + t_2 = 14$, or $\dfrac{45}{7-c} + \dfrac{45}{7+c} = 14$

Carry out. We solve the equation. Multiply both sides by the LCD, $(7+c)(7-c)$.

$$(7+c)(7-c)\left(\frac{45}{7+c} + \frac{45}{7-c}\right) =$$
$$(7+c)(7-c)14$$
$$45(7-c) + 45(7+c) = 14(49 - c^2)$$
$$315 - 45c + 315 + 45c = 686 - 14c^2$$
$$14c^2 - 56 = 0$$
$$14(c+2)(c-2) = 0$$
$$c + 2 = 0 \ \ or \ c - 2 = 0$$
$$c = -2 \ or \ \ \ c = 2$$

Check. Since speed cannot be negative in this problem, -2 cannot be a solution of the original problem. If the speed of the current is 2 km/h, the barge travels upriver at $7 - 2$, or 5 km/h. At this rate it takes $\dfrac{45}{5}$, or 9 hr, to travel 45 km. The barge travels downriver at $7 + 2$, or 9 km/h. At this rate it takes $\dfrac{45}{9}$, or 5 hr, to travel 45 km. The total travel time is $9 + 5$, or 14 hr. The answer checks.

State. The speed of the current is 2 km/h.

35. *Familiarize.* Let r = the speed at which the train actually traveled in mph, and let t = the actual travel time in hours. We organize the information in a table.

	Distance	Speed	Time
Actual speed	120	r	t
Faster speed	120	$r + 10$	$t - 2$

Translate. From the first row of the table we have $120 = rt$, and from the second row we have $120 = (r + 10)(t - 2)$. Solving the first equation for t, we have $t = \dfrac{120}{r}$. Substituting for t in the second equation, we have $120 = (r + 10)\left(\dfrac{120}{r} - 2\right)$.

Carry out. We solve the equation.

$$120 = (r+10)\left(\frac{120}{r} - 2\right)$$
$$120 = 120 - 2r + \frac{1200}{r} - 20$$
$$20 = -2r + \frac{1200}{r}$$
$$r \cdot 20 = r\left(-2r + \frac{1200}{r}\right)$$
$$20r = -2r^2 + 1200$$
$$2r^2 + 20r - 1200 = 0$$
$$2(r^2 + 10r - 600) = 0$$
$$2(r + 30)(r - 20) = 0$$
$$r = -30 \ or \ r = 20$$

Check. Since speed cannot be negative in this problem, -30 cannot be a solution of the original problem. If the speed is 20 mph, it takes $\dfrac{120}{20}$, or 6 hr, to travel 120 mi. If the speed is 10 mph faster, or 30 mph, it takes $\dfrac{120}{30}$, or 4 hr, to travel 120 mi. Since 4 hr is 2 hr less time than 6 hr, the answer checks.

State. The speed was 20 mph.

37. Write a proportion and then solve it.

$$\frac{b}{6} = \frac{7}{4}$$

$$b = \frac{7}{4} \cdot 6$$

$$b = \frac{42}{4}, \text{ or } 10.5$$

$\left(\text{Note that the proportions } \frac{6}{b} = \frac{4}{7}, \frac{b}{7} = \frac{6}{4},\right.$

$\left.\text{or } \frac{7}{b} = \frac{4}{6} \text{ could also be used.}\right)$

39. We write a proportion and then solve it.

$$\frac{4}{f} = \frac{6}{4}$$

$$4f \cdot \frac{4}{f} = 4f \cdot \frac{6}{4}$$

$$16 = 6f$$

$$\frac{8}{3} = f \qquad \text{Simplifying}$$

$\left(\text{One of the following proportions could}\right.$
$\left.\text{also be used:}\right.$

$$\frac{f}{4} = \frac{4}{6}, \frac{4}{f} = \frac{9}{6}, \frac{f}{4}$$

$$\left. = \frac{6}{9}, \frac{4}{9} = \frac{f}{6}, \frac{9}{4} = \frac{6}{f}\right)$$

41. We write a proportion and then solve it.

$$\frac{\text{Inches}}{\text{Feet}} \qquad \frac{P}{15} = \frac{1}{4}$$

$$P = \frac{1}{4} \cdot 15$$

$$P = \frac{15}{4}, \text{ or } 3\frac{3}{4} \text{ in.}$$

43. We write a proportion and then solve it.

$$\frac{\text{Inches}}{\text{Feet}} \qquad \frac{5}{r} = \frac{1}{4}$$

$$4r \cdot \frac{5}{r} = 4r \cdot \frac{1}{4}$$

$$20 = r$$

$$r = 20 \text{ ft}$$

45. We write a proportion and then solve it.

$$\frac{l}{10} = \frac{6}{4}$$

$$l = \frac{6}{4} \cdot 10$$

$$l = \frac{60}{4} = 15 \text{ ft}$$

47. We write a proportion and then solve it.

$$\frac{5}{7} = \frac{9}{r}$$

$$7r \cdot \frac{5}{7} = 7r \cdot \frac{9}{r}$$

$$5r = 63$$

$$r = \frac{63}{5}, \text{ or } 12.6$$

49. Write a proportion and then solve it.

$$\frac{n}{30} = \frac{384}{8}$$

$$n = \frac{384}{8} \cdot 30 = 96 \cdot 15$$

$$n = 1440 \text{ messages}$$

51. Let x = number of photos taken. Then:

$$\frac{234 \text{ photos}}{14 \text{ days}} = \frac{x \text{ photos}}{42 \text{ days}}$$

$$x = \frac{234 \cdot 42}{14} = 702 \text{ photos}$$

53. Let x = width of the wing in cm. Then:

$$\frac{24 \text{ cm}}{180 \text{ cm}} = \frac{x \text{ cm}}{200 \text{ cm}}$$

$$x = \frac{24 \cdot 200}{180} = \frac{24 \cdot 20\cancel{0}}{18\cancel{0}} = \frac{480}{18} = \frac{80}{3} = 26\frac{2}{3} \text{ cm}$$

55. Let x = number of defective drives.

Then: $\dfrac{7}{150} = \dfrac{x}{2700} \Rightarrow x = \dfrac{7 \cdot 2700}{150} = 126$

At the same rate 126 defective drives would be expected.

57. Let x = the ounces of water needed. Then:

$$\frac{12}{8} = \frac{x}{5} \Rightarrow x = \frac{12 \cdot 5}{8} = 7\frac{1}{2}$$

At the same rate the Bolognese would need $7\frac{1}{2}$ ounces of water.

59. Let x = the size of the pod. Then:

$$\frac{x}{27} = \frac{40}{12} \Rightarrow x = \frac{40 \cdot 27}{12} = 90$$

The pod has about 90 whales.

61. **Familiarize.** The ratio of the weight of an object on the moon to the weight of an object on Earth is 0.16 to 1.

  a) We wish to find how much a 12-ton rocket would weigh on the moon.

  b) We wish to find how much a 180-lb astronaut would weigh on the moon.

We can determine ratios.

$$\frac{0.16}{1} \quad \frac{R}{12} \quad \frac{A}{180}$$

Translate. Assuming the ratios are the same, we can translate to proportions.

  a) $\dfrac{\text{Weight on moon}}{\text{Weight on Earth}} \;\; \dfrac{0.16}{1} = \dfrac{R}{12}$

  b) $\dfrac{\text{Weight on moon}}{\text{Weight on Earth}} \;\; \dfrac{0.16}{1} = \dfrac{A}{180}$

Carry out. We solve each proportion.

  a) $\dfrac{0.16}{1} = \dfrac{R}{12}$

$$12(0.16) = R$$
$$1.92 = R$$

  b) $\dfrac{0.16}{1} = \dfrac{A}{180}$

$$120(0.16) = A$$
$$28.8 = A$$

Check. $\dfrac{0.16}{1} = 0.16, \quad \dfrac{1.92}{12} = 0.16,$

$$\text{and } \dfrac{28.8}{180} = 0.16.$$

The ratios are the same.

State.

  a) A 12-ton rocket would weigh 1.92 tons on the moon.

  b) A 180-lb astronaut would weigh 28.8 lb on the moon.

63. *Thinking and Writing Exercise.*

65. $a = \dfrac{b}{c}$

$$a \cdot c = \frac{b}{\cancel{c}} \cdot \cancel{c}$$

$$b = ac$$

67. $2x - 5y = 10$

$$2x \cancel{-5y} \cancel{+5y} - 10 = \cancel{10} + 5y \cancel{-10}$$

$$2x - 10 = 5y$$

$$\frac{1}{5} \cdot 2x - \frac{1}{5} \cdot 10 = \frac{1}{5} \cdot 5y$$

$$\frac{2}{5}x - 2 = y$$

69. $an + b = a$

$$\cancel{an} + b \cancel{- an} = a - an$$

$$b = a(1 - n)$$

$$b \cdot \frac{1}{1-n} = a \cancel{(1-n)} \cdot \frac{1}{\cancel{1-n}}$$

$$a = \frac{b}{1-n}$$

71. *Thinking and Writing Exercise.*

73. **Familiarize.** If the drainage gate is closed, $\dfrac{1}{9}$ of the bog is filled in 1 hr. If the bog is not being filled, $\dfrac{1}{11}$ of the bog is drained in 1 hr. If the bog is being filled with the drainage gate left open, $\dfrac{1}{9} - \dfrac{1}{11}$ of the bog is filled in 1 hr. Let t = the time it takes to fill the bog with the drainage gate left open.

Translate. We want to find t such that

$$t\left(\frac{1}{9} - \frac{1}{11}\right) = 1, \text{ or } \frac{t}{9} - \frac{t}{11} = 1.$$

Carry out. We solve the equation. First we multiply by the LCD, 99.

$$99\left(\frac{t}{9} - \frac{t}{11}\right) = 99 \cdot 1$$

$$11t - 9t = 99$$

$$2t = 99$$

$$t = \frac{99}{2}$$

Check. In $\frac{99}{2}$ hr, we have

$$\frac{99}{2}\left(\frac{1}{9} - \frac{1}{11}\right) = \frac{11}{2} - \frac{9}{2} = \frac{2}{2} = 1 \text{ full bog.}$$

State. It will take $\frac{99}{2}$, or $49\frac{1}{2}$ hr, to fill the

bog.

75. First let $t =$ the time in hours it takes Julia and Tristan working together to complete one batch:

$$\frac{t}{3} + \frac{t}{4} = 1$$
$$4t + 3t = 12$$
$$7t = 12$$
$$t = \frac{12}{7}$$

In $\frac{12}{7}$ of an hour, Julia completes:

$$\frac{1}{3} \cdot \frac{12}{7} = \frac{4}{7} \text{ of the batch. } \frac{4}{7} \approx 0.571 = 57.1\%$$

77. **Familiarize.** Let $p =$ the number of people per hour moved by the 60-cm-wide escalator. Then $2p =$ the number of people per hour moved by the 100-cm-wide escalator. We convert 1575 people per 14 minutes to people per hour:

$$\frac{1575 \text{ people}}{14 \text{ min}} \cdot \frac{60 \text{ min}}{1 \text{ hr}} = 6750 \text{ people/hr}$$

Translate. We use the information that together the escalators move 6750 people per hour to write an equation.

$$p + 2p = 6750$$

Carry out. We solve the equation.

$$p + 2p = 6750$$
$$3p = 6750$$
$$p = 2250$$

Check. If the 60 cm-wide escalator moves 2250 people per hour, then the 100 cm-wide escalator moves $2 \cdot 2250$, or 4500 people per hour. Together, they move $2250 + 4500$, or 6750 people per hour. The answer checks.
State. The 60 cm-wide escalator moves 2250 people per hour.

79. **Familiarize.** Let $d =$ the distance, in miles, the paddleboat can cruise upriver before it is time to turn around. The boat's speed upriver is $12 - 5$, or 7 mph, and its speed downriver is

$12 + 5$, or 17 mph. We organize the information in a table.

	Distance	Speed	Time
Upriver	d	7	t_1
Downriver	d	17	t_2

Translate. Using the formula Time = Distance/Rate we see that

$t_1 = \frac{d}{7}$ and $t_2 = \frac{d}{17}$. The time upriver and

back is 3 hr, so $t_1 + t_2 = 3$, or $\frac{d}{7} + \frac{d}{17} = 3$

Carry out. We solve the equation.

$$7 \cdot 17\left(\frac{d}{7} + \frac{d}{17}\right) = 7 \cdot 17 \cdot 3$$
$$17d + 7d = 357$$
$$24d = 357$$
$$d = \frac{119}{8}$$

Check. Traveling $\frac{119}{8}$ mi upriver at a speed

of 7 mph takes $\frac{119/8}{7} = \frac{17}{8}$ hr. Traveling

$\frac{119}{8}$ mi downriver at a speed of 17 mph takes

$\frac{119/8}{17} = \frac{7}{8}$ hr. The total time is

$\frac{17}{8} + \frac{7}{8} = \frac{24}{8} = 3$ hr. The answer checks.

State. The pilot can go $\frac{119}{8}$, or $14\frac{7}{8}$ mi

upriver before it is time to turn around.

81. **Familiarize.** The Admissions Office printer

can do $\frac{1}{50}$ of the job in 1 min, and the

Business Office printer can do $\frac{1}{40}$ of the job

in 1 min. Let $t =$ the time they work together.
Translate. The work of the Admission Office

printer is $t \cdot \frac{1}{50}$, and the work of the Business

Office printer is $t \cdot \frac{1}{40}$. Working together, to

do one entire job gives us the equation:

$$t \cdot \frac{1}{50} + t \cdot \frac{1}{40} = 1, \text{ or } \frac{t}{50} + \frac{t}{40} = 1$$

Carry out. We solve the equation.

$$\frac{t}{50} + \frac{t}{40} = 1, \text{ LCD is } 200$$

$$200\left(\frac{t}{50} + \frac{t}{40}\right) = 200 \cdot 1$$

$$4t + 5t = 200$$

$$9t = 200$$

$$t = \frac{200}{9} = 22\frac{2}{9}$$

It will require $22\frac{2}{9}$ min to complete the job, if they work together. To determine on what page they will meet, we will find the amount of work done by the faster Business Office machine. (Note: we could also find the amount of work done by the slower machine and do a similar computation.) The Business Office machine does $\frac{200}{9} \cdot \frac{1}{40}$ or $\frac{5}{9}$ of the job. To determine the page number, we take $\frac{5}{9}$ of $500 \approx 277.8$, or page 278, since this machine begins on page 1.

Check. We already have determined that the Business Office machine does $\frac{5}{9}$ of the job. Similarly, the Admissions Office machine does $\frac{200}{9} \cdot \frac{1}{50}$, or $\frac{4}{9}$ of the job. Since $\frac{5}{9} + \frac{4}{9} = 1$, the number checks.

State. Working together, the machines will meet on page 278.

83. **Familiarize.** Express the position of the hands in terms of minute units on the face of the clock. At 10:30 the hour hand is at $\frac{10.5}{12}$ hr $\times \frac{60 \text{ min}}{1 \text{ hr}}$, or 52.5 minutes, and the minute hand is at 30 minutes. The rate of the minute hand is 12 times the rate of the hour hand. (When the minute hand moves 60 minutes, the hour hand moves 5 minutes.) Let t = the number of minutes after 10:30 that the hands will first be perpendicular. After t minutes the minute hand has moved t units, and the hour hand has moved $\frac{t}{12}$ units. The position of the hour hand will be 15 units

"ahead" of the position of the minute hand when they are first perpendicular.

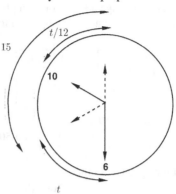

Translate.

Position of hour hand after t min	is	position of minute hand after t min	plus	15 min
↓	↓	↓	↓	↓
$52.5 + \dfrac{t}{12}$	=	$30 + t$	+	15

Solve. We solve the equation.

$$52.5 + \frac{t}{12} = 30 + t + 15$$

$$52.5 + \frac{t}{12} = 45 + t, \text{ LCM is } 12$$

$$12\left(52.5 + \frac{t}{12}\right) = 12(45 + t)$$

$$630 + t = 540 + 12t$$

$$90 = 11t$$

$$\frac{90}{11} = t \text{ or } t = 8\frac{2}{11}$$

Check. At $\frac{90}{11}$ min after 10:30, the position of the hour hand is at $52.5 + \frac{90/11}{12}$, or $53\frac{2}{11}$ min. The minute hand is at $30 + \frac{90}{11}$, or $38\frac{2}{11}$ min. The hour hand is 15 minutes ahead of the minute hand so the hands are perpendicular. The answer checks.

State. After 10:30 the hands of a clock will first be perpendicular in $8\frac{2}{11}$ min. The time is $10:38\frac{2}{11}$, or $21\frac{9}{11}$ min before 11:00.

85. *Familiarize.* Let r = the speed in mph Liam would have to travel for the last half of the trip in order to average a speed of 45 mph for the entire trip. We organize the information in a table.

	Distance	Speed	Time
First half	50	40	t_1
Last half	50	r	t_2

The total distance is 50 + 50, or 100 mi. The total time is $t_1 + t_2$, or $\dfrac{50}{40} + \dfrac{50}{r}$, or $\dfrac{5}{4} + \dfrac{50}{r}$.

The average speed is 45 mph.

Translate. Average speed = $\dfrac{\text{Total distance}}{\text{Total time}}$

$$45 = \frac{100}{\dfrac{5}{4} + \dfrac{50}{r}}$$

Carry out. We solve the equation.

$$45 = \frac{100}{\dfrac{5}{4} + \dfrac{50}{r}}$$

$$45 = \frac{100}{\dfrac{5r + 200}{4r}}$$

$$45 = 100 \cdot \frac{4r}{5r + 200}$$

$$45 = \frac{400r}{5r + 200}$$

$$(5r + 200)(45) = (5r + 200)\frac{400r}{5r + 200}$$

$$225r + 9000 = 400r$$

$$9000 = 175r$$

$$\frac{360}{7} = r$$

Check. Traveling 50 mi at 40 mph takes $\dfrac{50}{40}$, or $\dfrac{5}{4}$ hr. Traveling 50 mi at $\dfrac{360}{7}$ mph takes $\dfrac{50}{360/7}$, or $\dfrac{35}{36}$ hr. Then the total time is $\dfrac{5}{4} + \dfrac{35}{36} = \dfrac{80}{36} = \dfrac{20}{9}$ hr and the average speed is $\dfrac{100}{20/9} = 45$ mph. The answer checks.

State. Liam would have to travel at a speed of $\dfrac{360}{7}$, or $51\dfrac{3}{7}$ mph for the last half of the trip so that the average speed for the entire trip would be 45 mph.

Exercise Set 6.6

1. quotient

3. dividend

5. $\dfrac{32x^5 - 24x}{8} = \dfrac{32x^5}{8} - \dfrac{24x}{8}$

$$= \frac{32}{8}x^5 - \frac{24}{8}x$$

$$= 4x^5 - 3x$$

To check, we multiply the quotient by 8:

$(4x^5 - 3x)8 = 32x^5 - 24x.$

The answer checks.

7. $\dfrac{u - 2u^2 + u^7}{u} = \dfrac{u}{u} - \dfrac{2u^2}{u} + \dfrac{u^7}{u}$

$$= 1 - 2u + u^6$$

Check: We multiply.

$$
\begin{array}{r}
1 \;-\; 2u \;+\; u^6 \\
u \\
\hline
u \;-\; 2u^2 \;+\; u^7
\end{array}
$$

9. $(15t^3 - 24t^2 + 6t) \div (3t)$

$$= \frac{15t^3 - 24t^2 + 6t}{3t}$$

$$= \frac{15t^3}{3t} - \frac{24t^2}{3t} + \frac{6t}{3t}$$

$$= 5t^2 - 8t + 2$$

Check: We multiply.

$$
\begin{array}{r}
5t^2 \;-\; 8t \;+\; 2 \\
3t \\
\hline
15t^3 \;-\; 24t^2 \;+\; 6t
\end{array}
$$

11. $\left(24t^5 - 40t^4 + 6t^3\right) \div \left(4t^3\right)$

$$= \frac{24t^5}{4t^3} - \frac{40t^4}{4t^3} + \frac{6t^3}{4t^3}$$

$$= 6t^2 - 10t + \frac{3}{2}$$

Check: We multiply.

$$
\begin{array}{r}
6t^2 \quad - \quad 10t \quad + \quad \dfrac{3}{2} \\
4t^3 \\
\hline
24t^5 \quad - \quad 40t^4 \quad + \quad 6t^3
\end{array}
$$

13. $\left(15x^7 - 21x^4 - 3x^2\right) \div \left(-3x^2\right)$

$$= \frac{15x^7}{-3x^2} + \frac{-21x^4}{-3x^2} + \frac{-3x^2}{-3x^2}$$

$$= -5x^5 + 7x^2 + 1$$

Check. We multiply

$$
\begin{array}{r}
-5x^5 \quad + \quad 7x^2 \quad + \quad 1 \\
- \quad 3x^2 \\
\hline
15x^7 \quad - \quad 21x^4 \quad - \quad 3x^2
\end{array}
$$

15. $\dfrac{8x^2 - 10x + 1}{2x} = \dfrac{8x^2}{2x} - \dfrac{10x}{2x} + \dfrac{1}{2x}$

$$= 4x - 5 + \frac{1}{2x}$$

Check. We multiply.

$$
\begin{array}{r}
4x \quad - \quad 5 + \dfrac{1}{2x} \\
2x \\
\hline
8x^2 \quad - \quad 10x \quad + \quad 1
\end{array}
$$

17. $\dfrac{9r^2s^2 + 3r^2s - 6rs^2}{-3rs} = \dfrac{9r^2s^2}{-3rs} + \dfrac{3r^2s}{-3rs} - \dfrac{6rs^2}{-3rs}$

$$= -3rs - r + 2s$$

Check: We multiply.

$$
\begin{array}{r}
-3rs \quad - \quad r \quad + \quad 2s \\
- \quad 3rs \\
\hline
9r^2s^2 \quad + \quad 3r^2s \quad - \quad 6rs^2
\end{array}
$$

19. $\left(10x^5y^2 + 15x^2y^2 - 5x^2y\right) \div \left(5x^2y\right)$

$$= \frac{10x^5y^2}{5x^2y} + \frac{15x^2y^2}{5x^2y} - \frac{5x^2y}{5x^2y}$$

$$= 2x^3y + 3y - 1$$

Check. Multiply

$$
\begin{array}{r}
2x^3y \quad + \quad 3y \quad - \quad 1 \\
5x^2y \\
\hline
10x^5y^2 \quad + \quad 15x^2y^2 \quad - \quad 5x^2y
\end{array}
$$

21. $\left(x^2 + 10x + 21\right) \div (x + 7) = \dfrac{(x+7)(x+3)}{x+7}$

$$= \frac{(\cancel{x+7})(x+3)}{\cancel{x+7}}$$

$$= x + 3$$

Check. We multiply.

$$(x+3)(x+7) = x^2 + 7x + 3x + 21$$

$$= x^2 + 10x + 21$$

23.

$$
\begin{array}{r}
a - 12 \\
a + 4 \overline{\smash{\big)}\, a^2 - 8a - 16} \\
\underline{a^2 + 4a} \\
-12a - 16 \quad \left(a^2 - 8a\right) - \left(a^2 + 4a\right) = -12a \\
\underline{-12a - 48} \\
32 \quad (-12a - 16) - (-12a - 48) = 32
\end{array}
$$

The answer is $a - 12$, R 32, or $a - 12 + \dfrac{32}{a+4}$.

Check. We multiply

$$(a+4)\left[(a-12) + \frac{32}{a+4}\right]$$

$$= (a+4)(a-12) + (a+4)\left(\frac{32}{a+4}\right)$$

$$= a^2 - 12a + 4a - 48 + 32$$

$$= a^2 - 8a - 16$$

25.

$$
\begin{array}{r}
2x - 1 \\
x + 6 \overline{\smash{\big)}\, 2x^2 + 11x - 5} \\
\underline{-(2x^2 + 12x)} \\
-x - 5 \\
\underline{-(-x - 6)} \\
1
\end{array}
$$

The answer is $2x - 1$, R 1, or $2x - 1 + \dfrac{1}{x+6}$.

Check. Multiply.

$$(x+6)\left[2x - 1 + \frac{1}{x+6}\right]$$

$$= (x+6)(2x-1) + (x+6)\left(\frac{1}{x+6}\right)$$

$$= x^2 - x + 12x - 6 + 1$$

$$= x^2 + 11x - 5$$

27. $\left(y^2 - 25\right) \div (y+5) = \dfrac{y^2 - 25}{y+5}$

$\qquad\qquad\qquad\quad = \dfrac{(y+5)(y-5)}{y+5}$

$\qquad\qquad\qquad\quad = \dfrac{(\cancel{y+5})(y-5)}{\cancel{y+5}}$

$\qquad\qquad\qquad\quad = y - 5$

We could also find this quotient as follows.

$$\begin{array}{r} y-5 \\ y+5\overline{)y^2 + 0y - 25} \end{array}$$ Writing in the mising term

$$\begin{array}{r} \underline{y^2 + 5y} \\ -5y - 25 \\ \underline{-5y - 25} \\ 0 \end{array}$$

Check. We multiply.

$(y-5)(y+5) = y^2 - 5^2 = y^2 - 25$

29. $a+2\overline{)a^3 + 0a^2 + 0a + 8}$ Writing in the missing terms

$$\begin{array}{r} a^2 - 2a + 4 \\ \hline a^3 + 0a^2 + 0a + 8 \\ \underline{a^3 + 2a^2} \\ -2a^2 + 0a \\ \underline{-2a^2 - 4a} \\ 4a + 8 \\ \underline{4a + 8} \\ 0 \end{array}$$

Check. We multiply.

$$\begin{array}{r} a^2 - 2a + 4 \\ a + 2 \\ \hline 2a^2 - 4a + 8 \\ a^3 - 2a^2 + 4a \\ \hline a^3 \qquad\qquad + 8 \end{array}$$

31. $t-4\overline{)t^2 + 0t - 13}$

$$\begin{array}{r} t + 4 \\ \hline t^2 + 0t - 13 \\ \underline{t^2 - 4t} \\ 4t - 13 \\ \underline{4t - 16} \\ 3 \end{array}$$

The answer is $t+4$, R 3, or $t+4+\dfrac{3}{t-4}$

Check. We multiply.

$(t-4)\left[(t+4) + \dfrac{3}{t-4}\right]$

$= (t-4)(t+4) + (t-4)\left(\dfrac{3}{t-4}\right)$

$= t^2 - 4^2 + 3$

$= t^2 - 16 + 3$

$= t^2 - 13$

33. $2t-3\overline{)2t^3 - 9t^2 + 11t - 3}$

$$\begin{array}{r} t^2 - 3t + 1 \\ \hline 2t^3 - 9t^2 + 11t - 3 \\ \underline{2t^3 - 3t^2} \\ -6t^2 + 11t \\ \underline{-6t^2 + 9t} \\ 2t - 3 \\ \underline{2t - 3} \\ 0 \end{array}$$

Check. We multiply.

$$\begin{array}{r} t^2 \;-\; 3t \;+\; 1 \\ 2t \;-\; 3 \\ \hline -\,3t^2 \;+\; 9t \;-\; 3 \\ 2t^3 \;-\; 6t^2 \;+\; 2t \\ \hline 2t^3 \;-\; 9t^2 \;+\; 11t \;-\; 3 \end{array}$$

35. $5x+1\overline{)5x^2 - 14x}$

$$\begin{array}{r} x - 3 \\ \hline 5x^2 - 14x \\ \underline{-(5x^2 + x)} \\ -15x \\ \underline{-(-15x - 3)} \\ 3 \end{array}$$

The answer is $x-3$, R 3, or $x-3+\dfrac{3}{5x+1}$

Check. Multiply.

$(5x+1)\left[x - 3 + \dfrac{3}{5x+1}\right]$

$= (5x+1)(x-3) + (5x+1)\left(\dfrac{3}{5x+1}\right)$

$= 5x^2 - 15x + x - 3 + 3$

$= 5x^2 - 14x$

37. $\left(t^3 + t - t^2 - 1\right) \div (t + 1),$

   or $\left(t^3 - t^2 + t - 1\right) \div (t + 1)$

$$
\begin{array}{r}
t^2 - 2t + 3 \\
t + 1\overline{)t^3 - t^2 + t - 1} \\
\underline{t^3 + \ t^2} \\
-2t^2 + t \\
\underline{-2t^2 - 2t} \\
3t - 1 \\
\underline{3t + 3} \\
-4
\end{array}
$$

The answer is $t^2 - 2t + 3 + \dfrac{-4}{t+1}$.

Check. We multiply.

$(t + 1)\left[\left(t^2 - 2t + 3\right) + \dfrac{-4}{t+1}\right]$

$= t^3 - 2t^2 + 3t + t^2 - 2t + 3 - 4$

$= t^3 - t^2 + t - 1$

39. $t^2 + 5\overline{)t^4 + 0t^3 + 4t^2 + 3t - 6}$ with quotient $t^2 \quad -1$

$$
\begin{array}{r}
t^2 \qquad -1 \\
t^2 + 5\overline{)t^4 + 0t^3 + 4t^2 + 3t - 6} \\
\underline{t^4 \qquad + 5t^2} \\
-t^2 + 3t - 6 \\
\underline{-t^2 \qquad -5} \\
3t - 1
\end{array}
$$

The answer is $t^2 - 1 + \dfrac{3t - 1}{t^2 + 5}$.

Check. We multiply.

$\left(t^2 + 5\right)\left[\left(t^2 - 1\right) + \dfrac{3t - 1}{t^2 + 5}\right]$

$= \left(t^2 + 5\right)\left(t^2 - 1\right) + \left(t^2 + 5\right)\left(\dfrac{3t - 1}{t^2 + 5}\right)$

$= t^4 - t^2 + 5t^2 - 5 + 3t - 1$

$= t^4 + 4t^2 + 3t - 6$

41. $\left(4x^4 - 3 - x - 4x^2\right) \div \left(2x^2 - 3\right),$

   or $\left(4x^4 - 4x^2 - x - 3\right) \div \left(2x^2 - 3\right)$

$$
\begin{array}{r}
2x^2 \qquad + 1 \\
2x^2 - 3\overline{)4x^4 + 0x^3 - 4x^2 - x - 3} \\
\underline{4x^4 \qquad -6x^2} \\
2x^2 - x - 3 \\
\underline{2x^2 \qquad -3} \\
-x
\end{array}
$$

The answer is $2x^2 + 1 + \dfrac{-x}{2x^2 - 3}$.

Check. We multiply.

$\left(2x^2 - 3\right)\left[\left(2x^2 + 1\right) + \dfrac{-x}{2x^2 - 3}\right]$

$= \left(2x^2 - 3\right)\left(2x^2 + 1\right) + \left(2x^2 - 3\right)\left(\dfrac{-x}{2x^2 - 3}\right)$

$= 4x^4 + 2x^2 - 6x^2 - 3 - x$

$= 4x^4 - 4x^2 - x - 3$

43. $F(x) = \dfrac{f(x)}{g(x)} = \dfrac{6x^2 - 11x - 10}{3x + 2}$

$$
\begin{array}{r}
2x - 5 \\
3x + 2\overline{)6x^2 - 11x - 10} \\
\underline{6x^2 + \ 4x} \\
-15x - 10 \\
\underline{-15x - 10} \\
0
\end{array}
$$

Since $g(x)$ is 0 for $x = -\dfrac{2}{3}$, we have

$F(x) = 2x - 5,$ provided $x \neq -\dfrac{2}{3}$.

45. $F(x) = \dfrac{f(x)}{g(x)} = \dfrac{8x^3 - 27}{2x - 3} = \dfrac{(2x)^3 - 3^3}{2x - 3}$

$= \dfrac{(2x - 3)(4x^2 + 6x + 9)}{2x - 3}$

$= 4x^2 + 6x + 9$

The domain restriction is

$2x - 3 \neq 0,$ or $x \neq \dfrac{3}{2}$.

47. $F(x) = \dfrac{f(x)}{g(x)} = \dfrac{x^4 - 24x^2 - 25}{x^2 - 25}$

$$
\begin{array}{r}
x^2 + 1 \\
x^2 - 25\overline{)x^4 - 24x^2 - 25} \\
\underline{x^4 - 25x^2} \\
x^2 - 25 \\
\underline{x^2 - 25} \\
0
\end{array}
$$

Since $g(x)$ is 0 for $x = -5$ or $x = 5$, we have

$F(x) = x^2 + 1,$ provided $x \neq -5$ and $x \neq 5$.

49. We rewrite $f(x)$ in descending order.

$$F(x) = \frac{f(x)}{g(x)} = \frac{2x^5 - 3x^4 - 2x^3 + 8x^2 - 5}{x^2 - 1}$$

$$
\require{enclose}
\begin{array}{r}
2x^3 - 3x^2 + 5 \\
x^2-1 \enclose{longdiv}{2x^5 - 3x^4 - 2x^3 + 8x^2 - 5} \\
\underline{2x^5 - 2x^3} \\
-3x^4 + 8x^2 \\
\underline{-3x^4 + 3x^2} \\
5x^2 - 5 \\
\underline{5x^2 - 5} \\
0
\end{array}
$$

Since $g(x)$ is 0 for $x = -1$ or $x = 1$, we have $F(x) = 2x^3 - 3x^2 + 5$, provided $x \neq -1$ and $x \neq 1$.

51. *Thinking and Writing Exercise.*

53.

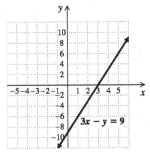

55.

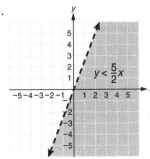

57.

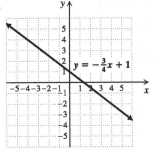

59. *Thinking and Writing Exercise.*

61.

$$
\require{enclose}
\begin{array}{r}
a^2 + ab \\
a^2+3ab+2b^2 \enclose{longdiv}{a^4 + 4a^3b + 5a^2b^2 + 2ab^3} \\
\underline{a^4 + 3a^3b + 2a^2b^2} \\
a^3b + 3a^2b^2 + 2ab^3 \\
\underline{a^3b + 3a^2b^2 + 2ab^3} \\
0
\end{array}
$$

The answer is $a^2 + ab$.

63.

$$
\require{enclose}
\begin{array}{r}
a^6 - a^5b + a^4b^2 - a^3b^3 + a^2b^4 - ab^5 + b^6 \\
a+b \enclose{longdiv}{a^7 + b^7} \\
\underline{a^7 + a^6b} \\
-a^6b \\
\underline{-a^6b - a^5b^2} \\
a^5b^2 \\
\underline{a^5b^2 + a^4b^3} \\
-a^4b^3 \\
\underline{-a^4b^3 - a^3b^4} \\
a^3b^4 \\
\underline{a^3b^4 + a^2b^5} \\
-a^2b^5 \\
\underline{-a^2b^5 - ab^6} \\
ab^6 + b^7 \\
\underline{ab^6 + b^7} \\
0
\end{array}
$$

The answer is
$a^6 - a^5b + a^4b^2 - a^3b^3 + a^2b^4 - ab^5 + b^6$.

65.

$$\begin{array}{r} x-5 \\ x+2\overline{)x^2-3x+2k} \\ \underline{x^2+2x} \\ -5x+2k \\ \underline{-5x-10} \\ 2k+10 \end{array}$$

The remainder is 7. Thus, we solve the following equation for k.

$$2k+10=7$$
$$2k=-3$$
$$k=-\frac{3}{2}$$

67. *Thinking and Writing Exercise.*

Exercise Set 6.7

1. True

3. True

5. True

7. $(x^3-4x^2-2x+5)\div(x-1)$

$$\begin{array}{r|rrrr} 1 & 1 & -4 & -2 & 5 \\ & & 1 & -3 & -5 \\ \hline & 1 & -3 & -5 & |\ 0 \end{array}$$

The answer is x^2-3x-5

9. $(a^2+8a+11)\div(a+3)=$

$$(a^2+8a+11)\div\left[a-(-3)\right]$$

$$\begin{array}{r|rrr} -3 & 1 & 8 & 11 \\ & & -3 & -15 \\ \hline & 1 & 5 & |\ -4 \end{array}$$

The answer is $a+5$, R -4, or $a+5+\dfrac{-4}{a+3}$.

11. $(2x^3-x^2-7x+14)\div(x+2)=$

$$(2x^3-x^2-7x+14)\div\left[x-(-2)\right]$$

$$\begin{array}{r|rrrr} -2 & 2 & -1 & -7 & 14 \\ & & -4 & 10 & -6 \\ \hline & 2 & -5 & 3 & |\ 8 \end{array}$$

The answer is $2x^2-5x+3$, R 8, or

$$2x^2-5x+3+\frac{8}{x+2}.$$

13. $(a^3-10a+12)\div(a-2)$

$$\begin{array}{r|rrrr} 2 & 1 & 0 & -10 & 12 \\ & & 2 & 4 & -12 \\ \hline & 1 & 2 & -6 & |\ 0 \end{array}$$

The answer is a^2+2a-6.

15. $(3y^3-7y^2-20)\div(y-3)$

$$\begin{array}{r|rrrr} 3 & 3 & -7 & 0 & -20 \\ & & 9 & 6 & 18 \\ \hline & 3 & 2 & 6 & |\ -2 \end{array}$$

The answer is $3y^2+2y+6$, R -2, or

$$3y^2+2y+6-\frac{2}{y-3}.$$

17. $(x^5-32)\div(x-2)=$

$$(x^5+0x^4+0x^3+0x^2+0x-32)\div(x-2)$$

$$\begin{array}{r|rrrrrr} 2 & 1 & 0 & 0 & 0 & 0 & -32 \\ & & 2 & 4 & 8 & 16 & 32 \\ \hline & 1 & 2 & 4 & 8 & 16 & |\ 0 \end{array}$$

The answer is $x^4+2x^3+4x^2+8x+16$.

19. $\left(3x^3+1-x+7x^2\right)\div\left(x+\dfrac{1}{3}\right)=$

$$\left(3x^3+7x^2-x+1\right)\div\left[x-\left(-\frac{1}{3}\right)\right]$$

$$\begin{array}{r|rrrr} -\dfrac{1}{3} & 3 & 7 & -1 & 1 \\ & & -1 & -2 & 1 \\ \hline & 3 & 6 & -3 & |\ 2 \end{array}$$

The answer is $3x^2+6x-3$, R 2, or

$$3x^2+6x-3+\frac{2}{x+\dfrac{1}{3}}.$$

21.

$$\begin{array}{r|rrrrr} -3 & 5 & 12 & 0 & 28 & 9 \\ & & -15 & 9 & -27 & -3 \\ \hline & 5 & -3 & 9 & 1 & |\ 6 \end{array}$$

The remainder tells us that $f(-3)=6$.

23.

$$
\begin{array}{r|rrrrr}
-3 & 2 & -1 & -7 & 1 & 2 \\
   &   & -6 & 21 & -42 & 123 \\
\hline
   & 2 & -7 & 14 & -41 & 125
\end{array}
$$

The remainder tells us that $P(-3) = 125$.

25.

$$
\begin{array}{r|rrrrr}
4 & 1 & -6 & 11 & -17 & 20 \\
  &   & 4 & -8 & 12 & -20 \\
\hline
  & 1 & -2 & 3 & -5 & 0
\end{array}
$$

The remainder tells us that $f(4) = 0$.

27. *Thinking and Writing Exercise.*

29. $\quad ac = b$

$$\frac{1}{a} \cdot ac = \frac{1}{a} \cdot b$$

$$c = \frac{b}{a}$$

31.

$$pq - rq = st$$

$$q(p - r) = st$$

$$q(p - r) \cdot \frac{1}{p - r} = st \cdot \frac{1}{p - r}$$

$$q = \frac{st}{p - r}$$

33.

$$ab - cd = 3b + d$$

$$ab - cd - 3b + cd = 3b + d - 3b + cd$$

$$ab - 3b = d + cd$$

$$b(a - 3) = d + cd$$

$$b(a - 3) \cdot \frac{1}{a - 3} = (d + cd) \cdot \frac{1}{a - 3}$$

$$b = \frac{d + cd}{a - 3}$$

35. *Thinking and Writing Exercise.*

37. a)   The degree of the remainder must be less than the degree of the divisor. Thus, the degree of the remainder must be 0, so R must be a constant.

    b) $P(x) = (x - r) \cdot Q(x) + R$

$$P(r) = (r - r) \cdot Q(r) + R$$

$$= 0 \cdot Q(r) + R = R$$

39. $f(x) = 4x^3 + 16x^2 - 3x - 45$

$$f(-3) = 4(-3)^3 + 16(-3)^2 - 3(-3) - 45$$

$$= 4(-27) + 16 \cdot 9 + 9 - 45$$

$$= -108 + 144 + 9 - 45$$

$$= 0$$

Since $f(-3) = 0$ we can use synthetic division to find the other two zeros.

$$
\begin{array}{r|rrrr}
-3 & 4 & 16 & -3 & -45 \\
   &   & -12 & -12 & 45 \\
\hline
   & 4 & 4 & -15 & 0
\end{array}
$$

The other two zeros are the solutions to

$$4x^2 + 4x - 15 = 0$$

$$(2x - 3)(2x + 5) = 0$$

$$x = \frac{3}{2}, -\frac{5}{2}$$

41. Graph $y_1 = 4x^3 + 16x^2 - 3x - 45$. Using TRACE, determine the y_1 value when $x = -3$. Determine all of the x values when $y_1 = 0$.

43. $f(x) = 4x^3 + 16x^2 - 3x - 45$

$$= x(4x^2 + 16x - 3) - 45$$

$$= x(x(4x + 16) - 3) - 45$$

$$f(-3) = -3(-3(4(-3) + 16) - 3) - 45$$

$$= -3(-3 \cdot 4 - 3) - 45$$

$$= -3(-15) - 45$$

$$= 45 - 45$$

$$= 0$$

Exercise Set 6.8

1. Clear denominators in an equation using the LCD. : d

3. $y = \dfrac{k}{x} \Rightarrow xy = k$ The product xy is constant. : e

5. If $y = kx$, then by definition, y varies directly as x. : a

7. Inverse

9. Direct

11. Inverse

13. $f = \dfrac{L}{d}$ Solve for d.

$df = L$ Multiplying by d

$d = \dfrac{L}{f}$ Dividing by f

15. $s = \dfrac{(v_1 + v_2)t}{2}$ Solve for v_1.

$2s = (v_1 + v_2)t$ Multiplying by 2

$\dfrac{2s}{t} = v_1 + v_2$ Dividing by t

$\dfrac{2s}{t} - v_2 = v_1$

This result can also be expressed as

$v_1 = \dfrac{2s - tv_2}{t}$.

17. $\dfrac{t}{a} + \dfrac{t}{b} = 1$ Solve for b.

$ab\left(\dfrac{t}{a} + \dfrac{t}{b}\right) = ab \cdot 1$ Multiplying by the LCD

$bt + at = ab$

$at = ab - bt$

$at = b(a - t)$ Factoring out b

$\dfrac{at}{a - t} = b$ Multiplying by $\dfrac{1}{a - t}$

19. $I = \dfrac{2V}{R + 2r}$ Solve for R.

$I(R + 2r) = \dfrac{2V}{R + 2r} \cdot (R + 2r)$ Multiplying by the LCD

$I(R + 2r) = 2V$

$R + 2r = \dfrac{2V}{I}$

$R = \dfrac{2V}{I} - 2r$, or $\dfrac{2V - 2Ir}{I}$

21. $R = \dfrac{gs}{g + s}$ Solve for g.

$(g + s) \cdot R = (g + s) \cdot \dfrac{gs}{g + s}$ Multiplying by the LCD

$Rg + Rs = gs$

$Rs = gs - Rg$

$Rs = g(s - R)$ Factoring out g

$\dfrac{Rs}{s - R} = g$ Multiplying by $\dfrac{1}{s - R}$

23. $\dfrac{1}{p} + \dfrac{1}{q} = \dfrac{1}{f}$ Solve for q.

$pqf\left(\dfrac{1}{p} + \dfrac{1}{q}\right) = pqf \cdot \dfrac{1}{f}$ Multiplying by the LCD

$qf + pf = pq$

$pf = pq - qf$

$pf = q(p - f)$

$\dfrac{pf}{p - f} = q$

25. $S = \dfrac{H}{m(t_1 - t_2)}$ Solve for t_1.

$(t_1 - t_2)S = \dfrac{H}{m}$ Multiplying by $t_1 - t_2$

$t_1 - t_2 = \dfrac{H}{Sm}$ Dividing by S

$t_1 = \dfrac{H}{Sm} + t_2$, or $\dfrac{H + Smt_2}{Sm}$

27. $\dfrac{E}{e} = \dfrac{R + r}{r}$ Solve for r

$er \cdot \dfrac{E}{e} = er \cdot \dfrac{R + r}{r}$ Mult. by LCD

$Er = e(R + r)$

$Er = eR + er$

$Er - er = eR$

$r(E - e) = eR$

$r = \dfrac{eR}{E - e}$

29. $S = \dfrac{a}{1-r}$ Solve for r.

$(1-r)S = a$ Multiplying by the LCD, $1-r$

$1 - r = \dfrac{a}{S}$ Dividing by S

$1 - \dfrac{a}{S} = r$ Adding r and $-\dfrac{a}{S}$

This result can also be expressed as

$r = \dfrac{S-a}{S}$.

31. $c = \dfrac{f}{(a+b)c}$ Solve for $(a+b)$.

$\dfrac{a+b}{c} \cdot c = \dfrac{a+b}{c} \cdot \dfrac{f}{(a+b)c}$

$a + b = \dfrac{f}{c^2}$

33. $P = \dfrac{A}{1+r}$ Solve for r.

$P(1+r) = \dfrac{A}{1+r} \cdot (1+r)$

$P(1+r) = A$

$1 + r = \dfrac{A}{P}$

$r = \dfrac{A}{P} - 1$, or $\dfrac{A-P}{P}$

35. $v = \dfrac{d_2 - d_1}{t_2 - t_1}$ Solve for t_2.

$(t_2 - t_1)v = (t_2 - t_1) \cdot \dfrac{d_2 - d_1}{t_2 - t_1}$

$(t_2 - t_1)v = d_2 - d_1$

$t_2 - t_1 = \dfrac{d_2 - d_1}{v}$

$t_2 = \dfrac{d_2 - d_1}{v} + t_1$, or $\dfrac{d_2 - d_1 + t_1 v}{v}$

37. $\dfrac{1}{t} = \dfrac{1}{a} + \dfrac{1}{b}$ Solve for t.

$tab \cdot \dfrac{1}{t} = tab\left(\dfrac{1}{a} + \dfrac{1}{b}\right)$

$ab = tb + ta$

$ab = t(b + a)$

$\dfrac{ab}{b+a} = t$

39. $A = \dfrac{2Tt + Qq}{2T + Q}$ Solve for Q.

$(2T + Q) \cdot A = (2T + Q) \cdot \dfrac{2Tt + Qq}{2T + Q}$

$2AT + AQ = 2Tt + Qq$

Adding $-2AT$ and $-Qq$

$AQ - Qq = 2Tt - 2At$

$Q(A - q) = 2Tt - 2AT$

$Q = \dfrac{2Tt - 2AT}{A - q}$

41. $p = \dfrac{-98.42 + 4.15c - 0.082w}{w}$

$w \cdot p = \cancel{w} \cdot \dfrac{-98.42 + 4.15c - 0.082w}{\cancel{w}}$

$pw = -98.42 + 4.15c - 0.082w$

$pw + 0.082w = -98.42 + 4.15c$

$w(p + 0.082) = 4.15c - 98.42$

$w = \dfrac{4.15c - 98.42}{p + 0.082}$

43. $y = kx$

$28 = k \cdot 4$ Substituting

$7 = k$

The variation constant is 7.

The equation of variation is $y = 7x$.

45. $y = kx$

$3.4 = k \cdot 2$ Substituting

$1.7 = k$

The variation constant is 1.7.

The equation of variation is $y = 1.7x$.

47. $y = kx$

    $2 = k \cdot \dfrac{1}{3}$ Substituting

    $6 = k$

    The variation constant is 6.
    The equation of variation is $y = 6x$.

49. $y = \dfrac{k}{x}$

    $3 = \dfrac{k}{20}$ Substituting

    $60 = k$

    The variation constant is 60.

    The equation of variation is $y = \dfrac{60}{x}$.

51. $y = \dfrac{k}{x}$

    $11 = \dfrac{k}{6}$ Substituting

    $66 = k$

    The variation constant is 66.

    The equation of variation is $y = \dfrac{66}{x}$.

53. $y = \dfrac{k}{x}$

    $27 = \dfrac{k}{\dfrac{1}{3}}$ Substituting

    $9 = k$

    The variation constant is 9.

    The equation of variation is $y = \dfrac{9}{x}$.

55. *Familiarize.* Because of the phrase "$d \ldots$ varies directly as $\ldots m$," we express the distance as a function of the mass. Thus we have $d(m) = km$. We know that $d(3) = 20$.

Translate. We find the variation constant and then find the equation of variation.

    $d(m) = km$

    $d(3) = k \cdot 3$ Replacing m with 3

    $20 = k \cdot 3$ Replacing $d(3)$ with 20

    $\dfrac{20}{3} = k$ Variation constant

The equation of variation is $d(m) = \dfrac{20}{3}m$.

Carry out. We compute $d(5)$.

    $d(m) = \dfrac{20}{3}m$

    $d(5) = \dfrac{20}{3} \cdot 5$ Replacing m with 5

    $d(5) = \dfrac{100}{3}$, or $33\dfrac{1}{3}$

Check. Reexamine the calculations. Note that the answer seems reasonable since

    $\dfrac{3}{20}$ and $\dfrac{5}{100/3}$ are equal.

State. The spring is stretched $33\dfrac{1}{3}$ cm by a hanging object with mass 5 kg.

57. *Familiarize.* Because T varies inversely as P, we write $T(P) = k/P$. We know that
$T(7) = 5$.

Translate. We find the variation constant and the equation of variation.

    $T(P) = \dfrac{k}{P}$

    $T(7) = \dfrac{k}{7}$ Replacing P with 7

    $5 = \dfrac{k}{7}$ Replacing $T(P)$ with 5

    $35 = k$ Variation constant

    $T(P) = \dfrac{35}{P}$ Equation of variation

Carry out. We find $T(10)$.

    $T(10) = \dfrac{35}{10}$

    $= 3.5$

Check. Reexamine the calculations.
State. It would take 3.5 hr for 10 volunteers to complete the job.

59.
$$W = kS$$
$$\frac{W}{S} = k$$
$$\frac{W_1}{S_1} = \frac{W_2}{S_2}$$
$$\frac{16.8 \text{ cm}}{150 \text{ cm}} = \frac{W_2}{500 \text{ in.}}$$
$$W_2 = \frac{16.8}{150} \cdot 500 \text{ in.} = 56 \text{ in.}$$

61. Since the number of kilograms of water varies directly as the mass, and a 96-kg person contains 64 kg of water, a 48-kg person (weighing 1/2 of 96 kg), would contain half

63.
$$F = \frac{k}{L}$$
$$FL = k$$
$$F_1L_1 = F_2L_2$$
$$F_2 = \frac{F_1L_1}{L_2}$$
$$\frac{F_1L_1}{L_2} = \frac{33 \text{ cm} \cdot 260 \text{ Hz}}{30 \text{ cm}} = 286 \text{ Hz}$$

65. *Familiarize.* Because of the phrase "t varies inversely as . . . u," we write $t = k/u$. We know that $70 = \dfrac{k}{4}$.

Translate. We find the variation constant and then we find the equation of variation.
$$70 = \frac{k}{4} \Rightarrow 70 \cdot 4 = k \Rightarrow 280 = k \Rightarrow t = \frac{280}{u}$$

Carry out. We find t when $u = 14$.
$$t = \frac{280}{14} = 20$$

Check. Reexamine the calculations. Note that, as expected, as the UV rating increases, the time it takes to burn goes down.

State. It will take 20 min to burn when the UV rating is 14.

67. *Familiarize.* Let A = the amount of carbon monoxide released, in tons, and P = the population. Then:
$$A = kP \Rightarrow \frac{A}{P} = k \Rightarrow \frac{A_1}{P_1} = \frac{A_2}{P_2} \Rightarrow A_2 = \frac{A_1}{P_1} \cdot P_2$$
$$= \frac{0.65 \text{ tons}}{2.6 \text{ people}} \cdot 308{,}000{,}000 \text{ people}$$
$$= 77{,}000{,}000 \text{ tons}$$

Check. Reexamine the calculations.
State. The U.S. released approximately 77 million tons of carbon monoxide in 1 year.

69.
$$y = kx^2$$
$$6 = k \cdot 3^2 \qquad \text{Substituting}$$
$$6 = 9k$$
$$\frac{6}{9} = k$$
$$\frac{2}{3} = k \qquad \text{Variation constant}$$
The equation of variation is $y = \dfrac{2}{3}x^2$.

71.
$$y = \frac{k}{x^2}$$
$$6 = \frac{k}{3^2} \qquad \text{Substituting}$$
$$6 = \frac{k}{9}$$
$$6 \cdot 9 = k$$
$$54 = k \qquad \text{Variation constant}$$
The equation of variation is $y = \dfrac{54}{x^2}$.

73.
$$y = kxz^2$$
$$105 = k \cdot 14 \cdot 5^2 \qquad \begin{array}{l}\text{Substituting 105 for } y, \\ \text{14 for } x, \text{ and 5 for } z.\end{array}$$
$$105 = 350k$$
$$\frac{105}{350} = k$$
$$0.3 = k \qquad \text{Variation constant}$$
The equation of variation is $y = 0.3xz^2$.

75.
$$y = k \cdot \frac{wx^2}{z}$$
$$49 = k \cdot \frac{3 \cdot 7^2}{12} \qquad \text{Substituting}$$
$$4 = k \qquad \text{Variation constant}$$
The equation of variation is $y = \dfrac{4wx^2}{z}$.

77. *Familiarize.* Because d varies directly as r^2, we write $d = kr^2$. We know that $d = 138$ ft when $r = 60$ mph.

Translate. Determine k and the equation of variation.

$$d = kr^2$$

$$138 = k \cdot 60^2$$

$$\frac{138}{3600} = k$$

$$\frac{23}{600} = k$$

$$d = \frac{23}{600} \cdot r^2 \quad \text{Equation of variation}$$

Carry out. Substitute 40 for r and solve for d.

$$d = \frac{23}{600} \cdot 40^2$$

$$d = 61\frac{1}{3}$$

Check. Reexamine the calculations.

State. A car traveling at 40 mph will require $61\frac{1}{3}$ ft to stop.

79. *Familiarize.* Because V varies directly as T and inversely as P, we write $V = \dfrac{kT}{P}$. We know that $V = 231$ when $T = 300$ and $P = 20$.

Translate. Determine k and the equation of variation.

$$V = \frac{kT}{P}$$

$$231 = \frac{k \cdot 300}{20}$$

$$\frac{20}{300} \cdot 231 = k$$

$$15.4 = k$$

$$V = \frac{15.4T}{P} \quad \text{Equation of variation}$$

Carry out. Substitute 320 for T and 16 for P to determine V.

$$V = \frac{15.4 \cdot 320}{16}$$

$$V = 308 \text{ cm}^3$$

Check. Reexamine the calculations.

State. The volume is 308 cm^3 when $T = 320°$K and $P = 16$ lb/cm^2.

81. *Familiarize.* The drag W varies jointly as the surface area A and velocity v, so we write $W = kAv$. We know that $W = 222$ when $A = 37.8$ and $v = 40$.

Translate. Find k.

$$W = kAv$$

$$222 = k(37.8)(40)$$

$$\frac{222}{37.8(40)} = k$$

$$\frac{37}{252} = k$$

$$W = \frac{37}{252}Av \quad \text{Equation of variation}$$

Carry out. Substitute 51 for A and 430 for W and solve for v.

$$430 = \frac{37}{252} \cdot 51 \cdot v$$

$$57.42 \text{ mph} \approx v$$

(If we had used the rounded value 0.1468 for k, the resulting speed would have been approximately 57.43 mph.)

Check. Reexamine the calculations.

State. The car must travel about 57.42 mph.

83. a) From the data table we see that as the population density increases the annual VMT per household decreases. Therefore the population P is inversely proportional to the VMT, V.

b) $P = \dfrac{k}{V} \Rightarrow 25 = \dfrac{k}{12{,}000}$

$$k = 25 \cdot 12{,}000 = 300{,}000$$

$$P = \frac{300{,}000}{V}$$

As a check, we note that $PV = k$ for all pairs of points (P, V).

$$50 \cdot 6000 = 300{,}000 \quad \checkmark$$

$$100 \cdot 3000 = 300{,}000 \quad \checkmark$$

$$200 \cdot 1500 = 300{,}000 \quad \checkmark$$

c) $P = \dfrac{300{,}000}{V}$

$$V = \frac{300{,}000}{P} = \frac{300{,}000}{10} = 30{,}000 \text{ VMT}$$

85. a) From the data table we see that, as the selling price increases, the seller's commission also increases. Therefore the commission C is directly proportional to the selling price P.

b) $C = kP \Rightarrow 111.85 = k \cdot 200$

$\Rightarrow k = \dfrac{111.85}{200} \approx 0.56$

$\Rightarrow C = 0.56P$

As a check, we check whether $\dfrac{C}{P} \approx k$ for

all pairs of points $(P,\ C)$.

$\dfrac{41.42}{75} \approx 0.552$

$\dfrac{55.50}{100} = 0.5550$

$\dfrac{111.85}{200} = 0.55925$

$\dfrac{240.55}{400} \approx 0.601$

We see that the "constant" of variation is not exactly constant. Over the range of selling prices, there is a rise of about 0.008 or 0.8% in the commission. However, we will use the equation determined above in accordance with the directions to use the point (200, 111.85).

c) $C = 0.56P = 0.56(150) = 84$ The seller receives approximately $84.00.

87. *Thinking and Writing Exercise.*

89.　　$f(x) = 4x - 7$

$f(a) + h = 4(a) - 7 + h$

$= 4a - 7 + h$

91. The denominator of the rational function $f(x)$ is zero when

$2x + 1 = 0$

$2x = -1$

$x = -\dfrac{1}{2}$

Therefore the domain of f is all real numbers

except $x = -\dfrac{1}{2}$. In set builder notation:

$\left\{ x \mid x \text{ is a real number, } x \neq -\dfrac{1}{2} \right\}$

93. The domain of f is all real numbers that satisfy the inequality

$2x + 8 \geq 0$

$2x \geq -8$

$x \geq -4$

In set builder notation:

$\left\{ x \mid x \text{ is a real number, } x \geq -4 \right\}$

95. *Thinking and Writing Exercise.*

97. Use the result of Example 2.

$h = \dfrac{2R^2 g}{V^2} - R$

We have $V = 6.5$ mi/sec, $R = 3960$ mi, and $g = 32.2$ ft/sec^2. We must convert 32.2 ft/sec^2 to mi/sec^2 so all units of length are the same.

$32.2 \dfrac{\cancel{ft}}{\sec^2} \cdot \dfrac{1\ \text{mi}}{5280\ \cancel{ft}} \approx 0.0060985 \dfrac{\text{mi}}{\sec^2}$

Now we substitute and compute.

$h = \dfrac{2(3960)^2 (0.0060985)}{(6.5)^2} - 3960$

$h \approx 567$

The satellite is about 567 mi from the surface of the earth.

99. $c = \dfrac{a}{a + 12} \cdot d$

$c = \dfrac{2a}{2a + 12} \cdot d$　　Doubling a

$= \dfrac{\cancel{2}a}{\cancel{2}(a + 6)} \cdot d$

$= \dfrac{a}{a + 6} \cdot d$　　Simplifying

The ratio of the larger dose to the smaller dose is

$\dfrac{\dfrac{a}{a + 6} \cdot d}{\dfrac{a}{a + 12} \cdot d} = \dfrac{\dfrac{ad}{a + 6}}{\dfrac{ad}{a + 12}}$

$= \dfrac{ad}{a + 6} \cdot \dfrac{a + 12}{ad}$

$= \dfrac{\cancel{ad}(a + 12)}{(a + 6)\cancel{ad}}$

$= \dfrac{a + 12}{a + 6}$

The amount by which the dosage increases is

$$\frac{a}{a+6} \cdot d - \frac{a}{a+12} \cdot d$$

$$= \frac{ad}{a+6} - \frac{ad}{a+12}$$

$$= \frac{ad}{a+6} \cdot \frac{a+12}{a+12} - \frac{ad}{a+12} \cdot \frac{a+6}{a+6}$$

$$= \frac{ad(a+12) - ad(a+6)}{(a+6)(a+12)}$$

$$= \frac{a^2 d + 12ad - a^2 d - 6ad}{(a+6)(a+12)}$$

$$= \frac{6ad}{(a+6)(a+12)}$$

Then the percent by which the dosage increases is

$$\frac{\dfrac{6ad}{(a+6)(a+12)}}{\dfrac{a}{a+12} \cdot d} = \frac{\dfrac{6ad}{(a+6)(a+12)}}{\dfrac{ad}{a+12}}$$

$$= \frac{6ad}{(a+6)(a+12)} \cdot \frac{a+12}{ad}$$

$$= \frac{6 \cdot \cancel{ad} \cdot \cancel{(a+12)}}{(a+6) \cancel{(a+12)} \cdot \cancel{ad}}$$

$$= \frac{6}{a+6}$$

This is a decimal representation for the percent of increase. To give the result in percent notation we multiply by 100 and use a percent symbol. We have

$$\frac{6}{a+6} \cdot 100\%, \text{ or } \frac{600}{a+6}\%$$

101. $$a = \frac{\dfrac{d_4 - d_3}{t_4 - t_3} - \dfrac{d_2 - d_1}{t_2 - t_1}}{t_4 - t_2}$$

$$a(t_4 - t_2) = \frac{d_4 - d_3}{t_4 - t_3} - \frac{d_2 - d_1}{t_2 - t_1}$$

Multiplying by $t_4 - t_2$

$$a(t_4 - t_2)(t_4 - t_3)(t_2 - t_1) =$$
$$(d_4 - d_3)(t_2 - t_1) - (d_2 - d_1)(t_4 - t_3)$$

Multiplying by $(t_4 - t_3)(t_2 - t_1)$

$$a(t_4 - t_2)(t_4 - t_3)(t_2 - t_1) - (d_4 - d_3)(t_2 - t_1) =$$
$$-(d_2 - d_1)(t_4 - t_3)$$

$$(t_2 - t_1)\left[a(t_4 - t_2)(t_4 - t_3) - (d_4 - d_3)\right] =$$
$$-(d_2 - d_1)(t_4 - t_3)$$

$$t_2 - t_1 = \frac{-(d_2 - d_1)(t_4 - t_3)}{a(t_4 - t_2)(t_4 - t_3) - (d_4 - d_3)}$$

$$t_2 + \frac{(d_2 - d_1)(t_4 - t_3)}{a(t_4 - t_2)(t_4 - t_3) + d_3 - d_4} = t_1$$

103. Write I as a function of w, the wattage and d, the distance, and then double both w and d.

$$I = \frac{w}{d^2}$$

$$I = \frac{2w}{(2d)^2}$$

$$I = \frac{2w}{4d^2}$$

$$I = \frac{1}{2} \cdot \frac{w}{d^2}$$

Comparing $I = \dfrac{w}{d^2}$ and $I = \dfrac{1}{2} \cdot \dfrac{w}{d^2}$, we see that the intensity is halved.

105. ***Familiarize.*** We write $T = kml^2 f^2$. We know that $T = 100$ when $m = 5$, $l = 2$, and $f = 80$.

Translate. Find k.

$$T = kml^2 f^2$$

$$100 = k(5)(2)^2 (80)^2$$

$$0.00078125 = k$$

$$T = 0.00078125 ml^2 f^2$$

Carry out. Substitute 72 for T, 5 for m, and 80 for f and solve for l.

$$72 = 0.00078125(5)l^2 (80)^2$$

$$2.88 = l^2$$

$$1.697 \approx l$$

Check. Recheck the calculations.

State. The string should be about 1.697 m long.

107. *Familiarize.* Because d varies inversely as s, we write $d(s) = k/s$. We know that $d(0.56) = 50$.

Translate.

$$d(s) = \frac{k}{s}$$

$$d(0.56) = \frac{k}{0.56} \quad \text{Replacing } s \text{ with } 0.56$$

$$50 = \frac{k}{0.56} \quad \text{Replacing } d(56) \text{ with } 50$$

$$28 = k$$

$$d(s) = \frac{28}{s} \quad \text{Equation of variation}$$

Carry out. Find $d(0.40)$.

$$d(0.40) = \frac{28}{0.40}$$
$$= 70$$

Check. Reexamine the calculations. Also serve that, as expected, when d decreases, then s increases.

State. The equation of variation is $d(s) = \frac{28}{s}$. The distance is 70 yd.

Chapter 6 Study Summary

1. $\dfrac{x^2 - 5x}{x^2 - 25} = \dfrac{x\,\cancel{(x-5)}}{(x+5)\,\cancel{(x-5)}} = \dfrac{x}{x+5}$ for $x \neq 5, -5$

2. $\dfrac{6x - 12}{2x^2 + 3x - 2} \cdot \dfrac{x^2 - 4}{8x - 8}$

   $= \dfrac{6(x-2)}{(2x-1)\,\cancel{(x+2)}} \cdot \dfrac{\cancel{(x+2)}\,(x-2)}{8(x-1)}$

   $= \dfrac{6(x-2)^2}{8(x-1)(2x-1)}$

   $= \dfrac{\cancel{2}\cdot 3(x-2)^2}{\cancel{2}\cdot 4(x-1)(2x-1)} = \dfrac{3(x-2)^2}{4(x-1)(2x-1)}$

3. $\dfrac{t-3}{6} \div \dfrac{t+1}{15} = \dfrac{t-3}{6} \cdot \dfrac{15}{t+1} = \dfrac{t-3}{\cancel{3}\cdot 2} \cdot \dfrac{\cancel{3}\cdot 5}{t+1}$

   $= \dfrac{5(t-3)}{2(t+1)}$

4. $\dfrac{5x+4}{x+3} + \dfrac{4x+1}{x+3} = \dfrac{(5x+4)+(4x+1)}{x+3} = \dfrac{9x+5}{x+3}$

5. $\dfrac{t}{t-1} - \dfrac{t-2}{t+1} = \dfrac{t}{t-1}\cdot\dfrac{t+1}{t+1} - \dfrac{t-2}{t+1}\cdot\dfrac{t-1}{t-1}$

   $= \dfrac{t(t+1) - (t-2)(t-1)}{(t+1)(t-1)} = \dfrac{t^2 + t - (t^2 - 2t - t + 2)}{(t+1)(t-1)}$

   $= \dfrac{\cancel{t^2} + t \,\cancel{- t^2} + 2t + t - 2}{(t+1)(t-1)} = \dfrac{4t - 2}{(t+1)(t-1)}$

6. $\dfrac{\dfrac{4}{x} - 4}{\dfrac{7}{x} - 7} = \dfrac{\dfrac{4}{x} - 4}{\dfrac{7}{x} - 7}\cdot\dfrac{x}{x} = \dfrac{\cancel{x}\cdot\dfrac{4}{\cancel{x}} - x\cdot 4}{\cancel{x}\cdot\dfrac{7}{\cancel{x}} - x\cdot 7} = \dfrac{4 - 4x}{7 - 7x}$

   $= \dfrac{4(1-x)}{7(1-x)} = \dfrac{4}{7}$

7. $\dfrac{3}{x+4} = \dfrac{1}{x-1}$

   $3(x-1) = 1(x+4)$

   $3x - 3 = x + 4$

   $3x - x = 4 + 3$

   $2x = 7$

   $x = \dfrac{7}{2}$

8. Jackson can sand $\dfrac{1}{12}$ of the floors per hour, and Charis can sand $\dfrac{1}{9}$. Let t = the time it takes for them to sand the floors working together. Then:

   $$\left(\frac{1}{12} + \frac{1}{9}\right)t = 1$$

   $$\frac{1}{12} + \frac{1}{9} = \frac{1}{t}$$

   $$36t\cdot\frac{1}{12} + 36t\cdot\frac{1}{9} = 36t\cdot\frac{1}{t}$$

   $$\cancel{12}\cdot 3t\cdot\frac{1}{\cancel{12}} + \cancel{9}\cdot 4t\cdot\frac{1}{\cancel{9}} = 36\cdot\cancel{t}\cdot\frac{1}{\cancel{t}}$$

   $$3t + 4t = 36$$

   $$7t = 36$$

   $$t = \frac{36}{7} = 5\frac{1}{7}$$

   They can do the job in $5\frac{1}{7}$ hours working together.

9. Let b = the speed of the boat in still water. We are told that the time going downstream is the same as the time going upstream. Thus:

$$t_{DOWNSTREAM} = t_{UPSTREAM}, \text{ or}$$

$$\left(\frac{d}{r}\right)_{DOWNSTREAM} = \left(\frac{d}{r}\right)_{UPSTREAM}$$

$$\frac{35}{b+4} = \frac{15}{b-4}$$

$$35(b-4) = 15(b+4)$$

$$35b - 140 = 15b + 60$$

$$35b - 15b = 60 + 140$$

$$20b = 200$$

$$b = \frac{200}{20} = 10$$

Drew's speed in still water is 10 mph.

10. $\frac{m}{16} = \frac{15}{10} = \frac{3}{2} \Rightarrow m = \frac{3}{2} \cdot 16 = \frac{3}{\cancel{2}} \cdot \cancel{2} \cdot 8 = 24$

11. $\left(32x^6 + 18x^5 - 27x^2\right) \div (6x^2)$

$$= \frac{32x^6 + 18x^5 - 27x^2}{6x^2}$$

$$= \frac{32x^6}{6x^2} + \frac{18x^5}{6x^2} - \frac{27x^2}{6x^2}$$

$$= \frac{2x^2 \cdot 16x^4}{2x^2 \cdot 3} + \frac{6x^2 \cdot 3x^3}{6x^2 \cdot 1} - \frac{3x^2 \cdot 9}{3x^2 \cdot 2}$$

$$= \frac{16}{3}x^4 + 3x^3 - \frac{9}{2}$$

12.
$$\begin{array}{r} x - 5 \\ x-4 \overline{)\, x^2 - 9x + 21} \\ -(\underline{x^2 - 4x}) \\ -5x + 21 \\ \underline{-(-5x + 20)} \\ 1 \end{array}$$

The answer is $x - 5$, R 1, or $x - 5 + \dfrac{1}{x-4}$.

13.
$$\begin{array}{r|rrrrr} 4 & 1 & -1 & -19 & 49 & -30 \\ & & 4 & 12 & -28 & 84 \\ \hline & 1 & 3 & -7 & 21 & 54 \end{array}$$

The remainder after applying synthetic division is 54. Therefore, by the Remainder Theorem, $f(4) = 54$.

14. If y varies directly as x, then $y = kx$. Therefore,

$$k = \frac{y}{x} = \frac{10}{0.2} = \frac{10 \cdot 10}{0.2 \cdot 10} = \frac{100}{2} = 50$$

$$y = 50x$$

15. If y varies inversely as x, then $y = \dfrac{k}{x}$.

Therefore,
$$k = xy = 5 \cdot 8 = 40$$

$$y = \frac{40}{x}$$

16. If y varies jointly as x and z, then $y = kxz$. Therefore,

$$y = kxz$$

$$k = \frac{y}{xz} = \frac{2}{5 \cdot 4} = \frac{2}{20} = \frac{\cancel{2}}{\cancel{2} \cdot 10} = \frac{1}{10}$$

$$y = \frac{1}{10}xz, \text{ or } \frac{xz}{10}$$

Chapter 6 Review Exercises

1. True

2. True

3. False

4. False

5. False

6. True

7. True

8. False

9. True

10. True

11. $f(t) = \dfrac{t^2 - 3t + 2}{t^2 - 9}$

a) $f(0) = \dfrac{0^2 - 3 \cdot 0 + 2}{0^2 - 9}$

$$= -\frac{2}{9}$$

b) $f(-1) = \dfrac{(-1)^2 - 3 \cdot (-1) + 2}{(-1)^2 - 9}$

$= \dfrac{6}{-8} = -\dfrac{3}{4}$

c) $f(2) = \dfrac{2^2 - 3 \cdot 2 + 2}{2^2 - 9}$

$= \dfrac{0}{-5} = 0$

12. $20x^3 = 2 \cdot 2 \cdot 5 \cdot x \cdot x \cdot x$

$24x^2 = 2 \cdot 2 \cdot 2 \cdot 3 \cdot x \cdot x$

$\text{LCM} = 2 \cdot 2 \cdot 2 \cdot 3 \cdot 5 \cdot x \cdot x \cdot x = 120x^3$

13. $x^2 + 8x - 20 = (x + 10)(x - 2)$

$x^2 + 7x - 30 = (x + 10)(x - 3)$

$\text{LCM} = (x + 10)(x - 2)(x - 3)$

14. $\dfrac{x^2}{x - 8} - \dfrac{64}{x - 8}$

$= \dfrac{x^2 - 64}{x - 8}$ Subtracting the numerators

$= \dfrac{(x + 8)(x - 8)}{x - 8}$ Factoring the numerator

$= \dfrac{(x + 8)\,\cancel{(x - 8)}}{\cancel{x - 8}}$ Removing the factor $\dfrac{x - 8}{x - 8} = 1$

$= x + 8$

15. $\dfrac{12a^2 b^3}{5c^3 d^2} \cdot \dfrac{25c^9 d^4}{9a^7 b}$

$= \dfrac{2 \cdot 2 \cdot 3 \cdot a^2 b^3 \cdot 5 \cdot 5 \cdot c^9 d^4}{5c^3 d^2 9a^7 b}$ Multiplying the numerators and the denominators

$= \dfrac{2 \cdot 2 \cdot \cancel{3} \cdot \cancel{a^2}\,\cancel{b} \cdot b^2 \cdot \cancel{5} \cdot 5\cancel{c^3} \cdot c^6\,\cancel{d^2} \cdot d^2}{\cancel{5}\,\cancel{c^3}\,\cancel{d^2} \cdot \cancel{3} \cdot \cancel{a^2} \cdot a^5\,\cancel{b}}$ Factoring out factors of 1

$= \dfrac{20b^2 c^6 d^2}{3a^5}$

16. $\dfrac{5}{6m^2 n^3 p} + \dfrac{7}{9mn^4 p^2}, \qquad \text{LCD is } 18m^2 n^4 p^2$

$= \dfrac{5}{6m^2 n^3 p} \cdot \dfrac{3np}{3np} + \dfrac{7}{9mn^4 p^2} \cdot \dfrac{2m}{2m}$

Multiplying by forms of 1

$= \dfrac{15np}{18m^2 n^4 p^2} + \dfrac{14m}{18m^2 n^4 p^2}$

$= \dfrac{15np + 14m}{18m^2 n^4 p^2}$ Adding the numerators

17. $\dfrac{x^3 - 8}{x^2 - 25} \cdot \dfrac{x^2 + 10x + 25}{x^2 + 2x + 4}$

$= \dfrac{(x - 2)(x^2 + 2x + 4)}{(x - 5)(x + 5)} \cdot \dfrac{(x + 5)^2}{x^2 + 2x + 4}$

Factoring the numerators and denominators

$= \dfrac{(x - 2)(x^2 + 2x + 4)(x + 5)(x + 5)}{(x - 5)(x + 5)(x^2 + 2x + 4)}$

Multiplying the numerators & denominators

$= \dfrac{(x - 2)\,\cancel{(x^2 + 2x + 4)}\,\cancel{(x + 5)}(x + 5)}{(x - 5)\,\cancel{(x + 5)}\,\cancel{(x^2 + 2x + 4)}}$

Removing factors of 1

$= \dfrac{(x - 2)(x + 5)}{x - 5}$

18. $\dfrac{x^2 - 4x - 12}{x^2 - 6x + 8} \div \dfrac{x^2 - 4}{x^3 - 64}$

$= \dfrac{x^2 - 4x - 12}{x^2 - 6x + 8} \cdot \dfrac{x^3 - 64}{x^2 - 4}$ Multiplying by the reciprocal of the divisor

$= \dfrac{(x - 6)(x + 2)}{(x - 4)(x - 2)} \cdot \dfrac{(x - 4)(x^2 + 4x + 16)}{(x + 2)(x - 2)}$

$= \dfrac{(x - 6)\,\cancel{(x + 2)}}{\cancel{(x - 4)}(x - 2)} \cdot \dfrac{\cancel{(x - 4)}(x^2 + 4x + 16)}{\cancel{(x + 2)}(x - 2)}$

$= \dfrac{(x - 6)(x^2 + 4x + 16)}{(x - 2)(x - 2)}$

$= \dfrac{(x - 6)(x^2 + 4x + 16)}{(x - 2)^2}$

19. $\dfrac{x}{x^2+5x+6} - \dfrac{2}{x^2+3x+2}$

$= \dfrac{x}{(x+3)(x+2)} - \dfrac{2}{(x+2)(x+1)}$

$\left[\text{LCD} = (x+3)(x+2)(x+1) \right]$

$= \dfrac{x}{(x+3)(x+2)} \cdot \dfrac{x+1}{x+1} - \dfrac{2}{(x+2)(x+1)} \cdot \dfrac{x+3}{x+3}$

$= \dfrac{x(x+1) - 2(x+3)}{(x+3)(x+2)(x+1)} = \dfrac{x^2+x-2x-6}{(x+3)(x+2)(x+1)}$

$= \dfrac{x^2-x-6}{(x+3)(x+2)(x+1)} = \dfrac{(x-3)(x+2)}{(x+3)(x+2)(x+1)}$

$= \dfrac{(x-3)\cancel{(x+2)}}{(x+3)\cancel{(x+2)}(x+1)} = \dfrac{x-3}{(x+3)(x+1)}$

20. $\dfrac{-4xy}{x^2-y^2} + \dfrac{x+y}{x-y} = \dfrac{-4xy}{(x+y)(x-y)} + \dfrac{x+y}{x-y}$

$\qquad$ LCD is $(x+y)(x-y)$

$= \dfrac{-4xy}{(x+y)(x-y)} + \dfrac{x+y}{x-y} \cdot \dfrac{x+y}{x+y}$

$= \dfrac{-4xy}{(x+y)(x-y)} + \dfrac{(x+y)^2}{(x+y)(x-y)}$

$= \dfrac{-4xy + x^2 + 2xy + y^2}{(x+y)(x-y)} = \dfrac{x^2 - 2xy + y^2}{(x+y)(x-y)}$

$= \dfrac{(x-y)^2}{(x+y)(x-y)} = \dfrac{(x-y)\cancel{(x-y)}}{(x+y)\cancel{(x-y)}} = \dfrac{x-y}{x+y}$

21. $\dfrac{5a^2}{a-b} + \dfrac{5b^2}{b-a} = \dfrac{5a^2}{a-b} - \dfrac{5b^2}{a-b}$ $\quad$ $b-a$ and $a-b$ are opposites

$= \dfrac{5a^2 - 5b^2}{a-b}$

$= \dfrac{5(a-b)(a+b)}{a-b}$

$= \dfrac{5\cancel{(a-b)}(a+b)}{\cancel{a-b}}$

$= 5(a+b)$

22. $\dfrac{3}{y+4} - \dfrac{y}{y-1} + \dfrac{y^2+3}{y^2+3y-4}$

$= \dfrac{3}{y+4} - \dfrac{y}{y-1} + \dfrac{y^2+3}{(y+4)(y-1)}$

$\left[\text{LCD is } (y+4)(y-1) \right]$

$= \dfrac{3}{y+4} \cdot \dfrac{y-1}{y-1} - \dfrac{y}{y-1} \cdot \dfrac{y+4}{y+4} + \dfrac{y^2+3}{(y+4)(y-1)}$

$= \dfrac{3(y-1) - y(y+4) + y^2+3}{(y+4)(y-1)}$

$= \dfrac{3y-3 - y^2 - 4y + y^2 + 3}{(y+4)(y-1)} = \dfrac{-y}{(y+4)(y-1)}$

23. $f(x) = \dfrac{4x-2}{x^2-5x+4} - \dfrac{3x+2}{x^2-5x+4}$

$= \dfrac{(4x-2) - (3x+2)}{x^2-5x+4}$

$= \dfrac{4x-2-3x-2}{x^2-5x+4}$

$= \dfrac{x-4}{(x-4)(x-1)}$ $\qquad$ Note $x \ne 4$ and $x \ne 1$

$= \dfrac{\cancel{(x-4)} \cdot 1}{\cancel{(x-4)}(x-1)}$

$= \dfrac{1}{x-1}$

$= \dfrac{1}{x-1}, x \ne 1, 4$

24. $f(x) = \dfrac{x+8}{x+5} \cdot \dfrac{2x+10}{x^2-64}$

$= \dfrac{x+8}{x+5} \cdot \dfrac{2(x+5)}{(x+8)(x-8)}$ $\quad$ Note: $x \ne -5,$ $x \ne -8, x \ne 8$

$= \dfrac{(x+8) \cdot 2(x+5)}{(x+5)(x+8)(x-8)}$

$= \dfrac{\cancel{(x+8)} \cdot 2\cancel{(x+5)}}{\cancel{(x+5)}\cancel{(x+8)}(x-8)}$

$= \dfrac{2}{x-8}$

$f(x) = \dfrac{2}{x-8}, x \ne -8, -5, 8$

25. $f(x) = \dfrac{9x^2 - 1}{x^2 - 9} \div \dfrac{3x + 1}{x + 3}$

$= \dfrac{(3x+1)(3x-1)}{(x+3)(x-3)} \div \dfrac{3x+1}{x+3}$

Note: $x \neq -3, x \neq 3,$ and $x \neq -\dfrac{1}{3}$

$= \dfrac{(3x+1)(3x-1)}{(x+3)(x-3)} \cdot \dfrac{x+3}{3x+1}$

$= \dfrac{(3x+1)(3x-1)(x+3)}{(x+3)(x-3)(3x+1)}$

$= \dfrac{\cancel{(3x+1)}(3x-1)\cancel{(x+3)}}{\cancel{(x+3)}(x-3)\cancel{(3x+1)}} = \dfrac{3x-1}{x-3}$

$= \dfrac{3x-1}{x-3}, x \neq -3, -\dfrac{1}{3}, 3$

26. $\dfrac{\dfrac{4}{x} - 4}{\dfrac{9}{x} - 9},$ LCD is x

$= \dfrac{\dfrac{4}{x} - 4}{\dfrac{9}{x} - 9} \cdot \dfrac{x}{x}$ Multiplying by 1 using the LCD

$= \dfrac{\dfrac{4}{x} \cdot x - 4x}{\dfrac{9}{x} \cdot x - 9x} = \dfrac{4 - 4x}{9 - 9x} = \dfrac{4(1-x)}{9(1-x)}$

$= \dfrac{4\cancel{(1-x)}}{9\cancel{(1-x)}} = \dfrac{4}{9}$

27. $\dfrac{\dfrac{5}{2x^2}}{\dfrac{3}{4x} + \dfrac{4}{x^3}} = \dfrac{\dfrac{5}{2x^2}}{\dfrac{3}{4x} + \dfrac{4}{x^3}} \cdot \dfrac{4x^3}{4x^3}$

$= \dfrac{4x^3 \cdot \dfrac{5}{2x^2}}{4x^3 \cdot \dfrac{3}{4x} + 4x^3 \cdot \dfrac{4}{x^3}}$

$= \dfrac{2x \cdot 5}{x^2 \cdot 3 + 4 \cdot 4} = \dfrac{10x}{3x^2 + 16}$

28. $\dfrac{\dfrac{y^2 + 4y - 77}{y^2 - 10y + 25}}{\dfrac{y^2 - 5y - 14}{y^2 - 25}}$

$= \dfrac{y^2 + 4y - 77}{y^2 - 10y + 25} \cdot \dfrac{y^2 - 25}{y^2 - 5y - 14}$

Multiplying by the reciprocal of the divisor

$= \dfrac{(y+11)(y-7)}{(y-5)^2} \cdot \dfrac{(y+5)(y-5)}{(y-7)(y+2)}$

$= \dfrac{(y+11)(y-7)(y+5)(y-5)}{(y-5)(y-5)(y-7)(y+2)}$

$= \dfrac{(y+11)\cancel{(y-7)}(y+5)\cancel{(y-5)}}{\cancel{(y-5)}(y-5)\cancel{(y-7)}(y+2)}$

$= \dfrac{(y+11)(y+5)}{(y-5)(y+2)}$

29. $\dfrac{\dfrac{5}{x^2 - 9} - \dfrac{3}{x+3}}{\dfrac{4}{x^2 + 6x + 9} + \dfrac{2}{x-3}}$

$= \dfrac{\dfrac{5}{(x+3)(x-3)} - \dfrac{3}{x+3}}{\dfrac{4}{(x+3)^2} + \dfrac{2}{x-3}}$

$\left[\text{LCD is } (x+3)^2(x-3)\right]$

$= \dfrac{(x+3)^2(x-3)}{(x+3)^2(x-3)} \cdot \dfrac{\dfrac{5}{(x+3)(x-3)} - \dfrac{3}{x+3}}{\dfrac{4}{(x+3)^2} + \dfrac{2}{x-3}}$

$= \dfrac{(x+3)^2(x-3) \cdot \left[\dfrac{5}{(x+3)(x-3)} - \dfrac{3}{x+3}\right]}{(x+3)^2(x-3) \cdot \left[\dfrac{4}{(x+3)^2} + \dfrac{2}{x-3}\right]}$

$= \dfrac{5(x+3) - 3(x+3)(x-3)}{4(x-3) + 2(x+3)^2}$

$$= \frac{5x+15-3\left(x^2-9\right)}{4x-12+2\left(x^2+6x+9\right)}$$

$$= \frac{5x+15-3x^2+27}{4x-12+2x^2+12x+18}$$

$$= \frac{-3x^2+5x+42}{2x^2+16x+6}$$

We can factor the numerator and denominator, but doing so leads to no further simplification.

30. $\dfrac{3}{x}+\dfrac{7}{x}=5$ LCD is $x; x \neq 0$

$x\left(\dfrac{3}{x}+\dfrac{7}{x}\right)=x\cdot 5$ Multiplying both sides by x

$3+7=5x$

$10=5x$

$2=x$

The solution is 2.

31. $\dfrac{5}{3x+2}=\dfrac{3}{2x}$

LCD is $2x(3x+2)$

$x \neq -\dfrac{2}{3}$ and $x \neq 0$

$2x(3x+2)\cdot\dfrac{5}{3x+2}=2x(3x+2)\cdot\dfrac{3}{2x}$

$2x\cdot 5=(3x+2)\cdot 3$

$10x=9x+6$

$x=6$

The solution is 6.

32. $\dfrac{4x}{x+1}+\dfrac{4}{x}+9=\dfrac{4}{x^2+x}$

$\dfrac{4x}{x+1}+\dfrac{4}{x}+9=\dfrac{4}{x(x+1)}$

$\left[\text{LCD is } x(x+1); x \neq 0, x \neq -1\right]$

$x(x+1)\left(\dfrac{4x}{x+1}+\dfrac{4}{x}+9\right)=x(x+1)\cdot\dfrac{4}{x(x+1)}$

$4x^2+4(x+1)+9x\cdot(x+1)=4$

$4x^2+4x+4+9x^2+9x=4$

$13x^2+13x=0$

$13x(x+1)=0$

$13x = 0 \; or \; x+1=0$

$x = 0 \; or \quad x = -1$

Since both $x = 0$ and $x = -1$ are excluded from the domain, the original equation has no solution.

33. $\dfrac{x+6}{x^2+x-6}+\dfrac{x}{x^2+4x+3}=\dfrac{x+2}{x^2-x-2}$

$\dfrac{x+6}{(x+3)(x-2)}+\dfrac{x}{(x+3)(x+1)}=$

$\dfrac{x+2}{(x-2)(x+1)}$

Note: $x \neq -3, x \neq 2, x \neq -1$

$\left[\text{LCD is } (x+3)(x-2)(x+1)\right]$

$(x+3)(x-2)(x+1)\left(\dfrac{x+6}{(x+3)(x-2)}+\right.$

$\left.\dfrac{x}{(x+3)(x+1)}\right)=$

$(x+3)(x-2)(x+1)\cdot\dfrac{x+2}{(x-2)(x+1)}$

$(x+1)(x+6)+(x-2)x=(x+3)(x+2)$

$x^2+7x+6+x^2-2x=x^2+5x+6$

$x^2=0$

$x=0$

34. $\dfrac{x}{x-3}-\dfrac{3x}{x+2}=\dfrac{5}{(x-3)(x+2)}$

Note: $x \neq 3, x \neq -2$

$\left[\text{LCD is } (x-3)(x+2)\right]$

$(x-3)(x+2)\left(\dfrac{x}{x-3}-\dfrac{3x}{x+2}\right)$

$=(x-3)(x+2)\dfrac{5}{(x-3)(x+2)}$

$x(x+2)-3x(x-3)=5$

$x^2+2x-3x^2+9x=5$

$-2x^2+11x-5=0$

$2x^2-11x+5=0$

$(2x-1)(x-5)=0$

$x=\dfrac{1}{2}, 5$

35. $f(x) = \dfrac{2}{x-1} + \dfrac{2}{x+2}$

$f(a) = \dfrac{2}{a-1} + \dfrac{2}{a+2}$ Substitute a for x

$1 = \dfrac{2}{a-1} + \dfrac{2}{a+2}$ Substitute 1 for $f(a)$

Note: $a \neq 1$ and $a \neq -2$

LCD is $(a-1)(a+2)$

$(a-1)(a+2) \cdot 1 = (a-1)(a+2)\left(\dfrac{2}{a-1} + \dfrac{2}{a+2}\right)$

$a^2 + a - 2 = 2(a+2) + 2(a-1)$

$a^2 + a - 2 = 2a + 4 + 2a - 2$

$a^2 - 3a - 4 = 0$

$(a-4)(a+1) = 0$

$a - 4 = 0 \ or \ a + 1 = 0$

$a = 4 \ or \qquad a = -1$

The solutions are -1 and 4.

36. **Familiarize.** Meg can do $\dfrac{1}{9}$ of the job in 1

hour, and Kelly can do $\dfrac{1}{12}$ of the same job in

1 hour. Let t = time worked together.
Translate. Rate × Time = Work. We have the equation:

$\dfrac{1}{9} \cdot t + \dfrac{1}{12} \cdot t = 1$, or $\dfrac{t}{9} + \dfrac{t}{12} = 1$

Carry out. Solve the equation.

$\dfrac{t}{9} + \dfrac{t}{12} = 1$, LCD = 36

$36\left(\dfrac{t}{9} + \dfrac{t}{12}\right) = 36 \cdot 1$

$36 \cdot \dfrac{t}{9} + 36 \cdot \dfrac{t}{12} = 36 \cdot 1$

$4t + 3t = 36$

$7t = 36$

$t = \dfrac{36}{7}$, or $5\dfrac{1}{7}$

Check. Meg's work is $\dfrac{1}{9} \cdot \dfrac{36}{7} = \dfrac{4}{7}$, and

Kelly's work is $\dfrac{1}{12} \cdot \dfrac{36}{7} = \dfrac{3}{7}$. Together they

do $\dfrac{4}{7} + \dfrac{3}{7}$, or 1 entire job. The number

checks.

State. Meg and Kelly, working together, can

arrange the books in $5\dfrac{1}{7}$ hours.

37. **Familiarize.** Let t = time for the Quad to process a file, then $t + 15$ = time for the Duo to process a file..

Translate. Using $R = \dfrac{W}{T}$, the rate of the

Quad is $\dfrac{1}{t}$ and the rate of the Duo is $\dfrac{1}{t+15}$.

Knowing it takes them 18 sec when they are working together, we write the equation.

$\dfrac{1}{t} \cdot 18 + \dfrac{1}{t+15} \cdot 18 = 1$, or $\dfrac{18}{t} + \dfrac{18}{t+15} = 1$

Carry out. We solve the equation.

$\dfrac{18}{t} + \dfrac{18}{t+15} = 1$, LCD is $t(t+15)$

$t(t+15)\left(\dfrac{18}{t} + \dfrac{18}{t+15}\right) = t(t+15) \cdot 1$

$18(t+15) + 18t = t(t+15)$

$18t + 270 + 18t = t^2 + 15t$

$0 = t^2 - 21t - 270$

$0 = (t-30)(t+9)$

$t - 30 = 0 \ or \ t + 9 = 0$

$t = 30 \ or \qquad t = -9$

Check. Since time cannot be negative, we

only check 30. The Quad does $\dfrac{18}{30}$, or $\dfrac{3}{5}$ of

the job, and the Duo does $\dfrac{18}{30+15} = \dfrac{18}{45}$, or $\dfrac{2}{5}$

of the job. Together they do $\dfrac{3}{5} + \dfrac{2}{5}$, or one

complete job.
State. The Quad can process the file in 30 sec and the Duo takes 45 sec.

38. **Familiarize.** The rate of the current is 6 mph. Let R = the speed of the boat in still water. The rate of the boat going downstream, with the current, is $R + 6$, and the rate of the boat going upstream, against the current, is $R - 6$.
Translate. Using the fact that the times are

the same and $T = \dfrac{D}{R}$, we have the equation

$\dfrac{50}{R+6} = \dfrac{30}{R-6}$

Carry out. We solve the equation.

$$\frac{50}{R+6} = \frac{30}{R-6}, \text{ LCD is } (R+6)(R-6)$$

$$(R+6)(R-6) \cdot \frac{50}{R+6} = (R+6)(R-6) \cdot \frac{30}{R-6}$$

$$50(R-6) = 30(R+6)$$

$$50R - 300 = 30R + 180$$

$$20R = 480$$

$$R = 24 \text{ mph}$$

Check. The time downstream is

$$\frac{50}{24+6} = \frac{50}{30}, \text{ or } \frac{5}{3} \text{ and the time upstream is}$$

$$\frac{30}{24-6} = \frac{30}{18}, \text{ or } \frac{5}{3}. \text{ The times are equal; the}$$

number checks.

State. The speed of the boat in still water is 24 mph.

39. *Familiarize.* Let R = speed of the motorcycle and $R + 8$ = speed of the car. The times are

equal. $T = \dfrac{D}{R}$

Translate. The time of the car is $\dfrac{105}{R+8}$, and

the time of the motorcycle is $\dfrac{93}{R}$. Since the

times are the same, we have the equation

$$\frac{105}{R+8} = \frac{93}{R}$$

Carry out. Solve the equation.

LCD = $(R+8)R$

$$(R+8)R \cdot \frac{105}{R+8} = (R+8)R \cdot \frac{93}{R}$$

$$105R = 93R + 744$$

$$12R = 744$$

$$R = 62$$

Check. The time of the car is

$$\frac{105}{62+8} = \frac{105}{70}, \text{ or } \frac{3}{2}; \text{ the time of the}$$

motorcycle is $\dfrac{93}{62}$, or $\dfrac{3}{2}$. The times are

equal; the number checks.

State. The speed of the motorcycle is 62 mph, and the speed of the car is 70 mph.

40. *Familiarize.* The ratio of the total number of tagged seals to the total seal population size ought to be close to the ratio of tagged seals in the sample to the sample size.

Translate. Let P = the seal population in the bay. Then build a proportion equation that reflects the equality of the two ratios:

$$\frac{33}{P} = \frac{24}{40}$$

Carry out. Solve the equation for P.

$$\frac{33}{P} = \frac{24}{40}$$

$$24P = 33 \cdot 40$$

$$P = \frac{33 \cdot 40}{24} = 55$$

Check.

$$\frac{33}{55} = \frac{24}{40}$$

$$0.6 = 0.6$$

The two ratios are equal. The answer checks.

State. There are approximately 55 harbor seals in Bristol Bay.

41. For similar triangles, the ratios of corresponding sides are equal.

$$\frac{x}{2.4} = \frac{8.5}{3.4}$$

$$\frac{x}{2.4} \cdot 2.4 = \frac{8.5}{3.4} \cdot 2.4$$

$$x = 6$$

42. $\left(30r^2s^3 + 25r^2s^2 - 20r^3s^3\right) \div 10r^2s$

$$= \frac{30r^2s^3}{10r^2s} + \frac{25r^2s^2}{10r^2s} - \frac{20r^3s^3}{10r^2s}$$

$$= 3s^2 + \frac{5}{2}s - 2rs^2$$

43. $\left(y^3 + 8\right) \div \left(y + 2\right)$

$$= \frac{y^3 + 2^3}{y+2} = \frac{(y+2)\left(y^2 - 2y + 4\right)}{y+2}$$

$$= \frac{\cancel{(y+2)}\left(y^2 - 2y + 4\right)}{\cancel{y+2}} = y^2 - 2y + 4$$

44. $\left(4x^3 + 3x^2 - 5x - 2\right) \div \left(x^2 + 1\right)$

$$
\begin{array}{r}
4x + 3 \\
x^2 + 1 \overline{)\ 4x^3 + 3x^2 - 5x - 2} \\
\underline{4x^3 \qquad\ + 4x} \\
3x^2 - 9x - 2 \\
\underline{3x^2 \qquad\ + 3} \\
-9x - 5
\end{array}
$$

The answer is $4x + 3, R - 9x - 5$, or

$4x + 3 + \dfrac{-9x - 5}{x^2 + 1}$

45. $\left(x^3 + 3x^2 + 2x - 6\right) \div \left(x - 3\right)$

$$
\begin{array}{r|rrrr}
3 & 1 & 3 & 2 & -6 \\
  &   & 3 & 18 & 60 \\
\hline
  & 1 & 6 & 20 & 54
\end{array}
$$

The answer is $x^2 + 6x + 20$, R 54,

or $x^2 + 6x + 20 + \dfrac{54}{x - 3}$

46. $f(x) = 4x^3 - 6x^2 - 9; f(5)$

$f(x) = 4x^3 - 6x^2 + 0x - 9$

$$
\begin{array}{r|rrrr}
5 & 4 & -6 & 0 & -9 \\
  &   & 20 & 70 & 350 \\
\hline
  & 4 & 14 & 70 & 341
\end{array}
$$

$f(5) = 341$

47. $C = \dfrac{Ag}{g + 12}$

$(g + 12)C = (g + 12) \cdot \dfrac{Ag}{g + 12}$

$gC + 12C = Ag$

$12C = Ag - Cg$

$12C = g(A - C)$

$\dfrac{12C}{A - C} = g$

48. $S = \dfrac{H}{m\left(t_1 - t_2\right)}$

$m \cdot S = m\left[\dfrac{H}{m\left(t_1 - t_2\right)}\right]$ Multiplying both sides of the equation by m

$mS = \dfrac{H}{t_1 - t_2}$

$\dfrac{1}{S} \cdot mS = \dfrac{1}{S} \cdot \dfrac{H}{t_1 - t_2}$ Multiplying both sides of the equation by $\dfrac{1}{S}$

$m = \dfrac{H}{S\left(t_1 - t_2\right)}$

49. $\dfrac{1}{ac} = \dfrac{2}{ab} - \dfrac{3}{bc}$, LCD is abc

$abc \cdot \dfrac{1}{ac} = abc\left(\dfrac{2}{ab} - \dfrac{3}{bc}\right)$

$\dfrac{abc}{ac} = \dfrac{abc \cdot 2}{ab} - \dfrac{abc \cdot 3}{bc}$

$b = 2c - 3a$

$b + 3a = 2c$

$\dfrac{b + 3a}{2} = c$

50. $T = \dfrac{A}{v\left(t_2 - t_1\right)}$

$\left(t_2 - t_1\right)T = \left(t_2 - t_1\right)\dfrac{A}{v\left(t_2 - t_1\right)}$

$Tt_2 - Tt_1 = \dfrac{A}{v}$

$-Tt_1 = \dfrac{A}{v} - Tt_2$

$-\dfrac{1}{T} \cdot -Tt_1 = -\dfrac{1}{T}\left(\dfrac{A}{v} - Tt_2\right)$

$t_1 = -\dfrac{A}{vT} + t_2$, or $\dfrac{-A + vTt_2}{vT}$

51. W varies directly as P, where W is waste and P is the number of people. We write $W = kP$. Using $W = 11.96$ and $P = 2.6$, we substitute to determine k and the equation of variation.

$$11.96 = k \cdot 2.6$$

$$4.6 = k \qquad \text{Variation constant}$$

$$W = 4.6P \qquad \text{Equation of variation}$$

Let $P = 5$ and determine W.

$$W = 4.6 \cdot 5$$

$$W = 23$$

A family of 5 will generate approximately 23 lb of waste daily.

52. The volume V of the dye used varies directly as the square of the diameter, or d^2. We write $V = kd^2$. Using $V = 4$ and $d = 10$, we substitute to determine k and the equation of variation.

$$4 = k \cdot 10^2$$

$$0.04 = k \qquad \text{Variation constant}$$

$$V = 0.04d^2$$

Let $d = 40$ and determine V.

$$V = 0.04 \cdot 40^2$$

$$V = 64$$

64 L of dye is needed for a 40-m wide circle.

53. $y = \dfrac{k}{x}$

Substitute $y = 3$ and $x = \dfrac{1}{4}$ to determine k.

$$3 = \frac{k}{\frac{1}{4}}$$

$$\frac{3}{4} = k \qquad \text{Variaton constant}$$

$$y = \frac{\frac{3}{4}}{x}, \text{ or } y = \frac{3}{4x} \qquad \text{Equation of variation}$$

54. a) Inverse: As the number of servings increases, the serving size decreases.

b) Using $(12, 2)$, $x = 12$ and $y = 2$

$$y = \frac{k}{x}$$

$$2 = \frac{k}{12}$$

$$24 = k \qquad \text{Variation constant}$$

$$y = \frac{24}{x} \qquad \text{Equation of variation}$$

c) $x = 8$

$$y = \frac{24}{8}$$

$$y = 3 \text{ oz}$$

55. *Thinking and Writing Exercise.*

56. *Thinking and Writing Exercise.*

57.
$$\frac{5}{x-13} - \frac{5}{x} = \frac{65}{x^2 - 13}$$

$$\frac{5}{x-13} - \frac{5}{x} = \frac{65}{x(x-13x)}$$

$$\text{Note: } x \neq 13 \text{ and } x \neq 0$$

$$\left[\text{LCD is } x(x-13) \right]$$

$$x(x-13)\left(\frac{5}{x-13} - \frac{5}{x} \right) = x(x-13) \cdot \frac{65}{x(x-13)}$$

$$5x - 5(x-13) = 65$$

$$5x - 5x + 65 = 65$$

$$65 = 65$$

Since 65 = 65 (Reflexive Property of Equality), we have an identity. We must remember to exclude the values which make the denominators equal zero; i.e., the solution is the domain. $\{x | x \text{ is a real number, and } x \neq 0 \text{ and } x \neq 13\}$

58. $\dfrac{\dfrac{x}{x^2-25}+\dfrac{2}{x-5}}{\dfrac{3}{x-5}-\dfrac{4}{x^2-10x+25}}=1$

Note: $x\neq\pm 5$

$\dfrac{x}{x^2-25}+\dfrac{2}{x-5}=\dfrac{3}{x-5}-\dfrac{4}{x^2-10x+25}$

Multiplying both sides of the equation by the denominator

$\left[\text{LCD is }(x-5)^2(x+5)\right]$

$(x-5)^2(x+5)\left[\dfrac{x}{(x-5)(x+5)}+\dfrac{2}{x-5}\right]=$

$\qquad (x-5)^2(x+5)\left[\dfrac{3}{x-5}-\dfrac{4}{(x-5)^2}\right]$

$x(x-5)+2(x-5)(x+5)=$

$\qquad 3(x-5)(x+5)-4(x+5)$

$x^2-5x+2x^2-50=3x^2-75-4x-20$

$\qquad\qquad 45=x$

The solution is 45.

59. $\dfrac{2a^2+5a-3}{a^2}\cdot\dfrac{5a^3+30a^2}{2a^2+7a-4}\div\dfrac{a^2+6a}{a^2+7a+12}$

$=\dfrac{(2a-1)(a+3)}{a^2}\cdot\dfrac{5a^2\,(a+6)}{(2a-1)\,(a+4)}$

$\quad\cdot\dfrac{(a+3)\,(a+4)}{a\,(a+6)}=\dfrac{5(a+3)^2}{a}$

60. $\dfrac{5(x-y)}{(x-y)(x+2y)}-\dfrac{5(x-3y)}{(x+2y)(x-3y)}$

$=\dfrac{5\,(x-y)}{(x-y)\,(x+2y)}-\dfrac{5\,(x-3y)}{(x+2y)\,(x-3y)}$

$=\dfrac{5}{x+2y}-\dfrac{5}{x+2y}=\dfrac{0}{x+2y}$

$=0$

Chapter 6 Test

1. $\dfrac{t+1}{t+3}\cdot\dfrac{5t+15}{4t^2-4}$

$=\dfrac{t+1}{t+3}\cdot\dfrac{5(t+3)}{4(t+1)(t-1)}$

$=\dfrac{(t+1)\cdot 5\cdot(t+3)}{(t+3)\cdot 4\cdot(t+1)(t-1)}$

$=\dfrac{5}{4(t-1)}$

2. $\dfrac{x^3+27}{x^2-16}\div\dfrac{x^2+8x+15}{x^2+x-20}$

$=\dfrac{x^3+27}{x^2-16}\cdot\dfrac{x^2+x-20}{x^2+8x+15}$

$=\dfrac{(x+3)(x^2-3x+9)}{(x+4)(x-4)}\cdot\dfrac{(x+5)(x-4)}{(x+5)(x+3)}$

$=\dfrac{(x+3)(x^2-3x+9)(x+5)(x-4)}{(x+4)(x-4)(x+5)(x+3)}$

$=\dfrac{x^2-3x+9}{x+4}$

3. $\dfrac{25x}{x+5}+\dfrac{x^3}{x+5}=\dfrac{25x+x^3}{x+5}$

4. $\dfrac{3a^2}{a-b}-\dfrac{3b^2-6ab}{b-a}$

$=\dfrac{3a^2}{a-b}+\dfrac{1}{-1}\cdot\dfrac{3b^2-6ab}{b-a}$

$=\dfrac{3a^2}{a-b}+\dfrac{3b^2-6ab}{a-b}$

$=\dfrac{3a^2-6ab+3b^2}{a-b}$

$=\dfrac{3(a-b)^2}{a-b}$

$=\dfrac{3(a-b)(a-b)}{a-b}$

$=3(a-b)$

5. $\dfrac{4ab}{a^2-b^2}+\dfrac{a^2+b^2}{a+b}$

$=\dfrac{4ab}{(a+b)(a-b)}+\dfrac{a^2+b^2}{a+b}$

LCD is $(a+b)(a-b)$

$=\dfrac{4ab}{(a+b)(a-b)}+\dfrac{a-b}{a-b}\cdot\dfrac{a^2+b^2}{a+b}$

$=\dfrac{4ab}{(a+b)(a-b)}+\dfrac{a^3+ab^2-a^2b-b^3}{(a+b)(a-b)}$

$=\dfrac{4ab+a^3+ab^2-a^2b-b^3}{(a+b)(a-b)}$

$=\dfrac{a^3-a^2b+4ab+ab^2-b^3}{(a+b)(a-b)}$

6. $\dfrac{6}{x^3-64}-\dfrac{4}{x^2-16}$

$=\dfrac{6}{(x-4)(x^2+4x+16)}-\dfrac{4}{(x+4)(x-4)}$

$\left[\text{LCD is }(x+4)(x-4)(x^2+4x+16)\right]$

$=\dfrac{x+4}{x+4}\cdot\dfrac{6}{(x-4)(x^2+4x+16)}-$

$\qquad\dfrac{x^2+4x+16}{x^2+4x+16}\cdot\dfrac{4}{(x+4)(x-4)}$

$=\dfrac{6(x+4)-4(x^2+4x+16)}{(x+4)(x-4)(x^2+4x+16)}$

$=\dfrac{6x+24-4x^2-16x-64}{(x+4)(x-4)(x^2+4x+16)}$

$=\dfrac{-4x^2-10x-40}{(x+4)(x-4)(x^2+4x+16)}$, or

$=\dfrac{-2(2x^2+5x+20)}{(x+4)(x-4)(x^2+4x+16)}$

7. $f(x)=\dfrac{4}{x+3}-\dfrac{x}{x-2}+\dfrac{x^2+4}{x^2+x-6}$

$=\dfrac{4}{x+3}-\dfrac{x}{x-2}+\dfrac{x^2+4}{(x+3)(x-2)}$

Note: $x\ne-3$ and $x\ne2$

$\left[\text{LCD is }(x+3)(x-2)\right]$

$=\dfrac{x-2}{x-2}\cdot\dfrac{4}{x+3}-\dfrac{x+3}{x+3}\cdot\dfrac{x}{x-2}+$

$\qquad\dfrac{x^2+4}{(x+3)(x-2)}$

$=\dfrac{4x-8-x^2-3x+x^2+4}{(x+3)(x-2)}$

$=\dfrac{x-4}{(x+3)(x-2)}$

$f(x)=\dfrac{x-4}{(x+3)(x-2)},x\ne-3,2$

8. $f(x)=\dfrac{x^2-1}{x+2}\div\dfrac{x^2-2x}{x^2+x-2}$

$=\dfrac{x^2-1}{x+2}\div\dfrac{x(x-2)}{(x+2)(x-1)}$

Note: $x\ne-2$ and $x\ne1$

$=\dfrac{(x+1)(x-1)}{x+2}\cdot\dfrac{(x+2)(x-1)}{x(x-2)}$

Note: $x\ne0$ and $x\ne2$

$=\dfrac{(x+1)(x-1)\,\cancel{(x+2)}\,(x-1)}{\cancel{(x+2)}\,x(x-2)}$

$f(x)=\dfrac{(x+1)(x-1)^2}{x(x-2)},x\ne-2,0,1,2$

9. $\dfrac{\dfrac{2}{a}+\dfrac{3}{b}}{\dfrac{5}{ab}+\dfrac{1}{a^2}}$ LCD is a^2b

$=\dfrac{a^2b}{a^2b}\cdot\left(\dfrac{\dfrac{2}{a}+\dfrac{3}{b}}{\dfrac{5}{ab}+\dfrac{1}{a^2}}\right)$

$=\dfrac{\dfrac{a^2b\cdot 2}{a}+\dfrac{a^2b\cdot 3}{b}}{\dfrac{a^2b\cdot 5}{ab}+\dfrac{a^2b}{a^2}}$

$=\dfrac{2ab+3a^2}{5a+b}$, or $\dfrac{a(2b+3a)}{5a+b}$

10. $\dfrac{\dfrac{x^2-5x-36}{x^2-36}}{\dfrac{x^2+x-12}{x^2-12x+36}}$

$=\dfrac{x^2-5x-36}{x^2-36}\div\dfrac{x^2+x-12}{x^2-12x+36}$

$=\dfrac{x^2-5x-36}{x^2-36}\cdot\dfrac{x^2-12x+36}{x^2+x-12}$

$=\dfrac{(x-9)(x+4)}{(x+6)(x-6)}\cdot\dfrac{(x-6)^2}{(x+4)(x-3)}$

$=\dfrac{(x-9)\cancel{(x+4)}\cancel{(x-6)}(x-6)}{(x+6)\cancel{(x-6)}\cancel{(x+4)}(x-3)}$

$=\dfrac{(x-9)(x-6)}{(x+6)(x-3)}$

11. $\dfrac{\dfrac{x}{8}-\dfrac{8}{x}}{\dfrac{1}{8}+\dfrac{1}{x}}=\dfrac{\dfrac{x}{8}-\dfrac{8}{x}}{\dfrac{1}{8}+\dfrac{1}{x}}\cdot\dfrac{8x}{8x}=\dfrac{8x\cdot\dfrac{x}{8}-8x\cdot\dfrac{8}{x}}{8x\cdot\dfrac{1}{8}+8x\cdot\dfrac{1}{x}}$

$=\dfrac{x^2-8^2}{x+8}=\dfrac{(x+8)(x-8)}{x+8}$

$=x-8$

12. $\dfrac{1}{t}+\dfrac{1}{3t}=\dfrac{1}{2}$ Note: $t\neq 0$

[LCD is $6t$]

$\dfrac{1}{t}\cdot 6t+\dfrac{1}{3t}\cdot 6t=\dfrac{1}{2}\cdot 6t$

$6+2=3t$

$8=3t$

$t=\dfrac{8}{3}$

The solution is $\dfrac{8}{3}$.

13. $\dfrac{t+11}{t^2-t-12}+\dfrac{1}{t-4}=\dfrac{4}{t+3}$ Note: $x\neq 4,-3$

$\left[\text{LCD is }(t-4)(t+3)\right]$

$(t-4)(t+3)\left[\dfrac{t+11}{(t-4)(t+3)}+\dfrac{1}{t-4}\right]=$

$(t-4)(t+3)\cdot\dfrac{4}{t+3}$

$t+11+t+3=4(t-4)$

$2t+14=4t-16$

$30=2t$

$15=t$

The solution is 15.

14. $\dfrac{15}{x}-\dfrac{15}{x-2}=-2$ Note: $x\neq 0,2$

$\left[\text{LCD is }x(x-2)\right]$

$x(x-2)\cdot\dfrac{15}{x}-x(x-2)\cdot\dfrac{15}{x-2}=-2\cdot x(x-2)$

$15(x-2)-15x=-2x^2+4x$

$15x-30-15x=-2x^2+4x$

$2x^2-4x-30=0$

$x^2-2x-15=0$

$(x-5)(x+3)=0$

The solutions are -3 and 5.

15. $f(x)=\dfrac{x+5}{x-1}$

$f(0)=\dfrac{0+5}{0-1}=\dfrac{5}{-1}=-5$

$f(-3)=\dfrac{-3+5}{-3-1}=\dfrac{2}{-4}=-\dfrac{1}{2}$

16. $f(a) = \dfrac{a+5}{a-1}$,   Note: $a \neq 1$

$10 = \dfrac{a+5}{a-1}$,   LCD is $a-1$

$(a-1) \cdot 10 = (a-1) \cdot \dfrac{a+5}{a-1}$

$10a - 10 = a + 5$

$9a = 15$

$a = \dfrac{15}{9} = \dfrac{5}{3}$

$f\left(\dfrac{5}{3}\right) = 10$

17. $\left(16a^4 b^3 c - 10a^5 b^2 c^2 + 12a^2 b^2 c\right) \div \left(4a^2 b\right)$

$= \dfrac{16a^4 b^3 c}{4a^2 b} - \dfrac{10a^5 b^2 c^2}{4a^2 b} + \dfrac{12a^2 b^2 c}{4a^2 b}$

$= 4a^2 b^2 c - \dfrac{5}{2} a^3 b c^2 + 3bc$

18.
$$\begin{array}{r} y - 14 \\ y-6 \overline{\smash{\big)}\ y^2 - 20y + 64} \\ \underline{y^2 - 6y} \\ -14y + 64 \\ \underline{-14y + 84} \\ -20 \end{array}$$

The answer is $y - 14$, $R - 20$,

or $y - 14 + \dfrac{-20}{y-6}$.

19.
$$\begin{array}{r} 6x^2 - 9 \\ x^2 + 2 \overline{\smash{\big)}\ 6x^4 + 3x^2 + 5x + 4} \\ \underline{6x^4 + 12x^2} \\ -9x^2 + 5x + 4 \\ \underline{-9x^2 \qquad -18} \\ 5x + 22 \end{array}$$

The answer is $6x^2 - 9$, $R\ 5x + 22$, or

$6x^2 - 9 + \dfrac{5x + 22}{x^2 + 2}$.

20. $\left(x^3 + 5x^2 + 4x - 7\right) \div (x - 2)$

$$\begin{array}{r|rrrr} 2 & 1 & 5 & 4 & -7 \\ & & 2 & 14 & 36 \\ \hline & 1 & 7 & 18 & \big|\ 29 \end{array}$$

The answer is $x^2 + 7x + 18 + \dfrac{29}{x-2}$

21.
$$\begin{array}{r|rrrrr} 4 & 3 & -5 & 0 & 2 & -7 \\ & & 12 & 28 & 112 & 456 \\ \hline & 3 & 7 & 28 & 114 & \big|\ 449 \end{array}$$

$f(4) = 449$

22. $R = \dfrac{gs}{g+s}$

$R(g + s) = gs$

$Rg + Rs = gs$

$Rg = gs - Rs$

$Rg = s(g - R)$

$s = \dfrac{Rg}{g - R}$

23. Let $x =$ the first integer and $x+1 =$ the second integer. The product of the reciprocals is $\dfrac{1}{110}$, so:

$\dfrac{1}{x} \cdot \dfrac{1}{1+x} = \dfrac{1}{110}$, LCD is $110x(x+1)$

$\dfrac{1}{x(x+1)} = \dfrac{1}{110}$

$110x(x+1) \cdot \dfrac{1}{x(x+1)} = 110x(x+1) \cdot \dfrac{1}{110}$

$110 = x(x+1)$

$0 = x^2 + x - 110$

$0 = (x+11)(x-10)$

$x + 11 = 0$ or $x - 10 = 0$

$x = -11$ or $x = 10$

If $x = -11$, $x+1 = -10$. $\dfrac{1}{-11} \cdot \dfrac{1}{-10} = \dfrac{1}{110}$

TRUE

If $x = 10$, $x+1 = 11$. $\dfrac{1}{10} \cdot \dfrac{1}{11} = \dfrac{1}{110}$ TRUE

Both numbers check.

The solutions are -11 and -10, or 10 and 11.

24. Ella can do $\frac{1}{5}$ of the job in one hour, and Sari

can do $\frac{1}{4}$ of the job in one hour.

Let t = time it takes them working together and write the equation.

$$t \cdot \frac{1}{5} + t \cdot \frac{1}{4} = 1$$

Solve the equation. LCD is 20.

$$20 \left(\frac{t}{5} + \frac{t}{4} \right) = 20 \cdot 1$$

$$4t + 5t = 20$$

$$9t = 20$$

$$t = \frac{20}{9} \text{ hr} = 2\frac{2}{9} \text{ hr}$$

Working together, they can install the

countertop in $2\frac{2}{9}$ hr.

25. Let r = speed of the wind. Terrel's speed with the wind is $12 + r$, and his speed against the wind is $12 - r$. Since the times are the same, we solve the equation $d = rt$ for time and write the equation.

$$d = rt \Rightarrow t = \frac{d}{r}$$

$$\frac{14}{12 + r} = \frac{8}{12 - r}$$

$$14(12 - r) = 8(12 + r)$$

$$168 - 14r = 96 + 8r$$

$$72 = 22r$$

$$r = \frac{72}{22} = \frac{36}{11} = 3\frac{3}{11}$$

The speed of the wind is $3\frac{3}{11}$ mph.

26. Let t_P = the time in hours it would take Pe'rez to mulch the flower beds working alone, and t_E = the time in hours it would take Ellia to mulch the flower beds working alone. Then:

$$t_P = t_E + 6 \quad (1)$$

$$\left(\frac{1}{t_P} + \frac{1}{t_E} \right) \left(\frac{20}{7} \right) = 1 \quad (2)$$

Substitute the expression for t_P in (1) into (2) and solve for t_E.

$$\left(\frac{1}{t_P} + \frac{1}{t_E} \right) \left(\frac{20}{7} \right) = 1 \quad (2)$$

$$\frac{1}{t_P} + \frac{1}{t_E} = \frac{7}{20}$$

$$\frac{1}{t_E + 6} + \frac{1}{t_E} = \frac{7}{20}$$

$$20t_E (t_E + 6) \left[\frac{1}{t_E + 6} + \frac{1}{t_E} \right] = 20t_E (t_E + 6) \left[\frac{7}{20} \right]$$

$$20t_E + 20(t_E + 6) = 7t_E (t_E + 6)$$

$$7t_E^2 + 2t_E - 120 = 0$$

$$(7t_E + 30)(t_E - 4) = 0$$

$t_E = 4$ since $t_E > 0$. Use (1) and the value for t_E to find t_P.

$$t_P = t_E + 6 \quad (1)$$

$$t_P = 4 + 6 = 10$$

It takes Ellia 4 hr working alone and Pe'rez 10 hr.

27. Let w = the amount of water needed for 6 cups of flour and build two ratios based on the fact that $3\frac{1}{2}c$ of flour are used with $1\frac{1}{4}c$ of water and set the two ratios equal to one another.

$$\frac{\text{Flour}}{\text{Water}} \quad \frac{3\frac{1}{2}}{1\frac{1}{4}} = \frac{6}{w}$$

Solve for w.

$$\frac{3\frac{1}{2}}{1\frac{1}{4}} = \frac{6}{w}$$

$$1\frac{1}{4} w \cdot \frac{3\frac{1}{2}}{1\frac{1}{4}} = 1\frac{1}{4} w \cdot \frac{6}{w}$$

$$3\frac{1}{2} w = 1\frac{1}{4} \cdot 6$$

$$\frac{7}{2} w = \frac{15}{2}$$

$$\frac{2}{7} \cdot \frac{7}{2} w = \frac{2}{7} \cdot \frac{15}{2}$$

$$w = \frac{15}{7}, \text{ or } 2\frac{1}{7}$$

Check. $\dfrac{3\frac{1}{2}}{1\frac{1}{4}} = 2.8$ $\dfrac{6}{2\frac{1}{7}} = 2.8$

The ratios are the same. $2\frac{1}{7}$ cups of warm water should be used with 6 cups of whole wheat flour.

28. n varies inversely as t, so $n = \dfrac{k}{t}$.

Let $n = 25$ and $t = 6$, and determine k, the variation constant and the equation of variation.

$$25 = \dfrac{k}{6}$$
$$150 = k \qquad \text{Variation constant}$$
$$n = \dfrac{150}{t} \qquad \text{Equation of variation}$$

Let $t = 5$ and solve for n.

$$n = \dfrac{150}{5}$$
$$n = 30$$

It will take 30 workers to clean the stadium in 5 hours.

29. The surface area, A, varies directly as the square of the radius, r^2.

The equation is $A = kr^2$

Let $A = 325$ and $r = 5$, and determine the variation constant and the equation of variation.

$$325 = k \cdot 5^2$$
$$13 = k \qquad \text{Variation constant}$$
$$A = 13r^2 \qquad \text{Equation of variation}$$

Let $r = 7$ to determine A for 7 in.

$$A = 13 \cdot 7^2$$
$$A = 13 \cdot 49$$
$$A = 637$$

When the radius is 7 in., the surface area of the balloon is 637 in.^{2}

30.
$$\dfrac{6}{x-15} - \dfrac{6}{x} = \dfrac{90}{x^2 - 15x}$$

Note: $x \neq 15$, $x \neq 0$.

$$\left[\text{LCD is } x(x-15) \right]$$

$$x(x-15)\left(\dfrac{6}{x-15} - \dfrac{6}{x} \right) = x(x-15) \cdot \dfrac{90}{x^2 - 15x}$$
$$6x - 6(x-15) = 90$$
$$6x - 6x + 90 = 90$$
$$90 = 90$$

Since $90 = 90$, the equation is an identity. Remember to exclude the values which make a denominator zero. (The solution is the domain of the equation.)

$$\{x \mid x \text{ is a real number, } x \neq 0, 15\}$$

31. $1 - \dfrac{1}{1 - \dfrac{1}{1 - \dfrac{1}{a}}} = 1 - \dfrac{1}{1 - \dfrac{1}{1 - \dfrac{1}{a}} \cdot \dfrac{a}{a}}$

$= 1 - \dfrac{1}{1 - \dfrac{a}{a-1}} = 1 - \dfrac{1}{1 - \dfrac{a}{a-1}} \cdot \dfrac{a-1}{a-1}$

$= 1 - \dfrac{a-1}{a-1-a} = 1 - \dfrac{a-1}{-1} = 1 + a - 1 = a$

32. The ratio of the number of lawns Andy mowed to the number of lawns Chad mowed is $\dfrac{4}{3}$. Let x = the number of lawns Andy mowed, then $98 - x$ = number of lawns Chad mowed. Set the ratios equal to one another and solve for x.

$$\dfrac{4}{3} = \dfrac{x}{98-x}, \text{ LCD is } 3(98-x)$$
$$3(98-x) \cdot \dfrac{4}{3} = 3(98-x) \cdot \dfrac{x}{98-x}$$
$$4(98-x) = 3x$$
$$392 - 4x = 3x$$
$$392 = 7x$$
$$56 = x$$

Substituting, $x = 56$, so $98 - 56 = 42$.

$\dfrac{56}{42} = \dfrac{4}{3}$ and $56 + 42 = 98$, so the numbers check. Andy mowed 56 lawns and Chad mowed 42 lawns.

Chapters 1 – 6
Cumulative Review

1. $\dfrac{2x - y^2}{x + y}$

   Let $x = 3$ and $y = -4$ and evaluate.

   $\dfrac{2(3) - (-4)^2}{3 + (-4)} = \dfrac{6 - 16}{-1} = 10$

2. The decimal point must be moved 8 places to the right. Therefore, the exponent of ten must be positive 8.

   $391,000,000 = 3.91 \times 10^8$.

3. $7x - 4y = 12$

   $-4y = -7x + 12$

   $y = \dfrac{7}{4}x - 3$

   $m = \dfrac{7}{4}$ (slope) and $b = -3$

   The y-intercept is $(0, -3)$.

4. Determine the slope:

   $m = \dfrac{7 - (-3)}{-1 - 4} = \dfrac{10}{-5} = -2$

   Use either point and the slope, and substitute them into the point-slope form for a line.

   $y - 7 = -2(x - (-1))$

   $y - 7 = -2(x + 1)$

   $y - 7 = -2x - 2$

   $y = -2x + 5$

5. $f(x) = \dfrac{x - 3}{x^2 - 11x + 30} = \dfrac{x - 3}{(x - 5)(x - 6)}$

   a) $f(3) = \dfrac{3 - 3}{(3 - 5)(3 - 6)} = \dfrac{0}{(-2)(-3)} = 0$

   b) The domain of f is the set of real numbers except those that make the denominator zero.

   $x - 5 = 0 \qquad x - 6 = 0$

   $x = 5 \qquad\quad x = 6$

The domain is the set
$\{x \mid x \text{ is a real number}, x \neq 5, 6\}$, or
$(-\infty, 5) \cup (5, 6) \cup (6, \infty)$.

6. Since the root is even, the domain of f is the set of real numbers that make the radicand nonnegative.

   $x - 9 \geq 0$

   $x \geq 9$

   The domain of f is
   $\{x \mid x \text{ is a real number}, x \geq 9\}$, or $[9, \infty)$.

7. Only two points are needed to graph a line. But it is okay to plot more.

x	$y = 5x$	(x, y)
-1	$5(-1) = -5$	$(-1, -5)$
0	$5 \cdot 0 = 0$	$(0, 0)$
1	$5 \cdot 1 = 5$	$(1, 5)$

8. Only two points are needed to graph a line. But it is okay to plot more.

   Solving the equation for y gives:

   $8y + 2x = 16$

   $8y = -2x + 16$

   $y = -\dfrac{1}{4}x + 2$

   To avoid needing to plot fractional numbers, choose multiples of 4 for x.

x	$y = -\dfrac{1}{4}x + 2$	(x, y)
-4	$-\dfrac{1}{4}(-4) + 2 = 1 + 2 = 3$	$(-4, 3)$
0	$-\dfrac{1}{4}(0) + 2 = 0 + 2 = 2$	$(0, 2)$
4	$-\dfrac{1}{4}(4) + 2 = -1 + 2 = 1$	$(4, 1)$

   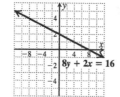

9. Begin by graphing the boundary line. Because the inequality is greater than *or equal to*, the line should be drawn as a solid line.

$$4x = 5y + 20$$

$$5y = 4x - 20$$

$$y = \frac{4}{5}x - 4$$

x	$y = \frac{4}{5}x - 4$	(x, y)
0	$\frac{4}{5}(0) - 4 = 0 - 4 = -4$	$(0, -4)$
5	$\frac{4}{5}(5) - 4 = 4 - 4 = 0$	$(5, 0)$
10	$\frac{4}{5}(10) - 4 = 8 - 4 = 4$	$(10, 4)$

Next use a test point not on the boundary line to determine which region to shade. (0, 0) is not on the boundary line.

$$4x \geq 5y + 20$$

$$4 \cdot 0 \geq 5 \cdot 0 + 20$$

$$0 \geq 20$$

Since $0 \geq 20$ is false, shade the region that does not include the point (0, 0).

10. Only two points are needed to graph a line. But it is okay to plot more.

x	$y = \frac{1}{3}x - 2$	(x, y)
-3	$\frac{1}{3}(-3) - 2 = -1 - 2 = -3$	$(-3, -3)$
0	$\frac{1}{3}(0) - 2 = 0 - 2 = -2$	$(0, -2)$
3	$\frac{1}{3}(3) - 2 = 1 - 2 = -1$	$(3, -1)$

11. $\left(8x^3 y^2\right)\left(-3xy^2\right)$

$= 8(-3)x^3 \cdot x \cdot y^2 \cdot y^2$

$= -24x^{3+1}y^{2+2}$

$= -24x^4 y^4$

12. $\left(5x^2 - 2x + 1\right)\left(3x^2 + x - 2\right)$

$$
\begin{array}{r}
3x^2 + x - 2 \\
5x^2 - 2x + 1 \\
\hline
3x^2 + x - 2 \quad \text{Mult. by } 1 \\
-6x^3 - 2x^2 + 4x \quad \text{Mult. by } -2 \\
15x^4 + 5x^3 - 10x^2 \quad \text{Mult. by } 5x^2 \\
\hline
15x^4 - x^3 - 9x^2 + 5x - 2 \quad \text{Adding}
\end{array}
$$

13. $\left(3x^2 + y\right)^2$

Applying the rule:

$$(a + b)^2 = a^2 + 2ab + b^2$$

$$\left(3x^2 + y\right)^2 = \left(3x^2\right)^2 + 2 \cdot 3x^2 \cdot y + y^2$$

$$= 9x^4 + 6x^2 y + y^2$$

14. $\left(2x^2 - 9\right)\left(2x^2 + 9\right)$

Applying the rule:

$$(a + b)(a - b) = a^2 - b^2$$

$$\left(2x^2 - 9\right)\left(2x^2 + 9\right) = \left(2x^2\right)^2 - 9^2$$

$$= 4x^4 - 81$$

15. $\left(-5m^3 n^2 - 3mn^3\right) + \left(-4m^2 n^2 + 4m^3 n^2\right) - \left(2mn^3 - 3m^2 n^2\right)$

$= \left(-5m^3 n^2 - 3mn^3\right) + \left(-4m^2 n^2 + 4m^3 n^2\right)$

$\quad - \left(2mn^3 - 3m^2 n^2\right)$

$= (-5 + 4)m^3 n^2 + (-4 + 3)m^2 n^2$

$\quad + (-3 + (-2))mn^3$

$= -m^3 n^2 - m^2 n^2 - 5mn^3$

16. $\dfrac{y^2-36}{2y+8}\cdot\dfrac{y+4}{y+6}$

$=\dfrac{(y+6)(y-6)(y+4)}{2(y+4)(y+6)}$

$=\dfrac{y-6}{2}$

17. $\dfrac{x^4-1}{x^2-x-2}\div\dfrac{x^2+1}{x-2}$

$=\dfrac{x^4-1}{x^2-x-2}\cdot\dfrac{x-2}{x^2+1}$

$=\dfrac{(x^2+1)(x+1)(x-1)}{(x-2)(x+1)}\cdot\dfrac{x-2}{x^2+1}$

$=\dfrac{(x^2+1)(x+1)(x-1)(x-2)}{(x-2)(x+1)(x^2+1)}$

$=x-1$

18. $\dfrac{5ab}{a^2-b^2}+\dfrac{a+b}{a-b}$

$=\dfrac{5ab}{(a+b)(a-b)}+\dfrac{a+b}{a-b}$

$\left[\text{LCD is }(a+b)(a-b)\right]$

$=\dfrac{5ab}{(a+b)(a-b)}+\dfrac{a+b}{a-b}\cdot\dfrac{a+b}{a+b}$

$=\dfrac{5ab+\left(a^2+2ab+b^2\right)}{(a+b)(a-b)}$

$=\dfrac{a^2+7ab+b^2}{(a+b)(a-b)}$

19. $\dfrac{2}{m+1}+\dfrac{3}{m-5}-\dfrac{m^2-1}{m^2-4m-5}$

$=\dfrac{2}{m+1}+\dfrac{3}{m-5}-\dfrac{m^2-1}{(m-5)(m+1)}$

$=\left[\text{LCD is }(m+1)(m-5)\right]$

$=\dfrac{2}{m+1}\cdot\dfrac{m-5}{m-5}+\dfrac{3}{m-5}\cdot\dfrac{m+1}{m+1}$

$\qquad-\dfrac{m^2-1}{(m-5)(m+1)}$

$=\dfrac{2(m-5)+3(m+1)-\left(m^2-1\right)}{(m+1)(m-5)}$

$=\dfrac{2m-10+3m+3-m^2+1}{(m+1)(m-5)}$

$=\dfrac{-m^2+5m-6}{(m+1)(m-5)},\text{ or }-\dfrac{(m-3)(m-2)}{(m+1)(m-5)}$

20. $y-\dfrac{2}{3y}$ LCD is $3y$.

$=y\cdot\dfrac{3y}{3y}-\dfrac{2}{3y}$

$=\dfrac{3y^2}{3y}-\dfrac{2}{3y}$

$=\dfrac{3y^2-2}{3y}$

21. $\dfrac{\dfrac{1}{x}-\dfrac{1}{y}}{x+y}$ LCD is xy.

$=\dfrac{\dfrac{1}{x}-\dfrac{1}{y}}{x+y}\cdot\dfrac{xy}{xy}$

$=\dfrac{\dfrac{1}{x}\cdot xy-\dfrac{1}{y}\cdot xy}{xy(x+y)}$

$=\dfrac{y-x}{xy(x+y)}$

22. $\left(9x^3+5x^2+2\right)\div(x+2)$

$=\left(9x^3+5x^2+0x+2\right)\div\left[x-(-2)\right]$

$\begin{array}{r|rrrr}
-2 & 9 & 5 & 0 & 2 \\
   &   & -18 & 26 & -52 \\
\hline
   & 9 & -13 & 26 & \underline{|-50} \\
\end{array}$

The answer is $9x^2-13x+26+\dfrac{-50}{x+2}$.

23. $4x^3+400x=4x\cdot x^2+4x\cdot100$

$\qquad\qquad\quad=4x\left(x^2+100\right)$

24. $x^2 + 8x - 84 = (x+14)(x-6)$

25. Applying the rule:

$$(a+b)(a-b) = a^2 - b^2$$

$$16y^2 - 25 = (4y)^2 - 5^2$$

$$= (4y+5)(4y-5)$$

26. $64x^3 + 8 = 8(8x^3 + 1)$

$$= 8(2x+1)(4x^2 - 2x + 1)$$

27. Applying the rule:

$$(a-b)^2 = a^2 - 2ab + b^2$$

$$t^2 - 16t + 64 = t^2 - 2 \cdot t \cdot 8 + 8^2$$

$$= (t-8)^2$$

28. $x^6 - x^2 = x^2(x^4 - 1)$

$$= x^2(x^2 + 1)(x^2 - 1)$$

$$= x^2(x^2 + 1)(x+1)(x-1)$$

29. Applying the rule:

$$x^3 - y^3 = (x-y)(x^2 + xy + y^y)$$

$$\frac{1}{8}b^3 - c^3 = \left(\frac{1}{2}b\right)^3 - c^3$$

$$= \left(\frac{1}{2}b - c\right)\left(\left(\frac{1}{2}b\right)^2 + \frac{1}{2}b \cdot c + c^2\right)$$

$$= \left(\frac{1}{2}b - c\right)\left(\frac{1}{4}b^2 + \frac{1}{2}bc + c^2\right)$$

30. Use the grouping method.
1. There is no common factor (other than 1 or –1).
2. We now factor the trinomial $3t^2 + 17t - 28$. Multiply the leading coefficient and the constant $3(-28) = -84$.
3. We want factors of –84 whose sum is 17. Since $21(-4) = -84$ and $21 + (-4) = 17$ we will split $17t$ into $21t$ and $-4t$
4. Factor by grouping:

$$3t^2 + 17t - 28 = 3t^2 + 21t - 4t - 28$$

$$= 3t(t+7) - 4(t+7)$$

$$= (t+7)(3t-4)$$

31. $x^5 - x^3y + x^2y - y^2$

$$= x^3(x^2 - y) + y(x^2 - y)$$

$$= (x^2 - y)(x^3 + y)$$

32. $8x = 1 + 16x^2$

$$0 = 16x^2 - 8x + 1$$

$$0 = (4x-1)^2$$

$$4x - 1 = 0$$

$$x = \frac{1}{4}$$

The solution is $\frac{1}{4}$.

33. $288 = 2y^2$

$$0 = 2y^2 - 288$$

$$0 = 2(y^2 - 144) = 2(y^2 - 12^2)$$

$$0 = 2(y+12)(y-12)$$

$$y + 12 = 0 \ or \ y - 12 = 0$$

$$y = -12 \ or \ y = 12$$

The solutions are $y = \pm 12$.

34. $\frac{1}{3}x - \frac{1}{5} \ge \frac{1}{5}x - \frac{1}{3}$ LCD is 15

$$15\left(\frac{1}{3}x - \frac{1}{5}\right) \ge 15\left(\frac{1}{5}x - \frac{1}{3}\right)$$

$$5x - 3 \ge 3x - 5$$

$$2x \ge -2$$

$$x \ge -1$$

The solution set is $\{x \mid x \ge -1\}$, or $[-1, \infty)$.

35. $-13 < 3x + 2 < -1$

$$-13 - 2 < 3x + 2 - 2 < -1 - 2$$

$$-15 < 3x < -3$$

$$\frac{1}{3} \cdot (-15) < \frac{1}{3} \cdot 3x < \frac{1}{3} \cdot (-3)$$

$$-5 < x < -1$$

The solution set is $\{x \mid -5 < x < -1\}$, or $(-5, -1)$.

36. $3x - 2 < -6$ or $x + 3 > 9$

$3x < -4$ or $x > 6$

$x < -\dfrac{4}{3}$ or $x > 6$

The solution set is $\left\{x \mid x < -\dfrac{4}{3} \text{ or } x > 6\right\}$, or

$\left(-\infty, -\dfrac{4}{3}\right) \cup (6, \infty)$.

37. $|x| > 6.4$

$x < -6.4$ or $x > 6.4$

The solution set is $\{x \mid x < -6.4 \text{ or } x > 6.4\}$,

or $(-\infty, -6.4) \cup (6.4, \infty)$.

38. $|3x - 2| \leq 14$

$-14 \leq 3x - 2 \leq 14$

$-14 + 2 \leq 3x - 2 + 2 \leq 14 + 2$

$-12 \leq 3x \leq 16$

$\dfrac{1}{3} \cdot (-12) \leq \dfrac{1}{3} \cdot 3x \leq \dfrac{1}{3} \cdot 16$

$-4 \leq x \leq \dfrac{16}{3}$

The solution set is $\left\{x \mid -4 \leq x \leq \dfrac{16}{3}\right\}$,

or $\left[-4, \dfrac{16}{3}\right]$.

39. $\dfrac{6}{x - 5} = \dfrac{2}{2x}$ Note: $x \neq 0, 5$

$\dfrac{6}{x - 5} = \dfrac{1}{x}$

$\left[\text{LCD is } x(x - 5)\right]$

$x(x - 5) \cdot \dfrac{6}{x - 5} = x(x - 5) \cdot \dfrac{1}{x}$

$6x = x - 5$

$5x = -5$

$x = -1$

The solution is $x = -1$.

40. $\dfrac{3x}{x - 2} - \dfrac{6}{x + 2} = \dfrac{24}{x^2 - 4}$

$\dfrac{3x}{x - 2} - \dfrac{6}{x + 2} = \dfrac{24}{(x + 2)(x - 2)}$ Note: $x \neq \pm 2$

$\left[\text{LCD is } (x + 2)(x - 2)\right]$

$(x + 2)(x - 2)\left(\dfrac{3x}{x - 2} - \dfrac{6}{x + 2}\right) =$

$(x + 2)(x - 2)\dfrac{24}{(x + 2)(x - 2)}$

$3x(x + 2) - 6(x - 2) = 24$

$3x^2 + 6x - 6x + 12 = 24$

$3x^2 - 12 = 0$

$3\left(x^2 - 4\right) = 0$

$3(x + 2)(x - 2) = 0$

$x + 2 = 0$ or $x - 2 = 0$

$x = -2$ or $x = 2$

Since $x \neq -2$ or $x \neq 2$, there is no solution.

41. $5x - 2y = -23$ (1)

$3x + 4y = 7$ (2)

Multiply equation (1) by 2 and add the result to equation (2).

$10x - 4y = -46$

$\underline{3x + 4y = 7}$

$13x = -39$

$x = -3$

Substitute –3 for x into any equation and solve for y. We choose equation (2).

$3(-3) + 4y = 7$

$4y = 16$

$y = 4$

The solution is $(-3, 4)$.

42. $-3x + 4y + z = -5$ (1)

$x - 3y - z = 6$ (2)

$2x + 3y + 5z = -8$ (3)

Add equations (1) and (2) to eliminate variable z.

$-3x + 4y + z = -5$ (1)

$\underline{x - 3y - z = 6}$ (2)

$-2x + y = 1$ (4)

Add 5 times equation (2) to equation (3).

$$5x - 15y - 5z = 30 \quad 5 \cdot (2)$$
$$\underline{2x + 3y + 5z = -8 \quad (3)}$$
$$7x - 12y = 22 \quad (5)$$

Multiply 12 times equation (4) and add it to equation (5).

$$-24x + 12y = 12 \quad 12 \cdot (4)$$
$$\underline{7x - 12y = 22 \quad (5)}$$
$$-17x \qquad = 34$$
$$x = -2$$

Substitute –2 for x in equation (4) and solve for y.

$$-2x + y = 1$$
$$-2(-2) + y = 1$$
$$4 + y = 1$$
$$y = -3$$

Substitute –2 for x and –3 for y in equation (2) and solve for z.

$$x - 3y - z = 6$$
$$-2 - 3(-3) - z = 6$$
$$-2 + 9 - 6 = z$$
$$1 = z$$

The solution is $(-2, -3, 1)$.

43.
$$P = \frac{4a}{a+b} \quad \text{LCD is } a+b$$

$$(a+b) \cdot P = (a+b) \cdot \frac{4a}{a+b}$$

$$Pa + Pb = 4a$$

$$Pb = 4a - Pa$$

$$Pb = a(4 - P)$$

$$\frac{Pb}{4 - P} = a$$

44. a) Let t = the year since 1986, and $r(t)$ = the revenue, in millions of dollars, for the Broadway season beginning in year t. Then the given information in the problem can be viewed as points on a line $(t, r(t))$, where $r(t) = mt + b$.

The two points are $(0, 20)$ and $(2008 - 1986, 943) = (22, 943)$. Use the two points to find m; $b = 20$.

$$m = \frac{r(t_2) - r(t_1)}{t_2 - t_1} = \frac{943 - 20}{22 - 0} = \frac{923}{22}$$

Therefore, $r(t) = \frac{923}{22}t + 20$.

b) In 2011, $t = 2011 - 1986 = 25$.

$$r(25) = \frac{923}{22}(25) + 20 \approx 1069$$

This corresponds to 1.069 billion, or $1 billion, 69 million.

c) 1.4 billion = 1400 million, so set $r(t) = 1400$ and solve for t.

$$\frac{923}{22}t + 20 = 1400$$

$$\frac{923}{22}t = 1380$$

$$t = \frac{22}{923} \cdot 1380 \approx 32.9$$

$$1986 + 32.9 = 2018.9$$

According to the model, the revenue will reach $1.4 billion in the 2018-2019 season.

45. **Familiarize.** The monthly rate of increase = total increase / number of months.
Translate. The total increase was $9424 - 7486 = 1938$; the number of months was the number of months from January 2006 to October 2010 = $4 \cdot 12 + (10 - 1) = 57$ (12 for each year from January 2006 to January 2010, plus the months in 2010 from January to October).
Carry out. The rate of increase =

$$\frac{1938 \text{ shows}}{57 \text{ month}} = 34 \text{ shows per month}$$

Check. $7486 + 34 \cdot 57 = 9424$. The number of shows in October 2010 equals the number of shows in January 2006 plus the monthly increase times the number of months from January 2006 to October 2010.
State. The monthly rate of increase over this time was 34 shows per month.

46. ***Familiarize.*** The problem states the percentage of nuts in two types of trail mixes, Himalayan Diamonds and Alpine Gold. The problems asks how much of each type must be used to create a mixture with a percentage of nuts in between those of the given two mixes. It also tells us the total amount of new mix desired.

Translate. Let x = the amount of Himalayan Diamonds mix needed, in pounds, and y = the amount of Alpine Gold mix needed, in pounds. Then the given information can be translated into two equations in x and y.

$$x + y = 20 \qquad (1)$$
$$.40x + .25y = .30 \cdot 20$$
$$40x + 25y = 30 \cdot 20$$
$$40x + 25y = 600$$
$$8x + 5y = 120 \qquad (2)$$

Note that the second equation was simplified by clearing decimals, and then dividing each term by the common factor 5.

Carry out. The system of equations can be solved using substitution. Solve equation (1) for y.

$$x + y = 20 \qquad (1)$$
$$y = 20 - x \qquad (3)$$

Substitute $20 - x$ for y in equation (2) and solve for x.

$$8x + 5y = 120 \qquad (2)$$
$$8x + 5(20 - x) = 120$$
$$8x + 100 - 5x = 120$$
$$3x = 20$$
$$x = \frac{20}{3} = 6\frac{2}{3}$$

Use the value of x and equation (3) to find y.

$$y = 20 - x \qquad (3)$$
$$y = 20 \cdot \frac{3}{3} - \frac{20}{3}$$
$$y = \frac{60}{3} - \frac{20}{3} = \frac{40}{3} = 13\frac{1}{3}$$

Check.

$x + y$	20
$\frac{20}{3} + \frac{40}{3}$	20
$\frac{60}{3}$	20
20	20

Equation (1) is satisfied.

$.40x + .25y$	$.30 \cdot 20$
$40 \cdot \frac{20}{3} + .25 \cdot \frac{40}{3}$	6
$\frac{8}{3} + \frac{10}{3}$	6
$\frac{18}{3}$	6
6	6

Equation (2) is satisfied. The answers check.

State. To create a mix that is 30% nuts, $6\frac{2}{3}$ lb of the Himalayan Diamond mix and $13\frac{1}{3}$ lb of the Alpine Gold mix are needed.

47. ***Familiarize.*** Let w = the width, in inches, of the quilt. Then $w + 4$ in = the quilt's length. The area of the quilt is given as 320 in^2. The area of a rectangle is lw. The perimeter of a rectangle is $2l + 2w$.

Translate.

$$\underbrace{\text{area of the quilt}}_{\displaystyle w(w+4)} \underset{=}{\text{ is }} \underset{320}{320} \text{ in}^2.$$

Carry out. Solve the equation:

$$w(w + 4) = 320$$
$$w^2 + 4w = 320$$
$$w^2 + 4w - 320 = 0$$
$$(w + 20)(w - 16) = 0$$
$$w + 20 = 0 \ \text{ or } \ w - 16 = 0$$
$$w = -20 \ \text{ or } \ w = 16$$

Since the width cannot be negative, exclude $w = -20.$ The dimensions are therefore 16 in and $16 + 4 = 20$ in.
The perimeter is:

$$P = 2l + 2w$$
$$= 2(20 \text{ in}) + 2(16 \text{ in})$$
$$= 40 \text{ in} + 32 \text{ in}$$
$$= 72 \text{ in}$$

Check. $(20 \text{ in})(16 \text{ in}) = 320 \text{ in}^2$. The dimensions check.

State. The perimeter of the quilt is 72 in.

48. **Familiarize.** The problem states the hours of driving delays in 2055 will be 250% of the hours of driving delays in 2005. The problem then states the predicted hours of delays in 2055 and asks for the hours of delays in 2005. To find the percent of a number, multiply the percent, in decimal form, by the number.

Translate. Let x = the hours of driving delays in 2005 and set 250% of x equal to the predicted hours of delays in 2055.

$$250\% \cdot x = \text{delays in 2005}$$
$$2.5x = 30 \text{ billion}$$

Carry out. $2.5x = 30 \text{ billion}$

$$x = \frac{30 \text{ billion}}{2.5} = 12 \text{ billion}$$

Check.

$250\% \cdot 12 \text{ billion} = 2.5 \cdot 12 \text{ billion} = 30 \text{ billion}$
The value checks.

State. There were 12 billion hours in driving delays in 2005.

49. **Familiarize.** The problem states that the time it takes Johann to drive to work is inversely related to his driving speed. Inverse relations are defined by the equation $y = \dfrac{k}{x}$, where k is constant of proportionality. The problem then gives the time required when traveling at 45 mph and asks what will be the time required at 40 mph.

Translate. Let t = the time, in minutes, required for Johann to drive to work, and s = his driving speed, in mph. Based on the inverse relationship between t and s, $t = \dfrac{k}{s}$.

Use the fact that it takes 20 min at 45 mph to find k, and then use k and the second driving speed to find the second time.

Carry out.

$t = \dfrac{k}{s} \Rightarrow k = ts$

$k = 45 \text{ mph} \cdot 20 \text{ min}$

$\quad = 900 \text{ mph} \cdot \text{min}$

Therefore,

$t = \dfrac{900 \text{ mph} \cdot \text{min}}{s}$

Find t when $s = 40$ mph

$t = \dfrac{900 \text{ mph} \cdot \text{min}}{s}$

$\quad = \dfrac{900 \text{ mph} \cdot \text{min}}{40 \text{ mph}}$

$\quad = 22.5 \text{ min}$

Check. Use $k = ts$ to check the solution.

$t_1 s_1$	$t_2 s_2$
45 mph · 20 min	40 mph · 22.5 min
900 mph · min	900 mph · min

The answer checks.

State. It will take Johann 22.5 minutes to drive to work at 40 mph.

50. **Familiarize.** The problem asks for the total number of rides and attractions at three theme parks. Note that the problem does not ask for the number of rides and the number of attractions separately. The problem then states three facts relating the numbers at each park. From these three facts we can build three equations and determine the number of rides and attractions at each park.

Translate. Let x = the number of rides and attractions at Magic Kingdom, y = the number of rides and attractions at Disneyland, and z = the number of rides and attractions at California Adventure. Since the total number of rides and attractions at all three parks is 133,

$x + y + z = 133$.

Half the total at California Adventure and Disneyland is $\dfrac{1}{2}(y + z)$. Subtract 5 to get 5 fewer than this number.

$x = \dfrac{1}{2}(y + z) - 5$

$2x = y + z - 10$

$2x - y - z = -10$

Three-fourths the total at Magic Kingdom and California Adventure is $\dfrac{3}{4}(x + z)$.

$y = \dfrac{3}{4}(x + z)$

$4y = 3(x + z)$

$4y = 3x + 3z$

$3x - 4y + 3z = 0$

We have a system of 3 equations:

$$x + y + z = 133 \quad (1)$$
$$2x - y - z = -10 \quad (2)$$
$$3x - 4y + 3z = 0 \quad (3)$$

Carry out. The system can be simplified rather quickly if one notices that adding equations (1) and (2) will eliminate both y and z and allow us to solve for x.

$$x + y + z = 133 \quad (1)$$
$$\underline{2x - y - z = -10 \quad (2)}$$
$$3x \qquad\quad = 123$$

$$x = \frac{123}{3} = 41$$

Substituting $x = 41$ into equations (1) and (3) and simplifying will give a 2x2 system in y and z.

$$x + y + z = 133 \quad (1)$$
$$3x - 4y + 3z = 0 \quad (3)$$
$$41 + y + z = 133 \quad (1)$$
$$3(41) - 4y + 3z = 0 \quad (3)$$
$$y + z = 92 \quad (1)$$
$$-4y + 3z = -123 \quad (3)$$

To solve this system, add 4 times equation (1) to equation (3) to eliminate y, and solve for z. Then use equation (1) and the value of z to find y.

$$4y + 4z = 368 \qquad 4 \cdot (1)$$
$$\underline{-4y + 3z = -123 \qquad (3)}$$
$$7z = 245$$

$$z = \frac{245}{7} = 35$$

$$y + z = 92 \quad (1)$$
$$y + 35 = 92$$
$$y = 92 - 35 = 57$$

Check.

$$\begin{array}{c|c}
x + y + z & 133 \\
\hline
41 + 57 + 35 & 133 \\
133 & 133
\end{array}$$

Condition (1) is satisfied.

$$\begin{array}{c|c}
x & \frac{1}{2}(y + z) - 5 \\
\hline
41 & \frac{1}{2}(57 + 35) - 5 \\
41 & \frac{1}{2}(92) - 5 \\
41 & 46 - 5 \\
41 & 41
\end{array}$$

Condition (2) is satisfied.

$$\begin{array}{c|c}
y & \frac{3}{4}(x + z) \\
\hline
57 & \frac{3}{4}(41 + 35) \\
57 & \frac{3}{4}(76) \\
57 & 57
\end{array}$$

Condition (3) is satisfied.
The values check.
State. There are 41 rides and attractions at Magic Kingdom, 57 at Disneyland, and 35 at California Adventure.

51. The data displayed in the graph shows both large increases and large decreases in the number of college student volunteers. In a linear function, growth or decline would be constant. This data could not be represented well by a linear function.

52. The data displayed in the graph shows nearly constant growth in college enrollment. Although the data is not exactly linear, the data could be modeled reasonably well by a linear function.

53.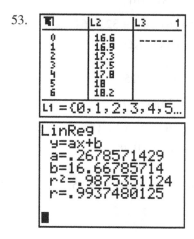

The function model is
$f(x) = 0.2679x + 16.6679$ where x is the
number of years after 2002.

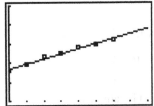

54. The graph appears to be the constant function
$y = 3$. The domain of this function is all real
numbers and the range is the set $\{3\}$.

55. The graph appears to be that of a line. Since
the line is neither horizontal, nor vertical, its
domain and range are both all real numbers.

56. The graph appears to be that of the absolute
value of a linear function that has been
shifted vertically by -3. The domain is all
real numbers, and the range is $\{y \mid y \geq -3\}$,
or $[-3, \infty)$.

57. The graph appears to be that of a parabola
that opens upward, and that has a vertex of
$(1, -4)$. The domain is all real numbers, and
the range is $\{y \mid y \geq -4\}$, or $[-4, \infty)$.

58. $(x-4)^3$
$= (x-4)(x-4)^2$
$= (x-4)(x^2 - 8x + 16)$
$= x^3 - 8x^2 + 16x - 4x^2 + 32x - 64$
$= x^3 - 12x^2 + 48x - 64$

59. $f(x) = x^4 - 34x^2 + 225$
$0 = (x^2 - 25)(x^2 - 9)$
$0 = (x+5)(x-5)(x+3)(x-3)$
$x+5 = 0 \; or \; x-5 = 0 \; or$
$x+3 = 0 \; or \; x-3 = 0$
The solutions are $-5, -3, 3,$ and 5.

60. $4 \leq |3 - x| \leq 6$
$4 \leq 3 - x \leq 6$ or $-6 \leq 3 - x \leq -4$
For $4 \leq 3 - x \leq 6$
$4 - 3 \leq 3 - x - 3 \leq 6 - 3$
$1 \leq -x \leq 3$
$1 \cdot (-1) \geq -x \cdot (-1) \geq 3 \cdot (-1)$
$-1 \geq x \geq -3$
$-3 \leq x \leq -1$
For $-6 \leq 3 - x \leq -4$
$-6 - 3 \leq 3 - x - 3 \leq -4 - 3$
$-9 \leq -x \leq -7$
$-9 \cdot (-1) \geq -x \cdot (-1) \geq -7 \cdot (-1)$
$9 \geq x \geq 7$
$7 \leq x \leq 9$
The solution is $\{x \mid -3 \leq x \leq -1 \text{ or } 7 \leq x \leq 9\}$,
or $[-3, -1] \cup [7, 9]$.

61. $\dfrac{18}{x-9} + \dfrac{10}{x+5} = \dfrac{28x}{(x-9)(x+5)}$ $x \neq -5, 9$

$\left[\text{LCD is } (x-9)(x+5) \right]$

$(x-9)(x+5) \left(\dfrac{18}{x-9} + \dfrac{10}{x+5} \right)$

$\qquad = (x-9)(x+5) \cdot \dfrac{28x}{(x-9)(x+5)}$

$18(x+5) + 10(x-9) = 28x$

$18x + 90 + 10x - 90 = 28x$

$28x = 28x$

Since this is an identity, the domain is all real
numbers in the domain of the equation.:
$\{x \mid x \text{ is a any real number and } x \neq -5, 9\}.\,.$

62. $16x^3 = x$
$16x^3 - x = 0$
$x(16x^2 - 1) = 0$
$x(4x - 1)(4x + 1) = 0$
$x = 0 \; or \; 4x - 1 = 0 \; or \; 4x + 1 = 0$
$x = 0 \; or \; x = \dfrac{1}{4} \; or \; x = -\dfrac{1}{4}$

The solutions are $-\dfrac{1}{4}$, 0, and $\dfrac{1}{4}$.

Chapter 7

Exponents and Radical Functions

1. Two

3. Positive

5. Irrational

7. Nonnegative

9. The square roots of 49 are 7 and –7, because $7^2 = 49$ and $(-7)^2 = 49$.

11. The square roots of 144 are 12 and –12, because $12^2 = 144$ and $(-12)^2 = 144$.

13. The square roots of 400 are 20 and –20, because $20^2 = 400$ and $(-20)^2 = 400$.

15. The square roots of 900 are 30 and –30, because $30^2 = 900$ and $(-30)^2 = 900$.

17. $\sqrt{49} = \sqrt{7 \cdot 7} = 7$

19. $-\sqrt{16} = -4$ since $\sqrt{16} = 4$.

21. $\sqrt{\dfrac{36}{49}} = \sqrt{\dfrac{6 \cdot 6}{7 \cdot 7}} = \dfrac{6}{7}$

23. $-\sqrt{\dfrac{16}{81}} = -\dfrac{4}{9}$ since $\sqrt{\dfrac{16}{81}} = \dfrac{4}{9}$.

25. $\sqrt{0.04} = 0.2$, since $(0.2)^2 = 0.04$.

27. $\sqrt{0.0081} = 0.09$

29. $5\sqrt{p^2} + 4$

The radicand is the expression written under the radical sign, p^2.
Since the index is not written, it is understood to be 2.

31. $xy\sqrt[5]{\dfrac{x}{y+4}}$

The radicand is the expression written under the radical sign, $\dfrac{x}{y+4}$.
The index is 5.

33. $f(t) = \sqrt{5t - 10}$

$f(3) = \sqrt{5 \cdot 3 - 10} = \sqrt{5}$

$f(2) = \sqrt{5 \cdot 2 - 10} = \sqrt{0} = 0$

$f(1) = \sqrt{5 \cdot 1 - 10} = \sqrt{-5}$

Since negative numbers do not have real-number square roots, $f(1)$ does not exist.

$f(-1) = \sqrt{5(-1) - 10} = \sqrt{-15}$

Since negative numbers do not have real-number square roots, $f(-1)$ does not exist.

35. $t(x) = -\sqrt{2x^2 - 1}$

$t(5) = -\sqrt{2(5)^2 - 1} = -\sqrt{49} = -7$

$t(0) = -\sqrt{2(0)^2 - 1} = -\sqrt{-1};$
  $t(0)$ does not exist

$t(-1) = -\sqrt{2(-1)^2 - 1} = -\sqrt{1} = -1$

$t\left(-\dfrac{1}{2}\right) = -\sqrt{2\left(-\dfrac{1}{2}\right)^2 - 1} = -\sqrt{-\dfrac{1}{2}};$
  $t\left(-\dfrac{1}{2}\right)$ does not exist

37. $f(t) = \sqrt{t^2 + 1}$

$f(0) = \sqrt{0^2 + 1} = \sqrt{1} = 1$

$f(-1) = \sqrt{(-1)^2 + 1} = \sqrt{2}$

$f(-10) = \sqrt{(-10)^2 + 1} = \sqrt{101}$

39. $\sqrt{64x^2} = \sqrt{(8x)^2} = |8x| = 8|x|$

Since x might be negative, absolute-value notation is necessary.

41. $\sqrt{(-4b)^2} = |-4b| = |-4| \cdot |b| = 4|b|$

Since b might be negative, absolute-value notation is necessary.

43. $\sqrt{(8-t)^2} = |8-t|$

Since $8-t$ might be negative, absolute-value notation is necessary.

45. $\sqrt{y^2+16y+64} = \sqrt{(y+8)^2} = |y+8|$

47. $\sqrt{4x^2+28x+49} = \sqrt{(2x+7)^2} = |2x+7|$

49. $-\sqrt[4]{256} = -4$ since $4^4 = 256$

51. $\sqrt[5]{-1} = -1$ since $-1^5 = -1$

53. $-\sqrt[5]{-\dfrac{32}{243}} = -\left(-\dfrac{2}{3}\right) = \dfrac{2}{3}$ since $\left(-\dfrac{2}{3}\right)^5 = -\dfrac{32}{243}$

55. $\sqrt[6]{x^6} = |x|$ The index is even. Use absolute-value notation since x could have a negative value.

57. $\sqrt[9]{t^9} = t$ The index is odd.

59. $\sqrt[4]{(6a)^4} = |6a| = 6|a|$ The index is even.

61. $\sqrt[10]{(-6)^{10}} = |-6| = 6$

63. $\sqrt[414]{(a+b)^{414}} = |a+b|$ The index is even.

65. $\sqrt{a^{22}} = |a^{11}|$ Since $\left(a^{11}\right)^2 = a^{22}$; a^{11} could have a negative value.

67. $\sqrt{-25}$ Cannot be simplified.

69. $\sqrt{16x^2} = \sqrt{(4x)^2} = 4x$ Assuming $x \geq 0$.

71. $-\sqrt{(3t)^2} = -3t$ Assuming $t \geq 0$.

73. $\sqrt{(a+1)^2} = a+1$ Assuming $a+1 \geq 0$.

75. $\sqrt{9t^2-12t+4} = \sqrt{(3t-2)^2} = 3t-2$

77. $\sqrt[3]{27a^3} = \sqrt[3]{(3a)^3} = 3a$

79. $\sqrt[4]{16x^4} = \sqrt[4]{(2x)^4} = 2x$

81. $\sqrt[5]{(x-1)^5} = x-1$

83. $-\sqrt[3]{-125y^3} = -\sqrt[3]{(-5y)^3} = -(-5y) = 5y$

85. $\sqrt{t^{18}} = \sqrt{(t^9)^2} = t^9$

87. $\sqrt{(x-2)^8} = \sqrt{\left[(x-2)^4\right]^2} = (x-2)^4$

89. $f(x) = \sqrt[3]{x+1}$

     $f(7) = \sqrt[3]{7+1} = \sqrt[3]{8} = 2$

     $f(26) = \sqrt[3]{26+1} = \sqrt[3]{27} = 3$

     $f(-9) = \sqrt[3]{-9+1} = \sqrt[3]{-8} = -2$

     $f(-65) = \sqrt[3]{-65+1} = \sqrt[3]{-64} = -4$

91. $g(t) = \sqrt[4]{t-3}$

     $g(19) = \sqrt[4]{19-3} = \sqrt[4]{16} = 2$

     $g(-13) = \sqrt[4]{-13-3} = \sqrt[4]{-16}$;

       $g(-13)$ does not exist

     $g(1) = \sqrt[4]{1-3} = \sqrt[4]{-2}$;

       $g(1)$ does not exist

     $g(84) = \sqrt[4]{84-3} = \sqrt[4]{81} = 3$

93. $f(x) = \sqrt{x-6}$

Since the index is even, the radicand, $x-6$, must be non-negative. We solve the inequality:

$$x - 6 \geq 0$$
$$x \geq 6$$

Domain of $f = \{x | x \geq 6\}$, or $[6, \infty)$

95. $g(t) = \sqrt[4]{t+8}$

Since the index is even, the radicand, $t+8$, must be non-negative. We solve the inequality:

$$t + 8 \geq 0$$
$$t \geq -8$$

Domain of $g = \{t | t \geq -8\}$, or $[-8, \infty)$

97. $g(x) = \sqrt[4]{2x-10}$

$$2x - 10 \geq 0$$
$$2x \geq 10$$
$$x \geq 5$$

Domain of $g = \{x | x \geq 5\}$, or $[5, \infty)$

99. $f(t) = \sqrt[5]{8-3t}$

Since the index is odd, the radicand can be any real number.

Domain of $f = \{t | t \text{ is a real number}\}$,

or $(-\infty, \infty)$

101. $h(z) = -\sqrt[6]{5z+2}$

$$5z + 2 \geq 0$$
$$5z \geq -2$$
$$z \geq -\frac{2}{5}$$

Domain of $h = \left\{z \middle| z \geq -\frac{2}{5}\right\}$, or $\left[-\frac{2}{5}, \infty\right)$

103. $f(t) = 7 + \sqrt[8]{t^8}$

Since we can compute $7 + \sqrt[8]{t^8}$ for any real number t, the domain is the set of real numbers, or $\{t | t \text{ is a real number}\}$,

or $(-\infty, \infty)$.

105. $f(x) = \sqrt{5-x}$

Find all values of x for which the radicand is non-negative.

$$5 - x \geq 0$$
$$5 \geq x$$

The domain is $\{x | x \leq 5\}$, or $(-\infty, 5]$.

We graph the function in the standard window.

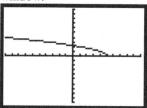

The range appears to be $\{y | y \geq 0\}$, or $[0, \infty)$.

107. $f(x) = 1 - \sqrt{x+1}$

Find all values of x for which the radicand is non-negative.

$$x + 1 \geq 0$$
$$x \geq -1$$

The domain is $\{x | x \geq -1\}$, or $[-1, \infty)$.

We graph the function in the standard window.

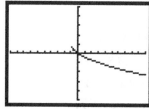

The range appears to be

$\{y | y \leq 1\}$, or $(-\infty, 1]$.

109. $g(x) = 3 + \sqrt{x^2 + 4}$

Since $x^2 + 4$ is positive for all values of x, the domain is $\{x | x \text{ is a real number}\}$,

or $(-\infty, \infty)$.

We graph the function in the standard window.

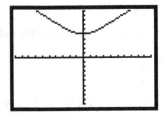

The range appears to be $\{y \mid y \ge 5\}$, or $[5, \infty)$.

111. For $f(x) = \sqrt{x-4}$, the domain is $[4, \infty)$ and all of the function values are non-negative. Graph (c) corresponds to this function.

113. For $h(x) = \sqrt{x^2 + 4}$, the domain is $(-\infty, \infty)$

Graph (d) corresponds to this function.

For problems 115 – 120, use a scatter plot of the data (for each problem, separately) and determine whether it could be modeled with a radical function.

115. Yes

117. Yes

119. No

121. $f(x) = 118.8\sqrt{x}$

$f(50) = 118.8\sqrt{50}$

$f(50) \approx 840$ GPM

$f(175) = 118.8\sqrt{175}$

$f(175) \approx 1572$ GPM

123. Thinking and Writing Exercise.

125. $(a^2 b)(a^4 b) = a^{2+4} b^{1+1} = a^6 b^2$

127. $(5x^2 y^{-3})^3 = 5^3 \cdot x^{2\cdot 3} y^{-3\cdot 3}$

$= 125 \cdot x^6 \cdot y^{-9} = \dfrac{125x^6}{y^9}$

129. $\left(\dfrac{10x^{-1} y^5}{5x^2 y^{-1}}\right)^{-1} = \left(\dfrac{2y^6}{x^3}\right)^{-1} = \dfrac{x^3}{2y^6}$

131. Thinking and Writing Exercise.

133. Thinking and Writing Exercise.

135. $N = 2.5\sqrt{A}$

a. $N = 2.5\sqrt{25} = 2.5(5) = 12.5 \approx 13$

b. $N = 2.5\sqrt{36} = 2.5(6) = 15$

c. $N = 2.5\sqrt{49} = 2.5(7) = 17.5 \approx 18$

d. $N = 2.5\sqrt{64} = 2.5(8) = 20$

137. $\{x \mid x \ge -5\}$, or $[-5, \infty)$

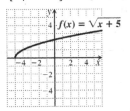

139. $\{x \mid x \ge 0\}$, or $[0, \infty)$

141. $g(x) = \dfrac{\sqrt[4]{5-x}}{\sqrt[6]{x+4}}$

In the numerator we must have $5 - x \ge 0$, or $x \le 5$, and in the denominator we must have $x + 4 > 0$, or $x > -4$. Thus we have $x \le 5$ and $x > -4$, so

Domain of $g = \{x \mid -4 < x \le 5\}$, or $(-4, 5]$.

143. Cubic

Exercise Set 7.2

1. g

3. e

5. a

7. b

9. $x^{1/6} = \sqrt[6]{x}$

11. $(16)^{1/2} = \sqrt{16} = 4$

13. $32^{1/5} = \sqrt[5]{32} = 2$

15. $9^{1/2} = \sqrt{9} = 3$

17. $(xyz)^{1/2} = \sqrt{xyz}$

19. $(a^2b^2)^{1/5} = \sqrt[5]{a^2b^2}$

21. $t^{2/5} = \sqrt[5]{t^2}$

23. $16^{3/4} = \sqrt[4]{16^3} = (\sqrt[4]{16})^3 = 2^3 = 8$

25. $27^{4/3} = \sqrt[3]{27^4} = (\sqrt[3]{3^3})^4 = 3^4 = 81$

27. $(81x)^{3/4} = \sqrt[4]{(81x)^3} = \sqrt[4]{81^3 x^3}$, or $\sqrt[4]{81^3} \cdot \sqrt[4]{x^3}$
$= (\sqrt[4]{81})^3 \cdot \sqrt[4]{x^3} = 3^3\sqrt[4]{x^3} = 27\sqrt[4]{x^3}$

29. $(25x^4)^{3/2} = \sqrt{(25x^4)^3} = \sqrt{25^3 \cdot x^{12}}$
$= \sqrt{25^3} \cdot \sqrt{x^{12}} = (\sqrt{25})^3 x^6$
$= 5^3 x^6 = 125x^6$

31. $\sqrt[3]{20} = 20^{1/3}$

33. $\sqrt{17} = 17^{1/2}$

35. $\sqrt{x^3} = x^{3/2}$

37. $\sqrt[5]{m^2} = m^{2/5}$

39. $\sqrt[4]{cd} = (cd)^{1/4}$ Parentheses are required.

41. $\sqrt[5]{xy^2z} = (xy^2z)^{1/5}$

43. $(\sqrt{3mn})^3 = (3mn)^{3/2}$

45. $(\sqrt[7]{8x^2y})^5 = (8x^2y)^{5/7}$

47. $\dfrac{2x}{\sqrt[3]{z^2}} = \dfrac{2x}{z^{2/3}}$

49. $8^{-1/3} = \dfrac{1}{8^{1/3}} = \dfrac{1}{(2^3)^{1/3}} = \dfrac{1}{2}$

51. $(2rs)^{-3/4} = \dfrac{1}{(2rs)^{3/4}}$

53. $\left(\dfrac{1}{16}\right)^{-3/4} = \dfrac{16^{3/4}}{1} = (2^4)^{3/4} = 2^3 = 8$

55. $\dfrac{2c}{a^{-3/5}} = 2c \cdot a^{3/5} = 2a^{3/5}c$

57. $5x^{-2/3}y^{4/5}z = 5 \cdot \dfrac{1}{x^{2/3}} \cdot y^{4/5}z = \dfrac{5y^{4/5}z}{x^{2/3}}$

59. $3^{-5/2}a^3b^{-7/3} = \dfrac{1}{3^{5/2}} \cdot a^3 \cdot \dfrac{1}{b^{7/3}} = \dfrac{a^3}{3^{5/2}b^{7/3}}$

61. $\left(\dfrac{2ab}{3c}\right)^{-5/6} = \left(\dfrac{3c}{2ab}\right)^{5/6}$

Find the reciprocal of the base and change the sign of the exponent.

63. $\dfrac{6a}{\sqrt[4]{b}} = \dfrac{6a}{b^{1/4}}$

65. $f(x) = \sqrt[4]{x+7} = (x+7)^{1/4}$

Enter $y = (x+7) \wedge (1/4)$, or
$y = (x+7) \wedge 0.25$.

Since the index is even, the domain of the function is the set of all x for which the radicand is non-negative, or $[-7, \infty)$. One good choice of a viewing window is $[-10, 25, -1, 5]$, Xscl $= 5$.

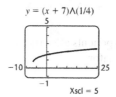

$y = (x+7)\wedge(1/4)$

67. $r(x) = \sqrt[7]{3x-2} = (3x-2)^{1/7}$

Enter $y = (3x-2) \wedge (1/7)$. Since the index is odd the domain of the function is $(-\infty, \infty)$. One good choice of a viewing window is $[-10, 10, -5, 5]$.

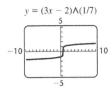
$y = (3x - 2) \wedge (1/7)$

69. $f(x) = \sqrt[6]{x^3} = (x^3)^{1/6} = x^{3/6}$

Enter $y = x \wedge (3/6)$. The function is defined only for non-negative values of x, so the domain is $[0, \infty)$. One good choice of a window is $[-5, 25, -1, 5]$, Xscl = 5.

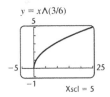
$y = x \wedge (3/6)$
Xscl = 5

71. $\sqrt[5]{9} = 9^{1/5} = 9 \wedge (1/5) \approx 1.552$

73. $\sqrt[4]{10} = 10^{1/4} = 10 \wedge (1/4) \approx 1.778$

75. $\sqrt[3]{(-3)^5} = (-3)^{5/3} = (-3) \wedge (5/3) \approx -6.240$

77. $7^{\frac{3}{4}} \cdot 7^{\frac{1}{8}} = 7^{\frac{3}{4}+\frac{1}{8}} = 7^{\frac{6}{8}+\frac{1}{8}} = 7^{\frac{7}{8}}$

We added exponents after finding their common denominator.

79. $\dfrac{3^{\frac{5}{8}}}{3^{-\frac{1}{8}}} = 3^{\frac{5}{8}-\left(-\frac{1}{8}\right)} = 3^{\frac{6}{8}} = 3^{\frac{3}{4}}$

81. $\dfrac{5.2^{-\frac{1}{6}}}{5.2^{-\frac{2}{3}}} = 5.2^{-\frac{1}{6}-\left(-\frac{2}{3}\right)} = 5.2^{-\frac{1}{6}+\frac{4}{6}} = 5.2^{\frac{3}{6}} = 5.2^{\frac{1}{2}}$

We subtracted exponents after finding a common denominator.

83. $\left(10^{\frac{3}{5}}\right)^{\frac{2}{5}} = 10^{\frac{3}{5}\cdot\frac{2}{5}} = 10^{\frac{6}{25}}$

85. $a^{\frac{2}{3}} \cdot a^{\frac{5}{4}} = a^{\frac{2}{3}+\frac{5}{4}} = a^{\frac{8}{12}+\frac{15}{12}} = a^{\frac{23}{12}}$

87. $\left(64^{\frac{3}{4}}\right)^{\frac{4}{3}} = 64^{\frac{3}{4}\cdot\frac{4}{3}} = 64^1 = 64$

89. $\left(m^{\frac{2}{3}}n^{-\frac{1}{4}}\right)^{\frac{1}{2}} = m^{\frac{2}{3}\cdot\frac{1}{2}}n^{-\frac{1}{4}\cdot\frac{1}{2}} = m^{\frac{1}{3}}n^{-\frac{1}{8}} = \dfrac{m^{\frac{1}{3}}}{n^{\frac{1}{8}}}$

91. $\sqrt[8]{x^4} = x^{4/8}$ Convert to exponential notation

     $= x^{1/2}$ Simplifying the exponent

     $= \sqrt{x}$ Returning to radical notation

93. $\sqrt[4]{a^{12}} = a^{12/4}$ Convert to exponential notation

     $= a^3$ Simplifying the exponent

95. $\sqrt[12]{y^8} = y^{8/12} = y^{2/3} = \sqrt[3]{y^2}$

97. $\left(\sqrt[7]{xy}\right)^{14} = (xy)^{14/7}$ Convert to exponential notation

     $= (xy)^2$ Simplifying the exponent

     $= x^2y^2$ Using the law of exponents

99. $\sqrt[4]{(7a)^2} = (7a)^{2/4}$ Convert to exponential notation

     $= (7a)^{1/2}$ Simplifying the exponent

     $= \sqrt{7a}$ Returing to radical notation

101. $\left(\sqrt[8]{2x}\right)^6 = (2x)^{6/8} = (2x)^{3/4}$

     $= \sqrt[4]{(2x)^3} = \sqrt[4]{8x^3}$

103. $\sqrt{\sqrt[5]{m}} = \sqrt{m^{1/5}}$ Convert to exponential notation

     $= (m^{1/5})^{1/2}$

     $= m^{1/10}$ Using a law of exponents

     $= \sqrt[10]{m}$ Returning to radical notation

105. $\sqrt[4]{(xy)^{12}} = (xy)^{12/4} = (xy)^3 = x^3 y^3$

107. $\left(\sqrt[5]{\left(a^2 b^4\right)}\right)^{15} = \left(a^2 b^4\right)^{15/5} = \left(a^2 b^4\right)^3 = a^6 b^{12}$

109. $\sqrt[3]{\sqrt[4]{xy}} = \sqrt[3]{(xy)^{1/4}} = \left[(xy)^{1/4}\right]^{1/3}$

$= (xy)^{1/12} = \sqrt[12]{xy}$

111. *Thinking and Writing Exercise.*

113. $(x+5)(x-5)$

$= x \cdot x + 5x - 5x + 5(-5)$

$= x^2 - 25$

115. $4x^2 + 20x + 25 = (2x)^2 + 2(2x)(5) + 5^2$

$= (2x+5)^2$

117. $5t^2 - 10t + 5 = 5\left(t^2 - 2t + 1\right)$

$= 5\left(t^2 - 2(t)(1) + 1^2\right)$

$= 5(t-1)^2$

119. *Thinking and Writing Exercise.*

121. $\sqrt{x\sqrt[3]{x^2}} = \sqrt{x \cdot x^{2/3}} = \left(x^{5/3}\right)^{1/2} = x^{5/6} = \sqrt[6]{x^5}$

123. $\sqrt[12]{p^2 + 2pq + q^2} = \sqrt[12]{(p+q)^2} = \left[(p+q)^2\right]^{1/12}$

$= (p+q)^{2/12} = (p+q)^{1/6}$

$= \sqrt[6]{p+q}$

125. $f(x) = k \cdot 2^{x/12}$

$f(24) = 440 \cdot 2^{24/12}$

$= 440 \cdot 2^2$

$= 1760 \text{ cycles per second}$

127. $2^{4/12} \approx 1.2599 \approx 1.25$ so the C sharp that is 4 half steps above concert A has a frequency that is 125% of, or 25% greater than, that of concert A.

129. $L = \dfrac{0.000169 d^{2.27}}{h}$

   a. $L = \dfrac{(0.000169)60^{2.27}}{1} \approx 1.8 \text{ m}$

   b. $L = \dfrac{(0.000169)75^{2.27}}{0.9906} \approx 3.1 \text{ m}$

   c. $L = \dfrac{(0.000169)80^{2.27}}{2.4} \approx 1.5 \text{ m}$

   d. $L = \dfrac{(0.000169)100^{2.27}}{1.1} \approx 5.3 \text{ m}$

131. $T = 0.936 d^{1.97} h^{0.85}$

$= 0.936 \cdot 3^{1.97} 80^{0.85}$

$\approx 338 \text{ ft}^3$

133. $y_1 = x^{1/2},\ y_2 = 3x^{2/5},$
$y_3 = x^{4/7},\ y_4 = \dfrac{1}{5}x^{3/4}$

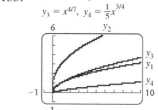

1. True

3. False

5. True

7. $\sqrt{5} \cdot \sqrt{7} = \sqrt{5 \cdot 7} = \sqrt{35}$

9. $\sqrt[3]{3} \cdot \sqrt[3]{2} = \sqrt[3]{3 \cdot 2} = \sqrt[3]{6}$

11. $\sqrt[4]{6} \cdot \sqrt[4]{3} = \sqrt[4]{6 \cdot 3} = \sqrt[4]{18}$

13. $\sqrt{2x} \cdot \sqrt{13y} = \sqrt{2x \cdot 13y} = \sqrt{26xy}$

15. $\sqrt[5]{8y^3} \cdot \sqrt[5]{10y} = \sqrt[5]{8y^3 \cdot 10y} = \sqrt[5]{80y^4}$

17. $\sqrt{y-b} \cdot \sqrt{y+b} = \sqrt{(y-b)(y+b)} = \sqrt{y^2 - b^2}$

19. $\sqrt[3]{0.7y} \cdot \sqrt[3]{0.3y} = \sqrt[3]{0.7y \cdot 0.3y} = \sqrt[3]{0.21y^2}$

21. $\sqrt[5]{x-2} \cdot \sqrt[5]{(x-2)^2} = \sqrt[5]{(x-2)(x-2)^2}$

$\qquad\qquad\qquad\qquad\quad = \sqrt[5]{(x-2)^3}$

23. $\sqrt{\dfrac{3}{t}} \cdot \sqrt{\dfrac{7s}{11}} = \sqrt{\dfrac{3}{t} \cdot \dfrac{7s}{11}} = \sqrt{\dfrac{21s}{11t}}$

25. $\sqrt[7]{\dfrac{x-3}{4}} \cdot \sqrt[7]{\dfrac{5}{x+2}} = \sqrt[7]{\dfrac{x-3}{4} \cdot \dfrac{5}{x+2}} = \sqrt[7]{\dfrac{5x-15}{4x+8}}$

27. $\sqrt{18}$

$\quad = \sqrt{9 \cdot 2}$ 9 is the greatest perfect
$\qquad\qquad\qquad$ square of 18

$\quad = \sqrt{9} \cdot \sqrt{2}$

$\quad = 3\sqrt{2}$

29. $\sqrt{27}$

$\quad = \sqrt{9 \cdot 3}$ 9 is the greatest perfect
$\qquad\qquad\qquad$ square of 27

$\quad = \sqrt{9} \cdot \sqrt{3}$

$\quad = 3\sqrt{3}$

31. $\sqrt{8x^9} = \sqrt{4x^8 \cdot 2x} = \sqrt{4x^8} \cdot \sqrt{2x} = 2x^4\sqrt{2x}$

33. $\sqrt{120} = \sqrt{4 \cdot 30} = \sqrt{4} \cdot \sqrt{30} = 2\sqrt{30}$

35. $\sqrt{36a^4 b}$

$\quad = \sqrt{36a^4 \cdot b}$ $36a^4$ is a perfect square

$\quad = \sqrt{36a^4} \cdot \sqrt{b}$ Factoring into 2 radicals

$\quad = 6a^2\sqrt{b}$ Taking the square root of $36a^4$

37. $\sqrt[3]{8x^3 y^2}$

$\quad = \sqrt[3]{8x^3 \cdot y^2}$ $8x^3$ is a perfect cube

$\quad = \sqrt[3]{8x^3} \cdot \sqrt[3]{y^2}$ Factoring into 2 radicals

$\quad = 2x\sqrt[3]{y^2}$ Taking the cube root of $8x^3$

39. $\sqrt[3]{-16x^6}$

$\quad = \sqrt[3]{-8x^6 \cdot 2}$ $-8x^6$ is a perfect cube

$\quad = \sqrt[3]{-8x^6} \cdot \sqrt[3]{2}$

$\quad = -2x^2\sqrt[3]{2}$ Taking the cube root of $-8x^6$

41. $f(x) = \sqrt[3]{125x^5}$

$\qquad = \sqrt[3]{125x^3 \cdot x^2}$

$\qquad = \sqrt[3]{125x^3} \cdot \sqrt[3]{x^2}$

$\qquad = 5x\sqrt[3]{x^2}$

43. $f(x) = \sqrt{49(x-3)^2}$ $49(x-3)^2$ is a
$\qquad\qquad\qquad\qquad\qquad$ perfect square

$\qquad = |7(x-3)|$, or $7|x-3|$

45. $f(x) = \sqrt{5x^2 - 10x + 5}$

$\qquad = \sqrt{5(x^2 - 2x + 1)}$

$\qquad = \sqrt{5(x-1)^2}$

$\qquad = \sqrt{(x-1)^2} \cdot \sqrt{5}$

$\qquad = |x-1|\sqrt{5}$

47. $\sqrt{a^6 b^7}$

$\quad = \sqrt{a^6 \cdot b^6 \cdot b}$ Identifying the greatest
$\qquad\qquad\qquad\qquad$ even powers of a and b

$\quad = \sqrt{a^6} \cdot \sqrt{b^6} \cdot \sqrt{b}$ Factoring into radicals

$\quad = a^3 b^3 \sqrt{b}$

49. $\sqrt[3]{x^5 y^6 z^{10}}$

$\quad = \sqrt[3]{x^3 \cdot x^2 \cdot y^6 \cdot z^9 \cdot z}$ Identifying the
$\quad$ greatest perfect-cube powers of x,
$\quad$ y, and z.

$\quad = \sqrt[3]{x^3} \cdot \sqrt[3]{y^6} \cdot \sqrt[3]{z^9} \cdot \sqrt[3]{x^2 z}$

$\quad$ Factoring into radicals

$\quad = xy^2 z^3 \sqrt[3]{x^2 z}$

51. $\sqrt[4]{16x^5 y^{11}}$

$\quad = \sqrt[4]{2^4 \cdot x^4 \cdot x \cdot y^8 \cdot y^3}$

$\quad = \sqrt[4]{2^4} \cdot \sqrt[4]{x^4} \cdot \sqrt[4]{y^8} \cdot \sqrt[4]{xy^3}$

$\quad = 2xy^2 \sqrt[4]{xy^3}$

53. $\sqrt[5]{x^{13}y^8z^{17}}$

$= \sqrt[5]{x^{10} \cdot x^3 \cdot y^5 \cdot y^3 \cdot z^{15} \cdot z^2}$

$= \sqrt[5]{x^{10}} \cdot \sqrt[5]{y^5} \cdot \sqrt[5]{z^{15}} \cdot \sqrt[5]{x^3y^3z^2}$

$= x^2yz^3\sqrt[5]{x^3y^3z^2}$

55. $\sqrt[3]{-80a^{14}} = \sqrt[3]{-8 \cdot 10 \cdot a^{12}a^2}$

$= \sqrt[3]{(-2)^3} \cdot \sqrt[3]{a^{12}} \cdot \sqrt[3]{10a^2}$

$= -2a^4\sqrt[3]{10a^2}$

57. $\sqrt{6} \cdot \sqrt{3} = \sqrt{18} = \sqrt{9 \cdot 2} = 3\sqrt{2}$

59. $\sqrt{10} \cdot \sqrt{14} = \sqrt{140} = \sqrt{4 \cdot 35} = 2\sqrt{35}$

61. $\sqrt[3]{9} \cdot \sqrt[3]{3} = \sqrt[3]{27} = \sqrt[3]{3^3} = 3$

63. $\sqrt{18a^3} \cdot \sqrt{18a^3} = \left(\sqrt{18a^3}\right)^2 = 18a^3$

65. $\sqrt[3]{5a^2} \cdot \sqrt[3]{2a} = \sqrt[3]{10a^3} = \sqrt[3]{a^3 \cdot 10} = a\sqrt[3]{10}$

67. $3\sqrt{2x^5} \cdot 4\sqrt{10x^2} = 12\sqrt{20x^7} = 12\sqrt{4x^6 \cdot 5x}$

$== 12 \cdot 2x^3\sqrt{5x} = 24x^3\sqrt{5x}$

69. $\sqrt[3]{s^2t^4} \cdot \sqrt[3]{s^4t^6} = \sqrt[3]{s^6t^{10}} = \sqrt[3]{s^6t^9t} = s^2t^3\sqrt[3]{t}$

71. $\sqrt[3]{(x+5)^2} \cdot \sqrt[3]{(x+5)^4} = \sqrt[3]{(x+5)^6} = (x+5)^2$

73. $\sqrt[4]{20a^3b^7} \cdot \sqrt[4]{4a^2b^5} = \sqrt[4]{80a^5b^{12}}$

$= \sqrt[4]{16 \cdot 5 \cdot a^4ab^{12}}$

$= 2ab^3\sqrt[4]{5a}$

75. $\sqrt[5]{x^3(y+z)^6} \cdot \sqrt[5]{x^3(y+z)^4} = \sqrt[5]{x^6(y+z)^{10}}$

$= \sqrt[5]{x^5(y+z)^{10} \cdot x}$

$= x(y+z)^2\sqrt[5]{x}$

77. *Thinking and Writing Exercise.*

79. $\dfrac{15a^2x}{8b} \cdot \dfrac{24b^2x}{5a} = \dfrac{15a^2x \cdot 24b^2x}{8b \cdot 5a}$

$= \dfrac{15a^2x}{5a} \cdot \dfrac{24b^2x}{8b}$

$= \dfrac{3a \cdot 5a \cdot x}{5a} \cdot \dfrac{3b \cdot 8b \cdot x}{8b}$

$= 3ax \cdot 3bx$

$= 9abx^2$

81. $\dfrac{x-3}{2x-10} - \dfrac{3x-5}{x^2-25} = \dfrac{x-3}{2(x-5)} - \dfrac{3x-5}{(x+5)(x-5)}$

LCD is $2(x+5)(x-5)$

$\dfrac{x-3}{2x-10} - \dfrac{3x-5}{x^2-25}$

$= \dfrac{(x-3)(x+5)}{2(x-5)(x+5)} - \dfrac{2(3x-5)}{2(x+5)(x-5)}$

$= \dfrac{(x-3)(x+5) - 2(3x-5)}{2(x-5)(x+5)}$

$= \dfrac{x^2-3x+5x-15-6x+10}{2(x-5)(x+5)}$

$= \dfrac{x^2-4x-5}{2(x-5)(x+5)}$

$= \dfrac{(x-5)(x+1)}{2(x-5)(x+5)}$

$= \dfrac{(x+1)}{2(x+5)}$

83. $\dfrac{a^{-1}+b^{-1}}{ab} = \dfrac{\dfrac{1}{a}+\dfrac{1}{b}}{ab}$

$= \dfrac{ab\left(\dfrac{1}{a}+\dfrac{1}{b}\right)}{ab(ab)}$

$= \dfrac{b+a}{a^2b^2}$

85. *Thinking and Writing Exercise.*

87. $R(x) = \dfrac{1}{2}\sqrt[4]{\dfrac{x \cdot 3.0 \times 10^6}{\pi^2}}$

$R(5 \times 10^4) = \dfrac{1}{2}\sqrt[4]{\dfrac{(5 \times 10^4) \cdot (3.0 \times 10^6)}{\pi^2}}$

≈ 175.6 mi

89. $T_w = 33 - \dfrac{\left(10.45 + 10\sqrt{v} - v\right)\left(33 - T\right)}{22}$

    a. $T_w = 33 - \dfrac{\left(10.45 + 10\sqrt{8} - 8\right)\left(33 - 7\right)}{22}$

       $\approx -3.3°C$

    b. $T_w = 33 - \dfrac{\left(10.45 + 10\sqrt{12} - 12\right)\left(33 - 0\right)}{22}$

       $\approx -16.6°C$

    c. $T_w = 33 - \dfrac{\left(10.45 + 10\sqrt{14} - 14\right)\left(33 - -5\right)}{22}$

       $\approx -25.5 \ °C$

    d. $T_w = 33 - \dfrac{\left(10.45 + 10\sqrt{15} - 15\right)\left(33 - -23\right)}{22}$

       $\approx -54.0 \ °C$

91. $\left(\sqrt[3]{25x^4}\right)^4 = \sqrt[3]{\left(25x^4\right)^4} = \sqrt[3]{25^4 x^{16}}$

             $= \sqrt[3]{25^3 \cdot 25 \cdot x^{15} \cdot x}$

             $= \sqrt[3]{25^3} \cdot \sqrt[3]{x^{15}} \cdot \sqrt[3]{25x}$

             $= 25x^5 \sqrt[3]{25x}$

93. $\left(\sqrt{a^3 b^5}\right)^7 = \sqrt{\left(a^3 b^5\right)^7} = \sqrt{a^{21} b^{35}}$

            $= \sqrt{a^{20} \cdot a \cdot b^{34} \cdot b}$

            $= a^{10} b^{17} \sqrt{ab}$

95.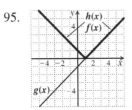

We see that $f(x) = h(x)$ and
$f(x) \neq g(x)$.

97. $g(x) = \sqrt{x^2 - 6x + 8}$

We must have $x^2 - 6x + 8 \geq 0$, or
$(x - 2)(x - 4) \geq 0$. We graph
$y = x^2 - 6x + 8$.

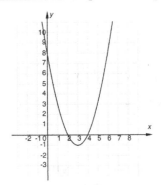

From the graph we see that $y \geq 0$ for $x \leq 2$
or $x \geq 4$, so the domain of g is
$\{x | x \leq 2 \ or \ x \geq 4\}$, or $(-\infty, 2] \cup [4, \infty)$.

99. $\sqrt[5]{4a^{3k+2}} \ \sqrt[5]{8a^{6-k}} = 2a^4$

      $\sqrt[5]{4a^{3k+2} \cdot 8a^{6-k}} = 2a^4$

        $\sqrt[5]{32a^{2k+8}} = 2a^4$

       $2\sqrt[5]{a^{2k+8}} = 2a^4$

         $\sqrt[5]{a^{2k+8}} = a^4$

       $\left(a^{2k+8}\right)^{1/5} = a^4$

          $a^{\frac{2k+8}{5}} = a^4$

Since the base is the same, the exponents
must be equal. We have:

    $\dfrac{2k + 8}{5} = 4$

    $2k + 8 = 20$

      $2k = 12$

       $k = 6$

101. *Thinking and Writing Exercise.*

Exercise Set 7.4

1.   e

3.   f

5.   h

7.   a

9. $\sqrt{\dfrac{36}{25}} = \dfrac{\sqrt{36}}{\sqrt{25}} = \dfrac{6}{5}$

11. $\sqrt[3]{\dfrac{64}{27}} = \dfrac{\sqrt[3]{64}}{\sqrt[3]{27}} = \dfrac{4}{3}$

13. $\sqrt{\dfrac{49}{y^2}} = \dfrac{\sqrt{49}}{\sqrt{y^2}} = \dfrac{7}{y}$

15. $\sqrt{\dfrac{36y^3}{x^4}} = \dfrac{\sqrt{36y^3}}{\sqrt{x^4}} = \dfrac{\sqrt{36y^2 \cdot y}}{\sqrt{x^4}} = \dfrac{6y\sqrt{y}}{x^2}$

17. $\sqrt[3]{\dfrac{27a^4}{8b^3}} = \dfrac{\sqrt[3]{27a^3 \cdot a}}{\sqrt[3]{8b^3}} = \dfrac{3a\sqrt[3]{a}}{2b}$

19. $\sqrt[4]{\dfrac{32a^4}{2b^4c^8}} = \sqrt[4]{\dfrac{16a^4}{b^4c^8}} = \dfrac{\sqrt[4]{16a^4}}{\sqrt[4]{b^4c^8}} = \dfrac{2a}{bc^2}$

21. $\sqrt[4]{\dfrac{a^5b^8}{c^{10}}} = \dfrac{\sqrt[4]{a^4 \cdot b^8 \cdot a}}{\sqrt[4]{c^8 \cdot c^2}} = \dfrac{ab^2\sqrt[4]{a}}{c^2\sqrt[4]{c^2}}$, or $\dfrac{ab^2}{c^2}\sqrt[4]{\dfrac{a}{c^2}}$

23. $\sqrt[5]{\dfrac{32x^6}{y^{11}}} = \dfrac{\sqrt[5]{32x^5 \cdot x}}{\sqrt[5]{y^{10} \cdot y}} = \dfrac{2x\sqrt[5]{x}}{y^2\sqrt[5]{y}}$, or $\dfrac{2x}{y^2}\sqrt[5]{\dfrac{x}{y}}$

25. $\sqrt[6]{\dfrac{x^6y^8}{z^{15}}} = \dfrac{\sqrt[6]{x^6y^6 \cdot y^2}}{\sqrt[6]{z^{12} \cdot z^3}} = \dfrac{xy\sqrt[6]{y^2}}{z^2\sqrt[6]{z^3}}$,

$\qquad$ or $\dfrac{xy}{z^2}\sqrt[6]{\dfrac{y^2}{z^3}}$

27. $\dfrac{\sqrt{18y}}{\sqrt{2y}} = \sqrt{\dfrac{18y}{2y}} = \sqrt{9} = 3$

29. $\dfrac{\sqrt[3]{26}}{\sqrt[3]{13}} = \sqrt[3]{\dfrac{26}{13}} = \sqrt[3]{2}$

31. $\dfrac{\sqrt{40xy^3}}{\sqrt{8x}} = \sqrt{\dfrac{40xy^3}{8x}} = \sqrt{5y^3} = \sqrt{y^2 \cdot 5y}$
$\qquad = \sqrt{y^2}\sqrt{5y} = y\sqrt{5y}$

33. $\dfrac{\sqrt[3]{96a^4b^2}}{\sqrt[3]{12a^2b}} = \sqrt[3]{\dfrac{96a^4b^2}{12a^2b}} = \sqrt[3]{8a^2b} = 2\sqrt[3]{a^2b}$

35. $\dfrac{\sqrt{100ab}}{5\sqrt{2}} = \dfrac{1}{5}\dfrac{\sqrt{100ab}}{\sqrt{2}} = \dfrac{1}{5}\sqrt{\dfrac{100ab}{2}}$
$\qquad = \dfrac{1}{5}\sqrt{50ab} = \dfrac{1}{5}\sqrt{25 \cdot 2ab}$
$\qquad = \dfrac{1}{5} \cdot 5\sqrt{2ab} = \sqrt{2ab}$

37. $\dfrac{\sqrt[4]{48x^9y^{13}}}{\sqrt[4]{3xy^{-2}}} = \sqrt[4]{\dfrac{48x^9y^{13}}{3xy^{-2}}} = \sqrt[4]{16x^8y^{15}}$
$\qquad = \sqrt[4]{16x^8y^{12} \cdot y^3} = 2x^2y^3\sqrt[4]{y^3}$

39. $\dfrac{\sqrt[3]{x^3 - y^3}}{\sqrt[3]{x-y}} = \sqrt[3]{\dfrac{x^3 - y^3}{x-y}}$

$\qquad = \sqrt[3]{\dfrac{(x-y)(x^2 + xy + y^2)}{x-y}}$

$\qquad = \sqrt[3]{\dfrac{\cancel{(x-y)}(x^2 + xy + y^2)}{\cancel{x-y}}}$

$\qquad = \sqrt[3]{x^2 + xy + y^2}$

41. $\sqrt{\dfrac{3}{2}} = \sqrt{\dfrac{3}{2} \cdot \dfrac{2}{2}} = \sqrt{\dfrac{6}{4}} = \dfrac{\sqrt{6}}{\sqrt{4}} = \dfrac{\sqrt{6}}{2}$

43. $\dfrac{2\sqrt{5}}{7\sqrt{3}} = \dfrac{2\sqrt{5}}{7\sqrt{3}} \cdot \dfrac{\sqrt{3}}{\sqrt{3}} = \dfrac{2\sqrt{15}}{7 \cdot 3} = \dfrac{2\sqrt{15}}{21}$

45. $\sqrt[3]{\dfrac{5}{4}} = \sqrt[3]{\dfrac{5}{2^2} \cdot \dfrac{2}{2}} = \sqrt[3]{\dfrac{10}{2^3}} = \dfrac{\sqrt[3]{10}}{\sqrt[3]{2^3}} = \dfrac{\sqrt[3]{10}}{2}$

47. $\dfrac{\sqrt[3]{3a}}{\sqrt[3]{5c}} = \dfrac{\sqrt[3]{3a}}{\sqrt[3]{5c}} \cdot \dfrac{\sqrt[3]{5^2c^2}}{\sqrt[3]{5^2c^2}} = \dfrac{\sqrt[3]{75ac^2}}{\sqrt[3]{5^3c^3}} = \dfrac{\sqrt[3]{75ac^2}}{5c}$

49. $\dfrac{\sqrt[4]{5y^6}}{\sqrt[4]{9x}} = \dfrac{y\sqrt[4]{5y^2}}{\sqrt[4]{3^2x^1}} \cdot \dfrac{\sqrt[4]{3^2x^3}}{\sqrt[4]{3^2x^3}} = \dfrac{y\sqrt[4]{45x^3y^2}}{\sqrt[4]{3^4x^4}}$
$\qquad = \dfrac{y\sqrt[4]{45x^3y^2}}{3x}$

51. $\sqrt[3]{\dfrac{2}{x^2y}} = \sqrt[3]{\dfrac{2}{x^2y} \cdot \dfrac{xy^2}{xy^2}} = \sqrt[3]{\dfrac{2xy^2}{x^3y^3}}$
$\qquad = \dfrac{\sqrt[3]{2xy^2}}{\sqrt[3]{x^3y^3}} = \dfrac{\sqrt[3]{2xy^2}}{xy}$

53. $\sqrt{\dfrac{7a}{18}} = \sqrt{\dfrac{7a}{18} \cdot \dfrac{2}{2}} = \sqrt{\dfrac{14a}{36}} = \dfrac{\sqrt{14a}}{6}$

55. $\sqrt{\dfrac{9}{20x^2 y}} = \sqrt{\dfrac{9}{20x^2 y} \cdot \dfrac{5y}{5y}}$

$= \sqrt{\dfrac{9 \cdot 5y}{100x^2 y^2}} = \dfrac{3\sqrt{5y}}{10xy}$

57. $\sqrt{\dfrac{10ab^2}{72a^3 b}} = \sqrt{\dfrac{5b}{36a^2}} = \dfrac{\sqrt{5b}}{6a}$

59. $\sqrt{\dfrac{5}{11}} = \dfrac{\sqrt{5}}{\sqrt{11}} = \dfrac{\sqrt{5}}{\sqrt{11}} \cdot \dfrac{\sqrt{5}}{\sqrt{5}} = \dfrac{\sqrt{25}}{\sqrt{55}} = \dfrac{5}{\sqrt{55}}$

61. $\dfrac{2\sqrt{6}}{5\sqrt{7}} = \dfrac{2\sqrt{6} \cdot \sqrt{6}}{5\sqrt{7} \cdot \sqrt{6}} = \dfrac{2 \cdot 6}{5\sqrt{42}} = \dfrac{12}{5\sqrt{42}}$

63. $\dfrac{\sqrt{8}}{2\sqrt{3x}} = \dfrac{\cancel{2}\sqrt{2}}{\cancel{2}\sqrt{3x}} \cdot \dfrac{\sqrt{2}}{\sqrt{2}} = \dfrac{2}{\sqrt{6x}}$

65. $\dfrac{\sqrt[3]{7}}{\sqrt[3]{2}} = \dfrac{\sqrt[3]{7}}{\sqrt[3]{2}} \cdot \dfrac{\sqrt[3]{7^2}}{\sqrt[3]{7^2}} = \dfrac{\sqrt[3]{7^3}}{\sqrt[3]{98}} = \dfrac{7}{\sqrt[3]{98}}$

67. $\sqrt{\dfrac{7x}{3y}} = \sqrt{\dfrac{7x \cdot 7x}{3y \cdot 7x}} = \dfrac{\sqrt{(7x)^2}}{\sqrt{21xy}} = \dfrac{7x}{\sqrt{21xy}}$

69. $\sqrt[3]{\dfrac{2a^5}{5b}} = \sqrt[3]{\dfrac{2a^5}{5b} \cdot \dfrac{4a}{4a}} = \sqrt[3]{\dfrac{8a^6}{20ab}} = \dfrac{2a^2}{\sqrt[3]{20ab}}$

71. $\sqrt{\dfrac{x^3 y}{2}} = \sqrt{\dfrac{x^3 y}{2} \cdot \dfrac{xy}{xy}} = \sqrt{\dfrac{x^4 y^2}{2xy}}$

$= \dfrac{\sqrt{x^4 y^2}}{\sqrt{2xy}} = \dfrac{x^2 y}{\sqrt{2xy}}$

73. *Thinking and Writing Exercise.*

75. $3x - 8xy + 2xz = x(3 - 8y + 2z)$

77. $(a+b)(a-b) = a^2 - b^2$

79. $(8+3x)(7-4x)$

$= 8 \cdot 7 - 8 \cdot 4x + 3x \cdot 7 - 3x \cdot 4x$

$= 56 - 32x + 21x - 12x^2$

$= 56 - 11x - 12x^2$

81. *Thinking and Writing Exercise.*

83. a. $T = 2\pi\sqrt{\dfrac{65}{980}} \approx 1.62 \, \text{sec}$

b. $T = 2\pi\sqrt{\dfrac{98}{980}} \approx 1.99 \, \text{sec}$

c. $T = 2\pi\sqrt{\dfrac{120}{980}} \approx 2.20 \, \text{sec}$

85. $\dfrac{\left(\sqrt[3]{81mn^2}\right)^2}{\left(\sqrt[3]{mn}\right)^2} = \dfrac{\sqrt[3]{\left(81mn^2\right)^2}}{\sqrt[3]{\left(mn\right)^2}} = \dfrac{\sqrt[3]{6561m^2 n^4}}{\sqrt[3]{m^2 n^2}}$

$= \sqrt[3]{\dfrac{6561m^2 n^4}{m^2 n^2}} = \sqrt[3]{6561n^2}$

$= \sqrt[3]{729 \cdot 9n^2} = \sqrt[3]{729}\sqrt[3]{9n^2}$

$= 9\sqrt[3]{9n^2}$

87. $\sqrt{a^2 - 3} - \dfrac{a^2}{\sqrt{a^2 - 3}}$

$= \sqrt{a^2 - 3} - \dfrac{a^2}{\sqrt{a^2 - 3}} \cdot \dfrac{\sqrt{a^2 - 3}}{\sqrt{a^2 - 3}}$

$= \sqrt{a^2 - 3} - \dfrac{a^2 \sqrt{a^2 - 3}}{a^2 - 3}$

$= \sqrt{a^2 - 3} \cdot \dfrac{a^2 - 3}{a^2 - 3} - \dfrac{a^2 \sqrt{a^2 - 3}}{a^2 - 3}$

$= \dfrac{a^2 \sqrt{a^2 - 3} - 3\sqrt{a^2 - 3} - a^2 \sqrt{a^2 - 3}}{a^2 - 3}$

$= \dfrac{-3\sqrt{a^2 - 3}}{a^2 - 3}, \text{ or } \dfrac{-3}{\sqrt{a^2 - 3}}$

89. Step 1: $\sqrt[n]{a} = a^{1/n}$, by definition;

Step 2: $\left(\dfrac{a}{b}\right)^{1/n} = \dfrac{a^{1/n}}{b^{1/n}}$, raising a quotient to a power;

Step 3: $a^{1/n} = \sqrt[n]{a}$, by definition.

91. $f(x) = \sqrt{18x^3}, g(x) = \sqrt{2x}$

$(f/g)(x) = \dfrac{f(x)}{g(x)} = \dfrac{\sqrt{18x^3}}{\sqrt{2x}}$

$= \sqrt{\dfrac{18x^3}{2x}} = \sqrt{9x^2} = 3x$

$\sqrt{2x}$ is defined for $2x \geq 0$, or $x \geq 0$. To avoid division by 0, we must exclude 0 from the domain. Thus, the domain of $f/g = \{x | x \text{ is a real number and } x > 0\}$, or $(0, \infty)$.

93. $f(x) = \sqrt{x^2 - 9}, g(x) = \sqrt{x-3}$

$(f/g)(x) = \dfrac{f(x)}{g(x)} = \dfrac{\sqrt{x^2-9}}{\sqrt{x-3}}$

$= \sqrt{\dfrac{x^2-9}{x-3}}$

$= \sqrt{\dfrac{(x+3)(x-3)}{x-3}}$

$= \sqrt{x+3}$

$\sqrt{x-3}$ is defined for $x-3 \geq 0$, or $x \geq 3$. To avoid division by 0, we must exclude 3 from the domain. Thus, the domain of $f/g = \{x | x \text{ is a real number and } x > 3\}$, or $(3, \infty)$.

Exercise Set 7.5

1. Radicands; indices

3. Bases

5. Numerator; conjugate

7. $2\sqrt{5} + 7\sqrt{5} = (2+7)\sqrt{5} = 9\sqrt{5}$

9. $7\sqrt[3]{4} - 5\sqrt[3]{4} = (7-5)\sqrt[3]{4} = 2\sqrt[3]{4}$

11. $\sqrt[3]{y} + 9\sqrt[3]{y} = (1+9)\sqrt[3]{y} = 10\sqrt[3]{y}$

13. $8\sqrt{2} - \sqrt{2} + 5\sqrt{2} = (8-1+5)\sqrt{2} = 12\sqrt{2}$

15. $9\sqrt[3]{7} - \sqrt{3} + 4\sqrt[3]{7} + 2\sqrt{3}$

$= (9+4)\sqrt[3]{7} + (-1+2)\sqrt{3}$

$= 13\sqrt[3]{7} + \sqrt{3}$

17. $4\sqrt{27} - 3\sqrt{3} = 4\sqrt{9 \cdot 3} - 3\sqrt{3}$

$= 4 \cdot 3\sqrt{3} - 3\sqrt{3}$

$= 12\sqrt{3} - 3\sqrt{3}$

$= (12-3)\sqrt{3} = 9\sqrt{3}$

19. $3\sqrt{45} - 8\sqrt{20} = 3\sqrt{9 \cdot 5} - 8\sqrt{4 \cdot 5}$

$= 3 \cdot 3\sqrt{5} - 8 \cdot 2\sqrt{5}$

$= 9\sqrt{5} - 16\sqrt{5} = -7\sqrt{5}$

21. $3\sqrt[3]{16} + \sqrt[3]{54} = 3\sqrt[3]{8 \cdot 2} + \sqrt[3]{27 \cdot 2}$

$= 3 \cdot 2\sqrt[3]{2} + 3\sqrt[3]{2} = 6\sqrt[3]{2} + 3\sqrt[3]{2}$

$= 9\sqrt[3]{2}$

23. $\sqrt{a} + 3\sqrt{16a^3} = \sqrt{a} + 3\sqrt{16a^2 \cdot a}$

$= \sqrt{a} + 3 \cdot 4a\sqrt{a}$

$= \sqrt{a} + 12a\sqrt{a}$

$= (1 + 12a)\sqrt{a}$

25. $\sqrt[3]{6x^4} - \sqrt[3]{48x} = \sqrt[3]{x^3 \cdot 6x} - \sqrt[3]{8 \cdot 6x}$

$= x\sqrt[3]{6x} - 2\sqrt[3]{6x}$

$= (x-2)\sqrt[3]{6x}$

27. $\sqrt{4a-4} + \sqrt{a-1} = \sqrt{4 \cdot (a-1)} + \sqrt{a-1}$

$= 2\sqrt{a-1} + \sqrt{a-1}$

$= 3\sqrt{a-1}$

29. $\sqrt{x^3 - x^2} + \sqrt{9x-9} = \sqrt{x^2(x-1)} + \sqrt{9(x-1)}$

$= x\sqrt{x-1} + 3\sqrt{x-1}$

$= (x+3)\sqrt{x-1}$

31. $\sqrt{3}(4 + \sqrt{3}) = \sqrt{3} \cdot 4 + \sqrt{3} \cdot \sqrt{3} = 4\sqrt{3} + 3$

33. $3\sqrt{5}(\sqrt{5} - \sqrt{2}) = 3\sqrt{5} \cdot \sqrt{5} - 3\sqrt{5} \cdot \sqrt{2}$

$= 3 \cdot 5 - 3\sqrt{10} = 15 - 3\sqrt{10}$

35. $\sqrt{2}\left(3\sqrt{10}-2\sqrt{2}\right)=\sqrt{2}\cdot3\sqrt{10}-\sqrt{2}\cdot2\sqrt{2}$

$\qquad = 3\sqrt{20}-2\cdot2 = 3\sqrt{4\cdot5}-4$

$\qquad = 3\cdot2\sqrt{5}-4 = 6\sqrt{5}-4$

37. $\sqrt[3]{3}\left(\sqrt[3]{9}-4\sqrt[3]{21}\right)=\sqrt[3]{3}\cdot\sqrt[3]{9}-\sqrt[3]{3}\cdot4\sqrt[3]{21}$

$\qquad = \sqrt[3]{27}-4\sqrt[3]{63} = 3-4\sqrt[3]{63}$

39. $\sqrt[3]{a}\left(\sqrt[3]{a^2}+\sqrt[3]{24a^2}\right)=\sqrt[3]{a}\cdot\sqrt[3]{a^2}+\sqrt[3]{a}\cdot\sqrt[3]{24a^2}$

$\qquad = \sqrt[3]{a^3}+\sqrt[3]{24a^3}$

$\qquad = a+\sqrt[3]{8a^3\cdot3}$

$\qquad = a+2a\sqrt[3]{3}$

41. $\left(2+\sqrt{6}\right)\left(5-\sqrt{6}\right)$

$\qquad = 2\cdot5-2\sqrt{6}+5\sqrt{6}-\sqrt{6}\cdot\sqrt{6}$

$\qquad = 10+3\sqrt{6}-6 = 4+3\sqrt{6}$

43. $\left(\sqrt{2}+\sqrt{7}\right)\left(\sqrt{3}-\sqrt{7}\right)$

$\qquad = \sqrt{2}\cdot\sqrt{3}-\sqrt{2}\cdot\sqrt{7}+\sqrt{7}\cdot\sqrt{3}-\sqrt{7}\cdot\sqrt{7}$

$\qquad = \sqrt{6}-\sqrt{14}+\sqrt{21}-7$

45. $\left(3-\sqrt{5}\right)\left(3+\sqrt{5}\right)=3^2-\left(\sqrt{5}\right)^2 = 9-5 = 4$

47. $\left(\sqrt{6}+\sqrt{8}\right)\left(\sqrt{6}-\sqrt{8}\right)=\left(\sqrt{6}\right)^2-\left(\sqrt{8}\right)^2$

$\qquad = 6-8 = -2$

49. $\left(3\sqrt{7}+2\sqrt{5}\right)\left(2\sqrt{7}-4\sqrt{5}\right)$

$\qquad = 3\sqrt{7}\cdot2\sqrt{7}-3\sqrt{7}\cdot4\sqrt{5}+2\sqrt{5}\cdot2\sqrt{7}-$

$\qquad\qquad\qquad\qquad 2\sqrt{5}\cdot4\sqrt{5}$

$\qquad = 3\cdot2\cdot7-12\sqrt{35}+4\sqrt{35}-2\cdot4\cdot5$

$\qquad = 42-8\sqrt{35}-40 = 2-8\sqrt{35}$

51. $\left(2+\sqrt{3}\right)^2=2^2+2\cdot2\cdot\sqrt{3}+\left(\sqrt{3}\right)^2$

$\qquad = 4+4\sqrt{3}+3 = 7+4\sqrt{3}$

53. $\left(\sqrt{3}-\sqrt{2}\right)^2=\left(\sqrt{3}\right)^2-2\sqrt{3}\sqrt{2}+\left(\sqrt{2}\right)^2$

$\qquad = 3-2\sqrt{6}+2 = 5-2\sqrt{6}$

55. $\left(\sqrt{2t}+\sqrt{5}\right)^2=\left(\sqrt{2t}\right)^2+2\sqrt{2t}\sqrt{5}+\left(\sqrt{5}\right)^2$

$\qquad = 2t+2\sqrt{10t}+5$

$\qquad = 2t+5+2\sqrt{10t}$

57. $\left(3-\sqrt{x+5}\right)^2=3^2-2\cdot3\cdot\sqrt{x+5}+\left(\sqrt{x+5}\right)^2$

$\qquad = 9-6\sqrt{x+5}+x+5$

$\qquad = 14+x-6\sqrt{x+5}$

59. $\left(2\sqrt[4]{7}-\sqrt[4]{6}\right)\left(3\sqrt[4]{9}+2\sqrt[4]{5}\right)$

$\qquad = 2\sqrt[4]{7}\cdot3\sqrt[4]{9}+2\sqrt[4]{7}\cdot2\sqrt[4]{5}-\sqrt[4]{6}\cdot3\sqrt[4]{9}-$

$\qquad\qquad\qquad\qquad\qquad\qquad \sqrt[4]{6}\cdot2\sqrt[4]{5}$

$\qquad = 6\sqrt[4]{63}+4\sqrt[4]{35}-3\sqrt[4]{54}-2\sqrt[4]{30}$

61. $\dfrac{6}{3-\sqrt{2}}=\dfrac{6}{3-\sqrt{2}}\cdot\dfrac{3+\sqrt{2}}{3+\sqrt{2}}=\dfrac{18+6\sqrt{2}}{9-2}$

$\qquad = \dfrac{18+6\sqrt{2}}{7}$

63. $\dfrac{2+\sqrt{5}}{6+\sqrt{3}}=\dfrac{2+\sqrt{5}}{6+\sqrt{3}}\cdot\dfrac{6-\sqrt{3}}{6-\sqrt{3}}$

$\qquad = \dfrac{12-2\sqrt{3}+6\sqrt{5}-\sqrt{15}}{36-3}$

$\qquad = \dfrac{12-2\sqrt{3}+6\sqrt{5}-\sqrt{15}}{33}$

65. $\dfrac{\sqrt{a}}{\sqrt{a}+\sqrt{b}}=\dfrac{\sqrt{a}}{\sqrt{a}+\sqrt{b}}\cdot\dfrac{\sqrt{a}-\sqrt{b}}{\sqrt{a}-\sqrt{b}}$

$\qquad = \dfrac{\sqrt{a}\left(\sqrt{a}-\sqrt{b}\right)}{\left(\sqrt{a}+\sqrt{b}\right)\left(\sqrt{a}-\sqrt{b}\right)}=\dfrac{a-\sqrt{ab}}{a-b}$

67. $\dfrac{\sqrt{7}-\sqrt{3}}{\sqrt{3}-\sqrt{7}}=\dfrac{-1\left(\sqrt{3}-\sqrt{7}\right)}{\sqrt{3}-\sqrt{7}}$

$\qquad = -1\cdot\dfrac{\sqrt{3}-\sqrt{7}}{\sqrt{3}-\sqrt{7}}=-1\cdot1 = -1$

69. $\dfrac{3\sqrt{2}-\sqrt{7}}{4\sqrt{2}+2\sqrt{5}} = \dfrac{3\sqrt{2}-\sqrt{7}}{4\sqrt{2}+2\sqrt{5}} \cdot \dfrac{4\sqrt{2}-2\sqrt{5}}{4\sqrt{2}-2\sqrt{5}}$

$= \dfrac{24 - 6\sqrt{10} - 4\sqrt{14} + 2\sqrt{35}}{32 - 20}$

$= \dfrac{\cancel{2}\left(12 - 3\sqrt{10} - 2\sqrt{14} + \sqrt{35}\right)}{\cancel{2} \cdot 6}$

$= \dfrac{12 - 3\sqrt{10} - 2\sqrt{14} + \sqrt{35}}{6}$

71. $\dfrac{\sqrt{5}+1}{4} = \dfrac{\sqrt{5}+1}{4} \cdot \dfrac{\sqrt{5}-1}{\sqrt{5}-1} = \dfrac{5-1}{4\left(\sqrt{5}-1\right)}$

$= \dfrac{4}{4\left(\sqrt{5}-1\right)} = \dfrac{1}{\sqrt{5}-1}$

73. $\dfrac{\sqrt{6}-2}{\sqrt{3}+7} = \dfrac{\sqrt{6}-2}{\sqrt{3}+7} \cdot \dfrac{\sqrt{6}+2}{\sqrt{6}+2}$

$= \dfrac{6-4}{\sqrt{18} + 2\sqrt{3} + 7\sqrt{6} + 14}$

$= \dfrac{2}{3\sqrt{2} + 2\sqrt{3} + 7\sqrt{6} + 14}$

75. $\dfrac{\sqrt{x}-\sqrt{y}}{\sqrt{x}+\sqrt{y}} = \dfrac{\sqrt{x}-\sqrt{y}}{\sqrt{x}+\sqrt{y}} \cdot \dfrac{\sqrt{x}+\sqrt{y}}{\sqrt{x}+\sqrt{y}}$

$= \dfrac{x-y}{x + 2\sqrt{xy} + y}$

77. $\dfrac{\sqrt{a+h}-\sqrt{a}}{h} = \dfrac{\sqrt{a+h}-\sqrt{a}}{h} \cdot \dfrac{\sqrt{a+h}+\sqrt{a}}{\sqrt{a+h}+\sqrt{a}}$

$= \dfrac{(a+h)-a}{h\left(\sqrt{a+h}+\sqrt{a}\right)}$

$= \dfrac{\cancel{h} \cdot 1}{\cancel{h}\left(\sqrt{a+h}+\sqrt{a}\right)}$

$= \dfrac{1}{\sqrt{a+h}+\sqrt{a}}$

79. $\sqrt[3]{a}\sqrt[6]{a}$

$= a^{1/3} \cdot a^{1/6}$ Converting to exponential notation

$= a^{3/6}$ Adding exponents

$= a^{1/2}$ Simplifying exponent

$= \sqrt{a}$ Returning to radical notation

81. $\sqrt[5]{b^2}\sqrt{b^3}$

$= b^{2/5} \cdot b^{3/2}$ Converting to exponential notation

$= b^{19/10}$ Adding exponents

$= b^{1+9/10}$ Writing 19/10 as a mixed number

$= b \cdot b^{9/10}$ Factoring

$= b\sqrt[10]{b^9}$ Returning to radical notation

83. $\sqrt{xy^3}\sqrt[3]{x^2 y} = \left(xy^3\right)^{1/2}\left(x^2 y\right)^{1/3}$

$= \left(xy^3\right)^{3/6}\left(x^2 y\right)^{2/6}$

$= \left[\left(xy^3\right)^3\left(x^2 y\right)^2\right]^{1/6}$

$= \sqrt[6]{x^3 y^9 \cdot x^4 y^2}$

$= \sqrt[6]{x^7 y^{11}}$

$= \sqrt[6]{x^6 y^6 \cdot xy^5}$

$= xy\sqrt[6]{xy^5}$

85. $\sqrt[4]{9ab^3}\sqrt{3a^4 b} = \left(9ab^3\right)^{1/4}\left(3a^4 b\right)^{1/2}$

$= \left(9ab^3\right)^{1/4}\left(3a^4 b\right)^{2/4}$

$= \left[\left(9ab^3\right)\left(3a^4 b\right)^2\right]^{1/4}$

$= \sqrt[4]{9ab^3 \cdot 9a^8 b^2}$

$= \sqrt[4]{81a^9 b^5}$

$= \sqrt[4]{81a^8 b^4 \cdot ab}$

$= 3a^2 b\sqrt[4]{ab}$

87. $\sqrt{a^4b^3c^4}\,\sqrt[3]{ab^2c} = \left(a^4b^3c^4\right)^{1/2}\left(ab^2c\right)^{1/3}$

$= \left(a^4b^3c^4\right)^{3/6}\left(ab^2c\right)^{2/6}$

$= \left[\left(a^4b^3c^4\right)^3\left(ab^2c\right)^2\right]^{1/6}$

$= \sqrt[6]{a^{12}b^9c^{12}\cdot a^2b^4c^2}$

$= \sqrt[6]{a^{14}b^{13}c^{14}}$

$= \sqrt[6]{a^{12}b^{12}c^{12}\cdot a^2bc^2}$

$= a^2b^2c^2\sqrt[6]{a^2bc^2}$

89. $\dfrac{\sqrt[3]{a^2}}{\sqrt[4]{a}}$

$= \dfrac{a^{2/3}}{a^{1/4}}$　　Converting to exponential notation

$= a^{2/3-1/4}$　Subtracting exponents

$= a^{5/12}$　　Converting back

$= \sqrt[12]{a^5}$　　to radical notation

91. $\dfrac{\sqrt[4]{x^2y^3}}{\sqrt[3]{xy}}$

$= \dfrac{\left(x^2y^3\right)^{1/4}}{\left(xy\right)^{1/3}}$　　Converting to exponential notation

$= \dfrac{x^{2/4}y^{3/4}}{x^{1/3}y^{1/3}}$　　Using the power and product rule

$= x^{2/4-1/3}y^{3/4-1/3}$　Subtracting exponents

$= x^{2/12}y^{5/12}$　　Converting back

$= \sqrt[12]{x^2y^5}$　　to radical notation

93. $\dfrac{\sqrt{ab^3}}{\sqrt[5]{a^2b^3}}$

$= \dfrac{\left(ab^3\right)^{1/2}}{\left(a^2b^3\right)^{1/5}}$　　Converting to exponential notation

$= \dfrac{a^{1/2}b^{3/2}}{a^{2/5}b^{3/5}}$　　Using the power rule

$= a^{1/10}b^{9/10}$　Subtracting exponents

$= \left(ab^9\right)^{1/10}$　　Converting back

$= \sqrt[10]{ab^9}$　　to radical notation

95. $\dfrac{\sqrt{(7-y)^3}}{\sqrt[3]{(7-y)^2}}$

$= \dfrac{(7-y)^{3/2}}{(7-y)^{2/3}}$　　Converting to exponential notation

$= (7-y)^{3/2-2/3}$　Subtracting exponents

$= (7-y)^{5/6}$　　Converting back

$= \sqrt[6]{(7-y)^5}$　　to radical notation

97. $\dfrac{\sqrt[4]{(5+3x)^3}}{\sqrt[3]{(5+3x)^2}}$

$= \dfrac{(5+3x)^{3/4}}{(5+3x)^{2/3}}$　　Converting to exponential notation

$= (5+3x)^{3/4-2/3}$　Subtracting exponents

$= (5+3x)^{1/12}$　　Converting back

$= \sqrt[12]{5+3x}$　　to radical notation

99. $\sqrt[3]{x^2y}\left(\sqrt{xy}-\sqrt[5]{xy^3}\right)$

$= \left(x^2y\right)^{1/3}\left[\left(xy\right)^{1/2}-\left(xy^3\right)^{1/5}\right]$

$= x^{2/3}y^{1/3}\left(x^{1/2}y^{1/2}-x^{1/5}y^{3/5}\right)$

$= x^{2/3}y^{1/3}x^{1/2}y^{1/2}-x^{2/3}y^{1/3}x^{1/5}y^{3/5}$

$= x^{2/3+1/2}y^{1/3+1/2}-x^{2/3+1/5}y^{1/3+3/5}$

$= x^{7/6}y^{5/6}-x^{13/15}y^{14/15}$

$= x^{1\frac{1}{6}}y^{\frac{5}{6}}-x^{13/15}y^{14/15}$

Writing a mixed numeral

$= x\cdot x^{\frac{1}{6}}y^{\frac{5}{6}}-x^{13/15}y^{14/15}$

$= x\left(xy^5\right)^{1/6}-\left(x^{13}y^{14}\right)^{1/15}$

$= x\sqrt[6]{xy^5}-\sqrt[15]{x^{13}y^{14}}$

101. $\left(m+\sqrt[3]{n^2}\right)\left(2m+\sqrt[4]{n}\right)$

$= \left(m+n^{2/3}\right)\left(2m+n^{1/4}\right)$

  Converting to exponential notation

$= 2m^2 + mn^{1/4} + 2mn^{2/3} + n^{2/3}n^{1/4}$

  Using FOIL

$= 2m^2 + mn^{1/4} + 2mn^{2/3} + n^{2/3+1/4}$

  Adding exponents

$= 2m^2 + mn^{1/4} + 2mn^{2/3} + n^{11/12}$

$= 2m^2 + m\sqrt[4]{n} + 2m\sqrt[3]{n^2} + \sqrt[12]{n^{11}}$

  Converting back to radical notation

103. $f(x) = \sqrt[4]{x}, \; g(x) = 2\sqrt{x} - \sqrt[3]{x^2}$

$(f \cdot g)(x) = \sqrt[4]{x}\left(2\sqrt{x} - \sqrt[3]{x^2}\right)$

$= \sqrt[4]{x}\left(2\sqrt{x}\right) - \left(\sqrt[4]{x}\right)\sqrt[3]{x^2}$

$= 2x^{\frac{1}{4}+\frac{1}{2}} - x^{\frac{1}{4}+\frac{2}{3}} = 2x^{\frac{3}{4}} - x^{\frac{11}{12}}$

$= 2\sqrt[4]{x^3} - \sqrt[12]{x^{11}}$

105. $f(x) = x+\sqrt{7}, \; g(x) = x-\sqrt{7}$

$(f \cdot g)(x) = \left(x+\sqrt{7}\right)\left(x-\sqrt{7}\right)$

$= x^2 - \left(\sqrt{7}\right)^2$

$= x^2 - 7$

107. $f(x) = x^2$

$f\left(5+\sqrt{2}\right) = \left(5+\sqrt{2}\right)^2$

$= 5^2 + 2 \cdot 5\sqrt{2} + \left(\sqrt{2}\right)^2$

$= 25 + 10\sqrt{2} + 2$

$= 27 + 10\sqrt{2}$

109. $f(x) = x^2$

$f\left(\sqrt{3}-\sqrt{5}\right) = \left(\sqrt{3}-\sqrt{5}\right)^2$

$= \left(\sqrt{3}\right)^2 - 2 \cdot \sqrt{3} \cdot \sqrt{5} + \left(\sqrt{5}\right)^2$

$= 3 - 2\sqrt{15} + 5$

$= 8 - 2\sqrt{15}$

111. *Thinking and Writing Exercise.*

113. $3x-1 = 125$

$3x = 125+1 = 126$

$x = \dfrac{126}{3} = 42$

115. $x^2 + 2x + 1 = 22 - 2x$

$x^2 + 2x + 2x - 22 = 0$

$x^2 + 4x - 21 = 0$

$(x+7)(x-3) = 0$

$x+7 = 0 \quad or \quad x-3 = 0$

$x = -7 \quad or \quad x = 3$

117. $\dfrac{1}{x} + \dfrac{1}{2} = \dfrac{1}{6}$

$6x \cdot \dfrac{1}{x} + 6x \cdot \dfrac{1}{2} = 6x \cdot \dfrac{1}{6}$

$6 + 3x = x$

$2x = -6$

$x = -3$

119. *Thinking and Writing Exercise.*

121. $f(x) =$

$= \sqrt{x^3 - x^2} + \sqrt{9x^3 - 9x^2} - \sqrt{4x^3 - 4x^2}$

$= \sqrt{x^2(x-1)} + \sqrt{9x^2(x-1)} - \sqrt{4x^2(x-1)}$

$= x\sqrt{x-1} + 3x\sqrt{x-1} - 2x\sqrt{x-1}$

$= 2x\sqrt{x-1}$

123. $f(x) = \sqrt[4]{x^5 - x^4} + 3\sqrt[4]{x^9 - x^8}$

$= \sqrt[4]{x^4(x-1)} + 3\sqrt[4]{x^8(x-1)}$

$= \sqrt[4]{x^4} \cdot \sqrt[4]{x-1} + 3\sqrt[4]{x^8} \cdot \sqrt[4]{x-1}$

$= x\sqrt[4]{x-1} + 3x^2\sqrt[4]{x-1}$

$= \left(x + 3x^2\right)\sqrt[4]{x-1}$

125. $7xy\sqrt{(x+y)^3} - 5xy\sqrt{x+y} - 2y\sqrt{(x+y)^3}$

$= 7xy\sqrt{(x+y)^2(x+y)} - 5xy\sqrt{x+y} -$
$\qquad 2y\sqrt{(x+y)^2(x+y)}$

$= 7x(x+y)\sqrt{x+y} - 5xy\sqrt{x+y} -$
$\qquad 2y(x+y)\sqrt{x+y}$

$= \left[7x(x+y) - 5xy - 2y(x+y)\right]\sqrt{x+y}$

$= \left(7x^2 + 7xy - 5xy - 2xy - 2y^2\right)\sqrt{x+y}$

$= \left(7x^2 - 2y^2\right)\sqrt{x+y}$

127. $\sqrt{8x(y+z)^5}\sqrt[3]{4x^2(y+z)^2}$

$= \left[8x(y+z)^5\right]^{1/2}\left[4x^2(y+z)^2\right]^{1/3}$

$= \left[8x(y+z)^5\right]^{3/6}\left[4x^2(y+z)^2\right]^{2/6}$

$= \left\{\left[2^3 x(y+z)^5\right]^3\left[2^2 x^2(y+z)^2\right]^2\right\}^{1/6}$

$= \sqrt[6]{2^9 x^3 (y+z)^{15} \cdot 2^4 x^4 (y+z)^4}$

$= \sqrt[6]{2^{13} x^7 (y+z)^{19}}$

$= \sqrt[6]{2^{12} x^6 (y+z)^{18} \cdot 2x(y+z)}$

$= 2^2 x(y+z)^3 \sqrt[6]{2x(y+z)}$, or

$4x(y+z)^3 \sqrt[6]{2x(y+z)}$

129. $\dfrac{\dfrac{1}{\sqrt{w}} - \sqrt{w}}{\dfrac{\sqrt{w}+1}{\sqrt{w}}} = \dfrac{\dfrac{1}{\sqrt{w}} - \sqrt{w}}{\dfrac{\sqrt{w}+1}{\sqrt{w}}} \cdot \dfrac{\sqrt{w}}{\sqrt{w}} = \dfrac{1-w}{\sqrt{w}+1}$

$= \dfrac{1-w}{\sqrt{w}+1} \cdot \dfrac{\sqrt{w}-1}{\sqrt{w}-1}$

$= \dfrac{\sqrt{w}-1-w\sqrt{w}+w}{w-1}$

$= \dfrac{(w-1)-\sqrt{w}(w-1)}{w-1}$

$= \dfrac{(w-1)(1-\sqrt{w})}{w-1} = 1-\sqrt{w}$

131. $x-5 = \left(\sqrt{x}\right)^2 - \left(\sqrt{5}\right)^2 = \left(\sqrt{x}+\sqrt{5}\right)\left(\sqrt{x}-\sqrt{5}\right)$

133. $x-a = \left(\sqrt{x}\right)^2 - \left(\sqrt{a}\right)^2$

$= \left(\sqrt{x}+\sqrt{a}\right)\left(\sqrt{x}-\sqrt{a}\right)$

135. $\left(\sqrt{x+2}-\sqrt{x-2}\right)^2$

$= \left(\sqrt{x+2}\right)^2 - 2\sqrt{x+2}\sqrt{x-2} + \left(\sqrt{x-2}\right)^2$

$= x+2 - 2\sqrt{(x+2)(x-2)} + x-2$

$= 2x - 2\sqrt{x^2-4}$

Mid-Chapter Review

Guided Solutions

1. $\sqrt{6x^9} \cdot \sqrt{2xy} = \sqrt{6x^9 \cdot 2xy}$

$= \sqrt{12x^{10} y}$

$= \sqrt{4x^{10} \cdot 3y}$

$= \sqrt{4x^{10}} \cdot \sqrt{3y}$

$= 2x^5 \sqrt{3y}$

2. $\sqrt{12} - 3\sqrt{75} + \sqrt{8} = 2\sqrt{3} - 3\cdot 5\sqrt{3} + 2\sqrt{2}$

$= 2\sqrt{3} - 15\sqrt{3} + 2\sqrt{2}$

$= -13\sqrt{3} + 2\sqrt{2}$

Mixed Review

1. $\sqrt{81} = \sqrt{9\cdot 9} = 9$

2. $-\sqrt{\dfrac{9}{100}} = -\sqrt{\dfrac{3\cdot 3}{10\cdot 10}} = -\dfrac{3}{10}$

3. $\sqrt{64t^2} = \sqrt{(8t)^2} = |8t| = 8|t|$

4. $\sqrt[5]{x^5} = x$

5. $f(x) = \sqrt[3]{12x-4}$

$f(-5) = \sqrt[3]{12(-5)-4}$

$= \sqrt[3]{-60-4}$

$= \sqrt[3]{-64} = -4$

6. $g(x) = \sqrt[4]{10 - x}$

Since the index is even, the radicand, $10 - x$, must be non-negative.

$$10 - x \geq 0$$
$$10 \geq x$$
$$x \leq 10$$

Domain of $f = \{x | x \leq 10\}$, or $(-\infty, 10]$

7. $8^{2/3} = \left(\sqrt[3]{8}\right)^2 = 2^2 = 4$

8. $\sqrt[6]{\sqrt{a}} = \sqrt[6]{a^{1/2}} = \left(a^{1/2}\right)^{1/6} = a^{1/12} = \sqrt[12]{a}$

9. $\sqrt[3]{y^{24}} = y^{24/3} = y^8$

10. $\sqrt{(t+5)^2} = t + 5 \ (\text{for } t + 5 \geq 0)$

11. $\sqrt[3]{-27a^{12}} = \sqrt[3]{\left(-3a^4\right)^3} = -3a^4$

12. $\sqrt{6x}\sqrt{15x} = \sqrt{6x \cdot 15x}$
$$= \sqrt{90x^2}$$
$$= \sqrt{9x^2 \cdot 10}$$
$$= 3x\sqrt{10} \ (\text{for } x \geq 0)$$

13. $\dfrac{\sqrt{20y}}{\sqrt{45y}} = \sqrt{\dfrac{20y}{45y}} = \sqrt{\dfrac{4 \cdot 5y}{9 \cdot 5y}} = \sqrt{\dfrac{4}{9}} = \dfrac{2}{3}$

14. $\sqrt{15t} + 4\sqrt{15t} = (1 + 4)\sqrt{15t} = 5\sqrt{15t}$

15. $\sqrt[5]{a^5 b^{10} c^{11}} = \sqrt[5]{a^5} \sqrt[5]{b^{10}} \sqrt[5]{c^{11}}$
$$= a\sqrt[5]{\left(b^2\right)^5} \sqrt[5]{c^{10} \cdot c}$$
$$= ab^2 \sqrt[5]{\left(c^2\right)^5} \sqrt[5]{c}$$
$$= ab^2 c^2 \sqrt[5]{c}$$

16. $\sqrt{6}\left(\sqrt{10} - \sqrt{33}\right) = \sqrt{6} \cdot \sqrt{10} - \sqrt{6} \cdot \sqrt{33}$
$$= \sqrt{60} - \sqrt{198}$$
$$= \sqrt{4 \cdot 15} - \sqrt{9 \cdot 22}$$
$$= 2\sqrt{15} - 3\sqrt{22}$$

17. $\dfrac{\sqrt{t}}{\sqrt[8]{t^3}} = \dfrac{t^{1/2}}{t^{3/8}} = t^{4/8 - 3/8} = t^{1/8} = \sqrt[8]{t}$

18. $\sqrt[5]{\dfrac{3a^{12}}{96a^2}} = \sqrt[5]{\dfrac{a^{10}}{32}} = \sqrt[5]{\dfrac{\left(a^2\right)^5}{2^5}} = \dfrac{a^2}{2}$

19. $2\sqrt{3} - 5\sqrt{12} = 2\sqrt{3} - 5\sqrt{4 \cdot 3}$
$$= 2\sqrt{3} - 5 \cdot 2\sqrt{3}$$
$$= 2\sqrt{3} - 10\sqrt{3}$$
$$= -8\sqrt{3}$$

20. $\left(\sqrt{5} + 3\right)\left(\sqrt{5} - 3\right) = \left(\sqrt{5}\right)^2 - 3^2 = 5 - 9 = -4$

21. $\left(\sqrt{15} + \sqrt{10}\right)^2 = \left(\sqrt{15}\right)^2 + 2\sqrt{15}\sqrt{10} + \left(\sqrt{10}\right)^2$
$$= 15 + 2\sqrt{150} + 10$$
$$= 25 + 2\sqrt{25 \cdot 6}$$
$$= 25 + 2 \cdot 5\sqrt{6}$$
$$= 25 + 10\sqrt{6}$$

22. $\sqrt{25x - 25} - \sqrt{9x - 9}$
$$= \sqrt{25(x-1)} - \sqrt{9(x-1)}$$
$$= 5\sqrt{x-1} - 3\sqrt{x-1}$$
$$= 2\sqrt{x-1}$$

23. $\sqrt{x^3 y} \sqrt[5]{xy^4} = x\sqrt{xy} \sqrt[5]{xy^4} = x \cdot x^{\frac{1}{2}} y^{\frac{1}{2}} \cdot x^{\frac{1}{5}} y^{\frac{4}{5}}$
$$= x \cdot x^{\frac{1}{2} + \frac{1}{5}} y^{\frac{1}{2} + \frac{4}{5}} = x \cdot x^{\frac{7}{10}} y^{\frac{13}{10}}$$
$$= x \cdot x^{\frac{7}{10}} \cdot y^1 \cdot y^{\frac{3}{10}} = xy\sqrt[10]{x^7 y^3}$$

24. $\sqrt[3]{5000} + \sqrt[3]{625} = \sqrt[3]{1000 \cdot 5} + \sqrt[3]{125 \cdot 5}$
$$= 10\sqrt[3]{5} + 5\sqrt[3]{5}$$
$$= 15\sqrt[3]{5}$$

25. $\sqrt[3]{12x^2 y^5} \sqrt[3]{18x^7 y} = y\sqrt[3]{12x^2 y^2} \cdot x^2 \sqrt[3]{18xy}$
$$= x^2 y\sqrt[3]{\left(2^2 \cdot 3x^2 y^2\right)\left(2 \cdot 3^2 xy\right)} = x^2 y\sqrt[3]{2^3 3^3 x^3 y^3}$$
$$= x^2 y \cdot 2 \cdot 3 \cdot xy = 6x^3 y^2$$

Exercise Set 7.6

1. False

3. True

5. True

7. $\sqrt{5x+1} = 4$

   $\left(\sqrt{5x+1}\right)^2 = 4^2$ Principle of powers

                        (squaring)

   $5x+1 = 16$

   $5x = 15$

   $x = \dfrac{15}{5} = 3$

   Check:

   $$\begin{array}{c|c}
   \sqrt{5x+1} = 4 & \\
   \hline
   \sqrt{5 \cdot 3 + 1} & 4 \\
   \sqrt{15+1} & \\
   \sqrt{16} & \\
   4 & 4 \quad \text{TRUE}
   \end{array}$$

   The solution is 3.

9. $\sqrt{3x} + 1 = 6$

   $\sqrt{3x} = 5$ Sub. to isolate the radical

   $\left(\sqrt{3x}\right)^2 = 5^2$ Principle of powers (squaring)

   $3x = 25$

   $x = \dfrac{25}{3}$

   Check:

   $$\begin{array}{c|c}
   \sqrt{3x} + 1 = 6 & \\
   \hline
   \sqrt{3 \cdot \dfrac{25}{3} + 1} & 6 \\
   5+1 & \\
   6 & 6 \quad \text{TRUE}
   \end{array}$$

   The solution is $\dfrac{25}{3}$.

11. $\sqrt{y+1} - 5 = 8$

   $\sqrt{y+1} = 13$

              Adding to isolate the radical

   $\left(\sqrt{y+1}\right)^2 = 13^2$

              Principle of powers (squaring)

   $y+1 = 169$

   $y = 168$

   Check:

   $$\begin{array}{c|c}
   \sqrt{y+1} - 5 = 8 & \\
   \hline
   \sqrt{168+1} - 5 & 8 \\
   13 - 5 & \\
   8 & 8 \quad \text{TRUE}
   \end{array}$$

   The solution is 168.

13. $\sqrt{8-x} + 7 = 10$

   $\sqrt{8-x} = 3$ Isolate the radical

   $\left(\sqrt{8-x}\right)^2 = 3^2$ Principle of powers

                         (squaring)

   $8 - x = 9$

   $-x = 1$

   $x = -1$

   Check:

   $$\begin{array}{c|c}
   \sqrt{8-x} + 7 = 10 & \\
   \hline
   \sqrt{8-(-1)} + 7 & 10 \\
   \sqrt{9} + 7 & \\
   3 + 7 & \\
   10 & 10 \quad \text{TRUE}
   \end{array}$$

   The solution is -1.

15. $\sqrt[3]{x+5} = 2$

   $\left(\sqrt[3]{x+5}\right)^3 = 2^3$

   $x+5 = 8$

   $x = 3$

   Check:

   $$\begin{array}{c|c}
   \sqrt[3]{x+5} = 2 & \\
   \hline
   \sqrt[3]{3+5} & 2 \\
   \sqrt[3]{8} & \\
   2 & 2 \quad \text{TRUE}
   \end{array}$$

   The solution is 3.

17. $\sqrt[4]{y-1} = 3$

$\left(\sqrt[4]{y-1}\right)^4 = 3^4$

$y - 1 = 81$

$y = 82$

Check:

$\sqrt[4]{y-1} = 3$		
$\sqrt[4]{82-1}$	3	
$\sqrt[4]{81}$		
3	3	TRUE

The solution is 82.

19. $3\sqrt{x} = x$

$\left(3\sqrt{x}\right)^2 = x^2$

$9x = x^2$

$0 = x^2 - 9x$

$0 = x(x-9)$

$x = 0 \text{ or } x = 9$

Check:

For 0:

$3\sqrt{x} = x$		
$3\sqrt{0}$	0	
$3 \cdot 0$		
0	0	TRUE

For 9:

$3\sqrt{x} = x$		
$3\sqrt{9}$	9	
$3 \cdot 3$		
9	9	TRUE

The solutions are 0 and 9.

21. $2y^{1/2} - 13 = 7$

$2\sqrt{y} - 13 = 7$

$2\sqrt{y} = 20$

$\sqrt{y} = 10$

$\left(\sqrt{y}\right)^2 = 10^2$

$y = 100$

Check:

$2y^{1/2} - 13 = 7$		
$2 \cdot 100^{1/2} - 13$	7	
$2 \cdot 10 - 13$		
7	7	TRUE

The solution is 100.

23. $\sqrt[3]{x} = -3$

$\left(\sqrt[3]{x}\right)^3 = (-3)^3$

$x = -27$

Check:

$\sqrt[3]{x} = -3$		
$\sqrt[3]{-27}$	-3	
-3	-3	TRUE

The solution is -27.

25. $z^{1/4} + 8 = 10$

$z^{1/4} = 2$

$\left(z^{1/4}\right)^4 = 2^4$

$z = 16$

Check:

$z^{1/4} + 8 = 10$		
$16^{1/4} + 8$	10	
$2 + 8$		
10	10	TRUE

The solution is 16.

27. $\sqrt{n} = -2$

This equation has no solution, since the principal square root is never negative.

29. $\sqrt[4]{3x+1} - 4 = -1$

$\sqrt[4]{3x+1} = 3$

$\left(\sqrt[4]{3x+1}\right)^4 = 3^4$

$3x + 1 = 81$

$3x = 80$

$x = \dfrac{80}{3}$

Check:

$$
\begin{array}{c|c}
\sqrt[4]{3x+1}-4=-1 & \\
\hline
\sqrt[4]{3\cdot\frac{80}{3}+1}-4 & -1 \\
\sqrt[4]{81}-4 & \\
3-4 & \\
-1 & -1 \quad \text{TRUE}
\end{array}
$$

The solution is $\frac{80}{3}$.

31.
$$(21x+55)^{1/3}=10$$
$$\left[(21x+55)^{1/3}\right]^3=10^3$$
$$21x+55=1000$$
$$21x=945$$
$$x=\frac{945}{21}=45$$

Check:

$$
\begin{array}{c|c}
(21x+55)^{1/3}=10 & \\
\hline
(21\cdot45+55)^{1/3} & 10 \\
(945+55)^{1/3} & \\
1000^{1/3} & \\
10 & 10 \quad \text{TRUE}
\end{array}
$$

The solution is 45.

33.
$$\sqrt[3]{3y+6}+7=8$$
$$\sqrt[3]{3y+6}=1$$
$$\left(\sqrt[3]{3y+6}\right)^3=1^3$$
$$3y+6=1$$
$$3y=-5$$
$$y=-\frac{5}{3}$$

Check:

$$
\begin{array}{c|c}
\sqrt[3]{3y+6}+7=8 & \\
\hline
\sqrt[3]{3\left(-\frac{5}{3}\right)+6}+7 & 8 \\
\sqrt[3]{1}+7 & \\
1+7 & \\
8 & 8 \quad \text{TRUE}
\end{array}
$$

The solution is $-\frac{5}{3}$.

35.
$$\sqrt{3t+4}=\sqrt{4t+3}$$
$$\left(\sqrt{3t+4}\right)^2=\left(\sqrt{4t+3}\right)^2$$
$$3t+4=4t+3$$
$$4=t+3$$
$$1=t$$

Check:

$$
\begin{array}{c|c}
\sqrt{3t+4}=\sqrt{4t+3} & \\
\hline
\sqrt{3\cdot1+4} & \sqrt{4\cdot1+3} \\
\sqrt{7} & \sqrt{7} \quad \text{TRUE}
\end{array}
$$

The solution is 1.

37.
$$3(4-t)^{1/4}=6^{1/4}$$
$$\left[3(4-t)^{1/4}\right]^4=\left(6^{1/4}\right)^4$$
$$81(4-t)=6$$
$$324-81t=6$$
$$-81t=-318$$
$$t=\frac{106}{27}$$

The number $\frac{106}{27}$ checks and is the solution.

39.
$$3+\sqrt{5-x}=x$$
$$\sqrt{5-x}=x-3$$
$$\left(\sqrt{5-x}\right)^2=(x-3)^2$$
$$5-x=x^2-6x+9$$
$$0=x^2-5x+4$$
$$0=(x-1)(x-4)$$
$$x-1=0 \ or \ x-4=0$$
$$x=1 \ or \quad x=4$$

Check:
For 1:

$$
\begin{array}{c|c}
3+\sqrt{5-x}=x & \\
\hline
3+\sqrt{5-1} & 1 \\
3+\sqrt{4} & \\
3+2 & \\
5 & 1 \quad \text{FALSE}
\end{array}
$$

For 4: $\dfrac{3+\sqrt{5-x}=x}{3+\sqrt{5-4}\;\Big|\;4}$

$$3+\sqrt{1}$$
$$3+1$$
$$4\;\Big|\;4 \quad\text{TRUE}$$

Since 4 checks but 1 does not, the solution is 4.

41. $\sqrt{4x-3}=2+\sqrt{2x-5}$ One radical is already isolated.

$\left(\sqrt{4x-3}\right)^2=\left(2+\sqrt{2x-5}\right)^2$ Squaring both sides

$$4x-3=4+4\sqrt{2x-5}+2x-5$$

$$2x-2=4\sqrt{2x-5}$$

$$x-1=2\sqrt{2x-5}$$

$$x^2-2x+1=8x-20$$

$$x^2-10x+21=0$$

$$(x-7)(x-3)=0$$

$$x-7=0 \;or\; x-3=0$$

$$x=7 \;or\;\;\; x=3$$

Both numbers check. The solutions are 7 and 3.

43. $\sqrt{20-x}+8=\sqrt{9-x}+11$

$\sqrt{20-x}=\sqrt{9-x}+3$ Isolating one radical

$\left(\sqrt{20-x}\right)^2=\left(\sqrt{9-x}+3\right)^2$ Squaring both sides

$$20-x=9-x+6\sqrt{9-x}+9$$

$2=6\sqrt{9-x}$ Isolating the remaining Radical

$1=3\sqrt{9-x}$ Multiplying by $\dfrac{1}{2}$

$1^2=\left(3\sqrt{9-x}\right)^2$ Squaring both sides

$$1=9(9-x)$$

$$1=81-9x$$

$$-80=-9x$$

$$\frac{80}{9}=x$$

The number $\dfrac{80}{9}$ checks and is the solution.

45. $\sqrt{x+2}+\sqrt{3x+4}=2$

$$\sqrt{x+2}=2-\sqrt{3x+4}$$

Isolating one radical

$$\left(\sqrt{x+2}\right)^2=\left(2-\sqrt{3x+4}\right)^2$$

$$x+2=4-4\sqrt{3x+4}+3x+4$$

$$-2x-6=-4\sqrt{3x+4}$$

Isolating the remaining radical

$$x+3=2\sqrt{3x+4}$$

Multiplying by $-\dfrac{1}{2}$

$$(x+3)^2=\left(2\sqrt{3x+4}\right)^2$$

$$x^2+6x+9=4(3x+4)$$

$$x^2+6x+9=12x+16$$

$$x^2-6x-7=0$$

$$(x-7)(x+1)=0$$

$$x-7=0 \;or\; x+1=0$$

$$x=7 \;or\;\;\; x=-1$$

Check:

For 7: $\dfrac{\sqrt{x+2}+\sqrt{3x+4}=2}{\dfrac{\sqrt{7+2}+\sqrt{3\cdot7+4}}{\sqrt{9}+\sqrt{25}}\;\Big|\;2}$

$$8\;\Big|\;2 \quad\text{FALSE}$$

For -1: $\dfrac{\sqrt{x+2}+\sqrt{3x+4}=2}{\dfrac{\sqrt{-1+2}+\sqrt{3\cdot(-1)+4}}{\sqrt{1}+\sqrt{1}}\;\Big|\;2}$

$$2\;\Big|\;2 \quad\text{TRUE}$$

Since -1 checks but 7 does not, the solution is -1.

47. We must have $f(x) = 1$ or $\sqrt{x} + \sqrt{x-9} = 1$.

$$\sqrt{x} + \sqrt{x-9} = 1$$

$$\sqrt{x-9} = 1 - \sqrt{x} \qquad \text{Isolating one radical term.}$$

$$\left(\sqrt{x-9}\right)^2 = \left(1 - \sqrt{x}\right)^2$$

$$x - 9 = 1 - 2\sqrt{x} + x$$

$$-10 = -2\sqrt{x} \qquad \text{Isolating the remaining radical term}$$

$$5 = \sqrt{x}$$

$$25 = x$$

This value does not check. There is no solution, so there is no value of x for which $f(x) = 1$.

49. We must have $f(t) = -3$ or

$$\sqrt{t-2} - \sqrt{4t+1} = -3.$$

$$\sqrt{t-2} - \sqrt{4t+1} = -3$$

$$\sqrt{t-2} = \sqrt{4t+1} - 3$$

$$\left(\sqrt{t-2}\right)^2 = \left(\sqrt{4t+1} - 3\right)^2$$

$$t - 2 = 4t + 1 - 6\sqrt{4t+1} + 9$$

$$-3t - 12 = -6\sqrt{4t+1}$$

$$t + 4 = 2\sqrt{4t+1}$$

$$(t+4)^2 = \left(2\sqrt{4t+1}\right)^2$$

$$t^2 + 8t + 16 = 4(4t+1)$$

$$t^2 + 8t + 16 = 16t + 4$$

$$t^2 - 8t + 12 = 0$$

$$(t-2)(t-6) = 0$$

$$t - 2 = 0 \text{ or } t - 6 = 0$$

$$t = 2 \text{ or } \quad t = 6$$

Both numbers check, so we have $f(t) = -3$ when $t = 2$ and when $t = 6$.

51. We must have $\sqrt{2x-3} = \sqrt{x+7} - 2$.

$$\sqrt{2x-3} = \sqrt{x+7} - 2$$

$$\left(\sqrt{2x-3}\right)^2 = \left(\sqrt{x+7} - 2\right)^2$$

$$2x - 3 = x + 7 - 4\sqrt{x+7} + 4$$

$$x - 14 = -4\sqrt{x+7}$$

$$(x-14)^2 = \left(-4\sqrt{x+7}\right)^2$$

$$x^2 - 28x + 196 = 16(x+7)$$

$$x^2 - 28x + 196 = 16x + 112$$

$$x^2 - 44x + 84 = 0$$

$$(x-2)(x-42) = 0$$

$$x = 2 \text{ or } x = 42$$

Since 2 checks but 42 does not, we have $f(x) = g(x)$ when $x = 2$.

53.

$$4 - \sqrt{t-3} = (t+5)^{1/2}$$

$$\left(4 - \sqrt{t-3}\right)^2 = \left[(t+5)^{1/2}\right]^2$$

$$16 - 8\sqrt{t-3} + t - 3 = t + 5$$

$$-8\sqrt{t-3} = -8$$

$$\sqrt{t-3} = 1$$

$$\left(\sqrt{t-3}\right)^2 = 1^2$$

$$t - 3 = 1$$

$$t = 4$$

The number 4 checks, so we have $f(t) = g(t)$ when $t = 4$.

55. *Thinking and Writing Exercise.*

57. *Familiarize.* We let x represent the width of the sign, and $13x + 5$ represent the length. Recall that the perimeter is given by the formula $P = 2w + 2l$.

Translate.

$$\underbrace{\text{The perimeter}} \quad \text{is} \quad 430 \text{ ft.}$$
$$\downarrow \qquad\qquad\quad \downarrow \quad\; \downarrow$$
$$2x + 2(13x + 5) \quad = \quad 430$$

Carry out. We solve the equation.

$$2x + 2(13x + 5) = 430$$
$$2x + 26x + 10 = 430$$
$$28x = 420 \quad \text{Combining like terms}$$
$$x = 15 \quad \text{Dividing by 28}$$

If the width of the sign is 15 ft, then the length is $13 \cdot 15 + 5 = 195 + 5 = 200$ ft.

Check The length of the sign 200 ft. is 13 times the width, 15 ft, plus 5 ft. The perimeter is $2 \cdot 200 + 2 \cdot 15 = 400 + 30 = 430$ ft. These results check.

State. The width of the sign is 15 ft., and its length is 200 ft.

59. *Familiarize.* Let x = the width of the photograph and $x + 4$ the length.

Translate.

$$\underbrace{\text{Area of photograph}} \quad \text{is} \quad 140 \text{ in}^2$$
$$\downarrow \qquad\qquad\qquad \downarrow \quad\; \downarrow$$
$$x(x + 4) \qquad = \qquad 140$$

Carry out. We solve the equation.

$$x(x + 4) = 140$$
$$x^2 + 4x - 140 = 0$$
$$(x + 14)(x - 10) = 0$$
$$x + 14 = 0 \quad or \quad x - 10 = 0$$
$$x = -14 \; or \quad x = 10$$

Check. Since length and width cannot be negative, we only check $x = 10$. The length of the photograph is $10 + 4 = 14$ in^2. The area of the photograph is $10 \cdot 14 = 140$ in^2. These results check.

State. The width is 10 inches and the length is 14 inches.

61. *Familiarize.* Let x represent the base of the right triangle, $x + 2$ represent the other leg, and $x + 4$ represent the hypotenuse. Thus, each side represents one of three consecutive integers. Recall: Pythagorean Theorem $a^2 + b^2 = c^2$.

Translate.

$$a^2 + \quad b^2 \quad = \quad c^2$$
$$\downarrow \qquad \downarrow \qquad\quad \downarrow$$
$$x^2 + (x + 2)^2 = (x + 4)^2$$

Carry out. We solve the equation:

$$x^2 + (x + 2)^2 = (x + 4)^2$$
$$x^2 + x^2 + 4x + 4 = x^2 + 8x + 16$$
$$x^2 + 4x - 8x + 4 - 16 = 0$$
$$x^2 - 4x - 12 = 0$$
$$(x - 6)(x + 2) = 0$$
$$x - 6 = 0 \quad or \quad x + 2 = 0$$
$$x = 6 \quad or \qquad x = -2$$

Check. Since measure cannot be negative, we know -2 is not a solution. If $x = 6$, $x + 2 = 8$, and $x + 4 = 10$.

$$6^2 + 8^2 = 10^2$$
$$36 + 64 = 100$$
$$100 = 100$$

The results check.

State. The lengths of the sides are 6, 8, and 10 units.

63. *Thinking and Writing Exercise.*

65. $v(p) = 12.1\sqrt{p}$

Substitute 100 for $v(p)$ and solve for p.

$$100 = 12.1\sqrt{p}$$
$$\sqrt{p} = \frac{100}{12.1}$$
$$p = \left(\frac{100}{12.1}\right)^2$$
$$p \approx 68.3$$

The pressure is about 68 psi.

67. $f(T) = k\sqrt{T}$

To find the value of k, substitute 260 for $f(T)$ and 28 for T.

$260 = k\sqrt{28}$

$k = \dfrac{260}{\sqrt{28}} = \dfrac{130}{\sqrt{7}}$

When $T = 32$ N,

$f(T) = \dfrac{130}{\sqrt{7}}\sqrt{32} = 260\sqrt{\dfrac{8}{7}} \approx 277.95$

The new frequency is about 278 Hz.

69. $S(t) = 1087.7\sqrt{\dfrac{9t + 2617}{2457}}$

Substitute 1880 for $S(t)$ and solve for t.

$1880 = 1087.7\sqrt{\dfrac{9t + 2617}{2457}}$

$1.7284 \approx \sqrt{\dfrac{9t + 2617}{2457}}$

$2.9874 \approx \dfrac{9t + 2617}{2457}$

$7340.0418 \approx 9t + 2617$

$4723.0418 \approx 9t$

$524.7824 \approx t$

The temperature is about 524.8°C.

71. $S(t) = 1087.7\sqrt{\dfrac{9t + 2617}{2457}}$

$\dfrac{S}{1087.7} = \sqrt{\dfrac{9t + 2617}{2457}}$

$\left(\dfrac{S}{1087.7}\right)^2 = \left(\sqrt{\dfrac{9t + 2617}{2457}}\right)^2$

$\dfrac{S^2}{1087.7^2} = \dfrac{9t + 2617}{2457}$

$\dfrac{2457 S^2}{1087.7^2} = 9t + 2617$

$\dfrac{2457 S^2}{1087.7^2} - 2617 = 9t$

$\dfrac{1}{9}\left(\dfrac{2457 S^2}{1087.7^2} - 2617\right) = t$

73. $d(n) = 0.75\sqrt{2.8n}$

Substitute 84 for $d(n)$ and solve for n.

$84 = 0.75\sqrt{2.8n}$

$112 = \sqrt{2.8n}$

$(112)^2 = \left(\sqrt{2.8n}\right)^2$

$12544 = 2.8n$

$4480 \approx n$

Approximately 4480 rpm will produce peak performance.

75. $v = \sqrt{2gr}\sqrt{\dfrac{h}{r + h}}$

$v^2 = 2gr \cdot \dfrac{h}{r + h}$ Squaring both sides

$v^2(r + h) = 2grh$ Multiplying by $r + h$

$v^2 r + v^2 h = 2grh$

$v^2 h = 2grh - v^2 r$

$v^2 h = r(2gh - v^2)$

$\dfrac{v^2 h}{2gh - v^2} = r$

77. $\dfrac{x + \sqrt{x+1}}{x - \sqrt{x+1}} = \dfrac{5}{11}$

$11\left(x + \sqrt{x+1}\right) = 5\left(x - \sqrt{x+1}\right)$

$11x + 11\sqrt{x+1} = 5x - 5\sqrt{x+1}$

$16\sqrt{x+1} = -6x$

$8\sqrt{x+1} = -3x$

$\left(8\sqrt{x+1}\right)^2 = (-3x)^2$

$64(x + 1) = 9x^2$

$64x + 64 = 9x^2$

$x = 9x^2 - 64x - 64$

$0 = (9x + 8)(x - 8)$

$9x + 8 = 0$ or $x - 8 = 0$

$9x = -8$ or $x = 8$

$x = -\dfrac{8}{9}$ or $x = 8$

Since $-\dfrac{8}{9}$ checks but 8 does not, the solution

is $-\dfrac{8}{9}$.

79. $\left(z^2+17\right)^{3/4}=27$

$\left[\left(z^2+17\right)^{3/4}\right]^{4/3}=\left(3^3\right)^{4/3}$

$z^2+17=3^4$

$z^2+17=81$

$z^2-64=0$

$(z+8)(z-8)=0$

$z=-8 \ or \ z=8$

Both -8 and 8 check. They are the solutions.

81. $\sqrt{8-b}=b\sqrt{8-b}$

$\left(\sqrt{8-b}\right)^2=\left(b\sqrt{8-b}\right)^2$

$(8-b)=b^2(8-b)$

$0=b^2(8-b)-(8-b)$

$0=(8-b)(b^2-1)$

$0=(8-b)(b+1)(b-1)$

$8-b=0 \ or \ b+1=0 \ or \ b-1=0$

$8=b \ or \quad b=-1 \ or \quad b=1$

Since the number 8 and 1 check but -1 does not, 8 and 1 are the solutions.

83. We find the values of x for which $g(x)=0$.

$6x^{1/2}+6x^{-1/2}-37=0$

$6\sqrt{x}+\dfrac{6}{\sqrt{x}}=37$

$\left(6\sqrt{x}+\dfrac{6}{\sqrt{x}}\right)^2=37^2$

$36x+72+\dfrac{36}{x}=1369$

$36x^2+72x+36=1369x$ Multiplying by x

$36x^2-1297x+36=0$

$(36x-1)(x-36)=0$

$36x-1=0 \ or \ x-36=0$

$36x=1 \quad or \qquad x=36$

$x=\dfrac{1}{36} \ or \qquad x=36$

Both numbers check. The x-intercepts are $\left(\dfrac{1}{36},0\right)$ and $(36,0)$.

1. d

3. e

5. f

7. $a^2+b^2=c^2$ Pythagorean equation

   $5^2+3^2=c^2$ Substituting

   $25+9=c^2$

   $34=c^2$

   $c=\sqrt{34}$ Exact answer

   $c\approx 5.831$ Approximation

9. Since $a=b$, this is an isosceles right triangle. The hypotenuse = length of a leg $\cdot\sqrt{2}$, or $9\sqrt{2}$, or approximately, 12.728.

11. $a^2+b^2=c^2$

   $a^2+12^2=13^2$

   $a^2+144=169$

   $a^2=25$

   $a=\sqrt{25}=5$

13. $a^2+b^2=c^2; c=8$, and $a=4\sqrt{3}$ (or $b\cdot 4\sqrt{3}$, since it can be the length of either leg.)

   $\left(4\sqrt{3}\right)^2+b^2=8^2$

   $48+b^2=64$

   $b^2=16$

   $b=\sqrt{16}$

   $b=4$ m

15. $a^2+b^2=c^2; c=\sqrt{20}, a=1$

   $1^2+b^2=\left(\sqrt{20}\right)^2$

   $1+b^2=20$

   $b^2=19$

   $b=\sqrt{19}$ in

   $b\approx 4.359$ in

17. $a^2 + b^2 = c^2; a = 1, c = \sqrt{2}$

$$1^2 + b^2 = \left(\sqrt{2}\right)^2$$
$$1 + b^2 = 2$$
$$b^2 = 1$$
$$b = 1 \text{ m}$$

Also, we might have observed that the hypotenuse is $\sqrt{2} \cdot$ length of the leg, 1. So, we know the right triangle is isosceles, thus $a = b = 1$.

19.

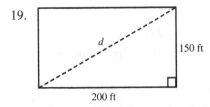

Let d = the distance, in feet, across the parking lot. A right triangle is formed in which the length of one leg is 200 ft., and the other is 150 ft.

We substitute these values into the Pythagorean equation to determine d.

$$d^2 = 200^2 + 150^2$$
$$d^2 = 40,000 + 22,500$$
$$d^2 = 62,500$$
$$d = \sqrt{62,500}$$
$$d = 250$$

The distance is 250 ft.

21.

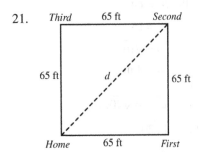

Let d = the distance, in feet, from home plate to second base.

Since the triangle formed is an isosceles right triangle, we know the hypotenuse, d, is $\sqrt{2} \cdot$ the length of a leg, 65. Thus,

$$d = 65\sqrt{2}$$
$$d \approx 91.924$$

The distance is $65\sqrt{2}$ ft, or approximately 91.924 ft.

23. Let h = height, in inches of the TV set. We substitute the given values into the Pythagorean equation, $w^2 + h^2 = d^2$, to determine h.

$$45^2 + h^2 = 51^2$$
$$2025 + h^2 = 2601$$
$$h^2 = 576$$
$$h = \sqrt{576} = 24$$

The height is 24 inches.

25. First, find the distance from corner to corner of the room.

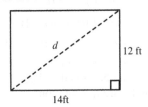

We use the Pythagorean equation to find d.

$$12^2 + 14^2 = d^2$$
$$144 + 196 = d^2$$
$$340 = d^2$$
$$d = \sqrt{340} = 2\sqrt{85}$$

Since 4 ft of slack is required on each end, we add $2 \cdot 4$, or 8 to d, or $2\sqrt{85} + 8$. The length of wire is $\left(2\sqrt{85} + 8\right)$ ft, or approximately 26.439 ft.

27.

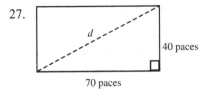

We use the Pythagorean equation to find d, the number of paces on the diagonal path.

$$d^2 = 40^2 + 70^2$$
$$d^2 = 1600 + 4900$$
$$d^2 = 6500$$
$$d = \sqrt{6500} = 10\sqrt{65}$$

If Marissa does not use the diagonal path, she will walk 40 + 70, or 110 paces. Marissa will save $\left(110-10\sqrt{65}\right)$ paces, or about 29.377 paces.

29. Since one acute angle is 45°, this is an isosceles right triangle with $b = 5$. Then $a = 5$ also. We substitute to find c.
$$c = a\sqrt{2}$$
$$c = 5\sqrt{2}$$
Exact answer: $a = 5; c = 5\sqrt{2}$
Approximation: $c \approx 7.071$

31. This is a 30-60-90 right triangle with $c = 14$.
$$c = 2a$$
$$14 = 2a$$
$$7 = a$$
$$b = a\sqrt{3}$$
$$b = 7\sqrt{3}$$
Exact answer: $a = 7, b = 7\sqrt{3}$
Approximation: $b \approx 12.124$

33. This is a 30-60-90 right triangle with $b = 15$.
$$b = a\sqrt{3}$$
$$15 = a\sqrt{3}$$
$$\frac{15}{\sqrt{3}} = a$$
$$\frac{15\sqrt{3}}{3} = a \quad \text{Rationalizing the denominator}$$
$$5\sqrt{3} = a \quad \text{Simplifying}$$
$$c = 2a$$
$$c = 2 \cdot 5\sqrt{3}$$
$$c = 10\sqrt{3}$$
Exact answer: $a = 5\sqrt{3}, c = 10\sqrt{3}$
Approximation: $a \approx 8.660, c \approx 17.321$

35. This is an isosceles right triangle with $c = 13$.
$$a = \frac{c\sqrt{2}}{2}$$
$$a = \frac{13\sqrt{2}}{2}$$
Since $a = b$, we have $b = \frac{13\sqrt{2}}{2}$ also.

Exact answer: $a = \dfrac{13\sqrt{2}}{2}, b = \dfrac{13\sqrt{2}}{2}$
Approximation $a \approx 9.192, b \approx 9.192$.

37. This is a 30-60-90 right triangle with $a = 14$.
$$b = a\sqrt{3} \qquad c = 2a$$
$$b = 14\sqrt{3} \qquad c = 2 \cdot 14$$
$$c = 28$$
Exact answer: $b = 14\sqrt{3}, c = 28$
Approximation: $b \approx 24.249$

39.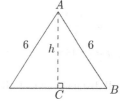

This is an equilateral triangle, so all the angles are 60°. The altitude bisects one angle and one side. Then triangle ABC is a 30-60-90 right triangle with the shorter leg of length 6/2, or 3, and hypotenuse of length 6. We substitute to find the length of the other leg.
$$b = a\sqrt{3}$$
$$h = 3\sqrt{3} \quad \text{Substituting } h \text{ for } b \text{ and 3 for } a$$
Exact answer: $h = 3\sqrt{3}$
Approximation: $h \approx 5.196$

41.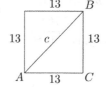

Triangle ABC is an isosceles right triangle with $a = 13$. We substitute to find c.
$$c = a\sqrt{2}$$
$$c = 13\sqrt{2}$$
Exact answer: $c = 13\sqrt{2}$
Approximation: $c \approx 18.385$

43.

Triangle ABC is an isosceles right triangle with $c = 19$. We substitute to find a.

$$a = \frac{c\sqrt{2}}{2}$$

$$a = \frac{19\sqrt{2}}{2}$$

Exact answer: $a = \dfrac{19\sqrt{2}}{2}$

Approximation: $a \approx 13.435$

45. We will express all distances in feet. Recall that 1 mi = 5280 ft.

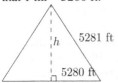

We use the Pythagorean equation to find h.

$$h^2 + (5280)^2 = (5281)^2$$

$$h^2 + 27{,}878{,}400 = 27{,}888{,}961$$

$$h^2 = 10{,}561$$

$$h = \sqrt{10{,}561}$$

$$h \approx 102.767$$

The height of the bulge is $\sqrt{10{,}561}$ ft., or about 102.767 ft.

47. The lodge is an equilateral triangle, so all the angles are 60°. The "height" bisects one angle and one side, so the the building can be depicted as two identical 30-60-90 right triangles, each with the shorter legs (a) of length 33/2 ft and hypotenuses (c) of length 33 ft. We substitute to find h, the length of the remaining side.

$$b = a\sqrt{3}$$

$$h = \left(\frac{33}{2}\right)\sqrt{3} \quad \text{Substituting } h \text{ for } b$$

$$\text{and } \frac{33}{2} \text{ for } a$$

The area of the lodge is the area of the equilateral triangle, which is $A = \dfrac{1}{2}ch$.

$$A = \frac{1}{2}(33)\left(\frac{33\sqrt{3}}{2}\right)$$

$$= \frac{1089\sqrt{3}}{4} \text{ ft}^2 \approx 471.551 \text{ ft}^2$$

49.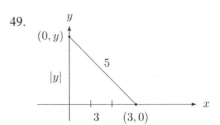

$$|y|^2 + 3^2 = 5^2$$

$$y^2 + 9 = 25$$

$$y^2 = 16$$

$$y = \pm 4$$

The points are $(0, -4)$ and $(0, 4)$.

51. Substitute the coordinates into the distance formula.

$$d = \sqrt{(x_2 - x_1)^2 + (y_2 - y_1)^2}$$

$$= \sqrt{(7 - 4)^2 + (1 - 5)^2}$$

$$= \sqrt{3^2 + (-4)^2}$$

$$= \sqrt{25}$$

$$= 5$$

53. Substitute the coordinates into the distance formula.

$$d = \sqrt{(x_2 - x_1)^2 + (y_2 - y_1)^2}$$

$$= \sqrt{(1 - 0)^2 + (-2 - (-5))^2}$$

$$= \sqrt{1^2 + 3^2}$$

$$= \sqrt{10}$$

$$\approx 3.162$$

55. Substitute the coordinates into the distance formula.
$$d = \sqrt{(x_2 - x_1)^2 + (y_2 - y_1)^2}$$
$$= \sqrt{(6-(-4))^2 + (-6-4)^2}$$
$$= \sqrt{10^2 + (-10)^2}$$
$$= \sqrt{200}$$
$$\approx 14.142$$

57. Substitute the coordinates into the distance formula.
$$d = \sqrt{(x_2 - x_1)^2 + (y_2 - y_1)^2}$$
$$= \sqrt{(-9.2-8.6)^2 + (-3.4-(-3.4))^2}$$
$$= \sqrt{(-17.8)^2 + 0^2}$$
$$= \sqrt{17.8^2}$$
$$= 17.8$$

59. Substitute the coordinates into the distance formula.
$$d = \sqrt{(x_2 - x_1)^2 + (y_2 - y_1)^2}$$
$$= \sqrt{\left(\frac{5}{6} - \frac{1}{2}\right)^2 + \left(-\frac{1}{6} - \frac{1}{3}\right)^2}$$
$$= \sqrt{\left(\frac{5}{6} - \frac{3}{6}\right)^2 + \left(-\frac{1}{6} - \frac{2}{6}\right)^2}$$
$$= \sqrt{\left(\frac{2}{6}\right)^2 + \left(-\frac{3}{6}\right)^2}$$
$$= \sqrt{\frac{4+9}{36}}$$
$$= \frac{\sqrt{13}}{6}$$
$$\approx 0.601$$

61. Substitute the coordinates into the distance formula.
$$d = \sqrt{(x_2 - x_1)^2 + (y_2 - y_1)^2}$$
$$= \sqrt{\left(0-(-\sqrt{6})\right)^2 + (0-\sqrt{6})^2}$$
$$= \sqrt{\left(\sqrt{6}\right)^2 + \left(-\sqrt{6}\right)^2}$$
$$= \sqrt{12}$$
$$\approx 3.464$$

63. Substitute the coordinates into the distance formula.
$$d = \sqrt{(x_2 - x_1)^2 + (y_2 - y_1)^2}$$
$$= \sqrt{(-2-(-1))^2 + (-40-(-30))^2}$$
$$= \sqrt{(-1)^2 + (-10)^2}$$
$$= \sqrt{101}$$
$$\approx 10.050$$

65. Substitute the coordinates into the midpoint formula.
$$\left(\frac{x_1 + x_2}{2}, \frac{y_1 + y_2}{2}\right)$$
$$\left(\frac{-2+8}{2}, \frac{5+3}{2}\right), \text{ or } \left(\frac{6}{2}, \frac{8}{2}\right), \text{ or } (3,4)$$

67. Substitute the coordinates into the midpoint formula.
$$\left(\frac{x_1 + x_2}{2}, \frac{y_1 + y_2}{2}\right)$$
$$\left(\frac{2+5}{2}, \frac{-1+8}{2}\right), \text{ or } \left(\frac{7}{2}, \frac{7}{2}\right)$$

69. Substitute the coordinates into the midpoint formula.
$$\left(\frac{x_1 + x_2}{2}, \frac{y_1 + y_2}{2}\right)$$
$$\left(\frac{-8+6}{2}, \frac{-5+(-1)}{2}\right), \text{ or } \left(\frac{-2}{2}, \frac{-6}{2}\right),$$
or $(-1, -3)$

71. Substitute the coordinates into the midpoint formula.
$$\left(\frac{x_1 + x_2}{2}, \frac{y_1 + y_2}{2}\right)$$
$$\left(\frac{-3.4+4.8}{2}, \frac{8.1+(-8.1)}{2}\right), \text{ or } \left(\frac{1.4}{2}, \frac{0}{2}\right),$$
or $(0.7, 0)$

73. Substitute the coordinates into the midpoint formula.

$$\left(\frac{x_1+x_2}{2},\frac{y_1+y_2}{2}\right)$$

$$\left(\frac{\frac{1}{6}+\left(-\frac{1}{3}\right)}{2},\frac{-\frac{3}{4}+\frac{5}{6}}{2}\right),\text{ or }\left(\frac{-\frac{1}{6}}{2},\frac{\frac{1}{12}}{2}\right),$$

$$\text{or }\left(-\frac{1}{12},\frac{1}{24}\right)$$

75. Substitute the coordinates into the midpoint formula.

$$\left(\frac{x_1+x_2}{2},\frac{y_1+y_2}{2}\right)$$

$$\left(\frac{\sqrt{2}+\sqrt{3}}{2},\frac{-1+4}{2}\right),\text{ or }\left(\frac{\sqrt{2}+\sqrt{3}}{2},\frac{3}{2}\right)$$

77. *Thinking and Writing Exercise.*

79. $y=2x-3$

Slope is 2, y-intercept is -3. Graph the point $(0,-3)$ and use the slope to determine a second point, from $(0,-3)$ go right 1 unit and up 2, or $(1,-1)$.

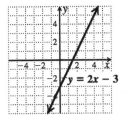

81. $8x-4y=8$

Rewrite the equation in slope-intercept form.

$8x-4y=8$

$8x-8=4y$

$2x-2=y$

Slope is 2, y-intercept is -2. Graph the point $(0,-2)$ and use the slope to determine a second point, from $(0,-2)$ go right 1 unit and up 2, to $(1,0)$.

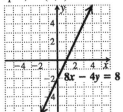

83. $x\geq 1$

Graph the boundary line $x=1$ as a solid line. Use a point of substitution (for instance, $(0, 1)$) to determine which half-plane to shade.

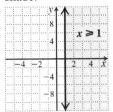

85. *Thinking and Writing Exercise.*

87. A regular hexagon has 6 sides of equal length, thus the length of each side is $72\div 6$, or 12. The shaded region has a base of 12, and using 1/2 of the shaded region, we determine the height (it is a $30°$-$60°$-$90°$ triangle). The height is $6\sqrt{3}$.

Area of triangle is	$\frac{1}{2}$	$\times$	length of the base	$\times$	length of the height
$\downarrow$	$\downarrow$ $\downarrow$		$\downarrow$		$\downarrow$
A	$=\frac{1}{2}$	$\cdot$	12	$\cdot$	$6\sqrt{3}$

$A=36\sqrt{3}$

The area of the shaded region is $36\sqrt{3}$ cm^2, or approximately 62.354 cm^2.

89.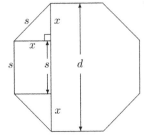

To determine x, use the Pythagorean equation on the right triangle, and solve.

$$x^2+x^2=s^2$$

$$2x^2=s^2$$

$$x^2=\frac{s^2}{2}$$

$$x=\sqrt{\frac{s^2}{2}}$$

$$x=\frac{s}{\sqrt{2}},\text{ or }\frac{s\sqrt{2}}{2}$$

Thus, $d = x + s + x$

$$d = 2x + s$$

$$d = 2\left(\frac{s\sqrt{2}}{2}\right) + s$$

$$d = s\sqrt{2} + s$$

91.

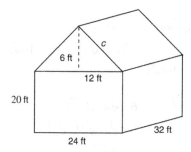

The area to be painted consists of two 20 ft by 24 ft rectangles, two 20 ft by 32 ft rectangles, and two triangles with height 6 ft and base 24 ft. The area of the two 20 ft by 24 ft rectangles is $2 \cdot 20$ ft $\cdot 24$ ft $= 960$ ft^2. The area of the two 20 ft by 32 ft rectangles is $2 \cdot 20$ ft $\cdot 32$ ft $= 1280$ ft^2. The area of the two triangles is $2 \cdot \frac{1}{2} \cdot 24$ ft $\cdot 6$ ft $= 144$ ft^2. Thus, the total area to be painted is 960 ft$^2 +$ 1280 ft$^2 + 144$ ft$^2 = 2384$ ft^2.

One gallon of paint covers 500 ft^2, so we divide to determine how many gallons of paint are required: $\frac{2384}{500} \approx 4.8$. Thus, 5 gallons of paint should be bought to paint the house.

93. First we find the radius of a circle with an area of 6160 ft^2.

$$A = \pi r^2$$

$$6160 = \pi r^2$$

$$\frac{6160}{\pi} = r^2$$

$$\sqrt{\frac{6160}{\pi}} = r$$

$$44.28 \approx r$$

Now we make a drawing. Let $s =$ the length of the side of the room.

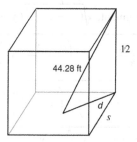

Using the Pythagorean equation, find d.

$$d^2 + 12^2 = 44.28^2$$

$$d^2 = 44.28^2 - 12^2$$

$$d \approx 42.63$$

$2 \cdot d$ is the length of the diagonal across the floor, and $s =$ the length of each side of the square room. Again, using the Pythagorean equations, we find s.

$$s^2 + s^2 = (2 \cdot 42.63)^2$$

$$2s^2 = 85.26^2$$

$$s^2 = \frac{85.26^2}{2}$$

$$s = \sqrt{\frac{85.26^2}{2}}$$

$$s \approx 60.28$$

The dimensions of the room are 60.28 ft by 60.28 ft.

95. i)

$$\sqrt{\left(\frac{x_1 + x_2}{2} - x_1\right)^2 + \left(\frac{y_1 + y_2}{2} - y_1\right)^2}$$

$$\overset{?}{=} \sqrt{\left(x_2 - \frac{x_1 + x_2}{2}\right)^2 + \left(y_2 - \frac{y_1 + y_2}{2}\right)^2}$$

$$\sqrt{\left(\frac{x_1 + x_2 - 2x_1}{2}\right)^2 + \left(\frac{y_1 + y_2 - 2y_1}{2}\right)^2}$$

$$\overset{?}{=} \sqrt{\left(\frac{2x_2 - x_1 - x_2}{2}\right)^2 + \left(\frac{2y_2 - y_1 - y_2}{2}\right)^2}$$

$$\sqrt{\left(\frac{x_2 - x_1}{2}\right)^2 + \left(\frac{y_2 - y_1}{2}\right)^2}$$

$$= \sqrt{\left(\frac{x_2 - x_1}{2}\right)^2 + \left(\frac{y_2 - y_1}{2}\right)^2}$$

ii) The sum of the lengths of the short segments must equal the length of the long segment, $\sqrt{(x_2 - x_1)^2 + (y_2 - y_1)^2}$.

$$\sqrt{\left(\frac{x_1 + x_2}{2} - x_1\right)^2 + \left(\frac{y_1 + y_2}{2} - y_1\right)^2}$$

$$+ \sqrt{\left(x_2 - \frac{x_1 + x_2}{2}\right)^2 + \left(y_2 - \frac{y_1 + y_2}{2}\right)^2}$$

$$= \sqrt{\left(\frac{x_1 + x_2 - 2x_1}{2}\right)^2 + \left(\frac{y_1 + y_2 - 2y_1}{2}\right)^2}$$

$$+ \sqrt{\left(\frac{2x_2 - x_1 - x_2}{2}\right)^2 + \left(\frac{2y_2 - y_1 - y_2}{2}\right)^2}$$

$$= \sqrt{\left(\frac{x_2 - x_1}{2}\right)^2 + \left(\frac{y_2 - y_1}{2}\right)^2}$$

$$+ \sqrt{\left(\frac{x_2 - x_1}{2}\right)^2 + \left(\frac{y_2 - y_1}{2}\right)^2}$$

$$= 2\sqrt{\left(\frac{x_2 - x_1}{2}\right)^2 + \left(\frac{y_2 - y_1}{2}\right)^2}$$

$$= 2\sqrt{\frac{1}{4}(x_2 - x_1)^2 + \frac{1}{4}(y_2 - y_1)^2}$$

$$= 2\left(\frac{1}{2}\right)\sqrt{(x_2 - x_1)^2 + (y_2 - y_1)^2}$$

$$= \sqrt{(x_2 - x_1)^2 + (y_2 - y_1)^2}$$

For the sum of the short segments to equal the long segment, the three points must be on the same line, and the short segments are equal in length. Thus, the midpoint formula is correct.

Exercise Set 7.8

1. False

3. True

5. False

7. False

9. $\sqrt{-100} = \sqrt{100 \cdot -1} = \sqrt{100} \cdot \sqrt{-1} = 10i$

11. $\sqrt{-13} = \sqrt{-1 \cdot 13} = \sqrt{-1} \cdot \sqrt{13}$
$= i\sqrt{13}$ or $\sqrt{13}i$

13. $\sqrt{-8} = \sqrt{4 \cdot 2 \cdot -1} = \sqrt{4} \cdot \sqrt{2} \cdot \sqrt{-1} = 2\sqrt{2}i$,
or $2i\sqrt{2}$

15. $-\sqrt{-3} = -\sqrt{3 \cdot -1} = -\sqrt{3}i$, or $-i\sqrt{3}$

17. $-\sqrt{-81} = -\sqrt{81 \cdot -1} = -9i$

19. $-\sqrt{-300} = -\sqrt{100 \cdot 3 \cdot -1}$
$= -\sqrt{100} \cdot \sqrt{3} \cdot \sqrt{-1} = -10\sqrt{3}i$,
or $-10i\sqrt{3}$

21. $6 - \sqrt{-84} = 6 - \sqrt{4 \cdot 21 \cdot -1}$
$= 6 - \sqrt{4} \cdot \sqrt{21} \cdot \sqrt{-1}$
$= 6 - 2\sqrt{21}i$,
or $6 - 2i\sqrt{21}$

23. $-\sqrt{-76} + \sqrt{-125} = -\sqrt{4 \cdot 19 \cdot -1} + \sqrt{25 \cdot 5 \cdot -1}$
$= -\sqrt{4} \cdot \sqrt{19} \cdot \sqrt{-1} +$
$\sqrt{25} \cdot \sqrt{5} \cdot \sqrt{-1}$
$= -2\sqrt{19}i + 5\sqrt{5}i$
$= \left(-2\sqrt{19} + 5\sqrt{5}\right)i$

25. $\sqrt{-18} - \sqrt{-100} = \sqrt{9 \cdot 2 \cdot -1} - \sqrt{100 \cdot -1}$
$= \sqrt{9} \cdot \sqrt{2} \cdot \sqrt{-1} - \sqrt{100} \cdot \sqrt{-1}$
$= 3\sqrt{2}i - 10i = \left(3\sqrt{2} - 10\right)i$

27. $(6 + 7i) + (5 + 3i) = (6 + 5) + (7 + 3)i$
$= 11 + 10i$

29. $(9 + 8i) - (5 + 3i) = (9 - 5) + (8 - 3)i = 4 + 5i$

31. $(7 - 4i) - (5 - 3i) = (7 - 5) + \left[-4 - (-3)\right]i$
$= 2 - i$

33. $(-5 - i) - (7 + 4i) = (-5 - 7) + (-1 - 4)i$
$= -12 - 5i$

35. $7i \cdot 6i = 42i^2 = -42 \quad \left[i^2 = -1\right]$

37. $(-4i)(-6i) = 24i^2 = -24$

39. $\sqrt{-36} \cdot \sqrt{-9} = \sqrt{36 \cdot -1} \cdot \sqrt{9 \cdot -1}$
$= 6i \cdot 3i = 18i^2 = -18$

41. $\sqrt{-5} \cdot \sqrt{-2} = \sqrt{5}i \cdot \sqrt{2}i = \sqrt{10}i^2 = -\sqrt{10}$

43. $\sqrt{-6} \cdot \sqrt{-21} = \sqrt{6}i \cdot \sqrt{21}i = \sqrt{126}i^2$
$= -\sqrt{9 \cdot 14} = -3\sqrt{14}$

45. $5i(2+6i)$
$= 5i \cdot 2 + 5i \cdot 6i \quad$ Distributive law
$= 10i + 30i^2$
$= 10i + 30(-1) \quad i^2 = -1$
$= -30 + 10i \quad$ Standard form

47. $-7i(3-4i)$
$= (-7i \cdot 3) + (-7i)(-4i)$
$= -21i + 28i^2$
$= -28 - 21i$

49. $(1+i)(3+2i)$
$= 1 \cdot 3 + 1 \cdot 2i + i \cdot 3 + i \cdot 2i \quad$ Using FOIL
$= 3 + 5i + 2i^2$
$= 3 + 5i - 2 \qquad\qquad i^2 = -1$
$= 1 + 5i$

51. $(6-5i)(3+4i)$
$= 18 + 9i - 20i^2$
$= 18 + 9i + 20$
$= 38 + 9i$

53. $(7-2i)(2-6i)$
$= 14 - 46i + 12i^2$
$= 14 - 46i - 12$
$= 2 - 46i$

55. $(3+8i)(3-8i) = 3^2 - (8i)^2$
$= 9 - 64i^2 = 9 - 64(-1) = 73$

57. $(-7+i)(-7-i) = (-7)^2 - (i)^2$
$= 49 - (-1) = 50$

59. $(4-2i)^2$
$= 4^2 - 2 \cdot 4 \cdot 2i + (2i)^2 \quad$ Squaring a binomial
$= 16 - 16i + 4i^2$
$= 12 - 16i$

61. $(2+3i)^2$
$= 2^2 + 2 \cdot 2 \cdot 3i + (3i)^2$
$= 4 + 12i + 9i^2$
$= 4 + 12i - 9$
$= -5 + 12i$

63. $(-2+3i)^2$
$= (-2^2) + 2 \cdot -2 \cdot 3i + (3i)^2$
$= 4 - 12i - 9$
$= -5 - 12i$

65. $\dfrac{10}{3+i}$

$= \dfrac{10}{3+i} \cdot \dfrac{3-i}{3-i} \quad$ Multiplying by 1, using the conjugate

$= \dfrac{30-10i}{9-i^2} \quad$ Multiplying

$= \dfrac{30-10i}{9-(-1)} \quad i^2 = -1$

$= \dfrac{30-10i}{10}$

$= \dfrac{30}{10} - \dfrac{10}{10}i \quad$ Standard form

$= 3 - i$

67. $\dfrac{2}{3-2i}$

$= \dfrac{2}{3-2i} \cdot \dfrac{3+2i}{3+2i} \quad$ Multiplying by 1, using the conjugate

$= \dfrac{6+4i}{9-4i^2} \quad$ Multiplying

$= \dfrac{6+4i}{9-4(-1)} \quad i^2 = -1$

$= \dfrac{6+4i}{13}$

$= \dfrac{6}{13} + \dfrac{4}{13}i \quad$ Standard form

69. $\dfrac{2i}{5+3i}$

$= \dfrac{2i}{5+3i} \cdot \dfrac{5-3i}{5-3i}$

$= \dfrac{10i - 6i^2}{25 - 9i^2}$

$= \dfrac{10i + 6}{25 + 9} \qquad i^2 = -1$

$= \dfrac{6 + 10i}{34} = \dfrac{\cancel{2}(3+5i)}{\cancel{2} \cdot 17}$

$= \dfrac{3}{17} + \dfrac{5}{17}i$

71. $\dfrac{5}{6i} = \dfrac{5}{6i} \cdot \dfrac{i}{i} = \dfrac{5i}{6i^2} = \dfrac{5i}{-6} = -\dfrac{5}{6}i$

73. $\dfrac{5-3i}{4i} = \dfrac{5-3i}{4i} \cdot \dfrac{i}{i} = \dfrac{5i - 3i^2}{4i^2}$

$= \dfrac{5i + 3}{-4} = -\dfrac{3}{4} - \dfrac{5}{4}i$

75. $\dfrac{7i+14}{7i} = \dfrac{7i}{7i} + \dfrac{14}{7i} = 1 + \dfrac{2}{i} = 1 + \dfrac{2}{i} \cdot \dfrac{i}{i}$

$= 1 + \dfrac{2i}{i^2} = 1 + \dfrac{2i}{-1} = 1 - 2i$

77. $\dfrac{4+5i}{3-7i} = \dfrac{4+5i}{3-7i} \cdot \dfrac{3+7i}{3+7i}$

$= \dfrac{12 + 28i + 15i + 35i^2}{9 - 49i^2}$

$= \dfrac{12 + 43i - 35}{9 + 49} = \dfrac{-23 + 43i}{58}$

$= -\dfrac{23}{58} + \dfrac{43}{58}i$

79. $\dfrac{2+3i}{2+5i} = \dfrac{2+3i}{2+5i} \cdot \dfrac{2-5i}{2-5i}$

$= \dfrac{4 - 10i + 6i - 15i^2}{4 - 25i^2}$

$= \dfrac{4 - 4i + 15}{4 + 25} = \dfrac{19 - 4i}{29}$

$= \dfrac{19}{29} - \dfrac{4}{29}i$

81. $\dfrac{3-2i}{4+3i} = \dfrac{3-2i}{4+3i} \cdot \dfrac{4-3i}{4-3i} = \dfrac{12 - 9i - 8i + 6i^2}{16 - 9i^2}$

$= \dfrac{12 - 17i - 6}{16 + 9} = \dfrac{6 - 17i}{25} = \dfrac{6}{25} - \dfrac{17}{25}i$

83. $i^7 = i^6 \cdot i = \left(i^2\right)^3 \cdot i = (-1)^3 \cdot i = -1 \cdot i = -i$

85. $i^{32} = \left(i^2\right)^{16} = (-1)^{16} = 1$

87. $i^{42} = \left(i^2\right)^{21} = (-1)^{21} = -1$

89. $i^9 = \left(i^2\right)^4 \cdot i = (-1)^4 \cdot i = 1 \cdot i = i$

91. $(-i)^6 = i^6 = \left(i^2\right)^3 = (-1)^3 = -1$

93. $(5i)^3 = 5^3 \cdot i^3 = 125 \cdot i^2 \cdot i$

$= 125(-1)(i) = -125i$

95. $i^2 + i^4 = -1 + \left(i^2\right)^2 = -1 + (-1)^2 = -1 + 1 = 0$

97. *Thinking and Writing Exercise.*

99. $\qquad x^2 - x - 6 = 0$

$(x-3)(x+2) = 0$

$\quad x - 3 = 0 \quad or \quad x + 2 = 0$

$\qquad x = 3 \quad or \qquad x = -2$

The solutions are -2 and 3.

101. $t^2 = 100$

$t = \pm\sqrt{100} = \pm 10$

The solutions are -10 and 10.

103. $\qquad\qquad 15x^2 = 14x + 8$

$\quad 15x^2 - 14x - 8 = 0$

$(5x + 2)(3x - 4) = 0 \qquad$ Factoring

$5x + 2 = 0 \quad or \quad 3x - 4 = 0$

$\quad 5x = -2 \quad or \qquad 3x = 4$

$\quad x = -\dfrac{2}{5} \ or \qquad x = \dfrac{4}{3}$

The solutions are $-\dfrac{2}{5}$ and $\dfrac{4}{3}$.

105. *Thinking and Writing Exercise.*

107.

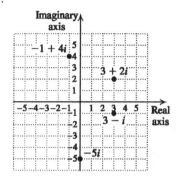

109. $|3+4i| = \sqrt{3^2 + 4^2} = \sqrt{9+16} = \sqrt{25} = 5$

111. $|-1+i| = \sqrt{(-1)^2 + 1^2} = \sqrt{1+1} = \sqrt{2}$

113. First we simplify $g(z)$.

$$g(z) = \frac{z^4 - z^2}{z-1} = \frac{z^2(z+1)\cancel{(z-1)}}{\cancel{z-1}}$$
$$= z^2(z+1) \text{ for } z \neq 1$$

Now we substitute.

$$g(3i) = (3i)^2(3i+1) = (9i^2)(3i+1)$$
$$= (-9)(3i+1) = -27i - 9$$
$$= -9 - 27i$$

115. We use the simplified form of $g(z)$ found in Exercise 113.

$$g(5i-1) = (5i-1)^2(5i-1+1)$$
$$= (25i^2 - 10i + 1)(5i)$$
$$= (-25 - 10i + 1)(5i)$$
$$= (-24 - 10i)(5i)$$
$$= -120i - 50i^2$$
$$= 50 - 120i$$

117. $\dfrac{1}{w - w^2} = \dfrac{1}{\left(\dfrac{1-i}{10}\right) - \left(\dfrac{1-i}{10}\right)^2} = \dfrac{1}{\left(\dfrac{1-i}{10}\right) - \dfrac{(1-i^2)}{100}}$

$$= \dfrac{1 \cdot 100}{\left(\dfrac{1-i}{10}\right) \cdot 100 - \dfrac{(1-i)^2}{100} \cdot 100}$$

$$= \dfrac{100}{(1-i)10 - (1-i)^2} = \dfrac{100}{10 - 10i - (1 - 2i + i^2)}$$

$$= \dfrac{100}{10 - 10i - (1 - 2i - 1)} = \dfrac{100}{10 - 8i} = \dfrac{50}{5 - 4i}$$

$$= \dfrac{50}{5-4i} \cdot \dfrac{5+4i}{5+4i} = \dfrac{50(5+4i)}{25+16}$$

$$= \dfrac{250 + 200i}{41} = \dfrac{250}{41} + \dfrac{200i}{41}$$

119. $(1-i)^3(1+i)^3$

$$= (1-i)(1+i) \cdot (1-i)(1+i) \cdot (1-i)(1+i)$$
$$= (1-i^2)(1-i^2)(1-i^2)$$
$$= (1+1)(1+1)(1+1) = 2 \cdot 2 \cdot 2 = 8$$

121. $\dfrac{6}{1 + \dfrac{3}{i}} = \dfrac{6}{\dfrac{i+3}{i}} = \dfrac{6i}{i+3} \cdot \dfrac{-i+3}{-i+3} = \dfrac{-6i^2 + 18i}{-i^2 + 9}$

$$= \dfrac{6 + 18i}{10} = \dfrac{6}{10} + \dfrac{18}{10}i = \dfrac{3}{5} + \dfrac{9}{5}i$$

123. $\dfrac{i - i^{38}}{1+i} = \dfrac{i - (i^2)^{19}}{1+i} = \dfrac{i - (-1)^{19}}{1+i}$

$$= \dfrac{i - (-1)}{1+i} = \dfrac{i+1}{i+1} = 1$$

Chapter 7 Study Summary

1. $-\sqrt{81} = -9$ since $\sqrt{81} = 9, -\sqrt{81} = -9$.

2. $\sqrt[3]{-1} = -1$ since $(-1)^3 = -1$

3. $\sqrt{36x^2} = \sqrt{(6x)^2} = |6x| = 6|x|$

    Since x might be negative, absolute-value notation is necessary.

4. $\sqrt[4]{x^4} = |x| = x$, for $x \geq 0$

5. $100^{-1/2} = \dfrac{1}{100^{1/2}} = \dfrac{1}{\left(10^2\right)^{1/2}} = \dfrac{1}{10}$

6. $\sqrt{7x} \cdot \sqrt{3y} = \sqrt{7x \cdot 3y} = \sqrt{21xy}$

7. $\sqrt{200x^5 y^{18}} = \sqrt{100 \cdot 2 \cdot x^4 \cdot x \cdot y^{18}}$
$= \sqrt{100x^4 y^{18} \cdot 2x}$
$= 10x^2 y^9 \sqrt{2x}$

The expression is undefined when $x < 0$, but y might be negative.

8. $\sqrt{\dfrac{12x^3}{25}} = \dfrac{\sqrt{4 \cdot 3 \cdot x^2 \cdot x}}{\sqrt{25}} = \dfrac{\sqrt{4x^2 \cdot 3x}}{5} = \dfrac{2x\sqrt{3x}}{5}$

Since the original expression is undefined when $x < 0$, there is no need to use the absolute value.

9. $\sqrt{\dfrac{2x}{3y^2}} = \dfrac{\sqrt{2x}}{y\sqrt{3}} \cdot \dfrac{\sqrt{3}}{\sqrt{3}} = \dfrac{\sqrt{6x}}{3y}$

10. $5\sqrt{8} - 3\sqrt{50} = 5\sqrt{4 \cdot 2} - 3\sqrt{25 \cdot 2}$
$= 5 \cdot 2\sqrt{2} - 3 \cdot 5\sqrt{2}$
$= 10\sqrt{2} - 15\sqrt{2}$
$= -5\sqrt{2}$

11. $\left(2 - \sqrt{3}\right)\left(5 - 7\sqrt{3}\right) =$
$= 2 \cdot 5 - 2 \cdot 7\sqrt{3} - \sqrt{3} \cdot 5 + 7\sqrt{3} \cdot \sqrt{3}$
$= 10 - 14\sqrt{3} - 5\sqrt{3} + 7 \cdot 3$
$= 31 - 19\sqrt{3}$

12. $\dfrac{\sqrt{15}}{3 + \sqrt{5}} = \dfrac{\sqrt{15}}{3 + \sqrt{5}} \cdot \dfrac{3 - \sqrt{5}}{3 - \sqrt{5}} = \dfrac{3\sqrt{15} - \sqrt{15} \cdot \sqrt{5}}{9 - 5}$
$= \dfrac{3\sqrt{15} - \sqrt{75}}{4} = \dfrac{3\sqrt{15} - \sqrt{25 \cdot 3}}{4}$
$= \dfrac{3\sqrt{15} - 5\sqrt{3}}{4}$

13. $\dfrac{\sqrt{x^5}}{\sqrt[3]{x}}$

$= \dfrac{x^{5/2}}{x^{1/3}}$ Converting to exponential notation

$= x^{5/2 - 1/3}$ Subtracting exponents

$= x^{13/6}$ Converting back

$= \sqrt[6]{x^{13}}$ to radical notation

14. $\sqrt{2x+3} = x$
$\left(\sqrt{2x+3}\right)^2 = x^2$
$2x + 3 = x^2$
$x^2 - 2x - 3 = 0$
$(x+1)(x-3) = 0$
$x + 1 = 0 \ \ or \ x - 3 = 0$
$x = -1 \ or \ \ \ \ x = 3$

Check:

For -1:
$$\begin{array}{c|c} \sqrt{2x+3} = x & \\ \hline \sqrt{-2+3} & -1 \\ \sqrt{1} & \\ 1 & -1 \end{array}$$ FALSE

For 3:
$$\begin{array}{c|c} \sqrt{2x+3} = x & \\ \hline \sqrt{6+3} & 3 \\ \sqrt{9} & \\ 3 & 3 \end{array}$$ TRUE

Since 3 checks but -1 does not, the solution is 3.

15. $a^2 + b^2 = c^2$ Pythagorean equation
$a^2 + 7^2 = 10^2$ Substituting
$a^2 + 49 = 100$
$a^2 = 100 - 49 = 51$
$a = \sqrt{51}$ Exact answer
$a \approx 7.141$ Approximation

16. Because this is an isosceles right triangle, the hypotenuse = length of a leg $\cdot \sqrt{2}$.
$6\sqrt{2} = a\sqrt{2}$
$6 = a$

17. For a 30-60-90 triangle, $c = 2a$, $b = \sqrt{3}a$
$a = 5$, $b = 5\sqrt{3} \approx 8.660$, $c = 2a = 10$

18. Substitute the coordinates into the distance formula.

$$d = \sqrt{(x_2 - x_1)^2 + (y_2 - y_1)^2}$$
$$= \sqrt{(6-(-2))^2 + (-10-1)^2}$$
$$= \sqrt{8^2 + (-11)^2}$$
$$= \sqrt{64 + 121}$$
$$= \sqrt{185}$$
$$\approx 13.601$$

19. Substitute the coordinates into the midpoint formula.

$$\left(\frac{x_1 + x_2}{2}, \frac{y_1 + y_2}{2}\right) = \left(\frac{-2+6}{2}, \frac{1+(-10)}{2}\right)$$
$$= \left(\frac{4}{2}, \frac{-9}{2}\right) = \left(2, -\frac{9}{2}\right)$$

20. $(5-3i) + (-8-9i) = (5-8) + (-3-9)i$
$$= -3 - 12i$$

21. $(2-i) - (-1+i) = 2 - i + 1 - i = 3 - 2i$

22. $(1-7i)(3-5i) = 1 \cdot 3 - 1 \cdot 5i - 7i \cdot 3 + 7 \cdot 5i^2$
$$= 3 - 5i - 21i - 35$$
$$= -32 - 26i$$

23. $\dfrac{1+i}{1-i} = \dfrac{1+i}{1-i} \cdot \dfrac{1+i}{1+i} = \dfrac{1+2i+i^2}{1-i^2} = \dfrac{1+2i+(-1)}{1-(-1)}$
$$= \dfrac{2i}{2} = i$$

Chapter 7 Review Exercises

1. True

2. False

3. False

4. True

5. True

6. True

7. True

8. False

9. $\sqrt{\dfrac{49}{9}} = \dfrac{\sqrt{49}}{\sqrt{9}} = \dfrac{\sqrt{7^2}}{\sqrt{3^2}} = \dfrac{7}{3}$

10. $-\sqrt{0.25} = -\sqrt{(0.5^2)} = -0.5$

11. $f(x) = \sqrt{2x-7}$
$$f(16) = \sqrt{2 \cdot 16 - 7} = \sqrt{32-7}$$
$$= \sqrt{25} = \sqrt{5^2} = 5$$

12. The domain of f are the values of x which make $2x - 7$ non-negative.
$$2x - 7 \geq 0$$
$$2x \geq 7$$
$$x \geq \frac{7}{2}$$
The domain is $\left\{x \,\middle|\, x \geq \dfrac{7}{2}\right\}$, or $\left[\dfrac{7}{2}, \infty\right)$

13. $M = -5 + \sqrt{6.7x - 444}$

   a. $x = 300$, so $M = -5 + \sqrt{6.7(300) - 444}$
   $$M \approx 34.6 \text{ lb}$$

   b. $x = 100$, so $M = -5 + \sqrt{6.7(100) - 444}$
   $$M \approx 10.0 \text{ lb}$$

   c. $x = 200$, so $M = -5 + \sqrt{6.7(200) - 444}$
   $$M \approx 24.9 \text{ lb}$$

   d. $x = 400$, so $M = -5 + \sqrt{6.7(400) - 444}$
   $$M \approx 42.3 \text{ lb}$$

14. $\sqrt{25t^2} = \sqrt{25} \cdot \sqrt{t^2} = 5|t|$

15. $\sqrt{(c+8)^2} = |c+8|$

16. $\sqrt{4x^2 + 4x + 1} = \sqrt{(2x+1)^2} = |2x+1|$

17. $\sqrt[5]{-32} = \sqrt[5]{(-2)^5} = -2$

18. $\left(\sqrt[3]{5ab}\right)^4 = (5ab)^{4/3}$

19. $\left(16a^6\right)^{3/4} = \left(2^4\right)^{3/4} \cdot \left(a^6\right)^{3/4} = 2^3 \cdot a^{9/2}$
$= 8 \cdot a^{\frac{8}{2}+\frac{1}{2}} = 8a^4\sqrt{a}$

20. $\sqrt{x^6 y^{10}} = \left(x^6 y^{10}\right)^{1/2} = \left(x^6\right)^{1/2}\left(y^{10}\right)^{1/2} = x^3 y^5$

21. $\left(\sqrt[6]{x^2 y}\right)^2 = \left[\left(x^2\right)^{1/6} y^{1/6}\right]^2 = \left[x^{1/3} y^{1/6}\right]^2$
$= \left(x^{1/3}\right)^2 \left(y^{1/6}\right)^2 = x^{2/3} y^{1/3}$
$= \left(x^2 y\right)^{1/3} = \sqrt[3]{x^2 y}$

22. $\left(x^{-2/3}\right)^{3/5} = x^{-2/3\cdot 3/5} = x^{-2/5} = \dfrac{1}{x^{2/5}}$

23. $\dfrac{7^{-1/3}}{7^{-1/2}} = 7^{-1/3-(-1/2)} = 7^{-2/6-(-3/6)} = 7^{1/6}$

24. $f(x) = \sqrt{25(x-6)^2} = \sqrt{5^2(x-6)^2}$
$f(x) = 5|x-6|$

25. $\sqrt[4]{16x^{20}y^8} = \sqrt[4]{\left(2x^5 y^2\right)^4} = 2x^5 y^2$

26. $\sqrt{250x^3 y^2} = \sqrt{25x^2 y^2 \cdot 10x} = 5xy\sqrt{10x}$

27. $\sqrt{2x}\cdot\sqrt{3y} = \sqrt{2x\cdot 3y} = \sqrt{6xy}$

28. $\sqrt[3]{3x^4 b}\sqrt[3]{9xb^2} = \sqrt[3]{3x^4 b \cdot 9xb^2} = \sqrt[3]{27x^5 b^3}$
$= \sqrt[3]{27x^3 b^3 \cdot x^2} = 3xb\sqrt[3]{x^2}$

29. $\sqrt[3]{-24x^{10}y^8}\cdot\sqrt[3]{18x^7 y^4} = \sqrt[3]{-432x^{17}y^{12}}$
$= \sqrt[3]{-216x^{15}y^{12}\cdot 2x^2}$
$= -6x^5 y^4\sqrt[3]{2x^2}$

30. $\dfrac{\sqrt[3]{60xy^3}}{\sqrt[3]{10x}} = \sqrt[3]{\dfrac{60xy^3}{10x}} = \sqrt[3]{6y^3} = y\sqrt[3]{6}$

31. $\dfrac{\sqrt{75x}}{2\sqrt{3}} = \dfrac{\sqrt{75x}}{2\sqrt{3}}\cdot\dfrac{\sqrt{3}}{\sqrt{3}} = \dfrac{\sqrt{225x}}{2\cdot 3} = \dfrac{15\sqrt{x}}{6} = \dfrac{5\sqrt{x}}{2}$

32. $\sqrt[4]{\dfrac{48a^{11}}{c^8}} = \dfrac{\sqrt[4]{48a^{11}}}{\sqrt[4]{c^8}} = \dfrac{\sqrt[4]{16a^8\cdot 3a^3}}{\sqrt[4]{c^8}} = \dfrac{2a^2\sqrt[4]{3a^3}}{c^2}$

33. $5\sqrt[3]{x} + 2\sqrt[3]{x} = (5+2)\sqrt[3]{x} = 7\sqrt[3]{x}$

34. $2\sqrt{75} - 9\sqrt{3} = 2\sqrt{25\cdot 3} - 9\sqrt{3}$
$= 2\cdot 5\sqrt{3} - 9\sqrt{3} = 10\sqrt{3} - 9\sqrt{3}$
$= (10-9)\sqrt{3} = \sqrt{3}$

35. $\sqrt[3]{8x^4} + \sqrt[3]{xy^6} = \sqrt[3]{8x^3\cdot x} + \sqrt[3]{y^6\cdot x}$
$= 2x\sqrt[3]{x} + y^2\sqrt[3]{x}$
$= \left(2x + y^2\right)\sqrt[3]{x}$

36. $\sqrt{50} + 2\sqrt{18} + \sqrt{32}$
$= \sqrt{25\cdot 2} + 2\sqrt{9\cdot 2} + \sqrt{16\cdot 2}$
$= 5\sqrt{2} + 2\cdot 3\sqrt{2} + 4\sqrt{2}$
$= (5+6+4)\sqrt{2} = 15\sqrt{2}$

37. $\left(3+\sqrt{10}\right)\left(3-\sqrt{10}\right) = (3)^2 - \left(\sqrt{10}\right)^2$
$= 9 - 10 = -1$

38. $\left(\sqrt{3}-3\sqrt{8}\right)\left(\sqrt{5}+2\sqrt{8}\right)$
$= \sqrt{3}\cdot\sqrt{5} + \sqrt{3}\cdot 2\sqrt{8} - 3\sqrt{8}\cdot\sqrt{5} - 3\sqrt{8}\cdot 2\sqrt{8}$
$= \sqrt{15} + 2\sqrt{24} - 3\sqrt{40} - 6\cdot 8$
$= \sqrt{15} + 2\sqrt{4\cdot 6} - 3\sqrt{4\cdot 10} - 48$
$= \sqrt{15} + 2\cdot 2\sqrt{6} - 3\cdot 2\sqrt{10} - 48$
$= \sqrt{15} + 4\sqrt{6} - 6\sqrt{10} - 48$

39. $\sqrt[4]{x}\cdot\sqrt{x} = x^{1/4}\cdot x^{1/2} = x^{1/4+1/2} = x^{1/4+2/4}$
$= x^{3/4} = \sqrt[4]{x^3}$

40. $\dfrac{\sqrt[3]{x^2}}{\sqrt[4]{x}} = \dfrac{x^{2/3}}{x^{1/4}} = x^{2/3-1/4} = x^{8/12-3/12} = x^{5/12} = \sqrt[12]{x^5}$

41. $f(x) = x^2$
$f\left(a-\sqrt{2}\right) = \left(a-\sqrt{2}\right)^2$
$= a^2 - 2\cdot a\cdot\sqrt{2} + \left(\sqrt{2}\right)^2$
$= a^2 - 2a\sqrt{2} + 2$

42. $\sqrt{\dfrac{x}{8y}} = \dfrac{\sqrt{x}}{2\sqrt{2y}} = \dfrac{\sqrt{x}}{2\sqrt{2y}} \cdot \dfrac{\sqrt{2y}}{\sqrt{2y}} = \dfrac{\sqrt{2xy}}{4y}$

43. $\dfrac{4\sqrt{5}}{\sqrt{2}+\sqrt{3}} = \dfrac{4\sqrt{5}}{\sqrt{2}+\sqrt{3}} \cdot \dfrac{\sqrt{2}-\sqrt{3}}{\sqrt{2}-\sqrt{3}}$

$\qquad = \dfrac{4\sqrt{5} \cdot \sqrt{2} - 4\sqrt{5} \cdot \sqrt{3}}{\left(\sqrt{2}\right)^2 - \left(\sqrt{3}\right)^2}$

$\qquad = \dfrac{4\sqrt{10} - 4\sqrt{15}}{2-3} = -4\sqrt{10} + 4\sqrt{15}$

44. $\dfrac{4\sqrt{5}}{\sqrt{2}+\sqrt{3}} = \dfrac{4\sqrt{5}}{\sqrt{2}+\sqrt{3}} \cdot \dfrac{\sqrt{5}}{\sqrt{5}}$

$\qquad = \dfrac{4 \cdot \left(\sqrt{5}\right)^2}{\sqrt{2} \cdot \sqrt{5} + \sqrt{3} \cdot \sqrt{5}}$

$\qquad = \dfrac{4 \cdot 5}{\sqrt{10}+\sqrt{15}} = \dfrac{20}{\sqrt{10}+\sqrt{15}}$

45. $\sqrt{y+6} - 2 = 3$

$\qquad \sqrt{y+6} = 5$

$\qquad \left(\sqrt{y+6}\right)^2 = 5^2$

$\qquad y+6 = 25$

$\qquad y = 19$

Check: $\sqrt{19+6} - 2 = \sqrt{25} - 2 = 5 - 2 = 3$
The solution is 19.

46. $\sqrt{x} = x - 6$

$\qquad \left(\sqrt{x}\right)^2 = (x-6)^2$

$\qquad x = x^2 - 12x + 36$

$\qquad 0 = x^2 - 13x + 36$

$\qquad 0 = (x-4)(x-9)$

$x - 4 = 0 \quad or \quad x - 9 = 0$

$\qquad x = 4 \quad or \qquad x = 9$

Check: $\sqrt{4} = 4 - 6$

$\qquad\qquad 2 = -2 \quad \text{FALSE}$

$\qquad\quad \sqrt{9} = 9 - 6$

$\qquad\qquad 3 = 3 \quad \text{TRUE}$

The solution is 9.

47. $(x+1)^{1/3} = -5$

$\qquad \left((x+1)^{1/3}\right)^3 = (-5)^3$

$\qquad x + 1 = -125$

$\qquad x = -126$

Check: $(-126+1)^{1/3} = (-125)^{1/3} = -5$
The solution is -126.

48. $1 + \sqrt{x} = \sqrt{3x-3}$

$\qquad \left(1+\sqrt{x}\right)^2 = \left(\sqrt{3x-3}\right)^2$

$\qquad 1 + 2\sqrt{x} + x = 3x - 3$

$\qquad 2\sqrt{x} = 2x - 4$

$\qquad 2\sqrt{x} = 2(x-2)$

$\qquad \sqrt{x} = x - 2$

$\qquad \left(\sqrt{x}\right)^2 = (x-2)^2$

$\qquad x = x^2 - 4x + 4$

$\qquad 0 = x^2 - 5x + 4$

$\qquad 0 = (x-4)(x-1)$

$x - 4 = 0 \ or \ x - 1 = 0$

$\qquad x = 4 \ or \qquad x = 1$

Check: $x = 4$:

$1 + \sqrt{4}$	$\sqrt{3(4)-3}$
$1 + 2$	$\sqrt{12-3}$
3	$\sqrt{9}$
3	3

TRUE

$x = 41$:

$1 + \sqrt{1}$	$\sqrt{3(1)-3}$
$1 + 1$	$\sqrt{3-3}$
2	$\sqrt{0}$
2	0

FALSE

The solution is 4.

49. $f(x) = \sqrt[4]{x+2}; f(a) = 2$. By substitution:

$\qquad 2 = \sqrt[4]{x+2}$

$\qquad 2^4 = \left(\sqrt[4]{x+2}\right)^4$

$\qquad 16 = x + 2$

$\qquad 14 = x$

Check: $\sqrt[4]{14+2} = \sqrt[4]{16} = \sqrt[4]{2^4} = 2$
The solution is 14.

50. Let x = the length of a side of the square.
Using the Pythagorean equations, solve for x.
$$x^2 + x^2 = 10^2$$
$$2x^2 = 100$$
$$x^2 = 50$$
$$x = \sqrt{50} = \sqrt{25 \cdot 2} = 5\sqrt{2}$$
$$x \approx 7.071$$
The length of a side of the square is $5\sqrt{2}$ cm, or approximately 7.071 cm.

51. Using the Pythagorean equation, determine b, the length of the base.
$$b^2 + 2^2 = 6^2$$
$$b^2 + 4 = 36$$
$$b^2 = 32$$
$$b = \sqrt{32} = \sqrt{16 \cdot 2} = 4\sqrt{2}$$
$$b \approx 5.657$$
The base is $4\sqrt{2}$ ft, or about 5.657 ft.

52. This is a 30°-60°-90° triangle. The length of short leg is 1/2 × the length of the hypotenuse, or $\frac{1}{2} \cdot 20 = 10$. The length of the long leg is $\sqrt{3} \times$ the length of the short leg, or $\sqrt{3} \cdot 10 = 10\sqrt{3} \approx 17.321$.

53. Substitute the coordinates into the distance formula.
$$d = \sqrt{(x_2 - x_1)^2 + (y_2 - y_1)^2}$$
$$= \sqrt{(-1-(-6))^2 + (5-4)^2}$$
$$= \sqrt{5^2 + 1^2}$$
$$= \sqrt{25+1}$$
$$= \sqrt{26}$$
$$\approx 5.099$$

54. Substitute the coordinates into the midpoint formula.
$$\left(\frac{x_1 + x_2}{2}, \frac{y_1 + y_2}{2} \right)$$
$$\left(\frac{-7+3}{2}, \frac{-2+(-1)}{2} \right), \text{ or } \left(\frac{-4}{2}, \frac{-3}{2} \right),$$
$$\text{or } \left(-2, -\frac{3}{2} \right)$$

55. $\sqrt{-45} = \sqrt{9 \cdot 5 \cdot -1} = \sqrt{9} \cdot \sqrt{5} \cdot \sqrt{-1} = 3\sqrt{5}i$, or $3i\sqrt{5}$

56. $(-4+3i)+(2-12i) = (-4+2)+[3+(-12)]i$
$$= -2 - 9i$$

57. $(9-7i)-(3-8i) = (9-3)+[-7-(-8)]i$
$$= 6 + i$$

58. $(2+5i)(2-5i) = 2^2 - (5i)^2 = 4 - 25i^2$
$$= 4 + 25 = 29$$

59. $i^{18} = (i^2)^9 = (-1)^9 = -1$

60. $(6-3i)(2-i)$
$$= 6 \cdot 2 + 6 \cdot -i + -3i \cdot 2 + -3i \cdot -i$$
$$= 12 - 6i - 6i + 3i^2$$
$$= 12 - 12i + 3(-1)$$
$$= 9 - 12i$$

61. $\dfrac{7-2i}{3+4i} \cdot \dfrac{3-4i}{3-4i} = \dfrac{21-34i+8i^2}{9-16i^2}$
$$= \frac{21-34i-8}{9+16} = \frac{13-34i}{25}$$
$$= \frac{13}{25} - \frac{34}{25}i$$

62. *Thinking and Writing Exercise.* A complex number $a+bi$ is real when $b = 0$; it is imaginary when $a = 0$.

63. *Thinking and Writing Exercise.* An absolute-value sign must be used to simplify $\sqrt[n]{x^n}$ when n is even, since x may be negative. If x is negative, while n is even, the radical expression cannot be simplified to x, since

$\sqrt[n]{x^n}$ represents the principal, or non-negative root. When n is odd, there is only one root, and it will be the same sign (positive or negative) as the radicand; thus, no absolute-value sign is needed when n is odd.

64.

$$\sqrt{11x+\sqrt{6+x}} = 6$$

$$\left(\sqrt{11x+\sqrt{6+x}}\right)^2 = 6^2$$

$$11x+\sqrt{6+x} = 36$$

$$\left(\sqrt{6+x}\right)^2 = (36-11x)^2$$

$$6+x = 1296-792x+121x^2$$

$$0 = 121x^2-793x+1290$$

$$0 = (121x-430)(x-3)$$

$$121x-430 = 0 \quad or \quad x-3=0$$

$$x = \frac{430}{121} \quad or \quad x = 3$$

Check: $x = \frac{430}{121}; \sqrt{11\left(\frac{430}{121}\right)+\sqrt{6+\frac{430}{121}}} \overset{?}{=} 6$

$$\sqrt{\frac{430}{11}+\sqrt{\frac{1156}{121}}}$$

$$\sqrt{\frac{430}{11}+\frac{34}{11}}$$

$$\sqrt{\frac{464}{11}} \neq 6$$

$x = 3; \sqrt{11(3)+\sqrt{6+3}} \overset{?}{=} 6$

$$\sqrt{33+\sqrt{9}}$$

$$\sqrt{33+3}$$

$$\sqrt{36}$$

$$6 = 6 \quad \text{True}$$

The solution is 3

65.

$$\frac{2}{1-3i} - \frac{3}{4+2i}$$

$$= \frac{2}{1-3i}\cdot\frac{1+3i}{1+3i} - \frac{3}{4+2i}\cdot\frac{4-2i}{4-2i}$$

$$= \frac{2+6i}{1+9} - \left(\frac{12-6i}{16+4}\right)$$

$$= \frac{2+6i}{10} - \left[\frac{2(6-3i)}{2\cdot 10}\right]$$

$$= \frac{2+6i-(6-3i)}{10}$$

$$= \frac{(2-6)+[6-(-3)]i}{10}$$

$$= \frac{-4+9i}{10} = \frac{-4}{10}+\frac{9}{10}i \text{ or } -\frac{2}{5}+\frac{9}{10}i$$

66. Variable answers. Possible form $\frac{ai}{bi}$ where a and b are integers (for instance, $\frac{2i}{3i}$).

67. The isosceles triangle has 2 legs of equal length, which equal the length of the hypotenuse $\div\sqrt{2}$, or $\frac{6}{\sqrt{2}}\cdot\frac{\sqrt{2}}{\sqrt{2}} = \frac{6\sqrt{2}}{2} = 3\sqrt{2}$. The area of the isosceles triangle is

$$\frac{1}{2}\cdot 3\sqrt{2}\cdot 3\sqrt{2} = 9.$$

The 30°-60°-90° triangle's shorter leg is $\frac{1}{2}$ the length of the hypotenuse, or $\frac{1}{2}\cdot 6 = 3$. The length of the longer leg is $\sqrt{3}\cdot$ the length of the shorter leg, or $3\sqrt{3}$. The area of the 30°-60°-90° triangle is

$$\frac{1}{2}\cdot 3\cdot 3\sqrt{3} = \frac{9}{2}\sqrt{3} \approx 7.794$$

The isosceles right triangle is greater by 9 – 7.794, or 1.206 ft^2.

Chapter 7 Test

1. $\sqrt{50} = \sqrt{25\cdot 2} = \sqrt{25}\cdot\sqrt{2} = 5\sqrt{2}$

2. $\sqrt[3]{-\dfrac{8}{x^6}} = \dfrac{\sqrt[3]{-8}}{\sqrt[3]{x^6}} = -\dfrac{2}{x^2}$

3. $\sqrt{81a^2} = \sqrt{81 \cdot a^2} = \sqrt{81} \cdot \sqrt{a^2} = 9|a|$

4. $\sqrt{x^2 - 8x + 16} = \sqrt{(x-4)^2} = |x-4|$

5. $\sqrt{7xy} = (7xy)^{1/2}$

6. $\left(4a^3b\right)^{5/6} = \sqrt[6]{\left(4a^3b\right)^5}$

7. $f(x) = \sqrt{2x - 10}$

The domain of f are those values of x which make $2x - 10$ non-negative.
$$2x - 10 \geq 0$$
$$2x \geq 10$$
$$x \geq 5$$
The domain of f is $\{x \mid x \geq 5\}$, or $[5, \infty)$

8. If $f(x) = x^2$, substitute to determine $f\left(5 + \sqrt{2}\right)$.
$$f\left(5 + \sqrt{2}\right) = \left(5 + \sqrt{2}\right)^2$$
$$= 5^2 + 2 \cdot 5 \cdot \sqrt{2} + \left(\sqrt{2}\right)^2$$
$$= 25 + 10\sqrt{2} + 2$$
$$= 27 + 10\sqrt{2}$$

9. $\sqrt[5]{32x^{16}y^{10}} = \sqrt[5]{32x^{15}y^{10} \cdot x} = 2x^3y^2\sqrt[5]{x}$

10. $\sqrt[3]{4w} \cdot \sqrt[3]{4v^2} = \sqrt[3]{4w \cdot 4v^2} = \sqrt[3]{16wv^2}$
$$= \sqrt[3]{8 \cdot 2wv^2} = 2\sqrt[3]{2wv^2}$$

11. $\sqrt{\dfrac{100a^4}{9b^6}} = \dfrac{\sqrt{100a^4}}{\sqrt{9b^6}} = \dfrac{10a^2}{3b^3}$

12. $\dfrac{\sqrt[5]{48x^6y^{10}}}{\sqrt[5]{16x^2y^9}} = \sqrt[5]{\dfrac{48x^6y^{10}}{16x^2y^9}} = \sqrt[5]{3x^4y}$

13. $\sqrt[4]{x^3}\sqrt{x} == x^{3/4} \cdot x^{1/2} = x^{3/4 + 1/2} = x^{5/4}$
$$= \sqrt[4]{x^5} = \sqrt[4]{x^4 \cdot x} = x\sqrt[4]{x}$$

14. $\dfrac{\sqrt{y}}{\sqrt[10]{y}} = \dfrac{y^{1/2}}{y^{1/10}} = y^{1/2 - 1/10} = y^{5/10 - 1/10} = y^{4/10}$
$$= y^{2/5} = \sqrt[5]{y^2}$$

15. $8\sqrt{2} - 2\sqrt{2} = (8 - 2)\sqrt{2} = 6\sqrt{2}$

16. $\sqrt{x^4y} + \sqrt{9y^3} = \sqrt{x^4 \cdot y} + \sqrt{9y^2 \cdot y}$
$$= x^2\sqrt{y} + 3y\sqrt{y}$$
$$= \left(x^2 + 3y\right)\sqrt{y}$$

17. $\left(7 + \sqrt{x}\right)\left(2 - 3\sqrt{x}\right)$
$$= 7 \cdot 2 + 7 \cdot \left(-3\sqrt{x}\right) + \sqrt{x} \cdot 2 + \sqrt{x} \cdot \left(-3\sqrt{x}\right)$$
$$= 14 - 21\sqrt{x} + 2\sqrt{x} - 3x$$
$$= 14 - 19\sqrt{x} - 3x$$

18. $\dfrac{\sqrt[3]{x}}{\sqrt[3]{4y}} = \dfrac{\sqrt[3]{x}}{\sqrt[3]{2^2 y}} = \dfrac{\sqrt[3]{x}}{\sqrt[3]{2^2 y}} \cdot \dfrac{\sqrt[3]{2y^2}}{\sqrt[3]{2y^2}} = \dfrac{\sqrt[3]{2xy^2}}{2y}$

19. $6 = \sqrt{x-3} + 5 \Rightarrow 1 = \sqrt{x-3}$
$$\Rightarrow 1^2 = \left(\sqrt{x-3}\right)^2$$
$$\Rightarrow 1 = x - 3 \Rightarrow x = 4$$
The value checks, $x = 4$.

20. $$x = \sqrt{3x+3} - 1$$
$$x + 1 = \sqrt{3x+3}$$
$$(x+1)^2 = \left(\sqrt{3x+3}\right)^2$$
$$x^2 + 2x + 1 = 3x + 3$$
$$x^2 - x - 2 = 0$$
$$(x+1)(x-2) = 0$$
$$x + 1 = 0 \quad or \quad x - 2 = 0$$
$$x = -1 \quad or \quad x = 2$$

Check: $x = -1$ $-1 \overset{?}{=} \sqrt{3(-1) + 3} - 1$
$$= \sqrt{-3+3} - 1 = -1$$

$x = 2$ $2 \overset{?}{=} \sqrt{3(2)+3} - 1$
$$= \sqrt{6+3} - 1 = \sqrt{9} - 1$$
$$= 3 - 1 = 2$$
The solutions are -1 and 2.

21.
$$\sqrt{2x} = \sqrt{x+1} + 1$$
$$\sqrt{2x} - 1 = \sqrt{x+1}$$
$$\left(\sqrt{2x} - 1\right)^2 = \left(\sqrt{x+1}\right)^2$$
$$2x - 2\sqrt{2x} + 1 = x + 1$$
$$x - 2\sqrt{2x} = 0$$
$$x = 2\sqrt{2x}$$
$$x^2 = \left(2\sqrt{2x}\right)^2$$
$$x^2 = 4 \cdot 2x$$
$$x^2 - 8x = 0$$
$$x(x-8) = 0$$
$$x = 0 \quad or \quad x - 8 = 0$$
$$x = 0 \quad or \quad\quad x = 8$$

Check: $x = 0$ $\quad \sqrt{2(0)} = 0 \overset{?}{=} \sqrt{0+1} + 1$
$$= \sqrt{1} + 1 = 1 + 1 = 2$$

The equation is false, so $x = 0$ is not a solution.

$x = 8$ $\quad\quad \sqrt{2(8)} \overset{?}{=} \sqrt{8+1} + 1$
$$\sqrt{16} \overset{?}{=} \sqrt{9} + 1$$
$$4 \overset{?}{=} 3 + 1$$
$$4 = 4$$

The solution is 8.

22. Let d = the diagonal distance. Using the given information and the Pythagorean equation, we solve for d.
$$d^2 = 50^2 + 90^2$$
$$d^2 = 2500 + 8100$$
$$d^2 = 10,600$$
$$d = \sqrt{10,600} = \sqrt{100 \cdot 106} = 10\sqrt{106}$$
$$d \approx 102.956$$
She jogged $10\sqrt{106}$ ft., or approximately 102.956 ft.

23. If the length of the hypotenuse is 10 cm, the length of the shorter leg is half as long (5 cm), and the length of the longer leg is $5\sqrt{3}$ cm, or approximately 8.660 cm.

24. Substitute the coordinates into the distance formula.
$$d = \sqrt{(x_2 - x_1)^2 + (y_2 - y_1)^2}$$
$$= \sqrt{(-1-3)^2 + (8-7)^2}$$
$$= \sqrt{(-4)^2 + 1^2}$$
$$= \sqrt{16+1}$$
$$= \sqrt{17}$$
$$\approx 4.123$$

25. Substitute the coordinates into the midpoint formula.
$$\left(\frac{x_1 + x_2}{2}, \frac{y_1 + y_2}{2}\right)$$
$$\left(\frac{2+1}{2}, \frac{-5+(-7)}{2}\right), \text{ or } \left(\frac{3}{2}, \frac{-12}{2}\right),$$
$$\text{or } \left(\frac{3}{2}, -6\right)$$

26. $\sqrt{-50} = \sqrt{50 \cdot -1} = \sqrt{25 \cdot 2 \cdot -1}$
$$= \sqrt{25} \cdot \sqrt{2} \cdot \sqrt{-1} = 5\sqrt{2}i$$
or $5i\sqrt{2}$

27. $(9+8i) - (-3+6i) = [9-(-3)] + [8-6]i$
$$= 12 + 2i$$

28. $\sqrt{-16} \cdot \sqrt{-36} = \sqrt{16 \cdot -1} \cdot \sqrt{36 \cdot -1}$
$$= 4i \cdot 6i = 24i^2 = -24$$

29. $(4-i)^2 = 4^2 - 2 \cdot 4 \cdot i + i^2 = 16 - 8i + i^2$
$$= 16 - 8i - 1 = 15 - 8i$$

30. $\dfrac{-2+i}{3-5i} = \dfrac{-2+i}{3-5i} \cdot \dfrac{3+5i}{3+5i} = \dfrac{-6-7i-5}{9+25}$
$$= \dfrac{-11-7i}{34} = -\dfrac{11}{34} - \dfrac{7}{34}i$$

31. $i^{37} = i^{36} \cdot i = \left(i^2\right)^{18} \cdot i = (-1)^{18} \cdot i = 1 \cdot i = i$

32. $\sqrt{2x-2}+\sqrt{7x+4}=\sqrt{13x+10}$

$\left(\sqrt{2x-2}+\sqrt{7x+4}\right)^2=\left(\sqrt{13x+10}\right)^2$

$(2x-2)+2\sqrt{2x-2}\cdot\sqrt{7x+4}+$
$\qquad\qquad (7x+4)=13x+10$

$9x+2+2\sqrt{(2x-2)(7x+4)}=13x+10$

$2\sqrt{14x^2-6x-8}=4x+8$

$2\sqrt{14x^2-6x-8}=2(2x+4)$

$\sqrt{14x^2-6x-8}=2x+4$

$\left(\sqrt{14x^2-6x-8}\right)^2=(2x+4)^2$

$14x^2-6x-8=4x^2+16x+16$

$10x^2-22x-24=0$

$2(5x^2-11x-12)=0$

$(5x+4)(x-3)=0$

$5x+4=0 \quad or \ x-3=0$

$\qquad x=-\dfrac{4}{5} \ or \qquad x=3$

Check: $x=-\dfrac{4}{5}$

$$\sqrt{2\left(-\dfrac{4}{5}\right)-2}=\sqrt{-\dfrac{8}{5}-\dfrac{10}{5}}=\sqrt{-\dfrac{18}{5}}$$

Since we cannot take the principle square of a

negative number, using real number, $-\dfrac{4}{5}$ is

not a solution.

Check: $x=3$

$$\sqrt{2(3)-2}+\sqrt{7(3)+4}\overset{?}{=}\sqrt{13(3)+10}$$

$\sqrt{6-2}+\sqrt{21+4}\qquad\sqrt{39+10}$

$\sqrt{4}+\sqrt{25}\qquad\qquad\sqrt{49}$

$\qquad\qquad 2+5=7$

The solution is 3.

33. $\dfrac{1-4i}{4i(1+4i)^{-1}}=\dfrac{1-4i}{4i\cdot\dfrac{1}{1+4i}}=\dfrac{1-4i}{\dfrac{4i}{1+4i}}$

$\qquad =(1-4i)\div\dfrac{4i}{1+4i}$

$\qquad =(1-4i)\cdot\dfrac{1+4i}{4i}$

$\qquad =\dfrac{(1-4i)(1+4i)}{4i}=\dfrac{1-16i^2}{4i}$

$=\dfrac{17}{4i}\cdot\dfrac{-i}{-i}=\dfrac{-17i}{-4i^2}=\dfrac{-17i}{4}$

$=\dfrac{-17}{4}i$

34. $D(h)=180\Rightarrow 1.2\sqrt{h}=180$

$\Rightarrow 1.2\sqrt{h}=180\Rightarrow\sqrt{h}=\dfrac{180}{1.2}=150$

$\Rightarrow\sqrt{h}^2=150^2\Rightarrow h=22{,}500$

The pilot must fly 22,500 ft above sea level.

Chapter 8

Quadratic Functions and Equations

1. $\sqrt{k}; -\sqrt{k}$

3. $t+3; t+3$

5. $25; 5$

7. There are 2 x-intercepts, so there are 2 real-number solutions.

9. There is 1 x-intercept, so there is 1 real-number solution.

11. There are no x-intercepts, so there are no real-number solutions.

13. $x^2 = 100 \Rightarrow x = \pm\sqrt{100} = \pm 10$

15. $p^2 - 50 = 0 \Rightarrow p^2 = 50 \Rightarrow p = \pm\sqrt{50}$
    $= \pm\sqrt{25 \cdot 2} = \pm\sqrt{25} \cdot \sqrt{2} = \pm 5\sqrt{2}$

17. $4x^2 = 20$

    $x^2 = 5$ Multiplying by $\dfrac{1}{4}$

    $x = \sqrt{5}$ or $x = -\sqrt{5}$ Using the principle of square roots

    The solutions are $\sqrt{5}$ and $-\sqrt{5}$, or $\pm\sqrt{5}$.

19. $x^2 = -4 \Rightarrow x = \pm\sqrt{-4} = \pm\sqrt{4}i = \pm 2i$

21. $9x^2 - 16 = 0 \Rightarrow 9x^2 = 16 \Rightarrow x^2 = \dfrac{16}{9}$

    $\Rightarrow x = \pm\sqrt{\dfrac{16}{9}} = \pm\dfrac{\sqrt{16}}{\sqrt{9}} = \pm\dfrac{4}{3}$

23. $5t^2 - 3 = 4 \Rightarrow 5t^2 = 7$

    $t^2 = \dfrac{7}{5}$

    $t = \sqrt{\dfrac{7}{5}}$ or $t = -\sqrt{\dfrac{7}{5}}$ Principle of square roots

    $t = \sqrt{\dfrac{7}{5} \cdot \dfrac{5}{5}}$ or $t = -\sqrt{\dfrac{7}{5} \cdot \dfrac{5}{5}}$ Rationalizing denominators

    $t = \dfrac{\sqrt{35}}{5}$ or $t = -\dfrac{\sqrt{35}}{5}$

    The solutions are $\pm\sqrt{\dfrac{7}{5}}$ or $\pm\dfrac{\sqrt{35}}{5}$.

25. $4d^2 + 81 = 0 \Rightarrow 4d^2 = -81 \Rightarrow d^2 = -\dfrac{81}{4}$

    $\Rightarrow d = \pm\sqrt{-\dfrac{81}{4}} = \pm\sqrt{\dfrac{81}{4}}i = \pm\dfrac{\sqrt{81}}{\sqrt{4}}i = \pm\dfrac{9}{2}i$

27. $(x-1)^2 = 49$

    $x - 1 = \pm 7$ Principle of square roots

    $x - 1 = 7$ or $x - 1 = -7$

    $x = 8$ or $\quad x = -6$

    The solutions are 8 and –6.

29. $(a-13)^2 = 18$

    $a - 13 = \pm\sqrt{18}$

    $a - 13 = -\sqrt{9 \cdot 2}$ or $a - 13 = \sqrt{9 \cdot 2}$

    $a = 13 - 3\sqrt{2}$ or $\quad a = 13 + 3\sqrt{2}$

    The solutions are $13 \pm 3\sqrt{2}$.

31. $(x+1)^2 = -9$

    $x + 1 = \pm\sqrt{-9}$

    $x + 1 = -\sqrt{9 \cdot -1}$ or $x + 1 = \sqrt{9 \cdot -1}$

    $x = -1 - 3i$ or $\quad x = -1 + 3i$

    The solutions are $-1 \pm 3i$.

33. $\left(y+\dfrac{3}{4}\right)^2 = \dfrac{17}{16}$

$y+\dfrac{3}{4} = \pm\sqrt{\dfrac{17}{16}}$

$y+\dfrac{3}{4} = -\dfrac{\sqrt{17}}{\sqrt{16}} \ or \ y+\dfrac{3}{4} = \dfrac{\sqrt{17}}{\sqrt{16}}$

$y = -\dfrac{3}{4}-\dfrac{\sqrt{17}}{4} \ or \ y = -\dfrac{3}{4}+\dfrac{\sqrt{17}}{4}$

$y = -\dfrac{3-\sqrt{17}}{4} \ or \ y = -\dfrac{3+\sqrt{17}}{4}$

The solutions are $\dfrac{-3\pm\sqrt{17}}{4}$.

35. $x^2 -10x+25 = 64$

$(x-5)^2 = 64$

$x-5 = \pm 8$

$x = 5\pm 8$

$x = 13 \ or \ x = -3$

The solutions are 13 and –3.

37. $f(x)=19 \Rightarrow x^2 = 19 \Rightarrow x = \pm\sqrt{19}$

39. $f(x)=16$

$(x-5)^2 = 16$ Substituting

$x-5 = 4 \ or \ x-5 = -4$

$x = 9 \ or \quad x = 1$

The solutions are 9 and 1.

41. $F(t)=13$

$(t+4)^2 = 13$ Substituting

$t+4 = \sqrt{13} \qquad or \ t+4 = -\sqrt{13}$

$t = -4+\sqrt{13} \ or \quad t = -4-\sqrt{13}$

The solutions are $-4+\sqrt{13}$ and $-4-\sqrt{13}$

or $-4\pm\sqrt{13}$.

43. $g(x) = x^2 +14x+49$

Observe first that $g(0) = 49$. Also observe

that when $x = -14$, then $x^2 +14x = (-14)^2$

$-(14)(14) = (14)^2 -(14)^2 = 0$, so

$g(-14) = 49$ as well. Thus, we have $x = 0$ or

$x = -14$. We can also do this problem as
follows.

$g(x)=49$

$x^2 +14x+49 = 49$ Substituting

$(x+7)^2 = 49$

$x+7 = 7 \ or \ x+7 = -7$

$x = 0 \ or \quad x = -14$

The solutions are 0 and –14.

45. $x^2 +16x$

Take half of the coefficient of x and square it:

half of 16 is 8 and $8^2 = 64$. Add 64.

$x^2 +16x+64, \ (x+8)^2$

47. $t^2 -10t$

Take half of the coefficient of t and square it:

half of –10 is –5, and $(-5)^2 = 25$. Add 25.

$t^2 -10t+25; \ (t-5)^2$

49. $t^2 -2t$

Take half of the coefficient of t and square it:

half of –2 is –1, and $(-1)^2 = 1$. Add 1.

$t^2 -2t+1; \ (t-1)^2$

51. $x^2 +3x$

Take half the coefficient of x and square it:

half of 3 is $\dfrac{3}{2}$, and $\left(\dfrac{3}{2}\right)^2 = \dfrac{9}{4}$. Add $\dfrac{9}{4}$.

$x^2 +3x+\dfrac{9}{4} = \left(x+\dfrac{3}{2}\right)^2$

53. $x^2 +\dfrac{2}{5}x$

$\dfrac{1}{2}\cdot\dfrac{2}{5} = \dfrac{1}{5}; \ \left(\dfrac{1}{5}\right)^2 = \dfrac{1}{25}; \ add \ \dfrac{1}{25}$

$x^2 +\dfrac{2}{5}x+\dfrac{1}{25}; \ \left(x+\dfrac{1}{5}\right)^2$

55. $t^2 - \dfrac{5}{6}t$

$\dfrac{1}{2} \cdot \dfrac{-5}{6} = \dfrac{-5}{12}; \; \left(\dfrac{-5}{12}\right)^2 = \dfrac{25}{144}; \;$ add $\dfrac{25}{144}$

$t^2 - \dfrac{5}{6}t + \dfrac{25}{144}; \; \left(t - \dfrac{5}{12}\right)^2$

57. $x^2 + 6x = 7$

$x^2 + 6x + 9 = 7 + 9$ Adding 9 to both sides
to complete the square

$(x+3)^2 = 16$ Factoring

$x + 3 = \pm 4$ Principle of square roots

$x = -3 \pm 4$

$x = -3 + 4 \; or \; x = -3 - 4$

$x = 1 \qquad or \; x = -7$

The solutions are 1 and –7.

59. $t^2 - 10t = -23$

$t^2 - 10t + 25 = -23 + 25$

$(t-5)^2 = 2$

$t - 5 = \pm\sqrt{2}$

$t = 5 \pm \sqrt{2}$

The solutions are $5 + \sqrt{2}$ and $5 - \sqrt{2}$.

61. $x^2 + 12x + 32 = 0$

$x^2 + 12x \quad = -32$

$x^2 + 12x + 36 = -32 + 36$

$(x+6)^2 = 4$

$x + 6 = \pm 2$

$x = -6 \pm 2$

$x = -6 + 2 \; or \; x = -6 - 2$

$x = -4 \quad or \; x = -8$

The solutions are –4 and –8.

63. $t^2 + 8t - 3 = 0$

$t^2 + 8t \quad = 3$

$t^2 + 8t + 16 = 3 + 16$

$(t+4)^2 = 19$

$t + 4 = \pm\sqrt{19}$

$t = -4 \pm \sqrt{19}$

The solutions are $-4 \pm \sqrt{19}$.

65. $f(x) = x^2 + 6x + 7$

To determine the x-intercepts, substitute $f(x) = 0$, and solve for x.

$x^2 + 6x + 7 = 0$

$x^2 + 6x \quad = -7$

$x^2 + 6x + 9 = -7 + 9$

$(x+3)^2 = 2$

$x + 3 = \pm\sqrt{2}$

$x = -3 \pm \sqrt{2}$

The intercepts are $\left(-3 - \sqrt{2}, 0\right)$ and $\left(-3 + \sqrt{2}, 0\right)$.

67. $x^2 + 9x - 25 = 0$

$x^2 + 9x \quad = 25$

$x^2 + 9x + \left(\dfrac{9}{2}\right)^2 = 25 + \left(\dfrac{9}{2}\right)^2 = 25 + \dfrac{81}{4}$

$\left(x + \dfrac{9}{2}\right)^2 = \dfrac{100}{4} + \dfrac{81}{4} = \dfrac{181}{4}$

$x + \dfrac{9}{2} = \pm\sqrt{\dfrac{181}{4}} = \pm\dfrac{\sqrt{181}}{2}$

$x = -\dfrac{9}{2} \pm \dfrac{\sqrt{181}}{2}$

The intercepts are $\left(\dfrac{-9 + \sqrt{181}}{2}, 0\right)$ and $\left(\dfrac{-9 - \sqrt{181}}{2}, 0\right)$.

69. $x^2 - 10x - 22 = 0$

$x^2 - 10x \quad = 22$

$x^2 - 10x + 25 = 22 + 25$

$(x-5)^2 = 47$

$x = 5 \pm \sqrt{47}$

The intercepts are $\left(5 - \sqrt{47}, 0\right)$ and $\left(5 + \sqrt{47}, 0\right)$.

71. $9x^2 + 18x = -8$

$$x^2 + 2x = -\frac{8}{9}$$

$$x^2 + 2x + 1 = -\frac{8}{9} + 1$$

$$(x+1)^2 = \frac{1}{9}$$

$$x + 1 = \pm\frac{1}{3}$$

$$x = -1 \pm \frac{1}{3}$$

$$x = -1 - \frac{1}{3} \ or \ x = -1 + \frac{1}{3}$$

$$x = -\frac{4}{3} \quad or \quad x = -\frac{2}{3}$$

The solutions are $-\frac{4}{3}$ and $-\frac{2}{3}$.

73. $3x^2 - 5x - 2 = 0$

$$3x^2 - 5x = 2$$

$$x^2 - \frac{5}{3}x = \frac{2}{3}$$

$$x^2 - \frac{5}{3}x + \frac{25}{36} = \frac{2}{3} + \frac{25}{36}$$

$$\left(x - \frac{5}{6}\right)^2 = \frac{49}{36}$$

$$x - \frac{5}{6} = \pm\frac{7}{6}$$

$$x = \frac{5}{6} \pm \frac{7}{6}$$

$$x = \frac{5}{6} - \frac{7}{6} \ or \ x = \frac{5}{6} + \frac{7}{6}$$

$$x = -\frac{2}{6} \quad or \quad x = \frac{12}{6}$$

$$x = -\frac{1}{3} \quad\quad or \quad x = 2$$

The solutions are 2 and $-\frac{1}{3}$.

75. $5x^2 + 4x - 3 = 0$

$5x^2 + 4x = 3$

$$x^2 + \frac{4}{5}x = \frac{3}{5}$$

$$x^2 + \frac{4}{5}x + \frac{4}{25} = \frac{3}{5} + \frac{4}{25}$$

$$\left(x + \frac{2}{5}\right)^2 = \frac{19}{25}$$

$$x = -\frac{2}{5} \pm \frac{\sqrt{19}}{5}$$

$$x = -\frac{2 \pm \sqrt{19}}{5}$$

The solutions are $\dfrac{-2 \pm \sqrt{19}}{5}$.

77. $4x^2 + 2x - 3 = 0$

$$x^2 + \frac{1}{2}x = \frac{3}{4}$$

$$x^2 + \frac{1}{2}x + \frac{1}{16} = \frac{3}{4} + \frac{1}{16}$$

$$\left(x + \frac{1}{4}\right)^2 = \frac{13}{16}$$

$$x + \frac{1}{4} = \pm\frac{\sqrt{13}}{4}$$

$$x = -\frac{1}{4} \pm \frac{\sqrt{13}}{4}$$

The x-intercepts are
$\left(\dfrac{-1-\sqrt{13}}{4}, 0\right)$ and $\left(\dfrac{-1+\sqrt{13}}{4}, 0\right)$.

79. $2x^2 - 3x - 1 = 0$

$$x^2 - \frac{3}{2}x = \frac{1}{2}$$

$$x^2 - \frac{3}{2}x + \frac{9}{16} = \frac{1}{2} + \frac{9}{16}$$

$$\left(x - \frac{3}{4}\right)^2 = \frac{17}{16}$$

$$x - \frac{3}{4} = \pm\frac{\sqrt{17}}{4}$$

$$x = \frac{3}{4} \pm \frac{\sqrt{17}}{4}; \ or \ \frac{3 \pm \sqrt{17}}{4}$$

The x-intercepts are
$\left(\dfrac{3-\sqrt{17}}{4}, 0\right)$ and $\left(\dfrac{3+\sqrt{17}}{4}, 0\right)$.

81. **Familiarize.** We are already familiar with the compound-interest formula.
Translate. We substitute into the formula.

$$A = P(1+r)^t$$

$$2420 = 2000(1+r)^2$$

Carry out. We solve for r.

$$2420 = 2000(1+r)^2$$

$$\frac{2420}{2000} = (1+r)^2$$

$$\frac{121}{100} = (1+r)^2$$

$$\pm\sqrt{\frac{121}{100}} = 1+r$$

$$\pm\frac{11}{10} = 1+r$$

$$-\frac{10}{10} \pm \frac{11}{10} = r$$

$$\frac{1}{10} = r \text{ or } -\frac{21}{10} = r$$

Check. Since the interest rate cannot be negative, we need only check $\frac{1}{10}$, or 10%. If $2000 were invested at 10% interest, compounded annually, then in 2 years it would grow to $2000(1.1)^2$, or $2420. The number 10% checks.
State. The interest rate is 10%.

83. **Familiarize.** We are already familiar with the compound-interest formula.
Translate. We substitute into the formula.

$$A = P(1+r)^t$$

$$6760 = 6250(1+r)^2$$

Carry out. We solve for r.

$$6760 = 6250(1+r)^2$$

$$\frac{6760}{6250} = (1+r)^2$$

$$\frac{676}{625} = (1+r)^2$$

$$\pm\frac{26}{25} = 1+r$$

$$-\frac{25}{25} \pm \frac{26}{25} = r$$

$$\frac{1}{25} = r \text{ or } -\frac{51}{25} = r$$

Check. Since the interest rate cannot be negative, we need only check $\frac{1}{25}$ or 4%. If $6250 were invested at 4% interest, compounded annually, then in 2 years it would grow to $6250(1.04)^2$, or $6760. The number 4% checks.
State. The interest rate is 4%.

85. Use $s = 16t^2$. We substitute 4000 for s, and solve for t.

$$4000 = 16t^2 \Rightarrow t^2 = \frac{4000}{16} = 250$$

$$t = +\sqrt{250} = \sqrt{25 \cdot 10} = 5\sqrt{10}$$

$$t \approx 15.8$$

It will take about 15.8 seconds.

87. Use $s = 16t^2$. We substitute 2063 for s, and solve for t.

$$2063 = 16t^2 \Rightarrow t^2 = \frac{2063}{16}$$

$$t = +\sqrt{\frac{2063}{16}} = \frac{\sqrt{2063}}{4}$$

$$t \approx 11.355$$

It will take about 11.4 seconds.

89. *Thinking and Writing Exercise.*

91. $b^2 - 4ac = 2^2 - 4(3)(-5) = 4 - (-60) = 64$

93. $\sqrt{200} = \sqrt{100 \cdot 2} = \sqrt{100}\sqrt{2} = 10\sqrt{2}$

95. $\sqrt{-4} = \sqrt{4}i = 2i$

97. $\sqrt{-8} = \sqrt{8}i = \sqrt{4}\sqrt{2}i = 2\sqrt{2}i$

99. *Thinking and Writing Exercise.*

101. In order for $x^2 + bx + 81$ to be a square, the following must be true:

$$\left(\frac{b}{2}\right)^2 = 81$$

$$\frac{b^2}{4} = 81$$

$$b^2 = 324$$

$$b = 18 \text{ or } b = -18$$

103. We see that x is a factor of each term, so x is also a factor of $f(x)$. We have

$$f(x) = x\left(2x^4 - 9x^3 - 66x^2 + 45x + 280\right).$$

Since $x^2 - 5$ is a factor of $f(x)$ it is also a factor of $2x^4 - 9x^3 - 66x^2 + 45x + 280$. We divide to find another factor.

$$
\begin{array}{r}
2x^2 - 9x - 56 \\
x^2 - 5 \overline{)\,2x^4 - 9x^3 - 66x^2 + 45x + 280} \\
\underline{2x^4 \qquad\; -10x^2} \\
-9x^3 - 56x^2 + 45x \\
\underline{-9x^3 \qquad\quad + 45x} \\
-56x^2 \qquad\; + 280 \\
\underline{-56x^2 \qquad\quad + 280} \\
0
\end{array}
$$

Then we have $f(x) = x(x^2 - 5)$ $(2x^2 - 9x - 56)$, or $f(x) = x(x^2 - 5)$ $(2x + 7)(x - 8)$. Now we find the values of a for which $f(a) = 0$.

$$f(a) = 0$$
$$a(a^2 - 5)(2a + 7)(a - 8) = 0$$
$$a = 0 \; or \; a^2 - 5 = 0$$
$$or \; 2a + 7 = 0 \; or \; a - 8 = 0$$
$$\Rightarrow a = 0 \; or \; a^2 = 5$$
$$or \; 2a = -7 \; or \; a = 8$$
$$\Rightarrow a = 0 \; or \; a = \pm\sqrt{5}$$
$$or \; a = -\frac{7}{2} \; or \; a = 8$$

The solutions are $0, \sqrt{5}, -\sqrt{5}, -\dfrac{7}{2}$, and 8.

105. **Familiarize.** It is helpful to list information in a chart and make a drawing. Let r represent the speed of the fishing boat. Then $r - 7$ represents the speed of the barge.

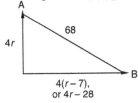

Boat	r	t	d
Fishing	r	4	$4r$
Barge	$r - 7$	4	$4(r - 7)$

Translate. We use the Pythagorean equation:
$$a^2 + b^2 = c^2$$
$$(4r - 28)^2 + (4r)^2 = 68^2$$

Carry out.
$$(4r - 28)^2 + (4r)^2 = 68^2$$
$$16r^2 - 224r + 784 + 16r^2 = 4624$$
$$32r^2 - 224r - 3840 = 0$$
$$r^2 - 7r - 120 = 0$$
$$(r + 8)(r - 15) = 0$$
$$r + 8 = 0 \; or \; r - 15 = 0$$
$$r = -8 \; or \qquad r = 15$$

Check. We check only 15 since the speeds of the boats cannot be negative. If the speed of the fishing boat is
15 km/h, then the speed of the barge is $15 - 7$, or 8 km/h, and the distances they travel are $4 \cdot 15$ (or 60) and $4 \cdot 8$ (or 32).
$$60^2 + 32^2 = 3600 + 1024 = 4624 = 68^2$$
The values check.

State. The speed of the fishing boat is 15 km/h, and the speed of the barge is 8 km/h.

Exercise Set 8.2

1. True

3. False

5. False

7. $2x^2 + 3x - 5 = 0 \Rightarrow (2x + 5)(x - 1) = 0$

$\Rightarrow 2x + 5 = 0 \Rightarrow 2x = -5 \Rightarrow x = -\dfrac{5}{2}$

or $x - 1 = 0 \Rightarrow x = 1$

The solutions are $x = -\dfrac{5}{2}$ and $x = 1$.

9. $u^2 + 2u - 4 = 0$

$a = 1,\ b = 2,\ c = -4$

$u = \dfrac{-2 \pm \sqrt{2^2 - 4 \cdot 1 \cdot (-4)}}{2 \cdot 1} = \dfrac{-2 \pm \sqrt{4 + 16}}{2}$

$= \dfrac{-2 \pm \sqrt{20}}{2} = \dfrac{-2 \pm \sqrt{4 \cdot 5}}{2} = \dfrac{-2 \pm 2\sqrt{5}}{2}$

$= \dfrac{\cancel{2}\left(-1 \pm \sqrt{5}\right)}{\cancel{2}} = -1 \pm \sqrt{5}$

The solutions are $u = -1 + \sqrt{5}$ and
$u = -1 - \sqrt{5}$.

11. $\qquad 3p^2 = 18p - 6$

$3p^2 - 18p + 6 = 0$

$\quad p^2 - 6p + 2 = 0 \quad$ Dividing by 3

$a = 1,\ b = -6,\ c = 2$

$p = \dfrac{-b \pm \sqrt{b^2 - 4ac}}{2a}$

$p = \dfrac{-(-6) \pm \sqrt{(-6)^2 - 4 \cdot 1 \cdot 2}}{2 \cdot 1} = \dfrac{6 \pm \sqrt{36 - 8}}{2}$

$p = \dfrac{6 \pm \sqrt{28}}{2} = \dfrac{6 \pm 2\sqrt{7}}{2}$

$p = \dfrac{2\left(3 \pm \sqrt{7}\right)}{2} = 3 \pm \sqrt{7}$

The solutions are $3 + \sqrt{7}$ and $3 - \sqrt{7}$.

13. $\qquad h^2 + 4 = 6h$

$h^2 - 6h + 4 = 0$

$a = 1,\ b = -6,\ c = 4$

$x = \dfrac{-(-6) \pm \sqrt{(-6)^2 - 4 \cdot 1 \cdot 4}}{2 \cdot 1} = \dfrac{6 \pm \sqrt{36 - 16}}{2}$

$x = \dfrac{6 \pm \sqrt{20}}{2} = \dfrac{6 \pm \sqrt{4 \cdot 5}}{2} = \dfrac{6 \pm 2\sqrt{5}}{2}$

$x = 3 \pm \sqrt{5}$

The solutions are $3 + \sqrt{5}$ and $3 - \sqrt{5}$.

15. $x^2 = 3x + 5 \Rightarrow x^2 - 3x - 5 = 0$

$\Rightarrow a = 1,\ b = -3,\ c = -5$

$x = \dfrac{-(-3) \pm \sqrt{(-3)^2 - 4 \cdot 1 \cdot (-5)}}{2 \cdot 1} = \dfrac{3 \pm \sqrt{9 + 20}}{2}$

$= \dfrac{3 \pm \sqrt{29}}{2} = \dfrac{3}{2} \pm \dfrac{\sqrt{29}}{2}$

The solutions are

$x = \dfrac{3}{2} + \dfrac{\sqrt{29}}{2}$ and $x = \dfrac{3}{2} - \dfrac{\sqrt{29}}{2}$.

17. $3t(t + 2) = 1 \Rightarrow 3t^2 + 6t - 1 = 0$

$\Rightarrow a = 3,\ b = 6,\ c = -1$

$t = \dfrac{-6 \pm \sqrt{6^2 - 4 \cdot 3 \cdot (-1)}}{2 \cdot 3} = \dfrac{-6 \pm \sqrt{36 + 12}}{6}$

$= \dfrac{-6 \pm \sqrt{48}}{6} = \dfrac{-6 \pm \sqrt{16 \cdot 3}}{6} = \dfrac{-6 \pm 4\sqrt{3}}{6}$

$= \dfrac{\cancel{2}\left(-3 \pm 2\sqrt{3}\right)}{\cancel{2} \cdot 3} = \dfrac{-3 \pm 2\sqrt{3}}{3}$

The solutions are

$t = \dfrac{-3 + 2\sqrt{3}}{3}$ and $t = \dfrac{-3 - 2\sqrt{3}}{3}$.

19. $\qquad \dfrac{1}{x^2} - 3 = \dfrac{8}{x} \qquad$ LCD is x^2

$x^2\left(\dfrac{1}{x^2} - 3\right) = x^2 \cdot \dfrac{8}{x}$

$\quad 1 - 3x^2 = 8x$

$\qquad 0 = 3x^2 + 8x - 1$

$a = 3,\ b = 8,\ c = -1$

$x = \dfrac{-8 \pm \sqrt{8^2 - 4 \cdot 3 \cdot (-1)}}{2 \cdot 3} = \dfrac{-8 \pm \sqrt{64 + 12}}{6}$

$x = \dfrac{-8 \pm \sqrt{76}}{6} = \dfrac{-8 \pm 2\sqrt{19}}{6}$

$x = \dfrac{2\left(-4 \pm \sqrt{19}\right)}{6} = \dfrac{-4 \pm \sqrt{19}}{3}$

The solutions are $-\dfrac{4}{3} \pm \dfrac{\sqrt{19}}{3}$.

21. $t^2 + 10 = 6t \Rightarrow t^2 - 6t + 10 = 0$

$\Rightarrow a = 1,\ b = -6,\ c = 10$

$t = \dfrac{-(-6) \pm \sqrt{(-6)^2 - 4 \cdot 1 \cdot 10}}{2 \cdot 1} = \dfrac{6 \pm \sqrt{36 - 40}}{2}$

$= \dfrac{6 \pm \sqrt{-4}}{2} = \dfrac{6 \pm \sqrt{4}i}{2} = \dfrac{6 \pm 2i}{2}$

$= \dfrac{\cancel{2}\left(3 \pm i\right)}{\cancel{2}} = 3 \pm i$

The solutions are $t = 3 + i$ and $t = 3 - i$.

23. $x^2 + 4x + 6 = 0 \Rightarrow a = 1, b = 4, c = 6$

$x = \dfrac{-4 \pm \sqrt{4^2 - 4 \cdot 1 \cdot 6}}{2 \cdot 1} = \dfrac{-4 \pm \sqrt{16 - 24}}{2}$

$= \dfrac{-4 \pm \sqrt{-8}}{2} = \dfrac{-4 \pm \sqrt{8}i}{2} = \dfrac{-4 \pm 2\sqrt{2}i}{2}$

$= \dfrac{\cancel{2}\left(-2 \pm \sqrt{2}i\right)}{\cancel{2}} = -2 \pm \sqrt{2}i$

The solutions are
$x = -2 + \sqrt{2}i$ and $x = -2 - \sqrt{2}i$.

25. $12t^2 + 17t = 40 \Rightarrow 12t^2 + 17t - 40 = 0$

$\Rightarrow a = 12, b = 17, c = -40$

$t = \dfrac{-17 \pm \sqrt{17^2 - 4 \cdot 12 \cdot (-40)}}{2 \cdot 12}$

$= \dfrac{-17 \pm \sqrt{289 + 1920}}{24}$

$= \dfrac{-17 \pm \sqrt{2209}}{24} = \dfrac{-17 \pm 47}{24}$

The solutions are
$t = \dfrac{-17 + 47}{24} = \dfrac{30}{24} = \dfrac{5}{4}$ and

$t = \dfrac{-17 - 47}{24} = \dfrac{-64}{24} = -\dfrac{8}{3}$.

27. $25x^2 - 20x + 4 = 0$

$(5x - 2)(5x - 2) = 0$

$5x - 2 = 0 \ or \ 5x - 2 = 0$

$x = \dfrac{2}{5} \ or \quad x = \dfrac{2}{5}$

The solution is $\dfrac{2}{5}$.

29. $7x(x + 2) + 5 = 3x(x + 1)$

$7x^2 + 14x + 5 = 3x^2 + 3x$

$4x^2 + 11x + 5 = 0$

$x = \dfrac{-11 \pm \sqrt{11^2 - 4 \cdot 4 \cdot 5}}{2 \cdot 4}$

$x = \dfrac{-11 \pm \sqrt{41}}{8}$

The solutions are
$\dfrac{-11 + \sqrt{41}}{8}$ and $\dfrac{-11 - \sqrt{41}}{8}$.

31. $14(x - 4) - (x + 2) = (x + 2)(x - 4)$

$14x - 56 - x - 2 = x^2 - 2x - 8$

$13x - 58 = x^2 - 2x - 8$

$0 = x^2 - 15x + 50$

$0 = (x - 10)(x - 5)$

$x - 10 = 0 \quad or \ x - 5 = 0$

$x = 10 \ or \qquad x = 5$

The solutions are 10 and 5.

33. $\qquad 5x^2 = 13x + 17$

$5x^2 - 13x - 17 = 0$

$x = \dfrac{-(-13) \pm \sqrt{(-13)^2 - 4(5)(-17)}}{2 \cdot 5}$

$x = \dfrac{13 \pm \sqrt{169 + 340}}{10} = \dfrac{13 \pm \sqrt{509}}{10}$

The solutions are $\dfrac{13 + \sqrt{509}}{10}$ and $\dfrac{13 - \sqrt{509}}{10}$.

35. $x(x - 3) = x - 9 \Rightarrow x^2 - 4x + 9 = 0$

$\Rightarrow a = 1, b = -4, c = 9$

$x = \dfrac{-(-4) \pm \sqrt{(-4)^2 - 4 \cdot 1 \cdot 9}}{2 \cdot 1} = \dfrac{4 \pm \sqrt{16 - 36}}{2}$

$= \dfrac{4 \pm \sqrt{-20}}{2} = \dfrac{4 \pm \sqrt{20}i}{2} = \dfrac{4 \pm \sqrt{4 \cdot 5}i}{2}$

$= \dfrac{4 \pm 2\sqrt{5}i}{2} = \dfrac{\cancel{2}\left(2 \pm \sqrt{5}i\right)}{\cancel{2}} = 2 \pm \sqrt{5}i$

The solutions are
$x = 2 + \sqrt{5}i$ and $x = 2 - \sqrt{5}i$.

37. $\qquad x^3 - 8 = 0$

$x^3 - 2^3 = 0$

$(x - 2)(x^2 + 2x + 4) = 0$

$x - 2 = 0 \ or \ x^2 + 2x + 4 = 0$

$x = 2 \ or \ x = \dfrac{-2 \pm \sqrt{2^2 - 4 \cdot 1 \cdot 4}}{2 \cdot 1}$

$x = 2 \ or \ x = \dfrac{-2 \pm \sqrt{-12}}{2} = \dfrac{-2 \pm 2i\sqrt{3}}{2}$

$x = 2 \ or \ x = -1 \pm i\sqrt{3}$

The solutions are 2, $-1 + i\sqrt{3}$ and $-1 - i\sqrt{3}$.

39.
$$g(x) = 0$$
$$4x^2 - 2x - 3 = 0 \text{ Substituting}$$
$$x = \frac{-(-2) \pm \sqrt{(-2)^2 - 4 \cdot 4 \cdot (-3)}}{2 \cdot 4} = \frac{2 \pm \sqrt{52}}{8}$$
$$x = \frac{2 \pm 2\sqrt{13}}{8} = \frac{1 \pm \sqrt{13}}{4}$$
The solutions are $\frac{1}{4} \pm \frac{\sqrt{13}}{4}$.

41.
$$g(x) = 1$$
$$\frac{2}{x} + \frac{2}{x+3} = 1 \text{ Substituting}$$
$$2x + 6 + 2x = x^2 + 3x$$
$$\qquad \text{Multiplying by } x(x+3)$$
$$0 = x^2 - x - 6$$
$$0 = (x-3)(x+2)$$
$$x - 3 = 0 \; or \; x + 2 = 0$$
$$x = 3 \; or \qquad x = -2$$
The solutions are 3 and –2.

43.
$$F(x) = G(x)$$
$$\frac{x+3}{x} = \frac{x-4}{3} \text{ Substituting}$$
$$3x\left(\frac{x+3}{x}\right) = 3x\left(\frac{x-4}{3}\right) \text{ Multiplying by the LCD}$$
$$3x + 9 = x^2 - 4x$$
$$0 = x^2 - 7x - 9$$
$$x = \frac{-(-7) \pm \sqrt{(-7)^2 - 4 \cdot 1 \cdot (-9)}}{2 \cdot 1}$$
$$x = \frac{7 \pm \sqrt{49 + 36}}{2} = \frac{7 \pm \sqrt{85}}{2}$$
The solutions are $\frac{7 + \sqrt{85}}{2}$ and $\frac{7 - \sqrt{85}}{2}$.

45. $x^2 + 4x - 7 = 0$
$$x = \frac{-4 \pm \sqrt{4^2 - 4 \cdot 1 \cdot (-7)}}{2 \cdot 1}$$
$$= \frac{-4 \pm \sqrt{44}}{2}$$
$$x = \frac{-4 + \sqrt{44}}{2} \approx 1.317 \text{ and}$$
$$x = \frac{-4 - \sqrt{44}}{2} \approx -5.317$$

47. $x^2 - 6x + 4 = 0$
$$x = \frac{-(-6) \pm \sqrt{(-6)^2 - 4 \cdot 1 \cdot 4}}{2 \cdot 1}$$
$$= \frac{6 \pm \sqrt{20}}{2}$$
$$x = \frac{6 + \sqrt{20}}{2} \approx 5.236 \text{ and}$$
$$x = \frac{6 - \sqrt{20}}{2} \approx 0.764$$

49. $2x^2 - 3x - 7 = 0$
$$x = \frac{-(-3) \pm \sqrt{(-3)^2 - 4 \cdot 2 \cdot (-7)}}{2 \cdot 2}$$
$$= \frac{3 \pm \sqrt{65}}{4}$$
$$x = \frac{3 + \sqrt{65}}{4} \approx 2.766 \text{ and}$$
$$x = \frac{3 - \sqrt{65}}{4} \approx -1.266$$

51. *Thinking and Writing Exercise.*

53. $(x - 2i)(x + 2i) = x^2 - (2i)^2 = x^2 - 4i^2$
$$= x^2 - 4(-1) = x^2 + 4$$

55. $(x - (2 - \sqrt{7}))(x - (2 + \sqrt{7}))$
$$= (x - 2 + \sqrt{7})(x - 2 - \sqrt{7})$$
$$= ((x-2) - \sqrt{7})((x-2) + \sqrt{7})$$
$$= (x-2)^2 - (\sqrt{7})^2 = (x^2 - 4x + 4) - 7$$
$$= x^2 - 4x - 3$$

57. $\dfrac{-6 \pm \sqrt{(-4)^2 - 4(2)(2)}}{2(2)} = \dfrac{-6 \pm \sqrt{16 - 16}}{4}$

$= \dfrac{-6 \pm 0}{4} = \dfrac{-6}{4} = -\dfrac{3}{2}$

59. *Thinking and Writing Exercise.*

61. $f(x) = \dfrac{x^2}{x-2} + 1$

To find the x-coordinates of the x-intercepts
of the graph of f, we solve $f(x) = 0$.

$\dfrac{x^2}{x-2} + 1 = 0$

$x^2 + x - 2 = 0$ Multiplying by $x - 2$

$(x+2)(x-1) = 0$

$x = -2 \ or \ x = 1$

The x-intercepts are $(-2, 0)$ and $(1, 0)$.

63. $f(x) = g(x)$

$\dfrac{x^2}{x-2} + 1 = \dfrac{4x-2}{x-2} + \dfrac{x+4}{2}$

Substituting

$2(x-2)\left(\dfrac{x^2}{x-2} + 1\right) = 2(x-2)\left(\dfrac{4x-2}{x-2} + \dfrac{x+4}{2}\right)$

Multiplying by the LCD

$2x^2 + 2(x-2) = 2(4x-2) + (x-2)(x+4)$

$2x^2 + 2x - 4 = 8x - 4 + x^2 + 2x - 8$

$2x^2 + 2x - 4 = x^2 + 10x - 12$

$x^2 - 8x + 8 = 0$

$a = 1, \ b = -8, \ c = 8$

$x = \dfrac{-(-8) \pm \sqrt{(-8)^2 - 4 \cdot 1 \cdot 8}}{2 \cdot 1} = \dfrac{8 \pm \sqrt{64 - 32}}{2}$

$x = \dfrac{8 \pm \sqrt{32}}{2} = \dfrac{8 \pm \sqrt{16 \cdot 2}}{2} = \dfrac{8 \pm 4\sqrt{2}}{2}$

$x = 4 \pm 2\sqrt{2}$

The solutions are $4 + 2\sqrt{2}$ and $4 - 2\sqrt{2}$.

65. $z^2 + 0.84z - 0.4 = 0$

$z = \dfrac{-0.84 \pm \sqrt{(0.84)^2 - 4 \cdot 1 \cdot (-0.4)}}{2 \cdot 1}$

$z = \dfrac{-0.84 \pm \sqrt{2.3056}}{2}$

$z = \dfrac{-0.84 + \sqrt{2.3056}}{2} \approx 0.339$

$z = \dfrac{-0.84 - \sqrt{2.3056}}{2} \approx -1.179$

67. $\sqrt{2}x^2 + 5x + \sqrt{2} = 0$

$x = \dfrac{-5 \pm \sqrt{5^2 - 4 \cdot \sqrt{2} \cdot \sqrt{2}}}{2\sqrt{2}} = \dfrac{-5 \pm \sqrt{17}}{2\sqrt{2}}$, or

$x = \dfrac{-5 \pm \sqrt{17}}{2\sqrt{2}} \cdot \dfrac{\sqrt{2}}{\sqrt{2}} = \dfrac{-5\sqrt{2} \pm \sqrt{34}}{4}$

The solutions are $\dfrac{-5\sqrt{2} \pm \sqrt{34}}{4}$.

69. $kx^2 + 3x - k = 0$

$k(-2)^2 + 3(-2) - k = 0$

Substituting -2 for

$4k - 6 - k = 0$

$3k = 6$

$k = 2$

$2x^2 + 3x - 2 = 0$

Substituting 2 for k

$(2x-1)(x+2) = 0$

$2x - 1 = 0 \ or \ x + 2 = 0$

$x = \dfrac{1}{2} \ or \ \ \ \ x = -2$

The other solution is $\dfrac{1}{2}$.

71. *Thinking and Writing Exercise.*

Exercise Set 8.3

1. Discriminant

3. Two

5. Rational

7. $a = 1$, $b = -7$, $c = 5$

$b^2 - 4ac = (-7)^2 - 4 \cdot 1 \cdot 5 = 29$

Since the discriminant is a positive number that is not a perfect square, there are two irrational solutions.

9. $a = 1$, $b = 0$, $c = 3$

$b^2 - 4ac = 0^2 - 4 \cdot 1 \cdot 3 = -12$

Since the discriminant is a negative number, there are two imaginary solutions.

11. $a = 1$, $b = 0$, $c = -5$

$b^2 - 4ac = 0^2 - 4 \cdot 1 \cdot (-5) = 20$

Two irrational solutions.

13. $a = 4$, $b = 8$, $c = -5$

$b^2 - 4ac = 8^2 - 4 \cdot 4 \cdot (-5) = 144$

Since the discriminant is a positive number that is a perfect square, there are two rational solutions.

15. $a = 1$, $b = 4$, $c = 6$

$b^2 - 4ac = 4^2 - 4 \cdot 1 \cdot 6 = -8$

Two imaginary solutions.

17. $a = 9$, $b = -48$, $c = 64$

$b^2 - 4ac = (-48)^2 - 4 \cdot 9 \cdot 64 = 0$

One rational solution.

19. Since $9t^2 - 3t$ is factorable, $3t(3t - 1)$, we know there are two rational solutions.

21. $x^2 + 4x - 8 = 0$ Standard form

$a = 1$, $b = 4$, $c = -8$

$b^2 - 4ac = 4^2 - 4 \cdot 1 \cdot (-8) = 48$

Two irrational solutions.

23. $2a^2 - 3a + 5 = 0$

$a = 2$, $b = -3$, $c = 5$

$b^2 - 4ac = (-3)^2 - 4 \cdot 2 \cdot 5 = -31$

Two imaginary solutions.

25. $7x^2 = 19x \Rightarrow 7x^2 - 19x = 0$

$\Rightarrow a = 7$, $b = -19$, $c = 0$

$\Rightarrow b^2 - 4ac = (-19)^2 - 4 \cdot 7 \cdot 0 = 19^2$

Two rational roots.

27. $y^2 - 4y + \dfrac{9}{4} = 0$

$a = 1$, $b = -4$, $c = \dfrac{9}{4}$

$b^2 - 4ac = (-4)^2 - 4 \cdot 1 \cdot \dfrac{9}{4} = 7$

Two irrational solutions.

29. The solutions are -7, 3

$x = -7$ or $x = 3$

$x + 7 = 0$ or $x - 3 = 0$

$(x + 7)(x - 3) = 0$

$x^2 + 4x - 21 = 0$

31. 3 is the only solution, so $x = 3$ "twice"

$x = 3$

$x - 3 = 0$

$(x - 3)^2 = 0$

$x^2 - 6x + 9 = 0$

Note: This is a perfect-square trinomial.

33. $x = -1$ or $x = -3$

$x + 1 = 0$ or $x + 3 = 0$

$(x + 1)(x + 3) = 0$

$x^2 + 4x + 3 = 0$

35. $x = 5$ or $x = \dfrac{3}{4}$

$x - 5 = 0$ or $x - \dfrac{3}{4} = 0$

$x - 5 = 0$ or $4x - 3 = 0$

$(x - 5)(4x - 3) = 0$

$4x^2 - 23x + 15 = 0$

37. $\left(x + \dfrac{1}{4}\right)\left(x + \dfrac{1}{2}\right) = 0$

$(4x + 1)(2x + 1) = 0$

$8x^2 + 6x + 1 = 0$

39. $(x - 2.4)(x + 0.4) = 0$

$x^2 - 2x - 0.96 = 0$

41. $\left(x + \sqrt{3}\right)\left(x - \sqrt{3}\right) = 0$

$x^2 - 3 = 0$

43. $\left(x-2\sqrt{5}\right)\left(x+2\sqrt{5}\right)=0$

$x^2-\left(2\sqrt{5}\right)^2=0$

$x^2-20=0$

45. $\left(x-4i\right)\left(x+4i\right)=0$

$x^2-\left(4i\right)^2=0$

$x^2-16i^2=0$

$x^2+16=0$

Reminder: $i^2=-1$

47. $x=2-7i \ or \qquad x=2+7i$

$x-2+7i=0 \qquad or \ x-2-7i=0$

$\left[x+\left(-2+7i\right)\right]\left[x+\left(-2-7i\right)\right]=0$

$x^2-2x-7xi-2x+4+14i$

$\qquad\qquad +7xi-14i-49i^2=0$

$x^2-4x+53=0$

49. $x=3-\sqrt{14} \qquad or \ x=3+\sqrt{14}$

$x-3+\sqrt{14}=0 \ or \ x-3-\sqrt{14}=0$

$\left[x-\left(3-\sqrt{14}\right)\right]\left[x-\left(3+\sqrt{14}\right)\right]=0$

$x^2-3x-x\sqrt{14}-3x+9+$

$\qquad 3\sqrt{14}+x\sqrt{14}-3\sqrt{14}-14=0$

$x^2-6x-5=0$

51.

$x=1-\dfrac{\sqrt{21}}{3} \ or \ x=1+\dfrac{\sqrt{21}}{3}$

$3x=3-\sqrt{21} \ or \ 3x=3+\sqrt{21}$

$3x-3+\sqrt{21}=0$

$or \ 3x-3-\sqrt{21}=0$

$\left(3x-3+\sqrt{21}\right)\left(3x-3-\sqrt{21}\right)=0$

$9x^2-9x-3x\sqrt{21}-9x+9+3\sqrt{21}+$

$\qquad\qquad 3x\sqrt{21}-3\sqrt{21}-21=0$

$\qquad\qquad\qquad 9x^2-18x-12=0$

$\qquad\qquad 3\left(3x^2-6x-4\right)=0$

$\qquad\qquad\qquad 3x^2-6x-4=0$

53. $\left(x+2\right)\left(x-1\right)\left(x-5\right)=0$

$\left(x^2+x-2\right)\left(x-5\right)=0$

$x^3+x^2-2x-5x^2-5x+10=0$

$x^3-4x^2-7x+10=0$

55. $\left(x+1\right)\left(x\right)\left(x-3\right)=0$

$\left(x^2+x\right)\left(x-3\right)=0$

$x^3-3x^2+x^2-3x=0$

$x^3-2x^2-3x=0$

57. *Thinking and Writing Exercise.*

59. $\dfrac{c}{d}=c+d \Rightarrow \cancel{d}\cdot\dfrac{c}{\cancel{d}}=d(c+d)$

$\Rightarrow c=cd+d^2 \Rightarrow c-cd=d^2$

$\Rightarrow c(1-d)=d^2 \Rightarrow c=\dfrac{d^2}{1-d}$

61. $x=\dfrac{3}{1-y} \Rightarrow x(1-y)=3 \Rightarrow x-xy=3$

$\Rightarrow x-3=xy \Rightarrow y=\dfrac{x-3}{x}$

63. Let $x=$ Jamal's walking speed in mph, and $y=$ Kade's walking speed in mph. Then:

$x=y+1.5$ (1)

$\dfrac{7}{x}=\dfrac{4}{y}$

or $7y=4x$ (2)

This system can be solved using substitution. Substitute $y+1.5$ for x in (2) and solve for y. Then use (1) and the value of y to find x.

$7y=4x$ (1)

$\Rightarrow 7y=4(y+1.5) \Rightarrow 7y=4y+6$

$\Rightarrow 3y=6 \Rightarrow y=2$ mph

$x=y+1.5$ (1)

$\Rightarrow x=2+1.5=3.5$ mph

Jamal walks at 3.5 mph, and Kade walks at 2 mph.

65. *Thinking and Writing Exercise.*

67. The graph includes the points
$(-3, 0), (0, -3)$, and $(1, 0)$. Substituting in
$y = ax^2 + bx + c$, we have three equations.
$$0 = 9a - 3b + c,$$
$$-3 = \phantom{9a - 3b + {}} c,$$
$$0 = a + b + c$$
The solution of this system of equations is
$a = 1$, $b = 2$, $c = -3$.

69. a) $kx^2 - 2x + k = 0$; one solution is -3
We first find k by substituting -3 for x.
$$k(-3)^2 - 2(-3) + k = 0$$
$$9k + 6 + k = 0$$
$$10k = -6$$
$$k = -\frac{6}{10}$$
$$k = -\frac{3}{5}$$

b) Now substitute $-\dfrac{3}{5}$ for k in the original
equation.
$$-\frac{3}{5}x^2 - 2x + \left(-\frac{3}{5}\right) = 0$$
$$3x^2 + 10x + 3 = 0$$
$$(3x + 1)(x + 3) = 0$$
$$x = -\frac{1}{3} \ or \ x = -3$$
The other solution is $-\dfrac{1}{3}$.

71. a) $x^2 - (6 + 3i)x + k = 0$; one solution is 3.
We first find k by substituting 3 for x.
$$3^2 - (6 + 3i)3 + k = 0$$
$$9 - 18 - 9i + k = 0$$
$$-9 - 9i + k = 0$$
$$k = 9 + 9i$$
b) Now we substitute $9 + 9i$ for k in the
original equation.
$$x^2 - (6 + 3i)x + (9 + 9i) = 0$$
$$x^2 - (6 + 3i)x + 3(3 + 3i) = 0$$
$$[x - (3 + 3i)][x - 3] = 0$$
$$x = 3 + 3i \ or \ x = 3$$
The other solution is $3 + 3i$

73. The solutions of $ax^2 + bx + c = 0$ are
$x = \dfrac{-b \pm \sqrt{b^2 - 4ac}}{2a}$. When there is just one
solution,
$b^2 - 4ac = 0$, so $x = \dfrac{-b \pm 0}{2a} = -\dfrac{b}{2a}$.

75. We substitute $(-3, 0), \left(\dfrac{1}{2}, 0\right)$, and $(0, -12)$ in
$f(x) = ax^2 + bx + c$ and get three equations.
$$0 = 9a - 3b + c,$$
$$0 = \frac{1}{4}a + \frac{1}{2}b + c,$$
$$-12 = c$$
The solution of this system of equations is
$a = 8$, $b = 20$, $c = -12$.

77. The only way an equation can have $\sqrt{2}$ as a
solution is if $x - \sqrt{2}$ is a factor of the factored
form of the equation $a(x - \sqrt{2})(x - k) = 0$. If
we select k so that the second factor is the
conjugate of $x - \sqrt{2}$, and a to be an integer,
then the left side of the equation will have
integer coefficients:
$$a(x - \sqrt{2})(x + \sqrt{2}) = a(x^2 - \sqrt{2}^2) = ax^2 - 2a.$$
Therefore any equation of the form:
$ax^2 - 2a = 0$, where a is an integer will work.
Letting $a = 1$ gives the simplest equation:
$x^2 - 2 = 0$.

79. If $1 - \sqrt{5}$ and $3 + 2i$ are two solutions, then
$1 + \sqrt{5}$ and $3 - 2i$ are also solutions. The
equation of lowest degree that has these
solutions is found as follows:
$$\left[x - (1 - \sqrt{5})\right]\left[x - (1 + \sqrt{5})\right]\left[x - (3 + 2i)\right]$$
$$\left[x - (3 - 2i)\right] = 0$$
$$(x^2 - 2x - 4)(x^2 - 6x + 13) = 0$$
$$x^4 - 8x^3 + 21x^2 - 2x - 52 = 0$$

Exercise Set 8.4

1. ***Familiarize.*** Let r represent the speed and t the time for the first part of the trip.

Trip	Distance	Speed	Time
1st part	120	r	t
2nd part	100	$r-10$	$4-t$

Translate. Using $r = \dfrac{d}{t}$, we get two equations from the table, $r = \dfrac{120}{t}$ and

$$r - 10 = \dfrac{100}{4-t}.$$

Carry out. We substitute $\dfrac{120}{t}$ for r in the second equation and solve for t.

$$\dfrac{120}{t} - 10 = \dfrac{100}{4-t}$$

$$\text{LCD is } t(4-t)$$

$$t(4-t)\left(\dfrac{120}{t} - 10\right) = t(4-t)\cdot\dfrac{100}{4-t}$$

$$120(4-t) - 10t(4-t) = 100t$$

$$480 - 120t - 40t + 10t^2 = 100t$$

$$10t^2 - 260t + 480 = 0$$

$$t^2 - 26t + 48 = 0$$

$$(t-24)(t-2) = 0$$

$$t = 24 \ or \ t = 2$$

Check. Since time cannot be negative (If $t = 24$, $4 - 24 = -20$), we check only 2. If $t = 2$, then $4 - 2 = 2$. The speed of the first part is $\dfrac{120}{2}$, or 60 mph. The speed of the second part is $\dfrac{100}{2}$, or 50 mph. The speed of the second part is 10 mph slower than the first   part. The value checks.
State. The speed of the first part was 60 mph, and the speed of the second part was 50 mph.

3. ***Familiarize.*** Let r represent the speed and t the time of the slower trip.

Trip	Distance	Speed	Time
Slower	200	r	t
Faster	200	$r+10$	$t-1$

Translate. Using $t = \dfrac{d}{r}$, we get two equations from the table, $t = \dfrac{200}{r}$ and

$$t - 1 = \dfrac{200}{r+10}.$$

Carry out. We substitute $\dfrac{200}{r}$ for t in the second equation and solve for r.

$$\dfrac{200}{r} - 1 = \dfrac{200}{r+10}$$

$$\text{LCD is } r(r+10)$$

$$r(r+10)\left(\dfrac{200}{r} - 1\right) = r(r+10)\cdot\dfrac{200}{r+10}$$

$$200(r+10) - r(r+10) = 200r$$

$$200r + 2000 - r^2 - 10r = 200r$$

$$0 = r^2 + 10r - 2000$$

$$0 = (r+50)(r-40)$$

$$r = -50 \ or \ r = 40$$

Check. Since speed cannot be negative, we check only 40 mph. If $r = 40$, then $r + 10 = 50$. The time for the slower trip is $\dfrac{200}{40}$, or 5 hrs. The time for the faster trip is $\dfrac{200}{50}$, or 4 hrs. The faster trip is 1 hr. less. The value checks.
State. The speed is 40 mph.

5. ***Familiarize.*** We let $r =$ the speed and $t =$ the time of the Cessna.

Plane	Distance	Speed	Time
Cessna	600	r	t
Beechcraft	1000	$r+50$	$t+1$

Translate. Using $t = d / r$, we get two equations from the table, $t = \dfrac{600}{r}$ and

$$t + 1 = \dfrac{1000}{r+50}.$$

Carry out. We substitute $\dfrac{600}{r}$ for t in the second equation and solve for r.

$$\frac{600}{r} + 1 = \frac{1000}{r+50}$$

$$\text{LCD is } r(r+50)$$

$$r(r+50)\left(\frac{600}{r}+1\right) = r(r+50)\cdot\frac{1000}{r+50}$$

$$600(r+50) + r(r+50) = 1000r$$

$$600r + 30,000 + r^2 + 50r = 1000r$$

$$r^2 - 350r + 30,000 = 0$$

$$(r-150)(r-200) = 0$$

$$r = 150 \ or \ r = 200$$

Check. If $r = 150$, then the Cessna's time is $\frac{600}{150}$, or 4 hr and the Beechcraft's time is $\frac{1000}{150+50}$ or $\frac{1000}{200}$, or 5 hr. If $r = 200$, then the Cessna's times is $\frac{600}{200}$, or 3 hr and the Beechcraft's time is $\frac{1000}{200+50}$, or $\frac{1000}{250}$, or 4 hr. Since the Beechcraft's time is 1 hr longer in each case, both values check. There are two solutions.

State. The speed of the Cessna is 150 mph and the speed of the Beechcraft is 200 mph; or the speed of the Cessna is 200 mph and the speed of the Beechcraft is 250 mph.

7. **Familiarize.** We let r represent the speed and t the time of the trip to Hillsboro.

Trip	Distance	Speed	Time
To Hillsboro	40	r	t
Return	40	$r-6$	$14-t$

Translate. Using $t = \frac{d}{r}$, we get two equations from the table,

$$t = \frac{40}{r} \text{ and } 14 - t = \frac{40}{r-6}.$$

Carry out. We substitute $\frac{40}{r}$ for t in the second equation and solve for r.

$$14 - \frac{40}{r} = \frac{40}{r-6}$$

$$\text{LCD is } r(r-6)$$

$$r(r-6)\left(14 - \frac{40}{r}\right) = r(r-6)\cdot\frac{40}{r-6}$$

$$14r(r-6) - 40(r-6) = 40r$$

$$14r^2 - 84r - 40r + 240 = 40r$$

$$14r^2 - 164r + 240 = 0$$

$$7r^2 - 82r + 120 = 0$$

$$(7r-12)(r-10) = 0$$

$$r = \frac{12}{7} \ or \ r = 10$$

Check. Since negative speed has no meaning in this problem (If $r = \frac{12}{7}$, then $r - 6 = -\frac{30}{7}$.). we check only 10 mph. If $r = 10$, then the time of the trip to Hillsboro is $\frac{40}{10}$, or 4 hr. The speed of the return trip is $10 - 6$, or 4 mph and the time is $\frac{40}{4}$, or 10 hr. The total time for the round trip is 4 hr + 10 hr, or 14 hr. The value checks.

State. Naoki's speed on the trip to Hillsboro was 10 mph and it was 4 mph on the return trip.

9. **Familiarize.** Let r represent the speed of the boat in still water and let t represent the time of the trip upriver.

Trip	Distance	Speed	Time
Upriver	60	$r-3$	t
Downriver	60	$r+3$	$9-t$

Translate. Using $t = \frac{d}{r}$, we get

$$t = \frac{60}{r-3} \text{ and } 9 - t = \frac{60}{r+3}.$$

Carry out. Substitute $\frac{60}{r-3}$ for t in the second equation and solve for r.

$$9 - \frac{60}{r-3} = \frac{60}{r+3}$$

LCD is $(r-3)(r+3)$

$$(r-3)(r+3)\left(9 - \frac{60}{r-3}\right)$$

$$= (r-3)(r+3) \cdot \frac{60}{r+3}$$

$$9(r-3)(r+3) - 60(r+3) = 60(r-3)$$

$$9r^2 - 81 - 60r - 180 = 60r - 180$$

$$9r^2 - 120r - 81 = 0$$

$$3r^2 - 40r - 27 = 0$$

We use the quadratic formula.

$$r = \frac{-(-40) \pm \sqrt{(-40)^2 - 4 \cdot 3 \cdot (-27)}}{2 \cdot 3}$$

$$r = \frac{40 \pm \sqrt{1924}}{6}$$

$r \approx 14$ or $r \approx -0.6$ Since speed cannot be negative, we check $r \approx 14$.

Check. Using the approximation, the speed upriver is $14 - 3$, or 11, and the time is about $\frac{60}{11}$, or 5.5 hrs. The speed downriver is $14 + 3$, or 17, and the time is about $\frac{60}{17}$, or 3.5. The total time is $5.5 + 3.5$, or 9 hrs. The value checks.

State. The speed of the boat in still water is approximately 14 mph.

11. **Familiarize.** Let x represent the time it takes the spring to fill the pool. Then $x - 6$ represents the time it takes the well to fill the pool. It takes them 4 hr to fill the pool when both wells are working together, so they can fill $\frac{1}{4}$ of the pool in 1 hr. The spring will fill $\frac{1}{x}$ of the pool in 1 hr, and the well will fill $\frac{1}{x-6}$ of the pool in 1 hr.

Translate. We have an equation.

$$\frac{1}{x} + \frac{1}{x-6} = \frac{1}{4}$$

Carry out. We solve the equation.
We multiply by the LCD, $4x(x-6)$.

$$4x(x-6)\left(\frac{1}{x} + \frac{1}{x-6}\right) = 4x(x-6) \cdot \frac{1}{4}$$

$$4(x-6) + 4x = x(x-6)$$

$$4x - 24 + 4x = x^2 - 6x$$

$$0 = x^2 - 14x + 24$$

$$0 = (x-2)(x-12)$$

$x = 2$ or $x = 12$

Check. Since negative time has no meaning in this problem, 2 is not a solution $(2 - 6 = -4)$. We check only 12 hr. This is the time it would take the spring working alone. Then the well would take $12 - 6$, or 6 hr working alone. The well would fill $4\left(\frac{1}{6}\right)$, or $\frac{2}{3}$, of the pool in 4 hr, and the spring would fill $4\left(\frac{1}{12}\right)$, or $\frac{1}{3}$, of the pool in 4 hr. Thus in 4 hr they would fill $\frac{2}{3} + \frac{1}{3}$ of the pool. This is all of it, so the numbers check.

State. It takes the spring, working alone, 12 hr to fill the pool.

13. **Familiarize.** We let r represent Antonio's speed in still water. Then $r - 2$ is the speed upstream and $r + 2$ is the speed downstream. Using $t = \frac{d}{r}$, we let $\frac{1}{r-2}$ represent the time upstream and $\frac{1}{r+2}$ represent the time downstream.

Trip	Distance	Speed	Time
Upriver	1	$r-2$	$\frac{1}{r-2}$
Downriver	1	$r+2$	$\frac{1}{r+2}$

Translate. The time for the round trip is 1 hour. We now have an equation.

$$\frac{1}{r-2} + \frac{1}{r+2} = 1$$

Carry out. We solve the equation. We multiply by the LCD, $(r-2)(r+2)$.

$$(r-2)(r+2)\left(\frac{1}{r-2}+\frac{1}{r+2}\right)$$

$$=(r-2)(r+2)\cdot 1$$

$$(r+2)+(r-2)=(r-2)(r+2)$$

$$2r=r^2-4$$

$$0=r^2-2r-4$$

$$r=\frac{-(-2)\pm\sqrt{(-2)^2-4\cdot 1(-4)}}{2\cdot 1}$$

$$r=\frac{2\pm\sqrt{4+16}}{2}=\frac{2\pm\sqrt{20}}{2}$$

$$r=\frac{2\pm 2\sqrt{5}}{2}=1\pm\sqrt{5}$$

$$1+\sqrt{5}\approx 1+2.236\approx 3.24$$

$$1-\sqrt{5}\approx 1-2.236=-1.24$$

Check. Since negative speed has no meaning in this problem, we check only 3.24 mph. If $r\approx 3.24$, then $r-2\approx 1.24$ and $r+2\approx 5.24$. The time it takes to travel upstream is approx. $\frac{1}{1.24}$, or 0.806 hr, and the time it takes to travel downstream is approx. $\frac{1}{5.24}$, or 0.191 hr. The total time is 0.997 which is approximately 1 hour. The value checks.
State. Antonio's speed in still water is approximately 3.24 mph.

15. $$A=4\pi r^2$$

$$\frac{A}{4\pi}=r^2 \quad \text{Dividing by } 4\pi$$

$$\frac{1}{2}\sqrt{\frac{A}{\pi}}=r \quad \text{Taking the positive square root}$$

17. $$A=2\pi r^2+2\pi rh$$

$$0=2\pi r^2+2\pi rh-A \quad \text{Standard form}$$

$$a=2\pi,\ b=2\pi h,\ c=-A$$

$$r=\frac{-2\pi h\pm\sqrt{(2\pi h)^2-4\cdot 2\pi\cdot(-A)}}{2\cdot 2\pi}$$

Using the quadratic formula

$$r=\frac{-2\pi h\pm\sqrt{4\pi^2 h^2+8\pi A}}{4\pi}$$

$$r=\frac{-2\pi h\pm 2\sqrt{\pi^2 h^2+2\pi A}}{4\pi}$$

$$r=\frac{-\pi h\pm\sqrt{\pi^2 h^2+2\pi A}}{2\pi}$$

Since taking the negative square root would result in a negative answer, we take the positive one.

$$r=\frac{-\pi h+\sqrt{\pi^2 h^2+2\pi A}}{2\pi}$$

19. $$F=\frac{Gm_1 m_2}{r^2}$$

$$Fr^2=Gm_1 m_2$$

$$r^2=\frac{Gm_1 m_2}{F}$$

$$r=\sqrt{\frac{Gm_1 m_2}{F}}$$

21. $$c=\sqrt{gH}\Rightarrow (c)^2=\left(\sqrt{gH}\right)^2\Rightarrow c^2=gH$$

$$\Rightarrow H=\frac{c^2}{g}$$

23. $$w=\frac{lg^2}{800}\Rightarrow\frac{800w}{l}=g^2\Rightarrow\sqrt{g^2}=\sqrt{\frac{800w}{l}}$$

$$|g|=\sqrt{\frac{800w}{l}}. \text{ However, if we assume } g>0,$$

we can drop the absolute value bars and get.

$$g=\sqrt{\frac{800w}{l}}.$$

25. $$a^2+b^2=c^2$$

$$b^2=c^2-a^2$$

$$b=\sqrt{c^2-a^2}$$

27. $s = v_0 t + \dfrac{gt^2}{2}$

$0 = \dfrac{gt^2}{2} + v_0 t - s$

$a = \dfrac{g}{2},\ b = v_0,\ c = -s$

$t = \dfrac{-v_0 \pm \sqrt{v_0^2 - 4\left(\dfrac{g}{2}\right)(-s)}}{2\left(\dfrac{g}{2}\right)}$

$t = \dfrac{-v_0 \pm \sqrt{v_0^2 + 2gs}}{g}$

Since taking the negative square root would result in a negative answer, we take the positive one.

$t = \dfrac{-v_0 + \sqrt{v_0^2 + 2gs}}{g}$

29. $N = \dfrac{1}{2}\left(n^2 - n\right)$

$N = \dfrac{1}{2}n^2 - \dfrac{1}{2}n$

$0 = \dfrac{1}{2}n^2 - \dfrac{1}{2}n - N$

$a = \dfrac{1}{2},\ b = -\dfrac{1}{2},\ c = -N$

$n = \dfrac{-\left(-\dfrac{1}{2}\right) \pm \sqrt{\left(-\dfrac{1}{2}\right)^2 - 4 \cdot \dfrac{1}{2} \cdot (-N)}}{2\left(\dfrac{1}{2}\right)}$

$n = \dfrac{1}{2} \pm \sqrt{\dfrac{1}{4} + 2N}$

$n = \dfrac{1}{2} \pm \sqrt{\dfrac{1 + 8N}{4}}$

$n = \dfrac{1}{2} \pm \dfrac{1}{2}\sqrt{1 + 8N}$

Since taking the negative square root would result in a negative answer, we take the positive one.

$n = \dfrac{1}{2} + \dfrac{1}{2}\sqrt{1 + 8N}$, or $\dfrac{1 + \sqrt{1 + 8N}}{2}$

31. $T = 2\pi\sqrt{\dfrac{l}{g}}$

$\dfrac{T}{2\pi} = \sqrt{\dfrac{l}{g}}$

$\dfrac{T^2}{4\pi^2} = \dfrac{l}{g}$

$gT^2 = 4\pi^2 l$

$g = \dfrac{4\pi^2 l}{T^2}$

33. $at^2 + bt + c = 0$
The quadratic formula gives the result.

$t = \dfrac{-b \pm \sqrt{b^2 - 4ac}}{2a}$

35. a) From example 4, we know

$t = \dfrac{-v_0 + \sqrt{v_0^2 + 19.6s}}{9.8}$

Substituting 500 for s and 0 for v_0, we

have $t = \dfrac{0 + \sqrt{0^2 + 19.6(500)}}{9.8}$

$t \approx 10.1$
It takes about 10.1 sec to reach the ground.

b) $t = \dfrac{-v_0 + \sqrt{v_0^2 + 19.6s}}{9.8}$

Substitute 500 for s and 30 for v_0.

$t = \dfrac{-30 + \sqrt{30^2 + 19.6(500)}}{9.8}$

$t \approx 7.49$
It takes about 7.49 sec to reach the ground.

c) We will use the formula in Example 4,
$s = 4.9t^2 + v_0 t$.

Substitute 5 for t and 30 for v_0.

$s = 4.9(5)^2 + 30(5) = 272.5$
The object will fall 272.5 m.

37. From Example 4, we know

$$t = \frac{-v_0 + \sqrt{v_0^2 + 19.6s}}{9.8}$$

Substituting 40 for s and 0 for v_0, we have

$$= \frac{0 + \sqrt{0^2 + 19.6(40)}}{9.8}$$

$t \approx 2.9$

He will be falling for about 2.9 sec.

39. From Example 3, we know $T = \dfrac{\sqrt{3V}}{12}$

Substituting 44 for V, we have

$$T = \frac{\sqrt{3 \cdot 44}}{12}$$

$T \approx 0.957$

LeBron's hang time is about 0.957 sec.

41. $s = 4.9t^2 + v_0 t$

Solve the formula for v_0

$s - 4.9t^2 = v_0 t$

$\dfrac{s - 4.9t^2}{t} = v_0$

Substitute 51.6 for s and 3 for t.

$$\frac{51.6 - 4.9(3)^2}{3} = v_0$$

$2.5 = v_0$

The initial velocity is 2.5 m/sec.

43. $A = P_1(1+r)^2 + P_2(1+r)$ where A = total amount in the account after 2 years, P is the amount of the original deposit, P_2 is the amount deposited at the beginning of the second year, and r is the annual interest rate. We are given $P_1 = 3000$, $P_2 = 1700$, and $A = 5253.70$.

Substitute to determine r.

$$r = -1 + \frac{-1700 + \sqrt{(1700)^2 + 4(3000)5253.70}}{2(3000)}$$

Using a calculator, we have $r = 0.07$. The annual interest rate is 0.07, or 7%.

45. *Thinking and Writing Exercise.*

47. $(m^{-1})^2 = m^{-1 \cdot 2} = m^{-2} = \dfrac{1}{m^2}$

49. $(y^{1/6})^2 = y^{2(1/6)} = y^{2/6} = y^{1/3}$

51. $t^{-1} = \dfrac{1}{2} \Rightarrow (t^{-1})^{-1} = \left(\dfrac{1}{2}\right)^{-1} \Rightarrow t^{(-1)(-1)} = \left(\dfrac{2}{1}\right)^1$

$\Rightarrow t = 2$

53. *Thinking and Writing Exercise.*

55. $A = 6.5 - \dfrac{20.4t}{t^2 + 36}$

$$(t^2 + 36)A = (t^2 + 36)\left(6.5 - \frac{20.4t}{t^2 + 36}\right)$$

$$At^2 + 36A = (t^2 + 36)(6.5) - (t^2 + 36)\left(\frac{20.4t}{t^2 + 36}\right)$$

$At^2 + 36A = 6.5t^2 + 234 - 20.4t$

$At^2 - 6.5t^2 + 20.4t + 36A - 234 = 0$

$(A - 6.5)t^2 + 20.4t + (36A - 234) = 0$

$a = A - 6.5, \ b = 20.4, \ c = 36A - 234$

$$t = \frac{-20.4 \pm \sqrt{(20.4)^2 - 4(A - 6.5)(36A - 234)}}{2(A - 6.5)}$$

$$t = \frac{-20.4 \pm \sqrt{416.16 - 144A^2 + 1872A - 6084}}{2(A - 6.5)}$$

$$t = \frac{-20.4 \pm \sqrt{-144A^2 + 1872A - 5667.84}}{2(A - 6.5)}$$

$$t = \frac{-20.4 \pm \sqrt{144(-A^2 + 13A - 39.36)}}{2(A - 6.5)}$$

$$t = \frac{-20.4 \pm 12\sqrt{-A^2 + 13A - 39.36}}{2(A - 6.5)}$$

$$t = \frac{2\left(-10.2 \pm 6\sqrt{-A^2 + 13A - 39.36}\right)}{2(A - 6.5)}$$

$$t = \frac{-10.2 \pm 6\sqrt{-A^2 + 13A - 39.36}}{A - 6.5}$$

57. Let a = the number. Then $a - 1$ is 1 less than a and the reciprocal of that number is $\dfrac{1}{a-1}$.

Also 1 more than the number is $a + 1$.

$$\dfrac{1}{(a-1)} = a+1$$

We solve the equation.

$$\dfrac{1}{a-1} = a+1, \text{ LCD is } a-1$$

$$(a-1)\cdot\dfrac{1}{a-1} = (a-1)(a+1)$$

$$1 = a^2 - 1$$

$$2 = a^2$$

$$\pm\sqrt{2} = a$$

Both numbers check. The numbers are $\sqrt{2}$ and $-\sqrt{2}$, or $\pm\sqrt{2}$.

59. $$\dfrac{w}{l} = \dfrac{l}{w+l}$$

$$l(w+l)\cdot\dfrac{w}{l} = l(w+l)\cdot\dfrac{l}{w+l}$$

$$w(w+l) = l^2$$

$$w^2 + lw = l^2$$

$$0 = l^2 - lw - w^2$$

Use the quadratic formula with $a = 1$, $b = -w$, and $c = -w^2$.

$$l = \dfrac{-(-w)\pm\sqrt{(-w)^2 - 4\cdot 1\cdot(-w^2)}}{2\cdot 1}$$

$$l = \dfrac{w\pm\sqrt{w^2+4w^2}}{2} = \dfrac{w\pm\sqrt{5w^2}}{2}$$

$$l = \dfrac{w\pm w\sqrt{5}}{2}$$

Since $\dfrac{w-w\sqrt{5}}{2}$ is negative we use the positive square root: $l = \dfrac{w+w\sqrt{5}}{2}$

61. $mn^4 - r^2pm^3 - r^2n^2 + p = 0$

Let $u = n^2$. Substitute and rearrange.

$$mu^2 - r^2u - r^2pm^3 + p = 0$$

$$a = m,\ b = -r^2,\ c = -r^2pm^3 + p$$

$$u = \dfrac{-(-r^2)\pm\sqrt{(-r^2)^2 - 4\cdot m(-r^2pm^3 + p)}}{2\cdot m}$$

$$u = \dfrac{r^2\pm\sqrt{r^4 + 4m^4r^2p - 4mp}}{2m}$$

$$n^2 = \dfrac{r^2\pm\sqrt{r^4 + 4m^4r^2p - 4mp}}{2m}$$

$$n = \sqrt{\dfrac{r^2\pm\sqrt{r^4 + 4m^4r^2p - 4mp}}{2m}}$$

63. Let s represent a length of a side of the cube, let S represent the surface area of the cube, let A represent the surface area of the sphere. Then the diameter of the sphere is s, so the radius r is $s/2$. From Exercise 15, we know, $A = 4\pi r^2$, so when $r = s/2$ we have $A = 4\pi\left(\dfrac{s}{2}\right)^2 = 4\pi\cdot\dfrac{s^2}{4} = \pi s^2$. From the formula for the surface area of a cube (See Exercise 16.) we know that $S = 6s^2$, so $\dfrac{S}{6} = s^2$ and then $A = \pi\cdot\dfrac{S}{6}$, or $A(S) = \dfrac{\pi S}{6}$.

Exercise Set 8.5

1. $x^6 = (x^3)^2$. Let $u = x^3$: f

3. $x^8 = (x^4)^2$. Let $u = x^4$: h

5. $x^{4/3} = (x^{2/3})^2$. Let $u = x^{2/3}$: g

7. $x^{-4/3} = (x^{-2/3})^2$. Let $u = x^{-2/3}$: e

9. $3p - 4\sqrt{p} + 6 = 3\left(\sqrt{p}\right)^2 - 4\sqrt{p} + 6$.

Let $u = \sqrt{p}$.

11. $(x^2 + 3)^2 + (x^2 + 3) - 7$

Let $u = x^2 + 3$.

13. $(1+t)^4 + (1+t)^2 + 4 = [(1+t)^2]^2 + (1+t)^2 + 4$

Let $u = (1+t)^2$.

15. $x^4 - 5x^2 + 4 = 0$

  Let $u = x^2$ and $u^2 = x^4$.

  $u^2 - 5u + 4 = 0$ Substituting u for x^2

  $(u - 1)(u - 4) = 0$

  $u - 1 = 0 \ or \ u - 4 = 0$

  $u = 1 \ or \qquad u = 4$

  Now replace u with x^2 and solve these equations.

  $x^2 = 1 \ or \ x^2 = 4$

  $x = \pm 1 \ or \ x = \pm 2$

  The numbers 1, –1, 2, and –2 check. They are the solutions.

17. $x^4 - 9x^2 + 20 = 0$

  Let $u = x^2$ and $u^2 = x^4$.

  $u^2 - 9u + 20 = 0$

  $(u - 4)(u - 5) = 0$

  $u = 4 \ or \ u = 5$

  $x^2 = 4 \ or \ x^2 = 5$

  $x = \pm 2 \ or \ x = \pm \sqrt{5}$

  All four numbers check.

19. $4t^4 - 19t^2 + 12 = 0$

  Let $u = t^2$ and $u^2 = t^4$.

  $4u^2 - 19u + 12 = 0$ Substituting

  $(4u - 3)(u - 4) = 0$

  $4u - 3 = 0 \ or \ u - 4 = 0$

  $u = \dfrac{3}{4} \ or \qquad u = 4$

  Now replace u with t^2 and solve these equations:

  $t^2 = \dfrac{3}{4} \quad or \ t^2 = 4$

  $t = \pm \dfrac{\sqrt{3}}{2} \ or \ t = \pm 2$

  The numbers $\dfrac{\sqrt{3}}{2}$, $-\dfrac{\sqrt{3}}{2}$, 2, and –2 check.

  They are the solutions.

21. $w + 4\sqrt{w} - 12 = 0 \Rightarrow \left(\sqrt{w}\right)^2 + 4\sqrt{w} - 12 = 0$

  $\Rightarrow u^2 + 4u - 12 = 0$ for $u = \sqrt{w}$

  $\Rightarrow (u + 6)(u - 2) = 0 \Rightarrow u = -6, \ or \ u = 2$

  $\Rightarrow \sqrt{w} = -6 \Rightarrow$ no solution

  $or \ \sqrt{w} = 2 \Rightarrow w = 2^2 = 4$

23. $\left(x^2 - 7\right)^2 - 3\left(x^2 - 7\right) + 2 = 0$

  Let $u = x^2 - 7$ and $u^2 = \left(x^2 - 7\right)^2$.

  $u^2 - 3u + 2 = 0$ Substituting

  $(u - 1)(u - 2) = 0$

  $u = 1 \ or \qquad u = 2$

  $x^2 - 7 = 1 \ or \ x^2 - 7 = 2 \quad$ Replacing u with $x^2 - 7$

  $x^2 = 8 \qquad or \qquad x^2 = 9$

  $x = \pm\sqrt{8} \ or \qquad x = \pm\sqrt{9}$

  $x = \pm 2\sqrt{2} \ or \qquad x = \pm 3$

  The numbers $2\sqrt{2}, \ -2\sqrt{2},$ 3, and –3 check. They are the solutions.

25. $r - 2\sqrt{r} - 6 = 0$

  Let $u = \sqrt{r}$ and $u^2 = r$

  $u^2 - 2u - 6 = 0$

  $u = \dfrac{-(-2) \pm \sqrt{(-2)^2 - 4 \cdot 1 \cdot (-6)}}{2 \cdot 1}$

  $u = \dfrac{2 \pm \sqrt{28}}{2} = \dfrac{2 \pm 2\sqrt{7}}{2}$

  $u = 1 \pm \sqrt{7}$

  $\sqrt{r} = 1 + \sqrt{7} \qquad or \ \sqrt{r} = 1 - \sqrt{7}$

  $r = 1 + 2\sqrt{7} + 7$ No solution: $1 - \sqrt{7} < 0$

  $r = 8 + 2\sqrt{7}$

  The number $8 + 2\sqrt{7}$ checks and is the solution.

27. $\left(1+\sqrt{x}\right)^2 + 5\left(1+\sqrt{x}\right) + 6 = 0$

Let $u = 1+\sqrt{x}$ and $u^2 = \left(1+\sqrt{x}\right)^2$.

$u^2 + 5u + 6 = 0$ Substituting

$(u+3)(u+2) = 0$

$u = -3 \ or \quad u = -2$

$1+\sqrt{x} = -3 \ or \ 1+\sqrt{x} = -2$ $\quad$ Replacing u with $1+\sqrt{x}$

$\sqrt{x} = -4 \ or \quad \sqrt{x} = -3$

Since the principal square root cannot be negative, this equation has no solution.

29. $x^{-2} - x^{-1} - 6 = 0$

Let $u = x^{-1}$ and $u^2 = x^{-2}$.

$u^2 - u - 6 = 0$ Substituting

$(u-3)(u+2) = 0$

$u = 3 \ or \ u = -2$

Now we replace u with x^{-1} and solve these equations.

$x^{-1} = 3 \ or \quad x^{-1} = -2$

$\dfrac{1}{x} = 3 \ or \quad \dfrac{1}{x} = -2$

$\dfrac{1}{3} = x \ or \quad -\dfrac{1}{2} = x$

Both $\dfrac{1}{3}$ and $-\dfrac{1}{2}$ check. They are the solutions.

31. $4y^{-2} - 3y^{-1} - 1 = 0$

Let $u = y^{-1}$, then $u^2 = (y^{-1})^2 = y^{-2}$.

$4u^2 - 3u - 1 = 0$

$(4u+1)(u-1) = 0$

$u = -\dfrac{1}{4} \ or \ u = 1$

Replace u with y^{-1} and solve the equations.

$y^{-1} = -\dfrac{1}{4} \ or \ y^{-1} = 1$

$\dfrac{1}{y} = -\dfrac{1}{4} \ or \ \dfrac{1}{y} = 1$

$y = -4 \ or \quad y = 1$

33. $t^{2/3} + t^{1/3} - 6 = 0$

Let $u = t^{1/3}$ and $u^2 = t^{2/3}$.

$u^2 + u - 6 = 0$ Substituting

$(u+3)(u-2) = 0$

$u = -3 \ or \ u = 2$

Now we replace u with $t^{1/3}$ and solve these equations.

$t^{1/3} = -3 \quad or \ t^{1/3} = 2$

$t = (-3)^3 \ or \quad t = 2^3$

$t = -27 \ or \quad t = 8$

Both –27 and 8 check. They are the solutions.

35. $y^{1/3} - y^{1/6} - 6 = 0$

Let $u = y^{1/6}$ and $u^2 = y^{1/3}$.

$u^2 - u - 6 = 0$ Substituting

$(u-3)(u+2) = 0$

$u = 3 \ or \ u = -2$

Now we replace u with $y^{1/6}$ and solve these equations.

$y^{1/6} = 3 \ or \ y^{1/6} = -2$

$\sqrt[6]{y} = 3 \ or \ \sqrt[6]{y} = -2$

$y = 3^6$ $\quad$ This equation has no solution since principal

$y = 729$ sixth roots are never negative.

The number 729 checks and is the solution.

37. $\quad t^{1/3} + 2t^{1/6} = 3$

$t^{1/3} + 2t^{1/6} - 3 = 0$

Let $u = t^{1/6}$ and $u^2 = t^{2/6} = t^{1/3}$.

$u^2 + 2u - 3 = 0$ Substituting

$(u+3)(u-1) = 0$

$u = -3 \ or \quad u = 1$

$t^{1/6} = -3 \ or \ t^{1/6} = 1$ Substituting $t^{1/6}$ for u

No solution $\quad t = 1$

The number 1 checks and is the solution.

39. $\left(10-\sqrt{x}\right)^2 - 2\left(10-\sqrt{x}\right) - 35 = 0$

Let $u = 10-\sqrt{x}$, then $u^2 = \left(10-\sqrt{x}\right)^2$.

$u^2 - 2u - 35 = 0 \Rightarrow (u-7)(u+5) = 0$

$\Rightarrow u = 7$ or $u = -5$

$\Rightarrow 10-\sqrt{x} = 7 \Rightarrow \sqrt{x} = 3 \Rightarrow x = 9$

or $10-\sqrt{x} = -5 \Rightarrow \sqrt{x} = 15 \Rightarrow x = 225$

The solutions are $x = 9$ and $x = 225$.

41. $16\left(\dfrac{x-1}{x-8}\right)^2 + 8\left(\dfrac{x-1}{x-8}\right) + 1 = 0$

Let $u = \dfrac{x-1}{x-8}$ and $u^2 = \left(\dfrac{x-1}{x-8}\right)^2$.

$16u^2 + 8u + 1 = 0$ Substituting

$(4u+1)(4u+1) = 0$

$$u = -\dfrac{1}{4}$$

Now we replace u with $\dfrac{x-1}{x-8}$ and solve this equation:

$\dfrac{x-1}{x-8} = -\dfrac{1}{4}$

$4x - 4 = -x + 8$ Multiplying by $4(x-8)$

$5x = 12$

$x = \dfrac{12}{5}$

The number $\dfrac{12}{5}$ checks and is the solution.

43. $x^4 + 5x^2 - 36 = 0$

$\Rightarrow u^2 + 5u - 36 = 0$ for $u = x^2$

$\Rightarrow (u+9)(u-4) = 0$

$\Rightarrow u = -9 \Rightarrow x^2 = -9 \Rightarrow x = \pm\sqrt{-9} = \pm 3i$

or $u = 4 \Rightarrow x^2 = 4 \Rightarrow x = \pm\sqrt{4} = \pm 2$

The solutions are $x = -2,\ 2,\ 3i,$ and $-3i$

45. $(n^2+6)^2 - 7(n^2+6) + 10 = 0$

$\Rightarrow u^2 - 7u + 10 = 0$ for $u = n^2 + 6$

$\Rightarrow (u-5)(u-2) = 0$

$\Rightarrow u = 5 \Rightarrow n^2 + 6 = 5 \Rightarrow n^2 = -1$

$\Rightarrow n = \pm\sqrt{-1} = \pm i$

or $u = 2 \Rightarrow n^2 + 6 = 2 \Rightarrow n^2 = -4$

$\Rightarrow n = \pm\sqrt{-4} = \pm 2i$

The solutions are $n = -i,\ i,\ -2i,$ and $2i$

47. The x-intercepts occur where $f(x) = 0$.

Thus, we must have $5x + 13\sqrt{x} - 6 = 0$.

Let $u = \sqrt{x}$ and $u^2 = x$.

$5u^2 + 13u - 6 = 0$

$(5u - 2)(u + 3) = 0$

$u = \dfrac{2}{5}$ or $u = -3$

Now replace u with $\sqrt{x}$ and solve these equations:

$\sqrt{x} = \dfrac{2}{5}$ or $\sqrt{x} = -3$

$x = \dfrac{4}{25}$ No Solution

The number $\dfrac{4}{25}$ checks. Thus, the

x-intercept is $\left(\dfrac{4}{25}, 0\right)$.

49. Solve: $\left(x^2 - 3x\right)^2 - 10\left(x^2 - 3x\right) + 24 = 0$

Let $u = x^2 - 3x$ and $u^2 = \left(x^2 - 3x\right)^2$.

$u^2 - 10u + 24 = 0$

$(u - 6)(u - 4) = 0$

$u = 6$ or $u = 4$

Now replace u with $x^2 - 3x$ and solve these equations:

$x^2 - 3x = 6$ or $x^2 - 3x = 4$

$x^2 - 3x - 6 = 0$ or $x^2 - 3x - 4 = 0$

$x = \dfrac{-(-3) \pm \sqrt{(-3)^2 - 4(1)(-6)}}{2 \cdot 1}$ or

$(x - 4)(x + 1) = 0$

$x = \dfrac{3 \pm \sqrt{33}}{2}$ or $x = 4$ or $x = -1$

All four numbers check. Thus, the x-intercepts are $\left(\dfrac{3+\sqrt{33}}{2},0\right)$, $\left(\dfrac{3-\sqrt{33}}{2},0\right)$, $(4,0)$, and $(-1,0)$

51. Solve: $x^{2/5}+x^{1/5}-6=0$

Let $u=x^{1/5}$ and $u^2=x^{2/5}$

$u^2+u-6=0$ Substituting

$(u+3)(u-2)=0$

$u=-3 \quad or \quad u=2$

$x^{1/5}=-3 \quad or \quad x^{1/5}=2$ Replacing u with $x^{1/5}$

$x=-243 \; or \quad x=32$ Raising to the fifth power

Both -243 and 32 check. Thus, the x-intercepts are $(-243,0)$ and $(32,0)$.

53. $f(x)=\left(\dfrac{x^2+2}{x}\right)^4+7\left(\dfrac{x^2+2}{x}\right)^2+5$

Observe that, for all real numbers x, each term is positive. Thus, there are no real-number values of x for which $f(x)=0$ and hence no x-intercepts.

55. *Thinking and Writing Exercise.*

57. $f(x)=x$

The function is linear. The graph is a line that goes through the points:

$(0,f(0))=(0,0)$ and $(2,f(2))=(2,2)$

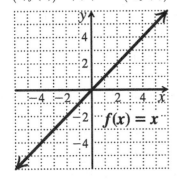

59. $h(x)=x-2$

The function is linear. The graph is a line that goes through the points:

$(0,f(0))=(0,-2)$ and $(2,f(2))=(2,0)$

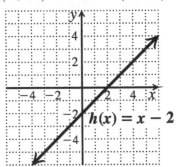

61. $g(x)=x^2+2$

The function is quadratic and the leading coefficient is positive. The graph is a parabola that opens upward with vertex at

$$\left(-\dfrac{b}{2a},f\left(-\dfrac{b}{2a}\right)\right)=\left(-\dfrac{0}{2\cdot1},f\left(-\dfrac{0}{2\cdot1}\right)\right)=(0,2)$$

and that goes through the points:

$(-2,f(-2))=(-2,6)$, $(2,f(2))=(2,6)$,

$(-1,f(-1))=(-1,3)$, and $(1,f(1))=(1,3)$

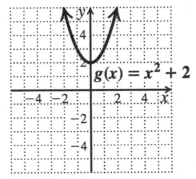

63. *Thinking and Writing Exercise.*

65. $5x^4 - 7x^2 + 1 = 0$

Let $u = x^2$ and $u^2 = x^4$

$5u^2 - 7u + 1 = 0$ Substituting

$$u = \frac{-(-7) \pm \sqrt{(-7)^2 - 4 \cdot 5 \cdot 1}}{2 \cdot 5}$$

$$u = \frac{7 \pm \sqrt{29}}{10}$$

$$x^2 = \frac{7 \pm \sqrt{29}}{10} \quad \text{Replacing } u \text{ with } x^2$$

$$x = \pm \sqrt{\frac{7 \pm \sqrt{29}}{10}}$$

All four numbers check and are the solutions.

67. $\left(x^2 - 4x - 2\right)^2 - 13\left(x^2 - 4x - 2\right) + 30 = 0$

Let $u = x^2 - 4x - 2$ and $u^2 = \left(x^2 - 4x - 2\right)^2$

$u^2 - 13u + 30 = 0$ Substituting

$(u - 3)(u - 10) = 0$

$\qquad u = 3 \ or \qquad\qquad u = 10$

$x^2 - 4x - 2 = 3 \ or \quad x^2 - 4x - 2 = 10$

$\qquad\qquad\qquad$ Replacing u with $x^2 - 4x - 2$

$x^2 - 4x - 5 = 0 \ or \quad x^2 - 4x - 12 = 0$

$(x - 5)(x + 1) = 0 \ or \ (x - 6)(x + 2) = 0$

$x = 5 \ or \ x = -1 \ or \ x = 6 \ or \ x = -2$

All four numbers check and are the solutions.

69. $\dfrac{x}{x-1} - 6\sqrt{\dfrac{x}{x-1}} - 40 = 0$

Let $u = \sqrt{\dfrac{x}{x-1}}$ and $u^2 = \dfrac{x}{x-1}$.

$u^2 - 6u - 40 = 0$ Substituting

$(u - 10)(u + 4)$

$\qquad u = 10 \quad or \qquad u = -4$

$\sqrt{\dfrac{x}{x-1}} = 10 \quad or \quad \sqrt{\dfrac{x}{x-1}} = -4$

$\dfrac{x}{x-1} = 100 \ or \quad \text{No solution}$

$\qquad x = 100x - 100$

$\qquad 100 = 99x$

$\qquad \dfrac{100}{99} = x$

The number $\dfrac{100}{99}$ checks. It is the solution.

71. $a^5\left(a^2 - 25\right) + 13a^3\left(25 - a^2\right) + 36a\left(a^2 - 25\right) = 0$

$a^5\left(a^2 - 25\right) - 13a^3\left(a^2 - 25\right) + 36a\left(a^2 - 25\right) = 0$

$a\left(a^2 - 25\right)\left(a^4 - 13a^2 + 36\right) = 0$

$a\left(a^2 - 25\right)\left(a^2 - 4\right)\left(a^2 - 9\right) = 0$

either $\quad a = 0,$

or $a^2 - 25 = 0 \qquad a^2 = 25, \quad a = \pm\sqrt{25} = \pm 5$

or $\quad a^2 - 4 = 0, \qquad a^2 = 4, \qquad a = \pm\sqrt{4} = \pm 2$

or $\quad a^2 - 9 = 0, \qquad a^2 = 9, \qquad a = \pm\sqrt{9} = \pm 3$

All seven numbers check. The solutions are $0, 5, -5, 2, -2, 3,$ and $-3.$

73. $x^6 - 28x^3 + 27 = 0$

Let $u = x^3$.

$u^2 - 28u + 27 = 0$

$(u - 27)(u - 1) = 0$

$u = 27 \ or \ u = 1$

$x^3 = 27 \ or \ x^3 = 1$

$\left(x^3 - 27\right) = (x - 3)\left(x^2 + 3x + 9\right)$

$\left(x^3 - 1\right) = (x - 1)\left(x^2 + x + 1\right)$

Using the quadratic formula on $x^2 + 3x + 9$ and $x^2 + x + 1$ gives

$-\dfrac{1}{2} \pm \dfrac{\sqrt{3}}{2}i$ and $-\dfrac{3}{2} \pm \dfrac{3\sqrt{3}}{2}i$, 1 and 3.

Mid-Chapter Review

Guided Solutions

1. $(x - 7)^2 = 5$

$\qquad x - 7 = \pm\sqrt{5}$

$\qquad\quad x = 7 \pm \sqrt{5}$

The solutions re $7 + \sqrt{5}$ and $7 - \sqrt{5}$.

2. $x^2 - 2x - 1 = 0$

$a = 1,\ b = -2,\ c = -1$

$$x = \frac{-(-2) \pm \sqrt{(-2)^2 - 4 \cdot 1 \cdot (-1)}}{2 \cdot 1}$$

$$= \frac{2 \pm \sqrt{8}}{2}$$

$$= \frac{2}{2} + \frac{2\sqrt{2}}{2}$$

$$= 1 \pm \sqrt{2}$$

The solutions are $1 + \sqrt{2}$ and $1 - \sqrt{2}$.

Mixed Review

1. $x^2 - 3x - 10 = 0 \Rightarrow (x - 5)(x + 2) = 0$

$\Rightarrow x - 5 = 0 \Rightarrow x = 5$

or $x + 2 = 0 \Rightarrow x = -2$

2. $x^2 = 121 \Rightarrow x = \pm\sqrt{121} = \pm 11$

3. $x^2 + 6x = 10 \Rightarrow x^2 + 6x - 10 = 0$

$\Rightarrow a = 1,\ b = 6,\ c = -10$

$$x = \frac{-6 \pm \sqrt{6^2 - 4 \cdot 1 \cdot (-10)}}{2 \cdot 1} = \frac{-6 \pm \sqrt{36 + 40}}{2 \cdot 1}$$

$$= \frac{-6 \pm \sqrt{76}}{2} = \frac{-6 \pm \sqrt{4 \cdot 19}}{2} = \frac{-6 \pm 2\sqrt{19}}{2}$$

$$= \frac{2(-3 \pm \sqrt{19})}{2} = -3 \pm \sqrt{19}$$

4. $x^2 + x - 3 = 0 \Rightarrow a = 1,\ b = 1,\ c = -3$

$$x = \frac{-1 \pm \sqrt{1^2 - 4 \cdot 1 \cdot (-3)}}{2 \cdot 1} = \frac{-1 \pm \sqrt{1 + 12}}{2}$$

$$= \frac{-1 \pm \sqrt{13}}{2} = -\frac{1}{2} \pm \frac{\sqrt{13}}{2}$$

5. $(x + 1)^2 = 2 \Rightarrow x + 1 = \pm\sqrt{2}$

$\Rightarrow x = -1 \pm \sqrt{2}$

6. $x^2 - 10x + 25 = 0 \Rightarrow (x - 5)^2 = 0$

$\Rightarrow x - 5 = 0 \Rightarrow x = 5$

7. $4t^2 = 11 \Rightarrow t^2 = \frac{11}{4} \Rightarrow t = \pm\sqrt{\frac{11}{4}} = \pm\frac{\sqrt{11}}{2}$

8. $2t^2 + 1 = 3t \Rightarrow 2t^2 - 3t + 1 = 0$

$\Rightarrow (2t - 1)(t - 1) = 0$

$\Rightarrow 2t - 1 = 0 \Rightarrow 2t = 1 \Rightarrow t = \frac{1}{2}$

or $t - 1 = 0 \Rightarrow t = 1$

9. $16c^2 = 7c \Rightarrow 16c^2 - 7c = 0 \Rightarrow c(16c - 7) = 0$

$\Rightarrow c = 0$

or $16c - 7 = 0 \Rightarrow 16c = 7 \Rightarrow c = \frac{7}{16}$

10. $y^2 - 2y + 8 = 0 \Rightarrow a = 1,\ b = -2,\ c = 8$

$$y = \frac{-(-2) \pm \sqrt{(-2)^2 - 4 \cdot 1 \cdot 8}}{2 \cdot 1} = \frac{2 \pm \sqrt{4 - 32}}{2}$$

$$= \frac{2 \pm \sqrt{-28}}{2} = \frac{2 \pm \sqrt{28}i}{2} = \frac{2 \pm 2\sqrt{7}i}{2}$$

$$= \frac{2(1 + \sqrt{7}i)}{2} = 1 \pm \sqrt{7}i$$

11. $x^4 - 10x^2 + 9 = 0$

$\Rightarrow u^2 - 10u + 9 = 0$ for $u = x^2$

$\Rightarrow (u - 9)(u - 1) = 0$

$\Rightarrow u = 9 \Rightarrow x^2 = 9 \Rightarrow x = \pm\sqrt{9} = \pm 3$

or $u = 1 \Rightarrow x^2 = 1 \Rightarrow x = \pm\sqrt{1} = \pm 1$

12. $x^4 - 8x^2 - 9 = 0 \Rightarrow u^2 - 8u - 9 = 0$ for $u = x^2$

$\Rightarrow (u - 9)(u + 1) = 0$

$\Rightarrow u = 9 \Rightarrow x^2 = 9 \Rightarrow x = \pm\sqrt{9} = \pm 3$

or $u = -1 \Rightarrow x^2 = -1 \Rightarrow x = \pm\sqrt{-1} = \pm i$

13. $(t + 4)(t - 3) = 18$

$\Rightarrow t^2 + 4t - 3t - 12 = 18$

$\Rightarrow t^2 + t - 30 = 0$

$\Rightarrow (t + 6)(t - 5) = 0$

$\Rightarrow t + 6 = 0 \Rightarrow t = -6$

or $t - 5 = 0 \Rightarrow t = 5$

14. $m^{-4} - 5m^{-2} + 6 = 0$

$\Rightarrow u^2 - 5u + 6 = 0$ for $u = m^{-2}$

$\Rightarrow (u - 3)(u - 2) = 0$

$\Rightarrow u = 3 \Rightarrow m^{-2} = 3 \Rightarrow \dfrac{1}{m^2} = 3 \Rightarrow m^2 = \dfrac{1}{3}$

$\Rightarrow m = \pm\sqrt{\dfrac{1}{3}} = \pm\dfrac{\sqrt{3}}{3}$

or $u = 2 \Rightarrow m^{-2} = 2 \Rightarrow \dfrac{1}{m^2} = 2 \Rightarrow m^2 = \dfrac{1}{2}$

$\Rightarrow m = \pm\sqrt{\dfrac{1}{2}} = \pm\dfrac{\sqrt{2}}{2}$

15. $x^2 - 8x + 1 = 0$

$\Rightarrow b^2 - 4ac = (-8)^2 - 4 \cdot 1 \cdot 1 = 60$

The coefficients of the quadratic equation are integers, the discriminant is positive, but not a perfect square. There are two irrational, real solutions.

16. $3x^2 = 4x + 7 \Rightarrow 3x^2 - 4x - 7 = 0$

$\Rightarrow b^2 - 4ac = (-4)^2 - 4 \cdot 3 \cdot (-7) = 100$

The coefficients of the quadratic equation are integers, the discriminant is a positive perfect square. There are two rational, real solutions.

17. $5x^2 - x + 6 = 0$

$\Rightarrow b^2 - 4ac = (-1)^2 - 4 \cdot 5 \cdot 6 = -119$

The coefficients of the quadratic equation are integers, the discriminant is negative. There are two imaginary solutions.

18. $F = \dfrac{Av^2}{400} \Rightarrow 400 \cdot F = \cancel{400} \cdot \dfrac{Av^2}{\cancel{400}}$

$\Rightarrow 400F = Av^2 \Rightarrow \dfrac{400F}{A} = \dfrac{\cancel{A}v^2}{\cancel{A}}$

$\Rightarrow v^2 = \dfrac{400F}{A} \Rightarrow \sqrt{v^2} = \sqrt{\dfrac{400F}{A}}$

$v = \sqrt{\dfrac{400F}{A}}$ or $20\sqrt{\dfrac{F}{A}}$

Since we are assuming all variables are ≥ 0, the negative square root is not shown.

19. $D^2 - 2Dd - 2hd = 0$

$\Rightarrow a = 1,\ b = -2d,\ c = -2hd$

$D = \dfrac{-(-2d) \pm \sqrt{(-2d)^2 - 4 \cdot 1 \cdot (-2hd)}}{2 \cdot 1}$

$= \dfrac{2d \pm \sqrt{4d^2 + 8hd}}{2} = \dfrac{2d \pm \sqrt{4(d^2 + 2hd)}}{2}$

$= \dfrac{2d \pm 2\sqrt{d^2 + 2hd}}{2} = \dfrac{\cancel{2}(d \pm \sqrt{d^2 + 2hd})}{\cancel{2}}$

$= d \pm \sqrt{d^2 + 2hd}$

Since $d^2 + 2hd \geq d$ when $h, d \geq 0$, we discard the difference and give the solution as

$D = d + \sqrt{d^2 + 2hd}$

20. Let x = Sophie's speed on the journey south, and y = Sophie's speed on the way back. Then:

$y = x - 30 \qquad (1)$

Also, $d = rt \Rightarrow t = \dfrac{d}{r}$

and $t_{SOUTH} + t_{BACK} = 8$

$\dfrac{225}{x} + \dfrac{225}{y} = 8 \quad (2)$

Substitute $x - 30$ for y in (2) and solve for x.

$\dfrac{225}{x} + \dfrac{225}{y} = 8 \quad (2)$

$\Rightarrow \dfrac{225}{x} + \dfrac{225}{x - 30} = 8$

$\Rightarrow x(x - 30) \cdot \left[\dfrac{225}{x} + \dfrac{225}{x - 30} \right] = x(x - 30) \cdot 8$

$\Rightarrow 225(x - 30) + 225x = 8x(x - 30)$

$\Rightarrow 225x - 6750 + 225x = 8x^2 - 240x$

$\Rightarrow 8x^2 - 690x + 6750 = 0$

$\Rightarrow 4x^2 - 345x + 3375 = 0$

$\Rightarrow a = 4,\ b = -345,\ c = 3375$

$x = \dfrac{-(-345) \pm \sqrt{(-345)^2 - 4 \cdot 4 \cdot 3375}}{2 \cdot 4}$

$= \dfrac{345 \pm \sqrt{119{,}025 - 54{,}000}}{8}$

$= \dfrac{345 \pm \sqrt{65{,}025}}{8}$

$= \dfrac{345 \pm 255}{8} = 75$ or $\cancel{11.25}$

We can rule out 11.25 since we know that x must be greater than 30 for y to come out positive. Thus Sophie drove south at 75 mph and back at $75 - 30 = 45$ mph.

Exercise Set 8.6

1. The quadratic expression is in vertex form. The vertex is located at $(1,3)$ and the leading coefficient is positive, so the parabola opens upward : h

3. The quadratic expression is in vertex form. The vertex is located at $(-1,3)$ and the leading coefficient is positive, so the parabola opens upward : f

5. The quadratic expression is in vertex form. The vertex is located at $(-1,3)$ and the leading coefficient is negative, so the parabola opens downward: b

7. The quadratic expression is in vertex form. The vertex is located at $(-1,-3)$ and the leading coefficient is positive, so the parabola opens upward : e

9. a) The parabola opens upward, so a is positive.
   b) The vertex is at $(3,1)$.
   c) The axis of symmetry is $x = 3$.
   d) The range is $[1,\infty)$.

11. a) The parabola opens downward, so a is negative.
   b) The vertex is $(-2,-3)$
   c) The axis of symmetry is $x = -2$
   d) The range is $(-\infty,-3]$

13. a) The parabola opens upward, so a is positive.
   b) The vertex is at $(-3,0)$
   c) The axis of symmetry is $x = -3$
   d) The range is $[0,\infty)$

15.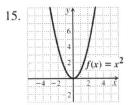

17.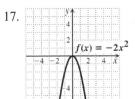

19.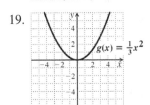

21.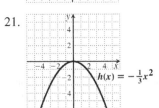

23.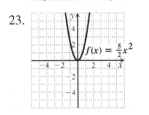

25. $g(x) = (x+1)^2 = [x-(-1)]^2$

   Vertex: $(-1,0)$, Axis of symmetry: $x = -1$

   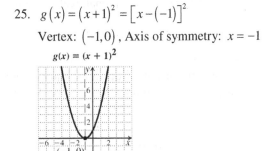

27. $f(x) = (x-2)^2$

   Vertex: $(2,0)$, Axis of symmetry: $x = 2$

   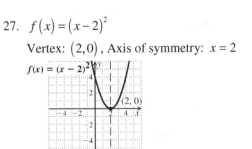

29. $f(x) = -(x+1)^2 = -\left[x-(-1)\right]^2$

Vertex: $(-1,0)$, Axis of symmetry: $x = -1$

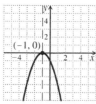

31. $g(x) = -(x-2)^2$

Vertex: $(2,0)$, Axis of symmetry: $x = 2$

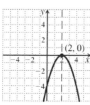

33. $f(x) = 2(x+1)^2 = 2\left[x-(-1)\right]^2$

Vertex: $(-1,0)$, Axis of symmetry: $x = -1$

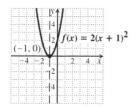

35. $g(x) = 3(x-4)^2$

Vertex: $(4,0)$, Axis of symmetry: $x = 4$

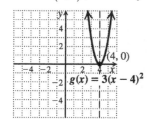

37. $h(x) = -\dfrac{1}{2}(x-4)^2$

Vertex: $(4,0)$, Axis of symmetry: $x = 4$

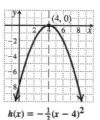

39. $f(x) = \dfrac{1}{2}(x-1)^2$

Vertex: $(1,0)$, Axis of symmetry: $x = 1$

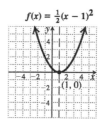

41. $f(x) = -2(x+5)^2 = -2\left[x-(-5)\right]^2$

Vertex: $(-5,0)$, Axis of symmetry: $x = -5$

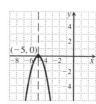

43. $h(x) = -3\left(x-\dfrac{1}{2}\right)^2$

Vertex: $\left(\dfrac{1}{2},0\right)$, Axis of symmetry: $x = \dfrac{1}{2}$

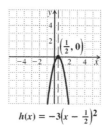

45. $f(x) = (x-5)^2 + 2$

     $a > 0$, so the function takes on a minimum at $x = 5$, of $f(5) = 2$, the y value of the vertex.

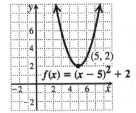

     The range of $f(x)$ is $[2, \infty)$.

47. $f(x) = -(x+2)^2 - 1$

     $a < 0$, so the function takes on a maximum at $x = -2$, of $f(-2) = -1$, the y value of the vertex.

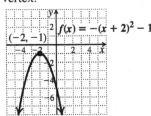

     The range of $f(x)$ is $(-\infty, -1]$.

49. $g(x) = \dfrac{1}{2}(x+4)^2 + 3$

     $a > 0$, so the function takes on a minimum at $x = -4$, of $g(-4) = 3$, the y value of the vertex.

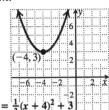

     The range of $g(x)$ is $[3, \infty)$.

51. $h(x) = -2(x-1)^2 - 3$

     $a < 0$, so the function takes on a maximum at $x = 1$, of $h(1) = -3$, the y value of the vertex.

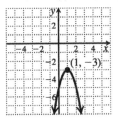

     The range of $h(x)$ is $(-\infty, -3]$.

53. $f(x) = (x+1)^2 - 3 = \left[x - (-1)\right]^2 + (-3)$

     Vertex: $(-1, -3)$

     Axis of symmetry: $x = -1$

     Minimum: -3

     Range: $[-3, \infty)$

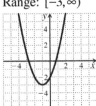

55. $g(x) = -(x+3)^2 + 5$

     Vertex: $(-3, 5)$

     Axis of symmetry: $x = -3$

     Maximum: 5

     Range: $(-\infty, 5]$

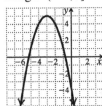

57. $f(x) = \frac{1}{2}(x-2)^2 + 1$

Vertex: $(2,1)$

Axis of symmetry: $x = 2$
Minimum: 1
Range: $[1,\infty)$

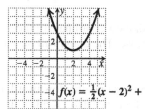

59. $h(x) = -2(x-1)^2 - 3$

Vertex: $(1,-3)$

Axis of symmetry: $x = 1$
Maximum: -3
Range: $(-\infty,-3]$

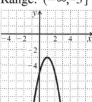

61. $f(x) = 2(x+4)^2 + 1$

Vertex: $(-4,1)$

Axis of symmetry: $x = -4$
Minimum: 1
Range: $[1,\infty)$

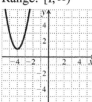

$f(x) = 2(x+4)^2 + 1$

63. $g(x) = -\frac{3}{2}(x-1)^2 + 4$

Vertex: $(1,4)$

Axis of symmetry: $x = 1$
Maximum: 4
Range: $(-\infty,4]$

$g(x) = -\frac{3}{2}(x-1)^2 + 4$

65. $f(x) = 6(x-8)^2 + 7$

Vertex: $(8,7)$

Axis of symmetry: $x = 8$
Minimum: 7

67. $h(x) = -\frac{2}{7}(x+6)^2 + 11 = -\frac{2}{7}\left[x-(-6)\right]^2 + 11$

Vertex: $(-6,11)$

Axis of symmetry: $x = -6$
Maximum: 11

69. $f(x) = \left(x-\frac{7}{2}\right)^2 - \frac{29}{4}$

Vertex: $\left(\frac{7}{2}, -\frac{29}{4}\right)$

Axis of symmetry: $x = \frac{7}{2}$

Minimum: $-\frac{29}{4}$

71. $f(x) = \sqrt{2}(x+4.58)^2 + 65\pi$
$= \sqrt{2}\left[x-(-4.58)\right]^2 + 65\pi$

Vertex: $(-4.58, 65\pi)$
Axis of symmetry: $x = -4.58$
Minimum: 65π

73. *Thinking and Writing Exercise.*

75. $8x - 6y = 24$
To find the x-intercept, let $y = 0$:
$8x - 6(0) = 24$
$8x = 24$
$x = 3$
To find the y-intercept, let $x = 0$:
$8(0) - 6y = 24$
$-6y = 24$
$y = -4$

The x-intercept is $(3, 0)$, and the y-intercept is $(0, -4)$.

77. $y = x^2 + 8x + 15$

To find the x-intercepts, let $y = 0$:

$0 = x^2 + 8x + 15$

$0 = (x + 3)(x + 5)$

$x = -5, -3$

The x-intercepts are $(-5, 0)$ and $(-3, 0)$.

79. $x^2 - 14x + \underline{} = (x - \underline{})^2$

$\left(\dfrac{14}{2}\right)^2 = 7^2 = 49$

$x^2 - 14x + 49 = (x - 7)^2$

81. *Thinking and Writing Exercise.*

83. The equation will be of the form

$f(x) = \dfrac{3}{5}(x - h)^2 + k$ with $h = 4$ and $k = 1$:

$f(x) = \dfrac{3}{5}(x - 4)^2 + 1$

85. The equation will be of the form

$f(x) = \dfrac{3}{5}(x - h)^2 + k$ with $h = 3$ and $k = -1$:

$f(x) = \dfrac{3}{5}(x - 3)^2 + (-1)$ or

$f(x) = \dfrac{3}{5}(x - 3)^2 - 1$

87. The equation will be of the form

$f(x) = \dfrac{3}{5}(x - h)^2 + k$ with $h = -2$ and $k = -5$:

$f(x) = \dfrac{3}{5}\left[x - (-2)\right]^2 + (-5)$ or

$f(x) = \dfrac{3}{5}(x + 2)^2 - 5$

89. Since there is a minimum at $(2, 0)$, the parabola will have the same shape as $f(x) = 2x^2$. It will be of the form

$f(x) = 2(x - h)^2 + k$ with $h = 2$ and $k = 0$:

$f(x) = 2(x - 2)^2$

91. Since there is a maximum at $(0, 3)$, the parabola will have the same shape as $g(x) = -2x^2$. It will be of the form

$g(x) = -2(x - h)^2 + k$ with $h = 0$ and $k = 3$:

$g(x) = -2x^2 + 3$

93. If h is increased, the x coordinate of the vertex will increase. The graph will be shifted right.

95. Since the parabola is opening downward, the original value of a is negative. Replacing a with $-a$ will then make the leading coefficient positive and the parabola will open upward, but the vertex will remain at (h, k).

97. The maximum value of $g(x)$ is 1 and occurs at the point $(5, 1)$, so for $F(x)$ we have $h = 5$ and $k = 1$. $F(x)$ has the same shape as $f(x)$ and has a minimum, so $a = 3$. Thus:

$F(x) = 3(x - 5)^2 + 1$.

99. The graph of $y = f(x - 1)$ looks like the graph of $y = f(x)$ moved 1 unit to the right.

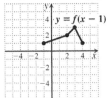

101. The graph of $y = f(x) + 2$ looks like the graph of $y = f(x)$ moved up 2 units.

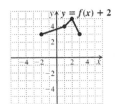

103. The graph of $y = f(x+3) - 2$ looks like the graph of $y = f(x)$ moved 3 units to the left and also moved down 2 units.

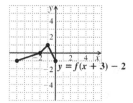

Exercise Set 8.7

1. True : The leading coefficient, 3, is positive.

3. True : The quadratic is in vertex form, $a(x-h)^2 + k$ with $(h,k) = (3,7)$.

5. False : The axis of symmetry is the line $x = h$. $g(x)$ is in vertex form with $h = \dfrac{3}{2}$.

7. False : The x-coordinate of the y-intercept is always 0. The graph goes through $(0, 7)$, not $(7, 0)$.

9. $x^2 - 8x + 2 = x^2 - 8x + \left(\dfrac{8}{2}\right)^2 + 2 - \left(\dfrac{8}{2}\right)^2$

   $= (x-4)^2 + 2 - 4^2 = (x-4)^2 + (-14)$

11. $x^2 + 3x - 5 = x^2 + 3x + \left(\dfrac{3}{2}\right)^2 - 5 - \left(\dfrac{3}{2}\right)^2$

    $= \left(x + \dfrac{3}{2}\right)^2 - 5 - \dfrac{9}{4} = \left(x - \left(-\dfrac{3}{2}\right)\right)^2 + \left(-\dfrac{29}{4}\right)$

13. $3x^2 + 6x - 2 = 3\left(x^2 + 2x + \left(\dfrac{2}{2}\right)^2\right) - 2 - 3 \cdot \left(\dfrac{2}{2}\right)^2$

    $= 3(x+1)^2 - 2 - 3 = 3(x - (-1))^2 + (-5)$

15. $-x^2 - 4x - 7 = -\left(x^2 + 4x + \left(\dfrac{4}{2}\right)^2\right) - 7 - (-1) \cdot \left(\dfrac{4}{2}\right)^2$

    $= -(x+2)^2 - 7 + 4 = -(x - (-2))^2 + (-3)$

17. $2x^2 - 5x + 10 = 2\left(x^2 - \dfrac{5}{2}x + \left(\dfrac{5}{4}\right)^2\right) + 10 - 2 \cdot \left(\dfrac{5}{4}\right)^2$

    $= 2\left(x - \dfrac{5}{4}\right)^2 + 10 - \dfrac{25}{8} = 2\left(x - \dfrac{5}{4}\right)^2 + \dfrac{55}{8}$

19. $f(x) = x^2 + 4x + 5$

    $= (x^2 + 4x + 4) + (-4 + 5)$

    $= (x+2)^2 + 1$

    Vertex: $(-2,1)$, Axis of symmetry: $x = -2$

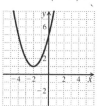

$f(x) = x^2 + 4x + 5$

21. $f(x) = x^2 + 8x + 20$

    $= (x^2 + 8x + 16) + (-16 + 20)$

    $= (x+4)^2 + 4$

    Vertex: $(-4,4)$, Axis of symmetry: $x = -4$

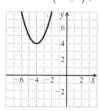

$f(x) = x^2 + 8x + 20$

23. $h(x) = 2x^2 - 16x + 25$

    $= 2(x^2 - 8x + 16) - 2 \cdot 16 + 25$

    $= 2(x-4)^2 - 7$

    Vertex: $(4,-7)$, Axis of symmetry: $x = 4$

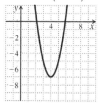

$h(x) = 2x^2 - 16x + 25$

25. $f(x) = -x^2 + 2x + 5$

$\qquad = -(x^2 - 2x - 5)$

$\qquad = -(x^2 - 2x + 1 - 1 - 5)$

$\qquad = -(x-1)^2 + 6$

Vertex: $(1,6)$, Axis of symmetry: $x = 1$

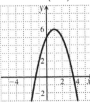

$f(x) = -x^2 + 2x + 5$

27. $g(x) = x^2 + 3x - 10$

$\qquad = \left(x^2 + 3x + \dfrac{9}{4}\right) - \dfrac{9}{4} - 10$

$\qquad = \left(x + \dfrac{3}{2}\right)^2 - \dfrac{49}{4}$

Vertex: $\left(-\dfrac{3}{2}, -\dfrac{49}{4}\right)$,

Axis of symmetry: $x = -\dfrac{3}{2}$

$g(x) = x^2 + 3x - 10$

29. $h(x) = x^2 + 7x$

$\qquad = \left(x^2 + 7x + \dfrac{49}{4}\right) - \dfrac{49}{4}$

$\qquad = \left(x + \dfrac{7}{2}\right)^2 - \dfrac{49}{4}$

Vertex: $\left(-\dfrac{7}{2}, -\dfrac{49}{4}\right)$

Axis of symmetry: $x = -\dfrac{7}{2}$

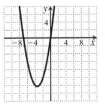

$h(x) = x^2 + 7x$

31. $f(x) = -2x^2 - 4x - 6$

$\qquad = -2(x^2 + 2x) - 6$

$\qquad = -2(x^2 + 2x + 1) - 2(-1) - 6$

$\qquad = -2(x+1)^2 - 4$

Vertex: $(-1, -4)$

Axis of symmetry: $x = -1$

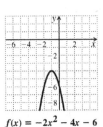

$f(x) = -2x^2 - 4x - 6$

33. a) $g(x) = x^2 - 6x + 13 \Rightarrow a = 1,\ b = -6$

$\qquad \Rightarrow -\dfrac{b}{2a} = -\dfrac{-6}{2 \cdot 1} = 3$

$\qquad g(3) = (3)^2 - 6(3) + 13$

$\qquad = 9 - 18 + 13 = 4$

Vertex : $(3, g(3)) = (3, 4)$

Axis of symmetry: $x = 3$

Minimum of g: $y = 4$

b)

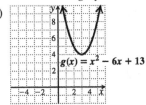

$g(x) = x^2 - 6x + 13$

35. a) $g(x) = 2x^2 - 8x + 3 \Rightarrow a = 2,\ b = -8$

$\Rightarrow -\dfrac{b}{2a} = -\dfrac{-8}{2 \cdot 2} = 2$

$g(2) = 2(2)^2 - 8(2) + 3$

$= 8 - 16 + 3 = -5$

Vertex : $(2, g(2)) = (2, -5)$

Axis of symmetry: $x = 2$

Minimum of g: $y = -5$

b)

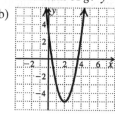

$g(x) = 2x^2 - 8x + 3$

37. a) $f(x) = 3x^2 - 24x + 50 \Rightarrow a = 3,\ b = -24$

$\Rightarrow -\dfrac{b}{2a} = -\dfrac{-24}{2 \cdot 3} = 4$

$f(4) = 3(4)^2 - 24(4) + 50$

$= 48 - 96 + 50 = 2$

Vertex : $(4, f(4)) = (4, 2)$

Axis of symmetry: $x = 4$

Minimum of f: $y = 2$

b)

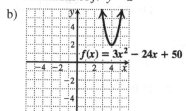

$f(x) = 3x^2 - 24x + 50$

39. a) $f(x) = -3x^2 + 5x - 2 \Rightarrow a = -3,\ b = 5$

$\Rightarrow -\dfrac{b}{2a} = -\dfrac{5}{2 \cdot (-3)} = \dfrac{5}{6}$

$f\left(\dfrac{5}{6}\right) = -3\left(\dfrac{5}{6}\right)^2 + 5\left(\dfrac{5}{6}\right) - 2$

$= -\dfrac{25}{12} + \dfrac{25}{6} - 2 = \dfrac{-25 + 2 \cdot 25 - 2 \cdot 12}{12} = \dfrac{1}{12}$

Vertex : $\left(\dfrac{5}{6}, f\left(\dfrac{5}{6}\right)\right) = \left(\dfrac{5}{6}, \dfrac{1}{12}\right)$

Axis of symmetry: $x = \dfrac{5}{6}$

Maximum of f: $y = \dfrac{1}{12}$

b)

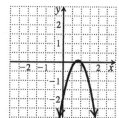

$f(x) = -3x^2 + 5x - 2$

41. a) $h(x) = \dfrac{1}{2}x^2 + 4x + \dfrac{19}{3} \Rightarrow a = \dfrac{1}{2},\ b = 4$

$\Rightarrow -\dfrac{b}{2a} = -\dfrac{4}{2 \cdot \left(\dfrac{1}{2}\right)} = -4$

$h(-4) = \dfrac{1}{2}(-4)^2 + 4(-4) + \dfrac{19}{3}$

$= 8 - 16 + \dfrac{19}{3} = \dfrac{-8 \cdot 3 + 19}{3} = -\dfrac{5}{3}$

Vertex : $(-4, h(-4)) = \left(-4, -\dfrac{5}{3}\right)$

Axis of symmetry: $x = -4$

Minimum of h: $y = -\dfrac{5}{3}$

b)

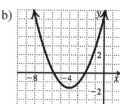

$h(x) = \dfrac{1}{2}x^2 + 4x + \dfrac{19}{3}$

43. $f(x) = x^2 + x - 6$

The coefficient of x^2 is positive so the graph opens upward and the function has a minimum value. Graph the function in a window that shows the vertex. The standard window is one good choice. Then use the Minimum feature from the CALC menu to find that the vertex is $(-0.5, -6.25)$.

45. $f(x) = 5x^2 - x + 1$

The coefficient of x^2 is positive so the graph opens upward and the function has a minimum value. Graph the function in a window that shows the vertex. The standard window is one good choice. Then use the Minimum feature from the CALC menu to find that the vertex is (0.1, 0.95).

47. $f(x) = -0.2x^2 + 1.4x - 6.7$

The coefficient of x^2 is negative so the graph opens downward and the function has a maximum value. Graph the function in a window that shows the vertex. The standard window is one good choice. Then use the Maximum feature from the CALC menu to find that the vertex is (3.5, –4.25).

49. $f(x) = x^2 - 6x + 3$

To find the x-intercepts, solve the equation $0 = x^2 - 6x + 3$.
Use the quadratic formula.

$$x = \frac{-(-6) \pm \sqrt{(-6)^2 - 4 \cdot 1 \cdot 3}}{2 \cdot 1}$$

$$x = \frac{6 \pm \sqrt{24}}{2} = \frac{6 \pm 2\sqrt{6}}{2} = 3 \pm \sqrt{6}$$

The x-intercepts are $\left(3 - \sqrt{6}, 0\right)$ and $\left(3 + \sqrt{6}, 0\right)$.

The y-intercept is $\left(0, f(0)\right)$, or $(0,3)$.

51. $g(x) = -x^2 + 2x + 3$

To find the x-intercepts, solve the equation $0 = -x^2 + 2x + 3$. We factor.

$0 = -x^2 + 2x + 3$
$0 = x^2 - 2x - 3$
$0 = (x - 3)(x + 1)$
$x = 3 \; or \; x = -1$

The x-intercepts are (–1,0) and (3,0). The y-intercepts are $\left(0, g(0)\right)$, or (0,3).

53. $f(x) = x^2 - 9x$

To find the x-intercepts, solve the equation.

$0 = x^2 - 9x$
$0 = x(x - 9)$
$x = 0 \; or \; x = 9$

The x-intercepts are (0,0) and (9,0). Since (0,0) is an x-intercept, we observe that (0,0) is also the y-intercept.

55. $h(x) = -x^2 + 4x - 4$

To find the x-intercepts, solve the equation.

$0 = -x^2 + 4x - 4$
$0 = x^2 - 4x + 4$
$0 = (x - 2)(x - 2)$
$x = 2 \; or \; x = 2$

The x-intercept is (2,0).

The y-intercept is $\left(0, h(0)\right)$, or (0,–4).

57. $g(x) = x^2 + x - 5$

To find the x-intercepts, solve the equation

$g(x) = 0 \Rightarrow x^2 + x - 5 = 0$
$\Rightarrow a = 1, \; b = 1, \; c = -5$

Use the quadratic formula.

$$x = \frac{-1 \pm \sqrt{1^2 - 4 \cdot 1 \cdot (-5)}}{2 \cdot 1} = \frac{-1 \pm \sqrt{21}}{2}$$

The x-intercepts are $\left(\frac{-1 - \sqrt{21}}{2}, 0\right)$ and $\left(\frac{-1 + \sqrt{21}}{2}, 0\right)$.

The y-intercept is $\left(0, g(0)\right) = (0,-5)$.

59. $f(x) = 2x^2 - 4x + 6$

To find the x-intercepts, solve the equation $0 = 2x^2 - 4x + 6$. We use the quadratic formula.

$$x = \frac{-(-4) \pm \sqrt{(-4)^2 - 4 \cdot 2 \cdot 6}}{2 \cdot 2}$$

$$x = \frac{4 \pm \sqrt{-32}}{4} = \frac{4 \pm 4i\sqrt{2}}{4} = 1 \pm i\sqrt{2}$$

There are no real-number solutions, so there is no x-intercept. The y-intercept is $\left(0, f(0)\right)$, or (0,6).

61. *Thinking and Writing Exercise.*

63. $x + y + z = 3$ (1)

  $x - y + z = 1$ (2)

  $-x - y + z = -1$ (3)

Adding (1) to (3) will generate an equation in y and z. Adding (2) to (3) will generate a second equation in y and z.

  $x + y + z = 3$ (1)

  $-x - y + z = -1$ (3)
  _____

  $2z = 2$ (4)

  $x - y + z = 1$ (2)

  $-x - y + z = -1$ (3)
  _____

  $-2y + 2z = 0$

  or $y - z = 0$

  or $y = z$ (5)

Solving (4) gives $z = 1$. (5) then gives $y = 1$ also. Use equation (1), and the values of y and z, to find x.

$x + y + z = 3$ (1)

$\Rightarrow x = 3 - y - z = 3 - 1 - 1 = 1$

$(x, y, z) = (1, 1, 1)$

65. $z = 8$ (1)

  $x + y + z = 23$ (2)

  $2x + y - z = 17$ (3)

Substitute 8 for z into (2) and (3) to generate a two-variable system in x and y.

  $x + y + z = 23$ (2) $x + y + 8 = 23$
                           $\Rightarrow$
  $2x + y - z = 17$ (3) $2x + y - 8 = 17$

  $\Rightarrow$ $x + y = 15$ (4)

   $2x + y = 25$ (5)

Add $-1 \cdot (4)$ and (5) to eliminate y and solve for x.

  $-x - y = -15$ $-1 \cdot (4)$

  $2x + y = 25$ (5)
  _____

  $x = 10$

Use (4) and the value of x to find y.

$x + y = 15$ (4)

$\Rightarrow y = 15 - x = 15 - 10 = 5$

$(x, y, z) = (10, 5, 8)$

67. $c = 1.5$ (1)

  $25a + 5b + c = 52.5$ (2)

  $4a + 2b + c = 7.5$ (3)

Substitute 1.5 for c into (2) and (3) to generate a two-variable system in a and b.

  $25a + 5b + c = 52.5$ (2) $25a + 5b + 1.5 = 52.5$
                               $\Rightarrow$
  $4a + 2b + c = 7.5$ (3) $4a + 2b + 1.5 = 7.5$

  $\Rightarrow 25a + 5b = 51$ (4)

   $4a + 2b = 6$

   or $2a + b = 3$ (5)

Add (4) to $-5 \cdot (5)$ to eliminate b and solve for a.

  $25a + 5b = 51$ (4)

  $-10a - 5b = -15$ $-5 \cdot (5)$
  _____

  $15a = 36$

$\Rightarrow x = \dfrac{36}{15} = \dfrac{12}{5} = 2.4$

Use (5) and the value of a to find b.

$2a + b = 3$ (5)

$\Rightarrow b = 3 - 2a = 3 - 2(2.4) = 3 - 4.8 = -1.8$

$(a, b, c) = (2.4, -1.8, 1.5)$

69. *Thinking and Writing Exercise.*

71. $f(x) = 2.31x^2 - 3.135x - 5.89$

a) The coefficient of x^2 is positive so the graph opens upward and the function has a minimum value. Graph the function in a window that shows the vertex. The standard window is one good choice. Then use the Minimum feature from the CALC menu to find that the minimum value is ≈ -6.95.

b) To find the first coordinates of the x-intercepts we use the Zero feature from the CALC menu to find the zeros of the function. They are about -1.06 and 2.41, so the x-intercepts are $(-1.06, 0)$ and $(2.41, 0)$. The y-intercept is $(0, f(0))$, or $(0, -5.89)$.

73. $g(x) = -1.25x^2 + 3.42x - 2.79$

   a) The coefficient of x^2 is negative so the graph opens downward and the function has a maximum value. Graph the function in a window that shows the vertex. The standard window is one good choice. Then use the Maximum feature from the CALC menu to find that the minimum value is about -0.45.

   b) The graph has no x-intercepts. The y-intercept is $(0, f(0))$, or $(0, -2.79)$.

75. $f(x) = x^2 - x - 6$

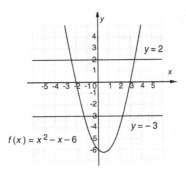

$f(x) = x^2 - x - 6$

   a) The solutions of $x^2 - x - 6 = 2$ are the first coordinates of the points of the intersection of the graphs of $f(x) = x^2 - x - 6$ and $y = 2$. From the graph we see that the solutions are approximately -2.4 and 3.4.

   b) The solutions $x^2 - x - 6 = -3$ are the first coordinates of the points of intersection of the graphs of $f(x) = x^2 - x - 6$ and $y = -3$. From the graph we see that the solutions are approximately -1.3 and 2.3.

77. $f(x) = mx^2 - nx + p$

$$= m\left(x^2 - \frac{n}{m}x\right) + p$$

$$= m\left(x^2 - \frac{n}{m}x + \frac{n^2}{4m^2} - \frac{n^2}{4m^2}\right) + p$$

$$= m\left(x - \frac{n}{2m}\right)^2 - \frac{n^2}{4m} + p$$

$$= m\left(x - \frac{n}{2m}\right)^2 + \frac{-n^2 + 4mp}{4m}, \text{ or}$$

$$m\left(x - \frac{n}{2m}\right)^2 + \frac{4mp - n^2}{4m}$$

79. Since the vertex is given as $(3, -5)$, the function must have the form:

$$f(x) = a(x - 3)^2 + (-5) = a(x - 3)^2 - 5$$

Use the point $(-1, f(-1)) = (-1, 0)$ to find a.

$$f(-1) = a(-1 - 3)^2 - 5 = 0$$

$$\Rightarrow a(-4)^2 = 5 \Rightarrow a = \frac{5}{16}$$

$$\Rightarrow f(x) = \frac{5}{16}(x - 3)^2 - 5$$

81. $f(x) = |x^2 - 1|$

We plot some points and draw the curve. Note that it will lie entirely on or above the x-axis since absolute value is never negative.

x	$f(x)$
-3	8
-2	3
-1	0
0	1
1	0
2	3
3	8

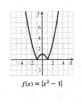

$f(x) = |x^2 - 1|$

83. $f(x) = \left|2(x-3)^2 - 5\right|$

We plot some points and draw the curve. Note that it will lie entirely on or above the x-axis since absolute value is never negative.

x	$f(x)$
-1	27
0	13
1	3
2	3
3	5
4	3
5	3
6	13

$f(x) = |2(x-3)^2 - 5|$

Exercise Set 8.8

1. e

3. c

5. d

7. $P(x) = 0.2x^2 - 2.8x + 9.8$

$\Rightarrow a = 0.2,\ b = -2.8,\ c = 9.8$

$P(x)$ is quadratic with a positive leading coefficient. Its minimum occurs at its vertex point.

$-\dfrac{b}{2a} = -\dfrac{-2.8}{2 \cdot 0.2} = 7$

7 corresponds to the month of July. The inches of precipitation is given by $P(7)$:

$P(7) = 0.2(7)^2 - 2.8(7) + 9.8$

$= 9.8 - 19.6 + 9.8 = 0$

It did not rain at all in the month of July.

9. Using $P(x) = R(x) - C(x)$, we have:

$P(x) = 1000x - x^2 - (3000 + 20x)$

$= -x^2 + 980x - 3000$

We determine the maximum of $P(x)$ by first

finding $-\dfrac{b}{2a}: -\dfrac{b}{2a} = -\dfrac{-980}{2 \cdot -1} = 490$

We now find the maximum value of the function $P(490)$.

$P(490) = -490^2 + 980(490) - 3000$

$= 237,100$

The maximum profit is \$237,100 when $x = 490$.

11. $P = 2l + 2w = 128$

$A = l \cdot w$

Solve the first equation for $l, l = 64 - w$ and substitute into the second equation.

$A = (64 - w)w$

$A = 64w - w^2$

$A = -(w^2 - 64w)$

Complete the square to get:

$A = -(w^2 - 64w + 1024) + 1024$

$A = -(w - 32)^2 + 1024$

The maximum function value is 1024, when $w = 32$, $l = 64-32$, or 32. The maximum area occurs when the dimensions are 32 in. by 32 in.

13. Since one side is the house, we have:

$P = l + 2w = 60$

$A = l \cdot w$

Solving P for l and substituting into A, we have:

$A = (60 - 2w)w$

$A = -2w^2 + 60w$

$A = -2(w^2 - 30w + 225) - (-2) \cdot 225$

$A = -2(w - 15)^2 + 450$

The maximum function value of 450 occurs when $w = 15$; $l = 60 - 2(15)$, or 30. Maximum area of 450 ft^2; dimensions 15 ft by 30 ft.

15. Let x = height of the file and y = width. We have two equations.
$$2x + y = 14$$
$$V = 8xy$$
Solve $2x + y = 14$ for y, $y = 14 - 2x$, and substitute into the second equation.
$$V = 8x(14 - 2x)$$
$$V = -16x^2 + 112x$$
$$V = -16(x^2 - 7x)$$
$$V = -16\left(x^2 - 7x + \frac{49}{4}\right) - (-16) \cdot \frac{49}{4}$$
$$V = -16\left(x - \frac{7}{2}\right)^2 + 196$$

The maximum of 196 occurs when $x = \frac{7}{2}$.

When $x = \frac{7}{2}, y = 14 - 2 \cdot \frac{7}{2} = 7$, the file should be $\frac{7}{2}$ in., or $3\frac{1}{2}$ in. tall.

17. Let x and y represent the numbers.
$$x + y = 18$$
$$P = xy$$
Solve the first equation for y and substitute into the second equation.
$$P = x(18 - x)$$
$$P = -x^2 + 18x$$
$$P = -(x^2 - 18x + 81) - (-81)$$
$$P = -(x - 9)^2 + 81$$
The maximum function value is 81 when $x = 9$. If $x = 9$, $y = 18 - 9$, or 9. The maximum product of 81 occurs for the numbers 9 and 9.

19. Let x and y represent the numbers.
$$x - y = 8$$
$$P = xy$$
Solve the first equation for x and substitute into the second equation.
$$P = (y + 8)y$$
$$P = y^2 + 8y$$
$$P = (y^2 + 8y + 16) - 16$$
$$P = (y + 4)^2 - 16$$
The minimum function value is -16 when

$y = -4$. If $y = -4$, then $x = -4 + 8$, or 4. The minimum product of -16 occurs for the numbers 4 and -4.

21. Let x and y represent the numbers.
$$x + y = -10$$
$$P = xy$$
Solve the first equation for y and substitute into the second equation.
$$P = x(-10 - x)$$
$$P = -x^2 - 10x$$
$$P = -(x^2 + 10x + 25) - (-25)$$
$$P = -(x + 5)^2 + 25$$
The maximum function value is 25 when $x = -5$. If $x = -5$, $y = -10 - (-5) = -5$. The maximum product of 25 occurs for the numbers $x = y = -5$.

23. The data points appear nearly linear.
$$f(x) = mx + b$$

25. The data points rise then fall. This appears to represent a parabola that opens downward.
$$f(x) = ax^2 + bx + c, a < 0$$

27. The data points fall then rise. This appears to represent a parabola that opens upward.
$$f(x) = ax^2 + bx + c, a > 0$$

29. The data points rise nonlinearly. This appears to represent a parabola that opens upward.
$$f(x) = ax^2 + bx + c, a > 0$$

31. The data appears linear over the years 1989-1195, but nonlinear after that. The points are neither linear or quadratic over the range 1989-2007.

33. Look for a function of the
    form $f(x) = ax^2 + bx + c$.
    Substituting the data points, we get
    $$4 = a(1)^2 + b(1) + c,$$
    $$-2 = a(-1)^2 + b(-1) + c,$$
    $$13 = a(2)^2 + b(2) + c,$$
       or
    $$4 = a + b + c,$$
    $$-2 = a - b + c,$$
    $$13 = 4a + 2b + c$$
    Solving this system, we get
    $a = 2$, $b = 3$, and $c = -1$.
    Therefore the function we are looking for is
    $$f(x) = 2x^2 + 3x - 1.$$

35. We look for a function of the form
    $f(x) = ax^2 + bx + c$. Substituting the data
    points, we get
    $$0 = a(2)^2 + b(2) + c,$$
    $$3 = a(4)^2 + b(4) + c,$$
    $$-5 = a(12)^2 + b(12) + c,$$
       or
    $$0 = 4a + 2b + c,$$
    $$3 = 16a + 4b + c,$$
    $$-5 = 144a + 12b + c$$
    Solving this system, we get
    $a = -\dfrac{1}{4}$, $b = 3$, and $c = -5$.
    Therefore the function we are looking for is
    $$f(x) = -\frac{1}{4}x^2 + 3x - 5.$$

37. a) $A(s) = as^2 + bs + c$, where $A(s)$
       represents the number of nighttime
       accidents (for every 200 million km) and s
       represents the travel speed (in
       km/h).

$$400 = a(60)^2 + b(60) + c,$$
$$250 = a(80)^2 + b(80) + c,$$
$$250 = a(100)^2 + b(100) + c,$$
   or
$$400 = 3600a + 60b + c,$$
$$250 = 6400a + 80b + c,$$
$$250 = 10,000a + 100b + c.$$
Solving the system of equations, we get
$$a = \frac{3}{16}, b = -\frac{135}{4}, c = 1750.$$
$$A(s) = \frac{3}{16}s^2 - \frac{135}{4}s + 1750 \text{ fits the data.}$$

   b) Find $A(50)$
$$A(50) = \frac{3}{16}(50)^2 - \frac{135}{4}(50) + 1750 - 531.25$$

   About 531 accidents occur at 50 km/h.

39. Think of a coordinate system placed on the
    drawing in the text with the origin at the point
    where the arrow is released. Then three
    points on the arrow's parabolic path are (0,0),
    (63,27), and (126,0). We look for a function
    of the form $h(d) = ad^2 + bd + c$, where
    $h(d)$ represents the arrow's height and d
    represents the distance the arrow has traveled
    horizontally.
    $$0 = a \cdot 0^2 + b \cdot 0 + c,$$
    $$27 = a \cdot 63^2 + b \cdot 63 + c,$$
    $$0 = a \cdot 126^2 + b \cdot 126 + c,$$
       or
    $$0 = c,$$
    $$27 = 3969a + 63b + c,$$
    $$0 = 15,876a + 126b + c.$$
    Solving the system of equations, we get
    $a \approx -0.0068$, $b \approx 0.8571$, and $c = 0$, and
    $$h(d) = -0.0068d^2 + 0.8571d.$$

41. a) Enter the data and use the quadratic
        regression feature. We have $D(x) = -0.0083x^2 + 0.8243x + 0.2122$

    b) $D(70) \approx 17.243$, so we estimate that the
        river is about 17.243 ft deep 70 ft from
        the left bank.

43. a) Enter the data and then use the quadratic regression feature. We have
$$t(x) = 18.125x^2 + 78.15x + 24,613$$

b) In 2017, $x = 2017 - 2005 = 12$
$$t(12) = 18.125(12)^2 + 78.15(12) + 24,613$$
$$\approx 28,161 \text{ teachers}$$

45. *Thinking and Writing Exercise.*

47. $\quad x^2 - 1 = 0$
$(x-1)(x+1) = 1$
$x = 1 \text{ or } x = -1$

49. $10x^3 - 30x^2 + 20x = 0$
$10x(x^2 - 3x + 2) = 0$
$10x(x-2)(x-1) = 0$
$x = 0 \text{ or } x - 2 = 0 \text{ or } x - 1 = 0$
$\qquad\qquad x = 2 \text{ or } \quad x = 1$
The solutions are 0, 1, and 2.

51. $\qquad \dfrac{x-3}{x+4} = 5$

$(x+4) \cdot \dfrac{x-3}{x+4} = (x+4) \cdot 5$

$\qquad x - 3 = 5x + 20$

$\qquad -23 = 4x$

$\qquad x = -\dfrac{23}{4}$

The solutions is $-\dfrac{23}{4}$.

53. $\dfrac{x}{(x-3)(x+7)} = 0$

$(x-3)(x+7) \cdot \dfrac{x}{(x-3)(x+7)}$

$\qquad = (x-3)(x+7) \cdot 0$

$\qquad\qquad x = 0$

The solutions is 0.

55. *Thinking and Writing Exercise.*

57. Position the bridge on a coordinate system as shown with the vertex of the parabola at (0, 30).

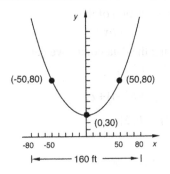

We find a function of the form $y = ax^2 + bx + c$ which represents the parabola. Since (0, 30), (–50, 80), and (50, 80) are on the parabola we know

$30 = a \cdot 0^2 + b \cdot 0 + c,$

$80 = a(-50)^2 + b(-50) + c,$

$80 = a(50)^2 + b(50) + c,$

or

$30 = c,$

$80 = 2500a - 50b + c,$

$80 = 2500a + 50b + c.$

Solving, we get $a = 0.02$, $b = 0$, $c = 30$. The function $y = 0.02x^2 + 30$ represents the parabola. The longest vertical cables occur at $x = -80$ and $x = 80$. For $x \pm 80$,

$y = 0.02(\pm 80)^2 + 30$

$\quad = 128 + 30$

$\quad = 158\, ft$

59.

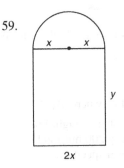

The perimeter of the semicircular portion of the window is $\dfrac{1}{2} \cdot 2\pi x$, or πx. The perimeter of the rectangular portion is $y + 2x + y$, or $2x + 2y$. The area of the semicircular portion of the window is $\dfrac{1}{2} \cdot \pi x^2$, or $\dfrac{\pi}{2} x^2$. The area of the rectangular portion is $2xy$.

We have two equations, one giving the perimeter of the window and the other giving the area.

$$\pi x + 2x + 2y = 24,$$

$$A = \frac{\pi}{2}x^2 + 2xy$$

Solve the first equation for y, and substitute into the second equation.

$$A = \frac{\pi}{2}x^2 + 2x\left(12 - \frac{\pi x}{2} - x\right)$$

$$A = \frac{\pi}{2}x^2 + 24x - \pi x^2 - 2x^2$$

$$A = -2x^2 - \frac{\pi}{2}x^2 + 24x$$

$$A = -\left(2 + \frac{\pi}{2}\right)x^2 + 24x$$

Completing the square, we get

$$A = -\left(2 + \frac{\pi}{2}\right)\left(x^2 + \frac{24}{-\left(2 + \frac{\pi}{2}\right)}x\right)$$

$$A = -\left(2 + \frac{\pi}{2}\right)\left(x^2 - \frac{48}{4 + \pi}x\right)$$

$$A = -\left(2 + \frac{\pi}{2}\right)\left(x - \frac{24}{4 + \pi}\right)^2 + \left(\frac{24}{4 + \pi}\right)^2$$

The maximum function value occurs when

$x = \frac{24}{4 + \pi}$. When $x = \frac{24}{4 + \pi}$,

$$y = 12 - \frac{\pi}{2}\left(\frac{24}{4 + \pi}\right) - \frac{24}{4 + \pi} =$$

$$\frac{48 + 12\pi}{4 + \pi} - \frac{12\pi}{4 + \pi} - \frac{24}{4 + \pi} = \frac{24}{4 + \pi}$$

The radius of the circular portion of the window and the height of the rectangular portion should each be $\frac{24}{4 + \pi}$ ft.

61. Let x represent the number of 25¢ increases in the admission price. Then $10 + 0.25x$ represents the admission price, and $80 - x$ represents the corresponding average attendance. Let R represent the total revenue.

$$R(x) = (10 + 0.25x)(80 - x)$$

$$= -0.25x^2 + 10x + 800$$

$$R(x) = -0.25(x - 20)^2 + 900$$

The maximum function value of 900 occurs when
$x = 20$. The owner should charge
$10 + \$0.25(20)$, or $15.

Exercise Set 8.9

1. True

3. True

5. False

7. We see that $p(x) = 0$ at $x = -4$ and $x = \frac{3}{2}$, and $p(x) < 0$ between -4 and $\frac{3}{2}$. The solution set of the inequality is $\left[-4, \frac{3}{2}\right]$, or $\left\{x \mid -4 \le x \le \frac{3}{2}\right\}$.

9. $$x^4 + 12x > 3x^3 + 4x^2$$
$$x^4 - 3x^3 - 4x^2 + 12x > 0$$
From the graph we see that $p(x) > 0$ on $(-\infty, -2) \cup (0, 2) \cup (3, \infty)$. This union, or $\{x \mid x < -2 \text{ or } 0 < x < 2 \text{ or } x > 3\}$, is the solution set of the inequality.

11. $$\frac{x-1}{x+2} < 3$$
$$\frac{x-1}{x+2} - 3 < 0$$
$$\frac{x-1}{x+2} - \frac{3(x+2)}{x+2} < 0$$
$$\frac{x-1-3(x+2)}{x+2} < 0$$
$$\frac{-2x-7}{x+2} < 0$$

$\frac{-2x-7}{x+2} < 0$ on $\left(-\infty, -\frac{7}{2}\right) \cup (-2, \infty)$. This union, or $\left\{x \mid x < -\frac{7}{2} \text{ or } x > -2\right\}$, is the solution set of the inequality.

13. $(x+4)(x-3)<0$

We solve the related equation.

$(x+4)(x-3)=0$

$x+4=0 \quad or \; x-3=0$

$\quad x=-4 \; or \quad\quad x=3$

The numbers –4 and 3 divide the number line into 3 intervals.

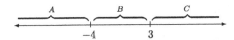

We graph $p(x)=(x+4)(x-3)$ in the window $[-10,10,-15,5]$ and determine the sign of the function in each interval.

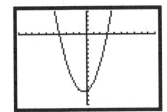

We see that $p(x)<0$ in interval B, or in $(-4,3)$. Thus, the solution set of the inequality is $(-4,3)$, or $\{x|-4<x<3\}$.

15. $(x+7)(x-2)\geq 0$

The solutions of $(x+7)(x-2)=0$ are –7 and 2. They divide the number line into three intervals as shown:

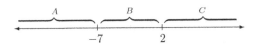

We graph $p(x)=(x+7)(x-2)$ in the window $[-10,10,-25,5]$, Yscl = 5.

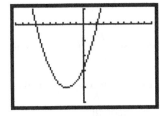

We see that $p(x)\geq 0$ in intervals A and C, or in $(-\infty,-7)\cup(2,\infty)$. We also know that

$p(-7)=0$ and $p(2)=0$. Thus, the solution set of the inequality is $(-\infty,-7]\cup[2,\infty)$, or $\{x|x\leq -7 \text{ or } x\geq 2\}$.

17. $\quad x^2-x-2>0$

$(x-2)(x+1)>0$

The solutions of $(x-2)(x+1)=0$ are 2 and –1. We graph $p(x)=(x-2)(x+1)$ in the standard window.

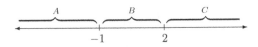

We see that $p(x)>0$ on $(-\infty,-1)$ and $(2,\infty)$. The solution set of the inequality is $(-\infty,-1)\cup(2,\infty)$, or $\{x|x<-1 \text{ or } x>2\}$.

19. $\quad x^2+4x+4<0$

$(x+2)^2<0$

Observe that $(x+2)^2\geq 0$ for all values of x. Thus, the solution set is $\varnothing$. The graph of $p(x)=x^2+4x+4$ confirms this.

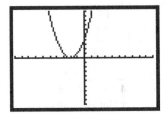

21. $\quad x^2-4x<10$

$x^2-4x-10<0$

The solutions of $x^2-4x-10<0$ are:

$x=\dfrac{-(-4)\pm\sqrt{(-4)^2-4\cdot 1\cdot(-10)}}{2\cdot 1}$

$=\dfrac{4\pm\sqrt{16+40}}{2}=\dfrac{4\pm\sqrt{56}}{2}$

$=\dfrac{4\pm 2\sqrt{14}}{2}=2\pm\sqrt{14}$

Graph $y=x^2-4x-10$ in the standard window.

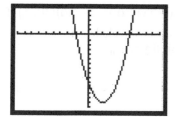

$y < 0$ on the interval $\left(2 - \sqrt{14}, 2 + \sqrt{14}\right)$, or $\left\{x \middle| 2 - \sqrt{14} < x < 2 + \sqrt{14}\right\}$. This is the solution set of the inequality.

23. $3x(x+2)(x-2) < 0$

The solutions of $3x(x+2)(x-2) = 0$ are 0, -2, and 2. Graph $p(x) = 3x(x+2)(x-2)$ in the window $[-5,5,-10,10]$.

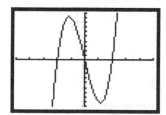

We see that $p(x) < 0$ on $(-\infty, -2) \cup (0, 2)$, or $\left\{x \middle| x < -2 \text{ or } 0 < x < 2\right\}$. This is the solution set for the inequality.

25. $(x-1)(x+2)(x-4) \geq 0$

The solutions of $(x-1)(x+2)(x-4) = 0$ are 1, -2, and 4. Graph $p(x) = (x-1)(x+2)(x-4)$ in the window $[-5,5,-12,12]$.

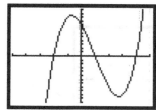

We see that $p(x) \geq 0$ on $[-2,1]$ and $[4,\infty)$. the solution set of the inequality includes the zeros; it is $[-2,1] \cup [4,\infty)$ or $\left\{x \middle| -2 \leq x \leq 1 \text{ or } x \geq 4\right\}$.

27. $4.32x^2 - 3.54x - 5.34 \leq 0$

Graph $p(x) = 4.32x^2 - 3.54x - 5.34$ in the window $[-5,5,-10,10]$.

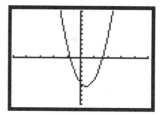

Using the Zero feature we find that $p(x) = 0$ when $x \approx -0.78$ and when $x \approx 1.59$. Also observe that $p(x) < 0$ in the interval $(-0.78, 1.59)$. Thus, the solution set of the inequality is $[-0.78, 1.59]$, or $\left\{x \middle| -0.78 \leq x \leq 1.59\right\}$.

29. $x^3 - 2x^2 - 5x + 6 < 0$

Graph $p(x) = x^3 - 2x^2 - 5x + 6$ in the window $[-5,5,-10,10]$.

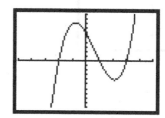

Using the Zero feature we find that $p(x) = 0$ when $x = -2$, when $x = 1$, and when $x = 3$. Then we see that $p(x) < 0$ on $(-\infty, -2) \cup (1, 3)$, or $\left\{x \middle| x < -2 \text{ or } 1 < x < 3\right\}$. This is the solution set of the inequality.

31. $f(x) \geq 3$

$7 - x^2 \geq 3$

$4 - x^2 \geq 0$

The solutions of $(x+2)(x-2) = 0$ are -2 and 2. Graph $p(x) = 4 - x^2$ in a standard window.

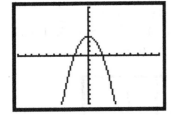

We see that $p(x) \geq 0$, or $f(x) \geq 3$, on $(-2,2)$. The solution set includes the zeros. $[-2,2]$ or $\{x \mid -2 \leq x \leq 2\}$.

33. $g(x) > 0$

$(x-2)(x-3)(x+1) > 0$

The solutions of $(x-2)(x-3)(x+1) = 0$ are 2, 3, and -1. Graph $g(x) = (x-2)(x-3)(x+1)$ in the window $[-5, 5, -10, 10]$.

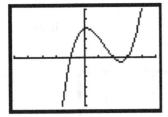

We see that $g(x) > 0$ on

$(-1,2) \cup (3, \infty)$, or

$\{x \mid -1 < x < 2 \text{ or } x > 3\}$.   This is

the solution set of the inequality.

35. $F(x) \leq 0$

$x^3 - 7x^2 + 10x \leq 0$

$x(x-5)(x-2) \leq 0$

The solutions of $x(x-5)(x-2) = 0$ are 0, 5, and 2. Graph $F(x) = x^3 - 7x^2 + 10x$.

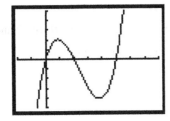

We see that $F(x) < 0$ on $(-\infty, 0) \cup (2,5)$. The solution set, which includes the zeros is

$(-\infty, 0] \cup [2,5]$, or

$\{x \mid x \leq 0 \text{ or } 2 \leq x \leq 5\}$.

37. $\dfrac{1}{x+5} < 0$

The related equation $\dfrac{1}{x+5} = 0$ has no

solution. Also $x + 5 = 0$ when $x = -5$.

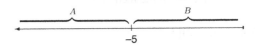

Choose a test number from each interval using substitution or a graphing calculator.

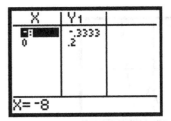

We see that $y < 0$ when $x < -5$, so the solution set is $(-\infty, -5)$, or $\{x \mid x < -5\}$.

39. $\dfrac{x+1}{x-3} \geq 0$

Find the values that make the denominator 0. $x - 3 = 0$ when $x = 3$

Graph $r(x) = \dfrac{x+1}{x-3}$ in the window $[-5, 15, -5, 5]$. Using the Zero feature we find that $r(x) = 0$ when $x = -1$. We include the -1 in the solution.

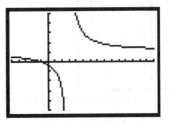

The solution is $(-\infty, -1] \cup (3, \infty)$ or

$\{x \mid x \leq -1 \text{ or } x > 3\}$.

41. $\dfrac{x+1}{x+6} \geq 1$

$\dfrac{x+1}{x+6} - 1 \geq 0$

If $r(x) = \dfrac{x+1}{x+6} - 1$, the solution set of the

inequality is all values of x for which

$r(x) \geq 0$.

First we solve $r(x) = 0$.

$$\dfrac{x+1}{x+6} - 1 = 0$$

$$(x+6)\left(\dfrac{x+1}{x+6} - 1\right) = (x+6)\cdot 0$$

$$(x+6)\left(\dfrac{x+1}{x+6}\right) - (x+6)\cdot 1 = 0$$

$$x + 1 - x - 6 = 0$$

$$-5 = 0$$

This equation has no solution. Find the
values that make the denominator 0.

$x + 6 = 0$

$x = -6$

Use -6 to divide the number line into
intervals.

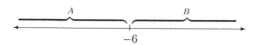

Enter $y = r(x)$ on a graphing calculator and
evaluate a test number in each interval. We
test -7 and 0.

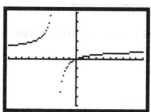

We see that $r(x) > 0$ in interval A. The

solution set is $(-\infty, -6)$, or $\{x | x < -6\}$.

43. $\dfrac{(x-2)(x+1)}{x-5} \leq 0$

Solve the related equation.

$$\dfrac{(x-2)(x+1)}{x-5} = 0$$

$$(x-2)(x+1) = 0$$

$x = 2 \ or \ x = -1$

Find the values that make the denominator 0.

$x - 5 = 0$

$x = 5$

Use the numbers 2, -1, and 5 to divide the
number line into intervals as shown:

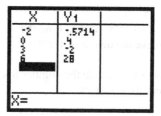

Enter $r(x) = \dfrac{(x-2)(x+1)}{x-5}$ and evaluate a test

number in each interval. We test -2, 0, 3, and
6.

We see that $r(x) < 0$ in intervals A and C.

From above we also know that $r(x) = 0$ when

$x = 2$ or $x = -1$. Thus, the solution set is

$(-\infty, -1] \cup [2,5)$, or $\{x | x \leq -1 \ or \ 2 \leq x < 5\}$.

45. $\dfrac{x}{x+3} \geq 0$

Graph $r(x) = \dfrac{x}{x+3}$ using DOT mode in the

window $[-10, 10, -5, 5]$.

Using the Zero feature we find that $r(x) = 0$

when $x = 0$. Also observe that $r(x) > 0$ in

the interval $(-\infty, -3)$ and in $(0, \infty)$. Then the

solution set is $(-\infty, -3) \cup [0, \infty)$, or

$\{x | x < -3 \ or \ x \geq 0\}$.

47. $\dfrac{x-5}{x} < 1$

$\dfrac{x-5}{x} - 1 < 0$

Let $r(x) = \dfrac{x-5}{x} - 1$ and solve $r(x) = 0$.

$\dfrac{x-5}{x} - 1 = 0$

$x\left(\dfrac{x-5}{x} - 1\right) = x \cdot 0$

$x\left(\dfrac{x-5}{x}\right) - x \cdot 1 = 0$

$x - 5 - x = 0$

$-5 = 0$

This equation has no solution. Find the values that make the denominator 0.

$x = 0$

Use the number 0 to divide the number line into two intervals as shown.

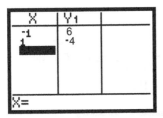

Enter $y = r(x)$ in a graphing calculator and evaluate a test number in each interval. We test -1 and 1.

X	Y1	
-1	6	
1	-4	
X=		

We see that $r(x) < 0$ in interval B. Thus, the solution set is $(0, \infty)$, or $\{x \mid x > 0\}$.

49. $\dfrac{x-1}{(x-3)(x+4)} \le 0$

Solve the related equation.

$\dfrac{x-1}{(x-3)(x+4)} = 0$

$x - 1 = 0$

$x = 1$

Find the values that make the denominator 0.

$(x-3)(x+4) = 0$

$x = 3 \text{ or } x = -4$

Use the numbers 1, 3, and -4 to divide the number line into intervals as shown.

Enter $r(x) = \dfrac{x-1}{(x-3)(x+4)}$ in a graphing calculator and evaluate a test point in each interval. We test -5, 0, 2, and 4.

X	Y1	
-5	-.75	
0	.08333	
2	-.1667	
4	.375	
X=		

We see that $r(x) < 0$ in intervals A and C. From above we also know that $r(x) = 0$ when $x = 1$. Thus, the solution set is

$(-\infty, -4) \cup [1, 3)$, or $\{x \mid x < -4 \text{ or } 1 \le x < 3\}$.

51. $\dfrac{5-2x}{4x+3} \ge 0$

Solve the related equation.

$\dfrac{5-2x}{4x+3} = 0$

$5 - 2x = 0$

$\dfrac{5}{2} = x$

Find the values that make the denominator 0.

$4x + 3 = 0$

$x = -\dfrac{3}{4}$

Now graph $r(x) = \dfrac{5-2x}{4x+3}$ in the window $[-10, 10, -5, 5]$.

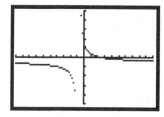

We see $r(x) > 0$ on $\left(-\dfrac{3}{4}, \dfrac{5}{2}\right)$. We include

the zero in the solution set:

$\left(-\dfrac{3}{4}, \dfrac{5}{2}\right]$ or $\left\{x \middle| -\dfrac{3}{4} < x \leq \dfrac{5}{2}\right\}$.

53. $\dfrac{1}{x-2} \leq 1$

$\dfrac{1}{x-2} - 1 \leq 0$

$\dfrac{1}{x-2} - \dfrac{x-2}{x-2} \leq 0$

$\dfrac{3-x}{x-2} \leq 0$

Solve the related equation.

$\dfrac{3-x}{x-2} = 0$

$3 - x = 0$

$3 = x$

Find the values that make the denominator 0.
$x - 2 = 0$ when $x = 2$.
Use these numbers to divide the number line into intervals.

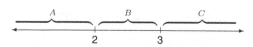

Choose a test point from each interval to determine where the original inequality is True i.e. the value is negative. We determine this to be intervals A and C.

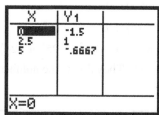

We include the zero in the solution.

$(-\infty, 2) \cup [3, \infty)$ or $\{x \mid x < 2 \text{ or } x \geq 3\}$

55. *Thinking and Writing Exercise.*

57. Make a table of values.

x	$f(x) = x^3 - 2$	(x, y)
-2	$(-2)^3 - 2 = -8 - 2 = -10$	$(-2, -10)$
-1	$(-1)^3 - 2 = -1 - 2 = -3$	$(-1) - 3$
0	$0^3 - 2 = 0 - 2 = -2$	$(0, -2)$
1	$1^3 - 2 = 1 - 2 = -1$	$(1, -1)$
2	$2^3 - 2 = 8 - 2 = 6$	$(2, 6)$

Plot the points and connect them with a smooth curve.

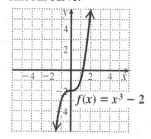

59. $f\left(\dfrac{1}{a^2}\right) = \dfrac{1}{a^2} + 7$

61. $g(2a + 5) = (2a + 5)^2 + 2$

$= 4a^2 + 2 \cdot 2a \cdot 5 + 25 + 2$

$= 4a^2 + 20a + 27$

63. *Thinking and Writing Exercise.*

65. $x^4 + x^2 < 0$
Note that when we raise any number to the fourth power or second power, the result is non-negative. If we add two non-negative numbers, the result is also non-negative. Thus $x^4 + x^2 \geq 0$, so $x^4 + x^2 < 0$ has no solution. The solution set is $\varnothing$.

67. $x^4 + 3x^2 \leq 0$

$x^2(x^2 + 3) \leq 0$

$x^2 = 0$ for $x = 0$, $x^2 > 0$ for $x \neq 0$, $x^2 + 3 > 0$ for all x. The solution set is $\{0\}$.

69. a) $-3x^2 + 630x - 6000 > 0$

$x^2 - 210x + 2000 < 0$ Multiply by $-\dfrac{1}{3}$

$(x - 200)(x - 10) < 0$

The solutions of

$f(x) = (x - 200)(x - 10) = 0$

are 200 and 10. They divide the number line as shown:

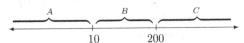

Enter $p(x) = x^2 - 210x + 2000$ in a graphing calculator and evaluate a test point in each interval. We test 9, 11, and 201. Note that only nonnegative values of x have meaning in this problem.

X	Y1	
9	191	
11	-189	
201	191	
X=		

We see that $p(x) < 0$ in interval B, so the company makes a profit for values of x such that $10 < x < 200$, or for values of x in the interval $(10, 200)$, or in the set $\{x | 10 < x < 200\}$.

   b)   See part (a). Keep in mind that x must be nonnegative since negative numbers have no meaning in this application.
The company loses money for values of x such that $0 \le x < 10$ or $x > 200$, or for values of x in the interval $[0, 10) \cup (200, \infty)$, or in the set $\{x | 0 \le x < 10 \text{ or } x > 200\}$.

71.   We find values of n such that $N \ge 66$ and $N \le 300$.
For $N \ge 66$

$\dfrac{n(n-1)}{2} \ge 66$

$n(n-1) \ge 132$

$n^2 - n - 132 \ge 0$

$(n - 12)(n + 11) \ge 0$

The solutions of $f(n) = (n - 12)(n + 11) = 0$

are 12 and -11. They divide the number line as shown:

However, only positive values of n have meaning in this exercise so we need only consider the intervals shown below:

Enter $p(x) = x^2 - x - 132$ in a graphing calculator and evaluate a test point in each interval. We test 1 and 13.

X	Y1	
1	-132	
13	24	
X=		

We see that $p(x) > 0$ in interval B. From above we also know that $p(12) = 0$, so the solution set for this inequality is $[12, \infty)$.
For $N \le 300$

$\dfrac{n(n-1)}{2} \le 300$

$n(n-1) \le 600$

$n^2 - n - 600 \le 0$

$(n - 25)(n + 24) \le 0$

The solutions of $f(n) = (n - 25)(n + 24) = 0$

are 25 and -24. They divide the number line as shown:

However, only positive values of n have meaning in this exercise so we need only consider the intervals shown below:

Enter $p(x) = x^2 - x - 600$ in a graphing calculator and evaluate a test point in each interval. We test 1 and 26.

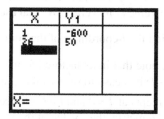

We see that $p(x) < 0$ in interval A. From above we also know that $p(25) = 0$, so the solution set for this inequality is $(0, 25]$. Then $66 \le N \le 300$ for $[12, \infty) \cap (0, 25]$, or $[12, 25]$. We can express the solution set as $\{n \mid n \text{ is an integer and } 12 \le n \le 25\}$.

73. From the graph we determine the following: $f(x)$ has no zeros.
The solutions of
$$f(x) < 0 \text{ are } (-\infty, 0) \text{ or } \{x \mid x < 0\}.$$
The solutions of
$$f(x) > 0 \text{ are } (0, \infty) \text{ or } \{x \mid x > 0\}.$$

75. From the graph we determine the following:
The solutions of $f(x) = 0$ are $-1, 0$
The solution of
$$f(x) < 0 \text{ is } (-\infty, -3) \cup (-1, 0) \text{ or}$$
$$\{x \mid x < -3 \text{ or } -1 < x < 0\}$$
The solution of
$$f(x) > 0 \text{ is } (-3, -1) \cup (0, 2) \cup (2, \infty) \text{ or}$$
$$\{x \mid -3 < x < -1 \text{ or } 0 < x < 2 \text{ or } x > 2\}$$

77. The domain of $f(x)$ are all x-values that make the expression under the square root symbol non-negative. Factoring the polynomial gives:
$$x^2 - 4x - 45 \ge 0$$
$$(x+5)(x-9) \ge 0$$
$x \le -5$ or $x \ge 9$. So the domain of f is $(-\infty, -5] \cup [9, \infty)$, or $\{x \mid x \le -5 \text{ or } x \ge 9\}$

79. The domain of $f(x)$ are all x-values that make the expression under the square root symbol non-negative. Factoring the polynomial gives:
$$x^2 + 8x \ge 0.$$
$$x(x+8) \ge 0$$
$x \le -8$ or $x \ge 0$. So the domain of f is $(-\infty, -8] \cup [0, \infty)$, or $\{x \mid x \le -8 \text{ or } x \ge 0\}$.

Chapter 8 Study Summary

1. $x^2 - 12x + 11 = 0$
   $(x-11)(x-1) = 0$
   $x = 1$, or $x = 11$

2. $x^2 - 18x + 81 = 5$
   $(x-9)^2 = 5$
   $x - 9 = \pm\sqrt{5} \Rightarrow x = 9 \pm \sqrt{5}$

3. $x^2 + 20x = 21$
   $x^2 + 20x + \left(\dfrac{20}{2}\right)^2 = 21 + \left(\dfrac{20}{2}\right)^2$
   $x^2 + 20x + (10)^2 = 21 + (10)^2$
   $(x+10)^2 = 121$
   $x + 10 = \pm\sqrt{121} = \pm 11$
   $x = -10 \pm 11$
   $x = -21$ or $x = 1$

4. $2x^2 - 3x - 9 = 0$
   $(2x+3)(x-3) = 0$
   $2x + 3 = 0$
   $2x = -3 \Rightarrow x = -\dfrac{3}{2}$
   or $x - 3 = 0 \Rightarrow x = 3$

5. $2x^2 + 5x + 9 = 0 \Rightarrow a = 2, \ b = 5, \ c = 9$

$b^2 - 4ac = 5^2 - 4 \cdot 2 \cdot 9 = 25 - 72 < 0$

The discriminant is negative. The equation has two imaginary solutions.

6. $a = n^2 + 1$

$n^2 = a - 1$

$n = \pm\sqrt{a - 1}$

7. $x - \sqrt{x} - 30 = 0$

$u^2 - u - 30 = 0$ for $u = \sqrt{x}$

$(u - 6)(u + 5) = 0$

$u = 6$ or $u = -5$

$\sqrt{x} = 6 \Rightarrow x = 6^2 = 36$

or $\sqrt{x} = -5$ which has no solution.

8. The function is quadratic with a positive leading coefficient. Its graph is a parabola opening upwards. It takes on a minimum value at its vertex point.

$f(x) = 2x^2 - 12x + 3 \Rightarrow a = 2, \ b = -12$

$\Rightarrow -\dfrac{b}{2a} = -\dfrac{-12}{2 \cdot 2} = 3$

$f(3) = 2(3)^2 - 12(3) + 3 = -15$

Vertex : $\big(3, f(3)\big) = (3, -15)$

Axis of symmetry: $x = 3$

Minimum of h: $y = -15$

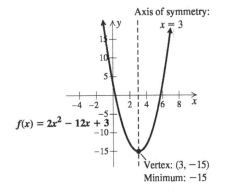

9. Let l = the length of the garden in feet, and w = its width, in feet. Then based on the amount of fencing she can afford, we have:
$2l + 2w = 120 \Rightarrow l + w = 60$

$\Rightarrow w = 60 - l$

Use this result to write the formula for the area in terms of one variable, l, then back substitute to find w.

$A = lw = l(60 - l) = -l^2 + 60l$

The formula for the area is quadratic with a negative leading coefficient. It takes on its maximum value at its vertex. We can use

$-\dfrac{b}{2a}$ to find the l-coordinate of the vertex.

However, note that in its factored form, $A = l(60 - l)$ it is easy to see that the graph has x-intercepts at $l = 0$ and $l = 60$. Since the graph is symmetric, the l-value of the vertex must be halfway between 0 and 60 at $l = 30$. Therefore the garden will have a maximum area when $l = 30$ ft, and $w = 60 - 30 = 30$ ft also. (This method of finding the vertex will work whenever $c = 0$ in $f(x) = ax^2 + bx + c$).

10. $x^2 - 11x - 12 < 0$

We find the zeros of the relation equation.

$x^2 - 11x - 12 = 0$

$(x + 1)(x - 12) = 0$

$x + 1 = 0$ or $x - 12 = 0$

$x = -1$ or $x = 12$

These zeros divide the number line into three intervals, A, B, and C.

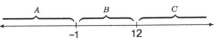

We choose a convenient test point from each interval to determine the sign of the interval. Using the table feature we see the signs on either side of the zeros.

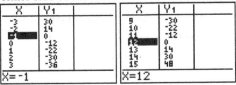

Since $x^2 - 11x - 12 < 0$, we want the interval where the sign is negative, or interval B.

The solution is $\{x \mid -1 < x < 12\}$ or $(-1, 12)$.

Chapter 8 Review Exercises

1. False: Some quadratics have repeated roots, or, equivalently, one solution.

2. False: Many quadratic equations have two imaginary-number solutions and no real-number solution.

3. True: The quadratic formula is derived by completing the square, so either method applies to all quadratic equations.

4. True: The discriminant is found under the square root term in the quadratic formula, and the square root of a negative number is imaginary.

5. True: The correct substitution can place many rational and radical equations into quadratic form.

6. False: The vertex is at $(-3, -4)$.

7. True: $b = 0 \Rightarrow -\dfrac{b}{2a} = 0$, so the axis of symmetry is $x = 0$.

8. True: When the leading coefficient is negative, the parabola opens downward, and, thus, takes on no minimum, but extends to $-\infty$.

9. True: Find the zeros by setting the function to zero. $x^2 - 9 = 0 \Rightarrow x = \pm 3$.

10. False: If a parabola's vertex is above the x-axis and it opens upward, or if its vertex is below the x-axis and it opens downward, then it will not cross the x-axis.

11. a) 2: The graph crosses the x-axis in two places.

    b) Positive : The parabola opens upward.

    c) -3: The minimum occurs at the vertex and is given by the y-coordinate.

12. $9x^2 - 2 = 0 \Rightarrow 9x^2 = 2 \Rightarrow x^2 = \dfrac{2}{9}$

    $\Rightarrow x = \pm\sqrt{\dfrac{2}{9}} = \pm\dfrac{\sqrt{2}}{3}$

    The solutions are $\pm\dfrac{\sqrt{2}}{3}$.

13. $8x^2 + 6x = 0$

    $2x(4x + 3) = 0$

    $2x = 0 \ or \ 4x + 3 = 0$

    $x = 0 \ or \qquad x = -\dfrac{3}{4}$

    The solutions are $-\dfrac{3}{4}$ and 0.

14. $x^2 - 12x + 36 = 9$

    $(x - 6)^2 = 9$

    $x - 6 = \pm\sqrt{9}$

    $x - 6 = -3 \ or \ x - 6 = 3$

    $x = 3 \quad or \qquad x = 9$

    The solutions are 3 and 9.

15. $x^2 - 4x + 8 = 0$

    $a = 1, b = -4, c = 8$

    $x = \dfrac{-(-4) \pm \sqrt{(-4)^2 - 4 \cdot 1 \cdot 8}}{2 \cdot 1}$

    $x = \dfrac{4 \pm \sqrt{16 - 32}}{2}$

    $x = \dfrac{4 \pm \sqrt{16 \cdot -1}}{2} = \dfrac{4 \pm 4i}{2}$

    $x = 2 \pm 2i$

    The solutions are $2 \pm 2i$.

16. $x(3x + 4) = 4x(x - 1) + 15$

    $3x^2 + 4x = 4x^2 - 4x + 15$

    $0 = x^2 - 8x + 15$

    $0 = (x - 5)(x - 3)$

    $x - 5 = 0 \ or \ x - 3 = 0$

    $x = 5 \ or \qquad x = 3$

    The solutions are 3 and 5.

17. $x^2 + 9x = 1$

$x^2 + 9x - 1 = 0$

$a = 1, \ b = 9, \ c = -1$

$x = \dfrac{-9 \pm \sqrt{9^2 - 4 \cdot 1 \cdot (-1)}}{2 \cdot 1}$

$x = \dfrac{-9 \pm \sqrt{81 + 4}}{2}$

$x = \dfrac{-9 \pm \sqrt{85}}{2}$

The solutions are $\dfrac{-9 \pm \sqrt{85}}{2}$, or $-\dfrac{9}{2} \pm \dfrac{\sqrt{85}}{2}$

18. $x^2 - 5x - 2 = 0$

$a = 1, \ b = -5, \ c = -2$

$x = \dfrac{-(-5) \pm \sqrt{(-5)^2 - 4 \cdot 1 \cdot (-2)}}{2 \cdot 1}$

$x = \dfrac{5 \pm \sqrt{25 + 8}}{2} = \dfrac{5 \pm \sqrt{33}}{2}$

$x = \dfrac{5 + \sqrt{33}}{2} \approx 5.3722813233$

$x = \dfrac{5 - \sqrt{33}}{2} \approx -0.3722813233$

The solutions rounded to 3 decimal places are 5.372 and –0.372.

19. Let $f(x) = 0$ and solve.

$0 = 4x^2 - 3x - 1$

$0 = (4x + 1)(x - 1)$

$4x + 1 = 0 \quad$ or $x - 1 = 0$

$x = -\dfrac{1}{4}$ or $\quad x = 1$

The solutions are $-\dfrac{1}{4}$ and 1.

20. $x^2 - 12x + \left(\dfrac{12}{2}\right)^2 = x^2 - 12x + 6^2$

$= (x - 6)^2$

21. $x^2 + \dfrac{3}{5}x + \left(\dfrac{1}{2} \cdot \dfrac{3}{5}\right)^2 = x^2 + \dfrac{3}{5}x + \left(\dfrac{3}{10}\right)^2$

$= \left(x + \dfrac{3}{10}\right)^2$

22. $x^2 - 6x + 1 = 0$

$x^2 - 6x + 3^2 = -1 + 3^2$

$(x - 3)^2 = 8$

$x - 3 = \pm\sqrt{8}$

$x = 3 \pm 2\sqrt{2}$

The solutions are $3 \pm 2\sqrt{2}$.

23. $A = P(1 + r)^t$

$2500(1 + r)^2 = 2704 \Rightarrow (1 + r)^2 = \dfrac{2704}{2500} = \dfrac{676}{625}$

$\Rightarrow 1 + r = \pm\sqrt{\dfrac{676}{625}} = \pm\dfrac{26}{25}$

Since r is positive, ignore the negative root.

$\Rightarrow r = -1 + \dfrac{26}{25} = \dfrac{1}{25} = 0.04 = 4\%$

24. $s = 16t^2 \Rightarrow 541 = 16t^2 \Rightarrow t^2 = \dfrac{541}{16}$

$\Rightarrow t = \pm\sqrt{\dfrac{541}{16}}$

Since the time to fall is positive, ignore the negative root.

$\Rightarrow t = \sqrt{\dfrac{541}{16}} = \dfrac{\sqrt{541}}{4} \approx 5.8$ seconds

25. $b^2 - 4ac = 3^2 - 4 \cdot 1 \cdot (-6) = 33$

Two irrational real solutions.

26. $b^2 - 4ac = 2^2 - 4 \cdot 1 + 5 = -16$

Two imaginary solutions.

27. $(x - 3i)(x + 3i) = 0 \Rightarrow x^2 - (3i)^2 = 0$

$\Rightarrow x^2 - 9i^2 = 0 \Rightarrow x^2 + 9 = 0$

28. $x = -4 \ or \ x = -4$

$(x + 4)^2 = 0$

$x^2 + 8x + 16 = 0$

29. Let x = the speed of the plane in still air, in mph. Then:

$$t_{\text{TO PLANT}} + t_{\text{BACK}} = 4 \Rightarrow \frac{300}{x+20} + \frac{300}{x-20} = 4$$

$$300(x+20) + 300(x-20)$$
$$= 4(x+20)(x-20)$$

$$300x \cancel{+6000} + 300x \cancel{-6000}$$
$$= 4x^2 - 1600$$

$$4x^2 - 600x - 1600 = 0$$

$$x^2 - 150x - 400 = 0$$

$$a = 1,\ b = -150,\ c = -400$$

$$x = \frac{-(-150) \pm \sqrt{(-150)^2 - 4 \cdot 1 \cdot (-400)}}{2 \cdot 1}$$

$$= \frac{150 \pm \sqrt{22,500 + 1600}}{2} = \frac{150 \pm \sqrt{24,100}}{2}$$

$$= \frac{150 \pm 10\sqrt{241}}{2} = 75 \pm 5\sqrt{241}$$

Since $75 - 5\sqrt{241} < 0$, the answer is:

$$75 + 5\sqrt{241} \approx 152.6 \text{ mph}$$

30. Let x = the time it takes Shawna to answer all the emails working alone, in hours, and y = the time it takes Erica to answer all the emails working alone, in hours. Then:

$$y = x + 6 \qquad (1)$$

$$\frac{1}{x} \cdot 4 + \frac{1}{y} \cdot 4 = 1 \quad (2)$$

Substitute $x + 6$ for y in (2) to find x.

$$\frac{1}{x} \cdot 4 + \frac{1}{y} \cdot 4 = 1 \quad (2)$$

$$\frac{4}{x} + \frac{4}{x+6} = 1$$

$$4(x+6) + 4x = 1 \cdot x(x+6)$$

$$4x + 24 + 4x = x^2 + 6x$$

$$x^2 - 2x - 24 = 0 \Rightarrow$$

$$(x-6)(x+4) = 0$$

$$x = -4 \text{ or } 6$$

Since x is greater than zero, Shawna can answer all the emails working alone in 6 hours.

31. Let $f(x) = 0$

$$0 = x^4 - 13x^2 + 36$$

$$0 = (x^2 - 9)(x^2 - 4)$$

$$0 = (x+3)(x-3)(x+2)(x-2)$$

$$x+3 = 0 \quad \text{or } x-3 = 0 \text{ or } x+2 = 0 \quad \text{or } x-2 = 0$$

$$x = -3 \text{ or} \quad x = 3 \text{ or} \quad x = -2 \text{ or} \quad x = 2$$

The x-intercepts are $(-3,0)$, $(-2,0)$, $(2,0)$, and $(3,0)$.

32. $15x^{-2} - 2x^{-1} - 1 = 0$

Let $u = x^{-1}$ $\left(\text{and } u^2 = x^{-2}\right)$

$$15u^2 - 2u - 1 = 0$$

$$(5u+1)(3u-1) = 0$$

$$5u+1 = 0 \quad \text{or } 3u-1 = 0$$

$$u = -\frac{1}{5} \text{ or} \quad u = \frac{1}{3}$$

$u = x^{-1}$; substitute and solve for x.

$$x^{-1} = -\frac{1}{5} \text{ or } x^{-1} = \frac{1}{3}$$

$$x = -5 \text{ or} \quad x = 3$$

The solutions are -5 and 3.

33. $\left(x^2 - 4\right)^2 - \left(x^2 - 4\right) - 6 = 0$

Let $u = x^2 - 4$ $\left[\text{and } u^2 = \left(x^2 - 4\right)^2\right]$

$$u^2 - u - 6 = 0$$

$$(u-3)(u+2) = 0$$

$$u = 3 \text{ or } u = -2$$

$u = x^2 - 4$; substitute and solve for x.

$$x^2 - 4 = 3 \quad \text{or } x^2 - 4 = -2$$

$$x^2 = 7 \quad \text{or} \quad x^2 = 2$$

$$x = \pm\sqrt{7} \text{ or} \quad x = \pm\sqrt{2}$$

All four values check. The solutions are $\pm\sqrt{7}$ and $\pm\sqrt{2}$.

34. a) $f(x) = -3(x+2)^2 + 4$

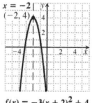

$f(x) = -3(x+2)^2 + 4$
Maximum: 4

b) Label the vertex (–2,4)
c) Draw the axis of symmetry $x = -2$
d) Maximum value is 4.

35. a) $f(x) = 2x^2 - 12x + 23$
$$= 2(x^2 - 6x) + 23$$
$$= 2(x^2 - 6x + 9) - 2 \cdot 9 + 23$$
$$= 2(x - 3)^2 + 5$$

Vertex: (3,5)
Axis of symmetry: $x = 3$

b)
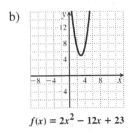

$f(x) = 2x^2 - 12x + 23$

36. $f(x) = x^2 - 9x + 14$

To determine the x-intercepts, let $f(x) = 0$
and solve the equations.
$$0 = x^2 - 9x + 14$$
$$0 = (x - 7)(x - 2)$$
$x = 7 \text{ or } x = 2$
The x-intercepts are (7,0) and (2,0); the
y-intercept is $(0, f(0))$, or (0,14).

37. $N = 3\pi\sqrt{\dfrac{1}{p}}$

$$\frac{N}{3\pi} = \sqrt{\frac{1}{p}}$$

$$\left(\frac{N}{3\pi}\right)^2 = \frac{1}{p}$$

$$\frac{9\pi^2}{N^2} = p$$

38. $2A + T = 3T^2$
$$0 = 3T^2 - T - 2A$$
This is a quadratic equation; use the quadratic
formula to solve. $a = 3$, $b = -1$, $c = -2A$

$$T = \frac{-(-1) \pm \sqrt{(-1)^2 - 4 \cdot 3 \cdot (-2A)}}{2 \cdot 3}$$

$$T = \frac{1 \pm \sqrt{1 + 24A}}{6}$$

39. The data looks linear over the period 2000 to
2004, but then deviates sharply from the line.
The data is neither quadratic nor linear.

40. The data points look like the right side of a
graph of a parabola which opens upward.
$$f(x) = ax^2 + bx + c, \ a > 0$$

41. The data points are almost linear.
$$f(x) = mx + b$$

42. Since we have two sides of fencing,
$30 = l + w$, where l is the length and w is the
width. The area of the rectangular area can
be expressed as the equation $A = l \cdot w$. Solve
the system of equations using substitution.
$30 = l + w \rightarrow l = 30 - w$

$$A = (30 - w)w$$

$$A = -w^2 + 30w$$

$$A = -(w^2 - 30w + 225) - (-225)$$

$$A = -(w - 15)^2 + 225$$

The maximum function value/area is $225\,ft^2$,
when $w = 15\,ft$ and $l = 30 - 15$, or $15\,ft$.

43. a) Find a function of the form
$f(x) = ax^2 + bx + c$ that satisfies the three
points listed. Substitute the given values for
x and $f(x)$ to get
$$a(0)^2 + b(0) + c = 1$$
$$\text{or } c = 1 \qquad (1)$$
$$a(20)^2 + b(20) + c = 1000$$
$$\text{or } 400a + 20b + c = 1000 \qquad (2)$$
$$a(60)^2 + b(60) + c = 32{,}000$$
$$\text{or } 3600a + 60b + c = 32{,}000 \quad (3)$$
Substitute $c = 1$ into (2) and (3) to generate
and two-variable system in a and b,

$$400a + 20b + 1 = 1000$$

$$\text{or } 400a + 20b = 999 \qquad (4)$$

$$3600a + 60b + 1 = 32{,}000$$

$$\text{or } 3600a + 60b = 31{,}999 \qquad (5)$$

Add $-3 \cdot (4)$ to (5) to eliminate b and find a.

$$-1200a - 60b = -2997 \qquad -3 \cdot (4)$$

$$\underline{3600a + 60b = 31{,}999 \qquad (5)}$$

$$2400a = 29{,}002$$

$$\Rightarrow a = \frac{29{,}002}{2400} = \frac{14{,}501}{1200}$$

Use (4) and the values of a to find b.

$$400a + 20b = 999 \quad (4)$$

$$\Rightarrow 20b = 999 - 400a = 999 - 400\left(\frac{14{,}501}{1200}\right)$$

$$= 999 - \frac{14{,}501}{3} = \frac{3 \cdot 999 - 14{,}501}{3}$$

$$\Rightarrow b = -\frac{11{,}504}{3 \cdot 20} = -\frac{\cancel{4} \cdot 2876}{3 \cdot \cancel{4} \cdot 5} = -\frac{2876}{15}$$

$$M(x) = \frac{14{,}501}{1200}x^2 - \frac{2876}{15}x + 1$$

b) In 2020, $x = 2020 - 1948 = 72$

$$M(72) = \frac{14{,}501}{1200}(72)^2 - \frac{2876}{15}(72) + 1$$

$$= \frac{14{,}501}{\cancel{24} \cdot \cancel{2} \cdot 25} \cdot \cancel{24} \cdot 3 \cdot \cancel{2} \cdot 36$$

$$- \frac{2876}{\cancel{3} \cdot 5} \cdot \cancel{3} \cdot 24 + 1$$

$$= \frac{14{,}501 \cdot 3 \cdot 36 - 5 \cdot 2876 \cdot 24 + 25 \cdot 1}{25}$$

$$= \frac{1{,}221{,}013}{25} \approx 48{,}841$$

44. a) $M(x) = 12.6207x^2 - 242.2557x + 706.6461$

　　b) $M(72)$

$$= 12.6207(72)^2 - 242.2557(72) + 706.6461$$

$$\approx 48{,}690$$

45. $\qquad x^3 - 3x > 2x^2$

$$x^3 - 2x^2 - 3x > 0$$

Solve the related equation.

$$x^3 - 2x^2 - 3x = 0$$

$$x(x^2 - 2x - 3) = 0$$

$$x(x - 3)(x + 1) = 0$$

$$x = 0 \quad \text{or} \quad x - 3 = 0 \quad \text{or} \quad x + 1 = 0$$

$$x = 0 \quad \text{or} \qquad x = 3 \quad \text{or} \qquad x = -1$$

The zeros are -1, 0, and 3. These zeros divide the number line into four intervals: A, B, C, and D.

We will select a convenient test point from each interval to determine the sign of $x^3 - 2x^2 - 3x$ for that interval.

Interval A: Let $x = -5$:

$$(-5)^3 - 2(-5)^2 - 3(-5)$$

$$= -125 - 2(25) - 3(-5)$$

$$= -125 - 50 + 15$$

$$= -160$$

Interval B: Let $x = -0.5$:

$$(-0.5)^3 - 2(-0.5)^2$$

$$- 3(-0.5)$$

$$= -0.125 - 2(0.25) - 3(-0.5)$$

$$= -0.125 - 0.5 + 1.5 = 0.875$$

Interval C: Let $x = 1$:

$$(1)^3 - 2(1)^2 - 3(1)$$

$$= 1 - 2(1) - 3(1)$$

$$= 1 - 2 - 3 = -4$$

Interval D: Let $x = 4$:

$$4^3 - 2 \cdot 4^2 - 3 \cdot 4$$

$$= 64 - 2 \cdot 16 - 3 \cdot 4$$

$$= 64 - 32 - 12 = 20$$

We indicate our results on the number line.

We are looking for the intervals where $x^3 - 2x^2 - 3x > 0$, or the $+$ intervals.

The solution set is $(-1, 0) \cup (3, \infty)$ or $\{x \mid -1 < x < 0 \text{ or } x > 3\}$

46. $\dfrac{x-5}{x+3} \le 0; \quad x \ne -3$

Solve the related equation.

$$\dfrac{x-5}{x+3} = 0 \qquad \text{LCD is } x+3$$

$$(x+3)\dfrac{x-5}{x+3} = (x+3) \cdot 0$$

$$x-5 = 0$$

$$x = 5$$

We must also find any values which make the denominator 0.

$$x+3 = 0$$

$$x = -3$$

Use 5 and –3 to divide the number line into intervals.

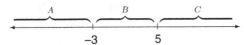

We will select a convenient test point from each interval to determine the sign of

$$\dfrac{x-5}{x+3}$$

Interval A: Let $x = -4$

$$\dfrac{-4-5}{-4+3} = \dfrac{-9}{-1} = 9$$

Interval B: Let $x = 0$

$$\dfrac{0-5}{0+3} = \dfrac{-5}{3}$$

Interval C: Let $x = 10$

$$\dfrac{10-5}{10+3} = \dfrac{5}{13}$$

We indicate our results on the number line.

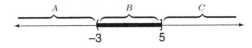

We are looking for intervals where $\frac{x-5}{x+3}$ is less than 0. We also want the value which makes the expression equal 0, or $x = 5$. The solution is $(-3, 5]$, or $\{x \mid -3 < x \le 5\}$.

47. *Thinking and Writing Exercise.* Completing the square was used to solve quadratic equations and to graph functions by rewriting the function in the vertex form:

$$f(x) = a(x-h)^2 + k.$$

48. *Thinking and Writing Exercise.* The model found in Exercise 44 predicts 4363 more restaurants in 2010 than the model from Exercise 43. The greater prediction seems to fit the pattern better.

49. *Thinking and Writing Exercise.* The most solutions a polynomial can have is equal to the degree of the polynomial. Thus the most solutions an equation of the form $ax^4 + bx^2 + c = 0$ can have is four. One way to see this is to remember that if k is a root, then $x - k$ must be a factor of the polynomial. Multiplying more than four linear factors of this type together would yield a polynomial of degree equal to the number of factors.

50. The x-intercepts are $(-3,0)$ and $(5,0)$, and the y-intercept is $(0,-7)$. Substituting these ordered pairs into the equation $f(x) = ax^2 + bx + c$ gives a system of equations.

$$0 = a \cdot (-3)^2 + b(-3) + c$$

$$0 = a \cdot 5^2 + b \cdot 5 + c$$

$$-7 = a \cdot 0 + b \cdot 0 + c$$

$$or$$

$$0 = 9a - 3b + c$$

$$0 = 25a + 5b + c$$

$$-7 = c$$

Solving this equation gives us:

$$a = \dfrac{7}{15}, b = -\dfrac{14}{15}, c = -7.$$

The equation is $f(x) = \dfrac{7}{15}x^2 - \dfrac{14}{15}x - 7$

51. It was shown that the sum of the solutions of $ax^2 + bx + c = 0$ is $-\dfrac{b}{a}$, and the product is $\dfrac{c}{a}$.

$$3x^2 - hx + 4k = 0$$

$$a = 3, \ b = -h, \text{ and } c = 4k$$

Substituting:

$$-\dfrac{b}{a} : \dfrac{-(-h)}{3} = 20, h = 60$$

$$\dfrac{c}{a} : \dfrac{4k}{3} = 80, k = 60$$

52. Let x and y represent two positive integers. Since one of the numbers is the square root of the other, we let $y = \sqrt{x}$. To find their average, we find their sum and divide by 2.

$$\frac{x + \sqrt{x}}{2} = 171$$

$$x + \sqrt{x} = 342$$

$$x + \sqrt{x} - 342 = 0$$

Let $u = \sqrt{x}$, (and $u^2 = x$).

$$u^2 + u - 342 = 0$$

$$(u + 19)(u - 18) = 0$$

$$u = -19 \ or \ u = 18$$

Substituting: $\sqrt{x} = -19$ or $\sqrt{x} = 18$

We use only $\sqrt{x} = 18$

$$x = 324$$

The numbers are 18 and 324.

Chapter 8 Test

1. a) 0 : The graph does not cross the x- axis.

   b) Negative : The parabola opens downward.

   c) -1 : The maximum value of the function is given by the y-coordinate of the vertex.

2. $25x^2 - 7 = 0 \Rightarrow 25x^2 = 7 \Rightarrow x^2 = \dfrac{7}{25}$

   $\Rightarrow x = \pm\sqrt{\dfrac{7}{25}} = \pm\dfrac{\sqrt{7}}{5}$

3. $4x(x - 2) - 3x(x + 1) = -18$

   $$4x^2 - 8x - 3x^2 - 3x = -18$$

   $$x^2 - 11x + 18 = 0$$

   $$(x - 9)(x - 2) = 0$$

   $$x - 9 = 0 \ or \ x - 2 = 0$$

   $$x = 9 \ or \quad x = 2$$

   The solutions are 9 and 2.

4. $x^2 + 2x + 3 = 0 \Rightarrow a = 1, \ b = 2, \ c = 3$

   $$x = \frac{-2 \pm \sqrt{2^2 - 4 \cdot 1 \cdot 3}}{2 \cdot 1} = \frac{-2 \pm \sqrt{-8}}{2}$$

   $$= \frac{-2 \pm 2\sqrt{2}i}{2} = \frac{\cancel{2}(-1 \pm \sqrt{2}i)}{\cancel{2}} = -1 \pm \sqrt{2}i$$

   The solutions are $-1 \pm \sqrt{2}i$.

5. $2x + 5 = x^2$

   $$0 = x^2 - 2x - 5$$

   $$a = 1, b = -2, c = 5$$

   $$x = \frac{-(-2) \pm \sqrt{(-2)^2 - 4 \cdot 1 \cdot (-5)}}{2 \cdot 1}$$

   $$x = \frac{2 \pm \sqrt{24}}{2} = \frac{2 \pm 2\sqrt{6}}{2} = \frac{2(1 \pm \sqrt{6})}{2}$$

   $$x = 1 \pm \sqrt{6}$$

   The solutions are $1 \pm \sqrt{6}$.

6. $x^{-2} - x^{-1} = \dfrac{3}{4}$

   Let $u = x^{-1}$ (and $u^2 = x^{-2}$)

   $$u^2 - u - \frac{3}{4} = 0$$

   $$4u^2 - 4u - 3 = 0$$

   $$(2u - 3)(2u + 1) = 0$$

   $$u = \frac{3}{2} \ or \ u = -\frac{1}{2}$$

   $u = x^{-1}$, so $x^{-1} = \dfrac{3}{2} \ or \ x^{-1} = -\dfrac{1}{2}$

   $$x = \frac{2}{3} \ or \quad x = -2$$

   The solutions are -2 and $\dfrac{2}{3}$.

7. $x^2 + 3x = 5$

   $$x^2 + 3x - 5 = 0$$

   $$a = 1, b = 3, c = -5$$

   $$x = \frac{-3 \pm \sqrt{3^2 - 4 \cdot 1 \cdot (-5)}}{2 \cdot 1} = \frac{-3 \pm \sqrt{29}}{2}$$

   $$x = \frac{-3 - \sqrt{29}}{2} \approx -4.193$$

   $$x = \frac{-3 + \sqrt{29}}{2} \approx 1.193$$

8. Let $f(x) = 0$ and solve for x.
$$0 = 12x^2 - 19x - 21$$
$$0 = (4x + 3)(3x - 7)$$
$$x = -\frac{3}{4} \text{ or } x = \frac{7}{3}$$

The solutions are $-\frac{3}{4}$ and $\frac{7}{3}$.

9. $x^2 - 20x + \left(\dfrac{20}{2}\right)^2 = x^2 - 20x + 10^2$
$$= (x - 10)^2$$

10. $x^2 + \dfrac{2}{7}x + \left(\dfrac{1}{2} \cdot \dfrac{2}{7}\right)^2 = x^2 + \dfrac{2}{7}x + \left(\dfrac{1}{7}\right)^2$
$$= \left(x + \dfrac{1}{7}\right)^2$$

11. $x^2 + 10x + 15 = 0$
$$x^2 + 10x = -15$$
$$x^2 + 10x + 25 = -15 + 25$$
$$(x + 5)^2 = 10$$
$$x + 5 = \pm\sqrt{10}$$
$$x = -5 \pm \sqrt{10}$$

The solutions are $-5 \pm \sqrt{10}$.

12. $x^2 + 2x + 5 = 0$
$$\Rightarrow b^2 - 4ac = 2^2 - 4 \cdot 1 \cdot 5 < 0$$
Because the discriminant is negative, the equation will have two imaginary solutions.

13. $(x - \sqrt{11})(x - (-\sqrt{11})) = 0$
$$\Rightarrow (x - \sqrt{11})(x + \sqrt{11}) = 0$$
$$\Rightarrow x^2 - 11 = 0$$

14. Let r = speed of the boat in still water. Since the rate of the river is 4 km/h, the rate upstream is $r - 4$ and the rate downstream is $r + 4$. Using $T = \dfrac{d}{r}$, we have time upstream is $\dfrac{60}{r - 4}$ and the time downstream is $\dfrac{60}{r + 4}$.
The total time is 8 hr. We have the equation:

$\dfrac{60}{r - 4} + \dfrac{60}{r + 4} = 8$ LCD is $(r - 4)(r + 4)$

$$(r - 4)(r + 4) \cdot \left(\dfrac{60}{r - 4} + \dfrac{60}{r + 4}\right) = (r - 4)(r + 4) \cdot 8$$

$$60(r + 4) + 60(r - 4) = 8(r^2 - 16)$$

$$60r + 240 + 60r - 240 = 8r^2 - 128$$

$$0 = 8r^2 - 120r - 128$$

$$0 = r^2 - 15r - 16$$

$$0 = (r - 16)(r + 1)$$

$r = 16$ or $r = -1$
Since the speed cannot be negative, the speed of the boat in still water is 16 km/h.

15. Let x = the number of hours it takes Dal alone
and
$x + 4$ = the number of hours it takes Kim.

Dal can do $\dfrac{1}{x}$ of the job in 1 hr, and Kim can

do $\dfrac{1}{x + 4}$. They work together for

$1\dfrac{1}{2}$ hrs., or $\dfrac{3}{2}$ hrs. We solve the equation

$\dfrac{1}{x} \cdot \dfrac{3}{2} + \dfrac{1}{x + 4} \cdot \dfrac{3}{2} = 1$, or $\dfrac{3}{2x} + \dfrac{3}{2(x + 4)} = 1$

for x
LCD is $2x(x + 4)$

$$2x(x + 4) \cdot \left(\dfrac{3}{2x} + \dfrac{3}{2(x + 4)}\right) = 2x(x + 4) \cdot 1$$

$$3(x + 4) + 3x = 2x^2 + 8x$$

$$3x + 12 + 3x = 2x^2 + 8x$$

$$0 = 2x^2 + 2x - 12$$

$$0 = x^2 + x - 6$$

$$0 = (x + 3)(x - 2)$$

$x = -3$ or $x = 2$
Since negative time has no meaning for this problem, Dal can assemble the swing set in 2 hrs.

16. $f(x) = 0 \Rightarrow x^4 - 15x^2 - 16 = 0$

$\Rightarrow u^2 - 15u - 16 = 0$ for $u = x^2$

$\Rightarrow (u - 16)(u + 1) = 0$

$\Rightarrow u = 16 \Rightarrow x^2 = 16 \Rightarrow x = \pm\sqrt{16} = \pm 4$

or $u = -1 \Rightarrow x^2 = -1 \Rightarrow x = \pm\sqrt{-1} = \pm i$

The function only has two real roots. Those correspond to the x-intercepts: $(-4, 0)$, $(4, 0)$.

17. a)

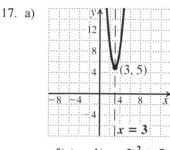

$$f(x) = 4(x - 3)^2 + 5$$
Minimum: 5

b) Label the vertex $(3,5)$

c) Draw the axis of symmetry $x = 3$

d) Minimum function value of 5.

18. a) $f(x) = 2x^2 + 4x - 6$

$= 2(x^2 + 2x) - 6$

$= 2(x^2 + 2x + 1) - 2 \cdot 1 - 6$

$= 2(x + 1)^2 - 8$

Vertex: $(-1, -8)$
Axis of symmetry: $x = -1$

b)

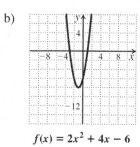

$$f(x) = 2x^2 + 4x - 6$$

19. To find the x-intercepts, set $f(x) = 0$.

$0 = x^2 - x - 6$

$0 = (x - 3)(x + 2)$

$x = 3$ or $x = -2$

The x-intercepts are $(-2,0)$ and $(3,0)$ and the y-intercept is $\left(0, f(0)\right)$, or $(0,-6)$.

20. $V = \frac{1}{3}\pi\left(R^2 + r^2\right) \Rightarrow \frac{3V}{\pi} = R^2 + r^2$

$\Rightarrow r^2 = \frac{3V}{\pi} - R^2 \Rightarrow r = \sqrt{\frac{3V}{\pi} - R^2}$

Ignore the negative root since we are assuming $r > 0$.

21. The data points rise then fall; this appears to represent a parabola which opens downward. A quadratic function.

22. $C(x) = 0.2x^2 - 1.3x + 3.4025$

$C(x) = 0.2\left(x^2 - 6.5x\right) + 3.4025$

$C(x) = 0.2\left(x^2 - 6.5x + 10.5625\right) - 0.2(10.5625) + 3.4025$

$C(x) = 0.2\left(x - 3.25\right)^2 + 1.29$

A minimum of $1.29 hundred, or $129 when 3.25 hundred, or 325 cabinets are built.

23. We look for a function of the form $f(x) = ax^2 + bx + c$. Substituting the data points, we have:

$35 = a(0)^2 + b(0) + c$

$310 = a(4)^2 + b(4) + c$

$200 = a(6)^2 + b(6) + c$

or

$35 = c$

$310 = 16a + 4b + c$

$200 = 36a + 6b + c$

Solving this system we get:

$a = -\frac{165}{8}, \ b = \frac{605}{4}, \ c = 35$

The function is $p(x) = -\frac{165}{8}x^2 + \frac{605}{4}x + 35$

24. Enter the data and then use the quadratic regression feature. We have

$p(x) = -23.5417x^2 + 162.2583x + 57.6167$

25. $x^2 + 5x \le 6$

$x^2 + 5x - 6 \le 0$

We find the zeros of the relation equation.

$x^2 + 5x - 6 = 0$

$(x + 6)(x - 1) = 0$

$x + 6 = 0$ or $x - 1 = 0$

$x = -6$ or $x = 1$

These zeros divide the number line into three intervals, A, B, and C.

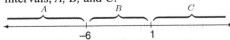

Choose a convenient test point from each interval to determine the sign of the interval.

X	Y1	
-10	44	
0	-6	
3	18	

X = -10

Since $x^2 + 5x - 6 \le 0$, we want the interval where the sign is negative, or interval B. The endpoints of the interval, where $x^2 + 5x - 6 = 0$, should also be included in the solution.

Our solution is $\{x \mid -6 \le x \le 1\}$ or $[-6, 1]$.

26. $x - \dfrac{1}{x} > 0$, $x \ne 0$

We find the zeros of the related equation.

$x - \dfrac{1}{x} = 0$, LCD is x

$x\left(x - \dfrac{1}{x}\right) = 0$

$x^2 - 1 = 0$

$(x + 1)(x - 1) = 0$

$x = -1$ or $x = 1$

As already noted, $x \ne 0$.

These three numbers divide the number line into four intervals: A, B, C, and D.

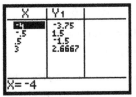

We choose a convenient test point from each interval to determine the sign of the interval.

X	Y1	
-4	-3.75	
-.5	1.5	
.5	-1.5	
3	2.6667	

X = -4

Since $x - \dfrac{1}{x} > 0$, we want the interval(s) where the sign is positive, so we choose intervals B and D.

Our solution is $\{x \mid -1 < x < 0, \text{ or } x > 1\}$, or $(-1, 0) \cup (1, \infty)$.

27. $kx^2 + 3x - k = 0$; one solution is -2.

We first find k by substituting -2 for x.

$k(-2)^2 + 3(-2) - k = 0$

$4k - 6 - k = 0$

$3k = 6$

$k = 2$

We now substitute 2 for k in the original equation.

$2x^2 + 3x - 2 = 0$

$(2x - 1)(x + 2) = 0$

$2x - 1 = 0$ or $x + 2 = 0$

$x = \dfrac{1}{2}$ or $x = -2$

The other solution is $\dfrac{1}{2}$.

28. If $2i$ is a solution, then $-2i$ must also be a solution since the imaginary solutions of $P(x) = 0$, where $P(x)$ is a polynomial, always come in conjugate pairs.

Thus $x - 2i$, $x + 2i$, and $x + \sqrt{3}$ must be factors of the polynomial. In order for the coefficients of the equation to be integers, $x - \sqrt{3}$ must also be a factor. Since the product of these factors will be a fourth degree polynomial, these are the only factors we need.

$(x - 2i)(x + 2i)(x + \sqrt{3})(x - \sqrt{3}) = 0$

$(x^2 + 4)(x^2 - 3) = 0$

$x^4 + x^2 - 12 = 0$

The left side of the equation could also be multiplied by any constant and retain the same roots.

29. $x^4 - 4x^2 - 1 = 0$

$u^2 - 4u - 1 = 0$ for $u = x^2$

$u = \dfrac{4 \pm \sqrt{16+4}}{2} = \dfrac{4 \pm 2\sqrt{5}}{2} = 2 \pm \sqrt{5}$

$x^2 = 2 + \sqrt{5} \Rightarrow x = \pm\sqrt{2+\sqrt{5}}$

or $x^2 = 2 - \sqrt{5} \Rightarrow x = \pm\sqrt{2-\sqrt{5}}$

However, since $\sqrt{5} > 2$, these last two solutions are imaginary. Using $2 - \sqrt{5} = -(\sqrt{5}-2)$, we can write these solutions in terms of i.

$x = \pm\sqrt{2-\sqrt{5}} = \pm\sqrt{-(\sqrt{5}-2)} = \pm\sqrt{\sqrt{5}-2}\,i$

Thus the four solutions are:

$x = \sqrt{2+\sqrt{5}},\ x = -\sqrt{2+\sqrt{5}}.\ x = \sqrt{\sqrt{5}-2}\,i$

and $x = -\sqrt{\sqrt{5}-2}\,i$

Chapter 9

Exponential Functions and Logarithmic Functions

1. True

3. False

5. False

7. True

9. a) $(f \circ g)(1) = f(g(1)) = f(1-3)$
   $= f(-2) = (-2)^2 + 1 = 5$

   b) $(g \circ f)(1) = g(f(1)) = g(1^2 + 1)$
   $= g(2) = 2 - 3 = -1$

   c) $(f \circ g)(x) = f(g(x)) = f(x-3)$
   $= (x-3)^2 + 1$
   $= x^2 - 6x + 9 + 1$
   $= x^2 - 6x + 10$

   d) $(g \circ f)(x) = g(f(x)) = g(x^2 + 1)$
   $= (x^2 + 1) - 3$
   $= x^2 + 1 - 3$
   $= x^2 - 2$

11. a) $(f \circ g)(1) = f(g(1)) = f(2 \cdot 1^2 - 7)$
    $= f(2-7) = f(-5)$
    $= 5(-5) + 1 = -25 + 1 = -24$

    b) $(g \circ f)(1) = g(f(1)) = g(5 \cdot 1 + 1)$
    $= g(6) = 2(6)^2 - 7$
    $= 2 \cdot 36 - 7$
    $= 72 - 7 = 65$

c) $(f \circ g)(x) = f(g(x)) = f(2x^2 - 7)$
   $= 5(2x^2 - 7) + 1$
   $= 10x^2 - 35 + 1$
   $= 10x^2 - 34$

d) $(g \circ f)(x) = g(f(x)) = g(5x + 1)$
   $= 2(5x + 1)^2 - 7$
   $= 2(25x^2 + 10x + 1) - 7$
   $= 50x^2 + 20x + 2 - 7$
   $= 50x^2 + 20x - 5$

13. a) $(f \circ g)(1) = f(g(1)) = f\left(\dfrac{1}{1^2}\right)$
    $= f(1) = 1 + 7 = 8$

    b) $(g \circ f)(1) = g(f(1)) = g(1+7)$
    $= g(8) = \dfrac{1}{8^2} = \dfrac{1}{64}$

    c) $(f \circ g)(x) = f(g(x))$
    $= f\left(\dfrac{1}{x^2}\right) = \dfrac{1}{x^2} + 7$

    d) $(g \circ f)(x) = g(f(x))$
    $= g(x+7) = \dfrac{1}{(x+7)^2}$

15. a) $(f \circ g)(1) = f(g(1)) = f(1+3)$
    $= f(4) = \sqrt{4} = 2$

    b) $(g \circ f)(1) = g(f(1)) = g(\sqrt{1})$
    $= g(1) = 1 + 3 = 4$

    c) $(f \circ g)(x) = f(g(x)) = f(x+3)$
    $= \sqrt{x+3}$

    d) $(g \circ f)(x) = g(f(x)) = g(\sqrt{x})$
    $= \sqrt{x} + 3$

17. a) $(f \circ g)(1) = f(g(1)) = f\left(\frac{1}{1}\right)$

$= f(1) = \sqrt{4 \cdot 1} = 2$

b) $(g \circ f)(1) = g(f(1)) = g(\sqrt{4 \cdot 1})$

$= g(2) = \frac{1}{2}$

c) $(f \circ g)(x) = f(g(x)) = f\left(\frac{1}{x}\right) = \sqrt{\frac{4}{x}}$

d) $(g \circ f)(x) = g(f(x)) = g(\sqrt{4x}) = \frac{1}{\sqrt{4x}}$

19. a) $(f \circ g)(1) = f(g(1)) = f(\sqrt{1-1})$

$= f(0) = 0^2 + 4 = 4$

b) $(g \circ f)(1) = g(f(1)) = g(1^2 + 4)$

$= g(5) = \sqrt{5-1} = \sqrt{4} = 2$

c) $(f \circ g)(x) = f(g(x)) = f(\sqrt{x-1})$

$= (\sqrt{x-1})^2 + 4$

$= x - 1 + 4 = x + 3$

d) $(g \circ f)(x) = g(f(x)) = g(x^2 + 4)$

$= \sqrt{x^2 + 4 - 1}$

$= \sqrt{x^2 + 3}$

21. Since $(y_1 \circ y_2)(-3) = y_1(y_2(-3))$, we first find $y_2(-3)$. Locate –3 in the x-column and then move across to the y_2-column to find that $y_2(-3) = 1$. Now we have $y_1(y_2(-3))$ $= y_1(1)$. Locate 1 in the x-column and then move across to the y_1-column to find that $y_1(1) = 8$. Thus, $(y_1 \circ y_2)(-3) = 8$.

23. Since $(y_1 \circ y_2)(-1) = y_1(y_2(-1))$, we first find $y_2(-1)$. Locate –1 in the x-column and then move across to the y_2-column to find that $y_2(-1) = -3$. Now we have $y_1(y_2(-1))$ $= y_1(-3)$. Locate –3 in the x-column and then move across to the y_1-column to find that $y_1(-3) = -4$.

Thus $(y_1 \circ y_2)(-1) = -4$

25. Since $(y_2 \circ y_1)(1) = y_2(y_1(1))$, we first find $y_1(1)$. Locate 1 in the x-column and then move across to the y_1-column to find that $y_1(1) = 8$. Now we have $y_2(y_1(1)) = y_2(8)$. However, y_2 is not defined for $x = 8$, so $(y_2 \circ y_1)(1)$ is not defined.

27. Since $(f \circ g)(2) = f(g(2))$, we first find $g(2)$. Locate 2 in the x-column and then move across to the $g(x)$-column to find that $g(2) = 5$. Now we have $f(g(2)) = f(5)$. Locate 5 in the x-column and then move across to the $f(x)$-column to find that $f(5) = 4$. Thus, $(f \circ g)(2) = 4$.

29. To find $f(g(3))$ we first find $g(3)$. Locate 3 in the x-column and then move across to the $g(x)$-column to find that $g(3) = 8$. Now we have $f(g(3)) = f(8)$. However, $f(x)$ is not defined for $x = 8$, so $f(g(3))$ is not defined.

31. $h(x) = (3x - 5)^4$

This is $3x - 5$ raised to the fourth power, so the two most obvious functions are

$f(x) = x^4$ and $g(x) = 3x - 5$.

33. $h(x) = \sqrt{2x + 7}$

We have $2x + 7$ and take the square root of this expression.

$f(x) = \sqrt{x}$ and $g(x) = 2x + 7$

35. $h(x) = \frac{2}{x - 3}$

This is 2 divided by $x - 3$, so we can use the functions $f(x) = \frac{2}{x}$ and $g(x) = x - 3$.

37. $h(x) = \frac{1}{\sqrt{7x + 2}}$

This is the reciprocal of the square root of $7x + 2$. The two functions can be

$f(x) = \frac{1}{\sqrt{x}}$ and $g(x) = 7x + 2$.

39. $h(x) = \dfrac{1}{\sqrt{3x}} + \sqrt{3x}$

This is the reciprocal of the square root of $3x$ plus the square root of $3x$. Two functions that can be used are

$f(x) = \dfrac{1}{x} + x$ and $g(x) = \sqrt{3x}$.

41. The graph of $f(x) = x - 5$ is shown below.

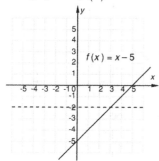

Since there is no horizontal line that crosses the graph more than once, the function is one-to-one.

43. The graph of $f(x) = x^2 + 1$ is shown below.

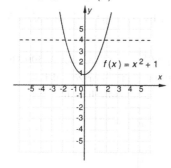

Observe that the graph of this function is a parabola that opens up. There are many horizontal lines that cross the graph more than once. In particular, the line $y = 4$ crosses the graph more than once. The function is not one-to-one.

45. Since no horizontal line crosses the graph more than once, the function is one-to-one.

47. Since we can draw at least one horizontal line that crosses the graph more than once, the function is not one-to-one.

49. a) The function $f(x) = x + 4$ is a linear function that is not constant, so it passes the horizontal-line test. Thus, f is one-to-one.

b) Replace $f(x)$ by y: $y = x + 4$
Interchange x and y: $x = y + 4$
Solve for y: $x - 4 = y$
Replace y by $f^{-1}(x)$: $f^{-1}(x) = x - 4$

51. a) The function $f(x) = 2x$ is a linear function that is not constant, so it passes the horizontal-line test. Thus, f is one-to-one.

b) Replace $f(x)$ by y: $y = 2x$
Interchange x and y: $x = 2y$
Solve for y: $\dfrac{x}{2} = y$
Replace y by $f^{-1}(x)$: $f^{-1}(x) = \dfrac{x}{2}$

53. a) The function $g(x) = 3x - 1$ is a linear function that is not constant, so it passes the horizontal-line test. Thus, g is one-to-one.

b) Replace $g(x)$ by y: $y = 3x - 1$
Interchange x and y: $x = 3y - 1$
Solve for y: $\dfrac{x + 1}{3} = y$
Replace y by $g^{-1}(x)$: $g^{-1}(x) = \dfrac{x + 1}{3}$

55. a) The function $f(x) = \frac{1}{2}x + 1$ is a linear function that is not constant, so it passes the horizontal-line test. Thus, f is one-to-one.

b) Replace $f(x)$ by y: $y = \dfrac{1}{2}x + 1$
Interchange x and y: $x = \dfrac{1}{2}y + 1$
Solve for y: $2x - 2 = y$
Replace y by $f^{-1}(x)$: $f^{-1}(x) = 2x - 2$

57. $g(x) = x^2 + 5$

a) The graph of this function is a parabola which opens upward. This does not pass the horizontal-line test. The function is not one-to-one.

59. a)   The function $h(x) = -10 - x$ is a linear

function that is not constant, so it passes
the horizontal-line test.
Thus, h his one-to-one.

b)   Replace $h(x)$ by y: $y = -10 - x$

Interchange x and y: $x = -10 - y$

Solve for y: $y = -x - 10$

Replace y by $h^{-1}(x)$: $h^{-1}(x) = -x - 10$,

or $h^{-1}(x) = -10 - x$

61. a)   The graph of $f(x) = \dfrac{1}{x}$ is shown below.

It passes the horizontal-line test, so the
function is one-to-one.

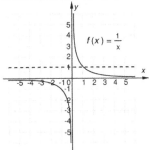

b)   Replace $f(x)$ by y: $y = \dfrac{1}{x}$

Interchange x and y: $x = \dfrac{1}{y}$

Solve for y: $xy = 1$

$$y = \dfrac{1}{x}$$

Replace y by $f^{-1}(x)$: $f^{-1}(x) = \dfrac{1}{x}$

63. $G(x) = 4$

a)   The graph of this function is a horizontal
line, so the function is not one-to-one.

65. a)   The function $f(x) = \dfrac{2x+1}{3} = \dfrac{2}{3}x + \dfrac{1}{3}$ is a

linear function that is not constant, so it
passes the horizontal-line test. Thus, f is
one-to-one.

b)   Replace $f(x)$ by y: $y = \dfrac{2x+1}{3}$

Interchange x and y: $x = \dfrac{2y+1}{3}$

Solve for y: $3x = 2y + 1$

$$3x - 1 = 2y$$

$$\dfrac{3x-1}{2} = y$$

Replace y by $f^{-1}(x)$: $f^{-1}(x) = \dfrac{3x-1}{2}$

67. a)   The graph of $f(x) = x^3 - 5$ is shown

below. It passes the horizontal-line test,
so the function is one-to-one.

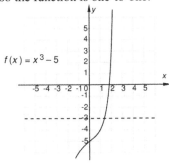

b)   Replace $f(x)$ by y: $y = x^3 - 5$

Interchange x and y: $x = y^3 - 5$

Solve for y: $x + 5 = y^3$

$$\sqrt[3]{x+5} = y$$

Replace y by $f^{-1}(x)$: $f^{-1}(x) = \sqrt[3]{x+5}$

69. a)   The graph of $g(x) = (x-2)^3$ is shown

below. It passes the horizontal-line test,
so the function is one-to-one.

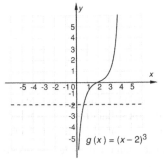

b)   Replace $g(x)$ by y: $y = (x-2)^3$

Interchange x and y: $x = (y-2)^3$

Solve for y: $\sqrt[3]{x} = y - 2$

$$\sqrt[3]{x} + 2 = y$$

Replace y by $g^{-1}(x)$: $g^{-1}(x) = \sqrt[3]{x} + 2$

71. a) The graph of $f(x) = \sqrt{x}$ is shown below. It passes the horizontal-line test, so the function is one-to-one.

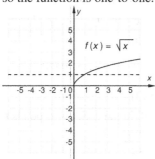

b) Replace $f(x)$ by y: $y = \sqrt{x}$

   $\left(\text{Note that } f(x) \geq 0.\right)$

   Interchange x and y: $x = \sqrt{y}$

   Solve for y: $x^2 = y$

   Replace y by $f^{-1}(x)$: $f^{-1}(x) = x^2$, $x \geq 0$

73. First graph $f(x) = \frac{2}{3}x + 4$. Then graph the inverse function by reflecting the graph of $f(x) = \frac{2}{3}x + 4$ across the line $y = x$. The graph of the inverse function can also be found by first finding a formula for the inverse, substituting to find function values, and then plotting points.

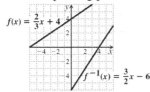

75. Follow the procedure in Exercise 73 to graph the function $f(x) = x^3 + 1$ and its inverse.

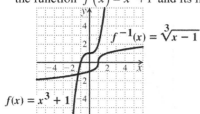

77. Follow the procedure in Exercise 73 to graph the function $g(x) = \frac{1}{2}x^3$ and its inverse.

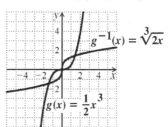

79. Follow the procedure in Exercise 73 to graph the function $F(x) = -\sqrt{x}$ and its inverse.

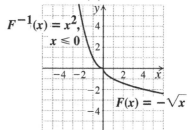

81. Follow the procedure in Exercise 73 to graph the function $f(x) = -x^2$, $x \geq 0$ and its inverse.

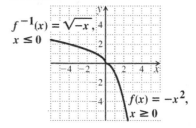

83. We check to see that $f^{-1} \circ f(x) = x$ and $f \circ f^{-1}(x) = x$.

$f^{-1} \circ f(x) = f^{-1}(f(x)) = f^{-1}\left(\sqrt[3]{x-4}\right)$

$\quad = \left(\sqrt[3]{x-4}\right)^3 + 4$

$\quad = x - 4 + 4 = x$

$f \circ f^{-1}(x) = f\left(f^{-1}(x)\right) = f\left(x^3 + 4\right)$

$\quad = \sqrt[3]{x^3 + 4 - 4} = \sqrt[3]{x^3} = x$

85. We check to see that $f^{-1} \circ f(x) = x$ and
$f \circ f^{-1}(x) = x$.

$$f^{-1} \circ f(x) = f^{-1}(f(x)) = f^{-1}\left(\frac{1-x}{x}\right)$$

$$= \frac{1}{\frac{1-x}{x}+1} = \frac{1}{\frac{1-x}{x}+1} \cdot \frac{x}{x}$$

$$= \frac{x}{1-x+x} = \frac{x}{1} = x$$

$$f \circ f^{-1}(x) = f(f^{-1}(x)) = f\left(\frac{1}{x+1}\right)$$

$$= \frac{1-\frac{1}{x+1}}{\frac{1}{x+1}} = \frac{1-\frac{1}{x+1}}{\frac{1}{x+1}} \cdot \frac{x+1}{x+1}$$

$$= \frac{x+1-1}{1} = \frac{x}{1} = x$$

87. Let $y_1 = f(x)$, $y_2 = g(x)$, $y_3 = y_1(y_2)$, and
$y_4 = y_2(y_1)$. A table of values shows that
$y_3 \neq x$ nor is $y_4 = x$, so $f(x)$ and $g(x)$ are
not inverses of each other.

89. Let $y_1 = f(x)$, $y_2 = g(x)$, $y_3 = y_1(y_2)$,
and $y_4 = y_2(y_1)$. A table of values shows
that $y_3 = x$ and $y_4 = x$ for any value of x, so
$f(x)$ and $g(x)$ are inverse of each other.

91. (1) C; (2) D; (3) B; (4) A

93. a. $f(8) = 8 + 32 = 40$
Size 40 in France corresponds to size 8 in
the U.S.
$f(10) = 10 + 32 = 42$
Size 42 in France corresponds to size 10
in the U.S.
$f(14) = 14 + 32 = 46$
Size 46 in France corresponds to size 14
in the U.S.
$f(18) = 18 + 32 = 50$
Size 50 in France corresponds to size 18
in the U.S.

b.   The function $f(x) = x + 32$ is a linear
function that is not constant, so it passes
the horizontal-line test. Thus, f is one-to-
one and has an inverse that is a function.
We now find a formula for the inverse.

Replace $f(x)$ by y: $y = x + 32$
Interchange x and y: $x = y + 32$
Solve for y: $x - 32 = y$
Replace y by $f^{-1}(x)$: $f^{-1}(x) = x - 32$

c. $f^{-1}(40) = 40 - 32 = 8$
Size 8 in the U.S. corresponds to size 40
in France.
$f^{-1}(42) = 42 - 32 = 10$
Size 10 in the U.S. corresponds to size 42
in France.
$f^{-1}(46) = 46 - 32 = 14$
Size 14 in the U.S. corresponds to size 46
in France.
$f^{-1}(50) = 50 - 32 = 18$
Size 18 in the U.S. corresponds to size 50
in France.

95. *Thinking and Writing Exercise.*

97. $2^{-3} = \left(\frac{1}{2}\right)^3 = \frac{1}{2} \cdot \frac{1}{2} \cdot \frac{1}{2} = \frac{1}{8}$

99. $4^{5/2} = \left(4^{1/2}\right)^5 = 2^5 = 32$

101. $y = x^3$

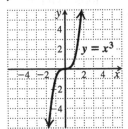

103. *Thinking and Writing Exercise.*

105.

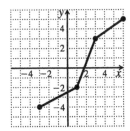

107. From Exercise 94(b), we know that a function that converts dress sizes in Italy to those in the United States is $g(x) = \dfrac{x}{2} - 12$. From Exercise 93(a), we know that a function that converts dress sizes in the United States to those in France is $f(x) = x + 32$. Then a function that converts dress sizes in Italy to those in France is

$$h(x) = (f \circ g)(x)$$
$$h(x) = f\left(\dfrac{x}{2} - 12\right)$$
$$h(x) = \dfrac{x}{2} - 12 + 32$$
$$h(x) = \dfrac{x}{2} + 20.$$

109. *Thinking and Writing Exercise.*

111. $\left(h \circ \left(g^{-1} \circ f^{-1}\right)\right)(x)$

$$= \left((f \circ g) \circ \left(g^{-1} \circ f^{-1}\right)\right)(x)$$
$$= \left(\left(f \circ \left(g \circ g^{-1}\right)\right) \circ f^{-1}\right)(x)$$
$$= \left((f \circ I) \circ f^{-1}\right)(x)$$
$$= \left(f \circ f^{-1}\right)(x) = x$$

Therefore, $\left(g^{-1} \circ f^{-1}\right)(x) = h^{-1}(x)$.

113. *Thinking and Writing Exercise.*

115. $(c \circ f)(n)$ represents the cost of mailing n copies of the book.

117. $R(10) \approx 18$ and $P(18) \approx 22$, so

$$P(R(10)) \approx 22 \text{ mm of mercury.}$$

119. Locate 20 on the vertical axis of the second graph, move across to the curve, and then move down to the horizontal axis to find that $P^{-1}(20) \approx 15$ liters per minute.

Exercise Set 9.2

1. True

3. True

5. False

7. The function values increase as x increases, so $a > 1$.

9. The function values decrease as x increases, so $0 < a < 1$.

11. Graph: $y = f(x) = 3^x$

We compute some function values and keep the results in a table.

$$f(0) = 3^0 = 1$$
$$f(1) = 3^1 = 3$$
$$f(2) = 3^2 = 9$$
$$f(-1) = 3^{-1} = \dfrac{1}{3^1} = \dfrac{1}{3}$$
$$f(-2) = 3^{-2} = \dfrac{1}{3^2} = \dfrac{1}{9}$$

x	y, or $f(x)$
0	1
1	3
2	9
-1	$\dfrac{1}{3}$
-2	$\dfrac{1}{9}$

Next we plot these points and connect them with a smooth curve.

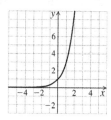

$y = f(x) = 3^x$

13. Graph: $y = 5^x$

We compute some function values, thinking of y as $f(x)$, and keep the results in a table.

$$f(0) = 5^0 = 1$$

$$f(1) = 5^1 = 5$$

$$f(2) = 5^2 = 25$$

$$f(-1) = 5^{-1} = \frac{1}{5^1} = \frac{1}{5}$$

$$f(-2) = 5^{-2} = \frac{1}{5^2} = \frac{1}{25}$$

x	y, or $f(x)$
0	1
1	5
2	25
−1	$\frac{1}{5}$
−2	$\frac{1}{25}$

Next we plot these points and connect them with a smooth curve.

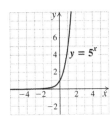

15. Graph: $y = 2^x + 3$

We compute some function values, thinking of y as $f(x)$, and keep the results in a table.

$$f(-4) = 2^{-4} + 3 = \frac{1}{2^4} + 3 = \frac{1}{16} + 3 = 3\frac{1}{16}$$

$$f(-2) = 2^{-2} + 3 = \frac{1}{2^2} + 3 = \frac{1}{4} + 3 = 3\frac{1}{4}$$

$$f(0) = 2^0 + 3 = 1 + 3 = 4$$

$$f(1) = 2^1 + 3 = 2 + 3 = 5$$

$$f(2) = 2^2 + 3 = 4 + 3 = 7$$

x	y, or $f(x)$
−4	$3\frac{1}{16}$
−2	$3\frac{1}{4}$
0	4
1	5
2	7

Next we plot these points and connect them with a smooth curve.

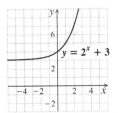

17. Graph: $y = 3^x - 1$

We compute some function values, thinking of y as $f(x)$, and keep the results in a table.

$$f(-3) = 3^{-3} - 1 = \frac{1}{3^3} - 1 = \frac{1}{27} - 1 = -\frac{26}{27}$$

$$f(-1) = 3^{-1} - 1 = \frac{1}{3} - 1 = -\frac{2}{3}$$

$$f(0) = 3^0 - 1 = 1 - 1 = 0$$

$$f(1) = 3^1 - 1 = 3 - 1 = 2$$

$$f(2) = 3^2 - 1 = 9 - 1 = 8$$

x	y, or $f(x)$
−3	$-\frac{26}{27}$
−1	$-\frac{2}{3}$
0	0
1	2
2	8

Next we plot these points and connect them with a smooth curve.

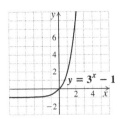

19. Graph: $y = 2^{x-3}$

We compute some function values, thinking of y as $f(x)$, and keep the results in a table.

$$f(-1) = 2^{-1-3} = 2^{-4} = \frac{1}{2^4} = \frac{1}{16}$$

$$f(0) = 2^{0-3} = 2^{-3} = \frac{1}{2^3} = \frac{1}{8}$$

$$f(1) = 2^{1-3} = 2^{-2} = \frac{1}{2^2} = \frac{1}{4}$$

$$f(2) = 2^{2-3} = 2^{-1} = \frac{1}{2^1} = \frac{1}{2}$$

$$f(3) = 2^{3-3} = 2^0 = 1$$

$$f(4) = 2^{4-3} = 2^1 = 2$$

$$f(5) = 2^{5-3} = 2^2 = 4$$

x	y, or $f(x)$
-1	$\frac{1}{16}$
0	$\frac{1}{8}$
1	$\frac{1}{4}$
2	$\frac{1}{2}$
3	1
4	2
5	4

Next we plot these points and connect them with a smooth curve.

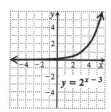

21. Graph: $y = 2^{x+3}$

We compute some function values, thinking of y as $f(x)$, and keep the results in a table.

$$f(-4) = 2^{-4+3} = 2^{-1} = \frac{1}{2}$$

$$f(-2) = 2^{-2+3} = 2$$

$$f(-1) = 2^{-1+3} = 2^2 = 4$$

$$f(0) = 2^{0+3} = 2^3 = 8$$

x	y, or $f(x)$
-4	$\frac{1}{2}$
-2	2
-1	4
0	8

Next we plot these points and connect them with a smooth curve.

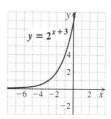

23. Graph: $y = \left(\frac{1}{5}\right)^x$

We compute some function values, thinking of y as $f(x)$, and keep the results in a table.

$$f(0) = \left(\frac{1}{5}\right)^0 = 1$$

$$f(1) = \left(\frac{1}{5}\right)^1 = \frac{1}{5}$$

$$f(2) = \left(\frac{1}{5}\right)^2 = \frac{1}{25}$$

$$f(-1) = \left(\frac{1}{5}\right)^{-1} = \frac{1}{\frac{1}{5}} = 5$$

$$f(-2) = \left(\frac{1}{5}\right)^{-2} = \frac{1}{\frac{1}{25}} = 25$$

x	y, or $f(x)$
0	1
1	$\dfrac{1}{5}$
2	$\dfrac{1}{25}$
−1	5
−2	25

Next we plot these points and connect them with a smooth curve.

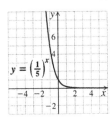

25. Graph: $y = \left(\dfrac{1}{10}\right)^x$

We compute some function values, thinking of y as $f(x)$, and keep the results in a table.

$$f(0) = \left(\dfrac{1}{10}\right)^0 = 1$$

$$f(1) = \left(\dfrac{1}{10}\right)^1 = \dfrac{1}{10}$$

$$f(2) = \left(\dfrac{1}{10}\right)^2 = \dfrac{1}{100}$$

$$f(-1) = \left(\dfrac{1}{10}\right)^{-1} = \dfrac{1}{\left(\frac{1}{10}\right)^1} = \dfrac{1}{\frac{1}{10}} = 10$$

$$f(-2) = \left(\dfrac{1}{10}\right)^{-2} = \dfrac{1}{\left(\frac{1}{10}\right)^2} = \dfrac{1}{\frac{1}{100}} = 100$$

x	y, or $f(x)$
0	1
1	$\dfrac{1}{10}$
2	$\dfrac{1}{100}$
−1	10
−2	100

Next we plot these points and connect them with a smooth curve.

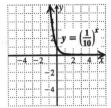

27. Graph: $y = 2^{x-3} - 1$

We compute some function values, thinking of y as $f(x)$, and keep the results in a table.

$$f(0) = 2^{0-3} - 1 = 2^{-3} - 1 = \dfrac{1}{8} - 1 = -\dfrac{7}{8}$$

$$f(1) = 2^{1-3} - 1 = 2^{-2} - 1 = \dfrac{1}{4} - 1 = -\dfrac{3}{4}$$

$$f(2) = 2^{2-3} - 1 = 2^{-1} - 1 = \dfrac{1}{2} - 1 = -\dfrac{1}{2}$$

$$f(3) = 2^{3-3} - 1 = 2^0 - 1 = 1 - 1 = 0$$

$$f(4) = 2^{4-3} - 1 = 2^1 - 1 = 2 - 1 = 1$$

$$f(5) = 2^{5-3} - 1 = 2^2 - 1 = 4 - 1 = 3$$

$$f(6) = 2^{6-3} - 1 = 2^3 - 1 = 8 - 1 = 7$$

x	y, or $f(x)$
0	$-\dfrac{7}{8}$
1	$-\dfrac{3}{4}$
2	$-\dfrac{1}{2}$
3	0
4	1
5	3
6	7

Next we plot these points and connect them with a smooth curve.

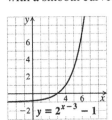

29. Graph: $y = 1.7^x$

We use a graphing calculator.

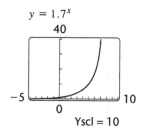

31. Graph: $y = 0.15^x$

We use a graphing calculator.

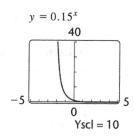

33. Graph: $x = 3^y$

We can find ordered pairs by choosing values for y and then computing values for x.

For $y = 0$, $x = 3^0 = 1$.

For $y = 1$, $x = 3^1 = 3$.

For $y = 2$, $x = 3^2 = 9$.

For $y = 3$, $x = 3^3 = 27$.

For $y = -1$, $x = 3^{-1} = \dfrac{1}{3^1} = \dfrac{1}{3}$.

For $y = -2$, $x = 3^{-2} = \dfrac{1}{3^2} = \dfrac{1}{9}$.

For $y = -3$, $x = 3^{-3} = \dfrac{1}{3^3} = \dfrac{1}{27}$.

x	y
1	0
3	1
9	2
27	3
$\dfrac{1}{3}$	-1
$\dfrac{1}{9}$	-2
$\dfrac{1}{27}$	-3

We plot the points and connect them with a smooth curve.

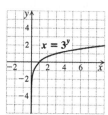

35. Graph: $x = 2^{-y} = \left(\dfrac{1}{2}\right)^y$

We can find ordered pairs by choosing values for y and then computing values for x. Then we plot these points and connect them with a smooth curve.

For $y = 0$, $x = \left(\dfrac{1}{2}\right)^0 = 1$.

For $y = 1$, $x = \left(\dfrac{1}{2}\right)^1 = \dfrac{1}{2}$.

For $y = 2$, $x = \left(\dfrac{1}{2}\right)^2 = \dfrac{1}{4}$.

For $y = 3$, $x = \left(\dfrac{1}{2}\right)^3 = \dfrac{1}{8}$.

For $y = -1$, $x = \left(\dfrac{1}{2}\right)^{-1} = \dfrac{1}{\frac{1}{2}} = 2$.

For $y = -2$, $x = \left(\dfrac{1}{2}\right)^{-2} = \dfrac{1}{\frac{1}{4}} = 4$.

For $y = -3$, $x = \left(\dfrac{1}{2}\right)^{-3} = \dfrac{1}{\frac{1}{8}} = 8$.

x	y
1	0
$\dfrac{1}{2}$	1
$\dfrac{1}{4}$	2
$\dfrac{1}{8}$	3
2	-1
4	-2
8	-3

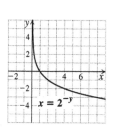

37. Graph: $x = 5^y$

We can ordered pairs by choosing values for y and then computing values for x. Then we plot these points and connect them with a smooth curve.

For $y = 0$, $x = 5^0 = 1.$

For $y = 1$, $x = 5^1 = 5.$

For $y = 2$, $x = 5^2 = 25.$

For $y = -1$, $x = 5^{-1} = \dfrac{1}{5}.$

For $y = -2$, $x = 5^{-2} = \dfrac{1}{25}.$

x	y
1	0
5	1
25	2
$\dfrac{1}{5}$	-1
$\dfrac{1}{25}$	-2

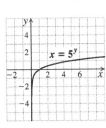

39. Graph: $x = \left(\dfrac{3}{2}\right)^y$

We can find ordered pairs by choosing values for y and then computing values for x. Then we plot these points and connect them with a smooth curve.

For $y = 0$, $x = \left(\dfrac{3}{2}\right)^0 = 1.$

For $y = 1$, $x = \left(\dfrac{3}{2}\right)^1 = \dfrac{3}{2}.$

For $y = 2$, $x = \left(\dfrac{3}{2}\right)^2 = \dfrac{9}{4}.$

For $y = 3$, $x = \left(\dfrac{3}{2}\right)^3 = \dfrac{27}{8}.$

For $y = -1$, $x = \left(\dfrac{3}{2}\right)^{-1} = \dfrac{1}{\frac{3}{2}} = \dfrac{2}{3}.$

For $y = -2$, $x = \left(\dfrac{3}{2}\right)^{-2} = \dfrac{1}{\frac{9}{4}} = \dfrac{4}{9}.$

For $y = -3$, $x = \left(\dfrac{3}{2}\right)^{-3} = \dfrac{1}{\frac{27}{8}} = \dfrac{8}{27}.$

x	y
1	0
$\dfrac{3}{2}$	1
$\dfrac{9}{4}$	2
$\dfrac{27}{8}$	3
$\dfrac{2}{3}$	-1
$\dfrac{4}{9}$	-2
$\dfrac{8}{27}$	-3

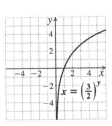

41. Graph $y = 3^x$ (see Exercise 11) and $x = 3^y$ (see Exercise 33) using the same set of axes.

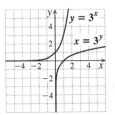

43. Graph $y = \left(\frac{1}{2}\right)^x$ and $x = \left(\frac{1}{2}\right)^y$ using the same set of axes.

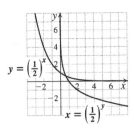

45. $y = \left(\frac{5}{2}\right)^x$ is an exponential function of the form $y = a^x$ with $a > 1$, so y-values will increase as x-values increase. Also, observe that when $x = 0$, $y = 1$. Thus, graph (d) corresponds to this equation.

47. For $x = \left(\frac{5}{2}\right)^y$, when $y = 0$, $x = 1$. The only graph that contains the point $(1, 0)$ is (f). This graph corresponds to the given equation.

49. $y = \left(\frac{2}{5}\right)^{x-2}$ is an exponential function of the form $y = a^x$ with $0 < a < 1$, so y-values will decrease as x-values increase. Also, observe that when $x = 2$, $y = 1$. Thus, graph (c) corresponds to the given equation.

51. a. In 2006, $t = 2006 - 2003 = 3$

$$M(3) = 0.353(1.244)^3 \approx 0.680$$

The number of tracks downloaded in 2006 was about 0.68 billion.
In 2008, $t = 2008 - 2003 = 5$

$$M(5) = 0.353(1.244)^5 \approx 1.052$$

The number of tracks downloaded in 2008 was about 1.052 billion.
In 2012, $t = 2012 - 2003 = 9$

$$M(9) = 0.353(1.244)^9 \approx 2.519$$

The number of tracks downloaded in 2012 was about 2.519 billion.

b.

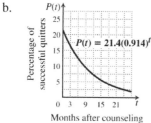

53. a. Since initial time is 0, we substitute the time(s) into the function

$$P(t) = 21.4(0.914)^t$$

$$P(1) = 21.4(0.914)^1 \approx 19.6$$

$$P(3) = 21.4(0.914)^3 \approx 16.3$$

$$P(12) = 21.4(0.914)^{12} \approx 7.3$$

Of the smokers who receive phone counseling, about 19.6% are able to quit for 1 month, about 16.3% are able to quit for 3 months, and about 7.3% are able to quit for 1 year.

b.

55. a. In 1930, $t = 1930 - 1900 = 30$.

$$P(t) = 150(0.960)^t$$

$$P(30) = 150(0.960)^{30}$$

$$\approx 44.079$$

In 1930, about 44.079 thousand, or 44,079, humpback whales were alive.
In 1960, $t = 1960 - 1900 = 60$.

$$P(t) = 150(0.960)^t$$

$$P(60) = 150(0.960)^{60}$$

$$\approx 12.953$$

In 1960, about 12.953 thousand, or 12,953, humpback whales were alive.

b. Plot the points found in part (a), (30, 44, 079) and (60, 12,953) and additional points as needed and graph the function.

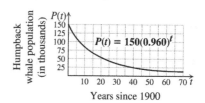

57. a. In 1992, $t = 1992 - 1982 = 10$

$$P(10) = 5.5(1.08)^{10} \approx 11.874$$

In 1992, there were about 11,874 humpback whales.
In 2006, $t = 2006 - 1982 = 24$

$$P(24) = 5.5(1.08)^{24} \approx 34.876$$

In 2006, there were about 34,876 humpback whales.

b.

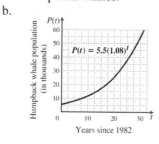

59. a. Since initial time is 0, we substitute the time(s) into the function

$$R(t) = 2(1.75)^t$$

For $t = 10$,

$$R(10) = 2(1.75)^{10} \approx 539$$

After 10 years, there will be about 539 ruffe in the lake.

For $t = 15$,

$$R(15) = 2(1.75)^{15} \approx 8843$$

After 15 years, there will be about 8843 ruffe in the lake.

b. Plot the points found in part (a), (10, 539) and (15, 8843) and additional points as needed and graph the function.

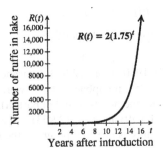

61. *Thinking and Writing Exercise.*

63. $3x^2 - 48 = 3(x^2 - 16) = 3(x^2 - 4^2)$
     $ = 3(x+4)(x-4)$

65. $6x^2 + x - 12$
     Use the FOIL method to factor the trinomial
     $(3x+)(2x+)$.
     Factor the last term, -12. The correct pair of factors is -4 and 3.
     $(3x-4)(2x+3)$
     $6x^2 + x - 12 = (3x-4)(2x+3)$

67. $6y^2 + 36y - 240 = 6(y^2 + 6y - 40)$

     Factoring $y^2 + 6y - 40$, we want two factors of -40, whose sum is 6. They are 10 and -4.
     $6y^2 + 36y - 240 = 6(y+10)(y-4)$

69. *Thinking and Writing Exercise.*

71. Since the bases are the same, the one with the larger exponent is the larger number. Thus $\pi^{2.4}$ is larger.

73. Graph: $y = 2^x + 2^{-x}$
     Construct a table of values, thinking of y as $f(x)$. Then plot these points and connect them with a curve.

     $$f(0) = 2^0 + 2^{-0} = 1 + 1 = 2$$

     $$f(1) = 2^1 + 2^{-1} = 2 + \frac{1}{2} = 2\frac{1}{2}$$

     $$f(2) = 2^2 + 2^{-2} = 4 + \frac{1}{4} = 4\frac{1}{4}$$

     $$f(3) = 2^3 + 2^{-3} = 8 + \frac{1}{8} = 8\frac{1}{8}$$

     $$f(-1) = 2^{-1} + 2^{-(-1)} = \frac{1}{2} + 2 = 2\frac{1}{2}$$

     $$f(-2) = 2^{-2} + 2^{-(-2)} = \frac{1}{4} + 4 = 4\frac{1}{4}$$

     $$f(-3) = 2^{-3} + 2^{-(-3)} = \frac{1}{8} + 8 = 8\frac{1}{8}$$

x	y, or $f(x)$
0	2
1	$2\frac{1}{2}$
2	$4\frac{1}{4}$
3	$8\frac{1}{8}$
-1	$2\frac{1}{2}$
-2	$4\frac{1}{4}$
-3	$8\frac{1}{8}$

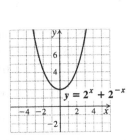

75. Graph: $y = \left| 2^x - 2 \right|$
     We construct a table of values, thinking of y as $f(x)$. Then plot these points and connect them with a curve.

$$f(0) = \left|2^0 - 2\right| = |1 - 2| = |-1| = 1$$

$$f(1) = \left|2^1 - 2\right| = |2 - 2| = |0| = 0$$

$$f(2) = \left|2^2 - 2\right| = |4 - 2| = |2| = 2$$

$$f(3) = \left|2^3 - 2\right| = |8 - 2| = |6| = 6$$

$$f(-1) = \left|2^{-1} - 2\right| = \left|\frac{1}{2} - 2\right| = \left|-\frac{3}{2}\right| = \frac{3}{2}$$

$$f(-3) = \left|2^{-3} - 2\right| = \left|\frac{1}{8} - 2\right| = \left|-\frac{15}{8}\right| = \frac{15}{8}$$

$$f(-5) = \left|2^{-5} - 2\right| = \left|\frac{1}{32} - 2\right| = \left|-\frac{63}{32}\right| = \frac{63}{32}$$

x	y, or $f(x)$
0	1
1	0
2	2
3	6
−1	$\frac{3}{2}$
−3	$\frac{15}{8}$
−5	$\frac{63}{32}$

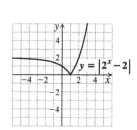

$y = |2^x - 2|$

77. Graph: $y = \left|2^{x^2} - 1\right|$

We construct a table of values, thinking of y as $f(x)$. Then we plot these points and connect them with a curve.

$$f(0) = \left|2^{0^2} - 1\right| = |1 - 1| = 0$$

$$f(1) = \left|2^{1^2} - 1\right| = |2 - 1| = 1$$

$$f(2) = \left|2^{2^2} - 1\right| = |16 - 1| = 15$$

$$f(-1) = \left|2^{(-1)^2} - 1\right| = |2 - 1| = 1$$

$$f(-2) = \left|2^{(-2)^2} - 1\right| = |16 - 1| = 15$$

x	y, or $f(x)$
0	0
1	1
2	15
−1	1
−2	15

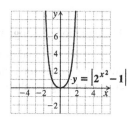

$y = \left|2^{x^2} - 1\right|$

79. Determine a table of values for each function and graph.

x	y
0	3
1	1
2	$\frac{1}{3}$
3	$\frac{1}{9}$
−1	9

$y = 3^{-(x-1)}$

x	y
3	0
1	1
$\frac{1}{3}$	2
$\frac{1}{9}$	3
9	−1

$x = 3^{-(y-1)}$

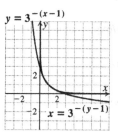

$y = 3^{-(x-1)}$

$x = 3^{-(y-1)}$

81. Enter the data points (0, 0.5), (4, 4) and (8, 50) and use the exponential regression feature to find an exponential function that models the data.

$$N(t) = 0.4642(1.7783)^t$$

In 2012, $x = 2012 - 2000 = 12$

$$N(12) = 0.4642(1.7783)^{12}$$

$$\approx 464.26$$

In 2012, about 464 million GPS systems will be in use.

83. *Thinking and Writing Exercise.*

85.

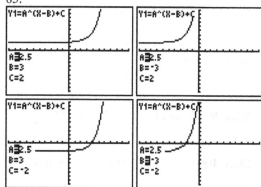

Exercise Set 9.3

1. g

3. a

5. b

7. e

9. $\log_{10} 1000$ is the power/exponent to which we raise 10 to get 1000.
    Since $10^3 = 1000$, $\log_{10} 1000 = 3$

11. $\log_2 16$ is the power/exponent to which we raise 2 to get 16.
    Since $2^4 = 16$, $\log_2 16 = 4$

13. Since $3^4 = 81$, $\log_3 81 = 4$

15. Since $4^{-2} = \dfrac{1}{16}$, $\log_4 \dfrac{1}{16} = -2$

17. Since $7^{-1} = \dfrac{1}{7}$, $\log_7 \dfrac{1}{7} = -1$

19. Since $5^4 = 625$, $\log_5 625 = 4$

21. Since $8^1 = 8$, $\log_8 8 = 1$

23. Since $8^0 = 1$, $\log_8 1 = 0$

25. Since $9^5 = 9^5$, $\log_9 9^5 = 5$

27. Since $10^{-2} = 0.01$, $\log_{10} 0.01 = -2$

29. Since $9^{\frac{1}{2}} = 3\left[\left(3^2\right)^{\frac{1}{2}} = 3\right]$, $\log_9 3 = \dfrac{1}{2}$

31. Since $9 = 3^2$ and $\left(3^2\right)^{3/2} = 27$, $\log_9 27 = \dfrac{3}{2}$

33. Since $1000 = 10^3$ and $\left(10^3\right)^{2/3} = 10^2 = 100$,
    $\log_{1000} 100 = \dfrac{2}{3}$

35. Since $\log_5 7$ is the power to which we raise 5 to get 7, then 5 raised to this power is 7. That is, $5^{\log_5 7} = 7$.

37. Graph: $y = \log_{10} x$

    The equation $y = \log_{10} x$ is equivalent to $10^y = x$. We can find ordered pairs by choosing values for y and computing the corresponding x-values.

    For $y = 0$, $x = 10^0 = 1$.
    For $y = 1$, $x = 10^1 = 10$.
    For $y = 2$, $x = 10^2 = 100$.
    For $y = -1$, $x = 10^{-1} = \dfrac{1}{10}$.
    For $y = -2$, $x = 10^{-2} = \dfrac{1}{100}$.

x, or 10^y	y
1	0
10	1
100	2
$\dfrac{1}{10}$	-1
$\dfrac{1}{100}$	-2

    (1) Select y.

    (2) Compute x.

    We plot the set of ordered pairs and connect the points with a smooth curve.

    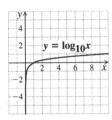

39. Graph: $y = \log_3 x$

    The equation $y = \log_3 x$ is equivalent to $3^y = x$. We can find ordered pairs by choosing values for y and computing the corresponding x-values.

For $y = 0$, $x = 3^0 = 1$.

For $y = 1$, $x = 3^1 = 3$.

For $y = 2$, $x = 3^2 = 9$.

For $y = -1$, $x = 3^{-1} = \dfrac{1}{3}$.

For $y = -2$, $x = 3^{-2} = \dfrac{1}{9}$.

x, or 3^y	y
1	0
3	1
9	2
$\dfrac{1}{3}$	-1
$\dfrac{1}{9}$	-2

For $y = 0$, $x = 2.5^0 = 1$.

For $y = 1$, $x = 2.5^1 = 2.5$.

For $y = 2$, $x = 2.5^2 = 6.25$.

For $y = 3$, $x = 2.5^3 = 15.625$.

For $y = -1$, $x = 2.5^{-1} = 0.4$.

For $y = -2$, $x = 2.5^{-2} = 0.16$.

x, or 2.5^y	y
1	0
2.5	1
6.25	2
15.625	3
0.4	-1
0.16	-2

We plot the set of ordered pairs and connect the points with a smooth curve.

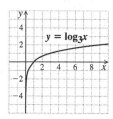

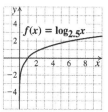

41.   Graph: $f(x) = \log_6 x$

Think of $f(x)$ as y. Then $y = \log_6 x$ is equivalent to $6^y = x$. We find ordered pairs by choosing values for y and computing the corresponding x-values. Then we plot the points and connect them with a smooth curve.

For $y = 0$, $x = 6^0 = 1$.

For $y = 1$, $x = 6^1 = 6$.

For $y = 2$, $x = 6^2 = 36$.

For $y = -1$, $x = 6^{-1} = \dfrac{1}{6}$.

For $y = -2$, $x = 6^{-2} = \dfrac{1}{36}$.

x, or 6^y	y
1	0
6	1
36	2
$\dfrac{1}{6}$	-1
$\dfrac{1}{36}$	-2

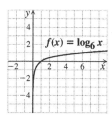

43.   Graph: $f(x) = \log_{2.5} x$

Think of $f(x)$ as y. Then $y = \log_{2.5} x$ is equivalent to $2.5^y = x$. We construct a table of values, plot these points and connect them with a smooth curve.

45.   Graph $f(x) = 3^x$ (see Exercise Set 9.2, Exercise 11) and $f^{-1}(x) = \log_3 x$ (see Exercise 39 above) on the same set of axes.

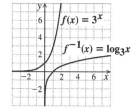

47. $\log 4 \approx 0.6021$

49. $\log 13,400 \approx 4.1271$

51. $\log 0.527 \approx -0.2782$

53. $10^{2.3} \approx 199.5262$

55. $10^{-2.9523} \approx 0.0011$

57. $10^{0.0012} \approx 1.0028$

59. $y = \log(x + 2)$

$y = \log(x + 2)$

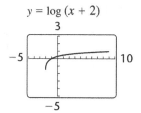

61. $y = \log(2x) - 3$

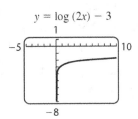

63. $y = \log\left(x^2\right)$

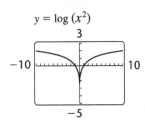

65.   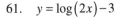 $x = \log_{10} 8 \Rightarrow 10^x = 8$

The base remains the same.

The logarithm is the exponent.

67. $\log_9 9 = 1 \Rightarrow 9^1 = 9$

The logarithm is the exponent.

The base remains the same.

69. $\log_{10} 0.1 = -1 \Rightarrow 10^{-1} = 0.1$

71. $\log_{10} 7 = 0.845 \Rightarrow 10^{0.845} = 7$

73. $\log_c m = 8 \Rightarrow c^8 = m$

75. $\log_t Q = r \Rightarrow t^r = Q$

77. $\log_e 0.25 = -1.3863 \Rightarrow e^{-1.3863} = 0.25$

79. $\log_r T = -x \Rightarrow r^{-x} = T$

81. $10^2 = 100 \Rightarrow 2 = \log_{10} 100$

The exponent is the logarithm.

The base remains the same.

83. $4^{-5} = \frac{1}{1024} \Rightarrow -5 = \log_4 \frac{1}{1024}$

The exponent is the logarithm.

The base remains the same.

85. $16^{3/4} = 8 \Rightarrow \frac{3}{4} = \log_{16} 8$

87. $10^{0.4771} = 3 \Rightarrow 0.4771 = \log_{10} 3$

89. $z^m = 6 \Rightarrow m = \log_z 6$

91. $p^m = V \Rightarrow m = \log_p V$

93. $e^3 = 20.0855 \Rightarrow 3 = \log_e 20.0855$

95. $e^{-4} = 0.0183 \Rightarrow -4 = \log_c 0.0183$

97. $\log_3 x = 2$

$3^2 = x$ Converting to an exponential equation

$9 = x$ Computing 3^2

99. $\log_5 125 = x$

$5^x = 125$ Converting to an exponential equation

$5^x = 5^3$

$x = 3$ The exponents must be equal/the same.

101. $\log_x 16 = 4$

$x^4 = 16$ Converting to an exponential equation

$x = \sqrt[4]{16}$

$x = 2$ Computing $\sqrt[4]{16}$

103. $\log_x 7 = 1$

$x^1 = 7$ Converting to an exponential equation

$x = 7$ Simplifying x^1

105. $\log_x 9 = \frac{1}{2}$

$\qquad x^{1/2} = 9$ Converting to an exponential equation

$\qquad x = 9^2$

$\qquad x = 81$ Computing 9^2

107. $\log_3 x = -2$

$\qquad 3^{-2} = x$ Converting to an exponential equation

$\qquad \left(\frac{1}{3}\right)^2 = x$

$\qquad \frac{1}{9} = x$ Computing $\left(\frac{1}{3}\right)^2$

109. $\log_{32} x = \frac{2}{5}$

$\qquad 32^{2/5} = x$ Converting to an exponential equation

$\qquad \left(2^5\right)^{2/5} = x$

$\qquad 4 = x$ Simplifying

111. *Thinking and Writing Exercise.*

113. $a^{12} \cdot a^6 = a^{12+6} = a^{18}$

115. $\dfrac{x^{12}}{x^4} = x^{12-4} = x^8$

117. $\left(y^3\right)^5 = y^{3\cdot5} = y^{15}$

119. $x^2 \cdot x^3 = x^{2+3} = x^5$

121. *Thinking and Writing Exercise.*

123. Graph: $y = \left(\dfrac{3}{2}\right)^x$ Graph: $y = \log_{3/2} x$, or

$$x = \left(\frac{3}{2}\right)^y$$

x	y or $\left(\dfrac{3}{2}\right)^x$
0	1
1	$\dfrac{3}{2}$
2	$\dfrac{9}{4}$
3	$\dfrac{27}{8}$
-1	$\dfrac{2}{3}$
-2	$\dfrac{4}{9}$

x, or $\left(\dfrac{3}{2}\right)^y$	y
1	0
$\dfrac{3}{2}$	1
$\dfrac{9}{4}$	2
$\dfrac{27}{8}$	3
$\dfrac{2}{3}$	-1
$\dfrac{4}{9}$	-2

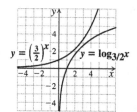

125. Graph: $y = \log_3 |x+1|$

Choose values of x and determine corresponding y-values. Plot these points and connect them with a smooth curve.

For $x = 0$, $y = \log_3 |0+1| = 0$

For $x = 2$, $y = \log_3 |2+1| = 1$

For $x = 8$, $y = \log_3 |8+1| = 2$

For $x = -2$, $y = \log_3 |-2+1| = 0$

For $x = -4$, $y = \log_3 |-4+1| = 1$

For $x = -10$, $y = \log_3 |-10+1| = 2$

x	y
0	0
2	1
8	2
-2	0
-4	1
-10	2

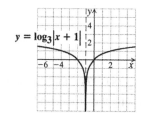

127. $\log_4(3x-2) = 2$
$$4^2 = 3x-2$$
$$16 = 3x-2$$
$$18 = 3x$$
$$6 = x$$

129. $\log_{10}(x^2 + 21x) = 2$
$$10^2 = x^2 + 21x$$
$$0 = x^2 + 21x - 100$$
$$0 = (x+25)(x-4)$$
$$x = -25 \quad \text{or} \quad x = 4$$

131. Let $\log_{1/5} 25 = x$. Then
$$\left(\frac{1}{5}\right)^x = 25$$
$$\left(5^{-1}\right)^x = 25$$
$$5^{-x} = 5^2$$
$$-x = 2$$
$$x = -2.$$
Thus, $\log_{1/5} 25 = -2$.

133. $\log_{10}\left(\log_4\left(\log_3 81\right)\right)$
$$= \log_{10}\left(\log_4 4\right) \quad \left(\log_3 81 = 4\right)$$
$$= \log_{10} 1 \quad\quad \left(\log_4 4 = 1\right)$$
$$= 0$$

135. Let $b = 0$, $x = 1$, and $y = 2$. Then $0^1 = 0^2$, but $1 \neq 2$. Let $b = 1, x = 1$, and $y = 2$. Then $1^1 = 1^2$, but $1 \neq 2$.

Exercise Set 9.4

1. e

3. a

5. c

7. $\log_3(81 \cdot 27) = \log_3 81 + \log_3 27$
Using the product rule.

9. $\log_4(64 \cdot 16) = \log_4 64 + \log_4 16$
Using the product rule.

11. $\log_c(rst) = \log_c r + \log_c s + \log_c t$
Using the product rule.

13. $\log_a 5 + \log_a 14 = \log_a(5 \cdot 14), \text{or} \log_a 70$
Using the product rule.

15. $\log_c t + \log_c y = \log_c(t \cdot y)$
Using the product rule.

17. $\log_a r^8 = 8 \log_a r$
Using the power rule.

19. $\log_2 y^{1/3} = \frac{1}{3}\log_2 y$
Using the power rule.

21. $\log_b C^{-3} = -3 \log_b C$
Using the power rule.

23. $\log_2 \dfrac{25}{13} = \log_2 25 - \log_2 13$
Using the quotient rule.

25. $\log_b \dfrac{m}{n} = \log_b m - \log_b n$
Using the quotient rule.

27. $\log_a 17 - \log_a 6 = \log_a \dfrac{17}{6}$
Using the quotient rule.

29. $\log_b 36 - \log_b 4 = \log_b \dfrac{36}{4} = \log_b 9$
Using the quotient rule.

31. $\log_a x - \log_a y = \log_a \dfrac{x}{y}$
Using the quotient rule.

33. $\log_a(xyz) = \log_a x + \log_a y + \log_a z$
Using the product rule.

35. $\log_a(x^3 z^4) = \log_a x^3 + \log_a z^4$
Using the product rule.
$$= 3\log_a x + 4\log_a z$$
Using the power rule.

37. $\log_a(x^2 y^{-2} z) = \log_a x^2 + \log_a y^{-2} + \log_a z$

Using the product rule
$= 2\log_a x - 2\log_a y + \log_a z$

Using the power rule.

39. $\log_a \dfrac{x^4}{y^3 z} = \log_a x^4 - \log_a y^3 z$

Using the quotient rule.
$= \log_a x^4 - (\log_a y^3 + \log_a z)$

Using the product rule.
$= 4\log_a x - 3\log_a y - \log_a z$

Using the power rule.

41. $\log_b \dfrac{xy^2}{wz^3} = \log_b xy^2 - \log_b wz^3$

Using the quotient rule.
$= \log_b x + \log_b y^2 - (\log_b w + \log_b z^3)$

Using the product rule.
$= \log_b x + 2\log_b y - \log_b w - 3\log_b z$

Using the power rule.

43. $\log_a \sqrt{\dfrac{x^7}{y^5 z^8}}$

$= \log_a \left(\dfrac{x^7}{y^5 z^5} \right)^{1/2}$

$= \dfrac{1}{2} \log_a \dfrac{x^7}{y^5 z^8}$ Using the power rule

$= \dfrac{1}{2} (\log_a x^7 - \log_a y^5 z^8)$ Using the quotient rule

$= \dfrac{1}{2} \Big[\log_a x^7 - (\log_a y^5 + \log_a z^8) \Big]$ Using the product rule

$= \dfrac{1}{2} (7\log_a x - 5\log_a y - 8\log_a z)$ Using the power rule

45. $\log_a \sqrt[3]{\dfrac{x^6 y^3}{a^2 z^7}}$

$= \log_a \left(\dfrac{x^6 y^3}{a^2 z^7} \right)^{1/3}$

$= \dfrac{1}{3} \log_a \left(\dfrac{x^6 y^3}{a^2 z^7} \right)$ Using the power rule

$= \dfrac{1}{3} \left(\log_a x^6 y^3 - \log_a a^2 z^7 \right)$ Using the quotient rule

$= \dfrac{1}{3} \Big[\log_a x^6 + \log_a y^3 - \left(\log_a a^2 + \log_a z^7 \right) \Big]$ Using the product rule

$= \dfrac{1}{3} \Big(\log_a x^6 + \log_a y^3 - 2 - \log_a z^7 \Big)$ 2 is the number to which we raise a to get a^2.

$= \dfrac{1}{3} \big(6\log_a x + 3\log_a y - 2 - 7\log_a z \big)$

47. $8\log_a x + 3\log_a z$

$= \log_a x^8 + \log_a z^3$ Using the power rule

$= \log_a x^8 z^3$ Using the product rule

49. $\log_a x^2 - 2\log_a \sqrt{x}$

$= \log_a x^2 - \log_a (\sqrt{x})^2$ Using the power rule

$= \log_a x^2 - \log_a x$ $(\sqrt{x})^2 = x$

$= \log_a \dfrac{x^2}{x}$ Using the quotient rule

$= \log_a x$ Simplifying

51. $\dfrac{1}{2}\log_a x + 5\log_a y - 2\log_a x$

$= \log_a x^{1/2} + \log_a y^5 - \log_a x^2$ Using the power rule

$= \log_a x^{1/2} y^5 - \log_a x^2$ Using the product rule

$= \log_a \dfrac{x^{1/2} y^5}{x^2}$ Using the quotient rule

The result can also be expressed as

$\log_a \dfrac{\sqrt{x} y^5}{x^2}$ or as $\log_a \dfrac{y^5}{x^{3/2}}$.

53. $\log_a(x^2-4)-\log_a(x+2)$

$\quad = \log_a \dfrac{x^2-4}{x+2}$ Using the quotient rule

$\quad = \log_a \dfrac{(\cancel{x+2})(x-2)}{(\cancel{x+2})}$ Simplifying

$\quad = \log_a(x-2)$

55. $\log_b 15 = \log_b(3 \cdot 5)$

$\quad\quad = \log_b 3 + \log_b 5$ Using the product rule

$\quad\quad = 0.792 + 1.161$

$\quad\quad = 1.953$

57. $\log_b \dfrac{3}{5} = \log_b 3 - \log_b 5$ Using the quotient rule

$\quad\quad\quad = 0.792 - 1.161$

$\quad\quad\quad = -0.369$

59. $\log_b \dfrac{1}{5} = \log_b 1 - \log_b 5$ Using the quotient rule

$\quad\quad\quad = 0 - 1.161$ $(\log 1 = 0)$

$\quad\quad\quad = -1.161$

61. $\log_b \sqrt{b^3} = \log_b b^{3/2} = \dfrac{3}{2}$ (log is the exponent)

63. $\log_b 8$ since 8 cannot be expressed using the numbers 1, 3, and 5 (as factors/quotients), we cannot calculate $\log_b 8$ using the given information.

65. $\log_t t^7 = 7$ 7 is the power/exponent/ logarithm

67. $\log_e e^m = m$ m is the power/exponent/ logarithm

69. $\log_5 125 = 3$ and $\log_5 625 = 4$, so

$\log_5(125 \cdot 625) = \log_5 125 + \log_5 625$

$\quad\quad\quad\quad = 3 + 4 = 7$

71. $\log_2 16 = 4$, so

$\log_2 16^5 = 5 \cdot \log_2 16 = 5 \cdot 4 = 20$

73. *Thinking and Writing Exercise.*

75. Graph $f(x) = \sqrt{x} - 3$.

We construct a table of values, plot points, and connect them with a smooth curve. Note that we must choose nonnegative values of x in order for $\sqrt{x}$ to be a real number.

x	$f(x)$
0	-3
1	-2
4	-1
9	0

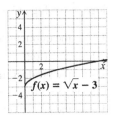

77. Graph: $g(x) = x^3 + 2$

We construct a table of values, plot points, and connect them with a smooth curve.

x	$f(x)$
-2	-6
-1	1
0	2
1	3
2	10

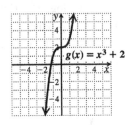

79. $f(x) = \dfrac{x-3}{x+7}$

$f(x)$ cannot be calculated for any x-value for which the denominator, $x+7$, is 0. To find the excluded values, we solve:

$x + 7 = 0$

$\quad x = -7$

The domain of f is $\{\,x\,|\,x$ is a real number *and* $x \neq -7\,\}$, or $(-\infty, -7) \cup (-7, \infty)$.

81. $g(x) = \sqrt{10-x}$

Since the index is even, the radicand, $10 - x$, must be non-negative. We solve the inequality:

$\quad\quad 10 - x \geq 0$

$\quad\quad\quad -x \geq -10$

$\quad\quad\quad\quad x \leq 10$

Domain of $g = \{x\,|\,x \leq 10\}$, or $(-\infty, 10]$.

83. *Thinking and Writing Exercise.*

85. $\log_a (x^8 - y^8) - \log_a (x^2 + y^2)$

$= \log_a \dfrac{x^8 - y^8}{x^2 + y^2}$

$= \log_a \dfrac{(x^4 + y^4)(x^2 + y^2)(x + y)(x - y)}{x^2 + y^2}$

$= \log_a [(x^4 + y^4)(x^2 - y^2)]$ Simplifying

$= \log_a (x^6 - x^4 y^2 + x^2 y^4 - y^6)$

87. $\log_a \sqrt{1 - s^2}$

$= \log_a (1 - s^2)^{1/2}$

$= \dfrac{1}{2} \log_a (1 - s^2)$

$= \dfrac{1}{2} \log_a [(1 - s)(1 + s)]$

$= \dfrac{1}{2} \log_a (1 - s) + \dfrac{1}{2} \log_a (1 + s)$

89. $\log_a \dfrac{\sqrt[3]{x^2 z}}{\sqrt[3]{y^2 z^{-2}}}$

$= \log_a \left(\dfrac{x^2 z^3}{y^2} \right)^{1/3}$

$= \dfrac{1}{3} (\log_a x^2 z^3 - \log_a y^2)$

$= \dfrac{1}{3} (2 \log_a x + 3 \log_a z - 2 \log_a y)$

$= \dfrac{1}{3} [2 \cdot 2 + 3 \cdot 4 - 2 \cdot 3]$

$= \dfrac{1}{3} (10)$

$= \dfrac{10}{3}$

91. $\log_a x = 2$, so $a^2 = x$.

Let $\log_{1/a} x = n$ and solve for n.

$\log_{1/a} a^2 = n$ Substituting a^2 for x

$\left(\dfrac{1}{a} \right)^n = a^2$

$(a^{-1})^n = a^2$

$a^{-n} = a^2$

$-n = 2$

$n = -2$

Thus, $\log_{1/a} x = -2$ when $\log_a x = 2$.

93. $\log_2 80 + \log_2 x = 5$

Evaluating $\log_2 80$,

$\log_2 80 = \log_2 (32 \cdot 2.5)$

$= \log_2 32 + \log_2 2.5$

$= 5 + \log_2 2.5$

Substituting this expression into the original equation,

$\log_2 80 + \log_2 x = 5$

$5 + \log_2 2.5 + \log_2 x = 5$

$\log_2 2.5 + \log_2 x = 5 - 5$

$\log_2 x = -\log_2 2.5$

$\log_2 x = \log_2 2.5^{-1}$

$x = 2.5^{-1}$

$x = \dfrac{1}{2.5} = \dfrac{2}{5}$

95. True; $\log_a (Q + Q^2) = \log_a [Q(1 + Q)]$

$= \log_a Q + \log_a (1 + Q) = \log_a Q + \log_a (Q + 1)$.

Mid-Chapter Review

Guided Solutions

1. $y = 2x - 5$

$x = 2y - 5$

$x + 5 = 2y$

$\dfrac{x + 5}{2} = y$

$f^{-1}(x) = \dfrac{x + 5}{2}$

2. $\log_4 x = 1 \Rightarrow x = 4^1$

$x = 4$

Mixed Review

1. $(f \circ g)(x) = f(g(x)) = f(x - 5)$

$= (x - 5)^2 + 1$

$= x^2 - 10x + 25 + 1$

$= x^2 - 10x + 26$

2. $h(x) = \sqrt{5x-3}$

    We have $5x-3$ and take the square root of this expression.

    $f(x) = \sqrt{x}$ and $g(x) = 5x - 3$

3. Replace $g(x)$ by y: $y = 6 - x$

    Interchange x and y: $x = 6 - y$

    Solve for y: $x - 6 = -y$

    $\qquad\qquad -x + 6 = y$

    Replace y by $g^{-1}(x)$: $g^{-1}(x) = -x + 6$

    or $g^{-1}(x) = 6 - x$

4. Graph: $y = 2^x + 3$

    We compute some function values, thinking of y as $f(x)$, and keep the results in a table.

    $f(-4) = 2^{-4} + 3 = \dfrac{1}{2^4} + 3 = \dfrac{1}{16} + 3 = 3\dfrac{1}{16}$

    $f(-2) = 2^{-2} + 3 = \dfrac{1}{2^2} + 3 = \dfrac{1}{4} + 3 = 3\dfrac{1}{4}$

    $f(0) = 2^0 + 3 = 1 + 3 = 4$

    $f(1) = 2^1 + 3 = 2 + 3 = 5$

    $f(2) = 2^2 + 3 = 4 + 3 = 7$

x	y, or $f(x)$
-4	$3\dfrac{1}{16}$
-2	$3\dfrac{1}{4}$
0	4
1	5
2	7

    Next we plot these points and connect them with a smooth curve.

    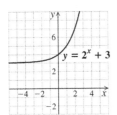

5. $\log_4 16$ is the power/exponent to which we raise 4 to get 16.

    Since $4^2 = 16$, $\log_4 16 = 2$

6. Since $5^{-1} = \dfrac{1}{5}$, $\log_5 \dfrac{1}{5} = -1$

7. $\log_{100} 10$ is the power/exponent to which we raise 100 to get 10.

    Since $100^{1/2} = 10$, $\log_{100} 10 = \dfrac{1}{2}$

8. $\log_b b^1 = 1 \cdot \log_b b = 1 \cdot 1 = 1$

9. $\log_8 8^{19} = 19 \cdot \log_8 8 = 19 \cdot 1 = 19$

10. $\log_t 1 = 0$

    0 is the power/exponent to which t is raised to yield 1 ($t^0 = 1$).

11. $\log_x 3 = m$

    $x^m = 3$, or $x = \sqrt[m]{3}$

12. $\log_2 1024 = 10$

    $2^{10} = 1024$

13. $e^t = x \Rightarrow t = \log_e x$

14. $64^{2/3} = 16 \Rightarrow \dfrac{2}{3} = \log_{64} 16$

15. $\log \sqrt{\dfrac{x^2}{yz^3}}$

    $= \log\left(\dfrac{x^2}{yz^3}\right)^{1/2}$

    $= \dfrac{1}{2}\log \dfrac{x^2}{yz^3}$ Using the power rule

    $= \dfrac{1}{2}(\log x^2 - \log yz^3)$ Using the quotient rule

    $= \dfrac{1}{2}\left[\log x^2 - \left(\log y + \log z^3\right)\right]$ Using the product rule

    $= \dfrac{1}{2}(2\log x - \log y - 3\log z)$ Using the power rule

    $= \log x - \tfrac{1}{2}\log y - \tfrac{3}{2}\log z$ Simplify

16. $\log a - 2\log b - \log c$

    $= \log a - \log b^2 - \log c$ Using the power rule

    $= \log \dfrac{a}{b^2} - \log c$ Using the quotient rule

    $= \log \dfrac{a}{b^2 c}$ Using the quotient rule

17. $\log_x 64 = 3$

    $x^3 = 64$ Converting to an exponential equation

    $x = \sqrt[3]{64}$

    $x = 4$ Computing $\sqrt[3]{64}$

18. $\log_3 x = -1$

    $3^{-1} = x$ Converting to an exponential equation

    $\dfrac{1}{3} = x$

19. $\log x = 5$

    $10^5 = x$ Converting to an exponential equation

    $100,000 = x$ Computing 10^5

20. $\log_x 2 = \frac{1}{2}$

    $x^{1/2} = 2$ Converting to an exponential equation

    $x = 2^2$

    $x = 4$ Computing 2^2

Exercise Set 9.5

1. True

3. True

5. True

7. True

9. True

11. 1.6094

13. −5.0832

15. 96.7583

17. 15.0293

19. 0.0305

21. 0.8451

23. 13.0014

25. −0.4260

27. 4.9459

29. We will use common logarithms for the conversion. Let $a = 10, b = 6,$ and $M = 92$ substitute in the change-of-base formula.

$$\log_b M = \frac{\log_a M}{\log_a b}$$

$$\log_6 92 = \frac{\log_{10} 92}{\log_{10} 6} \approx \frac{1.963787827}{0.7781512504} \approx 2.5237$$

31. We will use common logarithms for the conversion. Let $a = 10, b = 2,$ and $M = 100$ and substitute in the change-of-base formula.

$$\log_2 100 = \frac{\log_{10} 100}{\log_{10} 2}$$

$$\approx \frac{2}{0.3010299957}$$

$$\approx 6.6439$$

33. We will use natural logarithms for the conversion. Let $a = e, b = 0.5,$ and $M = 5$ and substitute in the change-of-base formula.

$$\log_{0.5} 5 = \frac{\ln 5}{\ln 0.5} \approx \frac{1.609437912}{-0.6931471806} \approx -2.3219$$

35. We will use common logarithms for the conversion. Let $a = 10, b = 2,$ and $M = 0.2$ and substitute in the change-of-base formula.

$$\log_2 0.2 = \frac{\log_{10} 0.2}{\log_{10} 2}$$

$$\approx \frac{-0.6989700043}{0.3010299957}$$

$$\approx -2.3219$$

37. We will use natural logarithms for the
conversion. Let $a = e, b = \pi$, and $M = 58$
and substitute in the change-of-base formula.

$$\log_\pi 58 = \frac{\ln 58}{\ln \pi} \approx \frac{4.060443011}{1.144729886} \approx 3.5471$$

39. Graph: $f(x) = e^x$

We find some function values with a
calculator. We use these values to plot points
and draw the graph.

x	e^x
0	1
1	2.7
2	7.4
3	20.1
−1	0.4
−2	0.1

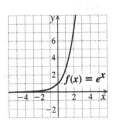

The domain is the set of real numbers and the
range is $(0, \infty)$.

41. Graph: $f(x) = e^x + 3$

We find some function values with a
calculator. We use these values to plot points
and draw the graph.

x	$e^x + 3$
0	4
1	5.72
2	10.39
−1	3.37
−2	3.14

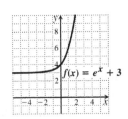

The domain is the set of real numbers and the
range is $(3, \infty)$.

43. Graph: $f(x) = e^x - 2$

We find some function values, plot points,
and draw the graph.

x	$e^x - 2$
0	−1
1	0.72
2	5.4
−1	−1.6
−2	−1.9

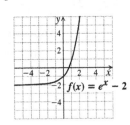

The domain is the set of real numbers and the
range is $(-2, \infty)$.

45. Graph: $f(x) = 0.5e^x$

We find some function values, plot points,
and draw the graph.

x	$0.5e^x$
0	0.5
1	1.36
2	3.69
−1	0.18
−2	0.07

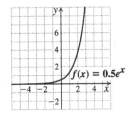

The domain is the set of real numbers and the
range is $(0, \infty)$.

47. Graph: $f(x) = 0.5e^{2x}$

We find some function values, plot points,
and draw the graph.

x	$0.5e^{2x}$
0	0.5
1	3.7
2	27.3
−1	0.07
−2	0.009

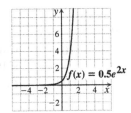

The domain is the set of real numbers and the
range is $(0, \infty)$.

49. Graph: $f(x) = e^{x-3}$

We find some function values, plot points,
and draw the graph.

x	e^{x-3}
5	7.4
3	1
0	0.05
−1	0.02

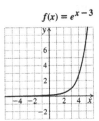

The domain is the set of real numbers
and the rage is $(0, \infty)$.

51. Graph: $f(x) = e^{x+2}$

We find some function values, plot points, and draw the graph.

x	e^{x+2}
1	20.1
0	7.4
−1	2.7
−2	1
−3	0.4

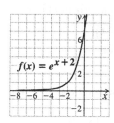

The domain is the set of real numbers and the range is $(0, \infty)$.

53. $f(x) = -e^x$

We find some function values, plot points, and draw the graph.

x	$-e^x$
0	−1
1	−2.7
2	−7.4
3	−20.1
−1	−0.4
−2	−0.1

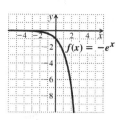

The domain is the set of real number and the range is $(-\infty, 0)$.

55. Graph: $g(x) = \ln x + 1$

Remember : $x > 0$

x	$\ln x + 1$
0.5	0.3
1	1
2	1.7
3	2.1
4	2.4

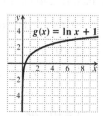

The domain is $(0, \infty)$ and the range is the set of real numbers.

57. Graph: $g(x) = \ln x - 2$

x	$\ln x - 2$
1	−2
2	−1.31
3	−0.90
4	−0.61
5	−0.39

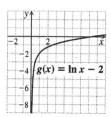

The domain is $(0, \infty)$ and the range is the set of real numbers.

59. Graph: $g(x) = 2 \ln x$

x	$2 \ln x$
0.5	−1.4
1	0
2	1.4
4	2.8
6	3.6

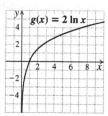

The domain is $(0, \infty)$ and the range is the set of real numbers.

61. Graph: $g(x) = -2 \ln x$

x	$-2 \ln x$
0.5	1.4
1	0
2	−1.4
3	−2.2
4	−2.8

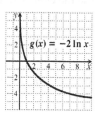

The domain is $(0, \infty)$ and the range is the set of real numbers.

63. Graph: $g(x) = \ln(x + 2)$

x	$\ln(x+2)$
−1.9	−2.3
−1	0
0	0.7
1	1.1
2	1.4

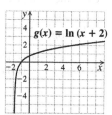

The domain is $(-2, \infty)$ and the range is the set of real numbers.

65. Graph: $g(x) = \ln(x-1)$

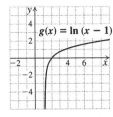

x	$\ln(x-1)$
1.1	−2.3
2	0
3	0.7
4	1.1
5	1.4

The domain is $(1, \infty)$ and the range is the set of real numbers.

67. We use the change-of-base formula:

$$f(x) = \frac{\log x}{\log 5} \text{ or } f(x) = \frac{\ln x}{\ln 5}$$

$y = \log(x)/\log(5)$, or
$y = \ln(x)/\ln(5)$

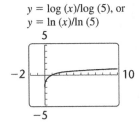

69. We use the change of base formula.

$$f(x) = \frac{\log(x-5)}{\log 2} \text{ or } f(x) = \frac{\ln(x-5)}{\ln 2}$$

$y = \log(x-5)/\log(2)$, or
$y = \ln(x-5)/\ln(2)$

71. We use the change of base formula.

$$f(x) = \frac{\log x}{\log 3} + x \text{ or } f(x) = \frac{\ln x}{\ln 3} + x$$

$y = \log(x)/\log(3) + x$, or
$y = \ln(x)/\ln(3) + x$

73. *Thinking and Writing Exercise.*

75. $x^2 - 3x - 28 = 0$
The trinomial can be factored by finding two numbers whose product is −28 and whose sum is −3. These factors are 4 and −7, and the factorization is $(x+4)(x-7)$.

$(x+4)(x-7) = 0$

$x+4 = 0 \quad or \quad x-7 = 0$

$x = -4 \quad or \quad\quad x = 7$

The solutions are −4 and 7.

77. $17x - 15 = 0$

$17x = 15$

$x = \dfrac{15}{17}$

The solution is $\dfrac{15}{17}$.

79. $(x-5) \cdot 9 = 11$

$9x - 45 = 11$

$9x = 45 + 11$

$9x = 56$

$x = \dfrac{56}{9}$

The solution is $\dfrac{56}{9}$.

81. $x^{1/2} - 6x^{1/4} + 8 = 0$
The trinomial can be factored by finding two numbers whose product is 8 and whose sum is −6. These factors are −2 and −4, and the factorization is $(x^{1/4} - 2)(x^{1/4} - 4)$.

$(x^{1/4} - 2)(x^{1/4} - 4) = 0$

$x^{1/4} - 2 = 0 \quad or \quad x^{1/4} - 4 = 0$

$x^{1/4} = 2 \quad or \quad\quad x^{1/4} = 4$

$x = 2^4 \quad or \quad\quad x = 4^4$

$x = 16 \quad or \quad\quad x = 256$

The solutions are 16 and 256.

83. *Thinking and Writing Exercise.*

85.   We use the change-of-base formula.

$$\log_6 81 = \frac{\log 81}{\log 6}$$

$$= \frac{\log 3^4}{\log(2 \cdot 3)}$$

$$= \frac{4 \log 3}{\log 2 + \log 3}$$

$$\approx \frac{4(0.477)}{0.301 + 0.477}$$

$$\approx 2.452$$

87.   We use the change-of-base formula.

$$\log_{12} 36 = \frac{\log 36}{\log 12}$$

$$= \frac{\log(2 \cdot 3)^2}{\log(2^2 \cdot 3)}$$

$$= \frac{2 \log(2 \cdot 3)}{\log 2^2 + \log 3}$$

$$= \frac{2(\log 2 + \log 3)}{2 \log 2 + \log 3}$$

$$\approx \frac{2(0.301 + 0.477)}{2(0.301) + 0.477}$$

$$\approx 1.442$$

89.   Use the change-of-base formula with $a = e$ and $b = 10$. We obtain

$$\log M = \frac{\ln M}{\ln 10}.$$

91. $\log(492x) = 5.728$

$$10^{5.728} = 492x$$

$$\frac{10^{5.728}}{492} = x$$

$$1086.5129 \approx x$$

93. $\log 692 + \log x = \log 3450$

$$\log x = \log 3450 - \log 692$$

$$\log x = \log \frac{3450}{692}$$

$$x = \frac{3450}{692}$$

$$x \approx 4.9855$$

95.   a. Domain: $\{x \mid x > 0\}$, or $(0, \infty)$; range: $\{y \mid y < 0.5135\}$, or $(-\infty, 0.5135)$;

b. $[-1, 5, -10, 5]$;

c.

$y = 3.4 \ln x - 0.25 e^x$

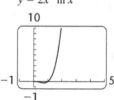

97.   a. Domain $\{x \mid x > 0\}$, or $(0, \infty)$; range: $\{y \mid y > -0.2453\}$, or $(-0.2453, \infty)$

b. $[-1, 5, -1, 10]$;

c.

$y = 2x^3 \ln x$

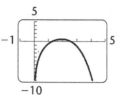

Exercise Set 9.6

1. e

3. f

5. b

7. g

9. $3^{2x} = 81 \Rightarrow 3^{2x} = 3^4 \Rightarrow 2x = 4 \Rightarrow x = 2$

11. $4^x = 32 \Rightarrow \left(2^2\right)^x = 2^5 \Rightarrow 2^{2x} = 2^5$

$$\Rightarrow 2x = 5 \Rightarrow x = \frac{5}{2}$$

13. $2^x = 10 \Rightarrow \log(2^x) = \log(10)$

$$\Rightarrow x \log(2) = 1 \Rightarrow x = \frac{1}{\log(2)} \approx 3.322$$

15. $2^{x+5} = 16 \Rightarrow 2^{x+5} = 2^4 \Rightarrow x + 5 = 4 \Rightarrow x = -1$

17. $8^{x-3} = 19 \Rightarrow \log(8^{x-3}) = \log(19)$

$\Rightarrow (x-3)\log(8) = \log(19)$

$\Rightarrow x-3 = \dfrac{\log(19)}{\log(8)} \Rightarrow x = \dfrac{\log(19)}{\log(8)} + 3 \approx 4.416$

19. $e^{t} = 50 \Rightarrow \ln(e^{t}) = \ln(50) \Rightarrow t\ln(e) = \ln(50)$

$\Rightarrow t = \ln(50) \approx 3.912$

21. $e^{-0.02t} = 8 \Rightarrow \ln(e^{-0.02t}) = \ln(8)$

$\Rightarrow -0.02t\ln(e) = \ln(8)$

$\Rightarrow -0.02t = \ln(8) \Rightarrow t = -\dfrac{\ln(8)}{0.02} \approx -103.972$

23. $5 = 3^{x+1}$

$\log 5 = \log 3^{x+1} \Rightarrow \log 5 = (x+1)\log 3$

$\Rightarrow \dfrac{\log 5}{\log 3} = \dfrac{(x+1)\log 3}{\log 3} \Rightarrow \dfrac{\log 5}{\log 3} = x+1$

$\Rightarrow x = \dfrac{\log 5}{\log 3} - 1 \approx 0.465$

25. $4.9^{x} - 87 = 0$

$4.9^{x} = 87$

$\log 4.9^{x} = \log 87$

$x\log 4.9 = \log 87$

$x = \dfrac{\log 87}{\log 4.9}$

$x \approx 2.810$

27. $19 = 2e^{4x}$

$\dfrac{19}{2} = \dfrac{2e^{4x}}{2}$

$\ln\left(\dfrac{19}{2}\right) = \ln e^{4x}$

$\ln\left(\dfrac{19}{2}\right) = 4x$

$\dfrac{\ln\left(\frac{19}{2}\right)}{4} = x$

$0.563 \approx x$

29. $7 + 3e^{5x} = 13$

$3e^{5x} = 6$

$e^{5x} = 2$

$\ln e^{5x} = \ln 2$

$5x = \ln 2$

$x = \dfrac{\ln 2}{5}$

$x \approx 0.139$

31. $\log_3 x = 4$

$x = 3^4$ Writing as an equivalent
                    exponential equation

$x = 81$

33. $\log_2 x = -3$

$x = 2^{-3}$ Writing as an equivalent
                        exponential equation

$x = \dfrac{1}{8}$

35. $\ln x = 5$

$x = e^5$ ($\ln x$ has base of e)

$x \approx 148.413$

37. $\ln 4x = 3$

$4x = e^3$

$x = \dfrac{e^3}{4}$

$x \approx 5.021$

39. $\log x = 2.5$

$x = 10^{2.5}$ ($\log x$ has base of 10)

$x \approx 316.228$

41. $\ln(2x+1) = 4$

$2x+1 = e^4$

$2x = e^4 - 1$

$x = \dfrac{e^4 - 1}{2}$

$x \approx 26.799$

43. $\ln x = 1$

$x = e^1$

$x \approx 2.718$

45. $5\ln x = -15$

$\ln x = -3$

$x = e^{-3}$

$x \approx 0.050$

47. $\log_2(8 - 6x) = 5$

$8 - 6x = 2^5$

$8 - 6x = 32$

$-6x = 24$

$x = -4$

The answer checks. The solution is -4.

49. $\log(x - 9) + \log x = 1$

$\log_{10}\left[(x - 9)(x)\right] = 1$

$x(x - 9) = 10^1$

$x^2 - 9x = 10$

$x^2 - 9x - 10 = 0$

$(x + 1)(x - 10) = 0$

$x = -1$ or $x = 10$

The number -1 does not check, because negative numbers do not have logarithms. The solution is 10.

51. $\log x - \log(x + 3) = 1$

$\log_{10}\dfrac{x}{x + 3} = 1$

$\dfrac{x}{x + 3} = 10^1$

$x = 10(x + 3)$

$x = 10x + 30$

$-9x = 30$

$x = -\dfrac{10}{3}$

The number $-\frac{10}{3}$ does not check. The equation has no solution.

53. $\log(2x + 1) = \log(5) \Rightarrow 2x + 1 = 5 \Rightarrow x = 2$

55. $\log_4(x + 3) = 2 + \log_4(x - 5)$

$\log_4(x + 3) - \log_4(x - 5) = 2$

$\log_4\dfrac{x + 3}{x - 5} = 2$

$\dfrac{x + 3}{x - 5} = 4^2$

$\dfrac{x + 3}{x - 5} = 16$

$x + 3 = 16(x - 5)$

$x + 3 = 16x - 80$

$83 = 15x$

$\dfrac{83}{15} = x$

The number $\frac{83}{15}$ checks. It is the solution.

57. $\log_7(x + 1) + \log_7(x + 2) = \log_7 6$

$\log_7\left[(x + 1)(x + 2)\right] = \log_7 6$

$\log_7(x^2 + 3x + 2) = \log_7 6$

$x^2 + 3x + 2 = 6$

$x^2 + 3x - 4 = 0$

$(x + 4)(x - 1) = 0$

$x = -4$ or $x = 1$.

Only the number 1 checks; it is the solution.

59. $\log_5(x + 4) + \log_5(x - 4) = \log_5 20$

$\log_5(x^2 - 16) = \log_5 20$

$x^2 - 16 = 20$

$x^2 = 36$

$x = \pm\sqrt{36} = \pm 6$

Only 6 checks; it is the solution.

61. $\ln(x + 5) + \ln(x + 1) = \ln 12$

$\ln(x^2 + 6x + 5) = \ln 12$

$x^2 + 6x + 5 = 12$

$x^2 + 6x - 7 = 0$

$(x + 7)(x - 1) = 0$

$x = -7$ or $x = 1$

Only 1 checks; it is the solution.

63. $\log_2(x-3)+\log_2(x+3)=4$

$\log_2(x^2-9)=4$

$x^2-9=2^4$

$x^2-9=16$

$x^2=25$

$x=\pm\sqrt{25}=\pm5$

Only 5 checks; it is the solution.

65. $\log_{12}(x+5)-\log_{12}(x-4)=\log_{12}3$

$\log_{12}\dfrac{x+5}{x-4}=\log_{12}3$

$\dfrac{x+5}{x-4}=3$

$x+5=3(x-4)$

$x+5=3x-12$

$17=2x$

$\dfrac{17}{2}=x$

The number $\frac{17}{2}$ checks and is the solution.

67. $\log_3(x-2)+\log_2 x=3$

$\log_2\big[(x-2)(x)\big]=3$

$x(x-2)=2^3$

$x^2-2x=8$

$x^2-2x-8=0$

$(x-4)(x+2)=0$

$x=4 \text{ or } x=-2$

The number 4 checks, but -2 does not. The solution is 4.

69. $e^{0.5x}-7=2x+6$

Graph $y_1=e^{0.5x}-7$ and $y_2=2x+6$ in a window that shows the points of intersection of the graphs. One good choice is $[-10,10,-10,25]$, Yscl = 5. Use Intersect to find the first coordinates of the points of intersection. They are the solutions of the given equation. They are about -6.480 and 6.519.

71. $\ln(3x)=3x-8$

Graph $y_1=\ln(3x)$ and $y_2=3x-8$ in a window that shows the points of intersection of the graphs. One good choice is $[-5,5,-15,5]$. When we use Intersect in this window we can find only the coordinates of the right-hand point of intersection. They are about $(3.445,2.336)$, so one solution of the equation is about 3.445. To find the coordinates of the left-hand point of intersection we make the window smaller. One window that is appropriate is $[-1,1,-15,5]$. Using Intersect again we find that the other solution of the equation is about 0.0001. (The answer approximated to the nearest thousandth is 0.000, so we express it to the nearest ten-thousandth.)

73. Solve $\ln x=\log x$.

Graph $y_1=\ln x$ and $y_2=\log x$ in a window that shows the point of intersection of the graphs. One good choice is $[-5,5,5,5]$. Use Intersect to find the first coordinate of the point of intersection. It is solution of the given equation. It is 1.

75. *Thinking and Writing Exercise.*

77. Let $x=$ the width of the rectangle, in feet. Then $x+6$ will equal the length, in feet. So:

$P=2l+2w\Rightarrow 2(x+6)+2x=26$

$\Rightarrow(x+6)+x=13\Rightarrow 2x+6=13$

$\Rightarrow 2x=7$

$\Rightarrow x=\dfrac{7}{2}$ ft = 3.5 ft

$\Rightarrow x+6=3.5+6=9.5$ ft

The width is 3.5 ft and the length is 9.5 ft.

79. Let $x=$ the amount, in pounds, of the 25% sunflower seed brand, and $y=$ the amount, in pounds, of the 40% sunflower seed brand. Then:

$x+y=50 \qquad\qquad (1)$

$0.25x+0.40y=0.33(50)$

or $25x+40y=33\cdot 50$

or $5x+8y=33\cdot 10=330 \quad (2)$

Solve (1) for y:

$x + y = 50$ (1)

$\Rightarrow y = 50 - x$ (3)

Substitute $50 - x$ for y in (2) and solve for x:

$5x + 8y = 330$ (2)

$\Rightarrow 5x + 8(50 - x) = 330$

$\Rightarrow 5x + 400 - 8x = 330$

$\Rightarrow -3x = 330 - 400 = -70$

$\Rightarrow x = \dfrac{-70}{-3} = 23\dfrac{1}{3}$

Use (3) and the value of x to find y:

$y = 50 - x$ (3)

$\Rightarrow y = 50 - 23\dfrac{1}{3} = 49\dfrac{3}{3} - 23\dfrac{1}{3} = 26\dfrac{2}{3}$

Joanna should use $23\tfrac{1}{3}$ lbs of the 25% sunflower seed mix and $26\tfrac{2}{3}$ lbs of the 40% sunflower seed mix.

81. Max's rate is $\dfrac{1 \text{ score}}{2 \text{ hr}}$. Miles' rate is $\dfrac{1 \text{ score}}{3 \text{ hr}}$.

When working together their rate is::

$\dfrac{1 \text{ score}}{2 \text{ hr}} + \dfrac{1 \text{ score}}{3 \text{ hr}} = \dfrac{3 \text{ score}}{6 \text{ hr}} + \dfrac{2 \text{ score}}{6 \text{ hr}} = \dfrac{5 \text{ score}}{6 \text{ hr}}$

Therefore,

$\dfrac{5 \text{ score}}{6 \text{ hr}} \cdot t = 1 \text{ score}$

$\Rightarrow t = 1 \text{ score} \cdot \dfrac{6 \text{ hr}}{5 \text{ score}}$

$= \dfrac{6}{5} \text{ hr} = 1\dfrac{1}{5} \text{hr} = 1 \text{ hour, 12 min}$

83. *Thinking and Writing Exercise.*

85. $27^x = 81^{2x-3}$

$\left(3^3\right)^x = \left(3^4\right)^{2x-3}$

$3^{3x} = 3^{8x-12}$

$3x = 8x - 12$

$12 = 5x$

$\dfrac{12}{5} = x$

The solution is $\frac{12}{5}$.

87. $\log_x\left(\log_3 27\right) = 3$

$\log_3 27 = x^3$

$3 = x^3$ $\left(\log_3 27 = 3\right)$

$\sqrt[3]{3} = x$

The solution is $\sqrt[3]{3}$.

89. $x \cdot \log\dfrac{1}{8} = \log 8$

$x \cdot \log 8^{-1} = \log 8$

$x\left(-\log 8\right) = \log 8$ Using the power rule

$x = -1$

The solution is -1.

91. $2^{x^2+4x} = \dfrac{1}{8}$

$2^{x^2+4x} = \dfrac{1}{2^3}$

$2^{x^2+4x} = 2^{-3}$

$x^2 + 4x = -3$

$x^2 + 4x + 3 = 0$

$\left(x+3\right)\left(x+1\right) = 0$

$x = -3 \text{ or } x = -1$

The solutions are -3 and -1.

93. $\log_5 |x| = 4$

$|x| = 5^4$

$|x| = 625$

$x = 625 \text{ or } x = -625$

The solutions are 625 and -625.

95. $\log \sqrt{2x} = \sqrt{\log 2x}$

$\log\left(2x\right)^{1/2} = \sqrt{\log 2x}$

$\dfrac{1}{2}\log 2x = \sqrt{\log 2x}$

$\dfrac{1}{4}\left(\log 2x\right)^2 = \log 2x$ Squaring both sides

$\dfrac{1}{4}\left(\log 2x\right)^2 - \log 2x = 0$

Let $u = \log 2x$

$\dfrac{1}{4}u^2 - u = 0$

$u\left(\dfrac{1}{4}u - 1\right) = 0$

$u = 0 \quad$ or $\quad \dfrac{1}{4}u - 1 = 0$

$u = 0 \quad$ or $\quad \dfrac{1}{4}u = 1$

$u = 0 \quad$ or $\quad u = 4$

$\log 2x = 0 \quad$ or $\quad \log 2x = 4 \qquad$ Replacing u with $\log 2x$

$2x = 10^0 \quad$ or $\quad 2x = 10^4$

$2x = 1 \quad$ or $\quad 2x = 10{,}000$

$x = \dfrac{1}{2} \quad$ or $\quad x = 5000$

Both numbers check. The solutions are $\frac{1}{2}$ and 5000.

97. $3^{x^2} \cdot 3^{4x} = \dfrac{1}{27}$

$3^{x^2 + 4x} = 3^{-3}$

$x^2 + 4x = -3$

$x^2 + 4x + 3 = 0$

$(x+3)(x+1) = 0$

$x = -3 \quad$ or $\quad x = -1$

Both numbers check. The solutions are -3 and -1.

99. $\log x^{\log x} = 25$

$\log x \left(\log x \right) = 25 \quad$ Using the power rule

$\left(\log x \right)^2 = 25$

$\log x = \pm 5$

$x = 10^5 \qquad$ or $\quad x = 10^{-5}$

$x = 100{,}000 \quad$ or $\quad x = \dfrac{1}{100{,}000}$

Both numbers check. The solutions are

$100{,}000$ and $\dfrac{1}{100{,}000}$.

101. $\left(81^{x-2}\right)\left(27^{x+1}\right) = 9^{2x-3}$

$\left[\left(3^4\right)^{x-2} \right]\left[\left(3^3\right)^{x+1} \right] = \left(3^2\right)^{2x-3}$

$\left(3^{4x-8}\right)\left(3^{3x+3}\right) = 3^{4x-6}$

$3^{7x-5} = 3^{4x-6}$

$7x - 5 = 4x - 6$

$3x = -1$

$x = -\dfrac{1}{3}$

The solution is $-\dfrac{1}{3}$.

103. $2^y = 16^{x-3}$ and $3^{y+2} = 27^x$

$2^y = \left(2^4\right)^{x-3} \quad$ and $\quad 3^{y+2} = \left(3^3\right)^x$

$y = 4x - 12 \quad$ and $\quad y + 2 = 3x$

$12 = 4x - y \quad$ and $\quad 2 = 3x - y$

Solving this system of equations we get $x = 10$ and $y = 28$. Then $x + y = 10 + 28 = 38$.

Exercise Set 9.7

---

1. a) $A(t) = 4000 \Rightarrow 77(1.283)^t = 4000$

$\Rightarrow 1.283^t = \dfrac{4000}{77}$

$\Rightarrow \log(1.283^t) = \log\left(\dfrac{4000}{77}\right)$

$\Rightarrow t \log(1.283) = \log\left(\dfrac{4000}{77}\right)$

$\Rightarrow t = \dfrac{\log\left(\dfrac{4000}{77}\right)}{\log(1.283)} \approx 15.85$

$1990 + 15.85 = 2005.85.$ About 2006.

b) $A(t) = 2 \cdot 77 = 77(1.283)^t \Rightarrow 1.283^t = 2$

$\Rightarrow t = \dfrac{\log(2)}{\log(1.283)} \approx 2.78$

The doubling time is approximately 2.8 years.

3. a) $S(t) = 100 \Rightarrow 180(0.97)^t = 100$

$\Rightarrow 0.97^t = \dfrac{100}{180} \Rightarrow \log(0.97^t) = \log\left(\dfrac{100}{180}\right)$

$\Rightarrow t\log(0.97) = \log\left(\dfrac{100}{180}\right)$

$\Rightarrow t = \dfrac{\log\left(\dfrac{100}{180}\right)}{\log(0.97)} \approx 19.30$

$1960 + 19.3 = 1979.3$. About 1979.

b) $S(t) = 25 \Rightarrow 180(0.97)^t = 25$

$\Rightarrow 0.97^t = \dfrac{25}{180} \Rightarrow \log(0.97^t) = \log\left(\dfrac{25}{180}\right)$

$\Rightarrow t\log(0.97) = \log\left(\dfrac{25}{180}\right)$

$\Rightarrow t = \dfrac{\log\left(\dfrac{25}{180}\right)}{\log(0.97)} \approx 64.81$

$1960 + 64.81 = 2024.81$. About 2025.

5. a) $A(t) = 29,000(1.03)^t$

Let $A(t) = 35,000,$ and solve for t.

$35,000 = 29,000(1.03)^t$

$\dfrac{35,000}{29,000} = (1.03)^t$

$\log 1.2069 \approx \log(1.03)^t$

$\log 1.2069 \approx t\log 1.03$

$\dfrac{\log 1.2069}{\log 1.03} \approx t$

$6.4 \approx t$

The amount after about 6.4 years will reach $35,000.

b) $2 \cdot P_0 = 2 \cdot 29,000 = 58,000$

Let $A(t) = 58,000,$ and solve for t.

$58,000 = 29,000(1.03)^t$

$\dfrac{58,000}{29,000} = (1.03)^t$

$\log 2 = \log 1.03^t$

$\log 2 = t\log 1.03$

$\dfrac{\log 2}{\log 1.03} = t$

$23.4 \approx t$

The doubling time is approximately 23.4 years.

7. a) $m(t) = 48 \Rightarrow 1507(0.94)^t = 48$

$\Rightarrow 0.94^t = \dfrac{48}{1507}$

$\Rightarrow \log(0.94^t) = \log\left(\dfrac{48}{1507}\right)$

$\Rightarrow t\log(0.94) = \log\left(\dfrac{48}{1507}\right)$

$\Rightarrow t = \dfrac{\log\left(\dfrac{48}{1507}\right)}{\log(0.94)} \approx 55.70$

About $56°$ Fahrenheit.

b) $m(t) = 15 \Rightarrow 1507(0.94)^t = 15$

$\Rightarrow 0.94^t = \dfrac{15}{1507}$

$\Rightarrow \log(0.94^t) = \log\left(\dfrac{15}{1507}\right)$

$\Rightarrow t\log(0.94) = \log\left(\dfrac{15}{1507}\right)$

$\Rightarrow t = \dfrac{\log\left(\dfrac{15}{1507}\right)}{\log(0.94)} \approx 74.50$

About $75°$ Fahrenheit.

9. a) Note that 1 billion = 1000 million.

$S(t) = 1000 \Rightarrow 2.05(1.8)^t = 1000$

$\Rightarrow 1.8^t = \dfrac{1000}{2.05} \Rightarrow \log(1.8^t) = \log\left(\dfrac{1000}{2.05}\right)$

$\Rightarrow t\log(1.8) = \log\left(\dfrac{1000}{2.05}\right)$

$\Rightarrow t = \dfrac{\log\left(\dfrac{1000}{2.05}\right)}{\log(1.8)} \approx 10.53$

$2002 + 10.53 = 2012.53$. About 2013.

b) $m(t) = 2 \cdot 2.05 \Rightarrow 2.05(1.8)^t \Rightarrow 1.8^t = 2$

$\Rightarrow t = \dfrac{\log(2)}{\log(1.8)} \approx 1.18$

The doubling time is approximately 1.2 years.

11. $\text{pH} = -\log\left[H^+\right]$

$= -\log\left[1.3 \times 10^{-5}\right]$

$\approx -(-4.886057)$

≈ 4.9

The pH of fresh-brewed coffee is about 4.9.

13. $\text{pH} = -\log\left[H^+\right]$

$7.0 = -\log\left[H^+\right]$

$-7.0 = \log\left[H^+\right]$

$10^{-7.0} = \left[H^+\right]$ Converting to an
                         exponential equation

10^{-7} moles per liter

15. $L = 10 \cdot \log\dfrac{I}{I_0} = 10 \cdot \log\dfrac{10}{10^{-12}}$

$= 10 \cdot \log\left(10^{13}\right) = 10 \cdot 13 = 130$

130 decibels.

17. $L = 128.8 \Rightarrow 10 \cdot \log\left(\dfrac{I}{I_0}\right) = 128.8$

$\Rightarrow \log\left(\dfrac{I}{I_0}\right) = \dfrac{128.8}{10} = 12.88 \Rightarrow \dfrac{I}{I_0} = 10^{12.88}$

$\Rightarrow I = 10^{12.88} I_0 == 10^{12.88} 10^{-12} \text{ W/m}^2$

$= 10^{0.88} \text{ W/m}^2 \approx 7.6 \text{ W/m}^2$

19. $M = 7.5 \Rightarrow \log\left(\dfrac{v}{1.34}\right) = 7.5 \Rightarrow \dfrac{v}{1.34} = 10^{7.5}$

$\Rightarrow v = 1.34 \cdot 10^{7.5} \approx 4.24 \cdot 10^7$

Approximately 42.4 million emails a day.

21. $k = 2.5\%; \ 0.025$

(a) Using the exponential growth formula,

we have $P(t) = P_0 e$

$P(t) = P_0 e^{0.025t}$

(b) $P_0 = \$5000$

After 1 year, $t = 1$.

$P(1) = 5000 e^{0.025 \cdot 1}$

$P(1) = 5000 e^{0.025}$

$\approx \$5126.58$

After 2 years, $t = 2$.

$P(2) = 5000 e^{0.025 \cdot 2}$

$= 5000 e^{0.05}$

$\approx \$5256.36$

(c) $2 \cdot 5,000 = 10,000$

$10,000 = 5000 e^{0.025t}$

$2 = e^{0.025t}$

$\ln 2 = \ln e^{0.025t}$

$\ln 2 = 0.025t$

$\dfrac{\ln 2}{0.025} = t$

$27.7 \approx t$

The investment will double in approximately 27.7 years.

23.   (a) $P_0 = 310$ million, $k = 1.0\%$, or 0.01

$P(t) = P_0 e^{kt}$

$P(t) = 310 e^{0.01t}$, where $t =$ the number

of years after 2010 and $P(t)$ is in

millions.

(b) In 2016, $t = 2016 - 2010 = 6$

$P(6) = 310 e^{0.01 \cdot 6}$

$P(6) = 310 e^{0.06}$

$P(6) \approx 329$

In 2016, the U.S. population will be approximately 329 million.

(c)

$310 e^{0.01t} = 350 \Rightarrow e^{0.01t} = \dfrac{350}{310} = \dfrac{35}{31}$

$\Rightarrow 0.01t = \ln\left(\dfrac{35}{31}\right) \Rightarrow t = \dfrac{1}{0.01}\ln\left(\dfrac{35}{31}\right)$

$\Rightarrow t \approx 12.1$

The population will reach 350 million about 12 years after 2010, or about 2022.

25. $k = 12\%$, or 0.12 The doubling rate is when

$P(t) = 2P_0 \Rightarrow P_0 e^{0.12t} = 2P_0 \Rightarrow e^{0.12t} = 2$

$\Rightarrow 0.12t = \ln(2) \Rightarrow t = \dfrac{\ln(2)}{0.12} \approx 5.78$

The number of online college students doubles approximately every 5.8 years.

27. (a) $Y(10) = 87 \ln\left(\dfrac{10}{6.1}\right) \approx 43.00$

    $2000 + 43 = 2043$

    According to the model, the world population will reach 10 billion in about 2043.

  (b) $Y(12) = 87 \ln\left(\dfrac{12}{6.1}\right) \approx 58.87$

    $2000 + 58.87 = 2058.87$

    According to the model, the world population will reach 12 billion in late 2058.

  (c) Using the values we computed in parts (a) and (b) and any others we wish to calculate, we sketch the graph:

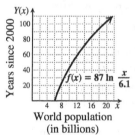

    World population (in billions)

29. (a) $S(0) = 68 - 20 \log(0+1)$

    $= 68 - 20 \log 1 = 68 - 20(0)$

    $= 68\%$

  (b) $S(4) = 68 - 20 \log(4+1)$

    $= 68 - 20 \log 5$

    $\approx 68 - 20(0.69897)$

    $\approx 54\%$

    $S(24) = 68 - 20 \log(24+1)$

    $= 68 - 20 \log 25$

    $\approx 68 - 20(1.39794)$

    $\approx 40\%$

  (c) Using the values we computed in parts (a) and (b) and any others we wish to calculate, we sketch the graph:

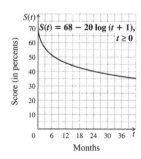

    Months

d. $50 = 68 - 20 \log(t+1)$

    $-18 = -20 \log(t+1)$

    $0.9 = \log(t+1)$

    $10^{0.9} = t+1$

    $7.9 \approx t+1$

    $6.9 \approx t$

    After about 6.9 months, the average score was 50.

31. (a) Let $t =$ the year since 1990, and P be the power generated by wind in thousands of megawatts. Then:

    $P_0 = P(0) = 2$ and $P(19) = 35$ So:

    $P(t) = P_0 e^{kt} = 2e^{kt}$

    $\Rightarrow P(19) = 2e^{k19} = 35 \Rightarrow e^{19k} = \dfrac{35}{2} = 17.5$

    $\Rightarrow 19k = \ln(17.5)$

    $\Rightarrow k = \dfrac{\ln(17.5)}{19} \approx 0.150642$

    $\Rightarrow P(t) = 2e^{0.150642t}$

  (b) $P(t) = 50 \Rightarrow 2e^{0.150642t} = 50$

    $\Rightarrow e^{0.150642t} = 25 \Rightarrow 0.150642t = \ln(25)$

    $\Rightarrow t = \dfrac{\ln(25)}{0.150642} \approx 21.37$

    $1990 + 21.37 = 2011.37$

    According to the model, wind-power capacity will 50,000 MW of power in 2011.

33. (a) Let $t =$ the year since 1997, and C be the cost per gigabit per second per mile in dollars. Then:

    $C_0 = C(0) = 8200$ and $C(10) = 500$ So:

    $C(t) = C_0 e^{kt} = 8200 e^{-kt}$

    $\Rightarrow C(10) = 8200 e^{-k10} = 500$

    $\Rightarrow e^{-10k} = \dfrac{500}{8200} = \dfrac{5}{82} \Rightarrow -10k = \ln\left(\dfrac{5}{82}\right)$

    $\Rightarrow k = -\dfrac{1}{10} \cdot \ln\left(\dfrac{5}{82}\right) \approx 0.2797$

    $\Rightarrow C(t) = 8200 e^{-0.2797t}$

  (b) $t = 2010 - 1997 = 13$

    $C(13) = 8200 e^{-0.2797 \cdot 13} \approx 216$

    According to the model, it cost approximately $216 per gigabit per second per mile to lay subsea cable in 2010.

(c) Use $C(t) = 1$.

$$8200e^{-0.2797t} = 1$$

$$e^{-0.2797t} = \frac{1}{8200}$$

$$-0.2797t = \ln\left(\frac{1}{8200}\right)$$

$$t = \frac{\ln\left(\frac{1}{8200}\right)}{-0.2797}$$

$$\approx 32.22$$

$$1997 + 32 = 2029$$

According to the model, it will cost \$1 per gigabit per second per mile in about 2029.

35. We will use the function derived in Example 7:
$$P(t) = P_0 e^{-0.00012t}$$

If the seed lost 21% of its carbon-14 from an initial amount P_0, then

$(100\% - 21\%)P_0 = 79\% P_0 = 0.79 P_0$ is the amount present. To find the age of the seed set $P(t) = 0.79 P_0$ and solve for t.

$$P(t) = 0.79 P_0 \Rightarrow P_0 e^{-0.00012t} = 0.79 P_0$$

$$\Rightarrow e^{-0.00012t} = 0.79 \Rightarrow -0.00012t = \ln(0.79)$$

$$\Rightarrow t = -\frac{\ln(0.79)}{0.00012} \approx 1964.35$$

The seed was about 1964 years old.

37. The function $P(t) = P_0 e^{-kt}, k > 0$, can be used to model decay. For iodine-131, $k = 9.6\%$, or 0.096. To find the half-life we substitute 0.096 for k and $\frac{1}{2}P_0$ for $P(t)$, and solve for t.

$$x = 2010 - 1980 = 30$$

$$\frac{1}{2}P_0 = P_0 e^{-0.096t}, \text{ or } \frac{1}{2} = e^{-0.096t}$$

$$\ln\frac{1}{2} = \ln e^{-0.096t} = -0.096t$$

$$t = \frac{\ln 0.5}{-0.096} \approx \frac{-0.6931}{-0.096} \approx 7.2 \text{ days}$$

39. (a) Use $P(t) = P_0 e^{-kt}$, $k > 0$ for exponential decay. To find k substitute $\frac{1}{2}P_0$ for $P(t)$ and 5 for t.

$$P_0 e^{-k \cdot 5} = \frac{1}{2}P_0 \Rightarrow e^{-k \cdot 5} = \frac{1}{2}$$

$$\Rightarrow -5k = \ln\left(\frac{1}{2}\right) = -\ln(2)$$

$$\Rightarrow k = \frac{\ln(2)}{5} \approx 0.1386 \text{ or } 13.86\%$$

$$\Rightarrow P(t) = P_0 e^{-0.1386t}$$

(b) If 95% has decayed then
$$P(t) = (1 - 0.95)P_0 = 0.05 P_0$$

$$P(t) = 0.05 P_0 \Rightarrow P_0 e^{-0.1386t} = 0.05 P_0$$

$$\Rightarrow e^{-0.1386t} = 0.05 \Rightarrow -0.1386t = \ln(0.05)$$

$$\Rightarrow t = -\frac{\ln(0.05)}{0.1386} \approx 21.6$$

It takes approximately 21.6 hours for 95% of the caffeine to expelled.

41. (a) In 1990, $t = 0$ and $V_0 = 9$ In 2010, $t = 20$ and $V(20) = 104.3$ Substitute into the exponential growth formula and solve for k.

$$V(t) = V_0 e^{kt} = 9 e^{kt}$$

$$\Rightarrow V(20) = 9 e^{20k} = 104.3$$

$$\Rightarrow e^{20k} = \frac{104.3}{9} \Rightarrow 20k = \ln\left(\frac{104.3}{9}\right)$$

$$\Rightarrow k = \frac{1}{20}\ln\left(\frac{104.3}{9}\right) \approx 0.1225$$

$$\Rightarrow V(t) = 9 e^{0.1225t}$$

(b) In 2020, $t = 2020 - 1990 = 30$
$$V(30) = 9 e^{0.1225 \cdot 30} = 9 e^{3.675} \approx 355$$

According to the model the sculpture will be worth approximately   \$355 million in 2020.

(c) $9 e^{0.1225t} = 2 \cdot 9 \Rightarrow e^{0.1225t} = 2$

$$\Rightarrow 0.1225t = \ln(2)$$

$$\Rightarrow t = \frac{\ln(2)}{0.1225} \approx 5.66$$

The doubling time is about 5.7 years.

(d) $9e^{0.1225t} = 1000 \Rightarrow e^{0.1225t} = \dfrac{1000}{9}$

$\Rightarrow 0.1225t = \ln\left(\dfrac{1000}{9}\right)$

$\Rightarrow t = \dfrac{1}{0.1225} \cdot \ln\left(\dfrac{1000}{9}\right) \approx 38.45$

According to the model, the sculpture's value will exceed $1 billion after about 38.3 years, or in 2028.

43. The number of transistors increases from 1974 to 2006 at a rate that makes it appear that an exponential function might fit the data.

45. The number of accidents increase then decrease at different rates. It does not appear that an exponential function would be a good model.

47. (a) Let t = the number of years after 1974, and n = the number of transistors per chip, in thousands. Then from the table, we have the points (0, 6), (4, 29), (8, 134), (15, 1200), (19, 3300), (25, 9500), and (32, 291,000). Enter the data on a graphing calculator and then use the exponential regression feature to get:

$n(t) = 7.8(1.3725)^t$

(b) $e^k = 1.3725$

$\Rightarrow k = \ln(1.3725) \approx 0.3166 = 31.66\%$

(c) For 2010, $t = 2010 - 1974 = 36$.

$n(36) = 7.8(1.3725)^{36} \approx 6.95888 \cdot 10^5$

Since n is in thousands, this corresponds to approximately 696 million transistors per chip.

49. (a) We enter the data from the table and then use the exponential regression feature. We have:

$f(x) = 20917152(0.87055)^x$

(b) $f(95) = 20917152(0.8705505633)^{95}$

≈ 4 hr.

51. *Thinking and Writing Exercise.*

53. $d^2 = (x_2 - x_1)^2 + (y_2 - y_1)^2$

$= (-3 - (-2))^2 + (7 - 6)^2 = 2$

$\Rightarrow d = \sqrt{2}$

55. $(x_M, y_M) = \left(\dfrac{x_1 + x_2}{2}, \dfrac{y_1 + y_2}{2}\right)$

$= \left(\dfrac{3 + 5}{2}, \dfrac{-8 + (-6)}{2}\right) = (4, -7)$

57. $x^2 + 8x = 1 \Rightarrow x^2 + 8x + \left(\dfrac{8}{2}\right)^2 = 1 + \left(\dfrac{8}{2}\right)^2$

$\Rightarrow (x + 4)^2 = 1 + 16 = 17$

$\Rightarrow x + 4 = \pm\sqrt{17} \Rightarrow x = -4 \pm \sqrt{17}$

59. $y = x^2 - 5x - 6 = (x - 6)(x + 1)$

The function is quadratic with a positive leading coefficient. Therefore its graph is a parabola opening upward.

By the factored form we can see the x-intercepts are at $(-1, 0)$ and $(6, 0)$. Since the graph is symmetric along the axis of symmetry, the x-coordinate of the vertex must be halfway between -1 and 6, at $\dfrac{6 + (-1)}{2} = \dfrac{5}{2} = 2\dfrac{1}{2}$. To find the y-coordinate of the vertex find the value of the function at

$x = \dfrac{5}{2}$. $y\left(\dfrac{52}{2}\right) = \left(\dfrac{5}{2}\right)^2 - 5\left(\dfrac{5}{2}\right) - 6$

$= \dfrac{25}{4} - \dfrac{25}{2} - 6 = \dfrac{25}{4} - \dfrac{50}{4} - \dfrac{24}{4}$

$= -\dfrac{49}{4} = -12\dfrac{1}{4}$

Since $y(0) = -6$, the y-intercept is at $(0, -6)$. Again, using the axis of symmetry and noting that $x = 0$ is $2\dfrac{1}{2}$ units to the left of the vertex, at $x = 5$, which is $2\dfrac{1}{2}$ units to the right of the vertex, y should also be -6.

Thus we have a parabola that opens upward, with vertex $\left(\dfrac{5}{2}, -\dfrac{49}{4}\right)$, x-intercepts of $(-1, 0)$ and $(6, 0)$, y-intercept of $(0, -6)$, and which passes through the point $(5, -6)$. This is plenty of information to plot the graph:

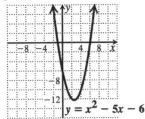

61. *Thinking and Writing Exercise.*

63. For continuous compounding, use the formula
$P(t) = P_0 e^{kt}$, with P_0 = the amount needed in
2008, in millions of dollars, to have $20 million
dollars in 2016, $k = 4\% = 0.04$, and
$t = 2016 - 2008 = 8$ years.
$$P(t) = 20 \Rightarrow P_0 e^{0.04 \cdot 8} = 20$$
$$\Rightarrow P_0 = \frac{20}{e^{0.32}} \approx 14.5$$

If one could find an institution that offered
continuous compounding for 8 years, $14.5
million dollars would need to be invested to
have $20 million 8 years later.

65. $m(I) = -(19 + 2.5 \cdot \log I)$.
(a) Let $I = 1390 \, \text{W/m}^2$ and solve.
$$m(I) = -(19 + 2.5 \log 1390)$$
$$\approx -(19 + 2.5 \cdot 3.1430)$$
$$\approx -26.9$$
(b) Let $m(I) = 23$ and solve for I.
$$23 = -(19 + 2.5 \cdot \log I)$$
$$23 = -19 - 2.5 \log I$$
$$42 = -2.5 \log I$$
$$\frac{42}{-2.5} = \log I$$
$$-16.8 = \log I \Rightarrow I = 10^{-16.8}$$
$$I \approx 1.58 \times 10^{-17} \, \text{W/m}^2$$

67. Since doubling time is the amount of time for
$P(t)$ to equal $2 \cdot P_0$, substitute and solve for t.
$$P(t) = P_0 e^{kt}$$
$$\frac{2P_0}{P_0} = \frac{P_0 e^{kt}}{P_0}$$
$$2 = e^{kt}$$
$$\ln 2 = \ln e^{kt}$$
$$\ln 2 = kt$$
$$\frac{\ln 2}{k} = t.$$

69. (a) Enter the data into a graphing calculator and
choose the logistic option in the STAT
CALC MENU to obtain:
$$f(x) = \frac{62.2245}{1 + 2.2661 e^{-0.4893x}}$$
(b) In 2010, $x = 2010 - 1997 = 13$
$$f(13) = \frac{62.2245}{1 + 2.2661 e^{-0.4893(13)}}$$
$$\approx 0.62 = 62\%$$

Chapter 9 Study Summary

1. $f(x) = 1 - 6x$, $g(x) = x^2 - 3$
$$\Rightarrow (f \circ g)(x) = f(g(x)) = 1 - 6(g(x))$$
$$= 1 - 6(x^2 - 3) = 1 - 6x^2 + 18$$
$$= 19 - 6x^2$$

2. By definition, *f(x)* is 1-1 if
$f(x) = f(y) \Rightarrow x = y$.
For $f(x) = 5x - 7$,
$$f(x) = f(y) \Rightarrow 5x - 7 = 5y - 7$$
$$\Rightarrow 5x = 5y \Rightarrow x = y \Rightarrow f \text{ is 1-1}$$

3. First show that $f^{-1}(x)$ exists by showing that
f(x) is one-to-one. For $f(x) = 5x + 1$,
$$f(x) = f(y) \Rightarrow 5x + 1 = 5y + 1$$
$$\Rightarrow 5x = 5y \Rightarrow x = y \Rightarrow f \text{ is one-to-one.}$$
Replace $f(x)$ by y: $y = 5x + 1$
Interchange x and y: $x = 5y + 1$
Solve for y: $y = \dfrac{x - 1}{5}$
Replace y by $f^{-1}(x)$: $f^{-1}(x) = \dfrac{x - 1}{5}$

4. Creating a table of values is convenient for graphing exponential functions.

x	$f(x) = 2^x$
-3	$2^{-3} = \dfrac{1}{2^3} = \dfrac{1}{8}$
-2	$2^{-2} = \dfrac{1}{2^2} = \dfrac{1}{4}$
-1	$2^{-1} = \dfrac{1}{2^1} = \dfrac{1}{2}$
0	$2^0 = 1$
1	$2^1 = 2$
2	$2^2 = 4$
3	$2^3 = 8$

From the table you can see that as x gets more negative 2^x gets smaller, but never becomes negative, or zero. (Note that a negative sign in an exponent does not make the expression negative, it just puts the base in the denominator when the sign of the exponent is changed.) This is equivalent to saying that as x goes to $-\infty$, 2^x approaches the line $y = 0$ (the x-axis) asymptotically.

As x gets more positive, however, 2^x grows without bound.

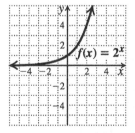

5. There are two ways to approach the graphs of log functions. One is to create a table of values, as in the previous exercise. The other is to recognize that $f(x) = \log_b(x)$ is the inverse function of the exponential function $g(x) = b^x$, and to remember that the graph of an inverse function is a reflection of the original function across the line $y = x$ (the line that goes through the points (x, x), such as $(-1, -1)$, $(0, 0)$, $(2, 2)$, and so on).

Therefore, since $f(x) = \log(x) = \log_{10}(x)$ is the inverse of $g(x) = 10^x$, we can plot the graph of 10^x, as the graph of 2^x was plotted in the previous

exercise, and then reflect the graph of 10^x across the line $y = x$.

As in exercise 4, create a table of values:

x	$g(x) = 10^x$
-3	$10^{-3} = \dfrac{1}{10^3} = \dfrac{1}{1000}$
-2	$10^{-2} = \dfrac{1}{10^2} = \dfrac{1}{100}$
-1	$10^{-1} = \dfrac{1}{10^1} = \dfrac{1}{10}$
0	$10^0 = 1$
1	$10^1 = 10$
2	$10^2 = 100$
3	$10^3 = 1000$

Again as x goes to $-\infty$, 10^x approaches the line $y = 0$ (the x-axis) asymptotically, and as x gets more positive 10^x grows without bound. Reflecting the resulting graph about the line $y = x$ gives the following graph for $f(x) = \log(x)$:

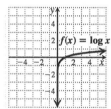

Note how the horizontal asymptote of $g(x) = 10^x$ has become a vertical asymptote for $f(x) = \log(x)$ upon reflection.

6. $5^4 = 625 \Rightarrow \log_5(5^4) = \log_5(625)$
$\Rightarrow 4\log_5(5) = \log_5(625) \Rightarrow 4 \cdot 1 = \log_5(625)$
$\Rightarrow \log_5(625) = 4$

7. $\log_9(xy) = \log_9(x) + \log_9(y)$

8. $\log_6\left(\dfrac{7}{10}\right) = \log_6(7) - \log_6(10)$

9. $\log(7^5) = 5\log(7)$

10. $\log_8(1) = 0$

11. $\log_7(7) = 1$

12. $\log_t(t^{12}) = 12\log_t(t) = 12 \cdot 1 = 12$

13. $\log(100) = \log_{10}(10^2) = 2\log_{10}(10) = 2$

14. $\ln(e) = \log_e(e) = 1$

15. $\log_2(5) = \dfrac{\log(5)}{\log(2)} \approx 2.3219$

16. $2^{3x} = 16 \Rightarrow 2^{3x} = 2^4 \Rightarrow 3x = 4 \Rightarrow x = \dfrac{4}{3}$

17. $e^{0.1x} = 10 \Rightarrow \ln(e^{0.1x}) = \ln(10)$

$\Rightarrow 0.1x\ln(e) = \ln(10) \Rightarrow 0.1x = \ln(10)$

$\Rightarrow x = \dfrac{\ln(10)}{0.1} \approx 23.02585$

18 (a) For exponential growth use the formula
$P(t) = P_0 e^{kt}$ where k is the growth rate, expressed as a decimal, and P_0 is the initial population (the population at $t = 0$). Since we are given $P_0 = 15{,}000$ and $k = 2.3\% = 0.023$, the function is: $P(t) = 15{,}000e^{0.023t}$

(b) To find the doubling time, set $P(t) = 2P_0$ and solve for t.

$P(t) = 2P_0 \Rightarrow 15{,}000e^{0.023t} = 2 \cdot 15{,}000$

$\Rightarrow e^{0.023t} = 2 \Rightarrow \ln(e^{0.023t}) = \ln(2)$

$\Rightarrow 0.023t = \ln(2) \Rightarrow t = \dfrac{\ln(2)}{0.023} \approx 30.1$

The doubling time is approximately 30.1 years.

19. For exponential decay use the formula
$P(t) = P_0 e^{-kt}$ where k is the decay rate, expressed as a decimal, and P_0 is the initial amount (the amount at $t = 0$). Since we are given $k = 1.98\% = 0.0198$, , the function is: $P(t) = P_0 e^{0.0198t}$
Even though the initial amount is unknown, the half-life can still be determined by setting $P(t) = 0.5P_0$ and solving for t.

$P(t) = 0.5P_0 \Rightarrow P_0 e^{-0.0198t} = 0.5P_0$

$\Rightarrow e^{-0.0198t} = 0.5 \Rightarrow \ln(e^{-0.0198t}) = \ln(0.5)$

$\Rightarrow -0.0198t = \ln(0.5) \Rightarrow t = -\dfrac{\ln(0.5)}{0.0198} \approx 35.0$

The half-life is approximately 35 days.

Chapter 9 Review Exercises

1. True

2. True

3. True

4. False

5. False

6. True

7. False

8. False

9. True

10. False

11. $f(x) = x^2 + 1, \qquad g(x) = 2x - 3$

$(f \circ g)(x) = f(g(x)) = f(2x - 3)$

$= (2x - 3)^2 + 1$

$= 4x^2 - 12x + 9 + 1$

$= 4x^2 - 12x + 10$

$(g \circ f)(x) = g(f(x)) = g(x^2 + 1)$

$= 2(x^2 + 1) - 3$

$= 2x^2 + 2 - 3$

$= 2x^2 - 1$

12. Possible answer: $h(x) = \sqrt{3 - x}$ is the square root of $(3 - x)$.

$f(x) = \sqrt{x}; \ g(x) = 3 - x.$

13. The graph of $f(x) = 4 - x^2$ is a parabola which opens downward and does not pass the horizontal line test. The function is not 1-1.

14. $f(x) = x - 8$: $f(x)$ is a linear function and is 1:1.

Replace $f(x)$ by y: $y = x - 8$

Interchange x and y: $x = y - 8$

Solve for y: $x + 8 = y$

Replace y by $f^{-1}(x)$: $f^{-1}(x) = x + 8$

15. $g(x) = \dfrac{3x+1}{2}$

$g(x)$ is one-to-one.

Replace $g(x)$ by y: $y = \dfrac{3x+1}{2}$

Interchange x and y: $x = \dfrac{3y+1}{2}$

Solve for y: $\dfrac{2x-1}{3} = y$

Replace y by $g^{-1}(x)$: $g^{-1}(x) = \dfrac{2x-1}{3}$

16. $f(x) = 27x^3$

$f(x)$ is one-to-one.

Replace $f(x)$ by y: $y = 27x^3$

Interchange x and y: $x = 27y^3$

Solve for y: $\sqrt[3]{\dfrac{x}{27}} = y$

Replace y by $f^{-1}(x)$: $f^{-1}(x) = \sqrt[3]{\dfrac{x}{27}}$ or

$$f^{-1}(x) = \dfrac{\sqrt[3]{x}}{3}$$

17. Graph: $f(x) = 3^x + 1$

Choose values of x and determine corresponding function values. Plot these points and connect them with a smooth curve.

$f(-2) = 3^{-2} + 1 = \dfrac{1}{9} + 1 = 1\dfrac{1}{9}$

$f(-1) = 3^{-1} + 1 = \dfrac{1}{3} + 1 = 1\dfrac{1}{3}$

$f(0) = 3^0 + 1 = 1 + 1 = 2$

$f(1) = 3^1 + 1 = 3 + 1 = 4$

$f(2) = 3^2 + 1 = 9 + 1 = 10$

x	$f(x)$, or $3^x + 1$
-2	$1\dfrac{1}{9}$
-1	$1\dfrac{1}{3}$
0	2
1	4
2	10

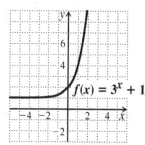

18. Graph: $x = \left(\dfrac{1}{4}\right)^y$

Choose values of y and determine corresponding x values. Plot these points and connect them with a smooth curve.

For $y = -2$, $x = \left(\dfrac{1}{4}\right)^{-2} = 16$

For $y = -1$, $x = \left(\dfrac{1}{4}\right)^{-1} = 4$

For $y = 0$, $x = \left(\dfrac{1}{4}\right)^{0} = 1$

For $y = 1$, $x = \left(\dfrac{1}{4}\right)^{1} = \dfrac{1}{4}$

For $y = 2$, $x = \left(\dfrac{1}{4}\right)^{2} = \dfrac{1}{16}$

x, or $\left(\dfrac{1}{4}\right)^y$	y
16	-2
4	-1
1	0
$\dfrac{1}{4}$	1
$\dfrac{1}{16}$	2

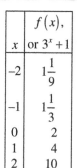

19. $y = \log_5 x$

The equation $y = \log_5 x$ is equivalent to $5^y = x$. Choose values of y and determine corresponding x values. Plot the points and connect them with a smooth curve.

For $y = -2$, $x = 5^{-2} = \dfrac{1}{25}$

For $y = -1$, $x = 5^{-1} = \dfrac{1}{5}$

For $y = 0$, $x = 5^0 = 1$

For $y = 1$, $x = 5^1 = 5$

For $y = 2$, $x = 5^2 = 25$

x, or 5^y	y
$\dfrac{1}{25}$	-2
$\dfrac{1}{5}$	-1
1	0
5	1
25	2

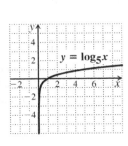

$y = \log_5 x$

20. $\log_3 9$

Since $9 = 3^2$, $\log_3 3^2 = 2$

21. $\log_{10} \dfrac{1}{100}$

Since $\dfrac{1}{100} = \dfrac{1}{10^2} = 10^{-2}$, $\log_{10} 10^{-2} = -2$

22. $\log_5 5^7 = 7$ The logarithm is the exponent.

23. $\log_9 3$

Since $9^{\frac{1}{2}} = 3$, $\left[\left(3^2\right)^{\frac{1}{2}} = 3 \right]$

$\log_9 3 = \dfrac{1}{2}$

24. $10^{-2} = \dfrac{1}{100} \Rightarrow \log_{10} \dfrac{1}{100} = -2$

The exponent is the logarithm

The base remains the same.

25. $25^{1/2} = 5 \Rightarrow \log_{25} 5 = \dfrac{1}{2}$

The exponent is the logarithm

The base remains the same.

26. $\log_4 16 = x \Rightarrow 16 = 4^x$

The exponent is the logarithm

The base remains the same

27. $\log_8 1 = 0 \Rightarrow 1 = 8^0$

The exponent is the logarithm

The base remains the same.

28. $\log_a x^4 y^2 z^3$

$= \log_a x^4 + \log_a y^2 + \log_a z^3$ Using the product rule

$= 4 \log_a x + 2 \log_a y + 3 \log_a z$ Using the power rule.

29. $\log_a \dfrac{x^5}{yz^2}$

$= \log_a x^5 - \log_a yz^2$ Using the quotient rule

$= \log_a x^5 - \left(\log_a y + \log_a z^2 \right)$ Using the product rule

$= 5 \log_a x - \log_a y - 2 \log_a z$ Using the power rule

30. $\log \sqrt[4]{\dfrac{z^2}{x^3 y}} = \log \left(\dfrac{z^2}{x^3 y} \right)^{\frac{1}{4}}$

$= \dfrac{1}{4} \log \dfrac{z^2}{x^3 y}$ Using the power rule

$= \dfrac{1}{4} \left(\log z^2 - \log x^3 y \right)$ Using the quotient rule

$= \dfrac{1}{4} \left[\log z^2 - \left(\log x^3 + \log y \right) \right]$ Using the product rule

$= \dfrac{1}{4} \left(2 \log z - 3 \log x - \log y \right)$ Using the power rule

31. $\log_a 7 + \log_a 8 = \log_a (7 \cdot 8)$ Using the

$= \log_a 56$ product rule

32. $\log_a 72 - \log_a 12 = \log_a \dfrac{72}{12}$ Using the

$= \log_a 6$ quotient rule

33. $\dfrac{1}{2}\log a - \log b - 2\log c$

   $= \log a^{\frac{1}{2}} - \log b - \log c^2$ Using the power rule

   $= \log a^{\frac{1}{2}} - \left(\log b + \log c^2\right)$ Distributive law

   $= \log a^{\frac{1}{2}} - \log bc^2$ Using the product rule

   $= \log \dfrac{a^{\frac{1}{2}}}{bc^2}$ Using the quotient rule

34. $\dfrac{1}{3}\left[\log_a x - 2\log_a y\right]$

   $= \dfrac{1}{3}\left[\log_a x - \log_a y^2\right]$ Using the power rule

   $= \dfrac{1}{3}\log_a \dfrac{x}{y^2}$ Using the quotient rule

   $= \log_a \left(\dfrac{x}{y^2}\right)^{\frac{1}{3}}$ Using the power rule

   $= \log_a \sqrt[3]{\dfrac{x}{y^2}}$

35. $\log_m m = 1$, since $m^1 = m$

36. $\log_m 1 = 0$, since $m^0 = 1$

37. $\log_m m^{17} = 17$; the logarithm is the exponent.

38. $\log_a 14$

   $= \log_a 2 + \log_a 7$ (Product rule)

   $= 1.8301 + 5.0999$

   $= 6.93$

39. $\log_a \dfrac{2}{7} = \log_a 2 - \log_a 7$ (Quotient rule)

   $= 1.8301 - 5.0999$

   $= -3.2698$

40. $\log_a 28 = \log_a 2^2 \cdot 7$ (Prime factorization)

   $= \log_a 2^2 + \log_a 7$ (Product rule)

   $= 2\log_a 2 + \log_a 7$ (Power rule)

   $= 2(1.8301) + 5.0999$

   $= 8.7601$

41. $\log_a 3.5 = \log_a \dfrac{7}{2}$

   $= \log_a 7 - \log_a 2$ (Quotient rule)

   $= 5.0999 - 1.8301$

   $= 3.2698$

42. $\log_a \sqrt{7} = \log_a 7^{\frac{1}{2}}$

   $= \dfrac{1}{2}\log_a 7$ (Power rule)

   $= \dfrac{1}{2}\cdot 5.0999$

   $= 2.54995$

43. $\log_a \dfrac{1}{4} = \log_a 4^{-1}$

   $= \log_a 2^{-2}$

   $= -2\log_a 2$ (Power rule)

   $= -2\cdot 1.8301$

   $= -3.6602$

44. $\log 75 \approx 1.8751$

45. $10^{1.789} \approx 61.5177$

46. $\ln 0.05 \approx -2.9957$

47. $e^{-0.98} \approx 0.3753$

48. Using common logarithms for the conversion, et $a = 10$, $b = 5$, and $m = 2$.

   $\log_b m = \dfrac{\log_a m}{\log_a b}$

   $\log_5 2 = \dfrac{\log_{10} 2}{\log_{10} 5}$

   $\approx \dfrac{0.3010299957}{0.6989700043}$

   ≈ 0.4307

49. Using common logarithms for the conversion, let $a = 10$, $b = 12$, and $m = 70$.

$$\log_b m = \frac{\log_a m}{\log_a b}$$

$$\log_{12} 70 = \frac{\log_{10} 70}{\log_{10} 12}$$

$$\approx \frac{1.84509804}{1.079181246}$$

$$\approx 1.7097$$

50. Graph: $f(x) = e^x - 1$

We find some function values, plot these points, and draw the graph.

x	$e^x - 1$
-2	-0.9
-1	-0.6
0	0
1	1.7
2	6.4

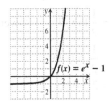

The domain is the set of real numbers and the range is $(-1, \infty)$.

51. Graph: $g(x) = 0.6 \ln x$

We find some function values, plot these points, and draw the graph.

x	$0.6 \ln x$
0.5	-1.4
1	0
2	0.4
3	0.7
10	1.4

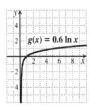

The domain is $(0, \infty)$. and the range is set of real numbers.

52. $2^x = 32$

$2^x = 2^5$

$x = 5$

53. $3^{2x} = \frac{1}{9} = 9^{-1} = 3^{-2} \Rightarrow 2x = -2$

$\Rightarrow x = \frac{-2}{2} = -1$

54. $\log_3 x = -4 \Rightarrow x = 3^{-4} \quad x = 3^{-4}$

$$x = \frac{1}{81}$$

55. $\log_x 16 = 4 \Rightarrow 16 = x^4$

$2^4 = x^4$

$2 = x$

56. $\log x = -3 \Rightarrow x = 10^{-3}$

$$x = \frac{1}{1000}$$

57. $3 \ln x = -6$

$\ln x^3 = -6$

$x^3 = e^{-6}$

$\left(x^3\right)^{1/3} = \left(e^{-6}\right)^{1/3}$

$x = e^{-2}$

$x \approx 0.1353$

58. $4^{2x-5} = 19$

$\log 4^{2x-5} = \log 19$

$(2x - 5) \log 4 = \log 19$

$2x \log 4 - 5 \log 4 = \log 19$

$2x \log 4 = \log 19 + 5 \log 4$

$\frac{2x \log 4}{2 \log 4} = \frac{\log 19 + 5 \log 4}{2 \log 4}$

$x = \frac{\log 19}{2 \log 4} + \frac{5}{2}$

$x \approx 3.5620$

59. $2^x = 12 \Rightarrow \log\left(2^x\right) = \log\left(12\right)$

$\Rightarrow x \log\left(2\right) = \log\left(12\right)$

$x = \frac{\log\left(12\right)}{\log\left(2\right)} \approx 3.585$

60. $e^{-0.1t} = 0.03$

$\ln e^{-0.1t} = \ln 0.03$

$-0.1t = \ln 0.03$

$t = \frac{\ln 0.03}{-0.1}$

$t \approx 35.0656$

61. $2\ln(x) = -6 \Rightarrow \ln(x) = -3$

    $\Rightarrow x = e^{-3} \approx 0.049787$

62. $\log(2x - 5) = 1 \Rightarrow 2x - 5 = 10^1$

    $2x = 15$

    $x = \dfrac{15}{2}$

63. $\log_4 x - \log_4(x - 15) = 2$

    $\Rightarrow \log_4\left[\dfrac{x}{x - 15}\right] = 2 \Rightarrow \dfrac{x}{x - 15} = 4^2 = 16$

    $\Rightarrow x = 16(x - 15) \Rightarrow x = 16x - 240$

    $\Rightarrow 15x = 240 \Rightarrow x = \dfrac{240}{15} = 16$

    The solution checks: $x = 16$.

64. $\log_3(x - 4) = 3 - \log_3(x + 4)$

    $\log_3(x - 4) + \log_3(x + 4) = 3$

    $\log_3\left(x^2 - 16\right) = 3$

    $x^2 - 16 = 3^3$

    $x^2 - 16 = 27$

    $x^2 = 43$

    $x = \pm\sqrt{43}$

    Only $\sqrt{43}$ checks and is the solution.

65. (a)

    $S(0) = 82 - 18\log(0 + 1)$

    $= 82 - 18\log 1$

    $= 82 - 18 \cdot 0 = 82$

    (b)

    $t = 6;\ 5(6) = 82 - 18\log(6 + 1)$

    $= 82 - 18\log 7 \approx 66.8$

    (c)

    Let $S(t) = 54$ and solve for t.

    $54 = 82 - 18\log(t + 1) \Rightarrow 18\log(t + 1) = 28$

    $\Rightarrow \log(t + 1) = \dfrac{28}{18} \Rightarrow t + 1 = 10^{\frac{14}{9}}$

    $\Rightarrow t + 1 \approx 35.938 \Rightarrow t \approx 34.938$

    Approximately 35 months.

66. (a) $V(t) = 900 \Rightarrow 1500(0.8)^t = 900$

    $\Rightarrow 0.8^t = \dfrac{900}{1500} = \dfrac{3}{5}$

    $\Rightarrow \ln\left(0.8^t\right) = \ln\left(\dfrac{3}{5}\right) = \ln(0.6)$

    $\Rightarrow t\ln(0.8) = \ln(0.6)$

    $\Rightarrow t = \dfrac{\ln(0.6)}{\ln(0.8)} \approx 2.29$

    The laptop will be worth \$900 in about 2.3 years.

    (b) $750 = 1500(0.8)^t$

    $\dfrac{750}{1500} = (0.8)^t$

    $\ln\dfrac{1}{2} = \ln 0.8^t$

    $\dfrac{\ln\dfrac{1}{2}}{\ln 0.8} = t$

    $3.11 \approx t$

    After about 3.1 years, the laptop will be worth half its original value.

67. For exponential growth use the formula:

    $A(t) = A_0 e^{kt}$

    (a) For 2007, $t = 0$, so $A_0 = 1.2$. To find k, use the value of A in 2012. In 2012, $t = 2012 - 2007 = 5$.

    Therefore:

    $A(5) = 2.1 \Rightarrow 1.2e^{5k} = 2.1$

    $\Rightarrow e^{5k} = \dfrac{2.1}{1.2} = 1.75 \Rightarrow \ln(e^{5k}) = \ln(1.75)$

    $\Rightarrow 5k = \ln(1.75) \Rightarrow k = \dfrac{\ln(1.75)}{5} \approx 0.112$

    $\Rightarrow A(t) = 1.2e^{0.112t}$

    (b) For 2015, $t = 2015 - 2007 = 8$.

    $A(8) = 1.2e^{0.112 \cdot 8} \approx 2.94$

    About \$2.94 billion will be spent in 2015.

(c) $A(t) = 4 \Rightarrow 1.2e^{0.112t} = 4 \Rightarrow e^{0.112t} = \dfrac{4}{1.2}$

$\Rightarrow \ln\left(e^{0.112t}\right) = \ln\left(\dfrac{4}{1.2}\right)$

$\Rightarrow 0.112t = \ln\left(\dfrac{4}{1.2}\right)$

$\Rightarrow t = \dfrac{\ln\left(\dfrac{4}{1.2}\right)}{0.112} \approx 10.75$

2007+10.75=2017.75
According to the model the amount spent on email marketing should reach \$4 billion in late 2017.

(d) The doubling time only depends on k, not A_0.

$A(t) = 2A_0 \Rightarrow A_0 e^{0.112t} = 2A_0$

$\Rightarrow e^{0.112t} = 2 \Rightarrow \ln\left(e^{0.112t}\right) = \ln(2)$

$\Rightarrow 0.112t = \ln(2) \Rightarrow t = \dfrac{\ln(2)}{0.112} \approx 6.2$

According to the model the amount spent on e-mail marketing doubles approximately every 6.2 years.

68. For exponential decay use the formula:

$M(t) = M_0 e^{-kt}$

(a) The value of k is given as 13.7% = 0.137, and, since we are letting $t = 0$ in 2005, $M_0 = M$ in 2005 $\Rightarrow M_0 = 3253$.

$\Rightarrow M(t) = 3253e^{-0.137t}$

(b) For 2012, $t = 2012 - 2005 = 7$.

$M(7) = 3253e^{-0.137 \cdot 7} \approx 1247$

In 2012, there will be 1247 spam messages per consumer.

(c) $M(t) = 100 \Rightarrow 3253e^{-0.137t} = 100$

$\Rightarrow e^{-0.137t} = \dfrac{100}{3253} \Rightarrow \ln\left(e^{-0.137t}\right) = \ln\left(\dfrac{100}{3253}\right)$

$\Rightarrow -0.137t = \ln\left(\dfrac{100}{3253}\right)$

$\Rightarrow t = -\dfrac{1}{0.137}\ln\left(\dfrac{100}{3253}\right) \approx 25.4$

2005+25.4=2030.4
According to the model the average number of spam messages per consumer should reach 100 in 2030.

69. (a)
Let x = the number of years after 1995, and f = the number of Hepatitis A cases in the U.S., in thousands. Then from the table, we have the points (0, 31.6), (5, 13.4), (8, 7.7), (9, 5.7), (10, 4.5), (11, 3.6), and (12, 3). Enter the data on a graphing calculator and then use the exponential regression feature to get:

$f(x) = 33.8684(0.8196)^x$

(b)
For 2010, $t = 2010 - 1995 = 15$.

$f(15) = 33.8684(0.8196)^{15} \approx 1.71$

According to the model, about 1.7 thousand cases, or 1700 cases, of Hepatitis A would occur in the U.S. in 2010.

(c)
$e^{-k} = 0.8196$

$\Rightarrow k = -\ln(0.8196) \approx 0.1989 = 19.89\%$

70. The value of the portfolio doubles in 6 years. Use this information to find k.

$2P_0 = P_0 e^{6k} \Rightarrow e^{6k} = 2$

$\Rightarrow \ln\left(e^{6k}\right) = \ln(2) \Rightarrow 6k = \ln(2)$

$\Rightarrow k = \dfrac{\ln(2)}{6} \approx 0.1155 = 11.55\%$

71. $2P_0 = 2 \cdot 7600;\ k = 4.2\% = .042,\ P_0 = 7600$.
Substitute into the exponential growth formula and solve for t.

$2 \cdot 7600 = 7600e^{0.042t}$

$2 = e^{0.042t}$

$\ln 2 = \ln e^{0.042t}$

$\ln 2 = 0.042t$

$\dfrac{\ln 2}{0.042} = t$

$16.5 \approx t$

\$7600 will double to \$13,200 in about 16.5 years.

72. If the skull has lost 34% of its carbon-14, $100\% - 34\% = 66\%$ remains. Substitute $66\% = 0.66$ into the decay formula and solve for t.

$$P(t) = P_0 e^{-0.00012t}$$
$$0.66 P_0 = P_0 e^{-0.00012t}$$
$$0.66 = e^{-0.00012t}$$
$$\ln 0.66 = \ln e^{-0.00012t}$$
$$\ln 0.66 = -0.00012t$$
$$\frac{\ln 0.66}{-0.00012} = t$$
$$3463 \approx t$$

The skull is about 3463 years-old.

73. $pH = -\log\left[H^+\right] = -\log\left[7.9 \cdot 10^{-6}\right] \approx 5.1$

74. Use $P(t) = P_0 e^{-kt}$, $k > 0$ for exponential decay. To find k substitute $\frac{1}{2}P_0$ for $P(t)$ and 5 for t.

$$P_0 e^{-k \cdot 24,360} = \frac{1}{2}P_0 \Rightarrow e^{-24,360k} = \frac{1}{2}$$

$$\Rightarrow -24,360k = \ln\left(\frac{1}{2}\right) = -\ln(2)$$

$$\Rightarrow k = \frac{\ln(2)}{24,360} \approx 2.8454 \times 10^{-5}$$

$$\Rightarrow P(t) = P_0 e^{-2.8454 \times 10^{-5} t}$$

If 90% has decayed then
$$P(t) = (1 - 0.90)P_0 = 0.1P_0$$
$$P(t) = 0.1P_0 \Rightarrow P_0 e^{-2.8454 \times 10^{-5} \cdot t} = 0.1P_0$$
$$\Rightarrow e^{-2.8454 \times 10^{-5} \cdot t} = 0.1$$
$$\Rightarrow -2.8454 \times 10^{-5} \cdot t = \ln(0.1)$$
$$\Rightarrow t = -\frac{\ln(0.1)}{2.8454 \times 10^{-5}} \approx 80,923$$

It takes approximately 80.923 years for Plutonium-239 to loose 90% of its radioactivity.

75. Let $I = 2.5 \cdot 10^{-1} = 0.25$ W/m^2 and solve the given equation.

$$L = 10 \cdot \log \frac{0.25}{10^{-12}} \approx 114$$

This sound level is approximately 114 dB.

76. *Thinking and Writing Exercise.* Negative numbers do not have logarithms because logarithm bases are positive, and there is no exponent to which a positive number can be raised to yield a negative number.

77. *Thinking and Writing Exercise.* Taking the logarithm on each side of an equation produces an equivalent equation because the logarithm function is one-to-one. If two quantities are equal, their logarithms must be equal, and if the logarithms of two quantities are equal, the quantities must be the same.

78. $\ln(\ln x) = 3 \Rightarrow \ln x = e^3$

$\ln x = e^3 \Rightarrow x = e^{e^3}$

79. $2^{x^2 + 4x} = \dfrac{1}{8}$

$$2^{x^2 + 4x} = 2^{-3}$$
$$x^2 + 4x = -3$$
$$x^2 + 4x + 3 = 0$$
$$(x + 3)(x + 1) = 0$$
$$x = -3 \quad \text{or} \quad x = -1$$

Both numbers check and are the solutions.

80. $5^{x+y} = 25 \Rightarrow 5^{x+y} = 5^2$

$x + y = 2$

$2^{2x-y} = 64 \Rightarrow 2^{2x-y} = 2^6$

$2x - y = 6$

Solve the system: $x + y = 2$

$2x - y = 6$

Add the equations: $3x = 8$

$$x = \frac{8}{3}$$

$$\frac{8}{3} + y = 2$$

$$y = \frac{-2}{3}$$

The solution is $\left(\dfrac{8}{3}, \dfrac{-2}{3}\right)$.

81. $P_0 e^{5.32k} = 2P_0 \Rightarrow e^{5.32k} = 2$

$$\Rightarrow \ln\left(e^{5.32k}\right) = \ln(2) \Rightarrow 5.32k = \ln(2)$$

$$\Rightarrow k = \frac{\ln(2)}{5.32} \approx 0.1303 = 13.03\%$$

Chapter 9 Test

1. $f(x) = x + x^2, g(x) = 2x + 1$

   $(f \circ g)(x) = f(g(x)) = f(2x+1)$
   $= (2x+1) + (2x+1)^2$
   $= 2x+1+4x^2+4x+1$
   $= 4x^2+6x+2$

   $(g \circ f)(x) = g(f(x)) = g(x+x^2)$
   $= 2(x+x^2)+1$
   $= 2x+2x^2+1$
   $= 2x^2+2x+1$

2. Possible answer. $h(x)$ is the multiplicative inverse of $2x^2+1$.

   $f(x) = \dfrac{1}{x}$ and $g(x) = 2x^2+1$

3. No. Example: $f(x) = 4$ when $x = 7$ and when $x = -1$. Also note, the graph of the function does not pass the horizontal line test.

4. $f(x) = 3x+4$

   $f(x)$ is one-to-one.

   Replace $f(x)$ by y: $y = 3x+4$

   Interchange x and y: $x = 3y+4$

   Replace y by $f^{-1}(x)$: $f^{-1}(x) = \dfrac{x-4}{3}$

5. $g(x) = (x+1)^3$

   $g(x)$ is one-to-one.

   Replace $g(x)$ by y: $y = (x+1)^3$

   Interchange x and y: $x = (y+1)^3$

   Solve for y: $\sqrt[3]{x} = y+1$

   $\sqrt[3]{x} - 1 = y$

   Replace y by $g^{-1}(x)$: $g^{-1}(x) = \sqrt[3]{x} - 1$

6. Graph: $f(x) = 2^x - 3$

   Choose valves of x and determine the corresponding function values. Plot these points and connect them with a smooth curve.

   $f(-2) = 2^{-2} - 3 = \dfrac{1}{4} - 3 = -2\dfrac{3}{4}$

   $f(-1) = 2^{-1} - 3 = \dfrac{1}{2} - 3 = -2\dfrac{1}{2}$

   $f(0) = 2^0 - 3 = 1 - 3 = -2$

   $f(1) = 2^1 - 3 = 2 - 3 = -1$

   $f(2) = 2^2 - 3 = 4 - 3 = 1$

   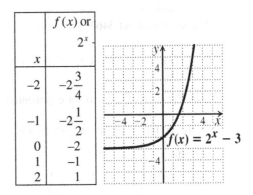

x	$f(x)$ or 2^x
-2	$-2\dfrac{3}{4}$
-1	$-2\dfrac{1}{2}$
0	-2
1	-1
2	1

7. Graph: $g(x) = \log_7 x$

   The equation $y = \log_7 x$ is equivalent to $7^y = x$. Choose values of y and determine corresponding x values. Plot the points and connect them with a smooth curve.

   For $y = -2$, $x = 7^{-2} = \dfrac{1}{49}$

   For $y = -2$, $x = 7^{-1} = \dfrac{1}{7}$

   For $y = 0$, $x = 7^0 = 1$

   For $y = 1$, $x = 7^1 = 7$

   For $y = 2$, $x = 7^2 = 49$

x, or 7^y	y
$\dfrac{1}{49}$	-2
$\dfrac{1}{7}$	-1
1	0
7	1
49	2

8. $\log_5 125$

   Since $125 = 5^3$, $\log_5 5^3 = 3$

9. $\log_{100} 10$

Since $10 = 100^{\frac{1}{2}}$, $\log_{100} 100^{\frac{1}{2}} = \dfrac{1}{2}$

10. $3^{\log_3 18} = 18$

11. $\log_n (n) = 1$

12. $\log_c (1) = 0$

13. $\log_a (a^{19}) = 19 \log_a (a) = 19 \cdot 1 = 19$

14.
$$5^{-4} = \frac{1}{625}$$

$$\log_5 (5^{-4}) = \log_5 \left(\frac{1}{625}\right)$$

$$(-4) \log_5 (5) = \log_5 \left(\frac{1}{625}\right)$$

$$(-4) \cdot 1 = \log_5 \left(\frac{1}{625}\right)$$

$$-4 = \log_5 \left(\frac{1}{625}\right)$$

15. $m = \log_2 \left(\dfrac{1}{2}\right)$

$$2^m = 2^{\log_2\left(\frac{1}{2}\right)}$$

$$2^m = \frac{1}{2}$$

16. $\log \dfrac{a^3 b^{1/2}}{c^2}$

$= \log(a^3 b^{1/2}) - \log c^2$ Using the quotient rule

$= \log a^3 + \log b^{1/2} - \log c^2$ Using the product rule

$= 3 \log a + \dfrac{1}{2} \log b - 2 \log c$ Using the power rule

17. $\dfrac{1}{3} \log_a x + 2 \log_a z$

$= \log_a x^{\frac{1}{3}} + \log_a z^2$ Using the power rule

$= \log_a x^{\frac{1}{3}} z^2$, or $\log_a z^2 \sqrt[3]{x}$ Using the product rule

18. $\log_a 14 = \log_a (2 \cdot 7)$

$= \log_a 2 + \log_2 7$ (Product Rule)

$= 0.301 + 0.845$

$= 1.146$

19. $\log_a 3 = \log_a \dfrac{6}{2}$

$= \log_a 6 - \log_a 2$ (Quotient Rule)

$= 0.778 - 0.301$

$= 0.477$

20. $\log_a 16 = \log_a 2^4$

$= 4 \log_a 2$ (Power Rule)

$= 4 \cdot 0.301$

$= 1.204$

21. $\log 12.3 \approx 1.0899$

22. $10^{-8} \approx 0.1585$

23. $\ln 0.4 \approx -0.9163$

24. $e^{4.8} \approx 121.5104$

25. Using common logarithms for the conversion, let $a = 10$, $b = 3$, and $m = 14$.

$$\log_b m = \frac{\log_a m}{\log_a b}$$

$$\log_3 14 = \frac{\log_{10} 14}{\log_{10} 3}$$

$$\approx \frac{1.146128036}{0.4771212547}$$

$$\approx 2.4022$$

26. Graph: $f(x) = e^x + 3$

We find some function values, plot these points, and draw the graph.

x	$f(x)$, or $e^x + 3$
-2	3.1
-1	3.4
0	4
1	5.7
2	10.4

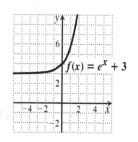

The domain is the set of real numbers and the range is $(3, \infty)$.

27. Graph: $g(x) = \ln(x-4)$

We find some function values, plot these points, and draw the graph

x	$f(x)$, or $\ln(x-4)$
4.1	-2.3
5	0
6	0.7
10	2.3

$g(x) = \ln(x-4)$

The domain is $(4, \infty)$ and the range is the set of real numbers.

28. $2^x = \dfrac{1}{32}$

$2^x = 2^{-5}$

$x = -5$

29. $\log_4 x = \dfrac{1}{2} \Rightarrow x = 4^{\frac{1}{2}} = (2^2)^{\frac{1}{2}} = 2$

30. $\log x = 4 \Rightarrow x = 10^4 = 10{,}000$

31. $5^{4-3x} = 87$

$\log 5^{4-3x} = \log 87$

$(4-3x)\log 5 = \log 87$

$4\log 5 - 3x\log 5 = \log 87$

$-3x\log 5 = \log 87 - 4\log 5$

$-\dfrac{1}{3}[-3x\log 5 = \log 87 - \log 5^4]$

$x\log 5 = -\dfrac{1}{3}(\log 87 - \log 5^4)$

$\dfrac{x\log 5}{\log 5} = \dfrac{-\dfrac{1}{3}(\log 87 - \log 5^4)}{\log 5}$

$x = -\dfrac{1}{3}\left(\dfrac{\log 87}{\log 5} - \dfrac{\log 5^4}{\log 5}\right)$

$x = -\dfrac{1}{3}\left(\dfrac{\log 87}{\log 5} - 4\right)$

$x \approx 0.4084$

32. $7^x = 1.2$

$\log 7^x = \log 1.2$

$x\log 7 = \log 1.2$

$x = \dfrac{\log 1.2}{\log 7}$

$x \approx 0.0937$

33. $\ln x = 3 \Rightarrow x = e^3 \approx 20.0855$

34. $\log(x-3) + \log(x+1) = \log 5$

$\log[(x-3)(x+1)] = \log 5$

$\log(x^2 - 2x - 3) = \log 5$

$x^2 - 2x - 3 = 5$

$x^2 - 2x - 8 = 0$

$(x-4)(x+2) = 0$

$x = 4 \text{ or } x = -2$

Only the number 4 checks, and it is the solution

35. (a) Find R when $P = 383$

$R = 0.37\ln P + 0.05$

$= 0.37\ln(383) + 0.05 \approx 2.25$

The average walking speed is approximately 2.25 ft/sec.

(b) Find P when $R = 3$

$R = 0.37\ln P + 0.05$

$\Rightarrow 0.37\ln(P) = R - 0.05$

$\Rightarrow \ln(P) = \dfrac{R - 0.05}{0.37}$

$\Rightarrow P = e^{\frac{R-0.05}{0.37}} = e^{\frac{3-0.05}{0.37}} \approx 2901$

According to the model, the population of San Diego is approximately 2,901,000.

36. (a) Let $t = 0$ in 2009. Then $P_0 = 8.2$ million.

The exponential growth rate is given as $k = 0.052\% = 0.00052$. Substitute this information into the formula for exponential growth.

$P(t) = P_0 e^{kt} = 8.2e^{0.00052t}$

(b) In 2020, $t = 2020 - 2009 = 11$

$P(11) = 8.2e^{0.00052 \cdot 11} \approx 8.247$

Population = 8,247,000.

In 2050, $t = 2050 - 2009 = 41$

$P(11) = 8.2e^{0.00052 \cdot 41} \approx 8.3767$

Population = 8,376,700

(c) $P(t) = 9 \Rightarrow 8.2e^{0.00052t} = 9$

$\Rightarrow e^{0.00052t} = \dfrac{9}{8.2} \Rightarrow \ln\left(e^{0.00052t}\right) = \ln\left(\dfrac{9}{8.2}\right)$

$\Rightarrow 0.00052t = \ln\left(\dfrac{9}{8.2}\right)$

$\Rightarrow t = \dfrac{1}{0.00052}\ln\left(\dfrac{9}{8.2}\right) \approx 179$

2009+179=2188

According to the model, the population will reach 9 million in about 2188.

(d) $P(t) = 2P_0 \Rightarrow 8.2e^{0.00052t} = 2 \cdot 8.2$

$\Rightarrow e^{0.00052t} = 2 \Rightarrow \ln\left(e^{0.00052t}\right) = \ln(2)$

$\Rightarrow 0.00052t = \ln(2)$

$\Rightarrow t = \dfrac{\ln(2)}{0.00052} \approx 1333$

The doubling time is approximately 1333 years.

37. (a) Let $t = 0$ in 2001. Then $P_0 = 21,855$.

Therefore, $P(t) = P_0 e^{kt} = 21,855e^{kt}$. In 2010, $t = 2010 - 2001 = 9$. To find k, set $P(9) = 35,600$ and solve for k.

$P(9) = 35,600 \Rightarrow 21,855e^{9k} = 35,600$

$\Rightarrow e^{9k} = \dfrac{35,600}{21,855} = \dfrac{7120}{4371}$

$\Rightarrow \ln\left(e^{9k}\right) = \ln\left(\dfrac{7120}{4371}\right) \Rightarrow 9k = \ln\left(\dfrac{7120}{4371}\right)$

$\Rightarrow k = \dfrac{1}{9}\ln\left(\dfrac{7120}{4371}\right) \approx 0.054$

$\Rightarrow P(t) = 21,855e^{0.0542t}$

(b) In 2015, $t = 2015 - 2001 = 14$

$P(14) = 21,855e^{0.054 \cdot 14} \approx 46,545$

Tuition = $46,545.

(c) $P(t) = 50,000 \Rightarrow 21,855e^{0.054t} = 50,000$

$\Rightarrow e^{0.054t} = \dfrac{50,000}{21,855} = \dfrac{10,000}{4371}$

$\Rightarrow \ln\left(e^{0.054t}\right) = \ln\left(\dfrac{10,000}{4371}\right)$

$\Rightarrow 0.054t = \ln\left(\dfrac{10,000}{4371}\right)$

$\Rightarrow t = \dfrac{1}{0.054}\ln\left(\dfrac{10,000}{4371}\right) \approx 15.33$

2001+15.33 = 2016.33

According to the model, tuition will reach $50,000 in 2016.

38. (a) Let x = the number of years after 2008 and $f(x)$ = the number of Kindle sales, in thousands. Then from the table, we have the points (0, 500), (1, 1027), and (2, 3533). Enter the data on a graphing calculator and then use the exponential regression feature to get:

$f(x) = 458.8188(2.6582)^x$

(b) For 2012, $x = 2012 - 2008 = 4$.

$f(4) = 458.8188(2.6582)^4 \approx 22,908$

According to the model, Kindle sales should reach approximately 22,908,000 in 2012.

39. P_0 is the initial investment, so $2P_0$ would be double that amount. $t = 15$. Use this information in the exponential growth formula to solve for k, the interest rate.

$2P_0 = P_0 \cdot e^{k \cdot 15}$

$2 = e^{15k}$

$\ln 2 = \ln e^{15k}$

$\ln 2 = 15k$

$\dfrac{\ln 2}{15} = k$

$0.046 \approx k$

$0.046 = 4.6\%$

The interest rate is 4.6%.

40. If 43% has been lost, $100\% - 43\%$, or 57% of the initial amount, P_0 remains.

$P(t) = P_0 e^{-0.00012t}$

$0.57P_0 = P_0 e^{-0.00012t}$

$0.57 = e^{-0.00012t}$

$\ln 0.57 = \ln e^{-0.00012t}$

$\ln 0.57 = -0.00012t$

$-\dfrac{\ln 0.57}{0.00012} = t$

$4684 \approx t$

The animal bone is approximately 4684 years-old.

41. $L = 10\log\left(\dfrac{I}{I_0}\right) = 10\log\left(\dfrac{I}{10^{-12}}\right)$

$L = 140 \Rightarrow 10\log\left(\dfrac{I}{10^{-12}}\right) = 140$

$\Rightarrow \log\left(\dfrac{I}{10^{-12}}\right) = \dfrac{140}{10} = 14$

$\dfrac{I}{10^{-12}} = 10^{14} \Rightarrow I = 10^{14}\cdot 10^{-12} = 10^{2}$

The intensity is 10^{2} W/m^2.

42. Let $[H+] = 1.0\times10^{-7}$ and solve for pH.

$pH = -\log\left[H^{+}\right]$

$pH = -\log\left(1.0\times10^{-7}\right)$

$pH = -(-7)$

$pH = 7$

The pH is 7.

43. $\log_5 |2x - 7| = 4 \Rightarrow |2x - 7| = 5^4$

$|2x - 7| = 625$

$2x - 7 = -625 \ \text{ or } \ 2x - 7 = 625$

$2x = -618 \ \text{ or } \ 2x = 632$

$x = -309 \ \text{ or } \ x = 316$

Both numbers are the solutions.

44. Express $\log_a \dfrac{\sqrt[3]{x^2 z}}{\sqrt[3]{y^2 z^{-1}}}$ using

the individual logarithms of x, y, and z.

$\log_a \dfrac{\sqrt[3]{x^2 z}}{\sqrt[3]{y^2 z^{-1}}}$

$= \log_a \sqrt[3]{\dfrac{x^2 z}{y^2 z^{-1}}}$

$= \log_a \sqrt[3]{\dfrac{x^2 z^2}{y^2}}$

$= \log_a \left(\dfrac{x^2 z^2}{y^2}\right)^{1/3}$

$= \dfrac{1}{3}\log_a \left(\dfrac{x^2 z^2}{y^2}\right)$ $\qquad$ Power Rule

$= \dfrac{1}{3}(\log_a x^2 z^2 - \log_a y^2)$ $\qquad$ Quotient Rule

$= \dfrac{1}{3}(\log_a x^2 + \log_a z^2 - \log_a y^2)$ $\quad$ Product Rule

$= \dfrac{1}{3}(2\log_a x + 2\log_a z - 2\log_a y)$ $\quad$ Power Rule

Substituting the given information, we have

$= \dfrac{1}{3}[2(2) + 2(4) - 2(3)]$

$= \dfrac{1}{3}[4 + 8 - 6]$

$= \dfrac{1}{3}\cdot 6$

$= 2$

Chapters 1 – 9

Cumulative Review

1. $\dfrac{x^0 + y}{-z} = \dfrac{6^0 + 9}{-(-5)} = \dfrac{1+9}{5} = \dfrac{10}{5} = 2$

2. $(-2x^2 y^{-3})^{-4} = (-2)^{-4}(x^2)^{-4}(y^{-3})^{-4}$

   $\qquad = \dfrac{1}{16} \cdot x^{-8} \cdot y^{12}$

   $\qquad = \dfrac{y^{12}}{16x^8}$

3. $(-5x^4 y^{-3} z^2)(-4x^2 y^2)$

   $\qquad = (-5)(-4)x^{4+2}y^{-3+2}z^2$

   $\qquad = 20x^6 y^{-1} z^2$

   $\qquad = \dfrac{20x^6 z^2}{y}$

4. $\dfrac{3x^4 y^6 z^{-2}}{-9x^4 y^2 z^3} = \dfrac{3}{-9}x^{4-4}y^{6-2}z^{-2-3}$

   $\qquad = \dfrac{-1}{3} \cdot x^0 y^4 z^{-5} \qquad (x^0 = 1)$

   $\qquad = -\dfrac{y^4}{3z^5}$

5. $(1.5 \times 10^{-3})(4.2 \times 10^{-12})$

   $= (1.5 \cdot 4.2) \cdot (10^{-3} \times 10^{-12})$

   $= 6.3 \times 10^{-15}$

6. $3^3 + 2^2 - (32 \div 4 - 16 \div 8)$

   $= 27 + 4 - (8 - 2)$

   $= 27 + 4 - 6 = 25$

7. $3(2x - 3) = 9 - 5(2 - x)$

   $\quad 6x - 9 = 9 - 10 + 5x$

   $\quad 6x - 5x = -1 + 9$

   $\qquad\quad x = 8$

8. $(1)\ 4x - 3y = 15$

   $(2)\ 3x + 5y = 4$

   Add: $5(1) + 3(2)$

   $20x - 15y = 75$

   $\underline{9x + 15y = 12}$

   $\quad 29x = 87$

   $\qquad x = 3$

   Substitute to determine x.

   $4 \cdot 3 - 3y = 15$

   $\qquad -3y = 3$

   $\qquad\quad y = -1$

   The solution is $(3, -1)$.

9. $x + y - 3z = -1 \ (1)$

   $2x - y + z = 4 \ (2)$

   $-x - y + z = 1 \ (3)$

   Add: $(1) + (2)$ to eliminate y.

   $x + y - 3z = -1$

   $\underline{2x - y + z = 4}$

   $3x \quad -2z = 3$

   Add: $(1) + (3)$ to eliminate y.

   $x + y - 3z = -1$

   $\underline{-x - y + z = 1}$

   $\qquad -2z = 0$

   $\qquad\quad z = 0$

   Substitute to determine x.

   $3x - 2 \cdot 0 = 3$

   $\qquad 3x = 3$

   $\qquad\ x = 1$

   Substitute to determine y.

   $1 + y - 3 \cdot 0 = -1$

   $\qquad\quad y = -2$

   The solution is $(1, -2, 0)$

10. $$x(x-3) = 70$$
$$x^2 - 3x = 70$$
$$x^2 - 3x - 70 = 0$$
$$(x-10)(x+7) = 0$$
$$x = 10 \text{ or } x = -7$$

11. $$\frac{7}{x^2 - 5x} - \frac{2}{x-5} = \frac{4}{x}$$
[LCD is $x^2 - 5x = x(x-5)$]
Note: $x \neq 0, 5$
$$x(x-5) \cdot \left(\frac{7}{x^2 - 5x} - \frac{2}{x-5} \right) = x(x-5) \cdot \frac{4}{x}$$
$$7 - 2x = 4(x-5)$$
$$\Rightarrow 7 - 2x = 4x - 20 \Rightarrow 27 = 6x$$
$$\Rightarrow x = \frac{27}{6} = \frac{9}{2}$$

12. $$\sqrt{4-5x} = 2x-1$$
$$\left(\sqrt{4-5x} \right)^2 = (2x-1)^2$$
$$4 - 5x = 4x^2 - 4x + 1$$
$$0 = 4x^2 + x - 3$$
$$0 = (4x-3)(x+1)$$
$$x = \frac{3}{4} \text{ or } x = -1$$

Only the number $\frac{3}{4}$ checks in the original

equation; $\frac{3}{4}$ is the solution.

13. $$\sqrt[3]{2x} = 1$$
$$\left(\sqrt[3]{2x} \right)^3 = 1^3$$
$$2x = 1$$
$$x = \frac{1}{2}$$

14. $$3x^2 + 48 = 0$$
$$3(x^2 + 16) = 0$$
$$x^2 + 16 = 0$$
$$x^2 = -16$$
$$x = \pm\sqrt{-16}$$
$$x = \pm 4i$$

15. $$x^4 - 13x^2 + 36 = 0$$
$$(x^2 - 9)(x^2 - 4) = 0$$
$$(x-3)(x+3)(x-2)(x+2) = 0$$
$$x = 3 \text{ or } x = -3 \text{ or } x = 2 \text{ or } x = -2$$
The solutions are $\pm 3, \pm 2$.

16. $$\log_x 81 = 2$$
$$x^{\log_x 81} = x^2$$
$$81 = x^2$$
$$x = \pm 9$$
However, we do not use negative numbers for base of logarithms. Therefore $x = 9$.

17. $$3^{5x} = 7$$
$$\log 3^{5x} = \log 7$$
$$5x \log 3 = \log 7$$
$$\frac{5x \log 3}{5 \log 3} = \frac{\log 7}{5 \log 3}$$
$$x = \frac{\log 7}{5 \log 3}$$
$$\approx 0.3542$$

18. $$\ln x - \ln(x-8) = 1$$
$$\ln \left(\frac{x}{x-8} \right) = 1$$
$$e^{\ln \left(\frac{x}{x-8} \right)} = e^1$$
$$\frac{x}{x-8} = e$$
$$(x-8) \cdot \frac{x}{x-8} = (x-8) \cdot e$$
$$x = ex - 8e$$
$$8e = ex - x$$
$$8e = x(e-1)$$
$$x = \frac{8e}{e-1} \approx 12.6558$$

19. $$x^2 + 4x > 5$$
$$x^2 + 4x - 5 > 0$$
$$(x+5)(x-1) > 0$$
Solve the related equation.
$$(x+5)(x-1) = 0$$
$$x = -5 \text{ or } x = 1$$
Choose a number from each interval
$(-\infty, -5), (-5, 1),$ and $(1, \infty)$ to determine

the solution.

The solution is $(-\infty, -5) \cup (1, \infty),$ or

$\{x \mid x < -5 \ or \ x > 1\},$ since the numbers

chosen from those intervals make

the original inequality true.

20. $f(x) = x^2 + 6x; \ f(a) = 11$

$f(a) = 11 \Rightarrow a^2 + 6a = 11 \Rightarrow x^2 + 6x - 11 = 0$

$a = \dfrac{-B \pm \sqrt{B^2 - 4AC}}{2A}$

$= \dfrac{-6 \pm \sqrt{6^2 - 4 \cdot 1 \cdot (-11)}}{2 \cdot 1}$

$= \dfrac{-6 \pm \sqrt{36 + 44}}{2}$

$= \dfrac{-6 \pm \sqrt{80}}{2} = \dfrac{-6 \pm 4\sqrt{5}}{2} = -3 \pm 2\sqrt{5}$

21. $f(x) = |2x - 3|; \ f(x) \ge 7$

$f(x) \ge 7 \Rightarrow |2x - 3| \ge 7$

$2x - 3 \le -7 \quad or \quad 2x - 3 \ge 7$

$2x \le -4 \quad or \quad 2x \ge 10$

$x \le -2 \quad or \quad x \ge 5$

$\{x \mid x \le -2 \ or \ x \ge 5\},$ or $(-\infty, -2] \cup [5, \infty)$

22. $D = \dfrac{ab}{b + a} \quad [LCD = b + a]$

$(b + a)D = (b + a)\dfrac{ab}{b + a}$

$Db + Da = ab$

$Db = ab - Da$

$Db = a(b - D)$

$\dfrac{Db}{b - D} = a$

23. $d = ax^2 + vx \Rightarrow ax^2 + vx - d = 0$

$x = \dfrac{-B \pm \sqrt{B^2 - 4AC}}{2A}$

$= \dfrac{-v \pm \sqrt{v^2 - 4 \cdot a \cdot (-d)}}{2 \cdot a} = \dfrac{-v \pm \sqrt{v^2 + 4ad}}{2a}$

24. $f(x) = \dfrac{x + 4}{3x^2 - 5x - 2}$

The domain in the set of real numbers, excluding values of x which make the denominator zero.

$3x^2 - 5x - 2 = 0$

$(3x + 1)(x - 2) = 0$

$x = -\dfrac{1}{3} \quad or \quad x = 2$

$\{x \mid x$ is a real number, $x \ne -\frac{1}{3}$ and $x \ne 2\}$

25. $(5p^2q^3 + 6pq - p^2 + p)$

$\quad - (2p^2q^3 + p^2 - 5pq - 9)$

$= 5p^2q^3 + 6pq - p^2 + p$

$\quad - 2p^2q^3 - p^2 + 5pq + 9$

$= 5p^2q^3 - 2p^2q^3 - p^2 - p^2$

$\quad + 6pq + 5pq + p + 9$

$= 3p^2q^3 - 2p^2 + 11pq + p + 9$

26. $(3x^2 - z^3)^2 = (3x^2)^2 - 2(3x^2)(z^3) + (z^3)^2$

$\qquad\qquad = 9x^4 - 6x^2z^3 + z^6$

27. $\dfrac{1 + \frac{3}{x}}{x - 1 - \frac{12}{x}} \quad [LCD = x]$

$\dfrac{x}{x} \cdot \dfrac{1 + \frac{3}{x}}{x - 1 - \frac{12}{x}} = \dfrac{x + 3}{x^2 - x - 12}$

$= \dfrac{\cancel{x + 3}}{(x - 4)(\cancel{x + 3})}$

$= \dfrac{1}{x - 4} \ for \ x \ne \{-3, 0, 4\}$

28. $\dfrac{a^2 - a - 6}{a^3 - 27} \cdot \dfrac{a^2 + 3a + 9}{6}$

$= \dfrac{(a - 3)(a + 2)}{(a - 3)(a^2 + 3a + 9)} \cdot \dfrac{a^2 + 3a + 9}{6}$

$= \dfrac{(a - 3)(a + 2)(a^2 + 3a + 9)}{(a - 3)(a^2 + 3a + 9) \cdot 6}$

$= \dfrac{(\cancel{a - 3})(a + 2)(\cancel{a^2 + 3a + 9})}{(\cancel{a - 3})(\cancel{a^2 + 3a + 9}) \cdot 6}$

$= \dfrac{a + 2}{6} \ for \ a \ne 3$

29. $\dfrac{3}{x+6} - \dfrac{2}{x^2-36} + \dfrac{4}{x-6}$

    [LCD is $x^2 - 36 = (x+6)(x-6)$]

$$= \frac{x-6}{x-6} \cdot \frac{3}{x+6} - \frac{2}{(x+6)(x-6)}$$

$$+ \frac{x+6}{x+6} \cdot \frac{4}{x-6}$$

$$= \frac{3x-18-2+4x+24}{(x+6)(x-6)}$$

$$= \frac{7x+4}{(x+6)(x-6)}$$

30. $\dfrac{\sqrt[3]{24xy^8}}{\sqrt[3]{3xy}} = \sqrt[3]{\dfrac{24xy^8}{3xy}} = \sqrt[3]{8y^7} = \sqrt[3]{2^3 y^6 \cdot y}$

$$= 2y^2 \sqrt[3]{y}$$

31. $\sqrt{x+5}\sqrt[5]{x+5} = (x+5)^{\frac{1}{2}}(x+5)^{\frac{1}{5}}$

$$= (x+5)^{\frac{1}{2}+\frac{1}{5}} = (x+5)^{\frac{7}{10}}, \text{ or } \sqrt[10]{(x+5)^7}$$

32. $(2 - i\sqrt{3})(6 + i\sqrt{3})$

$$= 12 + 2i\sqrt{3} - 6i\sqrt{3} - 3i^2$$

$$= 12 - 4i\sqrt{3} - 3(-1) \quad (i^2 = -1)$$

$$= 15 - 4i\sqrt{3}$$

33. Since the divisor is of the form $x - c$. we can use synthetic division.

$$\underline{3|}\ 1 \quad -8 \qquad 15 \qquad 1 \quad -3$$
$$\qquad\quad 3 \cdot 1 \quad 3 \cdot(-5) \quad 3 \cdot 0 \quad 3 \cdot 1$$
$$\overline{\quad 1 \quad -5 \qquad\ 0 \qquad\ 1 \qquad 0}$$

$$\Rightarrow (x^4 - 8x^3 + 15x^2 + x - 3) \div (x-3)$$

$$= x^3 - 5x^2 + 1$$

34. $xy + 2xz - xw$

$$= x \cdot y + x \cdot 2z - x \cdot w$$

$$= x(y + 2z - w)$$

35. $6x^2 + 8xy - 8y^2$

$$= 2(3x^2 + 4xy - 4y^2)$$

$$= 2(3x - 2y)(x + 2y)$$

36. $x^4 - 4x^3 + 7x - 28$

$$= (x^4 - 4x^3) + (7x - 28)$$

$$= x^3(x-4) + 7(x-4)$$

$$= (x^3 + 7)(x - 4)$$

37. $2m^2 + 12mn + 18n^2$

$$= 2(m^2 + 6mn + 9n^2)$$

$$= 2(m + 3n)^2$$

38. $x^4 - 16y^4 = (x^2)^2 - (4y^2)^2$

$$= (x^2 + 4y^2)(x^2 - 4y^2)$$

$$= (x^2 + 4y^2)[(x)^2 - (2y)^2]$$

$$= (x^2 + 4y^2)(x + 2y)(x - 2y)$$

39. $\dfrac{3 - \sqrt{y}}{2 - \sqrt{y}}$

    [The conjugate of the denominator is $2 + \sqrt{y}$]

$$\frac{3 - \sqrt{y}}{2 - \sqrt{y}} \cdot \frac{2 + \sqrt{y}}{2 + \sqrt{y}} = \frac{6 + \sqrt{y} - y}{4 - y}$$

40. $f(x) = 9 - 2x$

    $f(x)$ is one-to-one.

    Replace $f(x)$ by y: $y = 9 - 2x$

    Interchange x & y: $x = 9 - 2y$

    Solve for y: $\dfrac{x - 9}{-2} = y$,

       or $\dfrac{9 - x}{2} = y$

    Replace y with $f^{-1}(x)$: $f^{-1}(x) = \dfrac{9 - x}{2}$

41. Determine the slope of the line which contains $(0, -8)$ and $(-1, 2)$. Since $(0, -8)$ is the y-intercept use $y = mx + b$.

$$m = \frac{y_z - y_1}{x_2 - x_1} = \frac{2 - (-8)}{-1 - 0} = -10$$

$$y = -10x - 8$$

42. The slope of $2x + y = 6$ is

$$m = \frac{-A}{B} = \frac{-2}{1} = -2.$$ The slope of the line

perpendicular to $2x + y = 6$ is $\frac{1}{2}$, since

$l_1 \perp l_2 \Rightarrow m_1 \cdot m_2 = -1.$ Since $b = 5,$

$mx + b = \frac{1}{2}x + 5.$ The equation of the line is

$y = \frac{1}{2}x + 5.$

43. Graph: $5x = 15 + 3y$

Determine at least two points, plot these points, and graph the line.

x	y
0	-5
3	0

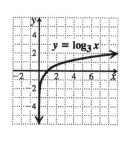

44. Graph: $y = \log_3 x \Rightarrow 3^y = x$

Choose values of y to determine corresponding values of x. Plot these points and connect with a smooth curve.

x_1 or 3^y	y
$3^{-2} = \frac{1}{9}$	-2
$3^{-1} = \frac{1}{3}$	-1
$3^0 = 1$	0
$3^1 = 3$	1
$3^2 = 9$	2

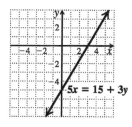

45. $-2x - 3y \leq 12$

Graph the boundary line: $-2x - 3y = 12$ as a solid line. The point, (0, 0) can be used to determine the correct half-plane to shade.
$-2(0) - 3(0) \leq 12$?

$\qquad 0 \leq 12$ True

Shade the half-plane which includes $(0,0)$.

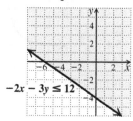

46. $f(x) = 2x^2 + 12x + 19$

(a) The vertex is at:

$$-\frac{b}{2a} = -\frac{12}{2 \cdot 2} = -3,$$
$$f(-3) = 2(-3)^2 + 12(-3) + 19$$
$$= 2 \cdot 9 - 39 + 19 = 1$$

Vertex: $\left(-\frac{b}{2a}, f\left(-\frac{b}{2a} \right) \right) = (-3, 1)$

(b) Axis of symmetry: $x = -3$

(c) The parabola open upward. Therefore the function takes on a minimum at its vertex of 1.

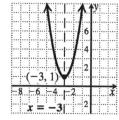

47. Graph: $f(x) = 2e^x$

Using values of x, determine ordered pairs of $f(x)$; plot these points and connect with a smooth curve.

x	$f(x)$, or $2e^x$
-2	0.3
-1	0.7
0	2
1	5.4
2	14.8

The domain is the set of real numbers and the range is $(0, \infty)$.

48. $3\log x - \dfrac{1}{2}\log y - 2\log z$

$= \log x^3 - \log y^{1/2} - \log z^2$ (Power Rule)

$= \log x^3 - (\log y^{1/2} + \log z^2)$

$= \log x^3 - \log y^{1/2}z^2$ (Product Rule)

$= \log\left(\dfrac{x^3}{y^{1/2}z^2}\right)$ (Quotient Rule)

49.   c

50.   b

51.   a

52.   d

53. Let x = the total volume of water carried by the Colorado River, in millions of acre-feet. Then the amount diverted for agricultural use is $0.90x$. Then:

$0.10x = 1.5 \Rightarrow x = \dfrac{1.5}{0.10} = 15$

$\Rightarrow 0.90x = 0.9 \cdot 15 = 13.5$

13.5 million acre-feet are diverted each year.

54. a) For exponential growth use the function:

$D(t) = D_0 e^{kt} = 15e^{kt}$. Then:

$D(17) = 55 \Rightarrow 15e^{17k} = 55 \Rightarrow e^{17k} = \dfrac{55}{15} = \dfrac{11}{3}$

$\Rightarrow \ln\left(e^{17k}\right) = \ln\left(\dfrac{11}{3}\right) \Rightarrow 17k = \ln\left(\dfrac{11}{3}\right)$

$\Rightarrow k = \dfrac{1}{17}\ln\left(\dfrac{11}{3}\right) \approx 0.0764$

$\Rightarrow D(t) = 15e^{0.0764t}$

b) In 2012, $t = 2012 - 1990 = 22$

$D(22) = 15e^{0.0764 \cdot 22} \approx 80.55$

According to the model 80.55 million m³ of water will be able to be desalinated per day by 2012.

c) $D(t) = 100 \Rightarrow 15e^{0.0764t} = 100$

$\Rightarrow e^{0.0764t} = \dfrac{100}{15} = \dfrac{20}{3}$

$\Rightarrow \ln\left(e^{0.0764t}\right) = \ln\left(\dfrac{20}{3}\right)$

$\Rightarrow 0.0764t = \ln\left(\dfrac{20}{3}\right)$

$\Rightarrow t = \dfrac{1}{0.0764}\ln\left(\dfrac{20}{3}\right) \approx 24.8$

$1990 + 24.8 = 2014.8$. Therefore, according to the model, the worldwide capacity will reach 100 million m³ in late 2014.

55. Anne can do $\frac{1}{10}$ of the task in 1 minute. And clay can do $\frac{1}{12}$. Let t = time working together; use $r \cdot t = w$ to write their work equation. Solve this equation for t.

$\dfrac{1}{10}t + \dfrac{1}{12}t = 1$ [LCD = 60]

$60\left(\dfrac{t}{10} + \dfrac{t}{12}\right) = 60 \cdot 1$

$6t + 5t = 60$

$11t = 60$

$t = \dfrac{60}{11}, \quad \text{or} \quad 5\dfrac{5}{11}$

Working together, it will take $5\frac{5}{11}$ min.

56. Let x = number of ounces of the 45% dressing and y = number of ounces of the 20% dressing.

Dressing	%	Number of ounces	Fat Calories
Thick & tasty	45%	x	$0.45x$
Light & lean	20%	y	$0.20x$
mix	30%	15	$0.30(15)$ $= 4.5$

Using the table, determine the equations
(1) $x + y = 15$ (total ounces) and

(2) $0.45x + 0.20 = 4.5$ (total fat calories).

Solve this system of equations using elimination

$$-x - y = -15$$
$$\underline{2.25x + y = 22.5}$$
$$1.25x = 7.5$$
$$x = 6$$
$$x + y = 15,$$
$$\text{so } 6 + y = 15$$
$$y = 9$$

To obtain the desired mix, use 6 oz of Thick and Tasty and 9 oz of Light and Lean.

57. Let x = the speed of the river in kph. The speed of the boat downstream is then $5 + x$, and the speed of the boat upstream is $5 - x$.

Use $d = rt \Rightarrow t = \dfrac{d}{r}$ to find x.

$$t_{\text{UPSTREAM}} = t_{\text{DOWNSTREAM}}$$

$$\frac{d_{\text{UPSTREAM}}}{r_{\text{UPSTREAM}}} = \frac{d_{\text{DOWNSTREAM}}}{r_{\text{DOWNSTREAM}}}$$

$$\frac{42}{5 + x} = \frac{12}{5 - x}$$
$$42(5 - x) = 12(5 + x)$$
$$210 - 42x = 60 + 12x$$
$$54x = 150$$

$$x = \frac{150}{54} = \frac{25}{9} = 2\frac{7}{9} \text{ km/h}$$

58. Since the function values increase at a steady rate, a linear function seems most appropriate.

59. Choose two ordered pairs and determine the slope, which signifies the rate of change.

$$m = \frac{y_2 - y_1}{x_2 - x_1}$$

We will use $(15, 18)$ and $(20, 23)$.

$$m = \frac{23 - 18}{20 - 15}$$
$$= \frac{5}{5}$$
$$= 1$$

The rate of change is $1/min.

60. From Exercise 2, we have $m = 1$. Choose an ordered pair and substitute to determine the equation.

$$y_2 - y_1 = m(x_2 - x_1)$$
$$y - 18 = 1(x - 15)$$
$$y = x + 3$$

Express the equation as a function of x.

$$f(x) = x + 3$$

61. Let $x = 10$.
$$f(10) = 10 + 3 = 13$$
A 10-min. would cost \$13.

62. $m = 1$ signifies the cost per minute, and $b = 3$ signifies the fixed cost or "startup" cost of each massage.

63. From the graph enter the data points $(35, 14.85), (40, 18.53), (50, 33.05),$ and $(55, 47.25)$. Use the ExpReg option in the STATCALC MENU to obtain the function.

$$m(x) = 1.8937(1.0596)^x$$

64. Let $x = 45$, and solve for m.
$$m(45) = 1.8937(1.0596)^{45}$$
$$m \approx 25.63$$
The monthly premium would be about \$25.63 for a 45-year-old-male.

65. $\dfrac{5}{3x-3}+\dfrac{10}{3x+6}=\dfrac{5x}{x^2+x-2}$

$\dfrac{5}{3(x-1)}+\dfrac{10}{3(x+2)}=\dfrac{5x}{(x+2)(x-1)}$

[LCD is $3(x+2)(x-1)$]

Note: $x \ne -2,1$

$3(x+2)(x-1)\left[\dfrac{5}{3(x-1)}+\dfrac{10}{3(x+2)}\right]$

$=3(x+2)(x-1)\left[\dfrac{5x}{(x+2)(x-1)}\right]$

$5(x+2)+10(x-1)=3\cdot 5x$

$5x+10+10x-10=15x$

$\qquad\qquad 15x=15x$

This is true for all values in the domain (reflexive property of equality) The solution is $\{x|x$ is a real number and $x \ne -2$ and $x \ne 1\}$.

66. $\log\sqrt{3x}=\sqrt{\log 3x}$

$\log(3x)^{1/2}=(\log 3x)^{1/2}$

$\dfrac{1}{2}\log 3x=(\log 3x)^{1/2}$

$\left(\dfrac{1}{2}\log 3x\right)^2=((\log 3x)^{1/2})^2$

$\dfrac{1}{4}(\log 3x)^2=\log 3x$

$\dfrac{1}{4}(\log 3x)^2-\log 3x=0$

$\log 3x\left(\dfrac{1}{4}\log 3x-1\right)=0$

$\log 3x=0 \quad$ or $\quad \dfrac{1}{4}\log 3x-1=0$

$3x=10^0 \quad$ or $\quad \dfrac{1}{4}\log 3x=1$

$x=\dfrac{1}{3} \quad$ or $\quad \log 3x=4$

$\qquad\qquad\qquad\qquad 3x=10^4$

$\qquad\qquad\qquad\qquad x=\dfrac{10,000}{3}$

Both numbers check and are the solutions.

67. Let $x=$ the original speed
and $x+5=$ the increased speed.

Using $T=\dfrac{D}{R}$, we determine the times.

Trips	Rate×	Time =	Distance
Original	x	$\dfrac{280}{x}$	280
Increased	$x+5$	$\dfrac{280}{x+5}$	280

The time for the trip is 1 hour greater. We have the equation:

$\dfrac{280}{x}=\dfrac{280}{x+5}+1$

[LCD $= x(x+5)$]

$x(x+5).\dfrac{280}{x}=x(x+5)\left[\dfrac{280}{x+5}+1\right]$

$280(x+5)=280x+x^2+5x$

$280x+1400=280x+x^2+5x$

$0=x^2+5x-1400$

$0=(x+40)(x-35)$

$x=-40$ or $x=35$

Since the speed cannot be negative, our solution is 35 mph.

Chapter 10
Conic Sections

Exercise Set 10.1

1. f

3. g

5. c

7. d

9.

11.

13.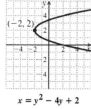

15.

17.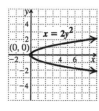

19.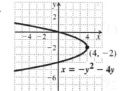

21.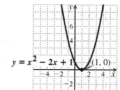

23.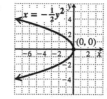

25.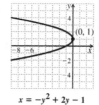

27.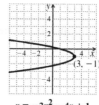

29. $(x-h)^2 + (y-k)^2 = r^2$ Standard form

$(x-0)^2 + (y-0)^2 = 6^2$ Substituting

$x^2 + y^2 = 36$ Simplifying

31. $(x-h)^2 + (y-k)^2 = r^2$ Standard form

$(x-7)^2 + (y-3)^2 = \left(\sqrt{5}\right)^2$ Substituting

$(x-7)^2 + (y-3)^2 = 5$

33. $(x-h)^2 + (y-k)^2 = r^2$

$\left[x-(-4)\right]^2 + (y-3)^2 = \left(4\sqrt{3}\right)^2$

$(x+4)^2 + (y-3)^2 = 48$

35. $(x-h)^2 + (y-k)^2 = r^2$

$\left[x-(-7)\right]^2 + \left[y-(-2)\right]^2 = \left(5\sqrt{2}\right)^2$

$(x+7)^2 + (y+2)^2 = 50$

37. Since the center is $(0,0)$, we have

$(x-0)^2 + (y-0)^2 = r^2$ or $x^2 + y^2 = r^2$. The

circle passes through $(-3,4)$. We find r^2 by

substituting –3 for x and 4 for y.

$(-3)^2 + 4^2 = r^2$

$9 + 16 = r^2$

$25 = r^2$

$x^2 + y^2 = 25$ is an equation of the circle.

39. Since the center is $(-4,1)$, we have

$\left[x-(-4)\right]^2 + (y-1)^2 = r^2$ or

$(x+4)^2 + (y-1)^2 = r^2$. The circle passes

through $(-2,5)$. We find r^2 by substituting

–2 for x and 5 for y.

$(-2+4)^2 + (5-1)^2 = r^2$

$4 + 16 = r^2$

$20 = r^2$

$(x+4)^2 + (y-1)^2 = 20$ is an equation of the

circle.

41. We write standard form

$(x-0)^2 + (y-0)^2 = 8^2$

Center: $(0,0)$; radius is 8.

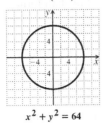

$x^2 + y^2 = 64$

43. We write standard form

$\left[x-(-1)\right]^2 + \left[y-(-3)\right]^2 = 6^2$

Center: $(-1,-3)$; radius is 6.

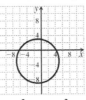

$(x+1)^2 + (y+3)^2 = 36$

45. We write standard form

$(x-4)^2 + \left[y-(-3)\right]^2 = \left(\sqrt{10}\right)^2$

Center: $(4,-3)$; radius is $\sqrt{10}$.

$(x-4)^2 + (y+3)^2 = 10$

47. We write standard form

$(x-0)^2 + (y-0)^2 = \left(\sqrt{10}\right)^2$

Center: $(0,0)$; radius is $\sqrt{10}$.

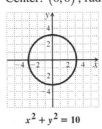

$x^2 + y^2 = 10$

49. We write standard form

$(x-5)^2 + (y-0)^2 = \left(\dfrac{1}{2}\right)^2$

Center: $(5,0)$; radius is $\dfrac{1}{2}$.

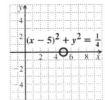

$(x-5)^2 + y^2 = \frac{1}{4}$

51. We write standard form
$$x^2 + 8x + y^2 - 6y = 15$$
$$\left(x^2 + 8x + 16\right) + \left(y^2 - 6y + 9\right) = 15 + 16 + 9$$
$$\left(x + 4\right)^2 + \left(y - 3\right)^2 = 40$$
$$\left[x - \left(-4\right)\right]^2 + \left(y - 3\right)^2 = \left(\sqrt{40}\right)^2$$

Center: $\left(-4, 3\right)$; radius is $\sqrt{40}$, or $2\sqrt{10}$.

$x^2 + y^2 + 8x - 6y - 15 = 0$

53.
$$x^2 - 8x + y^2 + 2y = -13$$
$$\left(x^2 - 8x + 16\right) + \left(y^2 + 2y + 1\right) = -13 + 16 + 1$$
$$\left(x - 4\right)^2 + \left(y + 1\right)^2 = 4$$
$$\left(x - 4\right)^2 + \left[y - \left(-1\right)\right]^2 = 2^2$$

Center: $\left(4, -1\right)$; radius is 2.

$x^2 + y^2 - 8x + 2y + 13 = 0$

55.
$$x^2 + y^2 + 10y = 75$$
$$x^2 + \left(y^2 + 10y + 25\right) = 75 + 25$$
$$\left(x - 0\right)^2 + \left(y + 5\right)^2 = 100$$
$$\left(x - 0\right)^2 + \left[y - \left(-5\right)\right]^2 = 10^2$$

Center: $\left(0, -5\right)$; radius is 10.

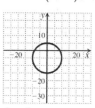

$x^2 + y^2 + 10y - 75 = 0$

57.
$$x^2 + 7x + y^2 - 3y = 10$$
$$\left(x^2 + 7x + \frac{49}{4}\right) + \left(y^2 - 3y + \frac{9}{4}\right) = 10 + \frac{49}{4} + \frac{9}{4}$$
$$\left(x + \frac{7}{2}\right)^2 + \left(y - \frac{3}{2}\right)^2 = \frac{98}{4}$$
$$\left[x - \left(-\frac{7}{2}\right)\right]^2 + \left(y - \frac{3}{2}\right)^2 = \left(\sqrt{\frac{98}{4}}\right)^2$$

Center: $\left(-\frac{7}{2}, \frac{3}{2}\right)$; radius is $\sqrt{\frac{98}{4}}$, or $\frac{7\sqrt{2}}{2}$.

$x^2 + y^2 + 7x - 3y - 10 = 0$

59.
$$36x^2 + 36y^2 = 1$$
$$\frac{1}{36} \cdot \left(36x^2 + 36y^2\right) = \frac{1}{36} \cdot 1$$
$$x^2 + y^2 = \left(\frac{1}{6}\right)^2$$

Center: $\left(0, 0\right)$; radius is $\frac{1}{6}$.

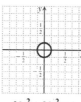

$36x^2 + 36y^2 = 1$

61. First we solve the equation for y.
$$x^2 + y^2 - 16 = 0$$
$$y^2 = 16 - x^2$$
$$y = \pm\sqrt{16 - x^2}$$

Then we graph $y_1 = \sqrt{16 - x^2}$ and $y_2 = -\sqrt{16 - x^2}$ on the same set of axes, choosing a squared window. We use $\left[-9, 9, -6, 6\right]$.

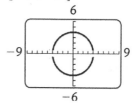

63. First we solve the equation for y. We can use the quadratic formula with $a = 1$, $b = -16$, and $c = x^2 + 14x + 54$ or we can complete the square on the y-terms and then proceed. We will complete the square.

$$x^2 + y^2 + 14x - 16y + 54 = 0$$

$$x^2 + 14x + y^2 - 16y + 64 - 64 + 54 = 0$$

$$x^2 + 14x + (y - 8)^2 - 10 = 0$$

$$(y - 8)^2 = 10 - x^2 - 14x$$

$$y = 8 \pm \sqrt{10 - x^2 - 14x}$$

Then we graph $y_1 = 8 + \sqrt{10 - x^2 - 14x}$ and $y_2 = 8 - \sqrt{10 - x^2 - 14x}$ on the same set of axes, choosing a squared window. We use $[-20, 7, -1, 17]$.

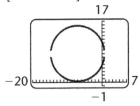

65. *Thinking and Writing Exercise.*

67. $$\frac{y^2}{16} = 1$$

$$y^2 = 16$$

$$y = \pm 4$$

69. $$\frac{(x-1)^2}{25} = 1$$

$$(x-1)^2 = 25$$

$$x - 1 = \pm 5$$

$$x = -4, 6$$

71. $$\frac{1}{4} + \frac{(y+3)^2}{36} = 1$$

$$\frac{(y+3)^2}{36} = \frac{3}{4}$$

$$(y+3)^2 = 27$$

$$y + 3 = \pm 3\sqrt{3}$$

$$y = -3 \pm 3\sqrt{3}$$

73. *Thinking and Writing Exercise.*

75. We make a drawing of the circle with center $(3, -5)$ and tangent to the y-axis.

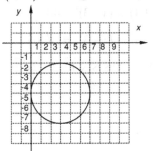

We see that the circle touches the y-axis at $(0, -5)$. Hence the radius is the distance between $(0, -5)$ and $(3, -5)$, or

$$\sqrt{(3-0)^2 + [-5-(-5)]^2}$$, or 3. Now we write the equation of the circle.

$$(x - h)^2 + (y - k)^2 = r^2$$

$$(x - 3)^2 + [y - (-5)]^2 = 3^2$$

$$(x - 3)^2 + (y + 5)^2 = 9$$

77. First we use the midpoint formula to find the center:

$$\left(\frac{7 + (-1)}{2}, \frac{3 + (-3)}{2}\right), \text{ or } \left(\frac{6}{2}, \frac{0}{2}\right), \text{ or } (3, 0)$$

The length of the radius is the distance between the center $(3, 0)$ and either endpoint of a diameter. We will use endpoint $(7, 3)$ in the distance formula:

$$r = \sqrt{(7 - 3)^2 + (3 - 0)^2} = \sqrt{25} = 5$$

Now we write the equation of the circle:

$$(x - h)^2 + (y - k)^2 = r^2$$

$$(x - 3)^2 + (y - 0)^2 = 5^2$$

$$(x - 3)^2 + y^2 = 25$$

79. Let $(0, y)$ be the point on the y-axis that is equidistant from $(2,10)$ and $(6,2)$. Then the distance between $(2,10)$ and $(0, y)$ is the same as the distance between $(6,2)$ and $(0, y)$.

$$\sqrt{(0-2)^2 + (y-10)^2} = \sqrt{(0-6)^2 + (y-2)^2}$$
$$(-2)^2 + (y-10)^2 = (-6)^2 + (y-2)^2$$
$$4 + y^2 - 20y + 100 = 36 + y^2 - 4y + 4$$
$$64 = 16y$$
$$4 = y$$

The number checks. The point is $(0,4)$.

81. The outer circle has a radius of $\dfrac{9}{2}$ (from $x^2 + y^2 = \dfrac{81}{4}$) and the inner edge of the red zone has a radius of 4 (from $x^2 + y^2 = 4^2$). The area of the red zone is the difference of their areas.

$$\pi \cdot \left(\frac{9}{2}\right)^2 - \pi \cdot (4^2) = \frac{81}{4}\pi - \frac{64}{4}\pi$$
$$= \frac{17}{4}\pi \approx 13.4 \text{ m}^2$$

83. The distance from the center of the circle at $(0, y)$ to the point at $(0, 23.5)$ is the same as the distance from the center to the point at $(580, 0)$. So, using the distance formula:

$$\sqrt{(0-0)^2 + (23.5 - y)^2} = \sqrt{(0-580)^2 + (0-y)^2}$$
$$\sqrt{y^2 - 47y + 552.25} = \sqrt{336,400 + y^2}$$
$$y^2 - 47y + 552.25 = 336,400 + y^2$$
$$-47y = 335,847.75$$
$$y \approx -7145.7$$

The radius is the distance from the center to $(0, 23.5)$, or $7145.7 + 23.5 \approx 7169$ mm.

85. a) When the circle is positioned on a coordinate system as shown in the text, the center lies on the y-axis. To find the center, we will find the point on the y-axis that is equidistant from $(-4,0)$ and $(0,2)$. Let $(0, y)$ be this point.

$$\sqrt{[0-(-4)]^2 + (y-0)^2} =$$
$$\sqrt{(0-0)^2 + (y-2)^2}$$

Squaring both sides
$$16 + y^2 = y^2 - 4y + 4$$
$$12 = -4y$$
$$-3 = y$$

The center of the circle is $(0,-3)$.

b) We find the radius of the circle.
Standard form:
$$(x-0)^2 + [y-(-3)]^2 = r^2$$
$$x^2 + (y+3)^2 = r^2$$

Substituting $(-4,0)$ for (x, y):
$$(-4)^2 + (0+3)^2 = r^2$$
$$16 + 9 = r^2$$
$$25 = r^2$$
$$5 = r$$

The radius is 5 ft.

87. We write the equation of a circle with center $(0, 30.6)$ and radius 24.3:
$$x^2 + (y - 30.6)^2 = 590.49$$

89.

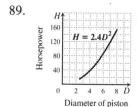

91. *Thinking and Writing Exercise.*

Exercise Set 10.2

1. True

3. False

5. True

7. True

9. $\dfrac{x^2}{1} + \dfrac{y^2}{9} = 1$

$\dfrac{x^2}{1^2} + \dfrac{y^2}{3^2} = 1$

The x-intercepts are $(1,0)$ and $(-1,0)$.

The y-intercepts are $(0,3)$ and $(0,-3)$.

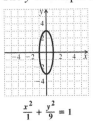

$\dfrac{x^2}{1} + \dfrac{y^2}{9} = 1$

11. $\dfrac{x^2}{25} + \dfrac{y^2}{9} = 1$

$\dfrac{x^2}{5^2} + \dfrac{y^2}{3^2} = 1$

The x-intercepts are $(5,0)$ and $(-5,0)$.

The y-intercepts are $(0,3)$ and $(0,-3)$.

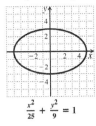

$\dfrac{x^2}{25} + \dfrac{y^2}{9} = 1$

13. $4x^2 + 9y^2 = 36$

$\dfrac{1}{36} \cdot \left(4x^2 + 9y^2\right) = \dfrac{1}{36} \cdot 36$

$\dfrac{x^2}{9} + \dfrac{y^2}{4} = 1$

$\dfrac{x^2}{3^2} + \dfrac{y^2}{2^2} = 1$

The x-intercepts are $(3,0)$ and $(-3,0)$.

The y-intercepts are $(0,2)$ and $(0,-2)$.

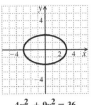

$4x^2 + 9y^2 = 36$

15. $16x^2 + 9y^2 = 144$

$\dfrac{1}{144} \cdot \left(16x^2 + 9y^2\right) = \dfrac{1}{144} \cdot 144$

$\dfrac{x^2}{9} + \dfrac{y^2}{16} = 1$

$\dfrac{x^2}{3^2} + \dfrac{y^2}{4^2} = 1$

The x-intercepts are $(3,0)$ and $(-3,0)$.

The y-intercepts are $(0,4)$ and $(0,-4)$.

$16x^2 + 9y^2 = 144$

17. $2x^2 + 3y^2 = 6$

$\dfrac{x^2}{3} + \dfrac{y^2}{2} = 1$

$\dfrac{x^2}{\left(\sqrt{3}\right)^2} + \dfrac{y^2}{\left(\sqrt{2}\right)^2} = 1$

The x-intercepts are $\left(\sqrt{3},0\right)$ and $\left(-\sqrt{3},0\right)$.

The y-intercepts are $\left(0,\sqrt{2}\right)$ and $\left(0,-\sqrt{2}\right)$.

$2x^2 + 3y^2 = 6$

19. $5x^2 + 5y^2 = 125$

$x^2 + y^2 = 5^2$

This is the equation of a circle with center $(0,0)$ and radius 5.

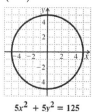

$5x^2 + 5y^2 = 125$

21. $3x^2 + 7y^2 - 63 = 0$

$3x^2 + 7y^2 = 63$

$\dfrac{x^2}{21} + \dfrac{y^2}{9} = 1$

$\dfrac{x^2}{\left(\sqrt{21}\right)^2} + \dfrac{y^2}{3^2} = 1$

The x-intercepts are $\left(\sqrt{21},0\right)$ and $\left(-\sqrt{21},0\right)$.

The y-intercepts are $(0,3)$ and $(0,-3)$.

$3x^2 + 7y^2 - 63 = 0$

23. $16x^2 = 16 - y^2$

$16x^2 + y^2 = 16$

$\dfrac{x^2}{1} + \dfrac{y^2}{16} = 1$

$\dfrac{x^2}{(1)^2} + \dfrac{y^2}{(4)^2} = 1$

The x-intercepts are $(1,0)$ and $(-1,0)$.

The y-intercepts are $(0,4)$ and $(0,-4)$.

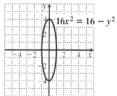

$16x^2 = 16 - y^2$

25. $16x^2 + 25y^2 = 1$

Note: $16 = \dfrac{1}{\frac{1}{16}}$ and $25 = \dfrac{1}{\frac{1}{25}}$

$\dfrac{x^2}{\frac{1}{16}} + \dfrac{y^2}{\frac{1}{25}} = 1$

$\dfrac{x^2}{\left(\frac{1}{4}\right)^2} + \dfrac{y^2}{\left(\frac{1}{5}\right)^2} = 1$

The x-intercepts are $\left(\dfrac{1}{4},0\right)$ and $\left(-\dfrac{1}{4},0\right)$.

The y-intercepts are $\left(0,\dfrac{1}{5}\right)$ and $\left(0,-\dfrac{1}{5}\right)$.

$16x^2 + 25y^2 = 1$

27. $\dfrac{(x-3)^2}{9} + \dfrac{(y-2)^2}{25} = 1$

$\dfrac{(x-3)^2}{3^2} + \dfrac{(y-2)^2}{5^2} = 1$

The center of the ellipse is $(3,2)$. Note that $a = 3$ and $b = 5$. We locate the center and then plot the points $(3+3,2)$, $(3-3,2)$, $(3,2+5)$ and $(3,2-5)$ or $(6,2)$, $(0,2)$, $(3,7)$ and $(3,-3)$.

$\dfrac{(x-3)^2}{9} + \dfrac{(y-2)^2}{25} = 1$

29. $\dfrac{(x+4)^2}{16} + \dfrac{(y-3)^2}{49} = 1$

$\dfrac{(x-(-4))^2}{4^2} + \dfrac{(y-3)^2}{7^2} = 1$

The center of the ellipse is $(-4,3)$. Note that $a = 4$ and $b = 7$. We locate the center and then plot the points $(-4+4,3)$, $(-4-4,3)$,

$(-4, 3+7)$ and $(-4, 3-7)$, or $(0,3)$, $(-8,3)$, $(-4,10)$ and $(-4,-4)$.

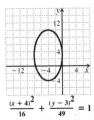

$$\frac{(x+4)^2}{16} + \frac{(y-3)^2}{49} = 1$$

31. $12(x-1)^2 + 3(y+4)^2 = 48$

$$\frac{(x-1)^2}{4} + \frac{(y+4)^2}{16} = 1$$

$$\frac{(x-1)^2}{2^2} + \frac{(y-(-4))^2}{4^2} = 1$$

The center of the ellipse is $(1,-4)$. Note that $a = 2$ and $b = 4$. We locate the center and then plot the points $(1+2, -4)$, $(1-2, -4)$, $(1, -4+4)$ and $(1, -4-4)$, or $(3,-4)$, $(-1,-4)$, $(1,0)$ and $(1,-8)$.

$12(x-1)^2 + 3(y+4)^2 = 48$

33. $4(x+3)^2 + 4(y+1)^2 - 10 = 90$

$$4(x+3)^2 + 4(y+1)^2 = 100$$

Observe that the x^2- and y^2-terms have the same coefficient. Dividing both sides by 4, we have

$(x+3)^2 + (y+1)^2 = 25$.

This is the equation of a circle with center $(-3,-1)$ and radius 5.

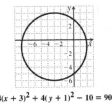

$4(x+3)^2 + 4(y+1)^2 - 10 = 90$

35. *Thinking and Writing Exercise.*

37. $x^2 - 5x + 3 = 0$

Use the quadratic formula:

$$\frac{-b \pm \sqrt{b^2 - 4ac}}{2a} = \frac{5 \pm \sqrt{(-5)^2 - 4(1)(3)}}{2(1)}$$

$$= \frac{5 \pm \sqrt{25 - 12}}{2}$$

$$= \frac{5 \pm \sqrt{13}}{2}$$

39. $\dfrac{4}{x+2} + \dfrac{3}{2x-1} = 2$

Note that $x+2$ is 0 when $x = -2$ and $2x-1$ is 0 when x is $\dfrac{1}{2}$, so -2 and $\dfrac{1}{2}$ cannot be solutions. We multiply by the LCD, $(x+2)(2x-1)$.

$$(x+2)(2x-1)\left(\frac{4}{x+2} + \frac{3}{2x-1}\right) =$$
$$(x+2)(2x-1)2$$
$$(2x-1) \cdot 4 + (x+2) \cdot 3 = 2(2x^2 + 3x - 2)$$
$$8x - 4 + 3x + 6 = 4x^2 + 6x - 4$$
$$11x + 2 = 4x^2 + 6x - 4$$
$$0 = 4x^2 - 5x - 6$$
$$0 = (4x+3)(x-2)$$
$$x = -\frac{3}{4}, 2$$

41. $x^2 = 11$

$x = \pm\sqrt{11}$

43. *Thinking and Writing Exercise.*

45. Plot the given points.

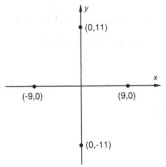

From the location of these points, we can see that the ellipse that contains them is centered at the origin with $a = 9$ and $b = 11$. We write the equation of the ellipse.

$$\frac{x^2}{9^2} + \frac{y^2}{11^2} = 1$$

$$\frac{x^2}{81} + \frac{y^2}{121} = 1$$

47. Plot the given points.

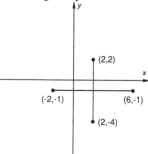

The vertical and horizontal segments intersect at $x = 2$ and $y = -1$. So the center of the ellipse is $(2, -1)$. Thus, the distance from $(-2, -1)$ to $(2, -1)$ is

$$\sqrt{\left[2 - (-2)\right]^2 + \left[-1 - (-1)\right]^2} = \sqrt{16} = 4,$$ so

$a = 4$. The distance from $(2, 2)$ to $(2, -1)$ is

$$\sqrt{(2 - 2)^2 + (-1 - 2)^2} = \sqrt{9} = 3,$$ so $b = 3$.

We write the equation of the ellipse.

$$\frac{(x - 2)^2}{4^2} + \frac{(y - (-1))^2}{3^2} = 1$$

$$\frac{(x - 2)^2}{16} + \frac{(y + 1)^2}{9} = 1$$

49. The soloist is in the center of the ellipse, or $(0, 0)$. If the ellipse is 6 ft wide, we have the points $(-3, 0)$ and $(3, 0)$, and if it is 10 ft long, we have the points $(0, 5)$ and $(0, -5)$.

Since $a = 3$ and $b = 5$, the equation is

$$\frac{x^2}{3^2} + \frac{y^2}{5^2} = 1, \text{ or } \frac{x^2}{9} + \frac{y^2}{25} = 1.$$

51. a) Let $F_1 = (-c, 0)$ and $F_2 = (c, 0)$. Then the sum of the distances from the foci to P is $2a$. By the distance formula,

$$\sqrt{(x + c)^2 + y^2} + \sqrt{(x - c)^2 + y^2} = 2a, \text{ or}$$

$$\sqrt{(x + c)^2 + y^2} = 2a - \sqrt{(x - c)^2 + y^2}.$$

Squaring, we get

$$(x + c)^2 + y^2 =$$

$$4a^2 - 4a\sqrt{(x - c)^2 + y^2} + (x - c)^2 + y^2$$

or

$$x^2 + 2cx + c^2 + y^2$$

$$= 4a^2 - 4a\sqrt{(x - c)^2 + y^2}$$

$$+ x^2 - 2cx + c^2 + y^2$$

Thus

$$-4a^2 + 4cx = -4a\sqrt{(x - c)^2 + y^2}$$

$$a^2 - cx = a\sqrt{(x - c)^2 + y^2}$$

Squaring again, we get

$$a^4 - 2a^2cx + c^2x^2 =$$

$$a^2\left(x^2 - 2cx + c^2 + y^2\right)$$

$$a^4 - 2a^2cx + c^2x^2 =$$

$$a^2x^2 - 2a^2cx + a^2c^2 + a^2y^2,$$

$$a^4 - a^2c^2 = a^2x^2 - c^2x^2 + a^2y^2$$

$$a^2\left(a^2 - c^2\right) = x^2\left(a^2 - c^2\right) + a^2y^2$$

$$\frac{x^2}{a^2} + \frac{y^2}{a^2 - c^2} = 1.$$

b) When P is at $(0, b)$, it follows that $b^2 = a^2 - c^2$.

Substituting, we have

$$\frac{x^2}{a^2} + \frac{y^2}{b^2} = 1.$$

53. For the given ellipse, $a = 6/2$, or 3, and $b = 2/2$, or 1. The patient's mouth should be at a distance of $2c$ from the light source, where the coordinates of the foci of the ellipse are $(-c, 0)$ and $(c, 0)$. From Exercise 52, we know that $b^2 = a^2 - c^2$. We use this to find c.

$$b^2 = a^2 - c^2$$
$$1^2 = 3^2 - c^2 \quad \text{Substituting}$$
$$c^2 = 8$$
$$c = \sqrt{8}$$

Then $2c = 2\sqrt{8} \approx 5.66$. The patient's mouth should be about 5.66 ft from the light source.

55.
$$x^2 - 4x + 4y^2 + 8y - 8 = 0$$
$$x^2 - 4x + 4y^2 + 8y = 8$$
$$x^2 - 4x + 4\left(y^2 + 2y\right) = 8$$
$$\left(x^2 - 4x + 4\right) + 4\left(y^2 + 2y + 1\right) = 8 + 4 + 4 \cdot 1$$
$$(x-2)^2 + 4(y+1)^2 = 16$$
$$\frac{(x-2)^2}{16} + \frac{(y+1)^2}{4} = 1$$
$$\frac{(x-2)^2}{4^2} + \frac{(y-(-1))^2}{2^2} = 1$$

The center of the ellipse is $(2, -1)$. Note that $a = 4$ and $b = 2$. We locate the center and then plot the points $(2+4, -1)$, $(2-4, -1)$, $(2, -1+2)$, $(2, -1-2)$, or $(6, -1)$, $(-2, -1)$, $(2, 1)$ and $(2, -3)$. Connect these points with an oval-shaped curve.

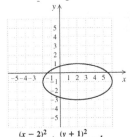

$$\frac{(x-2)^2}{16} + \frac{(y+1)^2}{4} = 1$$

Alternatively, we could write $x^2 - 4x + 4y^2 + 8y - 8 = 0$.

57. The sun is at the origin $(0, 0)$. Using the table of values, we see the maximum distance of the earth from the sun is 152.1 million km., or 152,100,000 km.

Exercise Set 10.3

---

1. d

3. h

5. g

7. c

9.
$$\frac{y^2}{16} - \frac{x^2}{16} = 1$$
$$\frac{y^2}{4^2} - \frac{x^2}{4^2} = 1$$

$a = 4$ and $b = 4$, so the asymptotes are $y = \frac{4}{4}x$ and $y = -\frac{4}{4}x$, or $y = x$ and $y = -x$. Replacing x with 0 and solving for y, we get $y = \pm 4$. The intercepts are $(0, 4)$ and $(0, -4)$.

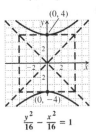

$$\frac{y^2}{16} - \frac{x^2}{16} = 1$$

11.
$$\frac{x^2}{4} - \frac{y^2}{25} = 1$$
$$\frac{x^2}{2^2} - \frac{y^2}{5^2} = 1$$

$a = 2$ and $b = 5$, so the asymptotes are $y = \frac{5}{2}x$ and $y = -\frac{5}{2}x$. Replacing y with 0 and solving for x, we get $x = \pm 2$. The intercepts are $(2, 0)$ and $(-2, 0)$.

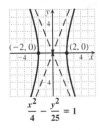

$$\frac{x^2}{4} - \frac{y^2}{25} = 1$$

13. $\dfrac{y^2}{36} - \dfrac{x^2}{9} = 1$

$\dfrac{y^2}{6^2} - \dfrac{x^2}{3^2} = 1$

$a = 3$ and $b = 6$, so the asymptotes are $y = \dfrac{6}{3}x$ and $y = -\dfrac{6}{3}x$, or $y = 2x$ and $y = -2x$. Replacing x with 0 and solving for y, we get $y = \pm 6$. The intercepts are $(0, 6)$ and $(0, -6)$.

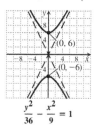

17. $25x^2 - 16y^2 = 400$

$\dfrac{x^2}{16} - \dfrac{y^2}{25} = 1$

$\dfrac{x^2}{4^2} - \dfrac{y^2}{5^2} = 1$

$a = 4$ and $b = 5$, so the asymptotes are $y = \dfrac{5}{4}x$ and $y = -\dfrac{5}{4}x$. Replacing y with 0 and solving for x, we get $x = \pm 4$. The intercepts are $(4, 0)$ and $(-4, 0)$.

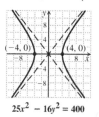

15. $y^2 - x^2 = 25$

$\dfrac{y^2}{25} - \dfrac{x^2}{25} = 1$

$\dfrac{y^2}{5^2} - \dfrac{x^2}{5^2} = 1$

$a = 5$ and $b = 5$, so the asymptotes are $y = \dfrac{5}{5}x$ and $y = -\dfrac{5}{5}x$, or $y = x$ and $y = -x$. Replacing x with 0 and solving for y, we get $y = \pm 5$. The intercepts are $(0, 5)$ and $(0, -5)$.

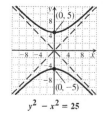

19. $xy = -6$

$y = -\dfrac{6}{x}$ Solving for y

We find some solutions, keeping the results in a table.

x	y
$\dfrac{1}{2}$	-12
1	-6
6	-1
12	$-\dfrac{1}{2}$
$-\dfrac{1}{2}$	12
-1	6
-6	1
-12	$\dfrac{1}{2}$

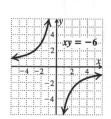

21. $xy = 4$

$y = \dfrac{4}{x}$ Solving for y

We find some solutions, keeping the results in a table.

x	y
$\dfrac{1}{2}$	8
1	4
4	1
8	$\dfrac{1}{2}$
$-\dfrac{1}{2}$	-8
-1	-4
-2	-2
-4	-1

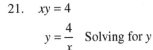

23. $xy = -2$

$y = -\dfrac{2}{x}$ Solving for y

We find some solutions, keeping the results in a table.

x	y
$\dfrac{1}{2}$	-4
1	-2
2	-1
4	$-\dfrac{1}{2}$
$-\dfrac{1}{2}$	4
-1	2
-2	1
-4	$\dfrac{1}{2}$

25. $xy = 1$

$y = \dfrac{1}{x}$ Solving for y

We find some solutions, keeping the results in a table.

x	y
$\dfrac{1}{4}$	4
$\dfrac{1}{2}$	2
1	1
2	$\dfrac{1}{2}$
4	$\dfrac{1}{4}$
$-\dfrac{1}{4}$	-4
$-\dfrac{1}{2}$	-2
-1	-1
-2	$-\dfrac{1}{2}$
-4	$-\dfrac{1}{4}$

27. $$x^2 + y^2 - 6x + 4y - 30 = 0$$
$$\left(x^2 - 6x \quad\right) + \left(y^2 + 4y \quad\right) = 30$$
$$\left(x^2 - 6x + 9\right) + \left(y^2 + 4y + 4\right) = 30 + 9 + 4$$
$$(x - 3)^2 + (y + 2)^2 = 43$$

Both variables are squared, so the graph is not a parabola. The plus sign between x^2 and y^2 indicates that we have either a circle or an ellipse. Since the coefficients of x^2 and y^2 are the same, the graph is a circle.

29. $$9x^2 + 4y^2 - 36 = 0$$
$$9x^2 + 4y^2 = 36$$
$$\dfrac{x^2}{4} + \dfrac{y^2}{9} = 1$$

Both variables are squared, so the graph is not a parabola. The plus sign between x^2 and y^2 indicates that we have either a circle or an ellipse. Since the coefficients of x^2 and y^2 are different, the graph is an ellipse.

31. $4x^2 - 9y^2 - 72 = 0$

        $4x^2 - 9y^2 = 72$

        $\dfrac{x^2}{18} - \dfrac{y^2}{8} = 1$

Both variables are squared, so the graph is not a parabola. The minus sign between x^2 and y^2 indicates that we have a hyperbola.

33. $y^2 = 20 - x^2$

    $x^2 + y^2 = 20$

Both variables are squared, so the graph is not a parabola. The plus sign between x^2 and y^2 indicates that we have either a circle or an ellipse. Since the coefficients of x^2 and y^2 are the same, the graph is a circle.

35. $x - 10 = y^2 - 6y$

     $x = y^2 - 6y + 10$

This equation has only one variable squared so we solve for the other variable. This is the equation for a parabola.

37. $x - \dfrac{8}{y} = 0$

       $x = \dfrac{8}{y}$

       $xy = 8$

We have the product of x and y which indicates that we have a hyperbola.

39. $y + 6x = x^2 + 5$

      $y = x^2 - 6x + 5$

This equation has only one variable squared so we solve for the other variable. This is the equation for a parabola.

41. $9y^2 = 36 + 4x^2$

   $9y^2 - 4x^2 = 36$

    $\dfrac{y^2}{4} - \dfrac{x^2}{9} = 1$

Both variables are squared, so the graph is not a parabola. The minus sign between x^2 and y^2 indicates that we have a hyperbola.

43. $3x^2 + y^2 - x = 2x^2 - 9x + 10y + 40$

   $x^2 + y^2 + 8x - 10y = 40$

Both variables are squared, so the graph is not a parabola. The plus sign between x^2 and y^2 indicates that we have either a circle or an ellipse. Since the coefficients of x^2 and y^2 are the same, the graph is a circle.

45. $16x^2 + 5y^2 - 12x^2 + 8y^2 - 3x + 4y = 568$

        $4x^2 + 13y^2 - 3x + 4y = 568$

Both variables are squared, so the graph is not a parabola. The plus sign between x^2 and y^2 indicates that we have either a circle or an ellipse. Since the coefficients of x^2 and y^2 are different, the graph is an ellipse.

47.   *Thinking and Writing Exercise.*

49. $5x + 2y = -3$

    $2x + 3y = 12$

Multiply the first equation by 3 and the second by –2, then add.

$15x + 6y = -9$

$\underline{-4x - 6y = -24}$

$11x = -33$

$x = -3$

Substitute into the first equation:

$5(-3) + 2y = -3$

       $2y = 12$

        $y = 6$

The solution is (–3, 6).

51. $\dfrac{3}{4}x^2 + x^2 = 7$

      $\dfrac{7}{4}x^2 = 7$

        $x^2 = 4$

        $x = \pm 2$

53. $x^2 - 3x - 1 = 0$

Use the quadratic formula:

$$\frac{-b \pm \sqrt{b^2 - 4ac}}{2a} = \frac{3 \pm \sqrt{(-3)^2 - 4(1)(-1)}}{2(1)}$$

$$= \frac{3 \pm \sqrt{9 + 4}}{2}$$

$$= \frac{3 \pm \sqrt{13}}{2}$$

55. *Thinking and Writing Exercise.*

57. Since the intercepts are $(0, 6)$ and $(0, -6)$, we know that the hyperbola is of the form

$\dfrac{y^2}{b^2} - \dfrac{x^2}{a^2} = 1$ and that $b = 6$. The equation of

the asymptotes tell us that $b / a = 3$, so

$$\frac{6}{a} = 3$$

$$a = 2$$

The equation is $\dfrac{y^2}{6^2} - \dfrac{x^2}{2^2} = 1$, or $\dfrac{y^2}{36} - \dfrac{x^2}{4} = 1$.

59. $\dfrac{(x-5)^2}{36} - \dfrac{(y-2)^2}{25} = 1$

$\dfrac{(x-5)^2}{6^2} - \dfrac{(y-2)^2}{5^2} = 1$

$h = 5, k = 2, a = 6, b = 5$

Center: $(5, 2)$

Vertices: $(5-6, 2)$ and $5 + 6, 2$ 9, or $(-1, 2)$

and $(11, 2)$

Asymptotes: $y - 2 = \dfrac{5}{6}(x - 5)$ and

$y - 2 = -\dfrac{5}{6}(x - 5)$

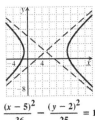

$\dfrac{(x-5)^2}{36} - \dfrac{(y-2)^2}{25} = 1$

61. $8(y + 3)^2 - 2(x - 4)^2 = 32$

$\dfrac{(y + 3)^2}{4} - \dfrac{(x - 4)^2}{16} = 1$

$\dfrac{(y - (-3))^2}{2^2} - \dfrac{(x - 4)^2}{4^2} = 1$

$h = 4, k = -3, a = 4, b = 2$

Center: $(4, -3)$

Vertices: $(4, -3 + 2)$ and $(4, -3 - 2)$, or

$(4, -1)$ and $(4, -5)$

Asymptotes: $y - (-3) = \dfrac{2}{4}(x - 4)$ and

$y - (-3) = -\dfrac{2}{4}(x - 4)$, or $y + 3 = \dfrac{1}{2}(x - 4)$

and $y + 3 = -\dfrac{1}{2}(x - 4)$

$8(y + 3)^2 - 2(x - 4)^2 = 32$

63. $4x^2 - y^2 + 24x + 4y + 28 = 0$

$4(x^2 + 6x) - (y^2 - 4y) = -28$

$4(x^2 + 6x + 9 - 9) - (y^2 - 4y + 4 - 4) = -28$

$4(x^2 + 6x + 9) - (y^2 - 4y + 4) = -28 + 4 \cdot 9 - 4$

$4(x + 3)^2 - (y - 2)^2 = 4$

$\dfrac{(x + 3)^2}{1} - \dfrac{(y - 2)^2}{4} = 1$

$\dfrac{(x - (-3))^2}{1^2} - \dfrac{(y - 2)^2}{2^2} = 1$

$h = -3, k = 2, a = 1, b = 2$

Center: $(-3, 2)$

Vertices: $(-3 - 1, 2)$ and $(-3 + 1, 2)$ or $(-4, 2)$

and $(-2, 2)$

Asymptotes: $y - 2 = \dfrac{2}{1}(x - (-3))$ and

$y - 2 = -\dfrac{2}{1}(x - (-3))$, or $y - 2 = 2(x + 3)$

and $y - 2 = -2(x + 3)$

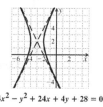

$4x^2 - y^2 + 24x + 4y + 28 = 0$

Mid-Chapter Review

Guided Solutions

1. $(x^2 - 4x) + (y^2 + 2y) = 6$

    $(x^2 - 4x + 4) + (y^2 + 2y + 1) = 6 + 4 + 1$

    $(x-2)^2 + (y+1)^2 = 11$

    The center of the circle is $(2, -1)$.

    The radius is $\sqrt{11}$.

2. $x^2 - \dfrac{y^2}{25} = 1$

    Is there both an x^2-term and a y^2-term? Yes
    Do both the x^2-term and the y^2-term have the same sign? No
    The graph of the equation is a hyperbola.

Mixed Review

1. $y = 3(x-4)^2 + 1$

    The equation of the parabola is already in the correct form. The vertex of the parabola is (h, k), or $(4, 1)$.
    Since the parabola has a vertical axis of symmetry, its equation is $x = h$, or $x = 4$.

2. $x = y^2 + 2y + 3$

    $x = (y^2 + 2y + 1) - 1 + 3$

    $x = (y+1)^2 + 2$

    1. Is there both an x^2-term and a y^2-term? Yes
    2. Do both the x^2-term and the y^2-term have the same sign? No
    3. The graph of the equation is a hyperbola.

3. $(x-3)^2 + (y-2)^2 = 5$

    Center: $(3, 2)$.

4. $x^2 + 6x + y^2 + 10y = 12$

    $x^2 + 6x + 9 + y^2 + 10y + 25 = 12 + 9 + 25$

    $(x+3)^2 + (y+5)^2 = 46$

    Center: $(-3, -5)$.

5. $\dfrac{x^2}{144} + \dfrac{y^2}{81} = 1$

    $\dfrac{x^2}{12^2} + \dfrac{y^2}{9^2} = 1$

    The x-intercepts are $(12, 0)$ and $(-12, 0)$.

    The y-intercepts are $(0, 9)$ and $(0, -9)$.

6. $\dfrac{x^2}{9} - \dfrac{y^2}{121} = 1$

    $\dfrac{x^2}{3^2} - \dfrac{y^2}{11^2} = 1$

    $a = 3$ and $b = 11$. Since the x term is positive, the vertices are $(3, 0)$ and $(-3, 0)$.

7. $4y^2 - x^2 = 4$

    $\dfrac{4y^2}{4} - \dfrac{x^2}{4} = \dfrac{4}{4}$

    $\dfrac{y^2}{1^2} - \dfrac{x^2}{2^2} = 1$

    $a = 2$ and $b = 1$. Since the y term is positive, the vertices are $(0, 1)$ and $(0, -1)$.

8. $\dfrac{y^2}{9} - \dfrac{x^2}{4} = 1$

    $\dfrac{y^2}{3^2} - \dfrac{x^2}{2^2} = 1$

    $a = 2$ and $b = 3$, so the asymptotes are $y = \dfrac{3}{2}x$ and $y = -\dfrac{3}{2}x$.

9. $x^2 + y^2 = 36$

    $x^2 + y^2 = 6^2$

    Both variables are squared, so the graph is not a parabola. The plus sign between x^2 and y^2 indicates that we have either a circle or an ellipse. Since the coefficients of x^2 and y^2 are the same, the graph is a circle. It has a center at $(0, 0)$ and a radius of 6.

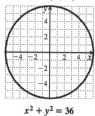

$x^2 + y^2 = 36$

10. $y = x^2 - 5$

This equation has only one variable squared. This is the equation for an upwards-opening parabola with a vertex at $(0, -5)$.

11. $\dfrac{x^2}{25} + \dfrac{y^2}{49} = 1$

$\dfrac{x^2}{5^2} + \dfrac{y^2}{7^2} = 1$

Both variables are squared, so the graph is not a parabola. The plus sign between x^2 and y^2 indicates that we have either a circle or an ellipse. Since the coefficients of x^2 and y^2 are different, the graph is an ellipse. Its intercepts are $(5, 0)$, $(-5, 0)$, $(0, 7)$, and $(0, -7)$.

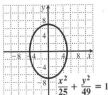

12. $\dfrac{x^2}{25} - \dfrac{y^2}{49} = 1$

$\dfrac{x^2}{5^2} - \dfrac{y^2}{7^2} = 1$

Both variables are squared, so the graph is not a parabola. The minus sign between x^2 and y^2 indicates that we have a hyperbola. Since the x-term is positive, it has a horizontal axis, with vertices at $(5, 0)$ and $(-5, 0)$ and asymptotes of $y = \pm\dfrac{7}{5}x$.

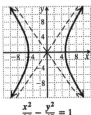

13. $x = (y + 3)^2 + 2$

This equation has only one variable squared. This is the equation for a rightwards-opening parabola with a vertex at $(2, -3)$.

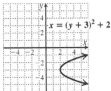

14. $4x^2 + 9y^2 = 36$

$\dfrac{4x^2}{36} + \dfrac{9y^2}{36} = 1$

$\dfrac{x^2}{3^2} + \dfrac{y^2}{2^2} = 1$

Both variables are squared, so the graph is not a parabola. The plus sign between x^2 and y^2 indicates that we have either a circle or an ellipse. Since the coefficients of x^2 and y^2 are different, the graph is an ellipse. Its intercepts are $(3, 0)$, $(-3, 0)$, $(0, 2)$, and $(0, -2)$.

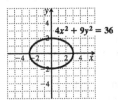

15. $xy = -4$

We have the product of x and y which indicates that we have a hyperbola. It passes through $(-4, 1)$, $(-2, 2)$, $(-1, 4)$, $(1, -4)$, $(2, -2)$, and $(4, -1)$.

16. $(x+2)^2 + (y-3)^2 = 1$

$(x+2)^2 + (y-3)^2 = 1^2$

Both variables are squared, so the graph is not a parabola. The plus sign between x^2 and y^2 indicates that we have either a circle or an ellipse. Since the coefficients of x^2 and y^2 are the same, the graph is a circle. It has a center at $(-2, 3)$ and a radius of 1.

17. $x^2 + y^2 - 8y - 20 = 0$

$x^2 + y^2 - 8y = 20$

$x^2 + y^2 - 8y + 16 = 20 + 16$

$x^2 + (y-4)^2 = 36$

$x^2 + (y-4)^2 = 6^2$

Both variables are squared, so the graph is not a parabola. The plus sign between x^2 and y^2 indicates that we have either a circle or an ellipse. Since the coefficients of x^2 and y^2 are the same, the graph is a circle. It has a center at $(0, 4)$ and a radius of 6.

18. $x = y^2 + 2y$

$x = y^2 + 2y + 1 - 1$

$x = (y+1)^2 - 1$

This equation has only one variable squared so we solve for the other variable. This is the equation for a rightwards-opening parabola with a vertex at $(-1, -1)$.

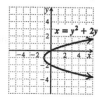

19. $16y^2 - x^2 = 16$

$\dfrac{16y^2}{16} - \dfrac{x^2}{16} = \dfrac{16}{16}$

$\dfrac{y^2}{1^2} - \dfrac{x^2}{4^2} = 1$

Both variables are squared, so the graph is not a parabola. The minus sign between x^2 and y^2 indicates that we have a hyperbola. Since the y-term is positive, it has a vertical axis, with vertices at $(0, 1)$ and $(0, -1)$ and asymptotes of $y = \pm \dfrac{1}{4} x$.

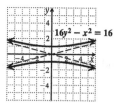

20. $x = \dfrac{9}{y}$

$xy = 9$

We have the product of x and y which indicates that we have a hyperbola. It passes through $(-9, -1)$, $(-3, -3)$, $(-1, -9)$, $(1, 9)$, $(3, 3)$, and $(9, 1)$.

Exercise Set 10.4

1. True

3. False

5. True

7. $x^2 + y^2 = 25$ (1)

   $y - x = 1$ (2)

First solve Eq. (2) for y.

$y = x + 1$ (3)

Then substitute $x + 1$ for y in Eq. (1) and solve for x.

$$x^2 + y^2 = 25$$

$$x^2 + (x+1)^2 = 25$$

$$x^2 + x^2 + 2x + 1 = 25$$

$$2x^2 + 2x + 1 = 25$$

$$2x^2 + 2x - 24 = 0$$

$$x^2 + x - 12 = 0$$

$$(x+4)(x-3) = 0$$

$$x+4 = 0 \ or \ x-3 = 0$$

$$x = -4 \ or \ x = 3$$

Now substitute these numbers in Eq. (3) and solve for y.

$$y = -4 + 1 = -3$$

$$y = 3 + 1 = 4$$

The pairs $(-4, -3)$ and $(3, 4)$ check, so they are the solutions.

9. $4x^2 + 9y^2 = 36$ (1)

    $3y + 2x = 6$ (2)

First solve Eq. (2) for y.

$$y = \frac{6-2x}{3} \quad (3)$$

Then substitute $\dfrac{6-2x}{3}$ for y in Eq. (1) and solve for x.

$$4x^2 + 9y^2 = 36$$

$$4x^2 + 9\left(\frac{6-2x}{3}\right)^2 = 36$$

$$4x^2 + 9\left(\frac{36 - 24x + 4x^2}{9}\right) = 36$$

$$4x^2 + 36 - 24x + 4x^2 = 36$$

$$8x^2 - 24x = 0$$

$$8x(x-3) = 0$$

$$8x = 0 \ or \ x-3 = 0$$

$$x = 0 \ or \ x = 3$$

Now substitute these numbers in Eq. (3) and solve for y.

$$y = \frac{6-2\cdot 0}{3} = 2$$

$$y = \frac{6-2\cdot 3}{3} = 0$$

The pairs $(0, 2)$ and $(3, 0)$ check, so they are the solutions.

11. $y^2 = x+3$ (1)

    $2y = x+4$ (2)

First solve Eq. (2) for x.

$$2y - 4 = x \quad (3)$$

Then substitute $2y - 4$ for x in Eq. (1) and solve for y.

$$y^2 = x+3$$

$$y^2 = (2y-4) + 3$$

$$y^2 = 2y - 1$$

$$y^2 - 2y + 1 = 0$$

$$(y-1)(y-1) = 0$$

$$y-1 = 0 \ or \ y-1 = 0$$

$$y = 1 \ or \ y = 1$$

Now substitute 1 for y in Eq. (3) and solve for x.

$$2\cdot 1 - 4 = x$$

$$-2 = x$$

The pair $(-2, 1)$ checks, so it is the solution.

13. $x^2 - xy + 3y^2 = 27$ (1)

    $x - y = 2$ (2)

First solve Eq. (2) for y.

$$x - 2 = y \quad (3)$$

Then substitute $x - 2$ for y in Eq. (1) and solve for x.

$$x^2 - xy + 3y^2 = 27$$

$$x^2 - x(x-2) + 3(x-2)^2 = 27$$

$$x^2 - x^2 + 2x + 3x^2 - 12x + 12 = 27$$

$$3x^2 - 10x - 15 = 0$$

$$x = \frac{-(-10) \pm \sqrt{(-10)^2 - 4(3)(-15)}}{2\cdot 3}$$

$$x = \frac{10 \pm \sqrt{100 + 180}}{6} = \frac{10 \pm \sqrt{280}}{6}$$

$$x = \frac{10 \pm 2\sqrt{70}}{6} = \frac{5 \pm \sqrt{70}}{3}$$

Now substitute these numbers in Eq. (3) and solve for y.

$$y = \frac{5+\sqrt{70}}{3} - 2 = \frac{-1+\sqrt{70}}{3}$$

$$y = \frac{5-\sqrt{70}}{3} - 2 = \frac{-1-\sqrt{70}}{3}$$

The pairs $\left(\dfrac{5+\sqrt{70}}{3}, \dfrac{-1+\sqrt{70}}{3} \right)$ and

$\left(\dfrac{5-\sqrt{70}}{3}, \dfrac{-1-\sqrt{70}}{3} \right)$ check, so they

are the solutions.

15. $x^2 + 4y^2 = 25 \quad (1)$
$\qquad x + 2y = 7 \quad (2)$

First solve Eq. (2) for x.

$x = -2y + 7 \quad (3)$

Then substitute $-2y + 7$ for x in Eq. (1) and
solve for y.

$$x^2 + 4y^2 = 25$$
$$(-2y+7)^2 + 4y^2 = 25$$
$$4y^2 - 28y + 49 + 4y^2 = 25$$
$$8y^2 - 28y + 24 = 0$$
$$2y^2 - 7y + 6 = 0$$
$$(2y-3)(y-2) = 0$$
$$2y-3 = 0 \ \ or \ \ y-2 = 0$$
$$y = \frac{3}{2} \ \ or \ \ y = 2$$

Now substitute these numbers in Eq. (3) and
solve for x.

$$x = -2 \cdot \frac{3}{2} + 7 = 4$$
$$x = -2 \cdot 2 + 7 = 3$$

The pairs $\left(4, \dfrac{3}{2} \right)$ and $(3, 2)$ check, so they
are the solutions.

17. $x^2 - xy + 3y^2 = 5 \quad (1)$
$\qquad x - y = 2 \quad (2)$

First solve Eq. (2) for y.

$x - 2 = y \quad (3)$

Then substitute $x - 2$ for y in Eq. (1) and
solve for x.

$$x^2 - xy + 3y^2 = 5$$
$$x^2 - x(x-2) + 3(x-2)^2 = 5$$
$$x^2 - x^2 + 2x + 3x^2 - 12x + 12 = 5$$
$$3x^2 - 10x + 7 = 0$$
$$(3x-7)(x-1) = 0$$
$$3x-7 = 0 \ \ or \ \ x-1 = 0$$
$$x = \frac{7}{3} \ \ or \ \ x = 1$$

Now substitute these numbers in Eq. (3) and
solve for y.

$$y = \frac{7}{3} - 2 = \frac{1}{3}$$
$$y = 1 - 2 = -1$$

The pairs $\left(\dfrac{7}{3}, \dfrac{1}{3} \right)$ and $(1, -1)$ check, so they
are the solutions.

19. $3x + y = 7 \quad (1)$
$\qquad 4x^2 + 5y = 24 \quad (2)$

First solve Eq. (1) for y.

$y = 7 - 3x \quad (3)$

Then substitute $7 - 3x$ for y in Eq. (2) and
solve for x.

$$4x^2 + 5y = 24$$
$$4x^2 + 5(7-3x) = 24$$
$$4x^2 + 35 - 15x = 24$$
$$4x^2 - 15x + 11 = 0$$
$$(4x-11)(x-1) = 0$$
$$4x-11 = 0 \ \ or \ \ x-1 = 0$$
$$x = \frac{11}{4} \ \ or \ \ x = 1$$

Now substitute these numbers in Eq. (3) and
solve for y.

$$y = 7 - 3 \cdot \frac{11}{4} = -\frac{5}{4}$$
$$y = 7 - 3 \cdot 1 = 4$$

The pairs $\left(\dfrac{11}{4}, -\dfrac{5}{4} \right)$ and $(1, 4)$ check, so they
are the solutions.

21. $a + b = 6$ (1)

    $ab = 8$ (2)

First solve Eq. (1) for a.

$a = 6 - b$ (3)

Then substitute $6 - b$ for a in Eq. (2) and solve for b.

$(6 - b)b = 8$

$6b - b^2 = 8$

$b^2 - 6b + 8 = 0$

$(b - 4)(b - 2) = 0$

        $b = 2, 4$

Now substitute these numbers in Eq. (3) and solve for a.

$a = 6 - b$

$a = 6 - 2$

$a = 4$

$a = 6 - b$

$a = 6 - 4$

$a = 2$

The pairs $(2, 4)$ and $(4, 2)$ check, so they are the solutions.

23. $2a + b = 1$ (1)

    $b = 4 - a^2$ (2)

Equation (2) is already solved for b.

Substitute $4 - a^2$ for b in Eq. (1) and solve for a.

$2a + 4 - a^2 = 1$

      $0 = a^2 - 2a - 3$

        $0 = (a - 3)(a + 1)$

$a - 3 = 0$ or $a + 1 = 0$

    $a = 3$ or $a = -1$

Now substitute these numbers in Eq. (2) and solve for b.

$b = 4 - 3^2 = -5$

$b = 4 - (-1)^2 = 3$

The pairs $(3, -5)$ and $(-1, 3)$ check, so they are the solutions.

25. $a^2 + b^2 = 89$ (1)

    $a - b = 3$ (2)

First solve Eq. (2) for a.

$a = b + 3$ (3)

Then substitute $b + 3$ for a in Eq. (1) and solve for b.

$(b + 3)^2 + b^2 = 89$

$b^2 + 6b + 9 + b^2 = 89$

$2b^2 + 6b - 80 = 0$

$b^2 + 3b - 40 = 0$

$(b + 8)(b - 5) = 0$

$b + 8 = 0$ or $b - 5 = 0$

    $b = -8$ or $b = 5$

Now substitute these numbers in Eq. (3) and solve for a.

$a = -8 + 3 = -5$

$a = 5 + 3 = 8$

The pairs $(-5, -8)$ and $(8, 5)$ check, so they are the solutions.

27. $y = x^2$ (1)

    $x = y^2$ (2)

Eq. (1) is already solved for y. Substitute x^2 for y in Eq. (2) and solve for x.

$x = y^2$

$x = \left(x^2\right)^2$

$x = x^4$

$0 = x^4 - x$

$0 = x\left(x^3 - 1\right)$

$0 = x(x - 1)\left(x^2 + x + 1\right)$

$x = 0$ or $x = 1$ or $x = \dfrac{-1 \pm \sqrt{1^2 - 4 \cdot 1 \cdot 1}}{2}$

$x = 0$ or $x = 1$ or $x = -\dfrac{1}{2} \pm \dfrac{\sqrt{3}}{2}i$

Now substitute these numbers in Eq. (1) and solve for y.

$y = \left(-\dfrac{1}{2} + \dfrac{\sqrt{3}}{2}i\right)^2 = -\dfrac{1}{2} - \dfrac{\sqrt{3}}{2}i$

$y = \left(-\dfrac{1}{2} - \dfrac{\sqrt{3}}{2}i\right)^2 = -\dfrac{1}{2} + \dfrac{\sqrt{3}}{2}i$

The pairs $(0,0)$, $(1,1)$, and

$\left(-\dfrac{1}{2}+\dfrac{\sqrt{3}}{2}i, -\dfrac{1}{2}-\dfrac{\sqrt{3}}{2}i\right)$ and

$\left(-\dfrac{1}{2}-\dfrac{\sqrt{3}}{2}i, -\dfrac{1}{2}+\dfrac{\sqrt{3}}{2}i\right)$ check, so they

are the solutions.

29. $x^2 + y^2 = 9$ (1)
$x^2 - y^2 = 9$ (2)

Here we use the elimination method.

$$\begin{array}{ll} x^2 + y^2 = 9 & (1) \\ \underline{x^2 - y^2 = 9} & (2) \\ 2x^2 \quad\ = 18 & \text{Adding} \\ \quad x^2 = 9 \\ \quad x = \pm 3 \end{array}$$

If $x = 3$, $x^2 = 9$, and if $x = -3$, $x^2 = 9$, so substituting 3 or –3 in Eq. (1) gives us

$$\begin{aligned} x^2 + y^2 &= 9 \\ 9 + y^2 &= 9 \\ y^2 &= 0 \\ y &= 0. \end{aligned}$$

The pairs $(3,0)$ and $(-3,0)$ check. They are the solutions.

31. $x^2 + y^2 = 25$ (1)
$xy = 12$ (2)

First we solve Eq. (2) for y.

$$\begin{aligned} xy &= 12 \\ y &= \frac{12}{x} \end{aligned}$$

Then we substitute $\dfrac{12}{x}$ for y in Eq. (1) and solve for x.

$$\begin{aligned} x^2 + y^2 &= 25 \\ x^2 + \left(\frac{12}{x}\right)^2 &= 25 \\ x^2 + \frac{144}{x^2} &= 25 \\ x^4 + 144 &= 25x^2 \quad \text{Multiplying by } x^2 \end{aligned}$$

$$\begin{aligned} x^4 - 25x^2 + 144 &= 0 \\ u^2 - 25u + 144 &= 0 \quad \text{Letting } u = x^2 \\ (u - 9)(u - 16) &= 0 \\ u = 9 \ \ &or \ \ u = 16 \end{aligned}$$

We now substitute x^2 for u and solve for x

$$\begin{aligned} x^2 &= 9 \ \ or \ \ x^2 = 16 \\ x &= \pm 3 \ \ or \ \ x = \pm 4 \end{aligned}$$

Since $y = 12/x$, if $x = 3$, $y = 4$; if $x = -3$, $y = -4$; if $x = 4$, $y = 3$; and if $x = -4$, $y = -3$. The pairs $(-4,-3)$, $(-3,-4)$, $(3,4)$ and $(4,3)$ check. They are the solutions.

33. $x^2 + y^2 = 9$ (1)
$25x^2 + 16y^2 = 400$ (2)

$$\begin{array}{ll} -16x^2 - 16y^2 = -144 & \text{Mult. (1) by } -16 \\ \underline{25x^2 + 16y^2 = \ \ 400} \\ 9x^2 \qquad\quad = 256 & \text{Adding} \end{array}$$

$$x = \pm\frac{16}{3}$$

$$\frac{256}{9} + y^2 = 9 \quad \text{Substituting in (1)}$$

$$y = \pm\frac{5\sqrt{7}}{3}i$$

The pairs $\left(\dfrac{16}{3}, \dfrac{5\sqrt{7}}{3}i\right)$, $\left(\dfrac{16}{3}, -\dfrac{5\sqrt{7}}{3}i\right)$,

$\left(-\dfrac{16}{3}, \dfrac{5\sqrt{7}}{3}i\right)$, and $\left(-\dfrac{16}{3}, -\dfrac{5\sqrt{7}}{3}i\right)$ check.

They are the solutions.

35. $x^2 + y^2 = 14$ (1)
$x^2 - y^2 = 4$ (2)

$$\begin{array}{ll} x^2 + y^2 = \ 14 \\ \underline{x^2 - y^2 = \ \ 4} \\ 2x^2 \qquad = \ 18 & \text{Adding} \\ x^2 = 9 \\ x = \pm 3 \\ 9 + y^2 = 14 & \text{Substituting in Eq. (1)} \\ \quad\ y^2 = 5 \\ \qquad y = \pm\sqrt{5} \end{array}$$

The pairs $\left(-3,-\sqrt{5}\right)$, $\left(-3,\sqrt{5}\right)$, $\left(3,-\sqrt{5}\right)$, and $\left(3,\sqrt{5}\right)$ check. They are the solutions.

37. $x^2 + y^2 = 20$ (1)
          $xy = 8$ (2)

First we solve Eq. (2) for y.

$$y = \frac{8}{x}$$

Then we substitute $\dfrac{8}{x}$ for y in Eq. (1) and solve for x.

$$x^2 + \left(\frac{8}{x}\right)^2 = 20$$

$$x^2 + \frac{64}{x^2} = 20$$

$$x^4 - 20x^2 + 64 = 0$$

$u^2 - 20u + 64 = 0$ Letting $u = x^2$

$(u-16)(u-4) = 0$

$u = 16$ or $u = 4$

$x^2 = 16$ or $x^2 = 4$

$x = \pm 4$ or $x = \pm 2$

$y = 8/x$, so if $x = 4$, $y = 2$; if $x = -4$,
$y = -2$; if $x = 2$, $y = 4$; if $x = -2$,
$y = -4$. The pairs $(4,2)$, $(-4,-2)$, $(2,4)$,
and $(-2,-4)$ check. They are the solutions.

39. $x^2 + 4y^2 = 20$ (1)
           $xy = 4$ (2)

First we solve Eq. (2) for y.

$$y = \frac{4}{x}$$

Then we substitute $\dfrac{4}{x}$ for y in Eq. (1) and solve for x.

$$x^2 + 4y^2 = 20$$

$$x^2 + 4\left(\frac{4}{x}\right)^2 = 20$$

$$x^2 + \frac{64}{x^2} = 20$$

$$x^4 + 64 = 20x^2$$

$x^4 - 20x^2 + 64 = 0$

$u^2 - 20u + 64 = 0$ Letting $u = x^2$

$(u-16)(u-4) = 0$

$u = 16$ or $u = 4$

We now substitute x^2 for u and solve for x.

$x^2 = 16$ or $x^2 = 4$

$x = \pm 4$ or $x = \pm 2$

$y = 4/x$, so if $x = 4$, $y = 1$; if $x = -4$,
$y = -1$; if $x = 2$, $y = 2$; and if $x = -2$,
$y = -2$. The pairs $(4,1)$, $(-4,-1)$, $(2,2)$,
and $(-2,-2)$ check. They are the solutions.

41. $2xy + 3y^2 = 7$ (1)
         $3xy - 2y^2 = 4$ (2)

$\quad 6xy + 9y^2 \doteq 21$ Mult. (1) by 3

$\underline{-6xy + 4y^2 = -8}$ Mult. (2) by -2

$\qquad\quad 13y^2 = 13$

$\qquad\quad\quad y^2 = 1$

$\qquad\quad\quad\; y = \pm 1$

Substitute for y in Eq. (1) and solve for x.

When $y = 1$: $2 \cdot x \cdot 1 + 3 \cdot 1^2 = 7$

$\qquad\qquad\qquad\qquad\quad 2x = 4$

$\qquad\qquad\qquad\qquad\quad\; x = 2$

When $y = -1$: $2 \cdot x \cdot (-1) + 3(-1)^2 = 7$

$\qquad\qquad\qquad\qquad\quad -2x = 4$

$\qquad\qquad\qquad\qquad\quad\;\; x = -2$

The pairs $(2,1)$ and $(-2,-1)$ check. They are the solutions.

43. $4a^2 - 25b^2 = 0$ (1)
         $2a^2 - 10b^2 = 3b + 4$ (2)

$\qquad 4a^2 - 25b^2 = \; 0$

$\underline{-4a^2 + 20b^2 = -6b - 8}$ Mult. (2) by -2

$\qquad\quad -5b^2 = -6b - 8$ Adding

$0 = 5b^2 - 6b - 8$

$0 = (5b+4)(b-2)$

Substitute for b in Eq. (1) and solve for y.

$$b = -\frac{4}{5}\ \ or\ \ b = 2$$

Substitute for b in Eq. (1) and solve for a.

When $b = -\dfrac{4}{5}$: $4a^2 - 25\left(-\dfrac{4}{5}\right)^2 = 0$

$$4a^2 = 16$$
$$a^2 = 4$$
$$a = \pm 2$$

When $b = 2$: $4a^2 - 25(2)^2 = 0$

$$4a^2 = 100$$
$$a^2 = 25$$
$$a = \pm 5$$

The pairs $\left(2, -\dfrac{4}{5}\right)$, $\left(-2, -\dfrac{4}{5}\right)$, $(5,2)$ and

$(-5,2)$ check. They are the solutions.

45. $ab - b^2 = -4$ (1)
$ab - 2b^2 = -6$ (2)

$$\begin{array}{l} ab - b^2 = -4 \\ \underline{-ab + 2b^2 = 6} \quad \text{Mult. (2) by } -1 \\ b^2 = 2 \\ b = \pm\sqrt{2} \end{array}$$

Substitute for b in Eq. (1) and solve for a.

When $b = \sqrt{2}$:

$$a\left(\sqrt{2}\right) - \left(\sqrt{2}\right)^2 = -4$$
$$a\sqrt{2} = -2$$
$$a = -\dfrac{2}{\sqrt{2}} = -\sqrt{2}$$

When $b = -\sqrt{2}$:

$$a\left(-\sqrt{2}\right) - \left(-\sqrt{2}\right)^2 = -4$$
$$-a\sqrt{2} = -2$$
$$a = \dfrac{-2}{-\sqrt{2}} = \sqrt{2}$$

The pairs $\left(-\sqrt{2}, \sqrt{2}\right)$ and $\left(\sqrt{2}, -\sqrt{2}\right)$ check.
They are the solutions.

47.   We let l and w represent the length and width, respectively.
The perimeter is 28 cm.
$2l + 2w = 28$, or $l + w = 14$
Using the Pythagorean theorem, we have another equation.
$l^2 + w^2 = 10^2$, or $l^2 + w^2 = 100$

We solve the system:
$l + w = 14$ (1)
$l^2 + w^2 = 100$ (2)
First solve Eq. 1 for w.
$w = 14 - l$ (3)
Then substitute $14 - l$ for w in Eq. (2) and solve for l.

$$l^2 + w^2 = 100$$
$$l^2 + \left(14 - l\right)^2 = 100$$
$$l^2 + 196 - 28l + l^2 = 100$$
$$2l^2 - 28l + 96 = 0$$
$$l^2 - 14l + 48 = 0$$
$$\left(l - 8\right)\left(l - 6\right) = 0$$
$$l = 8 \text{ or } l = 6$$

If $l = 8$, then $w = 14 - 8$, or 6. If $l = 6$, then $w = 14 - 6$, or 8. Since the length is usually considered to be longer than the width, we have the solution $l = 8$ and $w = 6$, or $(8,6)$.
The length is 8 cm and the width is 6 cm.

49.   Let l and w represent the length and width, respectively. We solve the system
$$lw = 2$$
$$2l + 2w = 6$$
We solve the first equation for l.
$$l = \dfrac{2}{w}$$

Then substitute the expression $\dfrac{2}{w}$ for l in the second equation.

$$2\left(\dfrac{2}{w}\right) + 2w = 6$$
$$4 + 2w^2 = 6w$$
$$2w^2 - 6w + 4 = 0$$
$$w^2 - 3w + 2 = 0$$
$$\left(w - 2\right)\left(w - 1\right) = 0$$
$$w = 2 \text{ or } w = 1$$

The solutions are $(1,2)$ and $(2,1)$. We choose the larger number to be the length, so the length is 2 yd and the width is 1 yd.

51. Let l equal the length and w equal the width of the cargo area.

The cargo area must be 60 ft^2, so we have one equation:

$lw = 60$

The Pythagorean equation gives us another equation:

$l^2 + w^2 = 13^2$, or $l^2 + w^2 = 169$

We solve the system of equations.

$$lw = 60 \quad (1)$$
$$l^2 + w^2 = 169 \quad (2)$$

First solve Eq. (1) for w:

$$lw = 60$$
$$w = \frac{60}{l} \quad (3)$$

Then substitute $60/l$ for w in Eq. (2) and solve for l.

$$l^2 + w^2 = 169$$
$$l^2 + \left(\frac{60}{l}\right)^2 = 169$$
$$l^2 + \frac{3600}{l^2} = 169$$
$$l^4 - 169l^2 + 3600 = 0$$

Let $u = l^2$ and $u^2 = l^4$ and substitute.

$$u^2 - 169u + 3600 = 0$$
$$(u - 144)(u - 25) = 0$$
$$u = 144 \ \text{ or } \ u = 25$$
$$l^2 = 144 \ \text{ or } \ l^2 = 25 \quad \text{Replacing } u \text{ with } l^2$$
$$l = \pm 12 \ \text{ or } \ l = \pm 5$$

Since the length cannot be negative, we consider only 12 and 5. We substitute in Eq. (3) to find w. When $l = 12$, $w = 60/12 = 5$; when $l = 5$, $w = 60/5 = 12$. Since we usually consider length to be longer than width, length is 12 ft and width is 5 ft.

53. Let x and y represent the numbers. Solve the system:

$$xy = 90$$
$$x^2 + y^2 = 261$$

The solutions are $(6, 15)$, $(-6, -15)$, $(15, 6)$, and $(-15, -6)$. The numbers are 6 and 15 or -6 and -15.

55. Let x equal the length of a side of one bed and y equal the length of a side of the other bed.

The area of the beds are x^2 and y^2, respectively.

Sum: $x^2 + y^2 = 832$

Difference: $x^2 - y^2 = 320$

Solve: Adding the two equations, we have:

$$2x^2 = 1152$$
$$x^2 = 576$$
$$x = \sqrt{576} = 24$$

Since the length cannot be negative, we use the positive value of x.

Substituting, we have

$$24^2 + y^2 = 832$$
$$y^2 = 256$$
$$y = 16 \ \ (16 \geq 0)$$

The length of the beds are 24 ft and 16 ft.

57. Let l equal the length of the rectangle and w equal the width.

Area: $lw = \sqrt{3} \ \ (1)$

From the Pythagorean theorem:

$$l^2 + w^2 = 2^2 \ \ (2)$$

We solve the system of equations.

We first solve Eq. (1) for w.

$$lw = \sqrt{3}$$
$$w = \frac{\sqrt{3}}{l}$$

Then we substitute $\dfrac{\sqrt{3}}{l}$ for w in Eq. (2) and solve for l.

$$l^2 + \left(\frac{\sqrt{3}}{l}\right)^2 = 4$$
$$l^2 + \frac{3}{l^2} = 4$$
$$l^4 + 3 = 4l^2$$
$$l^4 - 4l^2 + 3 = 0$$
$$u^2 - 4u + 3 = 0 \quad \text{Letting } u = l^2$$
$$(u - 3)(u - 1) = 0$$
$$u = 3 \ \text{ or } \ u = 1$$

We now substitute l^2 for u and solve for l.

$$l^2 = 3 \ \text{ or } \ l^2 = 1$$
$$l = \pm\sqrt{3} \ \text{ or } \ l = \pm 1$$

Measurements cannot be negative, so we only need to consider $l = \sqrt{3}$ and $l = 1$. Since $w = \sqrt{3}/l$, if $l = \sqrt{3}$, $w = 1$ and if $l = 1$, $w = \sqrt{3}$. Length is usually considered to be longer than width, so we have the solution $l = \sqrt{3}$ and $w = 1$, or $\left(\sqrt{3}, 1\right)$. The length is $\sqrt{3}$ m and the width is 1 m.

59. *Thinking and Writing Exercise.*

61. $(-1)^9 (-3)^2 = -1 \cdot 9 = -9$

63. $\dfrac{(-1)^k}{k-6} = \dfrac{(-1)^7}{7-6} = \dfrac{-1}{1} = -1$

65. $\dfrac{n}{2}(3+n) = \dfrac{11}{2}(3+11) = \dfrac{11}{2} \cdot 14 = 77$

67. *Thinking and Writing Exercise.*

69. Let (h,k) be a point on the line $5x + 8y = -2$ which is the center of a circle that passes through the points $(-2,3)$ and $(-4,1)$. The distance between (h,k) and $(-2,3)$ is the same as the distance between (h,k) and $(-4,1)$. This gives us one equation:

$$\sqrt{\left[h-(-2)\right]^2 + (k-3)^2} =$$
$$\sqrt{\left[h-(-4)\right]^2 + (k-1)^2}$$
$$(h+2)^2 + (k-3)^2 = (h+4)^2 + (k-1)^2$$
$$h^2 + 4h + 4 + k^2 - 6k + 9 =$$
$$h^2 + 8h + 16 + k^2 - 2k + 1$$
$$4h - 6k + 13 = 8h - 2k + 17$$
$$-4h - 4k = 4$$
$$h + k = -1$$

We get a second equation by substituting (h,k) in $5x + 8y = -2$.

$$5h + 8k = -2$$

We now solve the following system:

$$h + k = -1$$
$$5h + 8k = -2$$

The solution, which is the center of the circle, is $(-2,1)$.

Next, we find the length of the radius. We can find the distance between either $(-2,3)$ or $(-4,1)$ and the center $(-2,1)$. We use $(-2,3)$.

$$r = \sqrt{\left[-2-(-2)\right]^2 + (1-3)^2}$$
$$r = \sqrt{0^2 + (-2)^2}$$
$$r = \sqrt{4} = 2$$

We can write the equation of the circle with center $(-2,1)$ and radius 2.

$$(x-h)^2 + (y-k)^2 = r^2$$
$$\left[x-(-2)\right]^2 + (y-1)^2 = 2^2$$
$$(x+2)^2 + (y-1)^2 = 4$$

71. $p^2 + q^2 = 13$ (1)

    $\dfrac{1}{pq} = -\dfrac{1}{6}$ (2)

Solve Eq. (2) for p.

$$\dfrac{1}{q} = -\dfrac{p}{6}$$
$$-\dfrac{6}{q} = p$$

Substitute $-6/q$ for p in Eq. (1) and solve for q.

$$\left(-\dfrac{6}{q}\right)^2 + q^2 = 13$$
$$\dfrac{36}{q^2} + q^2 = 13$$
$$36 + q^4 = 13q^2$$
$$q^4 - 13q^2 + 36 = 0$$
$$u^2 - 13u + 36 = 0 \quad \text{Letting } u = q^2$$
$$(u-9)(u-4) = 0$$
$$u = 9 \ \ or \ \ u = 4$$
$$q^2 = 9 \ \ or \ \ q^2 = 4$$
$$q = \pm 3 \ \ or \ \ q = \pm 2$$

Since $p = -6/q$, if $q = 3$, $p = -2$; if $q = -3$, $p = 2$; if $q = 2$, $p = -3$; and if $q = -2$, $p = 3$. The pairs $(-2,3)$, $(2,-3)$, $(-3,2)$, and $(3,-2)$ check. They are the solutions.

73. Let l equal the length of a side of the fence and w equal the length of the other side.

$l + w = 100$ (Length of fencing)

$lw = 2475$ (Area of rectangle)

Solve the first equation for w.

$l + w = 100$

$w = 100 - l$

Substitute $100 - l$ for w in the second equation to determine:

$(100 - l)l = 2475$

$0 = l^2 - 100l + 2475$

$0 = (l - 55)(l - 45)$

$l = 55 \text{ or } l = 45$

Substituting:

$w = 100 - l = 100 - 55 = 45$

$w = 100 - l = 100 - 45 = 55$

We have the same pair of numbers for both solutions. We usually think of length as the greater of the two measures, so we have a rectangle with length of 55 ft and a width of 45 ft.

75. We let x and y represent the length and width of the base of the box, respectively. Make a drawing.

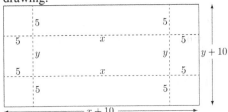

The dimensions of the metal sheet are $x + 10$ and $y + 10$. Solve the system:

$(x + 10)(y + 10) = 340$

$x \cdot y \cdot 5 = 350$

The solutions are $(10, 7)$ and $(7, 10)$.

Choosing the larger number as the length, we have the solution. The dimensions of the box are 10 in. by 7 in. by 5 in.

77. $\dfrac{\text{Length}}{\text{Height}} = \dfrac{16}{9}$

$l = \dfrac{16}{9}h$

Using the Pythagorean theorem, we have

$l^2 + h^2 = \text{Diagonal Measure}^2$

$l^2 + h^2 = 73^2$

$\left(\dfrac{16}{9}h\right)^2 + h^2 = 5329$

$\dfrac{337}{81}h^2 = 5329$

$h^2 \approx 1280.86$

$h \approx 35.8$

$l = \dfrac{16}{9}h$

$l = \dfrac{16}{9}(35.8)$

$l \approx 63.6$

The length is approximately 63.6 in., and the height is approximately 35.8 in.

79. $4xy - 7 = 0$

$x - 3y - 2 = 0$

Solve each equation for y.

$4xy - 7 = 0$

$4xy = 7$

$y = \dfrac{7}{4x}$

$x - 3y - 2 = 0$

$x - 2 = 3y$

$\dfrac{x - 2}{3} = y$

Using a graphing calculator, let

$y_1 = \dfrac{7}{4x}$ and $y_2 = \dfrac{x - 2}{3}$.

Use the INTERSECT feature to determine:

$(-1.50, -1.17)$ and $(3.50, 0.50)$.

These are the solutions.

Chapter 10 Study Summary

1. $x = y^2 + 6y + 7$

$x = y^2 + 6y + 9 - 9 + 7$

$x = (y + 3)^2 - 2$

This is a right-opening parabola, with a vertex at $(-2, -3)$.

2. $x^2 + y^2 - 6x + 5 = 0$

$x^2 - 6x + 9 + y^2 = -5 + 9$

$(x-3)^2 + y^2 = 2^2$

The center is (3, 0) and the radius is 2

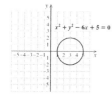

3. $\dfrac{x^2}{9} + y^2 = 1$

The intercepts are (3, 0), (–3, 0), (0, 1), and

(0, –1).

4. $\dfrac{y^2}{16} - \dfrac{x^2}{4} = 1$

This is a hyperbola with a vertical axis, with vertices at (0, 4) and (0, –4) and asymptotes of $y = \pm 2x$.

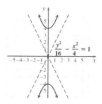

5. $x^2 + y^2 = 41$ (1)

$y - x = 1$ (2)

Solve (2) for y: $y = x + 1$. Then, substitute into (1):

$x^2 + y^2 = 41$

$x^2 + (x+1)^2 = 41$

$x^2 + x^2 + 2x + 1 = 41$

$2x^2 + 2x - 40 = 0$

$2(x+5)(x-4) = 0$

$x = -5, 4$

Substitute into (2):

$y - (-5) = 1$

$y = -4$

$y - 4 = 1$

$y = 5$

The solutions are (–5, –4) and (4, 5).

Chapter 10 Review Exercises

1. True

2. False

3. False

4. True

5. True

6. True

7. False

8. True

9. $(x+3)^2 + (y-2)^2 = 16$

$[x-(-3)]^2 + (y-2)^2 = (4)^2$

Center: $(-3, 2)$; radius is 4.

10. $(x-5)^2 + y^2 = 11$

$(x-5)^2 + (y-0)^2 = \left(\sqrt{11}\right)^2$

Center: $(5, 0)$; radius is $\sqrt{11}$.

11. $x^2 - 6x + y^2 - 2y = -1$

$(x^2 - 6x + 9) + (y^2 - 2y + 1) = -1 + 9 + 1$

$(x-3)^2 + (y-1)^2 = 3^2$

Center: $(3, 1)$; radius is 3.

12. $x^2 + 8x + y^2 - 6y = 20$

$(x^2 + 8x + 16) + (y^2 - 6y + 9) = 20 + 16 + 9$

$(x+4)^2 + (y-3)^2 = \left(\sqrt{45}\right)^2$

Center: $(-4, 3)$; radius is $3\sqrt{5}$.

13. $(x-h)^2 + (y-k)^2 = r^2$

$[x-(-4)]^2 + (y-3)^2 = (4)^2$

$(x+4)^2 + (y-3)^2 = 16$

14. $(x-h)^2 + (y-k)^2 = r^2$

$(x-7)^2 + [y-(-2)]^2 = (2\sqrt{5})^2$

$(x-7)^2 + (y+2)^2 = 20$

15. $5x^2 + 5y^2 = 80$

$x^2 + y^2 = 16$

Circle

$5x^2 + 5y^2 = 80$

16. $9x^2 + 2y^2 = 18$

$\dfrac{x^2}{2} + \dfrac{y^2}{9} = 1$

Ellipse

$9x^2 + 2y^2 = 18$

17. $y = -x^2 + 2x - 3$

Parabola

$y = -x^2 + 2x - 3$

18. $\dfrac{y^2}{9} - \dfrac{x^2}{4} = 1$

$\dfrac{y^2}{3^2} - \dfrac{x^2}{2^2} = 1$

Hyperbola

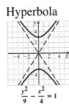

$\dfrac{y^2}{9} - \dfrac{x^2}{4} = 1$

19. $xy = 9$

$y = \dfrac{9}{x}$

Hyperbola

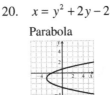

20. $x = y^2 + 2y - 2$

Parabola

$x = y^2 + 2y - 2$

21. $\dfrac{(x+1)^2}{3} + (y-3)^2 = 1$

Ellipse

$\dfrac{(x+1)^2}{3} + (y-3)^2 = 1$

22. $x^2 + y^2 + 6x - 8y - 39 = 0$

$(x^2 + 6x + 9) + (y^2 - 8y + 16) = 39 + 9 + 16$

$(x+3)^2 + (y-4)^2 = 64$

Circle

$x^2 + y^2 + 6x - 8y - 39 = 0$

23. $x^2 - y^2 = 21$ (1)

$x + y = 3$ (2)

Solve for y using Eq. (2) and substitute into

Eq. (1).

$x + y = 3 \rightarrow y = 3 - x$

$x^2 - (3-x)^2 = 21$

$x^2 - 9 + 6x - x^2 = 21$

$6x = 30$

$x = 5$

Substitute using $y = 3 - x$, $y = 3 - 5 = -2$.

This pair checks. Solution is $(5, -2)$.

24. $x^2 - 2x + 2y^2 = 8$ (1)

    $2x + y = 6$ (2)

Solve for y using Eq. (2) and substitute into

Eq. (1).

$2x + y = 6 \rightarrow y = 6 - 2x$

$\qquad x^2 - 2x + 2(6 - 2x)^2 = 8$

$x^2 - 2x + 72 - 48x + 8x^2 = 8$

$\qquad\qquad 9x^2 - 50x + 64 = 0$

$\qquad\qquad (9x - 32)(x - 2) = 0$

$x = \dfrac{32}{9}$ or $x = 2$

If $x = \dfrac{32}{9}$, $y = 6 - 2\left(\dfrac{32}{9}\right) = \dfrac{-10}{9}$.

If $x = 2$, $y = 6 - 2(2) = 2$.

We have $\left(\dfrac{32}{9}, \dfrac{-10}{9}\right)$ and $(2, 2)$. These pairs

check and are the solutions.

25. $x^2 - y = 5$ (1)

    $2x - y = 5$ (2)

Solve for y using Eq. (2) and substitute into

Eq. (1).

$2x - y = 5 \rightarrow y = 2x - 5$

$x^2 - (2x - 5) = 5$

$x^2 - 2x = 0$

$x(x - 2) = 0$

$x = 0$ or $x = 2$

If $x = 0$, $y = 2(0) - 5 = -5$.

If $x = 2$, $y = 2(2) - 5 = -1$.

We have $(0, -5)$ and $(2, -1)$. These pairs

check and are the solutions.

26. $x^2 + y^2 = 15$

    $x^2 - y^2 = 17$

Adding the two equations gives us:

$2x^2 = 32$

$\quad x^2 = 16$

$\quad\; x = \pm 4$

If $x = -4$, $(-4)^2 + y^2 = 15$

$\qquad\qquad\qquad y^2 = -1$

$\qquad\qquad\qquad\; y = \pm i$

If $x = 4$, $(4)^2 + y^2 = 15$

$\qquad\qquad\qquad y^2 = -1$

$\qquad\qquad\qquad\; y = \pm i$

We have $(-4, i)$, $(-4, -i)$, $(4, i)$, and

$(4, -i)$. These pairs check and are the

solutions.

27. $x^2 - y^2 = 3$

    $y = x^2 - 3$

Using substitution:

$x^2 - (x^2 - 3)^2 = 3$

$x^2 - (x^4 - 6x^2 + 9) = 3$

$0 = x^4 - 7x^2 + 12$

$0 = (x^2 - 4)(x^2 - 3)$

$x^2 - 4 = 0$ or $x^2 - 3 = 0$

$x = \pm 2$ or $x = \pm\sqrt{3}$

When $x = \pm 2$, $x^2 = 4$, so $y = 4 - 3 = 1$, and

we have the points $(2, 1)$ and $(-2, 1)$.

When $x = \pm\sqrt{3}$, $x^2 = 3$, so $3 - 3 = 0$, and we

have the points $\left(\sqrt{3}, 0\right)$ and $\left(-\sqrt{3}, 0\right)$.

These pairs check and are the solutions.

28. $x^2 + y^2 = 18$

    $2x + y = 3 \rightarrow y = 3 - 2x$

Using substitution:

$\qquad x^2 + (3 - 2x)^2 = 18$

$x^2 + 9 - 12x + 4x^2 = 18$

$\qquad\quad 5x^2 - 12x - 9 = 0$

$\qquad\quad (5x + 3)(x - 3) = 0$

$x = -\dfrac{3}{5}$ or $x = 3$

When $x = -\dfrac{3}{5}$, $y = 3 - 2\left(-\dfrac{3}{5}\right) = \dfrac{21}{5}$.

When $x = 3$, $y = 3 - 2(3) = -3$.

The pairs $\left(-\dfrac{3}{5}, \dfrac{21}{5}\right)$ and $(3, -3)$ check and

are the solutions.

29.　　$x^2 + y^2 = 100 \rightarrow x^2 = 100 - y^2$

$2x^2 - 3y^2 = -120$

Using substitution:

$2\left(100 - y^2\right) - 3y^2 = -120$

$200 - 2y^2 - 3y^2 = -120$

$320 = 5y^2$

$64 = y^2$

$\pm 8 = y$

$x^2 = 100 - 64$

$x^2 = 36$

$x = \pm 6$

All the pairs check. The solutions are $(6, 8)$, $(6, -8)$, $(-6, 8)$, and $(-6, -8)$.

30.　$x^2 + 2y^2 = 12$

$xy = 4 \rightarrow x = \dfrac{4}{y}$

Using substitution:

$\left(\dfrac{4}{y}\right)^2 + 2y^2 = 12$

$\dfrac{16}{y^2} + 2y^2 = 12$

$16 + 2y^4 = 12y^2$

$2y^4 - 12y^2 + 16 = 0$

$y^4 - 6y^2 + 8 = 0$

$\left(y^2 - 4\right)\left(y^2 - 2\right) = 0$

$y = \pm 2 \ or \ y = \pm\sqrt{2}$

When $y = -2$, $x = \dfrac{4}{-2} = -2$.

When $y = 2$, $x = \dfrac{4}{2} = 2$.

When $y = -\sqrt{2}$, $x = \dfrac{4}{-\sqrt{2}} = -2\sqrt{2}$.

When $y = \sqrt{2}$, $x = \dfrac{4}{\sqrt{2}} = 2\sqrt{2}$.

We have $(-2, -2)$, $(2, 2)$, $\left(-2\sqrt{2}, -\sqrt{2}\right)$, and $\left(2\sqrt{2}, \sqrt{2}\right)$. These pairs check and are the solutions.

31.　Using perimeter, we have $2l + 2w = 38$, or $l + w = 19$. Using area, we have $lw = 84$. Use substitution to solve.

$l = 19 - w$

$(19 - w)w = 84$

$19w - w^2 = 84$

$0 = w^2 - 19w + 84$

$0 = (w - 12)(w - 7)$

$w = 12 \ or \ w = 7$

$l = 19 - 12 = 7 \ or \ l = 19 - 7 = 12$

These are the same pair of numbers. We usually think of length as the greater measure, so length is 12 m and width is 7 m.

32.　Area: $lw = 108$

Using the Pythagorean theorem:

$l^2 + w^2 = 15^2$

Using substitution:

$l = \dfrac{108}{w}$

$\left(\dfrac{108}{w}\right)^2 + w^2 = 15^2$

$11,664 + w^4 = 225w^2$

$w^4 - 225w^2 + 11,664 = 0$

$\left(w^2 - 144\right)\left(w^2 - 81\right) = 0$

$w = \pm 12 \ or \ w = \pm 9$

Since measure cannot be negative, we disregard -12 and -9.

$w = 12$, $l = \dfrac{108}{12} = 9$

$w = 9$, $l = \dfrac{108}{9} = 12$

The length is 12 in. and the width is 9 in.

33.　Let x = length of side of lesser square and y = length of side of the greater. Using perimeter, we have

$4x + 12 = 4y$, or $x + 3 = y$.

Using area, we have

$x^2 + 39 = y^2$

Use substitution to solve the system of equations.

$x^2 + 39 = (x + 3)^2$

$x^2 + 39 = x^2 + 6x + 9$

$30 = 6x$

$5 = x$

$y = x + 3 = 5 + 3 = 8$

The perimeter of the lesser square is
$4x = 4 \cdot 5$, or 20 cm, and the perimeter of the
greater is $4y = 4 \cdot 8$, or 32 cm.

34. Let r_1 and r_2 represent the radii of the
circles. $A = \pi r^2$, so we have:

$$\pi r_1^2 + \pi r_2^2 = 130\pi$$
$$r_1^2 + r_2^2 = 130$$

$C = 2\pi r_1$, so we have:

$$2\pi r_1 - 2\pi r_2 = 16\pi$$
$$2\pi(r_1 - r_2) = 16\pi$$
$$r_1 - r_2 = 8$$

Solve the system of equations:

$$r_1^2 + r_2^2 = 130$$
$$r_1 - r_2 = 8 \rightarrow r_1 = r_2 + 8$$
$$(r_2 + 8)^2 + r_2^2 = 130$$
$$r_2^2 + 16r_2 + 64 + r_2^2 = 130$$
$$2r_2^2 + 16r_2 - 66 = 0$$
$$r_2^2 + 8r_2 - 33 = 0$$
$$(r_2 + 11)(r_2 - 3) = 0$$
$$r_2 = -11 \ or \ r_2 = 3$$

The radius cannot have negative length, so
$r_2 = 3$ and $r_1 = 3 + 8$, or 11. The radii have
lengths of 3 ft and 11 ft.

35. *Thinking and Writing Exercise.* The graph of
a parabola has one branch; whereas, the graph
of a hyperbola has two. A hyperbola has
asymptotes, but a parabola does not.

36. *Thinking and Writing Exercise.* Many of the
relations discussed are not functions.
Function notation could be used for vertical
parabolas and for hyperbolas that have the
axes as asymptotes.

37. $4x^2 - x - 3y^2 = 9 \quad (1)$
    $-x^2 + x + y^2 = 2 \quad (2)$

Multiply Eq. (2) by 3 and add to Eq. (1).

$$4x^2 - x - 3y^2 = 9$$
$$\underline{-3x^2 + 3x + 3y^2 = 6} \quad 3 \times \text{Eq. } (2)$$
$$x^2 + 2x \qquad = 15$$

$$x^2 + 2x - 15 = 0$$
$$(x + 5)(x - 3) = 0$$
$$x = -5 \ or \ x = 3$$

From Eq. (2), $y^2 = x^2 - x + 2$.

When $x = -5$, $y^2 = (-5)^2 - (-5) + 2$
$$y^2 = 25 + 5 + 2$$
$$y^2 = 32$$
$$y = \pm\sqrt{32} = \pm 4\sqrt{2}$$

When $x = 3$, $y^2 = 3^2 - 3 + 2$
$$y^2 = 9 - 3 + 2$$
$$y^2 = 8$$
$$y = \pm\sqrt{8} = \pm 2\sqrt{2}$$

The pairs check. The solutions are
$\left(-5, -4\sqrt{2}\right)$, $\left(-5, 4\sqrt{2}\right)$, $\left(3, -2\sqrt{2}\right)$, and
$\left(3, 2\sqrt{2}\right)$.

38. Let $(0, y)$ be the point on the y-axis that is
equidistant from $(-8, 0)$ and $(8, 0)$.
The distance is 10, so we have:

$$\sqrt{(-8 - 0)^2 + (y - 0)^2} = 10$$
$$\sqrt{64 + y^2} = 10$$
$$64 + y^2 = 100$$
$$y^2 = 36$$
$$y = \pm 6$$

The points are $(0, 6)$ and $(0, -6)$.

39. The three points are equidistant from the
center of the circle, (h, k). Using each of the
three points in the equation of a circle, we
have:

$$\left[x - (-2)\right]^2 + \left[y - (-4)\right]^2 = r^2$$
$$x^2 + 4x + 4 + y^2 + 8y + 16 = r^2$$
$$x^2 + 4x + y^2 + 8y + 20 = r^2 \quad (1)$$
$$(x - 5)^2 + \left[y - (-5)\right]^2 = r^2$$
$$x^2 - 10x + 25 + y^2 + 10y + 25 = r^2$$
$$x^2 - 10x + y^2 + 10y + 50 = r^2 \quad (2)$$
$$(x - 6)^2 + (y - 2)^2 = r^2$$
$$x^2 - 12x + y^2 - 4y + 40 = r^2 \quad (3)$$

$(1) = (2):$

$$x^2 + 4x + y^2 + 8y + 20$$
$$= x^2 - 10x + y^2 + 10y + 50$$

$$14x - 2y = 30$$

$$7x - y = 15 \ (4)$$

$(1) = (3):$

$$x^2 + 4x + y^2 + 8y + 20$$
$$= x^2 - 12x + y^2 - 4y + 40$$

$$16x + 12y = 20$$

$$4x + 3y = 5 \ (5)$$

Using (4) and (5), solve the system of equations.

$$3(7x - y = 15) = 21x - 3y = 45$$
$$\underline{4x + 3y = 5}$$
$$25x = 50$$
$$x = 2$$
$$y = -1$$

The center of the circle is $(2, -1)$.

$$(x - 2)^2 + (y + 1)^2 = r^2$$

Choose any of the three points on the circle to determine r^2.

$$(6 - 2)^2 + (2 + 1)^2 = r^2$$
$$4^2 + 3^2 = r^2$$
$$25 = r^2$$

The equation of the circle is

$$(x - 2)^2 + (y + 1)^2 = 25$$

40. From the x-intercepts, $(-9, 0)$ and $(9, 0)$, we know $a = 9$; from the y-intercepts, $(0, -5)$ and $(0, 5)$, we know $b = 5$. So, we have the equation:

$$\frac{x^2}{9^2} + \frac{y^2}{5^2} = 1$$

$$\frac{x^2}{81} + \frac{y^2}{25} = 1$$

41. Let $(x, 0)$ be the point on the x-axis that is equidistant from $(-3, 4)$ and $(5, 6)$.

$$\sqrt{[x - (-3)]^2 + (0 - 4)^2} = \sqrt{(x - 5)^2 + (0 - 6)^2}$$

$$x^2 + 6x + 9 + 16 = x^2 - 10x + 25 + 36$$

$$16x = 36$$

$$x = \frac{36}{16} = \frac{9}{4}$$

The point is $\left(\frac{9}{4}, 0\right)$.

Chapter 10 Test

1. $(h, k) = (3, -4)$ and $r = 2\sqrt{3}$.

$$(x - h)^2 + (y - k)^2 = r^2$$
$$(x - 3)^2 + (y - (-4))^2 = (2\sqrt{3})^2$$
$$(x - 3)^2 + (y + 4)^2 = 12$$

2. $$(x - 4)^2 + (y + 1)^2 = 5$$
$$[x - (4)]^2 + (y - (-1))^2 = (\sqrt{5})^2$$

Center: $(4, -1)$; radius is $\sqrt{5}$.

3. $$x^2 + y^2 + 4x - 6y + 4 = 0$$
$$x^2 + 4x + y^2 - 6y = -4$$
$$x^2 + 4x + 4 + y^2 - 6y + 9 = -4 + 4 + 9$$
$$(x + 2)^2 + (y - 3)^2 = 3^2$$

Center: $(-2, 3)$; radius is 3.

4. $y = x^2 - 4x - 1$

Parabola

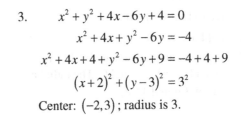

$$y = x^2 - 4x - 1$$

5. $x^2 + y^2 + 2x + 6y + 6 = 0$

Circle

$$x^2 + y^2 + 2x + 6y + 6 = 0$$

6. $\dfrac{x^2}{16} - \dfrac{y^2}{9} = 1$

Hyperbola

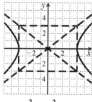

$$\dfrac{x^2}{16} - \dfrac{y^2}{9} = 1$$

7. $16x^2 + 4y^2 = 64$

Ellipse

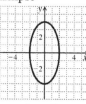

$$16x^2 + 4y^2 = 64$$

8. $xy = -5$

Hyperbola

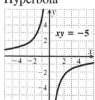

$xy = -5$

9. $x = -y^2 + 4y$

Parabola

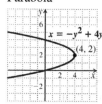

$x = -y^2 + 4y$

$(4, 2)$

10. $x^2 + y^2 = 36$ (1)

    $3x + 4y = 24$ (2)

Solve Eq. (2) for x and substitute into

Eq. (1).

$3x + 4y = 24$

$x = \dfrac{24 - 4y}{3}$

$x^2 + y^2 = 36$

$\left(\dfrac{24 - 4y}{3}\right)^2 + y^2 = 36$

$\dfrac{576 - 192y + 16y^2}{9} + y^2 = 36$

$576 - 192y + 25y^2 = 324$

$25y^2 - 192y + 252 = 0$

$(25y - 42)(y - 6) = 0$

$y = \dfrac{42}{25}, 6$

If $y = \dfrac{42}{25}$, $x = \dfrac{24 - 4\left(\frac{42}{25}\right)}{3} = \dfrac{144}{25}$.

If $y = 6$, $x = \dfrac{24 - 4(6)}{3} = 0$.

The pairs check.

The solutions are $\left(\dfrac{144}{25}, \dfrac{42}{25}\right)$ and $(0, 6)$.

11. $x^2 - y = 3$ (1)

    $2x + y = 5$ (2)

Add Eq. 2 and Eq. (1).

$\begin{array}{r} x^2 - y = 3 \\ 2x + y = 5 \\ \hline x^2 + 2x = 8 \end{array}$

$x^2 + 2x - 8 = 0$

$(x + 4)(x - 2) = 0$

$x = -4, 2$

$2x + y = 5$

$2(-4) + y = 5$

$y = 13$

$2(2) + y = 5$

$y = 1$

The pairs check. The solutions are $(-4, 13)$ and $(2, 1)$.

12. $x^2 - y^2 = 3$

    $xy = 2$

Solve $xy = 2$ for x and substitute into the first equation.

$$x = \frac{2}{y}$$

$$\left(\frac{2}{y}\right)^2 - y^2 = 3$$

$$\frac{4}{y^2} - y^2 = 3$$

$$4 - y^4 = 3y^2$$

$$0 = y^4 + 3y^2 - 4$$

$$0 = \left(y^2 + 4\right)\left(y^2 - 1\right)$$

$$y^2 + 4 = 0 \quad or \quad y^2 - 1 = 0$$

$$y = \pm\sqrt{-4} \quad or \quad y^2 = 1$$

$$y = \pm 2i \quad or \quad y = \pm 1$$

When $y = 2i$,

$$x = \frac{2}{2i} = \frac{1}{i} = \frac{i}{i^2} = -i.$$

When $y = -2i$,

$$x = \frac{2}{-2i} = \frac{1}{-i} = \frac{i}{-i^2} = i.$$

When $y = -1$, $x = \frac{2}{-1} = -2$.

When $y = 1$, $x = \frac{2}{1} = 2$.

The pairs check. The solutions are:
$(i, -2i)$, $(-i, 2i)$, $(2,1)$, and $(-2,-1)$.

13. $x^2 + y^2 = 10 \quad (1)$
     $x^2 = y^2 + 2 \quad (2)$

Using Eq. (2) substitute for x^2 into Eq. (1)
and solve.

$$\left(y^2 + 2\right) + y^2 = 10$$

$$2y^2 = 8$$

$$y^2 = 4$$

$$y = \pm 2$$

When $y = -2$, $y^2 = 4$, and when $y = 2$,
$y^2 = 4$.

$$x^2 = 4 + 2$$

$$x = \pm\sqrt{6}$$

The solutions are $\left(\sqrt{6}, 2\right)$, $\left(\sqrt{6}, -2\right)$,

$\left(-\sqrt{6}, 2\right)$, and $\left(-\sqrt{6}, -2\right)$.

14. Area: $lw = 22$

Pythagorean Theorem: $l^2 + w^2 = \left(5\sqrt{5}\right)^2$.

Solving the first equation for l, we have:

$$l = \frac{22}{w}.$$

Using substitution:

$$\left(\frac{22}{w}\right)^2 + w^2 = 125$$

$$\frac{484}{w^2} + w^2 = 125$$

$$484 + w^4 = 125w^2$$

$$w^4 - 125w^2 + 484 = 0$$

$$\left(w^2 - 121\right)\left(w^2 - 4\right) = 0$$

$$w^2 - 121 = 0 \quad or \quad w^2 - 4 = 0$$

$$w = \pm 11 \quad or \quad w = \pm 2$$

Measure cannot be negative, so we use only
$w = 11$ and $w = 2$.

If $w = 11$, then $l = \frac{22}{11} = 2$.

If $w = 2$, the $l = \frac{22}{2} = 11$.

We usually assign the greater measure to
length, so length is 11 inches and width is 2
inches.

15. Let x = the length of a side of the first square
and y = the length of a side of the second
square. The areas are x^2 and y^2,
respectively.

$$\text{Sum:} \quad x^2 + y^2 = 8$$

$$\text{Difference:} \quad \underline{x^2 - y^2 = 2}$$

$$2x^2 = 10 \quad \text{Add}$$

$$x^2 = 5$$

$$x = \pm\sqrt{5}$$

The length must be nonnegative.

If $x = \sqrt{5}$, $\quad x^2 + y^2 = 8$

$$\left(\sqrt{5}\right)^2 + y^2 = 8$$

$$y = \sqrt{3} \text{ (Positive Value)}$$

The lengths of the sides of the squares are
$\sqrt{5}$ m and $\sqrt{3}$ m.

16. Perimeter: $2l + 2w = 112$, or $l + w = 56$.
Pythagorean Theorem: $l^2 + w^2 = 40^2$.
From the first equation, $l = 56 - w$, using substitution we have:
$$(56 - w)^2 + w^2 = 40^2$$
$$3136 - 112w + w^2 + w^2 = 1600$$
$$2w^2 - 112w + 1536 = 0$$
$$w^2 - 56w + 768 = 0$$
$$(w - 32)(w - 24) = 0$$
$w = 32$ _or_ $w = 24$

Therefore, $l = 56 - 32$, or 24, and $l = 56 - 24$, or 32.
We usually express length as the greater measure. The length is 32 ft and the width is 24 ft.

17. $I = P \cdot R \cdot T$
For this problem, $T = 1$.
Brett invested P dollars at a rate of R.
$I = P \cdot R$

Erin invested $P + 240$ dollars at $\dfrac{5}{6}R$. They both had interest of $72. We have two equations:
$$72 = P \cdot R \qquad (1)$$
$$72 = (P + 240) \cdot \frac{5}{6}R \qquad (2)$$

Solve Eq. (1) for R and substitute into Eq. (2).
$$R = \frac{72}{P}$$
$$72 = (P + 240) \cdot \frac{5}{6} \cdot \frac{72}{P}$$
$$72 = (P + 240)\frac{60}{P}$$
$$72 = 60 + \frac{14,400}{P}$$
$$12 = \frac{14,400}{P}$$
$$P = \frac{14,400}{12}$$
$$P = 1200$$
$$R = \frac{72}{1200} = 0.06, \text{ or } 6\%$$
Brett invested $1200 at 6%.

18. Plot the given points. From the location of these points, we see that the ellipse which contains them is centered at $(6, 3)$.
The distance from $(1, 3)$ to $(6, 3)$ is $|6 - 1| = 5$, so $a = 5$. The distance from $(6, 6)$ to $(6, 3)$ is $|6 - 3| = 3$, so $b = 3$.
We write the equation of the ellipse.
$$\frac{(x - 6)^2}{5^2} + \frac{(y - 3)^2}{3^2} = 1$$
$$\frac{(x - 6)^2}{25} + \frac{(y - 3)^2}{9} = 1$$

19. Let $(0, y)$ be the point on the y-axis which is equidistant from $(-3, -5)$ and $(4, -7)$. We equate their distances and solve
$$\sqrt{[0 - (-3)]^2 + [y - (-5)]^2}$$
$$= \sqrt{(0 - 4)^2 + [y - (-7)]^2}$$
$$\sqrt{9 + y^2 + 10y + 25} = \sqrt{16 + y^2 + 14y + 49}$$
$$y^2 + 10y + 34 = y^2 + 14y + 65$$
$$-4y = 31$$
$$y = \frac{-31}{4}$$
The point is $\left(0, \dfrac{-31}{4}\right)$.

20. Let x and y represent the two numbers.
Sum: $x + y = 36$ (1)
Product: $xy = 4$ (2)
Solve Eq. (2) for y and substitute into Eq. (1).
$$y = \frac{4}{x}$$
$$x + \frac{4}{x} = 36$$
$$x^2 + 4 = 36x$$
$$x^2 - 36x + 4 = 0$$
We see that $a = 1$, $b = -36$, and $c = 4$.

$$x = \frac{-(-36) \pm \sqrt{(-36)^2 - 4 \cdot 1 \cdot 4}}{2 \cdot 1}$$

$$= \frac{36 \pm \sqrt{1280}}{2}$$

$$= \frac{36 \pm 16\sqrt{5}}{2}$$

$$= 18 \pm 8\sqrt{5}$$

When $x = 18 + 8\sqrt{5}$,

$$y = \frac{4}{18 + 8\sqrt{5}} \cdot \frac{18 - 8\sqrt{5}}{18 - 8\sqrt{5}}$$

$$= \frac{4(18 - 8\sqrt{5})}{324 - 320} = 18 - 8\sqrt{5}.$$

When $x = 18 - 8\sqrt{5}$,

$$y = \frac{4}{18 - 8\sqrt{5}} = 18 + 8\sqrt{5}.$$

These are the same pairs of numbers.
We want the sum of their reciprocals:

$$\frac{1}{18 + 8\sqrt{5}} + \frac{1}{18 - 8\sqrt{5}}$$

$$= \frac{1}{18 + 8\sqrt{5}} \cdot \frac{18 - 8\sqrt{5}}{18 - 8\sqrt{5}} + \frac{1}{18 - 8\sqrt{5}} \cdot \frac{18 + 8\sqrt{5}}{18 + 8\sqrt{5}}$$

$$= \frac{18 - 8\sqrt{5}}{324 - 320} + \frac{18 + 8\sqrt{5}}{324 - 320} = \frac{36}{4} = 9$$

The sum of the reciprocals is 9.

21. Let the actor be in the center at $(0,0)$. Using
    the information, we have $(-4,0)$ and $(4,0)$
    and $(0,-7)$ and $(0,7)$. Thus, $a = 4$ and
    $b = 7$. We write the equation of the ellipse.

$$\frac{x^2}{4^2} + \frac{y^2}{7^2} = 1$$

$$\frac{x^2}{16} + \frac{y^2}{49} = 1$$

Chapter 11

Sequences, Series, and the Binomial Theorem

1. f

3. d

5. c

7. $a_n = 2n - 3$
$a_8 = 2 \cdot 8 - 3 = 13$

9. $a_n = (3n + 1)(2n - 5)$
$a_9 = (3 \cdot 9 + 1)(2 \cdot 9 - 5) = 28 \cdot 13 = 364$

11. $a_n = (-1)^{n-1}(3.4n - 17.3)$
$a_{12} = (-1)^{12-1}(3.4 \cdot 12 - 17.3) = -23.5$

13. $a_n = 3n^2(9n - 100)$
$a_{11} = 3 \cdot 11^2 (9 \cdot 11 - 100)$
$= 363 \cdot (-1)$
$= -363$

15. $a_n = \left(1 + \dfrac{1}{n}\right)^2$
$a_{20} = \left(1 + \dfrac{1}{20}\right)^2 = \left(\dfrac{21}{20}\right)^2 = \dfrac{441}{400}$

17. $a_n = 2n + 3$
$a_1 = 2 \cdot 1 + 3 = 5$
$a_2 = 2 \cdot 2 + 3 = 7$
$a_3 = 2 \cdot 3 + 3 = 9$
$a_4 = 2 \cdot 4 + 3 = 11$
$a_{10} = 2 \cdot 10 + 3 = 23$
$a_{15} = 2 \cdot 15 + 3 = 33$

19. $a_n = n^2 + 2$
$a_1 = 1^2 + 2 = 3$
$a_2 = 2^2 + 2 = 6$
$a_3 = 3^2 + 2 = 11$
$a_4 = 4^2 + 2 = 18$
$a_{10} = 10^2 + 2 = 102$
$a_{15} = 15^2 + 2 = 227$

21. $a_n = \dfrac{n}{n+1}$
$a_1 = \dfrac{1}{1+1} = \dfrac{1}{2},$
$a_2 = \dfrac{2}{2+1} = \dfrac{2}{3},$
$a_3 = \dfrac{3}{3+1} = \dfrac{3}{4},$
$a_4 = \dfrac{4}{4+1} = \dfrac{4}{5};$
$a_{10} = \dfrac{10}{10+1} = \dfrac{10}{11};$
$a_{15} = \dfrac{15}{15+1} = \dfrac{15}{16}$

23. $a_n = \left(-\dfrac{1}{2}\right)^{n-1}$
$a_1 = \left(-\dfrac{1}{2}\right)^{1-1} = 1$
$a_2 = \left(-\dfrac{1}{2}\right)^{2-1} = -\dfrac{1}{2}$
$a_3 = \left(-\dfrac{1}{2}\right)^{3-1} = \dfrac{1}{4}$
$a_4 = \left(-\dfrac{1}{2}\right)^{4-1} = -\dfrac{1}{8}$
$a_{10} = \left(-\dfrac{1}{2}\right)^{10-1} = -\dfrac{1}{512}$
$a_{15} = \left(-\dfrac{1}{2}\right)^{15-1} = \dfrac{1}{16,384}$

25. $a_n = (-1)^n / n$

$a_1 = (-1)^1 / 1 = -1$

$a_2 = (-1)^2 / 2 = \dfrac{1}{2}$

$a_3 = (-1)^3 / 3 = -\dfrac{1}{3}$

$a_4 = (-1)^4 / 4 = \dfrac{1}{4}$

$a_{10} = (-1)^{10} / 10 = \dfrac{1}{10}$

$a_{15} = (-1)^{15} / 15 = -\dfrac{1}{15}$

27. $a_n = (-1)^n \left(n^3 - 1\right)$

$a_1 = (-1)^1 \left(1^3 - 1\right) = 0$

$a_2 = (-1)^2 \left(2^3 - 1\right) = 7$

$a_3 = (-1)^3 \left(3^3 - 1\right) = -26$

$a_4 = (-1)^4 \left(4^3 - 1\right) = 63$

$a_{10} = (-1)^{10} \left(10^3 - 1\right) = 999$

$a_{15} = (-1)^{15} \left(15^3 - 1\right) = -3374$

29. $-3, -1, 1, 3, 5$

31. $-1, -1, 1, 5, 11$

33. $\dfrac{1}{8}, \dfrac{4}{25}, \dfrac{1}{6}, \dfrac{8}{49}, \dfrac{5}{32}$

35. $2, 4, 6, 8, 10, \ldots$
These are even integers. $2n$

37. $1, -1, 1, -1, \ldots$
1 and -1 alternate, beginning with 1.
$(-1)^{n+1}$

39. $-1, 2, -3, 4, \ldots$
These are the first four natural numbers, but with alternating signs, beginning with a negative number. $(-1)^n \cdot n$

41. $3, 5, 7, 9, \ldots$
These are odd integers, beginning with 3.
$2n + 1$

43. $0, 3, 8, 15, 24, \ldots$
We can see a pattern if we write the sequence as
$0 \cdot 2, 1 \cdot 3, 2 \cdot 4, 3 \cdot 5, 4 \cdot 6$
$(n-1)(n+1),$ or $n^2 - 1$

45. $\dfrac{1}{2}, \dfrac{2}{3}, \dfrac{3}{4}, \dfrac{4}{5}, \dfrac{5}{6}, \ldots$
These are fractions in which the denominator is 1 greater than the numerator. Also, each numerator is 1 greater than the preceding numerator.
$\dfrac{n}{n+1}$

47. $5, 25, 125, 625, \ldots$
This sequence is powers of 5. 5^n

49. $-1, 4, -9, 16, \ldots$
This sequence is the squares of the first four natural numbers, but with alternating signs, beginning with a negative number.
$(-1)^n \cdot n^2$

51. $1, -2, 3, -4, 5, -6, \ldots;\ S_7$
$S_7 = 1 - 2 + 3 - 4 + 5 - 6 + 7 = 4$

53. $2, 4, 6, 8, \ldots;\ S_5$
$S_5 = 2 + 4 + 6 + 8 + 10 = 30$

55. $2, 3, \dfrac{11}{3}, \dfrac{25}{6}$

57. $-1, 3, -6, 10$

59. $\displaystyle\sum_{k=1}^{5} \dfrac{1}{2k} = \dfrac{1}{2 \cdot 1} + \dfrac{1}{2 \cdot 2} + \dfrac{1}{2 \cdot 3} + \dfrac{1}{2 \cdot 4} + \dfrac{1}{2 \cdot 5}$

$= \dfrac{1}{2} + \dfrac{1}{4} + \dfrac{1}{6} + \dfrac{1}{8} + \dfrac{1}{10}$

$= \dfrac{60}{120} + \dfrac{30}{120} + \dfrac{20}{120} + \dfrac{15}{120} + \dfrac{12}{120}$

$= \dfrac{137}{120}$

61. $\sum_{k=0}^{4} 10^k = 10^0 + 10^1 + 10^2 + 10^3 + 10^4$

$= 1 + 10 + 100 + 1000 + 10,000$

$= 11,111$

63. $\sum_{k=2}^{8} \frac{k}{k-1} = \frac{2}{2-1} + \frac{3}{3-1} + \frac{4}{4-1} + \frac{5}{5-1} + \frac{6}{6-1}$

$+ \frac{7}{7-1} + \frac{8}{8-1}$

$= 2 + \frac{3}{2} + \frac{4}{3} + \frac{5}{4} + \frac{6}{5} + \frac{7}{6} + \frac{8}{7}$

$= \frac{840}{420} + \frac{630}{420} + \frac{560}{420} + \frac{525}{420}$

$+ \frac{504}{420} + \frac{490}{420} + \frac{480}{42}$

$= \frac{4029}{420} = \frac{1343}{140}$

65. $\sum_{k=1}^{8} (-1)^{k+1} 2^k = (-1)^2 2^1 + (-1)^3 2^2 + (-1)^4 2^3$

$+ (-1)^5 2^4 + (-1)^6 2^5 + (-1)^7 2^6$

$+ (-1)^8 2^7 + (-1)^9 2^8$

$= 2 - 4 + 8 - 16 + 32 - 64$

$+ 128 - 256$

$= -170$

67. $\sum_{k=0}^{5} (k^2 - 2k + 3)$

$= (0^2 - 2 \cdot 0 + 3) + (1^2 - 2 \cdot 1 + 3)$

$+ (2^2 - 2 \cdot 2 + 3) + (3^2 - 2 \cdot 3 + 3)$

$+ (4^2 - 2 \cdot 4 + 3) + (5^2 - 2 \cdot 5 + 3)$

$= 3 + 2 + 3 + 6 + 11 + 18$

$= 43$

69. $\sum_{k=3}^{5} \frac{(-1)^k}{k(k+1)} = \frac{(-1)^3}{3(3+1)} + \frac{(-1)^4}{4(4+1)} + \frac{(-1)^5}{5(5+1)}$

$= \frac{-1}{3 \cdot 4} + \frac{1}{4 \cdot 5} + \frac{-1}{5 \cdot 6}$

$= -\frac{1}{12} + \frac{1}{20} - \frac{1}{30}$

$= -\frac{4}{60} = -\frac{1}{15}$

71. $\frac{2}{3} + \frac{3}{4} + \frac{4}{5} + \frac{5}{6} + \frac{6}{7}$

This is a sum of fractions in which the denominator is one greater than the numerator. Also, each numerator is 1 greater than the preceding numerator.

$\sum_{k=1}^{5} \frac{k+1}{k+2}$.

73. $1 + 4 + 9 + 16 + 25 + 36$

This is the sum of the squares of the first six natural numbers.

$\sum_{k=1}^{6} k^2$.

75. $4 - 9 + 16 - 25 + \cdots + (-1)^n n^2$

This is a sum of terms of the form $(-1)^k k^2$, beginning with $k = 2$ and continuing through $k = n$.

$\sum_{k=2}^{n} (-1)^k k^2$.

77. $5 + 10 + 15 + 20 + 25 + \ldots$

This is a sum of multiples of 5, and it is an infinite series.

$\sum_{k=1}^{\infty} 5k$.

79. $\frac{1}{1 \cdot 2} + \frac{1}{2 \cdot 3} + \frac{1}{3 \cdot 4} + \frac{1}{4 \cdot 5} + \ldots$

This is a sum of fractions in which the numerator is 1 and the denominator is a product of two consecutive integers. The greater integer in each product is the lesser integer in the succeeding product. It is an infinite series.

$\sum_{k=1}^{\infty} \frac{1}{k(k+1)}$.

81. $u = 3n + 1$

Yscl = 5

83. $u = (-1)^\wedge n(n^2)$

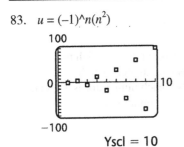

Yscl = 10

85. $u = 1/n$

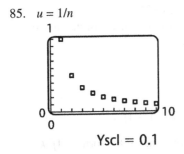

Yscl = 0.1

87. $u = (-1)^\wedge n / (n + 2)$

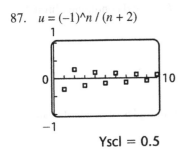

Yscl = 0.5

89. *Thinking and Writing Exercise.*

91. $\dfrac{7}{2}(a_1 + a_7) = \dfrac{7}{2}(8 + 20) = \dfrac{7}{2} \cdot 28 = 98$

93. $(a_1 + 3d) + d = a_1 + (3d + d)$
$= a_1 + 4d$

95. $(a_1 + a_n) + (a_1 + a_n) + (a_1 + a_n)$
$= (a_1 + a_1 + a_1) + (a_n + a_n + a_n)$
$= 3a_1 + 3a_n$, or $3(a_1 + a_n)$

97. *Thinking and Writing Exercise.*

99. $a_1 = 1,\ a_{n+1} = 5a_n - 2$
$a_1 = 1$
$a_2 = 5 \cdot 1 - 2 = 3$
$a_3 = 5 \cdot 3 - 2 = 13$
$a_4 = 5 \cdot 13 - 2 = 63$
$a_5 = 5 \cdot 63 - 2 = 313$
$a_6 = 5 \cdot 313 - 2 = 1563$

101. Find each term by multiplying the preceding term by 0.8:
$2500, $2000, $1600, $1280, $1024, $819.20, $655.36, $524.29, $419.43, $335.54

103. The sequence is $-1, 1, -1, 1, -1, 1, \ldots$
Therefore, $S_2 = -1 + 1 = 0$,
$S_4 = -1 + 1 + (-1) + 1 = 0$,
$S_6 = -1 + 1 + (-1) + 1 + (-1) + 1 = 0$,
and so on until
$S_{100} = -1 + 1 + \ldots + (-1) + 1 = 0$.
$S_{101} = S_{100} + (-1) = 0 - 1 = -1$

105. $a_n = i^n$
$a_1 = i^1 = i$
$a_2 = i^2 = -1$
$a_3 = i^3 = i^2 \cdot i = -1 \cdot i = -i$
$a_4 = i^4 = (i^2)^2 = (-1)^2 = 1$
$a_5 = i^5 = (i^2)^2 \cdot i = (-1)^2 \cdot i = 1 \cdot i = i$
$S_5 = i - 1 - i + 1 + i = i$

107. Enter $y_1 = x^5 - 14x^4 + 6x^3 + 416x^2$
$-655x - 1050$. Then scroll through a table of values. We see that $y_1 = 6144$ when $x = 11$, so the 11th term of the sequence is 6144.

Exercise Set 11.2

1. True

3. False

5. True

7. False

9. $2, 6, 10, 14, \ldots$

    $a_1 = 2 \quad d = 4$

11. $7, 3, -1, -5, \ldots$

    $a_1 = 7 \quad d = -4$

13. $\dfrac{3}{2}, \dfrac{9}{4}, 3, \dfrac{15}{4}, \ldots$

    $a_1 = \dfrac{3}{2} \quad d = \dfrac{3}{4}$

15. $\$5.12, \$5.24, \$5.36, \$5.48, \ldots$

    $a_1 = \$5.12 \quad d = \$0.12$

17. $7, 10, 13, \ldots$

    $a_1 = 7, \quad d = 3, \text{ and } n = 15$

    $a_n = a_1 + (n-1)d$

    $a_{15} = 7 + (15-1)3 = 7 + 14 \cdot 3 = 7 + 42 = 49$

19. $8, 2, -4, \ldots$

    $a_1 = 8, d = -6, \text{ and } n = 18$

    $a_n = a_1 + (n-1)d$

    $a_{18} = 8 + (18-1)(-6) = 8 + 17(-6)$

    $\quad\;\; = 8 - 102 = -94$

21. $\$1200, \$964.32, \$728.64, \ldots$

    $a_1 = \$1200, d = -\$235.68, \text{ and } n = 13$

    $a_n = a_1 + (n-1)d$

    $a_{13} = \$1200 + (13-1)(-\$235.68)$

    $\quad\;\; = \$1200 + 12(\$ - \$235.68)$

    $\quad\;\; = \$1200 - \$2828.16 = -\$1628.16$

23. $a_1 = 7, d = 3$

    $a_n = 7 + (n-1)3$

    Let $a_n = 82$

    $\quad 82 = 7 + (n-1)3$

    $\quad 82 = 7 + 3n - 3$

    $\quad 78 = 3n$

    $\quad 26 = n$

    26th term

25. $a_1 = 8, d = -6$

    $a_n = 8 + (n-1)(-6)$

Let $a_n = -328$

$\quad -328 = 8 - 6n + 6$

$\quad -342 = -6n$

$\quad\quad\; 57 = n$

57th term

27. $a_n = a_1 + (n-1)d$

    $a_{17} = 2 + (17-1)5$

    $\quad\;\; = 2 + 16 \cdot 5 \qquad$ Substituting 17 for n,

    $\quad\;\; = 2 + 80 \qquad\quad$ 2 for a_1, and 5 for d

    $\quad\;\; = 82$

29. $a_n = a_1 + (n-1)d$

    $33 = a_1 + (8-1)4$

    $33 = a_1 + 28 \qquad$ Substituting 33 for a_8,

    $\;\, 5 = a_1 \qquad\qquad$ 8 for n, and 4 for d

    (Note that this procedure is equivalent to subtracting d from a_8 seven times to get $a_1 : 33 - 7(4) = 33 - 28 = 5$)

31. $a_n = a_1 + (n-1)d$

    $-76 = 5 + (n-1)(-3)$

    $-76 = 5 - 3n + 3 \qquad$ Substituting

    $-76 = 8 - 3n \qquad\quad$ -76 for a_n, 5 for a_1,

    $-84 = -3n \qquad\quad\;$ and -3 for d

    $\;\; 28 = n$

33. We know that $a_{17} = -40$ and $a_{28} = -73$. We would have to add d eleven times to get from a_{17} to a_{28}. That is,

    $-40 + 11d = -73$

    $\quad\quad\;\; 11d = -33$

    $\quad\quad\quad\;\; d = -3.$

    Since $a_{17} = -40,$ we subtract d sixteen times to get to a_1.

    $a_1 = -40 - 16(-3) = -40 + 48 = 8$

    We write the first five terms of the sequence:

    $8, 5, 2, -1, -4$

35. $a_{13} = 13$ and $a_{54} = 54$

    Observe that for this to be true, $a_1 = 1$ and $d = 1.$

37. $1+5+9+13+\dots$

Note that $a_1 = 1, d = 4,$ and $n = 20.$ Before using the formula for $S_n,$ we find a_{20}:

$a_{20} = 1+(20-1)d$ Substituting 4 into
$\quad = 1+19\cdot 4$ the formula for d
$\quad = 77$

Then

$S_{20} = \dfrac{20}{2}(1+77)$
$\quad = 10(78)$ Using the formula
$\quad = 780.$ for S_n

39. The sum is $1+2+3+\dots+249+250.$ This is the sum of the arithmetic sequence for which $a_1 = 1, a_n = 250,$ and $n = 250.$ We use the formula for $S_n.$

$S_n = \dfrac{n}{2}(a_1+a_n)$

$S_{300} = \dfrac{250}{2}(1+250) = 125(251) = 31,375$

41. The sum is $2+4+6+\dots+98+100.$ This is the sum of the arithmetic sequence for which $a_1 = 2, a_n = 100,$ and $n = 50.$ Use the formula for $S_n.$

$S_n = \dfrac{n}{2}(a_1+a_n)$

$S_{50} = \dfrac{50}{2}(2+100) = 2550$

43. The sum is $6+12+18+\dots+96+102.$ This is the sum of the arithmetic sequence for which $a_1 = 6, a_n = 102,$ and $n = 17.$ We use the formula for $S_n.$

$S_n = \dfrac{n}{2}(a_1+a_n)$

$S_{17} = \dfrac{17}{2}(6+102) = \dfrac{17}{2}(108) = 918$

45. Before using the formula for $S_n,$ we find a_{20}:

$a_{20} = 4+(20-1)\cdot 5$ Substituting into
$\quad = 4+19\cdot 5 = 99$ the formula for a_n

Then

$S_{20} = \dfrac{20}{2}(4+99)$ Using the formula
$\quad = 10(103) = 1030.$ for S_n

47. We want to find the fifteenth term and the sum of an arithmetic sequence with $a_1 = 7, d = 2,$ and $n = 15.$

$a_n = a_1+(n-1)d$

$a_{15} = 7+(15-1)2 = 7+14\cdot 2 = 35$

$S_n = \dfrac{n}{2}(a_1+a_2)$

$S_{15} = \dfrac{15}{2}(7+35) = \dfrac{15}{2}\cdot 42 = 315$

35 musicians in the last row, and a total of 315 musicians.

49. We have an arithmetic sequence $36,32,28,\dots,8,4.$ $a_1 = 36,$ $d = -4,$ and $n = 9.$

$S_9 = \dfrac{9}{2}(36+4) = \dfrac{9}{2}(40) = 180$

There are 180 stones.

51. We have an arithmetic sequence $10,20,30,\dots,300,310$

$a_1 = 10, d = 10,$ and $n = 31$

$S_{31} = \dfrac{31}{2}(310+10) = \dfrac{31}{2}\cdot 320 = 4960$

There will be 4960¢, or $49.60.

53. We have the arithmetic sequence $20,22,24,\dots a_{19}$

Determine $a_{19}.$

$a_1 = 20, d = 2, n = 19$

$a_{19} = 20+(19-1)2 = 20+18\cdot 2 = 20+36$
$\quad = 56$

$S_{19} = \dfrac{19}{2}(20+56) = \dfrac{19}{2}(76) = 722$

There are 722 seats.

55. *Thinking and Writing Exercise.*

57. $m = \dfrac{1}{3}$ and $b = 10,$ so $y = \dfrac{1}{3}x+10$

59. $2x + y = 8$ is the same as $y = -2x + 8$, so
$m = -2$.
$$y - y_1 = m(x - x_1)$$
$$y - 0 = -2(x - 5)$$
$$y = -2x + 10$$

61. $$(x - h)^2 + (y - k)^2 = r^2$$
$$(x - 0)^2 + (y - 0)^2 = 4^2$$
$$x^2 + y^2 = 16$$

63. *Thinking and Writing Exercise.*

65. The frog is at the bottom of a 100 foot well, so $a_n = 100$. The frog climbs 4 ft. but slips down 1 ft. with each jump, so $d = 4 - 1$, or 3, and $a_1 = 3$
$$a_n = a_1 + (n - 1) \cdot 3$$
$$100 = 3 + (n - 1) \cdot 3$$
$$100 = 3 + 3n - 3$$
$$100 = 3n$$
$$33\tfrac{1}{3} = n$$
Since on the 33rd jump, the frog would be out of the well, he would not slip; thus, we disregard the $\frac{1}{3}$.
It takes the frog 33 jumps.

67. $a_1 = \$8760$
$a_2 = \$8760 + (-\$798.23) = \$7961.77$
$a_3 = \$8760 + 2(-\$798.23) = \$7163.54$
$a_4 = \$8760 + 3(-\$798.23) = \$6365.31$
$a_5 = \$8760 + 4(-\$798.23) = \$5567.08$
$a_6 = \$8760 + 5(-\$798.23) = \$4768.85$
$a_7 = \$8760 + 6(-\$798.23) = \$3970.62$
$a_8 = \$8760 + 7(-\$798.23) = \$3172.39$
$a_9 = \$8760 + 8(-\$798.23) = \$2374.16$
$a_{10} = \$8760 + 9(-\$798.23) = \$1575.93$

69. Let d = the common difference. Since $p, m,$ and q form an arithmetic sequence, $m = p + d$ and $q = p + 2d$. Then
$$\frac{p + q}{2} = \frac{p + (p + 2d)}{2} = p + d = m$$

71. Each integer from 501 through 750 is 500 more than the corresponding integer from 1 through 250. There are 250 integers from 501 through 750, so their sum is the sum of the integers from 1 to 250 plus $250 \cdot 500$. From Exercise 39, we know that the sum of the integers from 1 through 250 is 31,375. Thus, we have
$$31,375 + 250 \cdot 500, \text{ or } 156,375.$$

73. We graph the data points.

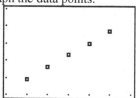

The points appear to be linear, so this could be the graph of an arithmetic sequence. The general term is $a_n = 0.7n + 68.3$, where $n = 1$ corresponds to a ball speed of 100 mph, $n = 2$ corresponds to a ball speed of 101 mph, and so on.

75. We graph the data points, and they do not appear to be linear. This is not the graph of an arithmetic sequence.

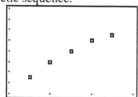

Exercise Set 11.3

1. Geometric sequence

3. Arithmetic sequence

5. Geometric series

7. Geometric series

9. $10, 20, 40, 80, \ldots$
$$\frac{20}{10} = 2, \frac{40}{20} = 2, \frac{80}{40} = 2$$
$$r = 2$$

11. $6, -0.6, 0.06, -0.006, \ldots$

$$\frac{-0.6}{6} = -0.1, \frac{0.06}{-0.6} = -0.1, \frac{-0.006}{0.06} = -0.1$$

$$r = -0.1$$

13. $\dfrac{1}{2}, -\dfrac{1}{4}, \dfrac{1}{8}, -\dfrac{1}{16}, \ldots$

$$\frac{\frac{-1}{4}}{\frac{1}{2}} = -\frac{1}{4} \cdot \frac{2}{1} = -\frac{1}{2}$$

$$\frac{\frac{1}{8}}{\frac{-1}{4}} = \frac{1}{8} \cdot \frac{-4}{1} = -\frac{1}{2}$$

$$\frac{\frac{-1}{16}}{\frac{1}{8}} = -\frac{1}{16} \cdot \frac{8}{1} = -\frac{1}{2}$$

$$r = -\frac{1}{2}$$

15. $75, 15, 3, \dfrac{3}{5}, \ldots$

$$\frac{15}{75} = \frac{1}{5}, \frac{3}{15} = \frac{1}{5}, \frac{\frac{3}{5}}{3} = \frac{1}{5}$$

$$r = \frac{1}{5}$$

17. $\dfrac{1}{m}, \dfrac{6}{m^2}, \dfrac{36}{m^3}, \dfrac{216}{m^4}$

$$\frac{\frac{6}{m^2}}{\frac{1}{m}} = \frac{6}{m^2} \cdot \frac{m}{1} = \frac{6}{m}$$

$$\frac{\frac{36}{m^3}}{\frac{6}{m^2}} = \frac{36}{m^3} \cdot \frac{m^2}{6} = \frac{6}{m}$$

$$\frac{\frac{216}{m^4}}{\frac{36}{m^3}} = \frac{216}{m^4} \cdot \frac{m^3}{36} = \frac{6}{m}$$

$$r = \frac{6}{m}$$

19. $3, 6, 12, \ldots$

$a_1 = 3, n = 7$, and $r = \dfrac{6}{3} = 2$

We use the formula $a_n = a_1 r^{n-1}$.

$a_7 = 3 \cdot 2^{7-1} = 3 \cdot 2^6 = 3 \cdot 64 = 192$

21. $\sqrt{3}, 3, 3\sqrt{3}, \ldots$

$a_1 = \sqrt{3}, n = 10, r = \dfrac{3}{\sqrt{3}} = \sqrt{3}$

$a_n = a_1 r^{n-1}$

$a_{10} = \sqrt{3} \cdot \sqrt{3}^{10-1} = \sqrt{3} \cdot \sqrt{3}^9 = \sqrt{3} \cdot 3^4 \left(\sqrt{3}\right)$

$a_{10} = 3^5 = 243$

23. $-\dfrac{8}{243}, \dfrac{8}{81}, -\dfrac{8}{27}, \ldots$

$a_1 = -\dfrac{8}{243}, n = 14$ and $r = \dfrac{\frac{8}{81}}{-\frac{8}{243}}$

$$= \frac{8}{81}\left(-\frac{243}{8}\right) = -3$$

$a_n = a_1 r^{n-1}$

$a_{14} = -\dfrac{8}{243} \cdot (-3)^{14-1}$

$$= -\frac{8}{243} \cdot (-3)^{13}$$

$$= -\frac{8}{243} \cdot -1,594,323$$

$$= 52,488$$

25. $a_1 = \$1000, n = 12$, and $r = \dfrac{\$1080}{\$1000} = 1.08$

$a_n = a_1 r^{n-1}$

$a_{12} = \$1000(1.08)^{12-1}$

$$\approx \$1000(2.331638997)$$

$$\approx \$2331.64$$

27. $1, 5, 25, 125, \ldots$

$a_1 = 1; \ r = \dfrac{5}{1} = 5$

$a_n = a_1 r^{n-1}$

$a_n = 1 \cdot 5^{n-1} = 5^{n-1}$

29. $1, -1, 1, -1, \ldots$

$a_1 = 1; \ r = \dfrac{-1}{1} = -1$

$a_n = a_1 r^{n-1}$

$a_n = 1 \cdot (-1)^{n-1} = (-1)^{n-1}$

31. $\dfrac{1}{x}, \dfrac{1}{x^2}, \dfrac{1}{x^3}, \ldots$

$a_1 = \dfrac{1}{x}; \ r = \dfrac{\frac{1}{x^2}}{\frac{1}{x}} = \dfrac{1}{x^2} \cdot \dfrac{x}{1} = \dfrac{1}{x}$

$a_n = a_1 r^{n-1}$

$a_n = \dfrac{1}{x} \cdot \left(\dfrac{1}{x}\right)^{n-1} = \left(\dfrac{1}{x}\right)^{1+n-1} = \left(\dfrac{1}{x}\right)^{n}$

$a_n = \left(\dfrac{1}{x}\right)^{n}$, or $a_n = x^{-n}$

33. $6 + 12 + 24 + \ldots$

$a_1 = 6, \ n = 9, \ r = \dfrac{12}{6} = 2$

$S_n = \dfrac{a_1\left(1 - r^n\right)}{1 - r}$

$S_9 = \dfrac{6\left(1 - 2^9\right)}{1 - 2} = \dfrac{6(-511)}{-1} = \dfrac{-3066}{-1}$

$S_9 = 3066$

35. $\dfrac{1}{18} - \dfrac{1}{6} + \dfrac{1}{2} - \ldots$

$a_1 = \dfrac{1}{18}, \ n = 7, \ \text{and} \ r = \dfrac{-\frac{1}{6}}{\frac{1}{18}} = -\dfrac{1}{6} \cdot \dfrac{18}{1} = -3$

$S_n = \dfrac{a_1\left(1 - r^n\right)}{1 - r}$

$S_7 = \dfrac{\frac{1}{18}\left[1 - (-3)^7\right]}{1 - (-3)} = \dfrac{\frac{1}{18}(1 + 2187)}{4} = \dfrac{\frac{1}{18}(2188)}{4}$

$= \dfrac{1}{18}(2188)\left(\dfrac{1}{4}\right) = \dfrac{547}{18}$

37. $1 + x + x^2 + x^3 + \ldots$

$a_1 = 1, \ n = 8, \ \text{and} \ r = \dfrac{x}{1}, \ \text{or} \ x$

$S_n = \dfrac{a_1\left(1 - r^n\right)}{1 - r}$

$S_8 = \dfrac{1\left(1 - x^8\right)}{1 - x} = \dfrac{\left(1 + x^4\right)\left(1 - x^4\right)}{1 - x}$

$= \dfrac{\left(1 + x^4\right)\left(1 + x^2\right)\left(1 - x^2\right)}{1 - x}$

$= \dfrac{\left(1 + x^4\right)\left(1 + x^2\right)\left(1 + x\right)\left(1 - x\right)}{1 - x}$

$= \left(1 + x^4\right)\left(1 + x^2\right)\left(1 + x\right)$

39. $\$200, \$200(1.06), \$200(1.06)^2, \ldots$

$a_1 = \$200, \ n = 16, \ \text{and} \ r = \dfrac{\$200(1.06)}{\$200}$

$= 1.06$

$S_n = \dfrac{a_1\left(1 - r^n\right)}{1 - r}$

$S_{16} = \dfrac{\$200\left[1 - (1.06)^{16}\right]}{1 - 1.06}$

$\approx \dfrac{\$200(1 - 2.540351685)}{-0.06}$

$\approx \$5134.51$

41. $18 + 6 + 2 + \ldots$

$|r| = \left|\dfrac{6}{18}\right| = \left|\dfrac{1}{3}\right| = \dfrac{1}{3}, \ \text{and since} \ |r| < 1, \ \text{the}$

series does have a limit.

$S_\infty = \dfrac{a_1}{1 - r} = \dfrac{18}{1 - \frac{1}{3}} = \dfrac{18}{\frac{2}{3}} = 18 \cdot \dfrac{3}{2} = 27$

43. $7 + 3 + \dfrac{9}{7} + \ldots$

$|r| = \left|\dfrac{3}{7}\right| = \dfrac{3}{7}, \ \text{and since} \ |r| < 1, \ \text{the series}$

does have a limit.

$S_\infty = \dfrac{a_1}{1 - r} = \dfrac{7}{1 - \frac{3}{7}} = \dfrac{7}{\frac{4}{7}} = 7 \cdot \dfrac{7}{4} = \dfrac{49}{4}$

45. $3 + 15 + 75 + \ldots$

$|r| = \left|\dfrac{15}{3}\right| = |5| = 5, \ \text{and since} \ |r| \nless 1 \ \text{the}$

series does not have a limit.

47. $4 - 6 + 9 - \dfrac{27}{2} + \ldots$

$|r| = \left| \dfrac{-6}{4} \right| = \left| -\dfrac{3}{2} \right| = \dfrac{3}{2}$, and since $|r| \not< 1$ the

series does not have a limit.

49. $0.43 + 0.0043 + 0.000043 + \ldots$

$|r| = \left| \dfrac{0.0043}{0.43} \right| = |0.01| = 0.01$, and since

$|r| < 1$, the series does have a limit.

$S_\infty = \dfrac{a_1}{1 - r} = \dfrac{0.43}{1 - 0.01} = \dfrac{0.43}{0.99} = \dfrac{43}{99}$

51. $\$500(1.02)^{-1} + \$500(1.02)^{-2}$

$\qquad + \$500(1.02)^{-3} + \ldots$

$|r| = \left| \dfrac{\$500(1.02)^{-2}}{\$500(1.02)^{-1}} \right| = |(1.02)^{-1}| = (1.02)^{-1},$

or $\dfrac{1}{1.02}$, and since $|r| < 1$, the series does

have a limit.

$S_\infty = \dfrac{a_1}{1 - r} = \dfrac{\$500(1.02)^{-1}}{1 - \left(\frac{1}{1.02}\right)} = \dfrac{\frac{\$500}{1.02}}{\frac{0.02}{1.02}}$

$\quad = \dfrac{\$500}{1.02} \cdot \dfrac{1.02}{0.02} = \$25,000$

53. $0.7777\ldots = 0.7 + 0.07 + 0.007 + 0.0007 + \ldots$

This is an infinite geometric series with

$a_1 = 0.7$.

$|r| = \left| \dfrac{0.07}{0.7} \right| = |0.1| = 0.1 < 1$, so the series has

a sum.

$S_\infty = \dfrac{a_1}{1 - r} = \dfrac{0.7}{1 - 0.1} = \dfrac{0.7}{0.9} = \dfrac{7}{9}$

Fractional notation for $0.7777\ldots$ is $\dfrac{7}{9}$.

55. $8.3838\ldots = 8.3 + 0.083 + 0.00083 + \ldots$

This is an infinite geometric series with

$a_1 = 8.3$.

$|r| = \left| \dfrac{0.083}{8.3} \right| = |0.01| = 0.01 < 1$, so the series

has a sum.

$S_\infty = \dfrac{a_1}{1 - r} = \dfrac{8.3}{1 - 0.01} = \dfrac{8.3}{0.99} = \dfrac{830}{99}$

Fractional notation for $8.3838\ldots$ is $\dfrac{830}{99}$.

57. $0.15151515\ldots = 0.15 + 0.0015$

$\qquad\qquad\qquad\quad + 0.000015 + \ldots$

This is an infinite geometric series with

$a_1 = 0.15$.

$|r| = \left| \dfrac{0.0015}{0.15} \right| = |0.01| = 0.01 < 1$, so the

series has a sum.

$S_\infty = \dfrac{a_1}{1 - r} = \dfrac{0.15}{1 - 0.01} = \dfrac{0.15}{0.99} = \dfrac{15}{99} = \dfrac{5}{33}$

Fractional notation for $0.15151515\ldots$ is $\dfrac{5}{33}$.

59. The rebound distances form a geometric

sequence:

$\dfrac{1}{4} \times 20, \left(\dfrac{1}{4}\right)^2 \times 20, \left(\dfrac{1}{4}\right)^3 \times 20, \ldots,$ or,

$5, \dfrac{1}{4} \times 5, \left(\dfrac{1}{4}\right)^2 \times 5, \ldots$

The height of the 6th rebound is the 6th term

of the sequence.

$a_n = a_n r^{n-1}$, with

$a_1 = 5, \quad r = \dfrac{1}{4}, \quad n = 6:$

$a_6 = 5\left(\dfrac{1}{4}\right)^{6-1}$

$a_6 = 5\left(\dfrac{1}{4}\right)^5 = \dfrac{5}{1024}$

It will rebound $\dfrac{5}{1024}$ ft the 6th time.

61. In one year, the population will be $100,000 +$

$0.03(100,000)$, or $(1.03)100,000$. In two

years, the population will be $(1.03)100,000 +$

$0.03(1.03)100,000$, or $(1.03)^2 100,000$.

Thus the populations form a geometric

sequence:

$100,000, (1.03)100,000, (1.03)^2 100,000, \ldots$

The population in 15 years will be the 16th

term of the sequence:

$a_n = a_1 r^{n-1}$

$a_1 = 100,000, \quad r = 1.03, \quad n = 16:$

$a_{16} = 100,000(1.03)^{16-1}$

$a_{16} \approx 155,797.$

In 15 years, the population will be about 155,797.

63. We have a geometric sequence:

$5000, 5000(0.96), 5000(0.96)^2,\ldots$

The number of fruit flies remaining alive after 15 minutes is given by the 16th term of the sequence.

$a_n = a_1 r^{n-1}$

$a_{16} = 5000 \cdot (0.96)^{16-1}$

$a_{16} \approx 2710$

There will be approx. 2710 fruit flies remaining after 15 minutes.

65. We have a geometric sequence.

$17; 17(1.04); 17(1.04)^2 \ldots 17(1.04)^{n-1}$

where n is the number of years after 2006.
$a_1 = 17, \quad r = 1.04,$ and $n = 9$ for 2015.

$S_n = \dfrac{a_1\left(1-r^n\right)}{1-r}$

$S_9 = \dfrac{17\left(1-1.04^9\right)}{1-1.04}$

$S_9 \approx 179.9$

Approximately 179.9 billion espresso-based coffees will be sold from 2007 through 2015.

67. The lengths of the falls form a geometric sequence:

$556, \left(\dfrac{3}{4}\right)556, \left(\dfrac{3}{4}\right)^2 556, \left(\dfrac{3}{4}\right)^3 556,\ldots$

The total length of the first 6 falls is the sum of the first six terms of this sequence. The heights of the rebounds also form a geometric sequence:

$\left(\dfrac{3}{4}\right)556, \left(\dfrac{3}{4}\right)^2 556, \left(\dfrac{3}{4}\right)^3 556,\ldots,$ or

$417, \left(\dfrac{3}{4}\right)417, \left(\dfrac{3}{4}\right)^2 417,\ldots$

When the ball hits the ground for the 6th time, it will have rebounded 5 times. Thus the

total length of the rebounds is the sum of the first five terms of this sequence.

We use the formula $S_n = \dfrac{a_1\left(1-r^n\right)}{1-r}$ twice,

once with $a_1 = 556, r = \frac{3}{4},$ and $n = 6$ and a second time with $a_1 = 417, r = \frac{3}{4},$ and $n = 5.$

$D = $ Length of falls $+$ length of rebounds

$= \dfrac{556\left[1-\left(\frac{3}{4}\right)^6\right]}{1-\frac{3}{4}} + \dfrac{417\left[1-\left(\frac{3}{4}\right)^5\right]}{1-\frac{3}{4}}.$

We use a calculator to obtain $D \approx 3100.35.$
The ball will have traveled about 3100.35 ft.

69. The heights of the stack form a geometric sequence:

$0.02, 0.02(2), 0.02\left(2^2\right),\ldots$

The height of the stack after it is doubled 10 times is given by the 11th term of this sequence.

$a_1 = 0.02, \quad r = 2, \quad n = 11.$

$a_n = a_1 r^{n-1}.$

$a_{11} = 0.02\left(2^{11-1}\right)$

$a_{11} = 0.02(1024) = 20.48$

The final stack will be 20.48 in. high.

71. The points lie on a straight line, so this is the graph of an arithmetic sequence.

73. The points lie on the graph of an exponential function, so this is the graph of a geometric sequence.

75. The points lie on the graph of an exponential function, so this is the graph of a geometric sequence.

77. *Thinking and Writing Exercise.*

79. $(x+y)^2 = (x+y)(x+y)$
$= x^2 + xy + xy + y^2$
$= x^2 + 2xy + y^2$

81. $(x-y)^3 = (x-y)(x-y)(x-y)$
$= (x^2 - 2xy + y^2)(x-y)$
$= x^3 - 2x^2 y + xy^2 - x^2 y + 2xy^2 - y^3$
$= x^3 - 3x^2 y + 3xy^2 - y^3$

83. $(2x+y)^3 = (2x+y)(2x+y)(2x+y)$

$$= (4x^2+4xy+y^2)(2x+y)$$

$$= 8x^3+8x^2y+2xy^2+4x^2y$$
$$\qquad\qquad +4xy^2+y^3$$

$$= 8x^3+12x^2y+6xy^2+y^3$$

85. *Thinking and Writing Exercise*

87. $\displaystyle\sum_{k=1}^{\infty} 6(0.9)^k$

$$a_1 = 6(0.9)^1 = 5.4$$

$$r = 0.9$$

$$S_\infty = \frac{a_1}{1-r} = \frac{5.4}{1-0.9} = \frac{5.4}{0.1} = 54$$

89. $x^2 - x^3 + x^4 - x^5 + \dots$

This is a geometric series with $a_1 = x^2$ and

$r = -x$.

$$S_n = \frac{a_1\left(1-r^n\right)}{1-r} = \frac{x^2\left[1-(-x)^n\right]}{1-(-x)}$$

$$= \frac{x^2\left[1-(-x)^n\right]}{1+x}$$

91. The length of a side of the first square is 16 cm. The length of a side of the next square is the length of the hypotenuse of a right triangle with legs 8 cm and 8 cm, or $8\sqrt{2}$ cm. The length of a side of the next square is the length of the hypotenuse of a right triangle with legs $4\sqrt{2}$ cm and $4\sqrt{2}$ cm, or 8 cm. The areas of the squares form a sequence:

$(16)^2, \left(8\sqrt{2}\right)^2, (8)^2, \dots,$ or

$256, 128, 64, \dots$

This is a geometric sequence with $a_1 = 256$ and $r = \frac{1}{2}$. We find the sum of the infinite geometric series $256+128+64+\dots$

$$S_\infty = \frac{256}{1-\frac{1}{2}} = \frac{256}{\frac{1}{2}} = 512 \text{ cm}^2$$

Mid-Chapter Review

Guided Solutions

1. $-6, -1, 4, 9, \dots$

$$a_n = a_1 + (n-1)d$$

$$n = 14, a_1 = -6, d = 5$$

$$a_{14} = -6 + (14-1)5$$

$$a_{14} = 59$$

2. $\dfrac{1}{9}, -\dfrac{1}{3}, 1, -3, \dots$

$$a_n = a_1 r^{n-1}$$

$$n = 7, a_1 = \frac{1}{9}, r = -3$$

$$a_7 = \frac{1}{9} \cdot (-3)^{7-1}$$

$$a_7 = 81$$

Mixed Review

1. $a_n = n^2 - 5n$

$$a_{20} = 20^2 - 5 \cdot 20$$

$$= 400 - 100$$

$$= 300$$

2. $\dfrac{1}{2}, \dfrac{1}{3}, \dfrac{1}{4}, \dfrac{1}{5}, \dots$

These are fractions in which each denominator is 1 greater than the preceding denominator, and the numerator is always 1.

$$\frac{1}{n+1}$$

3. $1, 2, 3, 4, \dots; \ S_{12}$

$$S_{12} = 1+2+3+4+5+6+7+$$
$$8+9+10+11+12 = 78$$

4. $\displaystyle\sum_{k=2}^{5} k^2 = 2^2 + 3^2 + 4^2 + 5^2$

$$= 4+9+16+25$$

$$= 54$$

5. $1-2+3-4+5-6$

This is a sum of terms of the form $(-1)^{k+1}k$, beginning with $k=1$ and continuing through $k=6$.

$$\sum_{k=1}^{6}(-1)^{k+1}k$$

6. $115, 112, 109, 106, \ldots$

$112 - 115 = -3$, $109 - 112 = -3$, $106 - 109 = -3$

$d = -3$

7. $10, 15, 20, 25, \ldots$

$a_1 = 10$, $d = 5$, and $n = 21$

$a_n = a_1 + (n-1)d$

$a_{21} = 10 + (21-1)5 = 10 + 20 \cdot 5 = 110$

8. $a_1 = 10$, $d = 0.2$

$a_n = 10 + (n-1)0.2$

Let $a_n = 22$

$22 = 10 + (n-1)0.2$

$22 = 10 + 0.2n - 0.2$

$12.2 = 0.2n$

$61 = n$

61st term

9. $a_n = a_1 + (n-1)d$

$a_1 = 9$, $d = -2$, and $n = 25$

$a_{25} = 9 + (25-1)(-2)$

$a_{25} = 9 + (24)(-2)$

$a_{25} = 9 - 48 = -39$

10. $a_n = a_1 + (n-1)d$

$a_5 = 65$, $d = 11$, and $n = 5$

$65 = a_1 + (5-1)11$

$65 = a_1 + (4)11$

$21 = a_1$

11. $a_n = a_1 + (n-1)d$

$0 = 5 + (n-1)\left(-\frac{1}{2}\right)$

$0 = 5 - \frac{1}{2}n + \frac{1}{2}$

$-\frac{11}{2} = -\frac{1}{2}n$

$11 = n$

Substituting 0 for a_n, 5 for a_1, and $-\frac{1}{2}$ for d

12. $2 + 12 + 22 + 32 + \ldots$

$a_1 = 2$, $d = 10$, and $n = 30$

$a_{30} = 2 + (30-1)(10)$

$= 2 + 29(10)$

$= 2 + 290 = 292$

$S_{30} = \frac{30}{2}[2 + 292]$

$= 15(294)$

$= 4410$

13. $\frac{1}{3}, -\frac{1}{6}, \frac{1}{12}, -\frac{1}{24}, \ldots$

$\frac{-\frac{1}{6}}{\frac{1}{3}} = -\frac{1}{2}$, $\frac{\frac{1}{12}}{-\frac{1}{6}} = -\frac{1}{2}$, $\frac{-\frac{1}{24}}{\frac{1}{12}} = -\frac{1}{2}$

$r = -\frac{1}{2}$

14. $5, 10, 20, 40, \ldots$

$a_1 = 5$, $n = 8$, and $r = \frac{10}{5} = 2$

We use the formula $a_n = a_1 r^{n-1}$.

$a_8 = 5 \cdot 2^{8-1} = 5 \cdot 2^7 = 5 \cdot 128 = 640$

15. $2, -2, 2, -2, \ldots$

$a_1 = 2$; $r = \frac{-2}{2} = -1$

$a_n = a_1 r^{n-1}$

$a_n = 2(-1)^{n-1}$ or $a_n = 2(-1)^{n+1}$

16. $\$100 + \$100(1.03) + \$100(1.03)^2 + \ldots$

$a_1 = \$100$, $n = 10$, $r = \frac{\$100(1.03)}{\$100} = 1.03$

$S_n = \frac{a_1(1-r^n)}{1-r}$

$S_{10} = \frac{100(1-(1.03)^{10})}{1-1.03}$

$\approx \$1146.39$

17. $0.9 + 0.09 + 0.009 + \ldots$

$|r| = \left|\frac{0.09}{0.9}\right| = |0.1| = 0.1 < 1$, so the series has a limit.

$$S_\infty = \frac{0.9}{1-0.1} = \frac{0.9}{0.9} = 1$$

18. $0.9 + 9 + 90 + \ldots$

$|r| = \left|\frac{9}{0.9}\right| = |10| = 10,$ and since $|r| \not< 1$ the

series does not have a limit.

19. We have the arithmetic sequence
$\$1, \$2, \$3, \$4 \ldots a_{30}$

Determine a_{30}

$a_1 = 1, d = 1, n = 30$

$a_{30} = 1 + (30-1)1 = 1 + 29 = 30$

$S_{30} = \frac{30}{2}(1+30) = 15(31)$

$\qquad = 465$

Renata earns \$465 in June.

20. We have a geometric sequence.
$\$1, \$2, \$4, \$8, \ldots a_{30}$

$a_1 = 1, \ r = 2,$ and $n = 30$

$S_n = \frac{a_1\left(1 - r^n\right)}{1 - r}$

$S_{30} = \frac{1\left(1 - 2^{30}\right)}{1 - 2}$

$S_{30} = 1,073,741,823$

Dwight earns \$1,073,741,823 in June.

Exercise Set 11.4

1. 2^5, or 32

3. 9

5. $\begin{pmatrix} 8 \\ 5 \end{pmatrix}$, or $\begin{pmatrix} 8 \\ 3 \end{pmatrix}$

7. $x^7 y^2$

9. $4! = 4 \cdot 3 \cdot 2 \cdot 1 = 24$

11. $11! = 11 \cdot 10 \cdot 9 \cdot 8 \cdot 7 \cdot 6 \cdot 5 \cdot 4 \cdot 3 \cdot 2 \cdot 1$
$\qquad = 39,916,800$

13. $\dfrac{8!}{6!} = \dfrac{8 \cdot 7 \cdot 6!}{6!} = 8 \cdot 7 = 56$

15. $\dfrac{9!}{4!5!} = \dfrac{9 \cdot 8 \cdot 7 \cdot 6 \cdot 5!}{4 \cdot 3 \cdot 2 \cdot 5!} = \dfrac{9 \cdot 8 \cdot 7 \cdot 6}{4 \cdot 3 \cdot 2} = 126$

17. $\begin{pmatrix} 7 \\ 4 \end{pmatrix} = \dfrac{7!}{3!4!} = \dfrac{7 \cdot 6 \cdot 5 \cdot 4!}{3!4!} = \dfrac{7 \cdot 6 \cdot 5}{3 \cdot 2} = 35$

19. $\begin{pmatrix} 9 \\ 9 \end{pmatrix} = \dfrac{9!}{9!0!} = \dfrac{1}{0!} = 1$

21. $\begin{pmatrix} 30 \\ 2 \end{pmatrix} = \dfrac{30!}{28!2!} = \dfrac{30 \cdot 29 \cdot 28!}{28!2!}$

$\qquad = \dfrac{30 \cdot 29}{2}$

$\qquad = 435$

23. $\begin{pmatrix} 40 \\ 38 \end{pmatrix} = \dfrac{40!}{2!38!} = \dfrac{40 \cdot 39 \cdot 38!}{2!38!} = \dfrac{40 \cdot 39}{2}$

$\qquad = 780$

25. Expand $(a-b)^4$

Form 1: The expansion of $(a-b)^4$ has $4+1$,
or 5 terms. The sum of the exponents in each
term is 4. The exponents of a begin with 4
and decrease to 0. The exponents of $-b$ begin
with 0 and increase to 4. We get the
coefficients from the 5th row of Pascal's
triangle.

1   4   6   4   1

$(a-b)^4 = 1 \cdot a^4 + 4a^3(-b) + 6a^2(-b)^2$

$\qquad + 4a(-b)^3 + (-b)^4$

$\qquad = a^4 - 4a^3b + 6a^2b^2 - 4ab^3 + b^4$

Form 2: We have $a = a, b = -b, n = 4$

$(a-b)^4 = \begin{pmatrix} 4 \\ 0 \end{pmatrix}a^4 + \begin{pmatrix} 4 \\ 1 \end{pmatrix}a^3(-b) + \begin{pmatrix} 4 \\ 2 \end{pmatrix}a^2(-b)^2$

$\qquad + \begin{pmatrix} 4 \\ 3 \end{pmatrix}a(-b)^3 + \begin{pmatrix} 4 \\ 4 \end{pmatrix}(-b)^4$

$\qquad = \dfrac{4!}{4!0!}a^4 + \dfrac{4!}{3!1!}a^3(-b)$

$\qquad + \dfrac{4!}{2!2!}a^2(-b)^2 + \dfrac{4!}{1!3!}a(-b)^3$

$\qquad + \dfrac{4!}{0!4!}(-b)^4$

$\qquad = a^4 - 4a^3b + 6a^2b^2 - 4ab^3 + b^4$

27. Expand $(p+q)^7$.

Form 1: We use the 8th row of Pascal's Triangle.

1   7   21   35   35   21   7   1

$(p+q)^7 = p^7 + 7p^6q + 21p^5q^2 + 35p^4q^3$
$\qquad + 35p^3q^4 + 21p^2q^5 + 7pq^6 + q^7$

Form 2: We have $a = p, b = q, n = 7$

$(p+q)^7 = \binom{7}{0}p^7 + \binom{7}{1}p^6q + \binom{7}{2}p^5q^2$

$\qquad + \binom{7}{3}p^4q^3 + \binom{7}{4}p^3q^4$

$\qquad + \binom{7}{5}p^2q^5 + \binom{7}{6}pq^6 + \binom{7}{7}q^7$

$\qquad = p^7 + 7p^6q + 21p^5q^2 + 35p^4q^3$
$\qquad + 35p^3q^4 + 21p^2q^5 + 7pq^6 + q^7$

29. Expand $(3c-d)^7$.

Form 1: We use the eighth row of Pascal's triangle.

1   7   21   35   35   21   7   1

$(3c-d)^7 = 1 \cdot (3c)^7 + 7(3c)^6(-d)$
$\qquad + 21(3c)^5(-d)^2 + 35(3c)^4(-d)^3$
$\qquad + 35(3c)^3(-d)^4 + 21(3c)^2(-d)^5$
$\qquad + 7(3c)(-d)^6 + 1 \cdot (-d)^7$
$\qquad = 2187c^7 - 5103c^6d + 5103c^5d^2$
$\qquad - 2835c^4d^3 + 945c^3d^4$
$\qquad - 189c^2d^5 + 21cd^6 - d^7$

Form 2: We have $a = 3c, b = -d, n = 7$

$(3c-d)^7 = \binom{7}{0}(3c)^7 + \binom{7}{1}(3c)^6(-d)$

$\qquad + \binom{7}{2}(3c)^5(-d)^2$

$\qquad + \binom{7}{3}(3c)^4(-d)^3$

$\qquad + \binom{7}{4}(3c)^3(-d)^4$

$\qquad + \binom{7}{5}(3c)^2(-d)^5$

$\qquad + \binom{7}{6}(3c)(-d)^6 + \binom{7}{7}(-d)^7$

$= 2187c^7 - 5103c^6d + 5103c^5d^2$
$\qquad - 2835c^4d^3 + 945c^3d^4$
$\qquad - 189c^2d^5 + 21cd^6 - d^7$

31. Expand $(t^{-2}+2)^6$.

Form 1: We use the 7th row of Pascal's triangle.

1   6   15   20   15   6   1

$(t^{-2}+2)^6 = 1 \cdot (t^{-2})^6 + 6(t^{-2})^5(2)$

$\qquad + 15(t^{-2})^4(2^2) + 20(t^{-2})^3(2^3)$

$\qquad + 15(t^{-2})^2(2^4)$

$\qquad + 6t^{-2}(2^5) + 1 \cdot 2^6$

$\qquad = t^{-12} + 12t^{-10} + 60t^{-8} + 160t^{-6}$

$\qquad + 240t^{-4} + 192t^{-2} + 64$

Form 2: $a = t^{-2}, b = 2, n = 6$

$(t^{-2}+2)^6 = \binom{6}{0}(t^{-2})^6 + \binom{6}{1}(t^{-2})^5(2)$

$\qquad + \binom{6}{2}(t^{-2})^4(2^2)$

$\qquad + \binom{6}{3}(t^{-2})^3(2^3)$

$\qquad + \binom{6}{4}(t^{-2})^2(2^4)$

$\qquad + \binom{6}{5}(t^{-2})(2^5) + \binom{6}{6}2^6$

$\qquad = t^{-12} + 12t^{-10} + 60t^{-8} + 160t^{-6}$

$\qquad + 240t^{-4} + 192t^{-2} + 64$

33. Expand $(x-y)^5$.

Form 1: We use the 6th row of Pascal's triangle.

1   5   10   10   5   1

$(x-y)^5 = 1 \cdot x^5 + 5x^4(-y) + 10x^3(-y)^2$

$\qquad + 10x^2(-y)^3 + 5x(-y)^4 + 1 \cdot (-y)^5$

$\qquad = x^5 - 5x^4y + 10x^3y^2 - 10x^2y^3$

$\qquad + 5xy^4 - y^5$

Form 2:

$$(x-y)^5 = \binom{5}{0}x^5 + \binom{5}{1}x^4(-y)$$

$$+ \binom{5}{2}x^3(-y)^2 + \binom{5}{3}x^2(-y)^3$$

$$+ \binom{5}{4}x(-y)^4 + \binom{5}{5}(-y)^5$$

$$= x^5 - 5x^4 y + 10x^3 y^2 - 10x^2 y^3$$

$$+ 5xy^4 - y^5$$

35. Expand $\left(3s + \dfrac{1}{t}\right)^9$.

Form 1: We use the tenth row of Pascal's triangle.

1 9 36 84 126 126 84 36 9 1

$$\left(3s + \frac{1}{t}\right)^9 = 1 \cdot (3s)^9 + 9(3s)^8 \left(\frac{1}{t}\right)$$

$$+ 36(3s)^7 \left(\frac{1}{t}\right)^2 + 84(3s)^6 \left(\frac{1}{t}\right)^3$$

$$+ 126(3s)^5 \left(\frac{1}{t}\right)^4 + 126(3s)^4 \left(\frac{1}{t}\right)^5$$

$$+ 84(3s)^3 \left(\frac{1}{t}\right)^6 + 36(3s)^2 \left(\frac{1}{t}\right)^7$$

$$+ 9(3s)\left(\frac{1}{t}\right)^8 + 1 \cdot \left(\frac{1}{t}\right)^9$$

$$= 19{,}683s^9 + \frac{59{,}049s^8}{t}$$

$$+ \frac{78{,}732s^7}{t^2} + \frac{61{,}236s^6}{t^3}$$

$$+ \frac{30{,}618s^5}{t^4} + \frac{10{,}206s^4}{t^5}$$

$$+ \frac{2268s^3}{t^6} + \frac{324s^2}{t^7}$$

$$+ \frac{27s}{t^8} + \frac{1}{t^9}$$

Form 2:

$$\left(3s + \frac{1}{t}\right)^9 = \binom{9}{0}(3s)^9 + \binom{9}{1}(3s)^8 \left(\frac{1}{t}\right)$$

$$+ \binom{9}{2}(3s)^7 \left(\frac{1}{t}\right)^2 + \binom{9}{3}(3s)^6 \left(\frac{1}{t}\right)^3$$

$$+ \binom{9}{4}(3s)^5 \left(\frac{1}{t}\right)^4 + \binom{9}{5}(3s)^4 \left(\frac{1}{t}\right)^5$$

$$+ \binom{9}{6}(3s)^3 \left(\frac{1}{t}\right)^6 + \binom{9}{7}(3s)^2 \left(\frac{1}{t}\right)^7$$

$$+ \binom{9}{8}(3s)\left(\frac{1}{t}\right)^8 + \binom{9}{9}\left(\frac{1}{t}\right)^9$$

$$= 19{,}683s^9 + \frac{59{,}049s^8}{t} + \frac{78{,}732s^7}{t^2}$$

$$+ \frac{61{,}236s^6}{t^3} + \frac{30{,}618s^5}{t^4}$$

$$+ \frac{10{,}206s^4}{t^5} + \frac{2268s^3}{t^6} + \frac{324s^2}{t^7}$$

$$+ \frac{27s}{t^8} + \frac{1}{t^9}$$

37. Expand $\left(x^3 - 2y\right)^5$

Form 1: We use the sixth row of Pascal's triangle.

1 5 10 10 5 1

$$\left(x^3 - 2y\right)^5 = 1 \cdot \left(x^3\right)^5 + 5\left(x^3\right)^4(-2y)$$

$$+ 10\left(x^3\right)^3(-2y)^2$$

$$+ 10\left(x^3\right)^2(-2y)^3 + 5\left(x^3\right)(-2y)^4$$

$$+ 1 \cdot (-2y)^5$$

$$= x^{15} - 10x^{12}y + 40x^9 y^2 - 80x^6 y^3$$

$$+ 80x^3 y^4 - 32y^5$$

Form 2:

$$\left(x^3 - 2y\right)^5 = \binom{5}{0}\left(x^3\right)^5 + \binom{5}{1}\left(x^3\right)^4(-2y)$$

$$+\binom{5}{2}\left(x^3\right)^3(-2y)^2$$

$$+\binom{5}{3}\left(x^3\right)^2(-2y)^3$$

$$+\binom{5}{4}\left(x^3\right)(-2y)^4 + \binom{5}{5}(-2y)^5$$

$$= x^{15} - 10x^{12}y + 40x^9y^2 - 80x^6y^3$$

$$+ 80x^3y^4 - 32y^5$$

39. Expand $\left(\sqrt{5} + t\right)^6$.

Form 1: We use the 7th row of Pascal's triangle.

1 6 15 20 15 6 1

$$\left(\sqrt{5} + t\right)^6 = 1 \cdot \left(\sqrt{5}\right)^6 + 6\left(\sqrt{5}\right)^5(t)$$

$$+ 15\left(\sqrt{5}\right)^4\left(t^2\right) + 20\left(\sqrt{5}\right)^3\left(t^3\right)$$

$$+ 15\left(\sqrt{5}\right)^2\left(t^4\right) + 6\left(\sqrt{5}\right)\left(t^5\right) + 1 \cdot t^6$$

$$= 125 + 150\sqrt{5}t + 375t^2 + 100\sqrt{5}t^3$$

$$+ 75t^4 + 6\sqrt{5}t^5 + t^6$$

Form 2:

$$\left(\sqrt{5} + t\right)^6 = \binom{6}{0}\left(\sqrt{5}\right)^6 + \binom{6}{1}\left(\sqrt{5}\right)^5(t)$$

$$+ \binom{6}{2}\left(\sqrt{5}\right)^4\left(t^2\right) + \binom{6}{3}\left(\sqrt{5}\right)^3\left(t^3\right)$$

$$+ \binom{6}{4}\left(\sqrt{5}\right)^2\left(t^4\right) + \binom{6}{5}\left(\sqrt{5}\right)\left(t^5\right)$$

$$+ \binom{6}{6}\left(t^6\right)$$

$$= 125 + 150\sqrt{5}t + 375t^2 + 100\sqrt{5}t^3$$

$$+ 75t^4 + 6\sqrt{5}t^5 + t^6$$

41. Expand $\left(\dfrac{1}{\sqrt{x}} - \sqrt{x}\right)^6$.

Form 1: We use the 7th row of Pascal's triangle:

1 6 15 20 15 6 1

$$\left(\frac{1}{\sqrt{x}} - \sqrt{x}\right)^6 = 1 \cdot \left(\frac{1}{\sqrt{x}}\right)^6 + 6\left(\frac{1}{\sqrt{x}}\right)^5(-\sqrt{x})$$

$$+ 15\left(\frac{1}{\sqrt{x}}\right)^4(-\sqrt{x})^2$$

$$+ 20\left(\frac{1}{\sqrt{x}}\right)^3(-\sqrt{x})^3$$

$$+ 15\left(\frac{1}{\sqrt{x}}\right)^2(-\sqrt{x})^4$$

$$+ 6\left(\frac{1}{\sqrt{x}}\right)(-\sqrt{x})^5 + 1 \cdot (-\sqrt{x})^6$$

$$= x^{-3} - 6x^{-2} + 15x^{-1} - 20$$

$$+ 15x - 6x^2 + x^3$$

Form 2:

$$\left(\frac{1}{\sqrt{x}} - \sqrt{x}\right)^6 = \binom{6}{0}\left(\frac{1}{\sqrt{x}}\right)^6$$

$$+ \binom{6}{1}\left(\frac{1}{\sqrt{x}}\right)^5(-\sqrt{x})$$

$$+ \binom{6}{2}\left(\frac{1}{\sqrt{x}}\right)^4(-\sqrt{x})^2$$

$$+ \binom{6}{3}\left(\frac{1}{\sqrt{x}}\right)^3(-\sqrt{x})^3$$

$$+ \binom{6}{4}\left(\frac{1}{\sqrt{x}}\right)^2(-\sqrt{x})^4$$

$$+ \binom{6}{5}\left(\frac{1}{\sqrt{x}}\right)(-\sqrt{x})^5$$

$$+ \binom{6}{6}(-\sqrt{x})^6$$

$$= x^{-3} - 6x^{-2} + 15x^{-1} - 20$$

$$+ 15x - 6x^2 + x^3$$

43. Find the 3rd term of $\left(a + b\right)^6$.

First, we note that $3 = 2 + 1$, $a = a$, $b = b$, and $n = 6$. The 3rd term of the expansion is

$$\binom{6}{2}a^{6-2}b^2, \text{ or } \frac{6!}{4!2!}a^4b^2, \text{ or } 15a^4b^2.$$

45. Find the 12th term of $(a-3)^{14}$.

First, we note that $12 = 11+1$, $a = a$, $b = -3$,

and $n = 14$. The 12th term of the expansion is

$$\binom{14}{11} a^{14-11} \cdot (-3)^{11} = \frac{14!}{3!11!} a^3 (-177,147)$$

$$= 364a^3 (-177,147)$$

$$= -64,481,508a^3$$

47. Find the 5th term of $\left(2x^3 + \sqrt{y}\right)^8$.

First, we note that $5 = 4+1$, $a = 2x^3$,

$b = \sqrt{y}$, and $n = 8$. The 5th term of the

expansion is

$$\binom{8}{4}\left(2x^3\right)^{8-4}\left(\sqrt{y}\right)^4 = \frac{8!}{4!4!}\left(2x^3\right)^4\left(\sqrt{y}\right)^4$$

$$= 70\left(16x^{12}\right)\left(y^2\right)$$

$$= 1120x^{12}y^2$$

49. The expansion of $\left(2u + 3v^2\right)^{10}$ has 11 terms

so the 6th term is the middle term. Note that

$6 = 5+1$, $a = 2u$, $b = 3v^2$, and $n = 10$. The

6th term of the expansion is

$$\binom{10}{5}\left(2u\right)^{10-5}\left(3v^2\right)^5 = \frac{10!}{5!5!}\left(2u\right)^5\left(3v^2\right)^5$$

$$= 252\left(32u^5\right)\left(243v^{10}\right)$$

$$= 1,959,552u^5v^{10}$$

51. The 9th term of $(x-y)^8$ is the last term, y^8.

53. *Thinking and Writing Exercise.*

55. $y = x^2 - 5$

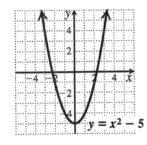

57. $y \geq x - 5$

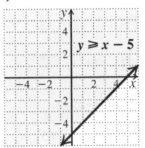

59. $f(x) = \log_5 x$

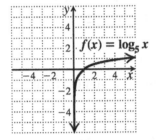

61. *Thinking and Writing Exercise.*

63. Consider a set of 5 elements, $\{a,b,c,d,e\}$.

List all the subsets of size 3:

$\{a,b,c\},\{a,b,d\},\{a,b,e\},\{a,c,d\},$

$\{a,c,e\},\{a,d,e\},\{b,c,d\},\{b,c,e\},$

$\{b,d,e\},\{c,d,e\}.$

There are exactly 10 subsets of size 3 and

$\binom{5}{3} = 10$, so there are exactly $\binom{5}{3}$ ways of

forming a subset of size 3 from a set of

5 elements.

65. $\binom{8}{5}\left(0.15\right)^3\left(0.85\right)^5 \approx 0.084$

67. Find and add the 7th through the 9th terms of

$\left(0.15 + 0.85\right)^8$:

$$\binom{8}{6}\left(0.15\right)^2\left(0.85\right)^6 + \binom{8}{7}\left(0.15\right)\left(0.85\right)^7$$

$$+ \binom{8}{8}\left(0.85\right)^8 \approx 0.89$$

69. Prove: $\dbinom{n}{r} = \dbinom{n}{n-r}$

Evaluate the right expression:

$$\dbinom{n}{n-r} = \frac{n!}{[n-(n-r)]!(n-r)!}$$

$$= \frac{n!}{r!(n-r)!} = \dbinom{n}{r}$$

71. $$\frac{\dbinom{5}{3}(p^2)^2\left(-\frac{1}{2}p\sqrt[3]{q}\right)^3}{\dbinom{5}{2}(p^2)^3\left(-\frac{1}{2}p\sqrt[3]{q}\right)^2} = \frac{-\frac{1}{8}p^7 q}{\frac{1}{4}p^8\sqrt[3]{q^2}}$$

$$= \frac{-\frac{1}{8}p^7 q}{\frac{1}{4}p^8 q^{2/3}}$$

$$= -\frac{1}{8}\cdot\frac{4}{1}\cdot p^{7-8}\cdot q^{1-2/3}$$

$$= -\frac{1}{2}p^{-1}q^{1/3} = -\frac{\sqrt[3]{q}}{2p}$$

73. $(x^2 + 2xy + y^2)(x^2 + 2xy + y^2)^2(x+y)$

$$= (x+y)^2(x+y)^{2\cdot 2}(x+y) = (x+y)^7$$

We can find the given product by finding the binomial expansion of $(x+y)^7$. It is

$x^7 + 7x^6 y + 21x^5 y^2 + 35x^4 y^3$

$\quad + 35x^3 y^4 + 21x^2 y^5 + 7xy^6 + y^7.$

(See Exercise 27).

Chapter 11 Study Summary

1. $a_n = n^2 - 1$

   $a_{12} = 12^2 - 1$

   $\quad = 144 - 1$

   $\quad = 143$

2. $-9, -8, -6, -3, 1, 6, 12$

   $S_5 = -9 + (-8) + (-6) + (-3) + 1$

   $\quad = -25$

3. $\displaystyle\sum_{k=0}^{3} 5k = 5(0) + 5(1) + 5(2) + 5(3)$

   $\quad = 0 + 5 + 10 + 15$

   $\quad = 30$

4. $6, 6.5, 7, 7.5, \ldots$

   $a_1 = 6, \quad d = 0.5, \text{ and } n = 20$

   $a_n = a_1 + (n-1)d$

   $a_{20} = 6 + (20-1)0.5 = 6 + 19 \cdot 0.5 = 15.5$

5. $6 + 6.5 + 7 + 7.5 + \ldots$

   $a_1 = 6, d = 0.5, \text{ and } n = 20$

   From Exercise 4, we know that $a_{20} = 15.5$.

   $S_{20} = \dfrac{20}{2}[6 + (15.5)]$

   $\quad = 10(21.5)$

   $\quad = 215$

6. $-5, -10, -20, \ldots$

   $a_1 = -5, n = 8, r = \dfrac{-10}{-5} = 2$

   $a_n = a_1 r^{n-1}$

   $a_8 = -5 \cdot 2^{8-1} = -5 \cdot 2^7 = -5 \cdot 128$

   $a_8 = -640$

7. $-5 - 10 - 20 - \ldots$

   $a_1 = -5, n = 12, r = \dfrac{-10}{-5} = 2$

   $S_n = \dfrac{a_1\left(1 - r^n\right)}{1 - r}$

   $S_{12} = \dfrac{-5(1 - 2^{12})}{1 - 2} = \dfrac{-5(-4095)}{-1}$

   $S_{12} = -20,475$

8. $20 - 5 + \dfrac{5}{4} - \ldots$

   $|r| = \left|\dfrac{-5}{20}\right| = \left|-\dfrac{1}{4}\right| = \dfrac{1}{4} < 1, \text{ so the series has a limit.}$

   $S_\infty = \dfrac{20}{1 - \left(-\frac{1}{4}\right)} = \dfrac{20}{\frac{5}{4}} = 20 \cdot \dfrac{4}{5} = 16$

9. $11! = 11 \cdot 10 \cdot 9 \cdot 8 \cdot 7 \cdot 6 \cdot 5 \cdot 4 \cdot 3 \cdot 2 \cdot 1$

   $\quad = 39,916,800$

10. $\dbinom{9}{3} = \dfrac{9!}{6!3!} = \dfrac{9 \cdot 8 \cdot 7 \cdot 6!}{3 \cdot 2 \cdot 6!} = \dfrac{9 \cdot 8 \cdot 7}{3 \cdot 2} = 84$

11. Expand $\left(x^2 - 2\right)^5$

Form 1: We use the 6th Row of Pascal's triangle.

1   5   10   10   5   1

$$\left(x^2 - 2\right)^5 = 1 \cdot \left(x^2\right)^5 + 5 \cdot \left(x^2\right)^4 \cdot (-2)$$
$$+ 10 \cdot \left(x^2\right)^3 \cdot (-2)^2$$
$$+ 10 \cdot \left(x^2\right)^2 \cdot (-2)^3$$
$$+ 5 \cdot \left(x^2\right) \cdot (-2)^4 + 1 \cdot (-2)^5$$
$$= x^{10} - 10x^8 + 40x^6 - 80x^4$$
$$+ 80x^2 - 32$$

Form 2: $a = x^2, b = -2, n = 5$

$$\left(x^2 - 2\right)^5 = \binom{5}{0}\left(x^2\right)^5 + \binom{5}{1}\left(x^2\right)^4 (-2)$$
$$+ \binom{5}{2}\left(x^2\right)^3 (-2)^2$$
$$+ \binom{5}{3}\left(x^2\right)^2 (-2)^3$$
$$+ \binom{5}{4}x^2 (-2)^4 + \binom{5}{5}(-2)^5$$
$$= x^{10} - 10x^8 + 40x^6 - 80x^4$$
$$+ 80x^2 - 32$$

12. Find the 4th term of $(t + 3)^{10}$.

First, we note that $4 = 3 + 1$, $a = t$, $b = 3$, and $n = 10$. The 4th term of the expansion is

$$\binom{10}{3}(t)^{10-3}(3)^3 = \frac{10!}{3!7!}t^7 \cdot 3^3$$
$$= 120 \cdot 27t^7$$
$$= 3240t^7$$

Chapter 11 Review Exercises

1. False

2. True

3. True

4. False

5. False

6. True

7. False

8. False

9. $a_n = 4n - 3$
$a_1 = 4 \cdot 1 - 3 = 1$
$a_2 = 4 \cdot 2 - 3 = 5$
$a_3 = 4 \cdot 3 - 3 = 9$
$a_4 = 4 \cdot 4 - 3 = 13$
$a_8 = 4 \cdot 8 - 3 = 29$
$a_{12} = 4 \cdot 12 - 3 = 45$

10. $a_n = \dfrac{n-1}{n^2 + 1}$
$a_1 = \dfrac{1-1}{1^2 + 1} = 0$
$a_2 = \dfrac{2-1}{2^2 + 1} = \dfrac{1}{5}$
$a_3 = \dfrac{3-1}{3^2 + 1} = \dfrac{2}{10} = \dfrac{1}{5}$
$a_4 = \dfrac{4-1}{4^2 + 1} = \dfrac{3}{17}$
$a_8 = \dfrac{8-1}{8^2 + 1} = \dfrac{7}{65}$
$a_{12} = \dfrac{12-1}{12^2 + 1} = \dfrac{11}{145}$

11. $-5, -10, -15, -20, \ldots$
These are multiples of -5, where
$a_1 = 1 \cdot (-5)$, $a_2 = 2 \cdot (-5)$, etc.
$a_n = -5n$

12. $-1, 3, -5, 7, -9, \ldots$
These could be odd counting/natural numbers with alternating signs, the first of which is negative.
$a_n = (-1)^n (2n - 1)$

13. $\displaystyle\sum_{k=1}^{5}(-2)^k = (-2)^1 + (-2)^2 + (-2)^3$

$\qquad\qquad + (-2)^4 + (-2)^5$

$\qquad\quad = -2 + 4 + (-8) + 16 + (-32)$

$\qquad\quad = -22$

14. $\displaystyle\sum_{k=2}^{7}(1-2k) = (1-2\cdot2) + (1-2\cdot3) + (1-2\cdot4)$

$\qquad\qquad\quad + (1-2\cdot5) + (1-2\cdot6)$

$\qquad\qquad\quad + (1-2\cdot7)$

$\qquad\qquad = -3 + (-5) + (-7) + (-9)$

$\qquad\qquad\quad + (-11) + (-13)$

$\qquad\qquad = -48$

15. $4 + 8 + 12 + 16 + 20$

$= 4\cdot1 + 4\cdot2 + 4\cdot3 + 4\cdot4 + 4\cdot5$

$\displaystyle = \sum_{k=1}^{5} 4k$

16. $\dfrac{-1}{2} + \dfrac{1}{4} + \dfrac{-1}{8} + \dfrac{1}{16} + \dfrac{-1}{32}$

$= -1\cdot\dfrac{1}{2^1} + \dfrac{1}{2^2} - 1\cdot\dfrac{1}{2^3} + \dfrac{1}{2^4} + -1\cdot\dfrac{1}{2^5}$

$\displaystyle = \sum_{k=1}^{5} \dfrac{1}{(-2)^k}$

17. $-6,\ 1,\ 8,\ \dots$

$\quad a_1 = -6,\ d = 7,\ n = 14$

$a_{14} = -6 + (14-1)\cdot7$

$\qquad = -6 + 13\cdot7$

$\qquad = -6 + 91$

$\qquad = 85$

18. $a_n = a_1 + (n-1)d$

$\quad a_1 = 11,\ a_{13} = 43$

$\quad 43 = 11 + (13-1)d$

$\quad 32 = 12d$

$\quad \dfrac{32}{12} = d$

$\quad \dfrac{8}{3} = d$

19. $a_n = a_1 + (n-1)d$

$20 = a_1 + (8-1)d,\ 20 = a_1 + 7d$

$100 = a_1 + (24-1)d,\ 100 = a_1 + 23d$

Solve the system of equations:

$\qquad 20 = a_1 + 7d$

$(-)\underline{100 = a_1 + 23d}$

$\qquad -80 = -16d$

$\qquad \dfrac{-80}{-16} = d$

$\qquad\quad 5 = d$

$\quad 20 = a_1 + 7\cdot5$

$\quad 20 - 35 = a_1$

$\quad -15 = a_1$

20. $-8 + (-11) + (-14) + \cdots$

$\quad d = -3,\ a_1 = -8,\ n = 17$

$\quad a_n = a_1 + (n-1)d$

$\quad a_{17} = -8 + (17-1)(-3)$

$\quad a_{17} = -8 + 16(-3)$

$\quad a_{17} = -56$

$\quad S_n = \dfrac{n}{2}(a_1 + a_n)$

$\qquad = \dfrac{17}{2}(-8 + -56)$

$\qquad = \dfrac{17}{2}(-64)$

$\qquad = -544$

21. $5, 10, 15, \dots, 500$

$\quad a_1 = 5,\ d = 5$

$\quad a_n = a_1 + (n-1)d$

$\quad 500 = 5 + (n-1)5$

$\quad 500 = 5 + 5n - 5$

$\quad 100 = n$

$\quad S_{100} = \dfrac{100}{2}(5 + 500)$

$\qquad = 50(505)$

$\qquad = 25,250$

22. $2, 2\sqrt{2}, 4, \cdots$

$a_1 = 2, r = \sqrt{2}, n = 20$

$a_{20} = 2 \cdot \sqrt{2}^{20-1}$

$\phantom{a_{20}} = 2\sqrt{2}^{19}$

$\phantom{a_{20}} = 2 \cdot 2^9 \sqrt{2}$

$\phantom{a_{20}} = 1024\sqrt{2}$

23. $40, 30, \dfrac{45}{2}, \cdots$

$\dfrac{30}{40} = \dfrac{3}{4}$

$\dfrac{\frac{45}{2}}{30} = \dfrac{45}{2} \cdot \dfrac{1}{30} = \dfrac{3}{4}$

$r = \dfrac{3}{4}$

24. $-2, 2, -2, \ldots$

These are negative and positive 2 alternating, beginning with –2.

$a_n = 2(-1)^n$

25. $3, \dfrac{3}{4}x, \dfrac{3}{16}x^2, \ldots$

$a_1 = 3$

$r = \dfrac{\frac{3x}{4}}{3} = \dfrac{x}{4}$

$a_n = 3\left(\dfrac{x}{4}\right)^{n-1}$

26. $3 + 12 + 48 + \cdots$

$a_1 = 3, n = 6, r = \dfrac{12}{3} = 4$

$S_n = \dfrac{a_1\left(1 - r^n\right)}{1 - r}$

$S_6 = \dfrac{3\left(1 - 4^6\right)}{1 - 4} = -\left(1 - 4^6\right)$

$ = 4095$

27. $3x - 6x + 12x - \ldots$

$a_1 = 3x, n = 12, r = \dfrac{-6x}{3x} = -2$

$S_n = \dfrac{a_1\left(1 - r^n\right)}{1 - r}$

$S_{12} = \dfrac{3x\left[1 - (-2)^{12}\right]}{1 - (-2)} = \dfrac{3x(1 - 4096)}{3}$

$\phantom{S_{12}} = -4095x$

28. $6 + 3 + 1.5 + 0.75 + \ldots$

$r = \dfrac{3}{6} = \dfrac{1}{2}; \left|\dfrac{1}{2}\right| < 1$

There is a limit.

$S_\infty = \dfrac{a_1}{1 - r}$

$S_\infty = \dfrac{6}{1 - \frac{1}{2}} = \dfrac{6}{\frac{1}{2}} = 6 \cdot 2 = 12$

29. $7 - 4 + \dfrac{16}{7} - \ldots$

$r = \dfrac{-4}{7}; \left|\dfrac{-4}{7}\right| < 1$

There is a limit.

$s_\infty = \dfrac{7}{1 - \frac{-4}{7}} = \dfrac{7}{\frac{11}{7}} = 7 \cdot \dfrac{7}{11} = \dfrac{49}{11}$

30. $2 + (-2) + 2 + (-2) + \ldots$

$r = \dfrac{-2}{2} = -1; |-1| \not< 1$

No; there is not a limit.

31. $0.04 + 0.08 + 0.16 + 0.32 + \ldots$

$r = \dfrac{0.08}{0.04} = 2; |2| \not< 1$

No limit.

32. $\$2000 + \$1900 + \$1805 + \$1714.75 + \ldots$

$r = \dfrac{1900}{2000} = 0.95; |0.95| < 1$

There is a limit.

$S_\infty = \dfrac{2000}{1 - 0.95} = \dfrac{2000}{0.05} = \$40,000$

33. $0.555555\ldots = 0.5 + 0.05 + 0.005 + \ldots$

This is an infinite geometric series with $a_1 = 0.5$.

$$|r| = \left|\frac{0.05}{0.5}\right| = 0.1$$

The series has a sum $(0.1 < 1)$.

$$S_\infty = \frac{0.5}{1 - 0.1} = \frac{0.5}{.9} = \frac{5}{9}$$

Fractional notation is $\frac{5}{9}$.

34. $1.39393939\ldots = 1 + 0.39393939\ldots$

$0.39393939\ldots$ is an infinite series with $a_1 = 0.39$, $a_2 = 0.0039$, etc.

$$|r| = \left|\frac{0.0039}{0.39}\right| = 0.01$$

The series has a sum.

$$S_\infty = \frac{0.39}{1 - 0.01} = \frac{0.39}{0.99} = \frac{39}{99} = \frac{13}{33}$$

Since $1 = \frac{33}{33}$, $1.39393939\ldots = \frac{33}{33} + \frac{13}{33}$

Fractional notation is $\frac{46}{33}$.

35. Adams starting wage is $a_1 = 11.50$.

$d = 0.4$, the 40¢ raise; he received 4 raises per year for 8 years, or 32 raises.

$n = 1 + 32 = 33$

$a_{33} = 11.50 + (33 - 1)0.4$

$\quad = 11.50 + 12.80$

$\quad = \$24.30$

At the end of 8 years, his hourly wage will be $24.30.

36. We have an arithmetic sequence
$42, 41, 40, \ldots, 1$.

$a_1 = 42$, $a_{42} = 1$, $n = 42$

$$S_{40} = \frac{42}{2}(42 + 1) = 21 \cdot 43 = 903$$

There are 903 poles.

37. $\$12,000, 1.04(\$12,000), 1.04^2(\$12,000)$

$\ldots 1.04^n(\$12,000)$.

The amount to be repaid at the end of 7 years is the amount owed at the beginning of the 8th year.

$a_1 = 12,000$, $r = 1.04$, $n = 8$

$a_8 = 12,000 \cdot 1.04^{8-1}$

$\quad = 12,000 \cdot 1.04^7$

$\quad \approx \$15,791.18$

38. Since $r = \frac{1}{3}$ and $\left|\frac{1}{3}\right| < 1$, we determine the sum.

$a_1 = 12$

$$S_\infty = \frac{12}{1 - \frac{1}{3}} = \frac{12}{\frac{2}{3}} = 12 \cdot \frac{3}{2} = 18$$

The total distance is 18, the fall distance is 12, so the rebound distance is $18 - 12$, or 6. The ball will rebound a total of 6 m.

39. $7! = 7 \cdot 6 \cdot 5 \cdot 4 \cdot 3 \cdot 2 \cdot 1 = 5040$

40. $\dbinom{8}{3} = \frac{8!}{5!3!} = \frac{8 \cdot 7 \cdot 6 \cdot 5!}{5!3!}$

$\quad = \frac{8 \cdot 7 \cdot 6}{3 \cdot 2} = 56$

41. $(a + b)^{20}$

Note: $3 = 2 + 1$, $a = a$, $b = b$, and $n = 20$.

$\dbinom{20}{2} a^{20-2} b^2 = 190 a^{18} b^2$

42. Expand: $(x - 2y)^4$

We use the 5th row of Pascal's triangle.

1 4 6 4 1

$(x - 2y)^4 = 1 \cdot x^4 + 4 \cdot x^3(-2y) + 6x^2(-2y)^2$

$\quad + 4x(-2y)^3 + (-2y)^4$

$\quad = x^4 - 8x^3 y + 24x^2 y^2 - 32xy^3 + 16y^4$

43. *Thinking and Writing Exercise.* For a geometric series with $|r| < 1$, as n increases, the absolute value of a_n decreases, since $|r|^n$ decreases.

44. *Thinking and Writing Exercise.* The first form uses Pascal's triangle to determine coefficients; the second form uses factorial notation. The second form does not require finding the preceding rows of Pascal's triangle, and is generally easier when only one term is needed. When several terms of an expansion are needed and n is "not large" (say, $n = 8$), it is often easier to use Pascal's triangle.

45. $1 - x + x^2 - x^3 + \ldots$

$$a_1 = 1; \; r = \frac{-x}{1} = -x$$

$$S_n = \frac{a_1(1 - r^n)}{1 - r}$$

$$S_n = \frac{1\left[1 - (-x)^n\right]}{1 - (-x)} = \frac{1 - (-x)^n}{x + 1}.$$

46. Expand $\left(x^{-3} + x^3\right)^5$

Use the 6th row of Pascal's triangle.

$$1 \quad 5 \quad 10 \quad 10 \quad 5 \quad 1$$

$$a = x^{-3}, b = x^3, n = 5$$

$$\left(x^{-3} + x^3\right)^5 = 1 \cdot \left(x^{-3}\right)^5 + 5\left(x^{-3}\right)^4\left(x^3\right)$$

$$+ 10\left(x^{-3}\right)^3\left(x^3\right)^2 + 10\left(x^{-3}\right)^2 \cdot$$

$$\left(x^3\right)^3 + 5\left(x^{-3}\right)\left(x^3\right)^4 + \left(x^3\right)^5$$

$$= x^{-15} + 5x^{-12}x^3 + 10x^{-9}x^6$$

$$+ 10x^{-6}x^9 + 5x^{-3}x^{12} + x^{15}$$

$$= x^{-15} + 5x^{-9} + 10x^{-3} + 10x^3$$

$$+ 5x^9 + x^{15}$$

Chapter 11 Test

1. $a_n = \dfrac{1}{n^2 + 1}$

$$a_1 = \frac{1}{1^2 + 1} = \frac{1}{2}$$

$$a_2 = \frac{1}{2^2 + 1} = \frac{1}{5}$$

$$a_3 = \frac{1}{3^2 + 1} = \frac{1}{10}$$

$$a_4 = \frac{1}{4^2 + 1} = \frac{1}{17}$$

$$a_5 = \frac{1}{5^2 + 1} = \frac{1}{26}$$

$$a_{12} = \frac{1}{12^2 + 1} = \frac{1}{145}$$

2. $\dfrac{4}{3}, \dfrac{4}{9}, \dfrac{4}{27}, \ldots$

$$\frac{4}{3^1}, \frac{4}{3^2}, \frac{4}{3^3}, \ldots, \frac{4}{3^n}$$

$$a_n = 4\left(\frac{1}{3}\right)^n$$

3. $\displaystyle\sum_{k=2}^{5}\left(1 - 2^k\right) = \left(1 - 2^2\right) + \left(1 - 2^3\right) + \left(1 - 2^4\right)$

$$+ \left(1 - 2^5\right)$$

$$= -3 + (-7) + (-15)$$

$$+ (-31)$$

$$= -56$$

4. $1 + (-8) + 27 + (-64) + 125$

$$1^3 + (-1) \cdot 2^3 + 3^3 + (-1) \cdot 4^3 + 5^3$$

$$\sum_{k=1}^{5}(-1)^{k+1} k^3$$

5. $\frac{1}{2}, 1, \frac{3}{2}, 2, \ldots$

$$a_1 = \tfrac{1}{2}, d = \tfrac{1}{2}, n = 13$$

$$a_n = a_1 + (n - 1)d$$

$$a_{13} = \tfrac{1}{2} + (13 - 1)\left(\tfrac{1}{2}\right)$$

$$= \tfrac{1}{2} + 12\left(\tfrac{1}{2}\right)$$

$$= \tfrac{13}{2}$$

6. $a_1 = 7, a_7 = -11$

Using substitution:

$a_n = a_1 + (n-1)d$

$a_7 = a_1 + (7-1)d$

$-11 = 7 + 6d$

$-18 = 6d$

$-3 = d$

7. Using substitution:

$16 = a_1 + (5-1)d, 16 = a_1 + 4d$

$-4 = a_1 + (10-1)d, -4 = a_1 + 9d$

Solve the system.

$16 = a_1 + 4d$

$(-)\ \underline{-4 = a_1 + 9d}$

$20 = \quad -5d$

$-4 = d$

$-4 = a_1 + 9(-4)$

$-4 = a_1 - 36$

$32 = a_1$

8. $24, 36, 48, \ldots, 240$

$a_1 = 24, d = 12, n = 19, a_{19} = 240$

$S_n = \dfrac{n}{2}(a_1 + a_n)$

$S_{19} = \dfrac{19}{2}(24 + 240) = \dfrac{19}{2}(264)$

$= 2508$

9. $-3, 6, -12, \ldots$

$a_1 = -3,\ n = 10,\ r = \dfrac{6}{-3} = -2$

$a_n = a \cdot r^{n-1}$

$a_{10} = -3 \cdot (-2)^{10-1} = -3 \cdot (-2)^9 = -3 \cdot -512$

$a_{10} = 1536$

10. $22\dfrac{1}{2}, 15, 10, \ldots$

$\dfrac{15}{22\dfrac{1}{2}} = 15 \cdot \dfrac{2}{45} = \dfrac{2}{3}$

$\dfrac{10}{15} = \dfrac{2}{3}$

$r = \dfrac{2}{3}$

11. $3, 9, 27, \ldots$

$3^1, 3^2, 3^3 \ldots 3^n$

The nth term is 3^n.

12. $11 + 22 + 44 + \ldots$

$a_1 = 11, n = 9, r = \dfrac{22}{11} = 2$

$S_n = \dfrac{a_1(1 - r^n)}{1 - r}$

$S_9 = \dfrac{(11)(1 - 2^9)}{1 - 2} = \dfrac{(11)(-511)}{-1}$

$S_9 = 5621$

13. $0.5 + 0.25 + 0.125 + \ldots$

$r = \dfrac{0.25}{0.5} = 0.5;\ |0.5| < 1$

There is a limit.

$S_\infty = \dfrac{a_1}{1 - r}$

$S_\infty = \dfrac{0.5}{1 - 0.5} = \dfrac{0.5}{0.5} = 1$

14. $0.5 + 1 + 2 + 4 + \ldots$

$r = \dfrac{1}{0.5} = 2; |2| \not< 1$

No limit.

15. $\$1000 + \$80 + \$6.40 + \ldots$

$r = \dfrac{\$80}{\$1000} = 0.08;\ |0.08| < 1$

There is a limit.

$S_\infty = \dfrac{a_1}{1 - r}$

$S_\infty = \dfrac{1000}{1 - 0.08} = \dfrac{1000}{0.92} \approx \$1086.96$

16. $0.85858585\ldots$

$= 0.85 + 0.0085 + 0.000085 + \ldots$

$a_1 = 0.85$

$r = \dfrac{0.0085}{0.85} = 0.01$

$S_\infty = \dfrac{a_1}{1 - r}$

$S_\infty = \dfrac{0.85}{1 - 0.01} = \dfrac{0.85}{0.99} = \dfrac{85}{99}$

17. This is an arithmetic sequence.
$31, 33, 35, \ldots a_{18}$

$a_1 = 31, d = 2, n = 17$

$a_{17} = 31 + (17 - 1)2$

$\quad = 31 + 16 \cdot 2$

$\quad = 63$

There are 63 seats in Row 17.

18. This is an arithmetic sequence.
$100, 200, 300, \ldots, 1800$

$a_1 = 100, \ a_{18} = 1800, \ n = 18$

$S_n = \dfrac{n}{2}(a_1 + a_2)$

$S_{18} = \dfrac{18}{2}(100 + 1800) = 9 \cdot 1900$

$\quad = 17{,}100$

Her uncle gave her a total of $17,100.

19. This is a geometric sequence.
$10{,}000; \ 10{,}000 \cdot 0.95; \ 10{,}000 \cdot 0.95^2, \ldots$

$a_1 = 10{,}000, \ n = 11, \ r = 0.95$

$a_n = a_1 \cdot r^{n-1}$

$a_{11} = 10{,}000 \cdot (0.95)^{11-1}$

$a_{11} = 10{,}000 \cdot (0.95)^{10}$

$a_{11} \approx \$5987.37$

After 10 weeks, the boat will cost $5987.37

20. Since $r = \dfrac{2}{3}$ and $\left|\dfrac{2}{3}\right| < 1$, we determine the

sum.

$a_1 = 18$

$S_\infty = \dfrac{18}{1 - \frac{2}{3}} = \dfrac{18}{\frac{1}{3}} = 18 \cdot 3 = 54$

The total distance is 54 m, the fall distance is 18 m, so the rebound distance is $54 - 18$, or 36 m.
The rebound total distance is 36 m.

21. $\dbinom{12}{9} = \dfrac{12!}{3!\,9!} = \dfrac{12 \cdot 11 \cdot 10 \cdot 9!}{3!\,9!} = \dfrac{12 \cdot 11 \cdot 10}{3 \cdot 2}$

$\quad = 220$

22. Expand $(x - 3y)^5$
We will use the 6th row of Pascal's triangle.

$1 \quad 5 \quad 10 \quad 10 \quad 5 \quad 1$

$a = x, b = -3y, n = 5$

$(x - 3y)^5 = 1 \cdot (x)^5 + 5(x)^4(-3y)$

$\qquad\qquad + 10(x)^3(-3y)^2$

$\qquad\qquad + 10(x)^2(-3y)^3$

$\qquad\quad + 5(x)(-3y)^4 + 1 \cdot (-3y)^5$

$= x^5 + 5x^4(-3y) + 10x^3 \cdot 9y^2$

$\quad + 10x^2(-27y^3)$

$\quad + 5x(81y^4) - 243y^5$

$= x^5 - 15x^4 y + 90x^3 y^2$

$\quad - 270x^2 y^3 + 405xy^4 - 243y^5$

23. $(a + x)^{12}$
Note: $4 = 3 + 1$

$a = a, b = x, n = 12$

$\dbinom{12}{3} a^{12-3} x^3 = 220a^9 x^3$

24. $2 + 4 + 6 + \ldots + 2n$

$a_1 = 2$

$S_n = \dfrac{n}{2}(2 + 2n)$

$S_n = \dfrac{n}{2} \cdot 2(1 + n)$

$S_n = n(n + 1)$

25. $1 + \dfrac{1}{x} + \dfrac{1}{x^2} + \dfrac{1}{x^3} + \ldots$

$a_1 = 1; \ r = \dfrac{\frac{1}{x}}{1} = \dfrac{1}{x}$

$S_n = \dfrac{1 \cdot \left(1 - \left(\frac{1}{x}\right)^n\right)}{1 - \frac{1}{x}} = \dfrac{1 - \left(\frac{1}{x}\right)^n}{1 - \frac{1}{x}}, \ \text{or}$

$\dfrac{x^n - 1}{x^{n-1}(x - 1)}$

Chapters 1 – 11

Cumulative Review

1. $\left|-\dfrac{2}{3}+\dfrac{1}{5}\right| = \left|-\dfrac{10}{15}+\dfrac{3}{15}\right|$

   $\qquad = \left|-\dfrac{7}{15}\right|$

   $\qquad = \dfrac{7}{15}$

2. $y-[3-4(5-2y)-3y]$

   $y-[3-20+8y-3y]$

   $y-[-17+5y]$

   $-4y+17$

3. $(10\cdot8-9\cdot7)^2-54\div9-3$

   $\qquad = (80-63)^2-6-3$

   $\qquad = 17^2-6-3$

   $\qquad = 289-6-3$

   $\qquad = 283-3$

   $\qquad = 280$

4. $(2.7\times10^{-24})(3.1\times10^{9})$

   $\qquad = (2.7\times3.1)(10^{-24}\times10^{9})$

   $\qquad = (8.37)(10^{-15})$

   $\qquad \approx 8.4\times10^{-15}$

5. $\dfrac{ab-ac}{bc} = \dfrac{(-2)3-(-2)(-4)}{3(-4)}$

   $\qquad = \dfrac{-6-8}{-12}$

   $\qquad = \dfrac{-14}{-12} = \dfrac{7}{6}$

6. $(5a^2-3ab-7b^2)-(2a^2+5ab+8b^2)$

   $\qquad = (5-2)a^2+[(-3)-5]\,ab+[(-7)-8]b^2$

   $\qquad = 3a^2-8ab-15b^2$

7. $(2a-1)(2a+1) = 4a^2+2a-2a-1$

   $\qquad = 4a^2-1$

8. $(3a^2-5y)^2 = (3a^2)^2-2(3a^2)(5y)+(5y)^2$

   $\qquad = 9a^4-30a^2y+25y^2$

9. $\dfrac{1}{x-2}-\dfrac{4}{x^2-4}+\dfrac{3}{x+2}$

   [LCD is $x^2-4=(x+2)(x-2)$]

   $= \dfrac{x+2}{x+2}\cdot\dfrac{1}{x-2}-\dfrac{4}{x^2-4}+\dfrac{x-2}{x-2}\cdot\dfrac{3}{x+2}$

   $= \dfrac{x+2-4+3(x-2)}{(x+2)(x-2)}$

   $= \dfrac{x+2-4+3x-6}{(x+2)(x-2)}$

   $= \dfrac{4x-8}{(x+2)(x-2)} = \dfrac{4(x-2)}{(x+2)(x-2)}$

   $= \dfrac{4}{x+2}$

10. $\dfrac{x^2-6x+8}{4x+12}\cdot\dfrac{x+3}{x^2-4}$

    $= \dfrac{(x-4)(x-2)}{4(x+3)}\cdot\dfrac{x+3}{(x-2)(x+2)}$

    $= \dfrac{(x-4)\cancel{(x-2)}\cancel{(x+3)}}{4\cancel{(x+3)}\cancel{(x-2)}(x+2)}$

    $= \dfrac{x-4}{4(x+2)}$

11. $\dfrac{3x+3y}{5x-5y}\div\dfrac{3x^2+3y^2}{5x^3-5y^3}$

    $= \dfrac{3x+3y}{5x-5y}\cdot\dfrac{5x^3-5y^3}{3x^2+3y^2}$

    $= \dfrac{3(x+y)}{5(x-y)}\cdot\dfrac{5(x-y)(x^2+xy+y^2)}{3(x^2+y^2)}$

    $= \dfrac{\cancel{3}(x+y)\cdot\cancel{5}\cancel{(x-y)}(x^2+xy+y^2)}{\cancel{5}\cancel{(x-y)}\cdot\cancel{3}(x^2+y^2)}$

    $= \dfrac{(x+y)(x^2+xy+y^2)}{x^2+y^2}$

12. $\dfrac{x-\frac{a^2}{x}}{1+\frac{a}{x}}$ [LCD is x]

$$\dfrac{x}{x}\cdot\dfrac{x-\frac{a^2}{x}}{1+\frac{a}{x}}=\dfrac{x^2-a^2}{x+a}$$

$$=\dfrac{(x+a)(x-a)}{x+a}$$

$$=x-a$$

13. $\sqrt{12a}\,\sqrt{12a^3b}=\sqrt{12^2\cdot a^4\cdot b}$

$$=12a^2\sqrt{b}$$

14. $(-9x^2y^5)(3x^8y^{-7})$

$$=-9(3)x^{2+8}y^{5+(-7)}$$

$$=-27x^{10}y^{-2}=-\dfrac{27x^{10}}{y^2}$$

15. $(125x^6y^{1/2})^{2/3}=(5^3)^{2/3}(x^6)^{2/3}(y^{1/2})^{2/3}$

$$=5^2x^4y^{1/3}$$

$$=25x^4y^{1/3}$$

16. $\dfrac{\sqrt[3]{x^2y^5}}{\sqrt[4]{xy^2}}=\dfrac{\sqrt[12]{x^8y^{20}}}{\sqrt[12]{x^3y^6}}=\sqrt[12]{x^5y^{14}}=y\sqrt[12]{x^5y^2}$

17. $(4+6i)(2-i)=4\cdot 2+4(-i)+6i\cdot 2+6i(-i)$

$$=8-4i+12i-6i^2$$

$$=8+8i+6$$

$$=14+8i$$

18. $4x^2-12x+9=(2x)^2-12x+3^2$

$$=(2x-3)^2$$

19. $27a^3-8=(3a)^3-(2)^3$

$$=(3a-2)[(3a)^2+3a\cdot 2+2^2]$$

$$=(3a-2)(9a^2+6a+4)$$

20. $12s^4-48t^2=12(s^4-4t^2)$

$$=12(s^2-2t)(s^2+2t)$$

21. $15y^4+33y^2-36=3(5y^4+11y^2-12)$

$$=3(5y^2-4)(y^2+3)$$

22. $(7x^4-5x^3+x^2-4)\div(x-2)$

$$2\underline{|7-5 \quad 1 \quad 0 \ -4}$$
$$\underline{\qquad 14 \ 18 \ 38 \ \ 76}$$
$$7 \ \ 9 \ \ 19 \ \ 38\ |72$$

$$7x^3+9x^2+19x+38+\dfrac{72}{x-2}$$

23. $f(x)=3x^2-4x$

$$f(-2)=3\cdot(-2)^2-4(-2)$$

$$=3\cdot 4+8$$

$$=20$$

24. $f(x)=\sqrt{2x-8}$

The domain of $f(x)$ is the set of values for which $2x-8\ge 0$.

$$2x-8\ge 0$$

$$2x\ge 8$$

$$x\ge 4$$

Domain: $[4,\infty)$; or $\{x\,|\,x\ge 4\}$

25. $g(x)=\dfrac{x-4}{x^2-10x+25}$

The domain of $g(x)$ is the set of values for which $x^2-10x+25\ne 0$.

Solve: $x^2-10x+25=0$, and exclude the value(s) from the domain.

$$x^2-10x+25=0$$

$$(x-5)^2=0$$

$$x=5$$

Domain: $\{x\,|\,x$ is a real number and $x\ne 5\}$,

$$\text{or } (-\infty,5)\cup(5,\infty)$$

26. $\dfrac{1-\sqrt{x}}{1+\sqrt{x}}$

[Conjugate of the denominator is $1-\sqrt{x}$]

$$\dfrac{1-\sqrt{x}}{1+\sqrt{x}}\cdot\dfrac{1-\sqrt{x}}{1-\sqrt{x}}=\dfrac{1-2\sqrt{x}+x}{1-x}$$

27. Parallel lines have equal slopes.

    $3x - y = 6$ has a slope of 3.

    $\left(m = \frac{-a}{b}, \frac{-3}{-1} = 3\right)$, as does the line with

    y-intercept of $(0, -8)$.

    $y = mx + b$

    $y = 3x + (-8)$

    $y = 3x - 8$

28. $x = 5\sqrt{2}$ or $x = -5\sqrt{2}$

    $x - 5\sqrt{2} = 0$ or $x + 5\sqrt{2} = 0$

    $(x - 5\sqrt{2})(x + 5\sqrt{2}) = 0$

    $x^2 - (5\sqrt{2})^2 = 0$

    $x^2 - 50 = 0$

29.
    $$x^2 + y^2 - 4x + 6y - 23 = 0$$
    $$x^2 - 4x + y^2 + 6y = 23$$
    $$(x^2 - 4x + 4) + (y^2 + 6y + 9) = 23 + 4 + 9$$
    $$(x - 2)^2 + (y + 3)^2 = 36$$
    $$(x - 2)^2 + [y - (-3)]^2 = 6^2$$

    Center is $(2, -3)$

    Radius is 6.

30. $\frac{2}{3}\log_a x - \frac{1}{2}\log_a y + 5\log_a z$

    $= \log_a x^{2/3} - \log_a y^{1/2} + \log_a z^5$

    $= \log_a \sqrt[3]{x^2} - \log_a \sqrt{y} + \log_a z^5$

    $= \log_a \frac{\sqrt[3]{x^2} \cdot z^5}{\sqrt{y}}$

31. $\log_a c = 5$

    $a^5 = c$

32. $\log 120 \approx 2.0792$

33. $\log_5 3 = \frac{\log 3}{\log 5}$

    ≈ 0.6826

34. $d = \sqrt{(x_2 - x_1)^2 + (y_2 - y_1)^2}$

    $d = \sqrt{[2 - (-1)]^2 + [-1 - (-5)]^2}$

    $d = \sqrt{3^2 + 4^2} = \sqrt{25} = 5$

35. $19, 12, 5, \ldots$

    $a_1 = 19, \quad d = -7, \quad n = 21$

    $a_{21} = a_1 + (n - 1)d$

    $a_{21} = 19 + (21 - 1) \cdot -7$

    $a_{21} = 19 + (-140)$

    $a_{21} = -121$

36. $-1 + 2 + 5 + \ldots$

    $a_1 = -1, \quad d = 3, \quad n = 25$

    $a_n = a_1 + (n - 1)d$

    $a_{25} = -1 + (25 - 1)3$

    $a_{25} = -1 + 72 = 71$

    $S_n = \frac{n}{2}(a_1 + a_n)$

    $S_{25} = \frac{25}{2}(-1 + 71)$

    $S_{25} = \frac{25}{2}(70) = 875$

37. $16, 4, 1, \ldots$

    $\frac{4}{16} = \frac{1}{4}, \frac{1}{4} = \frac{1}{4}$

    Thus, $r = \frac{1}{4}; \quad a_1 = 16$

    $a_n = a_1 r^{n-1}$

    $a_n = 16\left(\frac{1}{4}\right)^{n-1}$

38. $(a - 2b)^{10}; 7^{th}$ term

    $n = 10, \quad a = a, \quad b = -2b$

    $7 = 6 + 1, r = 6$

    $\binom{n}{r} a^{n-r} b^r$

    $\binom{10}{6} a^{10-6}(-2b)^6$

    $210 \cdot a^4 \cdot 64 b^6 = 13,440 a^4 b^6$

39. $4 + 6 + 9 + \ldots$

$a_1 = 4, \quad r = \dfrac{6}{4} = 1.5, \quad n = 9$

$S_n = \dfrac{a_1(1 - r^n)}{1 - r}$

$S_9 = \dfrac{4(1 - 1.5^9)}{1 - 1.5}$

$S_9 \approx 299.546875$

40. $8(x - 1) - 3(x - 2) = 1$

$8x - 8 - 3x + 6 = 1$

$5x - 2 = 1$

$5x = 3$

$x = \dfrac{3}{5}$

41. $\dfrac{6}{x} + \dfrac{6}{x + 2} = \dfrac{5}{2}$

LCD is $2x(x + 2)$

$2x(x + 2)\left[\dfrac{6}{x} + \dfrac{6}{x + 2}\right] = 2x(x + 2) \cdot \dfrac{5}{2}$

$2x(x + 2) \cdot \dfrac{6}{x} + 2x(x + 2) \cdot \dfrac{6}{x + 2}$

$\qquad\qquad = 2x(x + 2) \cdot \dfrac{5}{2}$

$12(x + 2) + 12x = 5x(x + 2)$

$12x + 24 + 12x = 5x^2 + 10x$

$0 = 5x^2 - 14x - 24$

$0 = (5x + 6)(x - 4)$

$5x + 6 = 0 \text{ or } x - 4 = 0$

$x = -\dfrac{6}{5} \text{ or } x = 4$

42. $2x + 1 > 5 \text{ or } x - 7 \le 3$

$2x > 4 \text{ or } x \le 10$

$x > 2 \text{ or } x \le 10$

All real numbers are either greater than 2, less than 10, or both.

$\mathbb{R}, \text{ or } (-\infty, \infty)$

43. $5x + 6y = -2 \quad (1)$

$3x + 10y = 2 \qquad (2)$

$-3 \cdot (1) + 5 \cdot (2)$

$-15x - 18y = 6$

$\underline{15x + 50y = 10}$

$32y = 16 \text{ (add)}$

$y = \dfrac{1}{2}$

$5x + 6\left(\dfrac{1}{2}\right) = -2$

$5x = -5$

$x = -1$

$\left(-1, \dfrac{1}{2}\right)$

44.

$x + y - z = 0 \;(1)$

$3x + y + z = 6 \;(2)$

$x - y + 2z = 5 \;(3)$

$(1) + (3) \quad x + y - z = 0$

$\underline{x - y + 2z = 5}$

$(4) \; 2x \quad\;\; + z = 5 \;\text{ (Add)}$

$(2) + (3) \quad 3x + y + z = 6$

$\underline{x - y + 2z = 5}$

$(5) \; 4x \quad\;\; + 3z = 11 \;\text{(Add)}$

Using (4) and (5) solve this system.

$2x + z = 5 \,(4)$

$4x + 3z = 11 \,(5)$

$-2 \cdot (4) + (5)$

$-4x - 2z = -10$

$\underline{4x + 3z = 11}$

$z = 1$

$2x + 1 = 5$

$2x = 4$

$x = 2$

$x + y - z = 0$

$2 + y - 1 = 0$

$y = -1$

$(2, -1, 1)$

45.
$$3\sqrt{x-1} = 5 - x$$
$$(3\sqrt{x-1})^2 = (5-x)^2$$
$$9(x-1) = 25 - 10x + x^2$$
$$9x - 9 = x^2 - 10x + 25$$
$$0 = x^2 - 19x + 34$$
$$0 = (x-2)(x-17)$$
$$x - 2 = 0 \quad \text{or} \quad x - 17 = 0$$
$$x = 2 \quad \text{or} \quad x = 17$$
Only the value 2 checks.
Solution is 2.

46. $x^4 - 29x^2 + 100 = 0$
Solve by Factoring.
$$(x^2 - 25)(x^2 - 4) = 0$$
$$(x-5)(x+5)(x-2)(x+2) = 0$$
$$x - 5 = 0 \quad \text{or} \quad x + 5 = 0 \quad \text{or} \quad x - 2 = 0$$
$$\text{or} \quad x + 2 = 0$$
$$x = 5 \quad \text{or} \quad x = -5 \quad \text{or} \quad x = 2$$
$$\text{or} \quad x = -2$$
Solutions are $\pm 2, \pm 5$.

47. $x^2 + y^2 = 8$
$$\underline{x^2 - y^2 = 2}$$
$$2x^2 \quad\;\; = 10 \;(\text{Add})$$
$$x^2 = 5$$
$$x = \pm\sqrt{5}$$
Substitute, to determine y.
Note:
$$\left(\sqrt{5}\right)^2 = 5$$
and
$$\left(-\sqrt{5}\right)^2 = 5$$
$$\left(\pm\sqrt{5}\right)^2 + y^2 = 8$$
$$5 + y^2 = 8$$
$$y^2 = 3$$
$$y = \pm\sqrt{3}$$
Solutions are $\left(\sqrt{5}, \sqrt{3}\right), \left(\sqrt{5}, -\sqrt{3}\right),$
$\left(-\sqrt{5}, \sqrt{3}\right)$ and $\left(-\sqrt{5}, -\sqrt{3}\right)$.

48. $4^x = 12$

$$\log 4^x = \log 12$$
$$x \log 4 = \log 12$$
$$x = \frac{\log 12}{\log 4}$$
$$x \approx 1.7925$$

49. $\log(x^2 - 25) - \log(x+5) = 3$
$$\log \frac{x^2 - 25}{x+5} = 3$$
$$\log(x - 5) = 3$$
$$10^3 = x - 5$$
$$1005 = x$$

50. $\log_5 x = -2$
$$x = 5^{-2}$$
$$x = \frac{1}{25}$$

51. $7^{2x+3} = 49$
$$7^{2x+3} = 7^2$$
$$2x + 3 = 2$$
$$2x = -1$$
$$x = -\frac{1}{2}$$

52. $|2x - 1| \le 5$
$$-5 \le 2x - 1 \le 5$$
$$-4 \le 2x \le 6$$
$$-2 \le x \le 3$$
$$\{x \,|\, -2 \le x \le 3\}, \;\text{or}\;\; [-2, 3]$$

53. $15x^2 + 45 = 0$
$$15x^2 = -45$$
$$x^2 = -3$$
$$x = \pm\sqrt{-3}$$
$$x = \pm i\sqrt{3}$$

54. $x^2 + 4x = 3$
$$x^2 + 4x + 4 = 3 + 4$$
$$(x + 2)^2 = 7$$
$$x + 2 = \pm\sqrt{7}$$
$$x = -2 \pm \sqrt{7}$$

55. $y^2 + 3y > 10$

Solve the related equation
$$y^2 + 3y - 10 = 0$$
$$(y + 5)(y - 2) = 0$$
$$y + 5 = 0 \quad \text{or} \quad y - 2 = 0$$
$$y = -5 \quad \text{or} \qquad y = 2$$

These values give us the intervals $(-\infty, -5), (-5, 2),$ and $(2, \infty)$. Using a test point from each interval, we determine the solution. Since $>$, we do not include the zeros.
$\{y \mid y < -5 \text{ or } y > 2\}$, or
$$(-\infty, -5) \cup (2, \infty).$$

56. $f(x) = x^2 - 2x; f(a) = 80$
$$80 = x^2 - 2x$$
$$0 = x^2 - 2x - 80$$
$$0 = (x + 8)(x - 10)$$
$$x + 8 = 0 \quad \text{or} \quad x - 10 = 0$$
$$x = -8 \quad \text{or} \qquad x = 10$$

57. $f(x) = \sqrt{-x + 4} + 3; \ g(x) = \sqrt{x - 2} + 3$

Let $f(a) = g(a)$
$$\sqrt{-a + 4} + 3 = \sqrt{a - 2} + 3$$
$$\sqrt{-a + 4} = \sqrt{a - 2}$$
$$-a + 4 = a - 2$$
$$6 = 2a$$
$$3 = a$$

Since we squared both sides of the equation, we must check our solution in the original equation. 3 checks and is the solution.

58.
$$V = P - Prt$$
$$V - P = -Prt$$
$$\frac{V - P}{-Pt} = r$$
or $\dfrac{P - V}{Pt} = r$

59.
$$I = \frac{R}{R + r}$$
$$I(R + r) = R$$
$$IR + Ir = R$$
$$IR - R = -Ir$$
$$R(I - 1) = -Ir$$
$$R = \frac{-Ir}{I - 1}$$
or $R = \dfrac{Ir}{1 - I}$

60. a. Linear
    b. $f(x) = 0$, when $x = 2$.

61. a. Logarithmic
    b. $f(x) = 0$, when $x = -1$.

62. a. Quadratic
    b. $f(x) = 0$, when $x = -1$ and when $x = 4$

63. a. Exponential
    b. The graph is asymptotic to the x-axis; no real zeros.

Problems 64 – 71 are graphing exercises.

64. $3x - y = 7$
$$m = \frac{-a}{b} = \frac{-3}{-1} = 3$$
$$b = \frac{c}{b} = \frac{7}{-1} = -7$$

Using $m = 3$ and the y-intercept $(0, -7)$, we graph the line.

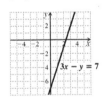

65. $x^2 + y^2 = 100$

$x^2 + y^2 = 10^2$

This graph is a circle whose center is $(0,0)$ and radius $= 10$.

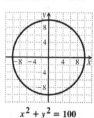

$$x^2 + y^2 = 100$$

66. $\dfrac{x^2}{36} - \dfrac{y^2}{9} = 1$

$\dfrac{x^2}{6^2} - \dfrac{y^2}{3^2} = 1$

$a = 6$ and $b = 3$. The asymptotes are

$y = \dfrac{3}{6}x$ and $y = -\dfrac{3}{6}x$, or

$y = \dfrac{1}{2}x$ and $y = -\dfrac{1}{2}x$

To help us sketch asymptotes and locate vertices, we use a and b, to form the pairs $(-6, 3), (6, 3), (-6, -3)$, and $(6, -3)$.

Plot these pairs and lightly sketch a rectangle. The asymptotes pass through the corners; since this is a horizontal hyperbola, the vertices are where the rectangle intersects the x-axis. Sketch the hyperbola.

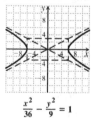

$$\frac{x^2}{36} - \frac{y^2}{9} = 1$$

67. $y = \log_2 x$

The graph contains the points $(1, 0)$ and $(2, 1)$, is increasing, and is asymptotic to the y-axis.

68. $f(x) = 2^x - 3$

$f(x) = 2^x$ contains the points $(0, 1)$ and $(1, 2)$, is increasing, and is asymptotic to the x-axis. We lightly sketch this function. To graph $f(x) = 2^x - 3$ we shift/translate the above graph 3 units downward.

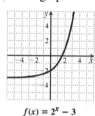

$$f(x) = 2^x - 3$$

69. $2x - 3y < -6$

Graph the related equation/line
$2x - 3y = -6$

You might use the intercepts, $(0, 2)$ and $(-3, 0)$. Use a dashed line, since $<$.

Choose a test point to determine the correct half-plane.

You might choose $(0, 0)$.

$2(0) - 3(0) < -6$

$0 < -6$

Since this is false, we shade the other half-plane.

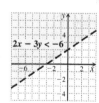

70. $f(x) = -2(x - 3)^2 + 1$

These are to be labeled on the graph.

a.   Vertex $(3, 1)$

b.   Axis of symmetry is $x = 3$

c. Since $a = -2$, the parabola opens downward. Thus, the maximum function value is 1, when $x = 3$.

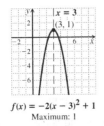

$$f(x) = -2(x - 3)^2 + 1$$
Maximum: 1

71.

$$y = \sqrt{(4 - x)}$$

From the graph we determine
Domain: $(-\infty, 4]$

Range: $[0, \infty)$

72. Let $w =$ the width and $200 - 2w =$ the length, as shown in the diagram.
Area = Length × Width

$$A = (200 - 2w)w$$

$$A = 200w - 2w^2$$

The graph of this function is a parabola which opens downward, so a maximum exists.

$$A = -2(w^2 - 100w)$$

$$= -2(w^2 - 100w + 2500) - (-2)(2500)$$

$$= -2(w - 50)^2 + 5000$$

The maximum area is 5000 ft^2.

73. $P = 2l + 2w$, or $34 = 2l + 2w$, or $17 = l + w$
Using the Pythagorean theorem,

$$a^2 + b^2 = c^2$$

$$l^2 + w^2 = 13^2$$

Substitute for l using $l = 17 - w$ and solve.

$$(17 - w)^2 + w^2 = 13^2$$

$$289 - 34w + w^2 + w^2 = 169$$

$$2w^2 - 34w + 120 = 0$$

$$2(w - 12)(w - 5) = 0$$

$w - 12 = 0$ or $w - 5 = 0$

$w = 12$ or $w = 5$

We usually think of length as the greater of the two measures, so $w = 5$ and $l = 17 - 5$ or 12.
The dimensions are 5 ft by 12 ft.

74. Let $x =$ number of movies rented. Limited members pay $40 plus $2.45 per movie, or $40 + \$2.45x$. Preferred members pay $60 plus $1.65 per movie, or $60 + \$1.65x$. We want the values of x which satisfy
$$\$60 + \$1.65x < \$40 + \$2.45x$$
We solve for x.
$$60 + 1.65x < 40 + 2.45x$$
$$20 < 0.8x$$
$$25 < x$$
It will be less expensive to be a preferred member when more than 25 movies are rented.

75. Let $x =$ the number of ounces of the herbs costing $2.68/$oz$. Let $y =$ the number of ounces of the herbs costing $4.60/$oz$. When mixed, they cost $3.80/$oz$, so
$$2.68x + 4.60y = 3.80(x + y).$$
We also know there are 24 oz in all, so
$$x + y = 24.$$
We solve the system using substitution.
$$2.68x + 4.60y = 3.80(x + y)$$
$$x + y = 24, \ x = 24 - y$$
$$2.68(24 - y) + 4.60y = 3.80(24)$$
$$64.32 - 2.68y + 4.60y = 91.2$$
$$1.92y = 26.88$$
$$y = 14$$
$x = 24 - y$, or $24 - 14 = 10$
The mix should contain 10 oz of the herbs costing $2.68/oz and 14 oz of the herbs costing $4.60/oz.

76. Let r = speed of the plane in still air,
$r + 30$ = speed of the plane with the wind,
and $r - 30$ = speed of the plane against the wind. Since the time for both is the same (equal), using $t = \frac{d}{r}$, we have:

$$\frac{190}{r+30} = \frac{160}{r-30}$$

LCD is $(r+30)(r-30)$

$$(r+30)(r-30) \cdot \frac{190}{r+30}$$

$$= (r+30)(r-30) \cdot \frac{160}{r-30}$$

$$(r-30) \cdot 190 = (r+30) \cdot 160$$

$$190r - 5700 = 160r + 4800$$

$$30r = 10,500$$

$$r = 350$$

The plane can fly 350 mph in still air.

77. Jack can tap the trees in 21 hours, so he can do $\frac{1}{21}$ of the job in an hour. Delia can tap the trees in 14 hours, so she can do $\frac{1}{14}$ of the job in an hour. Let t = number of hours they work together. Work = rate × time
Jack's work + Delia's work = 1

$$\frac{1}{21}t + \frac{1}{14}t = 1, \text{ or } \frac{t}{21} + \frac{t}{14} = 1$$

LCD is 42.

$$42\left(\frac{t}{21} + \frac{t}{14}\right) = 42 \cdot 1$$

$$2t + 3t = 42$$

$$5t = 42$$

$$t = \frac{42}{5} = 8\frac{2}{5} \text{ hr}$$

Working together, they can tap the trees in $8\frac{2}{5}$ hr, or 8 hr 24 min.

78. F varies directly as v^2 (velocity squared) and inversely as r (radius), so

$$F = k \cdot \frac{v^2}{r}$$

Substitute the given value to determine the constant of variation k.

$$8 = k \cdot \frac{1^2}{10}$$

$$80 = k$$

We have the variation equation

$$F = 80 \cdot \frac{v^2}{r}$$

Substitute the known values and solve for F.

$$F = 80 \cdot \frac{2^2}{16}$$

$$F = 80 \cdot \frac{4}{16}$$

$$F = 20$$

79. a. Determine the slope using (0, 173) and (4, 181).

$$m = \frac{181 - 173}{4 - 0}$$

$$m = \frac{8}{4}$$

$$m = 2$$

The average rate of change is 2 million card holders per year.

b. $m = 2$ and $b = 173$
Using slope-intercept form, we have
$$c = mt + b$$
$$c = 2t + 173, \text{ where}$$
c is in millions.

c. In 2015, $t = 9$
$$c = 2(9) + 173$$
$$= 18 + 173$$
$$= 191$$
There will be 191 million card holders in 2015.

d. Let $c = 250$:
$$250 = 2t + 173$$
$$77 = 2t$$
$$38.5 = t$$
$2006 + 38.5 = 2044.5$. There will be 250 million Americans with credit cards in 2044.

80. a. $P(t) = P_0 e^{kt}$
Using the given information,
$P_0 = 5$ billion. In 2006, $t = 27$ and
$P(27) = 60.$
Substitute to determine k.

$$60 = 5e^{k(27)}$$

$$\frac{60}{5} = e^{27k}$$

$$\ln 12 = 27k$$

$$\frac{\ln 12}{27} = k$$

$$0.092 \approx k$$

We have the exponential function $P(t) = 5e^{0.092t}$, where $P(t)$ is in billions of dollars.

b. In 2012, $t = 33$

$$P(33) = 5e^{0.092(33)}$$

$$P(33) \approx 104$$

In 2012, there will be approximately $104 billion in electronic payments.

c. $$200 = 5e^{0.092t}$$

$$40 = e^{0.092t}$$

$$\ln 40 = 0.092t$$

$$\frac{\ln 40}{0.092} = t$$

$$40.1 \approx t$$

There will be $200 billion in electronic payments in $1979 + 40 = 2019$.

81. This is a geometric sequence.
$$2000, 2000 \cdot 1.05, 2000 \cdot 1.05^2, \ldots 2000 \cdot 1.05^n$$
When Sarita is 62, $n = 40$.
$$2000 \cdot 1.05^{40} \approx \$14,079.98$$

82. $$\frac{9}{x} - \frac{9}{x+12} = \frac{108}{x^2+12x}$$
$$x \neq 0; \quad x \neq -12$$

LCD is $x^2 + 12x = x(x+12)$

$$x(x+12)\left(\frac{9}{x} - \frac{9}{x+12}\right) = (x^2+12x) \cdot \frac{108}{x^2+12x}$$

$$(x+12)9 - 9 \cdot x = 108$$

$$9x + 108 - 9x = 108$$

$$108 = 108$$

Since this is an identity (Reflexive Property of Equality), all values of x are solutions except 0 and -12.
$$\{x \mid x \in \mathbb{R}, x \neq 0, x \neq -12\}$$

83. $$\log_2\left(\log_3 x\right) = 2$$
$$\log_3 x = 2^2$$
$$\log_3 x = 4$$
$$x = 3^4$$
$$x = 81$$

84. y varies directly as the cube of x, so $y = kx^3$.
If x is multiplied by 0.5,
$$y = k\left(0.5x\right)^3, \text{ or } y = k \cdot 0.125x^3$$
y is divided by 8, since $0.125 = \frac{1}{8}$

85. Let x = the number of years Diaphantos lived, and y = number of years his son lived.

Since Diaphantos spent $(1/6)x$ as a child, $(1/12)x$ as an adolescent, and $(1/7)x$ as a bachelor, we have

$$\frac{1}{6}x + \frac{1}{12}x + \frac{1}{7}x =$$
$$\frac{14}{84}x + \frac{7}{12}x + \frac{12}{84}x = \frac{33}{84}x = \frac{11}{28}x$$

Five years after he was married, or $\frac{33}{84}x + 5$,

his son was born. His son died 4 years before his father (Diaphantos), so Diaphantos' years can be represented as

$$\frac{33}{84}x + 5 + y + 4 \text{ or } x$$

which gives us the equation

$$\frac{33}{84}x + 5 + y + 4 = x$$

$$y + 9 = \frac{51}{84}x$$

We also know the son lived half as long as Diaphantos, so we also have the equation

$y = \frac{1}{2}x$. We solve this system using substitution.

$$y + 9 = \frac{51}{84}x$$

$$\left(\frac{1}{2}x\right) + 9 = \frac{51}{84}x$$

$$9 = \frac{9}{84}x$$

$$84 = x$$

Diaphantos lived 84 years.